Conversion of SI Units to English Units

To Convert from	To	Multiply by:	
		Accurate	Common
joule (J)	foot-pound-force (ft-lbf)	$7.375\ 620 \times 10^{-1}$	0.737
kilogram (kg)	slug	$6.852\ 178 \times 10^{-2}$	0.0685
kilogram (kg)	pound-mass (lbm avoirdupois)	$2.204\ 622 \times 10^{0}$	2.20
kilogram (kg)	ton (short 2000 lbm)	$1.102\ 311 \times 10^{-3}$	0.001 10
kilometer (km)	mile (mi U.S. Statute)	$6.213\ 713 \times 10^{-1}$	0.621
kilowatt (kw)	horsepower (hp)	$1.341\ 022 \times 10^{0}$	1.34
meter (m)	foot (ft)	$3.280\ 840 \times 10^{0}$	3.28
meter (m)	inch (in)	$3.937\ 008 \times 10^{1}$	39.4
meter^3 (m^3)	gallon (gal U.S.)	$2.641\ 720 \times 10^{2}$	264.
newton (N)	pound-force (lbf avoirdupois)	$2.248\ 089 \times 10^{-1}$	0.225
pascal (Pa)	pound-force/foot^2 (lfb/ft^2)	$2.088\ 543 \times 10^{-2}$	0.0209
pascal (Pa)	pound-force/inch^2 (psi)	$1.450\ 370 \times 10^{-4}$	0.000 145
watt (W)	foot-pound-force/second (ft-lbf/s)	$7.375\ 620 \times 10^{-1}$	0.737

Mechanics
of Materials

Mechanics of Materials

Nelson R. Bauld, Jr.

CLEMSON UNIVERSITY

Brooks/Cole Engineering Division

Monterey, California

The cover illustration is a photograph of a small polymer scrap that was stretched, by hand, between two pins mounted on a glass slide. Under a low-power microscope, cross-polarized lighting created the various tones of color; these tones indicate the degrees of stress in the material.

Photo by Dante Petro.

Brooks/Cole Engineering Division
A Division of Wadsworth, Inc.

Printed in the United States of America

10 9 8 7 6 5 4 3 2 1

Library of Congress Cataloging in Publication Data
Bauld, Nelson R., 1931–
    Mechanics of materials.

    Includes index.
    1. Strength of materials. I. Title
TA405.B33   620.1'12   81-17072
ISBN 0-8185-0495-1   AACR2

Sponsoring Editor: Ray Kingman
Production Service: Mary Forkner, Publication
    Alternatives
Manuscript Editor: Ruth Cottrell
Interior Design and Art Direction: Al Burkhardt
Cover Design: Al Burkhardt
Illustrations: Innographics, Graphics House, Mary
    Burkhardt, Pat Rogondino, Maria Jedd, Camille
    Zawojski
Typesetting: Syntax International
Production Services Coordinator: Stacey C. Sawyer

Preface

This text is designed for a first course in strength of materials offered to engineering students in the sophomore or junior year. Normally courses in rigid body mechanics (statics) and in differential and integral calculus provide the mechanics and mathematic backgrounds necessary to a sound understanding of the material.

The first nine chapters contain the material usually covered in a three-semester hour course or a five-quarter hour course. The material in Chapters 10 through 14 is treated at a level consistent with that of students who have successfully completed an elementary course in strength of materials. It is rigorous enough to be of interest to exceptional students who wish to reach beyond the basic material covered in a first course or to be a source for advanced courses in strength of materials.

The mathematics required for an understanding of the topics contained in this book is nominal; however, mathematical techniques that are compatible with students' experiences are adopted where their use provides a sharper understanding of a topic or where their use is a necessity. Thus, mathematical techniques are used only when they enhance the understanding of a derivation or a computational procedure. A secondary objective of this book is to show the appropriateness and proper use for mathematics in strength of materials.

Considerable effort has been made in this book to define basic concepts in a direct but precise manner and to expose similarities in different developments where they arise. Important equations are emphasized in the text by shading, and a summary is included at the conclusion of each chapter that lists important formulas and conceptual ideas. Over 200 example problems are used to illustrate physical definitions and problem-solving procedures, and over 650 exercise problems provide a generous exposure to a variety of problems to which the strength of materials method can be applied. Over 950 diagrams are contained in the text and example problems, and nearly 650 additional diagrams accompany the exercise problems.

Exercise problems are placed at the conclusion of appropriate sections—that is, where the sections have covered enough material to warrant exercise problems. Otherwise, exercise problems are placed at the conclusion of a group of sections that collectively complete enough of a development to make the problems effective. Exercise problems also appear at the conclusion of each chapter, usually in the order in which the principles or procedures they emphasize occur in the text. "After-section" exercises usually focus on the notions introduced in the section that immediately precedes them, whereas exercises at the conclusions of the chapters include several concepts that have been covered in the chapter. Answers to odd-numbered problems are provided in an appendix so that students can assess their progress in problem solving, and a solutions manual containing detailed solutions for all exercise problems is available.

Any meaningful static structural theory must be based on three fundamental precepts—equilibrium, compatibility of deformations, and material behavior. This text emphasizes these precepts repeatedly because they occupy the central role in obtaining solutions to the plethora of problems common to the subject called *strength of materials* or *mechanics of deformable bodies*. This text not only demonstrates the methods by which these precepts are used to formulate solutions for problems, but it also nurtures thought processes that are invaluable in establishing new structural theories, improving existing theories, or extending existing theories beyond their common usages. Thus, these precepts serve as the common ground that characterizes the variety of problems usually covered in this course, and their importance in advanced courses in strength of materials, solid mechanics, and structural mechanics is undiminished.

An effort is made to elevate the modeling phase of the developments contained in this book to a level that reflects its importance in the study of strength of materials. Further, the emphasis on modeling is done in a way that is easily understood by students who have benefitted from the material that has preceded the introduction of a particular model.

Basic concepts of stress, strain, strain energy, and material properties are introduced early, whereas transformations of biaxial states of stress and strain are developed following a presentation of combined states of stress.

The design philosophy and the concept of statical indeterminate states of stress are integrated into the text as continuing themes, instead of treating them in separate chapters.

In this book the three computational procedures associated with the structural design process that students usually encounter first are considered different applications of the same equations. Therefore, determining appropriate member dimensions, calculating allowable loads for existing structures, or computing stresses and deflections induced in a structure by specified loads are shown to be different applications of the same basic equations. I have attempted to cultivate an appreciation for these fundamental computational design procedures by their repeated use in the example problems and exercises.

Although statically indeterminate problems are not elevated to a special level of importance, special care has been taken to explain the procedures used to solve them in text discussions and in example problems that emphasize the three fundamental precepts. The treatment of the statically indeterminate condition of a particular structural element usually occurs in the same unit or immediately following the unit in which the statically determinate analysis occurs. Thus, students can readily recognize the ramifications of statical indeterminacy and can easily identify the procedures required to obtain solutions for particular structural elements.

The material treated in this text is based predominately on linearly elastic behavior of isotropic materials. Stress-strain relations are developed for one-, two-, and three-dimensional states of stress and strain for isotropic materials. Inelastic stress-strain relations and nonlinear elastic stress-strain relations are discussed for uniaxial and pure torsional states of stress. This material is included to emphasize the limitations inherent in the use of a linearly elastic stress-strain relation, to establish a set of design formulas, and to indicate the role the three basic precepts play in establishing design formulas for inelastic and nonlinear elastic material behavior.

Both British gravitational units and SI units (*Systéme International Unité*) are used throughout this text in approximately equal proportions. This reflects my belief that transition from British gravitational units to SI units will evolve slowly in this country, requiring engineers to be familiar with both systems of units for several years. Generally, at least one example problem is presented in each system of units at the conclusion of each section.

Anyone who writes an elementary strength of materials book is indebted to innumerable persons. I must acknowledge the authors of strength of materials books from which I acquired my proficiency, professors who have provided the elusive quality inspiration, colleagues who have provided encouragement, and my family who permitted me to indulge my passion for the long task through their patience and understanding.

I also wish to acknowledge the assistance of Ms. Ingrid Jackson, who typed this manuscript, and Mr. Lih-Shyng Tzeng, who helped

viii   **Preface**

prepare the solutions manual. I also express my gratitude to Nuri Akkas, of Middle Eastern Technical University, Ankara, Turkey, for his helpful suggestions about the earliest version of this manuscript. Finally, I wish to acknowledge the many valuable comments and suggestions made by Milton E. Raville of Georgia Institute of Technology, Edwin R. Chubbuck of Louisiana State University, and the reviewers: Romesh C. Batra, University of Missouri at Rolla; Eric Becker, University of Texas; George Buchanan, Tennessee Technological Institute; William Clausen, The Ohio State University; Z. M. Elias, University of Washington; Jim Harp, University of Oklahoma; George Piotroski, University of Florida; and Edward Ting, Purdue University.

Nelson R. Bauld, Jr.

Contents

14

Failure Theories for Isotropic Materials   633

Appendices   663

Index

CHAPTER 1

Internal Forces

1-1 Equivalent Force-Couple Systems

An important principle of statics states that any system of forces and couples can be replaced by an equivalent force and couple. This equivalency is the result of the requirement that equivalent force systems have precisely the same external effect on a rigid body; that is, equivalent force systems tend to translate and rotate a rigid body in precisely the same manner. We emphasize that equivalent force systems are required to produce identical overall **external** effects on a body. The **internal** effects that such equivalent force systems produce may be quite different. Consequently, when determining internal quantities such as stresses, strains, and strain energy, which are to be defined soon, one must exercise care that they correspond to the actual forces that are applied to the body.

Figure 1-1 depicts a three-dimensional structure that is acted on by n concentrated forces, $\mathbf{F}_i$, and m couples, $\mathbf{C}_i$. In addition, forces that are distributed continuously over portions of the surface of the structure may be present.

An imaginary plane is perceived to separate the structure into two distinct portions as shown in Figure 1-1. One of these portions (either portion may be selected) is selected as a free-body diagram as shown in Figure 1-2a. The forces acting on this selected portion of the structure consist of both the external forces and couples that are applied directly to it and the distributed system of forces exerted on it by the portion that was removed. We assume that each distributed force system that acts on the surface of the portion that has been selected as a free-body diagram has been replaced by a statically equivalent force-couple system.

The distributed force system that acts on the cutting plane represents the molecular forces that material particles on either side

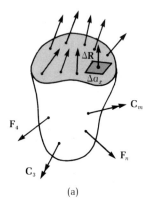

(a)

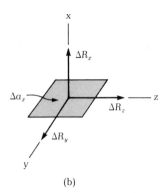

(b)

Figure 1-2   Irregular force distribution on an imaginary cutting plane.

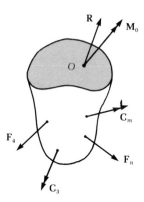

Figure 1-3   Equivalent force-couple system.

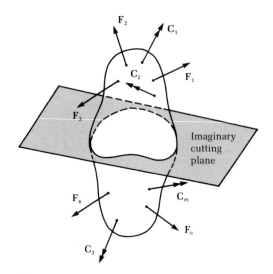

Figure 1-1

of the imaginary cutting plane exert on each other. The exact distribution of these molecular forces on the exposed plane of the free-body diagram is unknown. This distribution of forces must be such that, in combination with the external forces applied directly to the surface of the selected portion, it assures the equilibrium of this portion of the structure.

Equilibrium is guaranteed if the distributed force system acting on the imaginary plane is replaced by an equivalent force-couple system such as the one shown in Figure 1-3. Statical equivalency requires that

$$\mathbf{R} + \Sigma\mathbf{F}_i = 0$$

and   (1-1)

$$\mathbf{M}_0 + \Sigma\mathbf{r}_i \times \mathbf{F}_i + \Sigma\mathbf{C}_i = 0$$

Here $\mathbf{F}_i$ and $\mathbf{C}_i$ are the concentrated forces and couples that are applied directly to the surface of the portion of the structure that has been selected as a free-body diagram, and $\Sigma\mathbf{r}_i \times \mathbf{F}_i$ are the couples caused by the concentrated forces with respect to point O. Point O is an arbitrary point in the imaginary plane, but it is customarily chosen to coincide with the centroid of the plane area.

Ordinarily, it is convenient to decompose the force $\mathbf{R}$ and the couple $\mathbf{M}_0$ into rectangular components as depicted in Figures 1-4a and 1-4b. The origin of the xyz coordinate system is established at the centroid of the area, with one axis (the x axis) perpendicular to the area.

To connect the components R_x, R_y, and R_z and components M_x, M_y, and M_z with the actual force distribution that acts on the imaginary plane, we need the concept of stress—or force per unit area.

An increment of area Δa_x and the increment $\Delta \mathbf{R}$ of the force $\mathbf{R}$ that acts on it are shown in Figure 1-2a. The subscript x indicates that the area Δa_x is perpendicular to the x axis. Figure 1-2b depicts the components of $\Delta \mathbf{R}$ parallel to the coordinate axes in their positive senses. The ratios

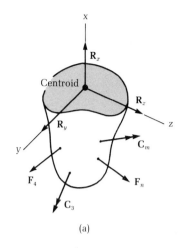

$$\lim_{\Delta a_x \to 0} \frac{\Delta R_x}{\Delta a_x}, \qquad \lim_{\Delta a_x \to 0} \frac{\Delta R_y}{\Delta a_x}, \qquad \text{and} \qquad \lim_{\Delta a_x \to 0} \frac{\Delta R_z}{\Delta a_x} \qquad (1\text{-}2)$$

are observed to possess units of pounds per square inch, newtons per square meter, or some similar convenient units. These ratios are called stresses. It is standard practice to denote these ratios by σ_x, τ_{xy}, and τ_{xz}, respectively. Accordingly, the **stress components** that act on a plane area perpendicular to the x axis are defined by the relations

$$\sigma_x = \lim_{\Delta a_x \to 0} \frac{\Delta R_x}{\Delta a_x}, \qquad \tau_{xy} = \lim_{\Delta a_x \to 0} \frac{\Delta R_y}{\Delta a_x}, \qquad \text{and} \qquad \tau_{xz} = \lim_{\Delta a_x \to 0} \frac{\Delta R_z}{\Delta a_x}$$

$$(1\text{-}3)$$

The placement of the subscripts serves to identify a particular stress component. For example, the first subscript identifies the area on which the stress component acts as perpendicular to the axis whose letter label appears in that position. The second subscript identifies the direction of the stress component. The stress component σ_x appears to be an exception to the rule; however, it is not really an exception because this notation is actually an abbreviated form of σ_{xx}.

The stress component σ_x that is perpendicular to the imaginary plane is called a **normal stress**, and it is either a **tensile** or a **compressive** normal stress depending on whether it pulls or pushes on the cutting plane. The stress components that lie in the cutting plane, τ_{xy} and τ_{xz}, are called shearing stresses.

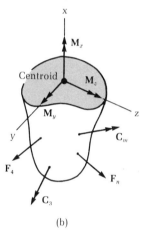

Figure 1-4

Having defined the stress components that act on the imaginary cutting plane, we can establish the equivalency between the force-couple system of Figure 1-4 and the force distribution that it replaces. The equivalency of the two-force systems is established through the relations

$$R_x = \int_{a_x} \sigma_x \, da_x, \qquad R_y = \int_{a_x} \tau_{xy} \, da_x, \qquad R_z = \int_{a_x} \tau_{xz} \, da_x$$

$$(1\text{-}4)$$

and

$$M_x = \int_{a_x} (y\tau_{xz} - z\tau_{xy}) \, da_x, \qquad M_y = \int_{a_x} \sigma_x z \, da_x,$$

$$(1\text{-}5)$$

$$M_z = -\int_{a_x} \sigma_x y \, da_x$$

The foregoing relations can be satisfied by an infinite number of stress distributions. Consequently, the stress components that satisfy

Eqs. (1-4) and (1-5) are not unique. The object of the strength-of-materials approach is to establish a reasonably simple stress distribution that will satisfy these equations and, at the same time, correspond to strains that are compatible with the deformations experienced by the structure. The requirement that the strain be compatible with the deformations is usually only approximately satisfied in the strength-of-materials approach and, therefore, laboratory tests are usually necessary to validate formulas that arise through this approach.

EXAMPLE 1-1

Consider a straight beam loaded as shown in Figure 1-5a. Calculate the components of the equivalent force-couple system that acts on the section C taken 11 ft from the left end of the beam.

SOLUTION

The distributed force system of 400 lb/ft is first replaced by an equivalent concentrated force as shown in Figure 1-5b. The reactions at the pin connection and at the roller are determined from equilibrium

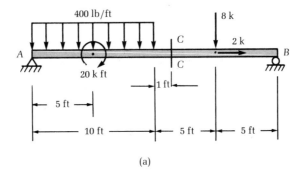

(a)

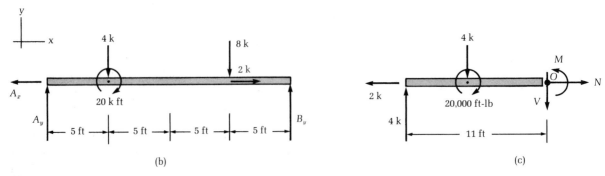

(b)

(c)

Figure 1-5

considerations. Thus, from Figure 1-5b,

$$\xrightarrow{+} \Sigma F_x = 0: \quad -A_x + 2 = 0, \qquad\qquad A_x =$$

$$\overset{\curvearrowright}{+} \Sigma M_A = 0: \quad 20B_y - 4(5) - 20 - 8(15) = 0, \qquad B_y = \quad ..$$

$$\overset{\curvearrowright}{+} \Sigma M_B = 0: \quad 20A_y - 4(15) + 20 - 8(5) = 0, \qquad A_y = 4 \text{ k}$$

Note that since A_x, A_y, and B_y each turned out to be positive, the senses assumed for these reactions (as shown in Figure 1-5b) are correct. As a check on the correctness of the calculations, observe that $\Sigma F_y = 0$ must be satisfied by the calculated reactions. This condition is an equilibrium condition, though not an independent one, that has not been used previously. Since

$$+\uparrow \Sigma F_y = 4 + 8 - 4 - 8 = 0$$

we conclude that the support reactions have been calculated correctly. The preferred technique is to calculate each unknown reaction independently of other unknown forces whenever possible, and to check these calculations when appropriate formulas are available.

The components of the equivalent force and couple that act on section C of the beam can now be calculated by considering the equilibrium of the free-body diagram depicted in Figure 1-5c. This free-body diagram has been obtained by imagining the beam to be cut into two portions by a plane passing through C perpendicular to the axis of the beam. Accordingly,

$$+\longrightarrow \Sigma F_x = 0: \quad N - 2 = 0 \qquad\qquad N = 2 \text{ k}$$

$$+\uparrow \; \Sigma F_y = 0: \quad -V + 4 - 4 = 0 \qquad\qquad V = 0$$

$$\overset{\curvearrowright}{+} \Sigma M_O = 0: \quad M - 4(11) + 4(6) - 20 = 0 \qquad M = 40 \text{ k ft}$$

These components of the equivalent force and couple are the only ones acting on the specified cross section. The remaining force and couple components are identically zero because the forces acting on the beam are coplanar.

EXAMPLE 1-2

Figure 1-6a depicts a step-shaft of circular cross section that is subjected to three couples as indicated. Determine the equivalent force-couple system that acts on a cross section between the 4000-N·m couple and the 5000-N·m couple.

SOLUTION

A free-body diagram that isolates the portion of the shaft to the right of an imaginary plane perpendicular to the axis of the shaft is shown in Figure 1-6b. The twisting moment about the x axis that acts on the section is denoted by T. Moment equilibrium with respect to the axis

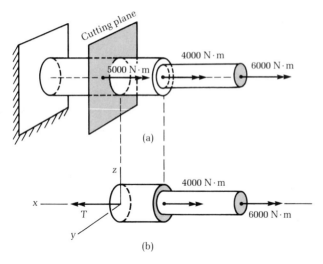

Figure 1-6

of the shaft gives

$$\Sigma M_x = 0: \quad T - 4000 - 6000 = 0, \qquad T = 10 \text{ kN} \cdot \text{m}$$

The components of the equivalent force on the section are zero because no external forces act on the shaft. The moments about the y and z axes, M_y and M_z, are also identically zero.

EXAMPLE 1-3

Figure 1-7a depicts a simple truss that is loaded at point A as shown. Determine the equivalent force-couple systems acting on planes perpendicular to the axes of members AE and AB.

SOLUTION

From the definition of a truss, each member must be a two-force member. Since members AE and AB are straight, it follows that only a force perpendicular to the cross section of each of these members can be nonzero. Thus the equivalent couple on the cross section is zero, and the equivalent force acts perpendicular to the cross section.

The reactions of the truss on the frictionless pulley are calculated from the free-body diagram shown in Figure 1-7b. Accordingly,

$$\xrightarrow{+} \Sigma F_x = 0: \quad -A_x + \tfrac{3}{5}(5000) = 0, \qquad A_x = 3000 \text{ lb}$$
$$+\uparrow \Sigma F_y = 0: \quad A_y - 5000 - \tfrac{4}{5}(5000) = 0, \qquad A_y = 9000 \text{ lb}$$

The reactions of the pulley on the truss are equal in magnitude to these reactions and opposite in sense. These reactions are shown on a free-body diagram of the pin at joint A of the truss in Figure 1-7c.

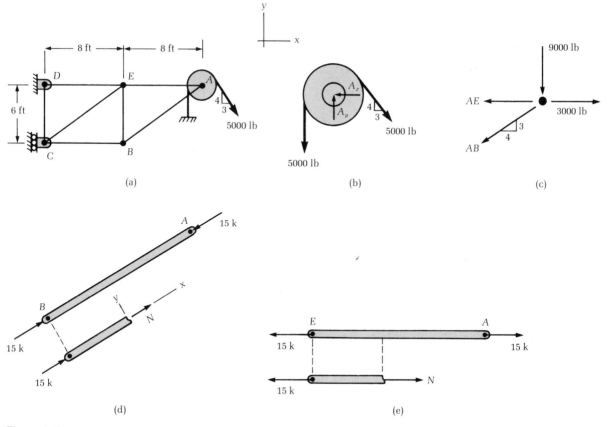

Figure 1-7

Equilibrium of the forces acting on the pin at A yields

$$+\uparrow\Sigma F_y = 0:\quad -\tfrac{3}{5}AB - 9000 = 0,\quad AB = 15\ \text{k(C)}$$
$$+\rightarrow\Sigma F_x = 0:\quad -AE + 3000 - \tfrac{4}{5}(-15000) = 0,$$
$$AE = 15\ \text{k}(T)$$

Figures 1-7d and 1-7e show that the required equivalent force-couple systems for members AB and AE are $N = -15\ \text{k} = 15\ \text{k(C)}$ (compression) and $N = 15\ \text{k}(T)$ (tension), respectively.

EXAMPLE 1-4

Figure 1-8a depicts a pin-connected plane frame supported at point C by a smooth pin and at point D by a smooth roller. It is loaded by a horizontal force of 15 kN at point A. Determine the equivalent force and couple that acts on a cross section of member AC 9 m from point C.

SOLUTION

The reactions at the pin connection and at the roller are determined from the free-body diagram of the entire frame as shown in Figure

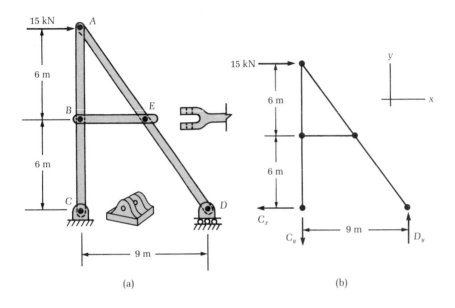

(a)

(b)

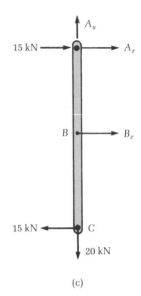

(c)

1-8b. Thus

$$\xrightarrow{+} \Sigma F_x = 0: \quad -C_x + 15 = 0, \qquad C_x = 15 \text{ kN}$$

$$\underset{+}{\curvearrowleft} \Sigma M_C = 0: \quad 9D_y - 15(12) = 0, \qquad D_y = 20 \text{ kN}$$

$$+\uparrow \Sigma F_y = 0: \quad -C_y + 20 = 0, \qquad C_y = 20 \text{ kN}$$

From the free-body diagram of member ABC shown in Figure 1-8c, we find the force exerted on ABC by the two-force member BE to be

$$\underset{+}{\curvearrowleft} \Sigma M_A = 0: \quad 6B_x - 15(12) = 0, \qquad B_x = 30 \text{ kN}$$

Finally, the required components of the equivalent force and couple at a section 9 m from point C are calculated from the free-body diagram shown in Figure 1-8d. Consequently,

$$\underset{+}{\curvearrowleft} \Sigma F_x = 0: \quad V = 15 \text{ kN}$$

$$\Sigma F_y = 0: \quad N = 20 \text{ kN}$$

$$\Sigma M_0 = 0: \quad M + 30(3) - 15(9) = 0, \qquad M = 45 \text{ kN} \cdot \text{m}$$

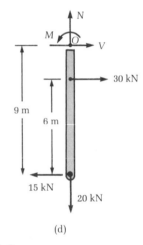

(d)

Figure 1-8

EXAMPLE 1-5

Figure 1-9a depicts a curved machine part that lies in the plane of the paper. A 700-lb force is applied at point A as shown. Determine the components of the equivalent force and couple that act on the cutting plane shown in the figure.

SOLUTION

Free-body diagrams that isolate the right portion of this machine element are shown in Figures 1-9b and 1-9c. To avoid cluttering, the

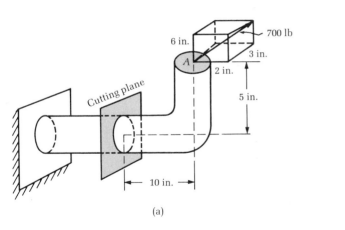

(a)

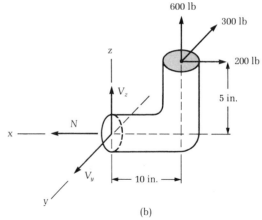

(b)

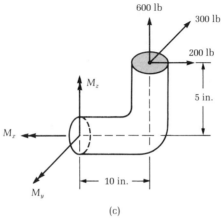

Figure 1-9

(c)

components of the equivalent force are shown in Figure 1-9b, and the components of the equivalent couple are shown in Figure 1-9c.

From equilibrium of forces acting on the free-body diagram of Figure 1-9b,

$$\Sigma F_x = 0: \quad N - 200 = 0, \qquad N = 200 \text{ lb}$$

$$\Sigma F_y = 0: \quad V_y - 300 = 0, \qquad V_y = 300 \text{ lb}$$

$$\Sigma F_z = 0: \quad V_z + 600 = 0, \qquad V_z = -600 \text{ lb}$$

Equilibrium of moments with respect to the x, y, and z axes of Figure 1-9c yields

$$\Sigma M_x = 0: \quad M_x + 300(5) = 0, \qquad\qquad M_x = -1500 \text{ in.-lb}$$

$$\Sigma M_y = 0: \quad M_y + 600(10) - 200(5) = 0, \qquad M_y = -5000 \text{ in.-lb}$$

$$\Sigma M_z = 0: \quad M_z + 300(10) = 0, \qquad\qquad M_z = -3000 \text{ in.-lb}$$

For this machine element, we can see that the equivalent force and couple each possesses three nonzero components.

EXAMPLE 1-6

Figure 1-10a depicts a single riveted butt-joint that is subjected to the end forces shown in the figure. The top view of the butt-joint is shown in Figure 1-10b. Determine the equivalent force system that acts on the cross-sectional planes of the rivet at the interfaces between the center plate and the two cover plates. Ignore friction forces between the plates.

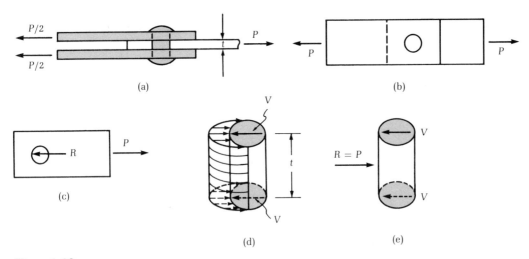

Figure 1-10

SOLUTION

Figure 1-10c shows a free-body diagram of the center plate with the rivet removed. R represents a concentrated force that is statically equivalent to the distribution of pressure that the rivet exerts on the cylindrical surface area of the hole. Figure 1-10d shows the reaction of the plate on the rivet, and Figure 1-10e shows the statically equivalent reaction R acting on the rivet. In both Figures 1-10d and 1-10e, only that portion of the rivet that is included in the center plate is shown.

Obviously, equilibrium of the center plate requires that $R = P$. Consequently, Figure 1-10e reveals that the shearing force acting in the cross section of the rivet must be $V = \frac{1}{2}R = \frac{1}{2}P$.

EXAMPLE 1-7

Figure 1-11a depicts a straight beam that rests on 100 mm × 100 mm blocks at its ends. Determine the reactions exerted by the blocks on the beam.

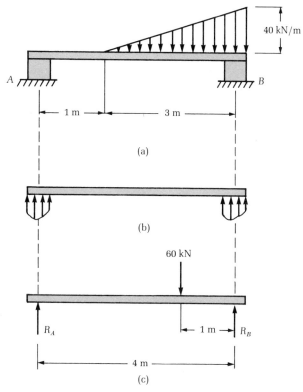

Figure 1-11

SOLUTION

The reactions exerted by blocks A and B on the beam are shown in Figure 1-11b as nonuniform distributions of pressure. Ordinarily, the exact distributions of these reactions are unknown. However, as far as external effects are concerned, these reactions can be replaced by statically equivalent concentrated forces such as the ones shown in Figure 1-11c. The magnitudes of these equivalent concentrated forces are equal to the area under the force distributions with which they are associated. Their action lines pass through the centroids of these distributions.

The exact locations of the centroids of these distributions are not known. However, because the dimension of either block parallel to the axis of the beam is small compared to the length of the beam, only a small error is introduced by assuming that the action lines of the equivalent concentrated forces act through the centers of the blocks.

The external distributed load is also replaced by an equivalent concentrated force shown in Figure 1-11c. In this case, the exact location of its action line is known.

The concentrated reactions R_A and R_B are calculated as follows:

$$\overset{\curvearrowright}{+}\ \Sigma M_A = 0: \quad 4R_B - 60(3) = 0, \qquad R_B = 45 \text{ kN}$$

$$\overset{\curvearrowleft}{+}\ \Sigma M_B = 0: \quad 4R_A - 60(1) = 0, \qquad R_A = 15 \text{ kN}$$

To check these computations, observe that the forces acting on the beam must satisfy the equilibrium requirement

$$+\uparrow \Sigma F_y = 0: \quad 15 + 45 - 60 = 0$$

PROBLEMS / Section 1-1

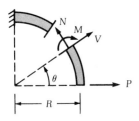

Figure P1-1

1-1   Determine formulas for the internal reactions at the cross section θ of the circular ring shown in Figure P1-1.

1-2   A steel bar is bent into the configuration shown in Figure P1-2. It is firmly attached to a rigid support at its left end. Establish formulas for the internal reactions at the section shown in the figure.

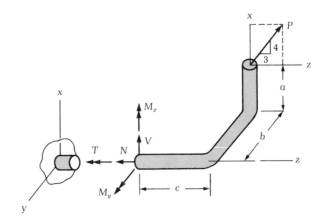

Figure P1-2

1-3   Establish formulas for the internal reactions acting at the base of the compression block shown in Figure P1-3.

1-4   Determine the shearing force acting on each rivet at the interface between the two plates of the riveted joint shown in Figure P1-4. Assume that the load applied to the joint divides equally among the rivets.

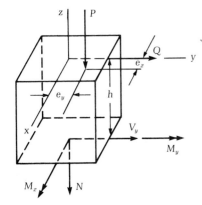

Figure P1-3

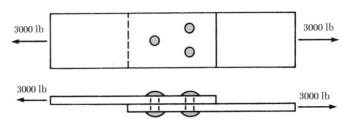

Figure P1-4

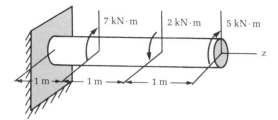

Figure P1-5

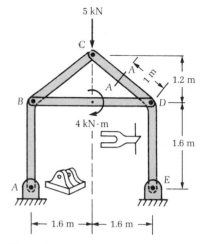

Figure P1-6

1-5   Determine the internal reactions acting on a cross section of the circular shaft shown in Figure P1-5 2.5 m from its free end.

1-6   Three members are connected by smooth pins at points B, C, and D as shown in Figure P1-6. The resulting structure is supported at points A and E by smooth pin connectors. Determine the internal reactions at section AA due to the 5-kN force and the 4-kN·m couple that acts in the plane of the structure.

1-7   Use the method joints to determine the internal reactions in members ED and EF of the pin-connected truss of Figure P1-7. Use the method of sections to determine the internal reactions in members FG, FC, and DC.

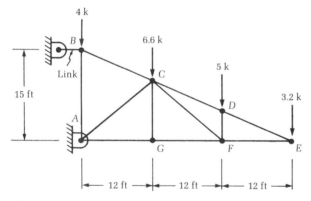

Figure P1-7

1-8   Determine the external reactions at the supports for the simply supported beam shown in Figure P1-8. Calculate the internal reactions at a section 12 ft from the left end of the beam.

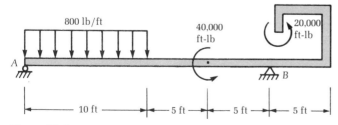

Figure P1-8

1-9 Determine the external reactions at supports A, C, and D of the beam shown in Figure P1-9. Determine the internal reactions at a section 5 m from the right end of the beam. The segments AB and BD are connected at point B by a smooth pin.

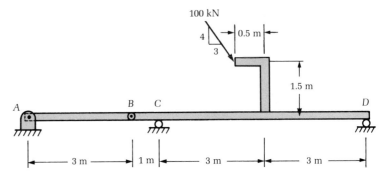

Figure P1-9

1-10 Determine the internal reactions at section AA of the engineering member shown in Figure P1-10.

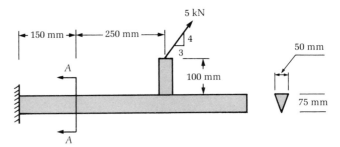

Figure P1-10

1-2 Average Normal, Shearing, and Bearing Stresses

After we determine the components of the equivalent force and couple that act on a particular cutting plane, the design process requires us to establish the stress distributions associated with them. Consequently, a major objective of this book is to develop approximate but reliable formulas that connect the components of the internal force and couple with the stress distribution to which they are equivalent.

In this chapter, we develop the concepts of average normal stress, average shearing stress, and average bearing stress.

Average normal stress. The normal force that acts on any cutting plane produces a *normal stress* (either tension or compression) that is statistically uniform over the area of the cutting plane, provided that the normal force acts at the centroid of the area. Consequently, if N denotes the component of **R** that is perpendicular to area A, then the normal stress that acts on A is calculated from the formula

$$\sigma_{ave} = \frac{N}{A} \tag{1-6}$$

Average bearing stress.   Bearing stress is actually a special form of normal stress. The normal stress that occurs as the result of two **distinct** structural elements bearing on, or pressing against, one another is called **bearing stress**. If N is the normal force acting between two contacting surfaces, and if A is the area of the contacting surfaces, then the average bearing stress between these surfaces is calculated from Eq. (1-6).

Average shearing stress.   Frequently, the only component of the equivalent force and couple acting on a cutting plane is a single shearing force. This situation exists in the rivets of riveted joints, pins of pin-connected trusses, and other machine members. Because the actual distribution of shearing stress is complicated, we use a simple approximate formula. Thus, if V is the shearing force acting on a bolt, pin, or rivet, the **average shearing stress** acting on a cross-sectional area A of either of these elements is given by

$$\tau_{ave} = \frac{V}{A} \tag{1-7}$$

The use of these formulas is illustrated in the following examples.

EXAMPLE 1-8

Calculate the average normal stress in the $\frac{3}{4}$-in. diameter plunger of the hydraulic cylinder shown in Figure 1-12a. The mechanism that the hydraulic cylinder supports weighs 6000 lb. Also calculate the average shearing stress in the $\frac{1}{2}$-in. diameter pin at A.

SOLUTION

A free-body diagram of member AB is shown in Figure 1-12b. Because the plunger acts as a two-force member, its reaction on the pin at B must be along the axis of the plunger. The components of this reaction are shown in Figure 1-12b.

From the free-body of Figure 1-12b, we find that

$$\curvearrowleft \Sigma M_A = 0: \quad 8\left(\frac{3}{\sqrt{13}}B\right) - 6\left(\frac{2}{\sqrt{13}}B\right) - 4(6) = 0$$

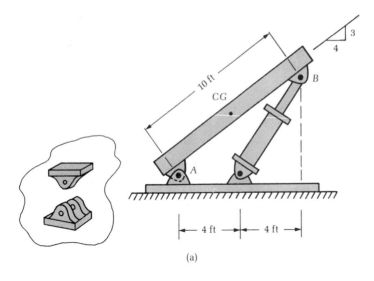

(a)

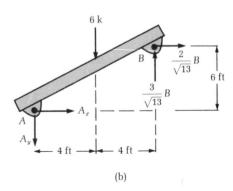

(b)

or

$$B = 2000\sqrt{13} \text{ lb}$$

$$\Sigma F_x = 0: \quad A_x + \frac{2}{\sqrt{13}}B = 0, \qquad A_x = -4000 \text{ lb}$$

$$\Sigma F_y = 0: \quad A_y - 6 + \frac{3}{\sqrt{13}}B = 0, \qquad A_y = 0$$

A free-body diagram of a portion of the plunger is shown in Figure 1-12c. Equilibrium of forces parallel to the axis of the plunger requires that $N = 2000\sqrt{13}$ lb. Accordingly, the average normal stress in the plunger is

$$\sigma_{ave} = \frac{N}{A} = \frac{2000\sqrt{13}}{0.442} = 16{,}315 \text{ psi(C)}$$

The details associated with the pin connection at A are shown in Figure 1-12a. A free-body diagram of the portion of the pin at A that passes through the part of the connection that is attached to member AB is shown in Figure 1-12d. From the free-body diagram, the shearing force $V = 2000$ lb. The average shearing stress in the pin is

$$\tau_{ave} = \frac{2000}{0.196} = 10{,}200 \text{ psi}$$

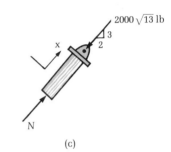

(c)

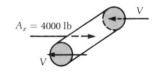

(d)

Figure 1-12

EXAMPLE 1-9

For the bolt shown in Figure 1-13a, determine (a) the average normal stress in the shank of the bolt, (b) the average shearing stress associated with the tendency of the shank to be sheared from the head of the bolt,

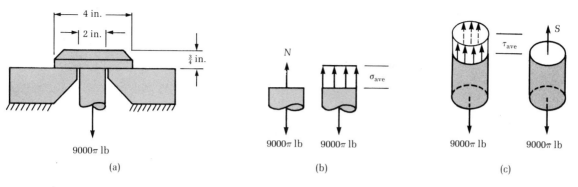

(a)   (b)   (c)

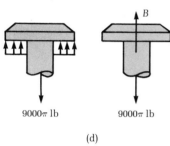

(d)

Figure 1-13

and (c) the average bearing stress between the head of the bolt and the base against which the head bears.

SOLUTION

(a) Figure 1-13b shows a free-body diagram of a portion of the shank. From the first part of this figure, we see that the internal normal force is

$$N = 9000\pi \text{ lb}$$

The average normal stress that corresponds to this internal force is

$$\sigma_{\text{ave}} = \frac{9000\pi}{\pi} = 9000 \text{ psi}(T)$$

and it is shown in the second part of this figure.

(b) The second part of Figure 1-13c is a free-body diagram of the shank of the bolt that is imagined to extend through the head of the bolt. The resultant shearing force S is seen to be

$$S = 9000\pi \text{ lb}$$

Accordingly, the average shearing stress on the shank that extends into the bolt head is

$$\tau_{\text{ave}} = \frac{9000\pi}{\pi(2)(\frac{3}{4})} = 6000 \text{ psi}$$

The distribution of this shearing stress is shown in the first part of this figure.

(c) From the second part of Figure 1-13d, one calculates the resultant bearing force

$$B = 9000\pi \text{ lb}$$

The average bearing stress between the bolt head and the base is

$$\sigma_{\text{ave}} = \frac{9000\pi}{(\pi/4)(4^2 - 2^2)} = 3000 \text{ psi}$$

This distribution is shown in the first part of Figure 1-13d.

EXAMPLE 1-10

The base of a concrete footing is 6 ft^2 and supports a force of 54π kips as shown in Figure 1-14a. The 54π-kips force is transmitted to the concrete footing through a 4-in. standard steel pipe and a 6-in. diameter bearing plate. Calculate (a) the bearing stress between the plate and the concrete and (b) the bearing stress between the concrete and the soil on which it rests.

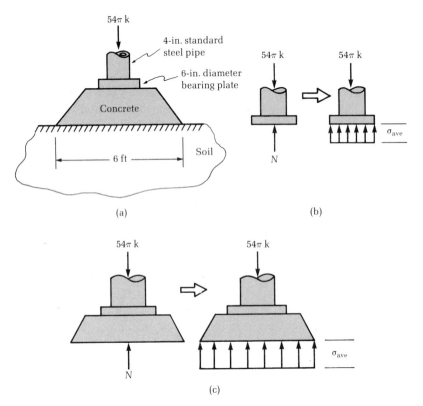

Figure 1-14

SOLUTION

(a) The first part of Figure 1-14b shows the resultant N of the bearing pressure exerted on the bearing plate by the concrete footing. Equilibrium requires that $N = 54\pi$ k. The average bearing stress between the plate and the concrete is

$$\sigma_{ave} = \frac{54\pi}{(\pi/4)(6)^2} = 6 \text{ ksi}$$

This bearing stress distribution is shown in the second part of Figure 1-14b.

(b) The first part of Figure 1-14c shows the resultant N of the bearing pressure exerted on the concrete footing by the soil. Again, equilibrium requires that $N = 54\pi$ k. Consequently, the average bearing pressure between the soil and the concrete footing is

$$\sigma_{\text{ave}} = \frac{54\pi}{72 \times 72} = 32.7 \text{ psi}$$

This bearing stress distribution is shown in the second part of Figure 1-14c.

EXAMPLE 1-11

A hole 25 mm in diameter is to be punched in an aluminum plate 2 mm thick by a punching mechanism such as the one shown in Figure 1-15a. Calculate the average shearing stress in the plate when the punching force is 5 kN.

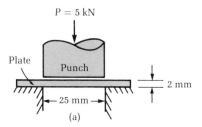

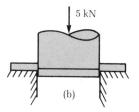

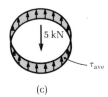

Figure 1-15

SOLUTION

Figure 1-15b indicates the shearing action of the punching mechanism of Figure 1-15a. The circular plug of material directly beneath the punch is removed from the aluminum plate by the shearing action along the cylindrical surface shown in Figure 1-15c. The area associated with this shearing action is

$$A = \pi d \cdot t = \pi(0.025)(0.002) = 1.57 \times 10^{-4} \text{ m}^2$$

The average shearing stress acting on this cylindrical area is

$$\tau_{\text{ave}} = \frac{5000}{1.57 \times 10^{-4}} = 31.9 \times 10^6 \text{ N/m}^2 = 31.9 \text{ MPa}$$

EXAMPLE 1-12

Calculate the shearing stress in the 6-mm diameter pin at the pulley shown in Figure 1-16a. Determine the average bearing stress between the pulley and the pin. The hub of the pulley is 12 mm thick.

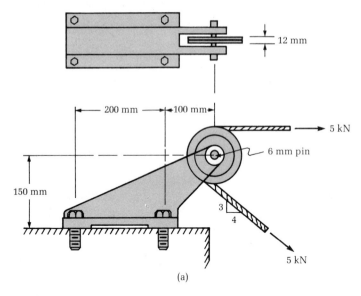

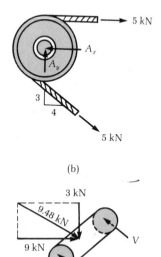

(a)

(b)

(c)

Figure 1-16

SOLUTION

A free-body diagram of the pulley with the pin removed is shown in Figure 1-16b. The reaction of the pin on the pulley is shown in component form. Equilibrium requires that

$$\overset{+}{\longleftarrow} \Sigma F_x = 0: \quad -A_x + 5 + 4 = 0, \qquad A_x = 9 \text{ kN}$$
$$\Sigma F_y = 0: \qquad A_y - 3 = 0, \qquad A_y = 3 \text{ kN}$$

Consequently, the pin reaction on the pulley has the magnitude

$$A = \sqrt{9^2 + 3^2} = 9.48 \text{ kN}$$

A free-body diagram of the portion of pin that is in contact with the inner surface of the hole in the pulley is shown in Figure 1-16c. The shearing force V that acts on a cross section of the pin is obtained from force equilibrium. Thus

$$2V = 9.48 \text{ kN}$$

or

$$V = 4.74 \text{ kN}$$

The average shearing in the pin is calculated from Eq. (1-7). It is

$$\tau_{\text{ave}} = \frac{4.74 \times 10^3}{9\pi \times 10^{-6}} = 167 \text{ MPa}$$

The average bearing stress between the pulley and the pin is

$$\sigma_{\text{ave}} = \frac{9.48 \times 10^3}{0.006(0.012)} = 131.6 \text{ MPa}$$

Note that, in calculating this average bearing stress, the bearing force is assumed to be uniformly distributed over the projected area of the pin (diameter of pin times the pulley thickness).

PROBLEMS / Section 1-2

1-11   The cross-sectional area of each member of the plane truss shown in Figure P1-11 is 100 mm². Determine the average normal stress acting in each member.

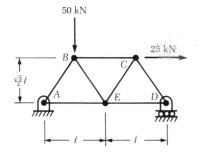

Figure P1-11

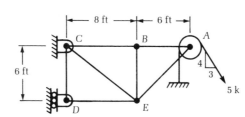

Figure P1-12

1-12   The cantilever truss shown in Figure P1-12 is fitted with a smooth pulley at *A*. Determine the average normal stress induced in members *BC*, *CE*, and *DE* when a 5000-lb force is applied to a rope draped around the pulley as shown in the figure. The cross-sectional area of each member is 0.785 in².

1-13   Two 1-in. diameter rods are connected by pins to form the mechanism shown in Figure P1-13. Determine the average normal stress in each rod.

1-14   Three metallic rods are connected at a common end by a ball-and-socket joint. The other three ends are attached to rigid supports, also by ball-and-socket joints as shown in Figure P1-14. If the cross-sectional area of each member of the assembly is 0.75 in², determine the average normal stress in each member.

1-15   Determine the shearing stress in the smooth pins at points *A* and *D* of the mechanism shown in Figure P1-15. Details of the connection at points *A* and *D* are also shown in the figure. The diameter of the pins at *A* and *D* is 25 mm.

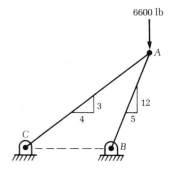

Figure P1-13

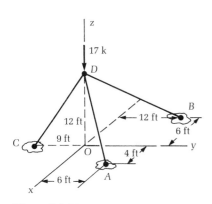

Figure P1-14

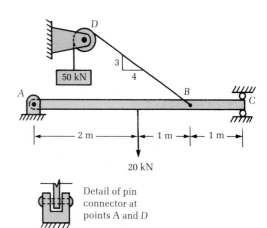

Figure P1-15

1-16   Determine the shearing stresses in the pins at B and C of the mechanism shown in Figure P1-16. These pins act in double shear as shown in the insert of Figure P1-15. The diameter of each pin is 20 mm.

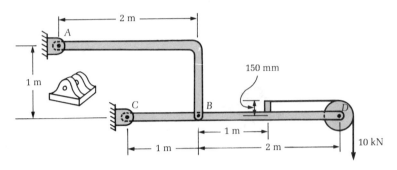

Figure P1-16

1-17   Determine the shearing stress in the pin at D and the tensile stress in the link AB of the bolt cutters shown in Figure P1-17. The diameter of the pin at D is 6 mm and the cross section of the link AB is a 5-mm × 20-mm rectangle. The pin at D acts in double shear.

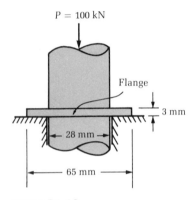

$P = 100$ kN

Flange

3 mm

28 mm

65 mm

Figure P1-18

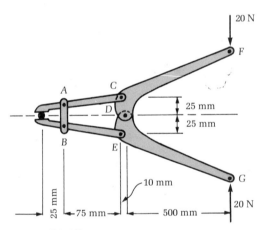

Figure P1-17

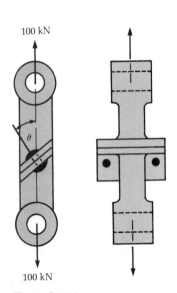

100 kN

θ

100 kN

Figure P1-19

1-18   Determine the shearing stress between the flange and the shaft of Figure P1-18. Also calculate the bearing stress between the flange and the support.

1-19   The coupling device shown in Figure P1-19 supports an axial force of 100 kN. The bolts used in the coupling are 15 mm in diameter at the root of their threads. Calculate the average normal and shearing stresses on a cross section of a bolt perpendicular to its axis when the flange angle $\theta = 0°$, 30°, and 90°.

1-20 Determine the normal stress in member *DE* (cross-sectional area = 125 mm²) and the shearing stress in the smooth pin at *C* of the plane frame shown in Figure P1-20. A detail of the pin connection at point *C* is shown in the insert near *C*. The diameter of the pin at *C* is 6 mm.

1-21 Determine the shearing stress in the $\frac{1}{2}$-in. diameter pin at *C* and the average normal stress in member *BD* of the plane frame shown in Figure P1-21. The cross-sectional area of member *BD* is 0.25 in², and the pin at *C* acts in double shear.

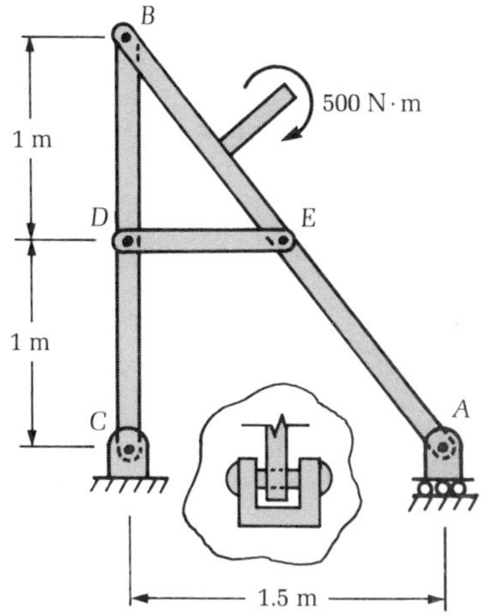

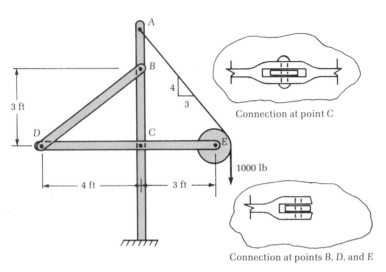

Connection at point *C*

1000 lb

Connection at points *B*, *D*, and *E*

Figure P1-21

500 N·m

1 m

1 m

1.5 m

Figure P1-20

1-3 SUMMARY

In this chapter, we used the methods of statics to express the components of the internal resultant force and the components of the internal couple in terms of externally applied forces and couples. We defined the average normal stress, the average bearing stress due to the normal component of the resultant force, and the average shearing stress due to the shearing component of the resultant.

PROBLEMS / CHAPTER 1

1-22 Determine the average normal stress in member *AB* and the average shearing stress in the pin at *C* of the mechanism shown in Figure P1-22. The cross-sectional area of member *AB* is 0.2 in², and the diameter of the pin at *C* is 0.5 in. Details of the pin connections at points *A*, *B*, and *C* are shown in Figure P1-22.

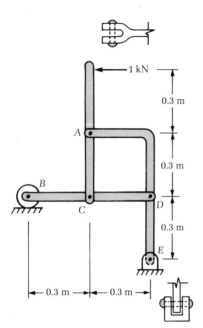

Figure P1-23

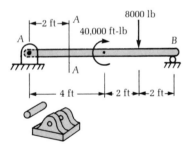

Figure P1-24

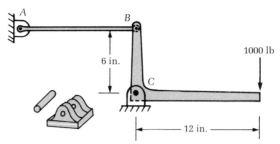

Figure P1-22

1-23 Determine the average shearing stress in the 10-mm diameter pin at A of the pin-connected frame shown in Figure P1-23.

1-24 Determine the average shearing stress in the 0.5-in. diameter pin at A of the beam shown in Figure P1-24. Determine the internal reactions at section AA.

1-25 Determine the internal reactions acting on a cross section of the drill bit shown in Figure P1-25.

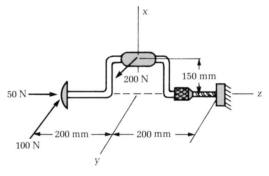

Figure P1-25

1-26 Determine the normal stresses in members AG, AB, and GC of the pin-connected truss shown in Figure P1-26 if the cross-sectional area of each member is 4909 mm^2.

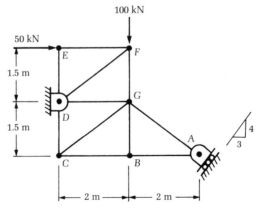

Figure P1-26

1-27 Determine the tensile stress in the 0.25-in. cables of the device shown in Figure P1-27. Also calculate the compressive stress in the spacer bar. The cross-sectional area of the spacer bar is 0.2 in^2.

1-28 Determine the internal reactions acting on a cross section perpendicular to the horizontal segment of the bar shown in Figure P1-28. Assume that the normal force acts at the centroid of the cross-sectional area.

1-29 Determine the internal reactions acting on the plane AA of the shaft shown in Figure P1-29.

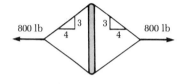

Figure P1-27

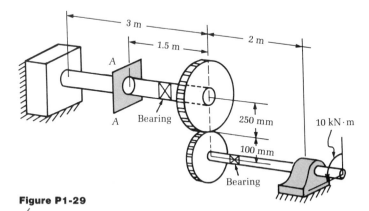

Figure P1-29

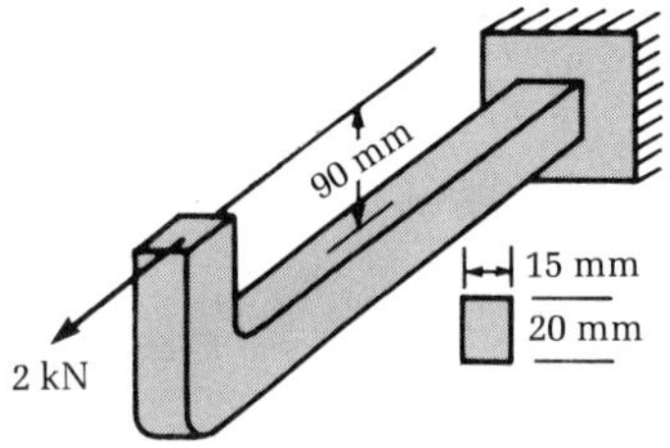

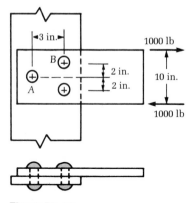

Figure P1-28

1-30 The cross-sectional area of each rivet of the riveted joint shown in Figure P1-30 is 0.2 in^2. If the force acting on any rivet because of the applied couple is assumed to vary linearly with its distance from the centroid of the rivet areas, determine the shearing force acting on rivets A and B. Calculate the shearing stress in these rivets and the bearing stresses in the rivet holes.

1-31 Determine the internal reactions acting at sections AA and BB of the standard steel pipe shown in Figure P1-31.

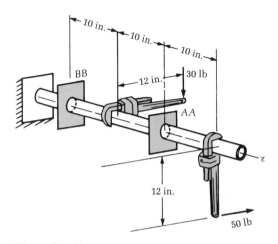

Figure P1-31

Figure P1-30

CHAPTER 2

Material Behavior—
Axially Loaded Members

2-1 Introduction

An engineering member is **designed for strength** when the maximum permissible load that can be applied to it causes stresses that do not exceed a specified limit. This limiting quantity has units of stress and is a property of the material from which the member is made. It is alternatively referred to as **allowable stress**, **working stress**, or **design stress**. Actually, it is desirable to distinguish between the stress induced in an engineering member because of the forces that are applied to it and the **strength of the material**. Allowable stress, working stress, or design stress refers to the strength of the material, not necessarily to the stress induced by the applied forces.

Frequently, the deflections of an engineering member must be restricted to acceptable values. In this case, the member is said to be **designed for stiffness**. A material property different from the strength properties described in the previous paragraph is required for this purpose.

In this chapter, we describe the traditional procedures used to establish the strength and stiffness properties of engineering materials.

The chapter closes with a detailed development of the equations pertinent to the design for strength and stiffness of axially loaded rods.

2-2 Engineering Stress-Engineering Strain

To effectively design an engineering member to perform a specified purpose requires certain parameters that characterize the response of the material from which the member is made to the loads it is expected to carry. Many important **material properties**, as these parameters are called, are obtained from graphs of stress versus strain.

Stress-strain diagrams are usually obtained from standard tension tests performed on a specimen made from the material whose properties are sought. The American Society for the Testing of Materials (ASTM) specifies precisely the dimensions and construction of standard tension specimens. Figure 2-1 depicts a standard circular specimen and indicates some of the more important dimension details.

The ordinate of a stress-strain diagram is the average stress acting on a cross section perpendicular to the longitudinal axis of the specimen. Accordingly,

$$\sigma_{\text{ave}} = \frac{N}{A_0} \tag{2-1}$$

where N is the force normal to a cross section of the test specimen, and A_0 is the area of the cross section prior to load application. We shall learn that the cross-sectional area changes as the applied load is increased. Accordingly, Eq. (2-1) is only an approximation, but a very satisfactory one for most engineering purposes. For this reason, the stress as given by Eq. (2-1) is referred to as engineering stress. Later we shall define the true stress acting on the cross-sectional area and determine the relationship between the engineering stress and the actual or true stress.

The abscissa of a stress-strain diagram is the average axial strain that occurs over a prescribed length of the test specimen. Points A

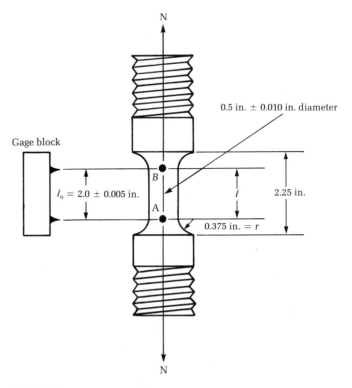

Figure 2-1

and B of Figure 2-1 represent marks made a prescribed distance apart on the test specimen. They are usually made by a so-called gage block that is nothing more than a heavy piece of steel with two hardened tips separated by the prescribed gage length. Most tests in the United States are made using a 2-in. gage length; however, an 8-in. length is also common. The average axial strain over the gage length of a test specimen is calculated from the formula

$$\epsilon_{ave} = \frac{\ell - \ell_0}{\ell_0} \tag{2-2}$$

where ℓ is the distance between the gage marks for the current value of applied load, and ℓ_0 is the initial or original gage length. The distance ℓ is customarily measured by an extensometer that can be either purely mechanical (a dial indicator and appropriate linkages) or electrical (displacement transducer). Because the average strain is based on the original gage length ℓ_0, in analogy to engineering stress, it is referred to as engineering strain. Later we shall define the true strain associated with the test specimen, and we shall obtain a relationship between true strain and engineering strain.

To obtain a stress-strain diagram for a specific material, we must prepare a test specimen made from this material according to ASTM standards for tension tests. Appropriate gage marks are inscribed, and an extensometer is attached to the specimen. The specimen is then placed in a testing machine. It is important for a specimen to be aligned in the testing machine so that the load is applied along its axis; otherwise the average stress as calculated from Eq. (2-1) will be erroneous because the eccentricity of the load will cause nonuniform stresses. Once the specimen has been properly aligned in the testing machine, the load is applied in convenient increments, and the distance between the gage marks is recorded for each successive load-level. For each load-level, the engineering stress and the engineering strain are calculated from Eqs. (2-1) and (2-2), respectively, and the corresponding engineering stress versus engineering strain diagram can be obtained for the material.

Figure 2-2 shows a typical hydraulic testing machine with a tensile specimen installed. An electronic extensometer is shown attached to the specimen. Figure 2-3 is a close-up view of the area around the tensile specimen showing the details of the extensometer and the specimen grips.

Figures 2-4 through 2-6 show the stress-strain diagrams for three different materials: ordinary structural steel, a magnesium alloy, and gray cast iron. The stress-strain diagrams for other engineering materials—such as aluminum, brass, copper, wood, glass, plastics, and concrete—are obtained in a similar manner.

With the use of the stress-strain diagrams, we can categorize several material properties as **strength properties** in the sense that they serve as limiting quantities that the induced stresses must not exceed.

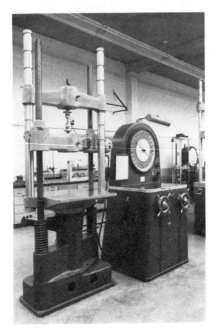

Figure 2-2

Figure 2-3

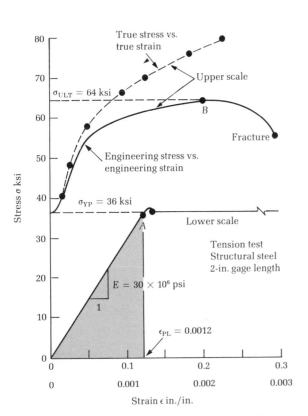

Figure 2-4

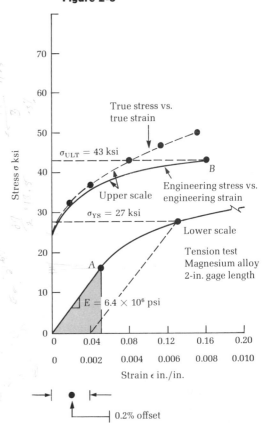

Figure 2-5

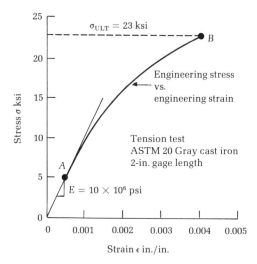

Figure 2-6

Proportional limit.   The proportional limit of a material is defined as the maximum stress for which stress remains proportional to strain, and it is usually denoted by σ_{PL}. It is the stress that corresponds to the last point on the initial linear portion of a stress-strain diagram. The proportional limit is signified by point A on each of the stress-strain diagrams shown in Figures 2-4 through 2-6.

Yield point.   The stress-strain diagrams for some materials possess a region for which increasing strain continues to occur with no apparent increase in stress. Structural steel exhibits this behavior. The stress that corresponds to this flat region of the stress-strain diagram that follows immediately after the initial linear portion is called the **yield point stress** for the material. It is signified by σ_{YP}. Figure 2-4 depicts this strength property. The flat region itself is called the **yield plateau**.

Yield strength.   The stress-strain diagrams for many materials do not exhibit a segment for which increasing strain continues to occur without an apparent increase in stress. Consequently, these materials do not possess a well-defined yield point. For these materials, an analogous strength property, called the **yield strength**, is defined; it serves in a capacity similar to that of yield stress. Yield strength is designated by σ_{YS}. Figures 2-5 and 2-6 depict materials for which no yield stress can be discerned. The yield strengths for these materials are obtained by the so-called **offset method**.

The offset method stipulates that the yield strength of a material be taken as the stress corresponding to a specified amount of permanent strain. The specified permanent strain is usually expressed as a percentage of strain. For example, the specified permanent strain for steel is usually about 0.2% strain. Accordingly, this amount of strain (0.002 in./in.) is measured along the abscissa of the stress-strain diagram of the material, and a straight line parallel to the initial straight line portion of the diagram is drawn through this abscissa. The stress corresponding to the intersection of this line with the graph of stress versus strain is taken as the yield strength of the material. This procedure is depicted in Figure 2-5 for a magnesium alloy. Permanent strain is discussed later in this section.

Ultimate strength. The engineering stress that corresponds to the highest point of the stress-strain diagram is called the **ultimate strength** of the material and is denoted by σ_{ULT}. The ultimate strength associated with structural steel, the magnesium alloy, or gray cast iron is designated as point B in Figures 2-4, 2-5, or 2-6, respectively.

Elastic limit. Another strength property frequently referred to in the literature is the **elastic limit stress**, which is signified by σ_{EL}. The elastic limit of a material is defined as the first point on the stress-strain diagram for which the material becomes inelastic. The elastic limit of a material does not technically correspond to the proportional limit. While the material must behave elastically to the elastic limit, the two points differ in that the material has ceased to be linear beyond the proportional limit. Between these two points, the material is said to behave nonlinearly elastically. Special care must be exercised to establish the elastic limit. Moreover, the magnitude of the stresses associated with the elastic and proportional limits do not differ by a significant amount for metals. For these reasons, there is no practical need to distinguish between the two stresses. Consequently, in tables listing the properties of materials, the terms **elastic limit stress**, **proportional limit stress**, and **elastic-proportional limit stress** are used interchangeably. The elastic limit is hardly ever determined as a design property.

The strength parameters (σ_{PL}, σ_{EL}, σ_{YP}, σ_{YS}, and σ_{ULT}) discussed in the foregoing paragraphs are all obtained from an engineering stress versus engineering strain diagram. It will soon be seen that some of these parameters do not relate well to the true stresses associated with them. Nevertheless, they are used extensively in the design and analysis of engineering members.

Elastic modulus. The modulus of elasticity of a material characterizes its capability to resist extensional deformations. The modulus of elasticity of a material is defined as the slope of the straight-line portion of the stress-strain diagram and is shown in Figures 2-4 through 2-6. The modulus of elasticity is denoted by E and is a measure of the stiffness of the material. Note that E has the same units as stress does since it is the ratio of stress to strain in the linear region of the stress-strain diagram. This linear relationship between stress and strain is known as *Hooke's law*. Symbolically,

$$\sigma = E\epsilon \tag{2-3}$$

Moreover, E is a relatively large number: 30×10^6 psi (210 GPa) for steel; 10×10^6 psi (70 GPa) for aluminum; 1.5×10^6 psi (10.5 GPa) for oak; and 0.5×10^6 psi (3.5 GPa) for plexiglass.

Modulus of resilience. The modulus of resilience of a material is the area under the portion of the stress-strain diagram that is de-limited by the proportional limit. Clearly, the modulus of resilience has units of energy per unit of volume (in.-lb/in^3). Consequently, the modulus of resilience is a measure of the capacity of the material to absorb energy without undergoing permanent deformations. The mod-uli of resilience for the structural steel shown in Figure 2-4 and for the magnesium alloy shown in Figure 2-5 are approximately 25 in.-lb/in^3 and 21 in.-lb/in^3, respectively. The modulus of resilience is usually used to compare the capacities of different materials to absorb elastic energy. The area associated with modulus of resilience for these ma-terials is shown by the shaded area in Figures 2-4 and 2-5.

Toughness. Toughness is defined to be the total area under the stress-strain diagram. It is a measure of the capability of a material to undergo large, permanent deformations prior to fracture. Its units are the same as those for the modulus of resilience.

Elasticity. If the stress in a tension specimen is increased to a magnitude not exceeding the elastic limit σ_{EL} of the material from which it is made, and if it is subsequently released, the material will unload along the loading path OA as shown in Figure 2-7. This be-havior, wherein the material returns to its original shape and di-mensions, is referred to as elastic behavior or, simply, **elasticity.**

On the other hand, if the stress exceeds the elastic limit of the material, such as occurs at point B in Figure 2-7, and if it is subse-quently released, the material unloads along a straight line BC that is essentially parallel to the initial straight line portion of the stress-strain diagram. The strain that remains after the stress has been reduced to zero is permanent deformation and is called **plastic** or **inelastic** deformation. It is usually designated as ϵ_P.

Strain hardening. When a tensile stress is reapplied, the ma-terial will load along line CB to point B and then proceed along line BD. The additional elastic strain that occurs during the reloading process between points C and B is designated as ϵ_E. Evidently, the proportional limit of the material has been increased by this so-called **strain-hardening** effect. Clearly, strain-hardening increases the elas-tic strength of the material (proportional limit) and its modulus of resilience; however, this increase has occurred at the expense of the toughness of the material.

Ductile behavior. A material is said to behave in a **ductile manner** if it can undergo large, permanent deformations before frac-ture. Thus toughness is a measure of the ductility of a material: larger

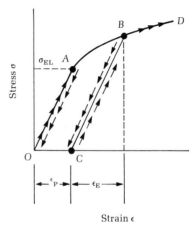

Figure 2-7

values of toughness signifying greater ductility, smaller values signifying diminishing amounts of ductility. Note that it is improper to refer to a material as a "ductile material." Whether a material behaves in a ductile manner depends on the temperature to which the material is subjected, the type of loading that is applied to it, and the rate at which the load is applied.

Brittle behavior.   A material is said to behave in a **brittle manner** if it undergoes only imperceptible inelastic deformations prior to fracture. This definition implies the lack of ductile behavior. Glass and ceramic materials are familiar examples of materials that behave in a brittle manner under ordinary conditions.

We can readily see that ductile behavior depends on the thermal environment in which the material is expected to function by observing that ordinary baling wire is extremely ductile at room temperature, but it becomes increasingly more brittle as the temperature approaches $-200°$ F. This behavior can be exhibited by placing a piece of baling wire in liquid nitrogen and subsequently attempting to bend it.

We can see that ductile behavior depends on the type of loading applied to the material by noting that cast iron or plexiglass behaves in an extremely ductile manner in compression, but it reacts in a brittle manner to tensile loads.

Finally, we can see that ductile behavior depends on the rate of loading by observing that "silly putty" undergoes very large permanent deformations when it is pulled slowly. If we want to break Silly Putty, we merely jerk it, thus applying stress rapidly.

Percent elongation and percent reduction in area.   There are two parameters besides toughness that are used to describe qualitatively the ductility of a material. They are **percent reduction in area** and **percent elongation**, and they are defined by the relations

$$\% \text{ reduction in area} = \frac{A_0 - A_f}{A_0} \times 100 \tag{2-4}$$

where A_f is the cross-sectional area of the test specimen after fracture has occurred, and

$$\% \text{ elongation} = \frac{\ell_f - \ell_0}{\ell_0} \times 100 \tag{2-5}$$

where ℓ_f is the distance between gage points when fracture occurs. This final length is usually obtained by fitting the separated parts of the test specimen together carefully and measuring the resulting distance between gage points. Since most of the inelastic deformation at fracture occurs over a very small part of the test length, it is essential to specify the gage length for which the percent elongation is calculated.

Isotropy.   A material is said to be isotropic if its elastic properties are the same in all directions. Most engineering materials are made from crystalline substances such as steel, brass, aluminum, copper, and titanium, or substances with small fibers such as wood. The dimensions of the crystals are small compared to the dimensions of the entire body, and they are randomly oriented so that the behavior of the material on the average is isotropic. Materials such as wood are not isotropic because their elastic properties differ significantly for directions parallel and perpendicular to the grain. Even metals that are ordinarily statistically isotropic can acquire directional properties through rolling or extruding processes because these processes tend to orient the crystals in preferred directions.

Homogeneity. A material is said to be homogeneous if its elastic properties are the same throughout the body. Most engineering materials can be considered to be homogeneous. Again, the crystals or fibers of a material are small compared to the dimensions of the body so that the elastic properties are statistically the same at each point in the body. Concrete is obviously not a homogeneous material because the elastic properties of the rock or gravel differ significantly from those of cement. Nevertheless, if a large enough sample of material surrounding a given point is considered, its elastic properties can be considered to be statistically homogeneous. We restrict developments in this text to homogeneous or nearly homogeneous materials.

PROBLEMS / Section 2-2

2-1 Data from a standard tension test of a 7075-T6 aluminum alloy are shown in Figure P2-1. The diameter of the tensile specimen was 0.5 in., and a gage length of 2 in. was used. Construct the stress-strain diagram for this material and from it determine (a) modulus of elasticity, (b) proportional limit stress and strain, (c) yield strength at 0.2% strain, (d) ultimate strength, (e) modulus of resilience, (f) toughness modulus, and (g) percent elongation.

2-2 The load-extension data obtained in a standard tension test is listed in Figure P2-2. The aluminum tension specimen had a diameter of 0.505 in. in the test section, and extensions were measured over an 8-in. gage length. Construct the stress-strain diagram for this material and determine (a) its modulus of elasticity, (b) its yield strength at 0.2% strain, and (c) its ultimate strength.

2-3 The load-extension data obtained for a structural steel test specimen is listed in Figure P2-3. The diameter of the test specimen was 17.5 mm and the extension measurements correspond to a 200-mm gage length. At fracture, the gage length was 256 mm and the diameter at the fracture section was 12 mm. Construct the stress-strain diagram for this material and determine (a) modulus of elasticity, (b) proportional limit stress, (c) yield point stress, (d) ultimate stress, (e) modulus of resilience, (f) toughness modulus, (g) percent elongation, and (h) percent reduction in area.

Load P (lb)	Change in gage length (in.)
0	0
1000	0.00102
2000	0.00204
3000	0.00306
4000	0.00408
5000	0.00510
6000	0.00612
7000	0.00712
8000	0.00816
9000	0.00918
10,000	0.01020
11,000	0.01122
12,000	0.01224
13,000	0.01327
14,000	0.01460
15,000	0.01900
16,000	0.2400
15,000	0.3240

Figure P2-1

Load (lb)	Extension (in.)
0	0.00258
400	0.00405
800	0.00574
1600	0.00870
2400	0.01165
3200	0.01537
4000	0.01940
4800	0.02499
5200	0.02980
5600	0.03780
6000	0.06910
6200	0.09310
6240	Fracture

Figure P2-2

Load (kN)	Extension (mm)	Load (kN)	Extension (mm)
6.3	0.025	68.5	0.500
12.9	0.050	68.5	2.5
19.1	0.075	86.1	5.0
24.7	0.100	100.7	10.0
31.1	0.125	107.4	15.0
36.9	0.150	110.8	20.0
43.1	0.175	112.5	25.0
48.9	0.200	113.9	30.0
55.2	0.225	114.3	35.0
60.7	0.250	114.3	40.0
66.7	0.275	113.4	45.0
71.8	0.300	105.0	50.0
71.2	0.325	81.4	55.0
69.4	0.350	79.2	Fracture
68.5	0.375		

Figure P2-3

2-3 Safety Factor-Allowable Stress

The existence of any engineering structure—whether it is an airplane, a bridge, a rifle, an elevator, a ship, a machine tool, or a lawn mower—raises the question of its potential to do harm if it should cease to function as it is intended. Safety, to a layperson, implies a relationship between personal well being and the functional behavior of the structure. To a designer, or analyst, safety is a concept that measures the **degree of probability** that the structure will perform its intended function. A designer attempts to design a structure so that the **probability of a failure** occurring for a stipulated **set of design conditions** is as low as is economically feasible; at the same time, it must be consistent with acceptable design procedures and structural esthetics.

The design of a structure requires the designer to establish the conditions the structure must satisfy. These conditions include the magnitude, direction, and type (static, dynamic, fatigue) of forces to which the structure is likely to be subjected, the strength and stiffness properties for each material to be used in the design, and the analytical formulas to be used to predict stresses and deflections in the members of the structure.

Each design condition injects an element of uncertainty into the design. The degree of uncertainty depends on how accurately the design condition can be established. For example, dead loads to which a bridge is subjected can be determined with a high degree of accuracy, but live loads such as wind and snow loads and dynamic effects of vehicular traffic cannot be predicted so accurately. Material strength and stiffness properties are only representative, and they may vary from those associated with the material used in the bridge. At best, design formulas predict only approximate values for stresses and displacements. Some predicted values are more reliable than others and, therefore, they inject uncertainties into a design in relation to their reliability.

To ensure an acceptable degree of reliability for a design, designers establish an allowable stress according to the collective degree of uncertainty they feel is inherent in the design by dividing one of the strength properties by a number called a **safety factor**. The magnitude of the safety factor depends on the degree of uncertainty inherent in the design. It is denoted by SF in this text. Accordingly, the allowable stress is calculated from the general formula

$$\text{Allowable stress} = \frac{\text{Strength property}}{\text{Safety factor}} \qquad (2\text{-}6)$$

When designers specify a safety factor, they must also specify the strength property to which it applies. For example, an allowable stress should be calculated using a safety factor of 3 based on the ultimate tensile strength of the material or a safety factor of 2 based on the yield strength at 0.2% offset. The magnitude of the safety factor depends on the amount of uncertainty associated with the various factors that enter into the design of a particular structure.

Frequently, the safety factor is applied to the loads that are applied to the structure. In this case, the **expected or actual loads** are multiplied by the safety factor and the allowable stress is taken to be equal to an appropriate strength parameter.

Application of the safety factor to a strength property or to the applied loads leads to identical designs whenever the stresses and the applied loads are linearly related. This condition prevails in all material in this text except for the buckling of columns in Chapter 9.

PROBLEMS / Section 2-3

2-4   A carbon hardened-steel ($\sigma_{\text{ULT}} = 66$ ksi) rod is required to carry an axial tensile force of 10,000 lb. Determine the required cross-sectional area if a safety factor of 3 with respect to fracture is required.

2-5   Determine the maximum permissible load that a structural steel bar 2 m long with a rectangular cross section (25 mm × 50 mm) can safely carry if the allowable tensile stress is 130 MPa and the elongation of the bar is not to exceed 1 mm ($E = 210$ GPa).

2-6   For the structural steel whose stress-strain diagram is shown in Figure 2-4 determine (a) the allowable stress corresponding to a safety factor of 2 with respect to the yield stress and (b) the allowable stress corresponding to a safety factor of 3 with respect to the ultimate strength.

2-7   For the magnesium alloy whose stress-strain diagram is shown in Figure 2-5, determine (a) the allowable stress corresponding to a safety factor of 3 with respect to its ultimate strength, (b) the allowable stress corresponding to a safety factor of 2 with respect to the yield strength at 0.2% strain, and (c) the allowable stress corresponding to a safety factor of 2 with respect to the proportional limit stress.

2-4 Poisson's Ratio

If we also measure the diameter of the specimen at each load-level during the tension test described in Section 2-2, we can calculate the strain perpendicular to the direction of the applied load. This lateral strain, signified by ϵ_l, is given by the formula

$$\epsilon_l = \frac{d - d_0}{d_0}, \tag{2-7}$$

where d is the diameter of the specimen for the current value of applied load, and d_0 is the initial or original diameter. Recall that the average axial strain (strain parallel to the direction of the load) is given by the relation

$$\epsilon_a = \frac{\ell - \ell_0}{\ell_0} \tag{2-8}$$

Figure 2-8 shows a graph of $|\epsilon_l|$ versus $|\epsilon_a|$. The relationship between $|\epsilon_l|$ and $|\epsilon_a|$ is found to be linear for most materials, and the proportionality factor is called **Poisson's ratio**. Consequently, $\epsilon_l = -\nu\epsilon_a$. In this book, it will be denoted by the Greek letter ν. Physically, it is observed that the strain perpendicular to the direction of load application is a contraction when the axial strain is **tensile**, an elongation when the axial strain is **compression**; that is, the lateral **strain** has a sign opposite that of the axial strain. Poisson's ratio for most engineering materials ranges from $\nu = 0.25$ to $\nu = 0.30$. The maximum value that Poisson's ratio can attain is 0.5, which corresponds to plastic deformation of metals.

It should be emphasized that, in the tension test described in Section 2-2, the axial tensile load causes lateral strain in addition to axial strain; however, the only stress occurring in the specimen is the one acting on a cross section perpendicular to its longitudinal axis. In other words, although a lateral strain occurs in the specimen during the test, no lateral stress develops. No lateral **stress** occurs in the specimen because the lateral **strain** is allowed to occur freely. If lateral deformation (shortening of the diameter of the tension specimen) is prevented, no lateral strain will occur, but lateral stresses will appear.

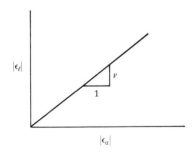

Figure 2-8

EXAMPLE 2-1

An axial force of 31,400 lb applied to a 2-in. diameter brass rod 30 in. long causes the length to increase 0.02 in. and the diameter to decrease

0.00033 in. Determine the modulus of elasticity E and Poisson's ratio for the brass.

SOLUTION

The axial and lateral strains are calculated from the basic definition of engineering strain. Therefore,

$$\epsilon_{axial} = \frac{0.02}{30} = 0.00067 \text{ in./in.}$$

and

$$\epsilon_{lateral} = -\frac{0.00033}{2} = -0.000165 \text{ in./in.}$$

Poisson's ratio for the material is

$$v = \left| \frac{\epsilon_{lateral}}{\epsilon_{axial}} \right| = \left| \frac{-0.000165}{0.00067} \right| = 0.246$$

To compute the modulus of elasticity, first calculate the normal stress corresponding to the applied load and, subsequently, divide this stress by the corresponding strain. Accordingly,

$$\sigma = \frac{31,400}{(\pi/4)(2)^2} = 10,000 \text{ psi}$$

and then

$$E = \frac{\sigma_{axial}}{\epsilon_{axial}} = \frac{10,000}{0.00067} = 14.9 \times 10^6 \text{ psi}$$

PROBLEMS / Section 2-4

2-8 A 2-in. diameter steel rod is subjected to a tensile force of 30,000 lb. Determine the change in diameter of the bar ($E = 30 \times 10^6$ psi and $v = 0.285$).

2-9 An aluminum block ($E = 70$ GPa, $v = 0.3$) with the dimensions shown in Figure P2-9 rests on a rigid surface and is subjected to an axial force of 200 kN through a rigid bearing plate. Assuming that the 200-kN force is uniformly distributed over the cross section of the block, calculate its dimensions after the force has been applied. Determine the change in volume of the block.

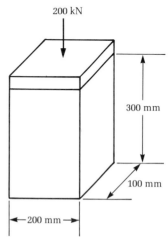

200 kN

300 mm

100 mm

200 mm

Figure P2-9

Figure 2-9

2-5 Necking of Tension Specimen

Tensile tests of materials that behave in a relatively ductile manner indicate that the strain in a specimen is reasonably uniform over the entire length of the test section up to the ultimate strength of the material. Figure 2-4 shows that the engineering stress decreases beyond the ultimate strength of the material and that rupture does not occur until a strain considerably in excess of the strain corresponding to the ultimate stress has been reached. The strain that occurs during this phase tends to be localized over a very short length of the test section leading to the **necking** phenomenon depicted in Figure 2-9. This necking is typical of materials that behave in a ductile manner. Figure 2-9 also shows a fractured tensile specimen of a material that behaves in a brittle manner. Notice that the necking phenomenon is not present for this type of material.

Note that, although the engineering stress at rupture is less than the ultimate strength of a material, the true stress, which is the topic of the next section, continues to increase throughout the tension test and, in fact, acquires its maximum value at rupture as seen in Figure 2-4.

Appendix A lists some pertinent mechanical properties for a few important engineering materials.

2-6 True Stress-True Strain

The stress-strain diagrams discussed in Section 2-2 are graphs of engineering stress versus engineering strain; that is, the ordinate of the stress-strain diagram was calculated using the original cross-sectional area, and the abscissa was obtained using the original gage length. Section 2-5 shows that the cross section of a test specimen decreases significantly under a sufficiently large tensile force so that the actual or so-called true stress is given by the relation

$$\bar{\sigma} = \frac{N}{A} \tag{2-9}$$

where A is the current area corresponding to the current load N.

Experimental evidence shows that the large inelastic strains that occur during the necking process do so in an incompressible manner; that is, they occur without a change in volume. Accordingly,

$$A_0 \ell_0 = A\ell \tag{2-10}$$

Consequently, Eq. (2-9) becomes

$$\bar{\sigma} = \frac{N}{A_0} \frac{\ell}{\ell_0} \tag{2-11}$$

and, since $\epsilon = (\ell - \ell_0)/\ell_0 = \ell/\ell_0 - 1$ and $\sigma = N/A_0$, it follows that

$$\bar{\sigma} = \sigma(1 + \epsilon). \tag{2-12}$$

Eq. (2-12) is a relationship between the true stress $\bar{\sigma}$ and the engineering stress σ and the engineering strain ϵ.

The true strain $\bar{\epsilon}$ is obtained by regarding the total strain as being the sum of a number of increments of strain. Such an increment is defined by the differential relationship

$$d\bar{\epsilon} = \frac{d\ell}{\ell} \tag{2-13}$$

so that the true strain associated with a gage length that changes from ℓ_0 to ℓ is

$$\bar{\epsilon} = \int_{\ell_0}^{\ell} \frac{d\ell}{\ell} = \ln \ell \bigg|_{\ell_0}^{\ell} = \ln \frac{\ell}{\ell_0} \tag{2-14}$$

or

$$\bar{\epsilon} = \ln(1 + \epsilon) \tag{2-15}$$

Eq. (2-15) relates the true strain $\bar{\epsilon}$ to the engineering strain ϵ. Because of the form of $\bar{\epsilon}$ as defined by Eq. (2-14), it is sometimes called **logarithmic strain**.

The true stress-true strain diagram for any material can be determined from the data used to construct the engineering stress-engineering strain diagram by means of Eqs. (2-12) and (2-15). The true stress-true strain diagrams for the structural steel and the magnesium alloy considered in Section 2-2 are shown as dashed curves in Figures 2-4 and 2-5. Note that the difference between the two diagrams is insignificant for stresses and strains up to and slightly beyond the proportional limit of the material; however, the difference becomes increasingly larger as the stress increases beyond this point.

PROBLEMS / Section 2-6

2-10 Construct the true stress-true strain diagram for the 7075-T6 aluminum alloy of Problem 2-1.

2-11 Construct the true stress-true strain diagram for the structural steel of Problem 2-3.

2-12 Construct the true stress-true strain diagram for the aluminum specimen of Problem 2-2.

2-7 Idealized Stress-Strain Diagrams

It is frequently necessary to represent only a portion of the stress-strain diagram that includes its initial part. Moreover, it is impractical to attempt to express the stress-strain curve by an exact mathematical equation because complicated stress-strain relations invariably lead to complicated formulas or theories. Consequently, since the strength-of-materials approach attempts to develop simplified theories that are reasonably reliable, stress-strain diagrams are usually idealized. Some such idealized stress-strain diagrams are illustrated in Figure 2-10. Figure 2-10a depicts an idealization of a material in which elastic deformations are not neglected and which possesses a well-defined yield point. For this material, strain-hardening in the plastic region is assumed to be negligible. Figure 2-10b depicts a so-called linear strain-hardening material that could be used to represent most of the stress-strain diagram for the magnesium alloy of Figure 2-5. Figure 2-10c represents an idealization that could be used to approximate the stress-strain diagram for the gray cast iron shown in Figure 2-6. Finally, Figure 2-10d exhibits the so-called rigidly elastic-ideally plastic material in which elastic deformations and strain-hardening in the plastic region are neglected. In extruding processes, the inelastic strains are frequently, extremely large compared to the elastic strains. Consequently, neglecting the elastic strains in this case is a reasonable approximation.

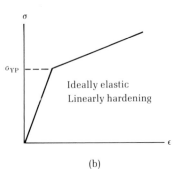

(a)

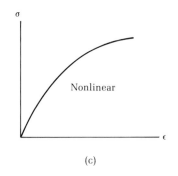

(b)

(c)

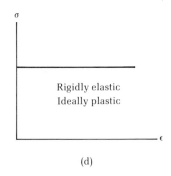

(d)

Figure 2-10

EXAMPLE 2-2

The stress-strain diagram for a certain material is represented closely by two straight lines as shown in Figure 2-11. Write the analytical expressions that relate stress and strain for the range of strain indicated in the figure.

SOLUTION

The stress-strain relation for the initial linear region, $0 \leq \epsilon \leq 0.001$, is

$$\sigma = 30{,}000\epsilon \text{ ksi}$$

The stress-strain relation for the second linear region, $0.001 \leq \epsilon \leq 0.003$, is

$$\sigma = 30 + 15{,}000(\epsilon - 0.001)$$

or

$$\sigma = 15 + 15{,}000\epsilon \text{ ksi}$$

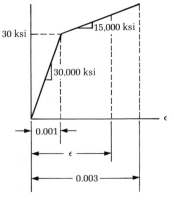

Figure 2-11

It is customary to write these separate relations in the collective form

$$\sigma = \begin{cases} 30,000\epsilon \text{ ksi} & 0 \leq \epsilon \leq 0.001 \\ 15 + 15,000\epsilon \text{ ksi} & 0.001 \leq \epsilon \leq 0.003 \end{cases}$$

EXAMPLE 2-3

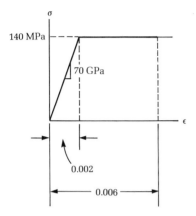

Figure 2-12

The stress-strain diagram for an ideally elastic-ideally plastic material is shown in Figure 2-12. Write the analytical expressions that relate stress and strain for the range of strain indicated in the figure.

SOLUTION

For this ideally elastic-ideally plastic material, we can see that the collective stress-strain relation is

$$\sigma = \begin{cases} 70,000\epsilon \text{ MPa} & 0 \leq \epsilon \leq 0.002 \\ 140 \text{ MPa} & 0.002 \leq \epsilon \leq 0.006 \end{cases}$$

2-8 Empirical Formulas

When a material is stressed beyond the elastic limit, such as to point B in Figure 2-7, the strain can be separated into two distinct parts. As described previously, the part that is recovered when the stress is released is elastic strain and is denoted by ϵ_E. The irrecoverable part of the total strain is called plastic strain or inelastic strain, and it is customarily denoted by ϵ_P. Accordingly, the total strain at point B (Figure 2-7) that is associated with curve $OABD$ is

$$\epsilon = \epsilon_E + \epsilon_P \tag{2-16}$$

The elastic part of the total strain is given by the uniaxial stress-strain relation

$$\epsilon_E = \frac{\sigma}{E} \tag{2-17}$$

where σ is the stress level corresponding to point B. A relationship between the plastic strain component and the stress producing it is usually obtained by empirical methods. The most common empirical formula used for this purpose is the power law

$$\epsilon_P = \left(\frac{\sigma}{K}\right)^n \tag{2-18}$$

where K and n are parameters chosen to obtain the best possible fit of the empirical formula (2-18) with a plot of stress versus plastic strain. The graph of plastic strain versus stress on log-log paper is usually nearly a straight line so that K and n are easily determined from such a graph. Substitution of Eqs. (2-17) and (2-18) in Eq. (2-16) gives the following expression for the total strain.

$$\epsilon = \frac{\sigma}{E} + \left(\frac{\sigma}{K}\right)^n \tag{2-19}$$

2-9 Thermal Strains

The effect of temperature on strain is twofold. First, the elastic constants of a material are functions of temperature; however, this effect is negligibly small for most engineering materials for a large range of temperatures around room temperature. Consequently, it will not be considered in this book. Second, temperature changes produce a strain even in the absence of stress. Temperature increases cause most engineering materials to expand, and temperature decreases cause them to contract. Suppose that gage points are marked on a standard tension specimen at a specified temperature, and that an extensometer is attached to it. If the distance between the gage points is recorded for a series of temperatures, a graph of thermal strain ($\epsilon_T = \Delta\ell/\ell_0$) versus ΔT can be constructed. Here ΔT is the difference between the current temperature of the specimen and the reference temperature T_R for which the gage points were marked on the specimen.

Figure 2-13 depicts a typical graph of thermal strain versus temperature. For most engineering materials, the graph is essentially linear for a large range of temperature around room temperature as a reference. The slope of this graph is denoted by α. It is called the linear coefficient of thermal expansion for the material. It has units in./in./°F. The linear coefficients of thermal expansion for several materials are given in Appendix A.

In the range of temperatures for which the graph of thermal strain versus temperature is linear, the thermal strain accompanying a temperature change is expressed as

$$\epsilon_T = \alpha\,\Delta T \tag{2-20}$$

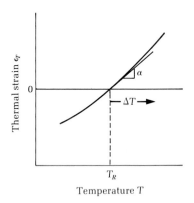

Figure 2-13

When the total strain at a point in a material results from combined mechanical (applied forces) and thermal effects, the mechanical and thermal strains can be added algebraically, provided that the mechanical strain, the thermal strain, and the algebraic sum of the two do not exceed the strain at the proportional limit of the material. Consequently, with the foregoing restrictions, the total strain is

$$\epsilon = \epsilon_{\text{mechanical}} + \epsilon_{\text{thermal}} \tag{2-21}$$

Note that, if one of the components of the total strain as given in this equation is the negative of the other, the total strain becomes zero although the individual components are nonzero. This case corresponds to the case of a rigidly restrained material subjected to a temperature change. This superposition of mechanical and thermal strains within the proportional limit of the material suggests that the mechanical strains can be calculated independently of the thermal strains and vice versa. This approach will be used to solve certain statically indeterminate problems that are subjected to the combined effects of mechanical and thermal loads.

2-10 Shearing Stress-Shearing Strain Diagrams

The mechanical properties of a material are occasionally determined from standard shearing tests. The most common test used to obtain shearing mechanical properties uses a thin-walled tube of thickness t as a test specimen. Such a test specimen is shown in Figure 2-14a. The tube is fitted with a twistometer that measures the relative angle of twist over a prescribed gage length ℓ_0. The twistometer consists of two stiff rings that are positioned at a prescribed distance ℓ_0 apart. The rings are fixed to the tube at sections A and B by a screw assembly such as the one shown in Figure 2-14b. The surface of ring B is graduated in degrees, and a pointer is rigidly attached to ring A as shown in Figure 2-14c. The movement of the pointer in response to twisting action is a measure of the relative rotation of the two rings

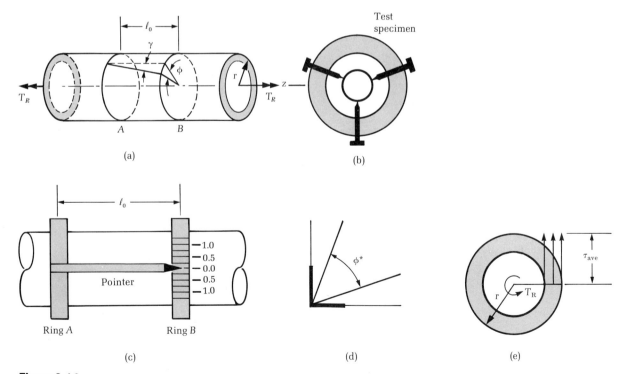

Figure 2-14

and, consequently, of the relative angle of twist of the ends of the shaft contained between the rings.

Figure 2-15 shows a typical torsion testing machine. The relative angle of twist between the sections where the pointers are attached is determined from the graduations on the circular surfaces of the pointers read relative to the thin horizontal wire.

Engineering shearing strain, which is designated by the symbol γ, is the change in angle (in radians) between two initially perpendicular line elements as shown in Figure 2-14d. Accordingly, engineering shearing strain is defined by the relation

$$\gamma = \frac{\pi}{2} - \varphi^* \tag{2-22}$$

where φ^* is the angle between the line elements after deformation. A decrease in the right angle usually represents a positive shearing strain.

Later we shall show that the shearing strain for a thin tube is

$$\gamma = \frac{r\varphi}{\ell_0} \tag{2-23}$$

where γ, r, φ, and ℓ_0 are defined pictorially in Figure 2-14a. Thus, as a torque T_R is applied in increments, the shearing strain can be calculated from Eq. (2-23).

Now, if the tube is thin, the shearing stress is nearly uniform over its thickness as shown in Figure 2-14e. Consequently, static equivalency of T_R and the stress distribution yields

$$T_R = \pi(r_o^2 - r_i^2)\tau_{ave}r_m \tag{2-24}$$

where r_m is the mean radius of the tube, and r_o and r_i are its outer and inner radii. Equation (2-24) can be written as

$$T_R = \pi(r_o + r_i)(r_o - r_i)\tau_{ave}r_m \tag{2-25}$$

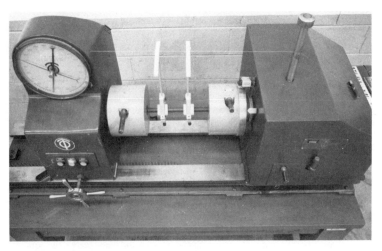

Figure 2-15

Since the tube is thin, the following approximate equation for T_R is obtained

$$T_R = 2\pi r_o^2 t \tau_{ave} \qquad (2\text{-}26)$$

Consequently,

$$\tau_{ave} = \frac{T_R}{2\pi r_o^2 t} \qquad (2\text{-}27)$$

Accordingly, the shearing stress can be calculated from Eq. (2-27) for any torque level.

Shearing stress-shearing strain diagrams are constructed using Eqs. (2-23) and (2-27). A typical shearing stress-shearing strain diagram is shown in Figure 2-16. Mechanical properties such as **shearing proportional limit** (τ_{PL}), **shearing yield strength** (τ_{YS}), **shearing yield point** (τ_{YP}), and **shearing ultimate strength** (τ_{ULT}) are defined in a manner completely analogous to the corresponding quantities for a simple tension test. **Ductile** and **brittle behavior** characteristics are measured by the **modulus of resilience** and **toughness**. Perhaps the most important mechanical property obtained from a simple shear test is the **modulus of rigidity** G. This modulus is sometimes called the **shearing modulus of elasticity**, and it is defined as the slope of the initially straight portion of the shearing stress-shearing strain diagram as shown in Figure 2-16. Note that G has the same units as stress since it is the ratio of stress to strain in the linear region of the diagram.

Ordinarily, the important properties that are needed to make design decisions are obtained from standard tension tests. Provided that Poisson's ratio is determined during the tension test, the modulus of rigidity can be calculated from the formula

$$G = \frac{E}{2(1 + \nu)} \qquad (2\text{-}28)$$

for isotropic materials. This relationship is derived in Chapter 7.

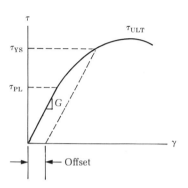

Figure 2-16

PROBLEMS / Section 2-10

2-13 A schematic of the friction braking mechanism of a bicycle is shown in Figure P2-13. Hard rubber friction blocks are pressed against the rim of the wheel as shown. The deformed configuration of the rubber blocks is indicated by the dotted lines. Determine the shearing strain in the rubber.

2-14 A 50-mm solid steel ($G = 84$ GPa) bar is subjected to a pure torque as shown in Figure P2-14b. Axial and circumferential lines PQ and PR are scribed on the surface as shown in P2-14a prior to loading. After application of the torque T, the lines PQ and PR become the lines P^*Q^* and P^*R^* shown in Figure P2-14b. The relative rotation of the ends of line PQ is observed to be $\Delta\varphi = 0.01$ rad. Calculate the shearing strain γ that occurs between lines PQ and PR as a result of the applied torque. Compute the corresponding shearing stress.

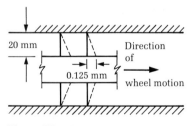

Figure P2-13

Torque (N·m)	Angle of twist (rad)
554	0.00051
893	0.00090
1290	0.00140
1695	0.00186
2085	0.00230
2335	0.00275
2800	0.00325
3125	0.00370
3407	0.00415
3770	0.00465
3927	0.00510
4190	0.00555
4385	0.00600
4555	0.00650
4750	0.00700
4910	0.00740
5005	0.00800
5200	0.00860
5385	0.00925
5530	0.01000
5705	0.01091
5765	0.01180
5930	0.01285

Figure P2-15

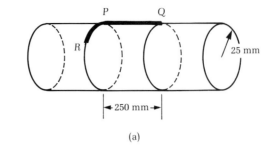

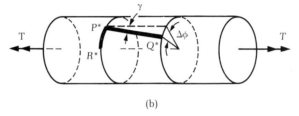

(b)

Figure P2-14

2-15 The torque/angle-of-twist data obtained from a torsion test of a hollow cast iron cylinder are listed in Figure P2-15. The inside and outside diameters of the test specimen were 81.5 mm and 86.5 mm, respectively. The data were obtained using a 100-mm gage length. Construct the stress-strain diagram and determine the shearing modulus of elasticity.

2-16 A thin aluminum plate is rigidly supported along its hypotenuse as shown in Figure P2-16. Determine the engineering shearing strain between line elements *PQ* and *PR* that results because of the deformation indicated by the dotted lines. Calculate the corresponding shearing stress.

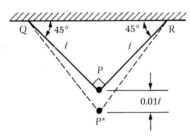

Figure P2-16

2-11 Deformation of Axially Loaded Rods

Consider a straight rod[†] of arbitrary cross section that is loaded by axial forces. The forces may be distributed along the length of the rod in some mathematically continuous way, or they may be a series

---

[†] The locus of centroids of cross sections along the rod forms a straight line.

of concentrated loads applied at different points along the axis of the rod.

Figure 2-17a depicts a rod in its unloaded state. Consider two cross sections A and B located by the axial coordinates x and $x + \Delta x$. In the loaded state shown in Figure 2-17b, these cross sections move to new positions signified by A^* and B^*, and they are located by the axial coordinates $x + u(x)$ and $x + \Delta x + u(x + \Delta x)$. Here $u(x)$ denotes the axial displacement of the centroid of the cross section located at coordinate x, and $u(x + \Delta x)$ denotes the axial displacement of the

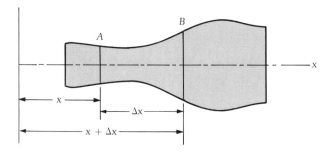

(a) Undeformed rod

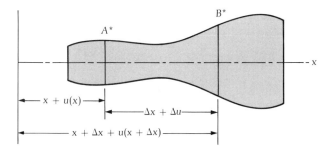

(b) Deformed rod

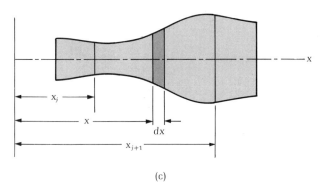

(c)

Figure 2-17

centroid at $x + \Delta x$. Note that $u(x + \Delta x) = u(x) + \Delta u$, where Δu is the change in the axial displacement between the two sections.

Evidently, the engineering strain associated with the length of rod between sections A and B is

$$\epsilon = \lim_{\Delta x \to 0} \frac{u(x + \Delta x) - u(x)}{\Delta x} = \lim_{\Delta x \to 0} \frac{\Delta u}{\Delta x} = \frac{du}{dx} \tag{2-29}$$

Suppose we require the displacement of a section at x_{j+1} of Figure 2-17c, relative to section at x_j. Furthermore, we restrict the analysis to linear elastic behavior. In other words, we assume the stresses occurring in the rod do not exceed σ_{PL} and, thus, the slope of the linear stress-strain diagram is given by E. Note that nothing has been said about the constancy of the modulus of elasticity of the material between the two sections under consideration. Therefore, in general, the modulus of elasticity may be a function of x; that is, $E = E(x)$. Then, through integration of Eq. (2-29),

$$\int_{u_j}^{u_{j+1}} du = u_{j+1} - u_j = \int_{x_j}^{x_{j+1}} \epsilon(x)\, dx = \int_{x_j}^{x_{j+1}} \frac{\sigma(x)}{E(x)}\, dx \tag{2-30}$$

Recall that $\sigma(x) = N(x)/A(x)$, where N and A are the axial force and the cross-sectional area of the rod, respectively, and are, in general, functions of x. Consequently, the relative displacement of a cross section at x_{j+1} with respect to a cross section at x_j is given by the formula

$$e_{j+1/j} = \int_{x_j}^{x_{j+1}} \frac{N(x)}{A(x)E(x)}\, dx \tag{2-31}$$

This equation is a general one giving the relative displacement between any two cross sections of a rod. It can be simplified in certain cases.

Equation (2-31) must be thought of as a sum of several integrals when the integrand N/AE contains discontinuities. Figure 2-18 depicts the N/AE diagram for a shaft for which N/AE is discontinuous at two points x_A and x_B. Discontinuities in the N/AE diagram are caused by abrupt changes in loading, abrupt changes in cross-sectional area, and/or abrupt changes in material. The integral in Eq. (2-31) becomes a sum of integrals—one for each region of the N/AE diagram for which N/AE is smooth and continuous.

Three important procedures are used to analyze axially loaded shafts for which the integrand N/AE contains discontinuities. They are (a) the direct integration method, (b) the discrete element method, and, (c) the superposition method.

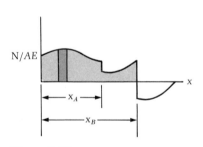

Figure 2-18

Direct integration method.   In this procedure, Eq. (2-30) is integrated directly. The internal force N at any section along the rod is obtained from an appropriate free-body diagram. Cross-sectional areas and material properties are obtained at the same section by inspection.

Discrete element method.   In this procedure, the rod is divided into a finite number of segments, for each of which N/AE is constant. Each of these segments becomes a uniform rod for which the relative displacement of its ends is given by Eq. (2-32). The relative displacement of two arbitrary sections along the rod is the sum of the relative displacements of the segments contained between them.

Superposition method.   The superposition principle for displacements of axially loaded rods stipulates that the relative displacement between two sections along the rod due to several applied forces acting simultaneously is equal to the **algebraic sum** of the relative displacements between the same two sections due to each external force acting separately. These computational procedures are illustrated in the example problems that follow.

Uniform rod.   A particularly important formula that can be obtained from Eq. (2-31) is a formula for the elongation of a uniform rod. A uniform rod is one for which neither the cross section nor the material properties vary along the axis of the rod. Moreover, the only forces applied to the rod are equal and opposite forces at its ends. Figure 2-19 shows a uniform rod of length ℓ and cross-sectional area A that is subjected to axial forces P at its ends. Since N/AE is now constant along the length of the rod, Eq. (2-31) yields

$$e = \frac{P\ell}{AE} \tag{2-32}$$

as the elongation of the rod due to the forces at its ends.

Equation (2-32) gives the shortening of the rod if it is subjected to compressive forces at its ends, provided that the modulus of elasticity to be used in the equation is the one obtained from a compression test rather than a tension test. Indeed, for most materials, the elastic moduli in tension and compression are equal.

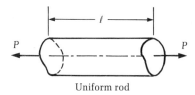

Uniform rod

Figure 2-19

EXAMPLE 2-4

An aluminum step-shaft is loaded by the two axial forces shown in Figure 2-20a. Determine the relative displacement of the free end at A with respect to the fixed end at D. The modulus of elasticity for aluminum is 10×10^6 psi. Solve this problem using (a) the direct integration method, (b) the discrete element method, and (c) the superposition method.

SOLUTION

Direct integration method.   In this method, the integrand of the general Eq. (2-31) is discontinuous at the abrupt change in cross section and again at the point of application of the 1500-lb force. The limits of the integral must be divided into three intervals for which the

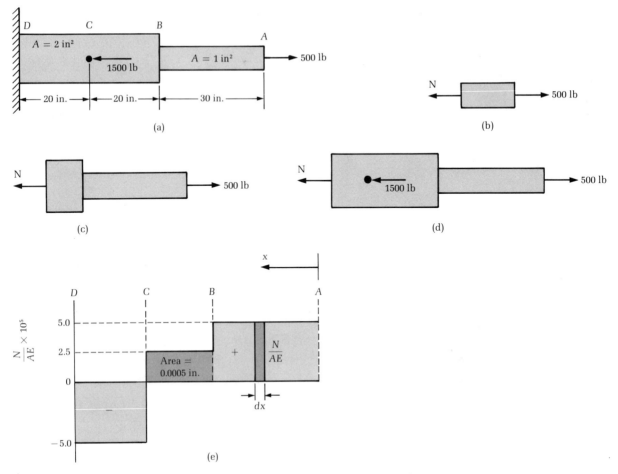

Figure 2-20

integrand is smooth and continuous. Thus

$$e_{A/D} = \int_0^{30} \frac{N}{AE}\, dx + \int_{30}^{50} \frac{N}{AE}\, dx + \int_{50}^{70} \frac{N}{AE}\, dx \tag{a}$$

The internal force N acting in each of the intervals $0 \le x \le 30$, $30 \le x \le 50$, and $50 \le x \le 70$ can be obtained from appropriate free-body diagrams as shown in Figures 2-20b through 2-20d. By direct substitution,

$$e_{A/D} = \int_0^{30} \frac{500}{(1)10 \times 10^6}\, dx + \int_{30}^{50} \frac{500}{(2)10 \times 10^6}\, dx$$

$$+ \int_{50}^{70} \frac{(-1000)}{(2)10 \times 10^6}\, dx \tag{b}$$

$$= 0.0015 + 0.0005 - 0.0010 = 0.001 \text{ in.}$$

Observe that the direct integration method merely entails determining the area under the N/AE diagram shown in Figure 2-20e between the sections whose relative displacement is sought. Thus the relative displacement of section B with respect to C is the shaded area in Figure 2-20e.

Discrete element method. The discrete element method requires the rod to be divided into finite segments for each of which N/AE is constant over its length. Consequently, the complete rod is viewed as a series of connected uniform rods. The relative elongation of the ends of each segment can then be calculated from the formula $e_i = N_i\ell_i/A_iE_i$. The relative displacement of a series of such segments is then given by the formula

$$e = \Sigma e_i \tag{c}$$

Figure 2-20e indicates that three discrete segments are required for the present problem. Consequently,

$$
\begin{aligned}
e_{A/D} &= e_{A/B} + e_{B/C} + e_{C/D} \\
&= \frac{500(30)}{(1)10 \times 10^6} + \frac{500(20)}{(2)10 \times 10^6} + \frac{(-1000)20}{(2)10 \times 10^6} \\
&= 0.0015 + 0.0005 - 0.001 = 0.001 \text{ in.}
\end{aligned}
\tag{d}
$$

This result is the same one that was obtained by the direct integration method.

Superposition method. The superposition principle asserts that the relative displacement between two sections of a straight rod due to several external loads acting simultaneously is equal to the algebraic sum of the relative displacement between the same two sections due to each load acting separately.

Figure 2-21a shows the change in the length of the rod caused by both external forces acting simultaneously. Figures 2-21b and 2-21c show the changes in length caused by each external force acting separately. From Figures 2-21b and 2-21c, calculate the relative displacement between sections A and D for each load acting separately. Accordingly,

$$e'_{A/D} = \frac{500(30)}{(1)10 \times 10^6} + \frac{500(40)}{(2)10 \times 10^6} = 0.0025 \text{ in.} \tag{e}$$

and

$$e''_{A/D} = -\frac{1500(20)}{(2)10 \times 10^6} = -0.0015 \text{ in.} \tag{f}$$

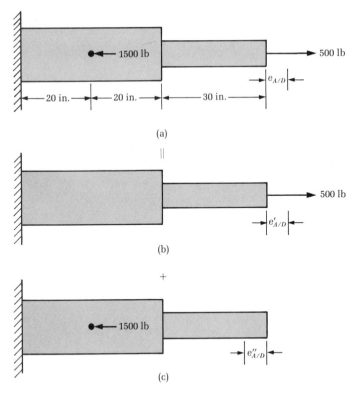

(a)

||

(b)

+

(c)

Figure 2-21

Consequently, the displacement of section A relative to section D when both forces act simultaneously is

$$e_{A/D} = e'_{A/D} + e''_{A/D}$$
$$= 0.0025 - 0.0015 = 0.001 \text{ in.} \qquad\qquad (g)$$

EXAMPLE 2-5

Find the deflection, caused by its own weight, of free end C of the solid truncated cone BC shown in Figure 2-22a. The weight density of material is $\gamma(N/m^3)$ and the modulus of elasticity is E (N/m^2). They are both constants. All the dimensions of the truncated cone are in meters.

SOLUTION

For this problem, the direct integration method is apparently the only method that will yield an exact solution. You can also analyze this problem by approximating the conical shape of the rod by a finite number of discrete elements as shown in Figure 2-22c. In this latter case, each discrete element will have a uniform cross-sectional area and, thus, the problem will be simplified; however, the solution will not be exact. Here we use the direct integration method.

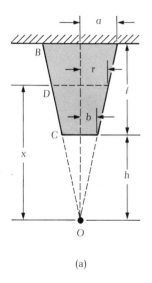

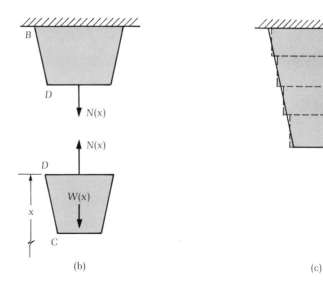

(a)

(b)

(c)

Figure 2-22

Consider the origin of the coordinate axes at the vertex of the extended cone as shown in Figure 2-22a. Since the extreme radii a and b and the length ℓ of the truncated cone are known, the distance h from the vertex to the smaller end can be determined easily. Thus

$$h = \frac{b\ell}{a - b} \tag{a}$$

In a similar fashion, the radius $r(x)$ of the cross-sectional area of the truncated cone at a distance x from the vertex can be determined. It is

$$r(x) = \frac{b}{h} x \tag{b}$$

The internal force $N(x)$ acting at a section x from the vertex can be obtained from an appropriate free-body diagram as shown in Figure 2-22b. $N(x)$ at a section D is simply equal to the weight $W(x)$ of the portion DC of the truncated cone. Accordingly,

$$N(x) = \tfrac{1}{3}\pi\gamma(r^2 x - b^2 h) \tag{c}$$

Finally, the cross-sectional area at D is

$$A(x) = \pi \frac{b^2}{h^2} x^2 \tag{d}$$

Now we have all the information we need for the use of Eq. (2-31). Thus

$$e_{C/B} = \int_h^{h+\ell} \frac{N(x)}{A(x)E} \, dx \tag{e}$$

which, in turn, yields

$$e_{C/B} = \frac{\gamma \ell^2 (\ell + 3h)}{6E(h + \ell)} \tag{f}$$

Eliminating h by using Eq. (a), we obtain

$$e_{C/B} = \frac{\gamma \ell^2 (a + 2b)}{6Ea}$$
(g)

If $a = b$, then $e_{C/B} = W_T \ell / 2AE$ where W_T is the total weight of the cylindrical rod, and A is its constant cross-sectional area.

PROBLEMS / Section 2-11

2-17 Determine the vertical displacement that point D of the mechanism shown in Figure P2-17 undergoes as a result of the 50-kN force applied at D. The 25-mm diameter rod AB is made of aluminum ($E = 70$ GPa), and the angle arm BCD is rigid. Consider CD to be horizontal in the undeformed configuration.

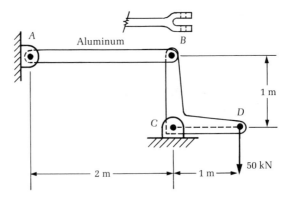

Figure P2-17

2-18 A steel bar of uniform cross-sectional area A weighing q lb/ft is suspended from a rigid support as shown in Figure P2-18. Determine a formula for the displacement of the lower end.

2-19 Determine the magnitude of the largest force Q that can be applied to the axially loaded composite step-shaft shown in Figure P2-19. The maximum permissible elongation is 0.025 mm, and the maximum permissible normal stress in either material is 2.5 MPa ($E_{ST} = 210$ GPa and $E_{AL} = 70$ GPa).

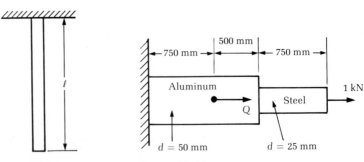

Figure P2-18   **Figure P2-19**

2-20 The cross-sectional areas of the steel ($E = 30 \times 10^6$ psi) and the aluminum ($E = 10 \times 10^6$ psi) segments of the step-shaft shown in Figure P2-20 are 1 in^2 and 3 in^2, respectively. Determine the displacement of the upper end of the shaft by (a) the direct integration method, (b) the discrete element method, and (c) the superposition method. Calculate the maximum compressive stress in the steel and the maximum tensile and compressive stresses in the aluminum. Justify the use of an elastic analysis.

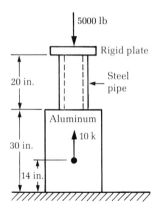

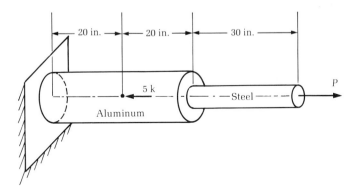

Figure P2-20

Figure P2-21

2-21 Determine the maximum force P that can be applied to the free end of the axially loaded step-shaft shown in Figure P2-21 if the axial elongation is not to exceed 0.01 in. and the maximum normal stress is not to exceed 10 ksi. The cross-sectional areas of the steel and aluminum segments are 1 in^2 and 2 in^2, respectively.

2-12 Statically Indeterminate Axially Loaded Members

An engineering structure is said to be externally statically indeterminate whenever the equilibrium conditions are insufficient to determine all unknown reactions acting on the structure. In such cases, the equilibrium equations must be augmented by equations that arise because certain points or planes in the structure are constrained to deform in prescribed ways. These equations are frequently referred to as **geometric compatibility equations**; they are also called **kinematic equations**.

In any static solid mechanics problem, there exist three basic ingredients: **equilibrium**, **geometric compatibility**, and **material behavior**. Equilibrium is concerned solely with the forces and/or stresses of the body, geometric compatibility is concerned only with its displacements and/or strains, and the stress-strain relations (material behavior, also called **constitutive equations**) provide a connection between the two. The following examples should clarify the simultaneous use of these three ingredients.

EXAMPLE 2-6

Determine the stresses induced in the aluminum and steel portions of the composite rod shown in Figure 2-23a due to a force of 7000 lb applied as shown. The cross-sectional areas of the steel and aluminum portions are 2 in^2 and 4 in^2, respectively, and their moduli of elasticity are 30×10^6 psi and 10×10^6 psi, respectively.

SOLUTION

Equilibrium.   Assume that the internal axial forces in both parts of the composite rod are tensile. For this relatively simple problem, it is obvious that the steel portion of the rod is actually in compression and the aluminum portion in tension. Our assumption is, therefore, incorrect; however, we shall keep it. A negative result in the solution will tell us that the corresponding portion is not in tension but in compression. In problem solving, it is a good practice to start with some assumptions about the senses of unknown forces rather than attempting to determine the correct senses. In more complex problems, determining the correct senses of unknown reactions may not be as trivial as it is in the present example. From the free-body diagram shown in Figure 2-23b, one, then, obtains

$$\overset{+}{\rightarrow} \Sigma F_x = 0: \quad -P_{ST} + P_{AL} - 7000 = 0 \tag{a}$$

Compatibility.   Since Eq. (a) is the only equilibrium equation among the unknowns P_{ST} and P_{AL}, we must augment it with an equation

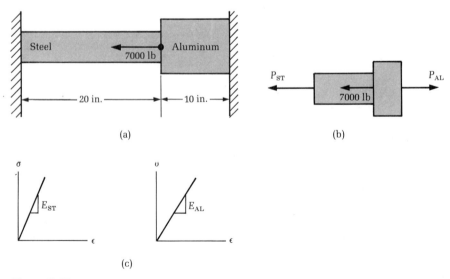

(a)   (b)

(c)

Figure 2-23

that expresses the fact that the total deformation of the composite rod from end to end is zero. Accordingly,

$$e = e_{ST} + e_{AL} = 0 \tag{b}$$

is the required compatibility equation.

Material behavior. The compatibility Eq. (b), in its present form, cannot be used together with Eq. (a) to solve for the unknown forces P_{ST} and P_{AL}; however, when we recall that the elongations are actually related to these unknown forces through force-deformation relations, the importance of the compatibility Eq. (b) becomes clear. Let both materials behave in a linear elastic manner, such as that exhibited by the stress-strain diagrams of Figure 2-20c. Then

$$e_{ST} = \frac{P_{ST}\ell_{ST}}{A_{ST}E_{ST}} \quad \text{and} \quad e_{AL} = \frac{P_{AL}\ell_{AL}}{A_{AL}E_{AL}} \tag{c}$$

From Eqs. (b) and (c), we obtain

$$P_{ST} = -\frac{\ell_{AL}}{\ell_{ST}} \frac{E_{ST}}{E_{AL}} \frac{A_{ST}}{A_{AL}} P_{AL} = -\frac{10}{20} \frac{30}{10} \frac{2}{4} P_{AL}$$

or

$$P_{ST} = -\tfrac{3}{4} P_{AL} \tag{d}$$

Solving Eqs. (a) and (d) simultaneously yields

$$P_{ST} = -3000 \text{ lb} \quad \text{and} \quad P_{AL} = 4000 \text{ lb} \tag{e}$$

from which we find the stresses to be

$$\sigma_{ST} = -\frac{3000}{4} = -1500 \text{ psi} = 1500 \text{ psi(C)}$$

and

$$\left. \vphantom{\frac{4000}{4}} \right\} \tag{f}$$

$$\sigma_{AL} = \frac{4000}{4} = 1000 \text{ psi(T)}$$

Since we assumed at the outset that both materials obeyed linear elastic stress-strain relations, we must validate this assumption by checking the calculated stresses (1500 psi for the steel and 1000 psi for the aluminum) against the proportional limit stresses for steel and aluminum. A quick glance at the table of properties given in Appendix A shows that these calculated stresses fall well within the proportional limits of their respective materials. Consequently, we can conclude that the assumption of linear elastic behavior is valid.

EXAMPLE 2-7

Suppose that the temperature is allowed to increase $10°$ F in Example 2-6. Determine the stress in each material due to the temperature change alone. The thermal expansion coefficients for steel and aluminum are 6.5×10^{-6} in./in./°F and 13×10^{-6} in./in./°F, respectively.

SOLUTION

Equilibrium. From the free-body diagram of Figure 2-24a, we obtain

$$\overset{+}{\rightarrow} \Sigma F_x = 0: \quad P_{ST} - P_{AL} = 0 \tag{a}$$

Here we assume that the unknown forces are both compressive.

Compatibility. Under the unconstrained thermal expansion indicated in Figure 2-24c, the composite rod would elongate an amount

$$\Delta = \Delta_{ST} + \Delta_{AL} \tag{b}$$

The elongations of the individual portions of the composite rod due to the temperature change can be determined by using Eq. (2-20).

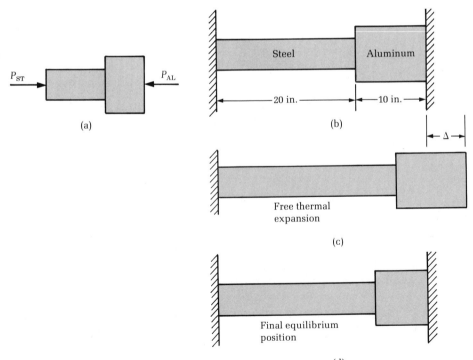

Figure 2-24

They are

$$\Delta_{ST} = \alpha_{ST} \ell_{ST} \Delta T \tag{c}$$

and

$$\Delta_{AL} = \alpha_{AL} \ell_{AL} \Delta T \tag{d}$$

Thus

$$\Delta = (6.5 \times 10^{-6})(20)(10) + (13 \times 10^{-6})(10)(10)$$
$$= 0.0013 + 0.0013 = 0.0026 \text{ in.} \tag{e}$$

Now the walls constrain the composite rod from expanding this amount. The net change in length of the composite rod is zero. Therefore, internal strains must be induced in the steel and the aluminum so that the sum of the deformations in the steel and the aluminum caused by thermal expansion and internal forces is zero.

$$\delta_{ST} + \delta_{AL} - 0.0026 = 0 \tag{f}$$

This is the compatibility equation.

Material behavior. Again let us assume linear elastic behavior for both materials. We have then

$$\delta_{ST} = \frac{P_{ST}\ell_{ST}}{A_{ST}E_{ST}} \quad \text{and} \quad \delta_{AL} = \frac{P_{AL}\ell_{AL}}{A_{AL}E_{AL}} \tag{g}$$

Eqs. (a), (f), and (g) yield

$$P_{ST}\left\{1 + \frac{E_{ST}}{E_{AL}} \frac{A_{ST}}{A_{AL}} \frac{\ell_{AL}}{\ell_{ST}}\right\} = \frac{E_{ST}A_{ST}}{\ell_{ST}} 0.0026$$

$$P_{ST}\left\{1 + \frac{30}{10} \frac{2}{4} \frac{10}{20}\right\} = \frac{30 \times 10^6 (2)}{20} 0.0026$$

$$\tfrac{7}{4} P_{ST} = 7800 \tag{h}$$

Consequently,

$$P_{ST} = P_{AL} = 4457 \text{ lb} \tag{i}$$

The corresponding stresses are

$$\sigma_{ST} = \frac{4457}{2} = 2228 \text{ psi(C)} \tag{j}$$

and

$$\sigma_{AL} = \frac{4457}{4} = 1114 \text{ psi(C)} \tag{k}$$

Since the forces (i) are positive, our initial assumptions regarding their proper senses are correct.

EXAMPLE 2-8

Suppose the composite rod of Examples 2-6 and 2-7 is subjected to an axial force of 7000 lb as shown in Figure 2-23a and a subsequent increase in temperature of 10° F. Determine the stress in the steel and aluminum portions of the rod due to the combined mechanical and thermal effects.

SOLUTION

The superposition principle applied to this problem asserts that the stress due to mechanical and thermal effects is equal to the algebraic sum of the stresses due to each effect acting separately. Accordingly, the stress in the steel and aluminum portions of the composite rod due to the combined effects of the 7000-lb axial force and the 10° F rise in temperature can be calculated by determining the stresses in each material due to the mechanical load and the thermal load independently, and then adding the results algebraically.

Accordingly, from Examples 2-6 and 2-7,

$$\sigma_{ST} = \sigma_{ST}^{mech} + \sigma_{ST}^{thermal} = -1500 - 2228 = 3728 \text{ psi(C)}$$
$$\sigma_{AL} = \sigma_{AL}^{mech} + \sigma_{AL}^{thermal} = 1000 - 1114 = 114 \text{ psi(C)}$$

Note that the mechanical stress and the thermal stress for the steel and for the aluminum are each less than the proportional limit stress for the corresponding material. Moreover, the sum of the mechanical stress and the thermal stress for each material is less than the corresponding proportional limit stress. The superposition procedure is therefore justified for this problem.

EXAMPLE 2-9

Figure 2-25a depicts an assembly consisting of a steel bolt and an aluminum collar. The pitch of the single-threaded bolt is 3 mm, and its cross-sectional area is 600 mm². The cross-sectional area of the collar is 900 mm². The nut is brought to a snug position and then given an additional $\frac{1}{8}$-in. turn. Determine the stresses in the aluminum collar and in the steel bolt.

SOLUTION

Equilibrium. Figure 2-25d shows a free-body diagram of the right-hand portion of assembly. P_{ST} denotes the axial force in the bolt, and P_{AL} denotes the force acting around the annular area of the collar. Equilibrium of forces along the axis of the bolt gives

$$\overset{+}{\rightarrow} \Sigma F_x = 0: \quad -P_{ST} + P_{AL} = 0 \tag{a}$$

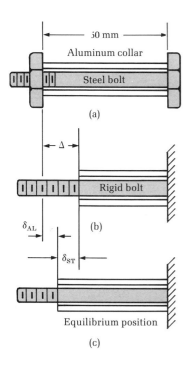

(a)

(b)

δ_{AL}

δ_{ST}

Equilibrium position

(c)

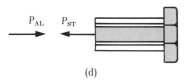

P_{AL} P_{ST}

(d)

Figure 2-25

Compatibility. To obtain an appropriate compatibility equation, we first imagine the bolt to be rigid and give the nut a $\frac{1}{8}$-in. turn. The distance traveled by the nut over the rigid bolt is

$$\Delta = \frac{1}{8}(0.003) \text{ m} \tag{b}$$

This situation is shown in Figure 2-25b.

Now imagine the bolt to possess elasticity so that the aluminum collar stretches the bolt until an equilibrium position such as the one shown in Figure 2-25c is attained. The stress-producing deformations in the aluminum and in the steel are shown in this figure. From the geometry of Figures 2-25b and 2-25c, we see immediately that the required compatibility equation is

$$\delta_{ST} + \delta_{AL} = \Delta \tag{c}$$

Material behavior. Assume that both the steel and the aluminum behave in a linear elastic manner. The stress-producing deformations in the aluminum and in the steel are given by the following force-deformation relations:

$$\delta_{ST} = \frac{P_{ST}\ell_{ST}}{A_{ST}E_{ST}} \quad \text{and} \quad \delta_{AL} = \frac{P_{AL}\ell_{AL}}{A_{AL}E_{AL}} \tag{d}$$

Equations (a), (c), and (d) combine to give

$$\frac{P_{ST}\ell_{ST}}{A_{ST}E_{ST}}\left\{1 + \frac{E_{ST}}{E_{AL}}\frac{A_{ST}}{A_{AL}}\frac{\ell_{AL}}{\ell_{ST}}\right\} = \frac{1}{8}(0.003) \tag{e}$$

or

$$P_{ST}\left\{1 + \frac{210}{70}\frac{600}{900}1\right\} = \frac{1}{8}(0.003)\frac{600 \times 10^{-6}(210 \times 10^9)}{0.050} \tag{f}$$

Consequently,

$$3P_{ST} = 945 \text{ kN}$$

or

$$P_{ST} = 315 \text{ kN(T)} \tag{g}$$

and

$$P_{AL} = 315 \text{ kN(C)} \tag{h}$$

The stresses corresponding to these forces are

$$\left. \begin{aligned} \sigma_{ST} &= \frac{P_{ST}}{A_{ST}} = \frac{315 \times 10^3}{600 \times 10^{-6}} = 525 \text{ MPa(T)} \\ \sigma_{AL} &= \frac{P_{AL}}{A_{AL}} = \frac{315 \times 10^3}{900 \times 10^{-6}} = 350 \text{ MPa(C)} \end{aligned} \right\} \tag{i}$$

EXAMPLE 2-10

Let the nut of the previous example be brought to a snug position and let the temperature increase 50° C. Determine the stresses induced in the aluminum collar and in the steel bolt. The bolt and collar assembly is repeated in Figure 2-26a to illustrate the relative positions of the bolt and the collar under free thermal expansion.

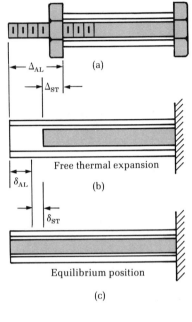

(a)

SOLUTION

Equilibrium. From the free-body diagram shown in Figure 2-26d, we find that

$$P_{ST} - P_{AL} = 0 \qquad (a)$$

and observe that the force in the steel bolt is tension and the force in the aluminum collar is compression.

Free thermal expansion

(b)

Compatibility. Figure 2-26b shows the positions the steel bolt and the aluminum collar assume under unconstrained thermal expansion, and Figure 2-26c shows the final equilibrium configuration of the assembly. Under unconstrained thermal expansion, the aluminum collar expands more than the steel bolt because the linear coefficient of thermal expansion for aluminum is larger than that for steel. However, such unequal expansion of the two parts of the collar-bolt assembly is physically impossible. The expansions of the two parts must be equal in the final equilibrium configuration as shown in Figure 2-26c. Thus the aluminum collar stretches the steel bolt, and the bolt in turn compresses the collar until it attains this final equilibrium configuration. We note from the geometry of Figures 2-26b and 2-26c that

$$\delta_{ST} + \delta_{AL} = \Delta_{AL} - \Delta_{ST} \qquad (b)$$

Equilibrium position

(c)

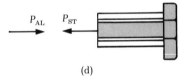

(d)

Figure 2-26

Material behavior. Assume that both materials behave in a linear elastic manner. Then, as before, the force-deformation relations are,

$$\delta_{ST} = \frac{P_{ST}\ell_{ST}}{A_{ST}E_{ST}} \quad \text{and} \quad \delta_{AL} = \frac{P_{AL}\ell_{AL}}{A_{AL}E_{AL}} \qquad (c)$$

Combining Eqs. (a), (b), and (c), and recalling that $\Delta_{AL} = \alpha_{AL}\ell_{AL}\,\Delta T$ and $\Delta_{ST} = \alpha_{ST}\ell_{ST}\,\Delta T$, we obtain

$$P_{ST}\left\{1 + \frac{E_{ST}}{E_{AL}}\frac{A_{ST}}{A_{AL}}\right\} = E_{ST}A_{ST}(\alpha_{AL} - \alpha_{ST})\,\Delta T \qquad (d)$$

Substituting the given data, we obtain

$$P_{ST}\left\{1 + \frac{210}{70}\frac{600}{900}\right\} = 210 \times 10^9(600 \times 10^{-6})(11.7 \times 10^{-6})50$$

or

$$P_{ST} = 24.57 \text{ kN}(T) \qquad \text{and} \qquad P_{AL} = 24.57 \text{ kN}(C) \qquad \text{(e)}$$

The corresponding stresses are

$$\sigma_{ST} = \frac{24,570}{600 \times 10^{-6}} = 41 \text{ MPa}(T) \qquad \text{(f)}$$

and

$$\sigma_{AL} = \frac{24,570}{900 \times 10^{-6}} = 27.3 \text{ MPa}(C) \qquad \text{(g)}$$

These stresses should be checked against the proportional limit stresses for steel and aluminum, respectively. If neither stress exceeds the proportional limit of its associated material, the assumption of elastic behavior is valid and the calculated stresses are correct. Otherwise, an analysis based on the appropriate material behavior must be undertaken.

EXAMPLE 2-11

Suppose that, in the bolt and collar assembly of the two previous examples, the nut is advanced to a snug position and then given an additional $\frac{1}{8}$-in. turn. The temperature of the assembly is subsequently increased 50° C. Determine the stresses induced in the steel and the aluminum by the combined effect.

SOLUTION

Employing the superposition principle, we first solve for the stresses due to the mechanical effect alone. This was accomplished in Example 2-9. We then calculate the stresses in the steel and the aluminum for the temperature increase as we did in Example 2-10.

According to the superposition principle, the stresses in the steel and the aluminum resulting from the combined mechanical and thermal effect are

$$\sigma_{ST} = \sigma_{ST}^{\text{mech}} + \sigma_{ST}^{\text{thermal}} = 525 + 41 = 566 \text{ MPa}(T)$$

and

$$\sigma_{AL} = \sigma_{AL}^{\text{mech}} + \sigma_{AL}^{\text{thermal}} = -350 - 27.3 = 377.3 \text{ MPa}(C)$$

These stresses are the correct ones provided they do not exceed the proportional limit stresses of their associated materials and that the mechanical and thermal components of these total stresses do not separately exceed the proportional limits of their respective materials.

PROBLEMS / Section 2-12

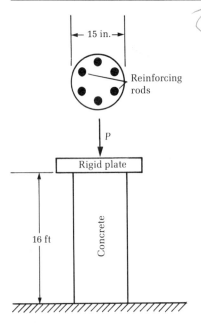

Figure P2-22

2-22 The reinforced concrete column of Figure P2-22 is required to support an axial force P. Six 1-in. diameter steel rods are used to reinforce the concrete. Determine the maximum permissible force P if the stress in the steel reinforcing rods is not to exceed 18 ksi, and the stress in the concrete is not to exceed 900 psi. Neglect the weight of the concrete and the steel and use $E_{ST} = 30 \times 10^6$ psi and $E_C = 3 \times 10^6$ psi.

2-23 A 4-in. standard steel pipe ($E = 30 \times 10^6$ psi) filled with concrete ($E = 3 \times 10^6$ psi) is used as a compression member as shown in Figure P2-23. Determine the average normal stress in the steel and in the concrete when the 10-k force acts along the geometric axis of the pipe. Neglect the weights of the concrete and the steel.

2-24 A rigid beam is suspended by two aluminum rods as shown in Figure P2-24. The rods are pin connected at points A, B, C, and D. The weight of the beam can be neglected in comparison with the 600-kN force applied to it. Calculate the displacements of the points A and B and the normal stress in each rod. The cross-sectional areas of rods AD and BC are 600 mm^2 and 1200 mm^2, respectively.

2-25 The composite bar shown in Figure P2-25 fits snugly between rigid supports. Calculate the normal stresses in the aluminum and the steel due to a temperature increase of 10° F. Pertinent material properties are $E_{ST} = 30 \times 10^6$ psi, $E_{AL} = 10 \times 10^6$ psi, $\alpha_{ST} = 6.5 \times 10^{-6}$ in./in./°F, and $\alpha_{AL} = 13 \times 10^{-6}$ in./in./°F. The cross-sectional areas of the steel and the aluminum segments are 1 in^2 and 2 in^2 respectively.

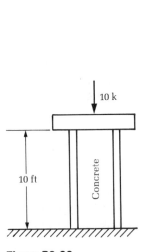

Figure P2-23

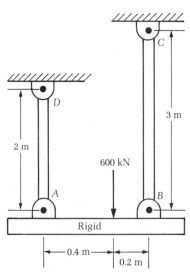

Figure P2-24

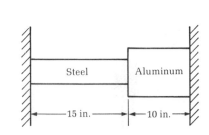

Figure P2-25

2-13 Stress Concentrations

Consider a thin rectangular plate with symmetrically located V notches under a uniform external tensile stress as shown in Figure 2-27a. Prior to the application of the external stress, the lines aa and bb are scribed on the plate a prescribed distance ℓ_0 apart. Due to the application of σ, points on the line bb displace to the positions indicated by the dotted line. Points on line aa do not displace because of symmetry considerations.

A rather crude distribution of the axial strain along line aa can be obtained experimentally by measuring the distances ℓ_f for a sufficient number of points with a set of micrometer calipers and, subsequently, by calculating the axial strain for the various points from the basic definition

$$\epsilon = \frac{\ell_f - \ell_0}{\ell_0} \tag{2-33}$$

The strain distribution along line aa is shown in Figure 2-27b. The strain distribution obtained in this manner will be rather crude; however, the essential effects of the symmetric notches on the strain distribution will be exhibited. More accurate strain distributions can be obtained using electrical resistance strain gages.

If we assume the material to behave elastically, the stress distribution along line aa is calculated from the strain distribution by multiplying the strain at a point on aa by the modulus of elasticity. The stress

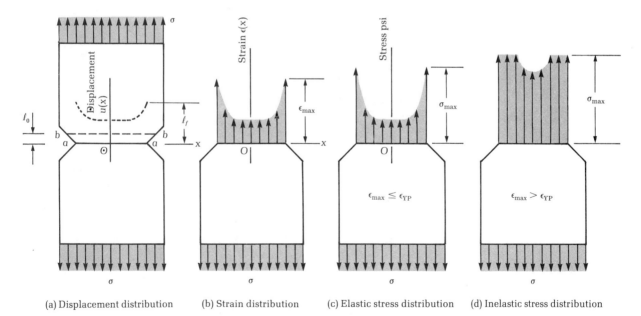

(a) Displacement distribution   (b) Strain distribution   (c) Elastic stress distribution   (d) Inelastic stress distribution

Figure 2-27

distribution for the thin rectangular plate with symmetric V notches is shown in Figure 2-27c. Observe that the maximum stress is significantly larger than the nominal stress as calculated from the relation P/A with A being the area of the net or reduced cross section. Also, note that this maximum stress is typically a local phenomena; that is, the maximum stress is concentrated in the vicinity of the root of the notch and rapidly diminishes to a value slightly less than P/A. For this reason, the V notch is called a **stress raiser**, and the region around the root of the notch where the stress reaches its maximum value is referred to as a **stress concentration**.

It is customary to express the maximum stress at a stress concentration in terms of the nominal or average stress P/A acting on the **reduced** cross-sectional area. Consequently, the maximum stress is usually written as

$$\sigma_{\max} = k_t \sigma_{\text{ave}} \tag{2-34}$$

where k_t is the so-called **theoretical stress concentration factor**. k_t is called a **theoretical stress concentration factor**, not because it is determined from mathematical models, but because the material is assumed to behave elastically so that k_t depends only on the geometry of the stress raiser. Most stress concentration factors are determined through the use of photoelastic methods, electrical resistance strain gages, or, more recently, the finite element method.

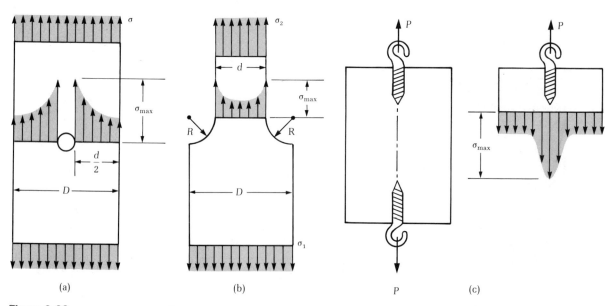

(a)   (b) P (c)

Figure 2-28

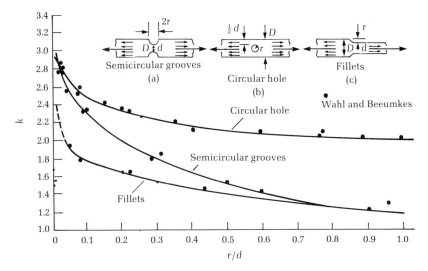

Figure 2-29   After M. M. Frocht, "Factors of Stress Concentration Photoelastically Determined," *Trans. ASME,* Vol. 57, 1935; by permission of the American Society of Mechanical Engineers.

Stress raisers are easy to locate. Any abrupt change in cross-sectional area such as a hole, a fillet, or a notch is a potential stress raiser. Moreover, stress raisers also occur at points of support, load application, and the intersection of one or several members. Some typical cases are shown in Figure 2-28.

Stress concentration factors are usually obtained by designers from graphs such as the ones shown in Figure 2-29.

Note that stress concentrations play an important role in the design of members made from materials that behave in a brittle manner. It is in the vicinity of a stress raiser that cracks are initiated, and, in the presence of time dependent loads, they are propagated until failure occurs. On the other hand, stress raisers are not so critical for members made from materials that behave in a ductile manner. For example, for a ideally elastic-ideally plastic material, the maximum stress at a stress raiser cannot exceed the yield stress. When the maximum strain at a stress raiser exceeds ϵ_{YP}, a redistribution of the stress occurs as shown in Figure 2-27d.

EXAMPLE 2-12

Determine the maximum normal stress in the 10-mm thick elastic bar shown in Figure 2-30a when the fillets' radii are 5 mm, 20 mm, and 100 mm.

SOLUTION

The maximum normal stress occurs at the section through the fillets. Figure 2-30b shows the stress distribution on this section.

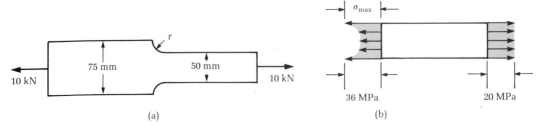

(a)   (b)

Figure 2-30

To obtain the appropriate stress concentration factors, calculate the ratios

$$\frac{r}{d} = \frac{5}{50}, \quad \frac{20}{50}, \quad \text{and} \quad \frac{100}{50}$$

Accordingly, the abscissas required to enter Figure 2-29 are 0.1, 0.4, and 2.0. The corresponding values of the stress concentration factor are $k = 1.8, 1.49,$ and 1.3. Consequently, the maximum normal stresses associated with fillets' radii of 5 mm, 20 mm, and 100 mm are

$$\sigma_{max} = 1.8(20 \text{ MPa}) = 36 \text{ MPa} \qquad (r = 5 \text{ mm})$$
$$\sigma_{max} = 1.49(20 \text{ MPa}) = 29.8 \text{ MPa} \qquad (r = 20 \text{ mm})$$
$$\sigma_{max} = 1.3(20 \text{ MPa}) = 26.0 \text{ MPa} \qquad (r = 100 \text{ mm})$$

The stress distribution for $r = 5$ mm is shown in Figure 2-30b.

Notice that increasing the fillet's radius decreases the magnitude of the maximum stress. Also, note that there is a limit to this reduction in magnitude of the stress concentration. Therefore, increasing the fillet's radius beyond 40–50 mm in this problem does not improve the situation significantly.

PROBLEMS / Section 2-13

2-26 Determine the maximum stress at each of the holes in the metallic bar shown in Figure P2-26 when $P = 30$ kN.

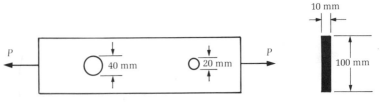

Figure P2-26

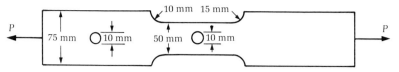

Figure P2-27

2-27 The metallic plate shown in Figure P2-27 is 10 mm thick and is subjected to an axial force $P = 37.5$ kN. Determine the maximum normal stress at the two holes and at the fillets.

2-14 SUMMARY

In this chapter, we explained the procedures used to construct engineering normal stress-normal strain diagrams and engineering shearing stress-shearing strain diagrams for isotropic materials, as well as defining tensile and shearing strength properties such as proportional limit stress, elastic limit stress, yield stress, yield strength, and ultimate stress. We also showed that the slope of the tensile stress-strain diagram is a measure of the capacity of the material to resist stretching or contraction, and that the slope of the shearing stress-shearing strain diagram is a measure of the capacity of a material to resist twisting deformations. We observed that moduli of resilience and toughness are a measure of the material to absorb energy elastically and inelastically, respectively. Percent elongation and percent reduction in a cross-sectional area characterize the ductility or brittleness of a material.

We discussed procedures used to construct true stress-true strain diagrams, and we developed formulas relating true stress with engineering normal stress and true strain with engineering normal strain. They are

$$\bar{\sigma} = \sigma(1 + \epsilon) \qquad \text{and} \qquad \bar{\epsilon} = \ln(1 + \epsilon)$$

Idealized stress-strain diagrams for materials called ideally elastic-ideally plastic, ideally elastic-linear hardening, nonlinear elastic, and rigidly elastic-ideally plastic were discussed. Procedures commonly used to represent these ideal materials mathematically were demonstrated.

Thermal strains in isotropic materials were discussed, and the important thermal property known as the coefficient of thermal expansion was described. The superposition principle for mechanical and thermal strains was discussed.

Procedures used to construct shearing stress-shearing strain diagram from the formulas

$$\tau_{ave} = \frac{T_R}{2\pi r_o^2 t} \quad \text{and} \quad \gamma_{ave} = \frac{r_o \varphi}{\ell_o}$$

were described. It was shown that mechanical properties analogous to those defined for engineering normal stress-engineering normal strain diagrams can be defined for shearing stress-shearing strain diagrams. Thus mechanical properties associated with shearing stress-shearing strain diagrams are proportional limit stress, elastic limit stress, yield stress, yield strength, ultimate stress, modulus of elasticity, modulus of resilience, and toughness. Percent elongation and reduction in area have no meaning for a shear test.

Three methods commonly used to determine axial deformations of axially loaded, elastic rods were demonstrated: the direct integration method, the discrete element method, and the superposition method. Formulas relevant to these methods are:

$$e_{j+1/j} = \int_{x_j}^{x_{j+1}} \frac{N(x)}{A(x)E(x)} \, dx \qquad \text{(direct integration method)}$$

$$e_i = \frac{N_i \ell_i}{A_i E_i} \quad \text{and} \quad e_{j+1/j} = \Sigma e_i \qquad \text{(discrete element method)}$$

$$e_{j+1/j} = e'_{j+1/j} + e''_{j+1/j} + \cdots \qquad \text{(superposition method)}$$

Procedures used to solve statically indeterminate, axially loaded elastic rods were described. The three ingredients characteristic of statically indeterminate problems were emphasized. They are equilibrium, geometric compatibility, and material behavior.

The concept of stress concentrations was established, and a procedure for calculating the maximum normal stress in the vicinity of a hole, notch, or fillet was described. The basic formula is

$$\sigma_{max} = k\sigma_{ave}$$

where k is the stress concentration factor associated with the hole, notch, or fillet and is obtained from an appropriate graph.

PROBLEMS / CHAPTER 2

2-28 The stress-strain diagram for a mild steel tensile specimen is shown in Figure P2-28. Determine (a) the modulus of elasticity, (b) the proportional limit stress, (c) the yield stress, (d) the ultimate strength, (e) the modulus of resilience, and (f) the toughness.

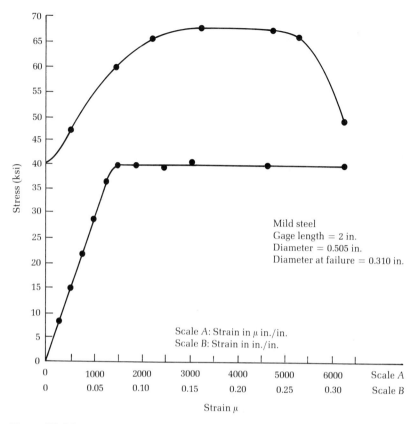

Figure P2-28

2-29 The stress-strain diagram for a certain material is listed in Figure P2-29. Determine (a) the modulus of elasticity, (b) the proportional limit, (c) the yield

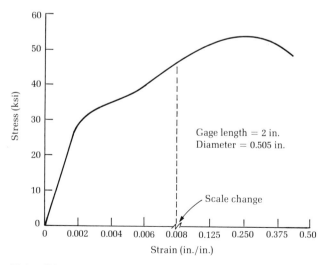

Figure P2-29

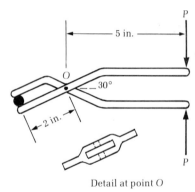

Detail at point O

Figure P2-30

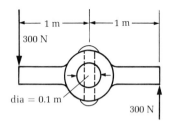

Figure P2-32

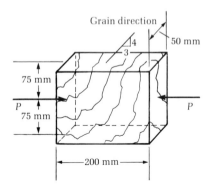

Figure P2-34

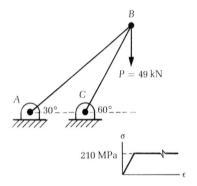

Figure P2-35

strength at 0.2% offset, and (d) the ultimate strength, the modulus of resilience, and the toughness.

2-30   The ultimate shearing stress for the 0.125-in. diameter pin in the pair of wire cutters shown in Figure P2-30 is 40 ksi. Determine the maximum load P that can be applied to the wire cutters if a safety factor of 2.5 with respect to the ultimate strength is required.

2-31   A 0.505-in. diameter bar is subjected to an axial tensile force P. At the proportional limit of the material, P = 5000 lb. The initial gage length has changed from 8 in. to 8.0125 in., and the diameter has decreased 0.00025 in. Determine (a) modulus of elasticity, (b) Poisson's ratio, and (c) proportional limit.

2-32   A reaming device is shown in Figure P2-32. Determine the minimum required diameter of the shear pin if the maximum shearing stress is not to exceed 60 MPa.

2-33   Determine the dimensions a and b for the wooden frame shown in Figure P2-33. The shearing stress parallel to the grain must not exceed 150 psi, and the compressive stress parallel to the grain must not exceed 1000 psi.

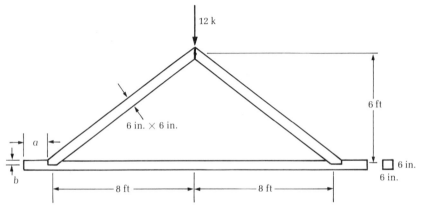

Figure P2-33

2-34   Determine the maximum force P that can be applied to the timber shown in Figure P2-34 if the shearing stress parallel to the grain is not to exceed 1050 kN/m². Assume that this force produces acceptable normal stress intensities.

2-35   An assemblage of rods is pin connected to form the mechanism shown in Figure P2-35. If the diameter of each member is 25 mm, determine the average normal stress acting in each member. Justify the use of any elastic analysis. The stress-strain diagram for material from which the mechanism is made is shown in the figure.

2-36   Two high-strength steel rods are attached to rigid supports at points A and B of Figure P2-36. The rods are connected to each other by a smooth pin at C. Determine the maximum permissible load W that the assembly can support if the ultimate strength of each rod is 1120 MPa and a safety factor of 4 is used. The cross-sectional areas of rods CA and CB are 125 mm² and 62.5 mm², respectively.

2-37   Determine the minimum permissible diameter of the hardened steel pins at A and B of the lock grips shown in Figure P2-37. The shearing ultimate strength of the material is 30 ksi, and a safety factor of 3 is required.

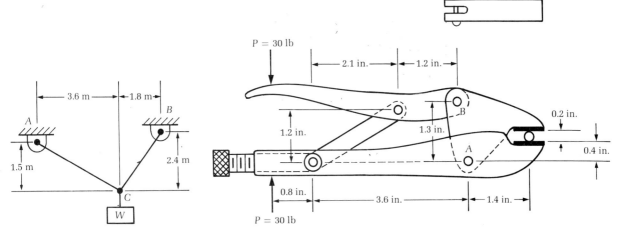

Figure P2-36   **Figure P2-37**

2-38 A steel bar 30-in. long with a 2-in. × 4-in. rectangular cross section is subjected to an axial tensile force. The length of the bar increases by 0.030 in., and the widths and thicknesses decrease by 0.001 in. and 0.0005 in., respectively. Determine Poisson's ratio for the material.

2-39 An aluminum block 2 in. × 2 in. × 15 in. long is subjected to a compressive axial load of 40,000 lb as shown in Figure P2-39. If $v = 0.3$ and $E = 10 \times 10^6$ psi, determine (a) the new dimensions of the block and (b) the unit change in volume.

2-40 The stress-strain data for a certain material were approximated by three straight line segments as shown in Figure P2-40. Show that the approximate stress-strain relations for this material can be written as

$$\sigma = \begin{cases} 60{,}000\epsilon \text{ (MPa)} & 0 \leq \epsilon \leq 0.002 \\ 90 + 15{,}000\epsilon \text{ (MPa)} & 0.002 \leq \epsilon \leq 0.006 \\ 150 + 5{,}000\epsilon \text{ (MPa)} & 0.006 \leq \epsilon \leq 0.008 \end{cases}$$

2-41 The shearing stress-shearing strain data for a certain material were approximated by two straight line segments as shown in Figure P2-41. Write

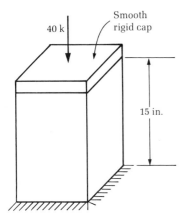

Figure P2-39

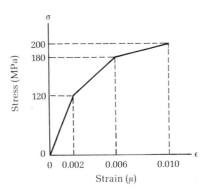

Figure P2-40   **Figure P2-41**

the relationships for stress and strain over the full range of strain indicated in the figure.

2-42 A thin aluminum tube with the dimensions shown in Figure P2-42a was tested in pure torsion. The data from this test are listed in Figure P2-42b. Plot the shearing stress-shearing strain diagram for the material. The gage length for which the data were acquired was 250 mm. Determine (a) the shearing modulus of elasticity, (b) the proportional limit stress, (c) the ultimate stress, (d) the yield strength at 0.2% offset, and (e) the modulus of resilience.

Torque (kN·m)	Angle of twist (Deg)
1.740	0.10
3.475	0.20
5.215	0.30
6.950	0.40
8.690	0.50
10.425	0.60
12.160	0.70
12.340	0.71
13.335	1.00
13.955	2.00
14.575	3.00
15.200	4.00
15.820	5.00
17.060	7.00
18.300	9.00
19.550	11.00

(a)   (b)

Figure P2-42

2-43 Use the superposition method to determine the displacement of point A of the axially loaded step-rod shown in Figure P2-43. Calculate the magnitudes of the maximum stress in the steel ($E = 210$ GPa), aluminum ($E = 70$ GPa), and brass ($E = 100$ GPa) segments. The cross-sectional areas of the two sections are 700 mm² and 1400 mm², respectively.

2-44 Use the discrete element method to determine the displacement of point A of the axially loaded, circular, aluminum rod shown in Figure P2-44. Calculate the magnitudes of the maximum tensile and compressive stresses in the rod.

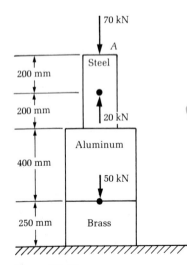

Figure P2-43

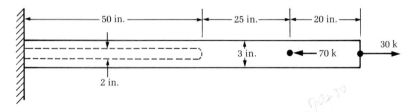

Figure P2-44

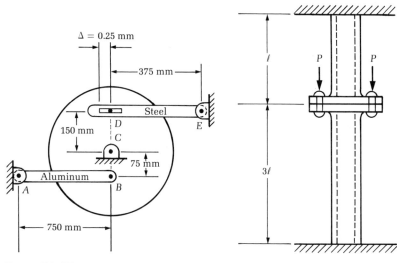

Figure P2-45

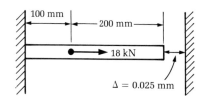

Figure P2-46

Figure P2-47

2-45 Determine the stresses induced in the aluminum and steel bars of the mechanism of Figure P2-45 due to an increase in temperature of 100 °F (E_{ST} = 210 GPa, E_{AL} = 70 GPa, α_{ST} = 6.5 × 10^{-6} in./in./°F, and α_{AL} = 13 × 10^{-6} in./in./°F. The cross-sectional areas of the aluminum and steel bars are equal. The circular plate can be considered rigid.

2-46 Two lengths of 2-in. standard steel pipe are connected by flanges as shown in Figure P2-46. The assembly is held firmly between two rigid supports. Determine (a) the reactions at the supports, (b) the maximum tensile and compressive stresses in the pipe if P = 10,000 lb, and (c) the axial tensile and compressive strain in the pipe.

2-47 An 18-kN force is applied to an aluminum rod as shown in Figure P2-47. The left end of the rod is attached to a rigid support. In the undeformed configuration, a gap (Δ = 0.025 mm) exists between the right end of the bar and another rigid support. If the cross-sectional area of the member is 500 mm^2, determine the reactions at the supports and the average tensile and compressive stresses.

2-48 Derive a formula for the reaction R at the right end of the rod shown in Figure P2-48.

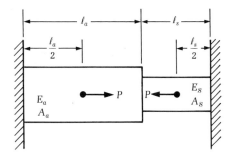

Figure P2-48

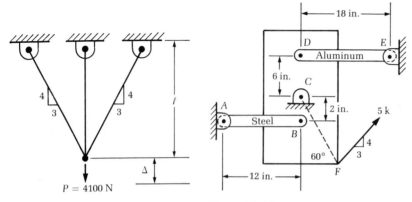

Figure P2-49   **Figure P2-50**

2-49 The cross-sectional area of each of the outside rods of the pin-connected assembly shown in Figure P2-49 is 10 mm^2. Determine the internal forces acting in each member and the corresponding average stress. (*Hint*: Note that the elongations of the outside rods are related to the displacement of joint A by the relations $e_1 = e_2 = \frac{4}{5}\Delta$.)

2-50 For the statically indeterminant mechanism shown in Figure P2-50, determine (a) the internal reactions acting in the aluminum and steel rods, (b) the average normal stress and strain associated with each bar, and (c) the elongation of each bar. Bars AB and DE are made of steel ($E = 30 \times 10^6$ psi) and aluminum ($E = 10 \times 10^6$ psi), respectively, and they have cross-sectional areas of 1 in^2 and 2 in^2, respectively. Consider the rectangular plate to be rigid and pin connected at C. The distance $CF = 8$ in.

CHAPTER 3

General States of Stress and Strain

3-1 Introduction

In Chapter 2, we defined engineering normal stress and normal strain for a uniaxial state of stress. We described in detail procedures used to establish the stress-strain diagram for an isotropic material subjected to a single tensile or compressive stress, and we explained how this diagram is used to obtain a relationship between stress and strain.

Chapter 2 also defined engineering shearing stress and shearing strain for a state of pure shear in a plane; it described the procedures used to establish shearing stress-shearing strain diagrams. Consequently, it showed how to obtain the stress-strain relation in pure shear.

In this chapter, we are concerned with establishing stress-strain relations for multiaxial states of stress. In many engineering members, normal stresses act in two or even three mutually perpendicular directions simultaneously. In addition, shearing stresses may act in two or even three mutually perpendicular planes. The most general state of stress consists of the three mutually perpendicular normal stresses and three shearing stresses acting simultaneously.

To analyze multiaxial states of stress, or to design an engineering member that experiences a multiaxial state of stress, we must establish analytical relationships that connect the stresses and the strains that occur in such a multiaxial state of stress. Moreover, it is desirable (for practical reasons) to determine material coefficients that appear in these stress-strain relations from the material coefficients obtained in simple tension and torsion tests. The chapter concludes with a detailed discussion of the elastic strain energy associated with uniaxial and multiaxial states of stress.

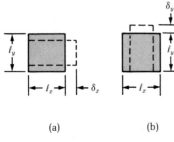

(a)   (b)

Figure 3-1

3-2 General Strain States

The engineering normal strain of a line element was defined in Section 2-2 as the ratio of its change in length divided by its original length. Since most structural members are inherently three dimensional, it is necessary to specify the normal strains of three mutually perpendicular line elements to characterize the volumetric change of a differential volume element. Figures 3-1a and 3-1b show the normal deformation caused by straining the differential element in the x direction and in the y direction independently. A third figure would show an analogous normal deformation for the z direction. The dimensions of the differential volume element in x, y, and z directions are denoted by ℓ_x, ℓ_y, and ℓ_z, respectively. According to the engineering definition, the normal strains of line elements that lie along the x, y, and z directions are

$$\left.\begin{aligned} \epsilon_x &= \frac{\delta_x}{\ell_x} \\ \epsilon_y &= \frac{\delta_y}{\ell_y} \\ \epsilon_z &= \frac{\delta_z}{\ell_z} \end{aligned}\right\} \tag{3-1}$$

where δ_x, δ_y, and δ_z are the normal deformations in the x, y, and z directions, respectively.

The three normal strains defined by Eq. (3-1) characterize the volumetric alterations of a differential volume element due to straining. A volumetric deformation is one for which the right angles of a differential element do not change. Deformations that cause alterations in the right angles of a differential volume element are called shearing deformations. Angular distortions of the volume element are characterized by three components of shearing strain.

Engineering shearing strain is customarily defined to be the change in angle between two initially perpendicular line elements. In Figure 3-2a, these line elements lie along the x and y axes and are shown as two sides of a differential volume element. After a pure shearing deformation denoted by δ_x in Figure 3-2a, the shape of the differential volume element is that shown by the dotted lines. Observe that the parallelopiped becomes a rhomboid. The shearing strain between the line element ℓ_x and the line element ℓ_y is

$$\gamma_{xy} = \frac{\delta_x}{\ell_y} \tag{3-2}$$

Shearing strains between line elements along the x and z axes and along the y and z axes are defined in a similar manner. Accordingly,

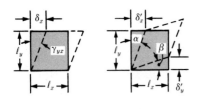

(a)   (b)

Figure 3-2

$$\gamma_{xz} = \frac{\delta_x}{\ell_z} \tag{3-3}$$

and

$$\gamma_{yz} = \frac{\delta_y}{\ell_z} \qquad (3\text{-}4)$$

Generally, a decrease in the original right angle between two line elements is considered a positive shearing strain.

Figure 3-2b shows a more symmetric way to define the components of shearing strain. In this figure, the dotted rhombus of Figure 3-2a is rotated so that angles α and β are equal. The shearing strains are now defined by the relations

$$\left. \begin{array}{l} \gamma_{xy} = \dfrac{\delta'_x}{\ell_y} + \dfrac{\delta'_y}{\ell_x} \\[2mm] \gamma_{xz} = \dfrac{\delta'_x}{\ell_z} + \dfrac{\delta'_z}{\ell_x} \\[2mm] \gamma_{yz} = \dfrac{\delta'_y}{\ell_z} + \dfrac{\delta'_z}{\ell_y} \end{array} \right\} \qquad (3\text{-}5)$$

Primes are used to emphasize that the individual shearing deformations of Figure 3-2b are not equal to those shown in Figure 3-2a. In the formulation of Eq. (3-5), we assume that angles α and β of Figure 3-2b are equal. However, a formulation with unequal angles will result in the same magnitudes for shearing strains because shearing strain is independent of the individual angles made with the coordinate axes.

The deformed shape of a differential volume element is completely defined by the three normal strains (ϵ_x, ϵ_y, ϵ_z) and the three shearing strains (γ_{xy}, γ_{xz}, γ_{yz}).

EXAMPLE 3-1

Determine the shearing strain between two line elements that were perpendicular before deformation if their relative orientation after deformation is 89° 54′.

SOLUTION

The shearing strain between two initially perpendicular line elements is defined as the change in the right angle that occurs because of the deformation. Therefore

$$\gamma = \frac{\pi}{2} - \frac{89.9(\pi)}{180} = 0.001745 = 1745\mu$$

The units on shearing strain are usually expressed as micro inches per inch (μ in./in.), micro meters per meter (μ m/m), or, frequently, as microns, since strain is dimensionless. It is usually convenient to retain the length dimension in expressing normal or shearing strain to remind us of their origins.

EXAMPLE 3-2

Three line elements OP, OQ, and OR shown in Figure 3-3 are mutually perpendicular in the undeformed state. In the deformed state, OQ and OR retain their orientations along the y and z axes, respectively; however, OP is now oriented as indicated in Figure 3-3. If $\varphi_{xy}^* = 89.95$ degrees and $\varphi_{xz}^* = 89.98$ degrees, determine the shearing strain between line elements OP and OQ and line elements OP and OR.

SOLUTION

By definition

$$\gamma_{xy} = \frac{\pi}{2} - \varphi_{xy}^* = \frac{\pi}{2} - \frac{89.95(\pi)}{180} = 872.7\mu$$

and

$$\gamma_{xz} = \frac{\pi}{2} - \varphi_{xz}^* = \frac{\pi}{2} - \frac{89.98(\pi)}{180} = 349.1\mu$$

Note that these shearing strains are positive because the right angles between line elements OP and OQ and line elements OP and OR decrease.

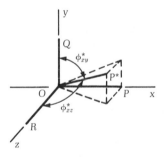

Figure 3-3

PROBLEMS / Section 3-2

3-1 A thin, cylindrical steel pressure vessel with the dimensions shown in Figure P3-1 is subjected to an internal pressure that causes both its length and its diameter to increase 0.238 mm and 0.417 mm, respectively. Determine the engineering strains associated with the line elements PQ and PR on the surface of the cylinder. Calculate the corresponding normal stresses and the shearing stress ($E = 210$ GPa and $v = 0.25$).

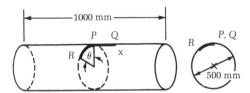

Figure P3-1

3-2 A thin membrane is stretched between two concentric hoops as shown in Figure P3-2. The inner hoop is kept from rotating. Determine the shearing strain that occurs between the line elements PQ and PR at the inner hoop when the outer hoop is rotated through a small angle φ.

3-3 A 2-in. square is scribed on a thin sheet of aluminum as shown in Figure P3-3a. A state of stress causes the diagonals PQ and RS of the square to attain the lengths 2.8298 in. and 2.8270 in., respectively, as shown in Figure P3-3b. Determine the shearing strain that occurs between the line elements RQ and QS and, hence, the shearing stress ($G = 3.8 \times 10^6$ psi).

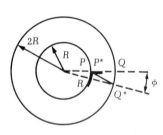

Figure P3-2

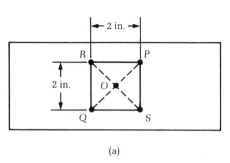

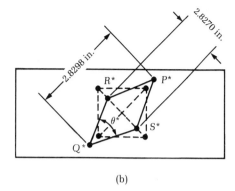

(a)   (b)

Figure P3-3

3-3 General Stress States

In Section 1-1, we defined the three components of stress that act on any cutting plane of a structural element. Figure 3-4a shows a three-dimensional structural element. The differential volume element shown in the figure has been obtained by passing appropriate cutting planes parallel to the coordinate axes through the structural element. The faces of the element whose outwardly directed normal vectors point in the directions of the positive coordinate axes are called the positive faces, and the opposite faces are called the negative faces. The stress components associated with the positive faces of this differential volume element are shown in Figure 3-4b. The corresponding stresses associated with the negative faces have opposite senses. We explained the placement of the subscripts of the stress components in Section 1-1. We observe that there appear to be nine independent stress components: the normal stresses σ_x, σ_y, σ_z, and the shearing stresses τ_{xy}, τ_{yx}, τ_{xz}, τ_{zx}, τ_{yz}, and τ_{zy}. However, the shearing stresses with identical letters as subscripts are equal. Thus, $\tau_{xy} = \tau_{yx}$, $\tau_{xz} = \tau_{zx}$, and $\tau_{yz} = \tau_{zy}$. These symmetry relations are easily verified by summing moments about the z, y, and x axes, respectively.

To demonstrate that the symmetry relations for shearing stress are indeed valid, we consider moments about the x axis. To facilitate the description a two-dimensional view of the differential volume element, Figure 3-4c shows only the forces that enter into the moment equation. Thus

$$\left\lfloor + \ \overset{\curvearrowleft}{\Sigma} M_x = 0: \quad \underbrace{(\tau_{yz}\,dx\,dz)}_{\text{force}} \ \underbrace{dy}_{\substack{\text{moment}\\\text{arm}}} \ - \ \underbrace{(\tau_{zy}\,dy\,dx)}_{\text{force}} \ \underbrace{dz}_{\substack{\text{moment}\\\text{arm}}} = 0 \right.$$

from which we obtain

$$\tau_{yz} = \tau_{zy} \tag{3-6}$$

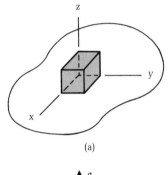

(a)

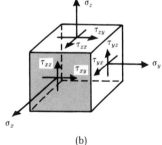

(b)

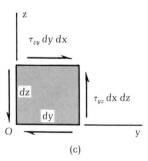

(c)

Figure 3-4

The other two symmetry relations can be verified in exactly the same manner.

As a result of the symmetry relations, six independent stress components completely characterize the state of stress at a given point in a structural member. Moreover, several of these components of stress are often identically zero.

3-4   Hooke's Law

An area of mechanics that receives a great deal of attention is the determination of functional relationships that connect the stress components $(\sigma_x, \sigma_y, \sigma_z, \tau_{xy}, \tau_{xz}, \tau_{yz})$ and the strain components $(\epsilon_x, \epsilon_y, \epsilon_z, \gamma_{xy}, \gamma_{xz}, \gamma_{yz})$ for various types of material behavior. Here we are concerned with the determination of suitable stress-strain relations for isotropic materials that behave linearly elastically. The most general linearly elastic material is called a **general Hookean material**, and it is defined by the linear relations

$$
\begin{aligned}
\epsilon_1 &= C_{11}\sigma_1 + C_{12}\sigma_2 + C_{13}\sigma_3 + C_{14}\sigma_4 + C_{15}\sigma_5 + C_{16}\sigma_6 \\
\epsilon_2 &= C_{21}\sigma_1 + C_{22}\sigma_2 + C_{23}\sigma_3 + C_{24}\sigma_4 + C_{25}\sigma_5 + C_{26}\sigma_6 \\
\epsilon_3 &= C_{31}\sigma_1 + C_{32}\sigma_2 + C_{33}\sigma_3 + C_{34}\sigma_4 + C_{35}\sigma_5 + C_{36}\sigma_6 \\
\epsilon_4 &= C_{41}\sigma_1 + C_{42}\sigma_2 + C_{43}\sigma_3 + C_{44}\sigma_4 + C_{45}\sigma_5 + C_{46}\sigma_6 \\
\epsilon_5 &= C_{51}\sigma_1 + C_{52}\sigma_2 + C_{53}\sigma_3 + C_{54}\sigma_4 + C_{55}\sigma_5 + C_{56}\sigma_6 \\
\epsilon_6 &= C_{61}\sigma_1 + C_{62}\sigma_2 + C_{63}\sigma_3 + C_{64}\sigma_4 + C_{65}\sigma_5 + C_{66}\sigma_6
\end{aligned}
\tag{3-7}
$$

In Eq. (3-7), we have used the notation

$$
\begin{aligned}
\epsilon_1 &= \epsilon_x \\
\epsilon_2 &= \epsilon_y \\
\epsilon_3 &= \epsilon_z \\
\epsilon_4 &= \gamma_{xz} \\
\epsilon_5 &= \gamma_{yz} \\
\epsilon_6 &= \gamma_{xy}
\end{aligned}
\quad\text{and}\quad
\begin{aligned}
\sigma_1 &= \sigma_x \\
\sigma_2 &= \sigma_y \\
\sigma_3 &= \sigma_z \\
\sigma_4 &= \sigma_{xz} \\
\sigma_5 &= \sigma_{yz} \\
\sigma_6 &= \sigma_{xy}
\end{aligned}
\tag{3-8}
$$

Notice that the subscripts 1, 2, and 3 replace the subscripts x, y, and z. Thus the x axis becomes the 1 axis, the y axis becomes the 2 axis, and the z axis becomes the 3 axis.

Observe that each strain component is a linear function of all six stress components. In general, a material of this type has different properties in different directions and is said to be **anisotropic**. The coefficients of the stress components C_{ij} are material properties and are called elastic constants or compliance coefficients.

Isotropic materials are characterized by the following features:

- The normal strains are independent of the shearing stresses.
- The shearing strains are independent of the normal stresses.

- A shearing strain depends only on the shearing stress that bears the same subscripts.
- The coefficients C_{ij} do not depend on the directions chosen for the x, y, and z axes.

Thus the stress-strain relations for an isotropic material reduce to

$$
\begin{aligned}
\epsilon_1 &= C_{11}\sigma_1 + C_{12}\sigma_2 + C_{13}\sigma_3 \\
\epsilon_2 &= C_{21}\sigma_1 + C_{22}\sigma_2 + C_{23}\sigma_3 \\
\epsilon_3 &= C_{31}\sigma_1 + C_{32}\sigma_2 + C_{33}\sigma_3 \\
\epsilon_4 &= C_{44}\sigma_4 \\
\epsilon_5 &= C_{55}\sigma_5 \\
\epsilon_6 &= C_{66}\sigma_6
\end{aligned}
\tag{3-9}
$$

The coefficients in the first three equations can be determined from simple tension tests. For example, if a tension specimen is cut from the material so that its center line is parallel to the x axis, and if the simple tension test shown in Figure 3-5 is performed, we have

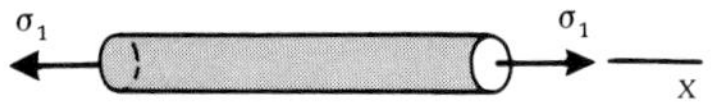

Figure 3-5

$$
\begin{aligned}
\epsilon_1 &= C_{11}\sigma_1 \\
\epsilon_2 &= C_{21}\sigma_1 \\
\epsilon_3 &= C_{31}\sigma_1
\end{aligned}
\tag{3-10}
$$

Recalling the definitions of the modulus of elasticity E from Section 2-2 and Poisson's ratio v from Section 2-4, we obtain

$$
\begin{aligned}
C_{11} &= \frac{\epsilon_1}{\sigma_1} = \frac{1}{E} \\[2mm]
C_{21} &= \frac{\epsilon_2}{\sigma_1} = -\frac{v\epsilon_1}{\sigma_1} = -\frac{v}{E} \\[2mm]
C_{31} &= \frac{\epsilon_3}{\sigma_1} = -\frac{v\epsilon_1}{\sigma_1} = -\frac{v}{E}
\end{aligned}
\tag{3-11}
$$

Similar tension tests on specimens that are cut from material with their center lines parallel to the y and z direction yield

$$
\begin{array}{lll}
C_{12} = -\dfrac{v}{E} & & C_{13} = -\dfrac{v}{E} \\[3mm]
C_{22} = \dfrac{1}{E} & \text{and} & C_{23} = -\dfrac{v}{E} \\[3mm]
C_{32} = -\dfrac{v}{E} & & C_{33} = \dfrac{1}{E}
\end{array}
\tag{3-12}
$$

We have used the fact that the material properties v and E are independent of direction for an isotropic material.

The last three equations of Eqs. (3-9) show that

$$C_{44} = \frac{\epsilon_4}{\sigma_4} = \frac{1}{G}$$

$$C_{55} = \frac{\epsilon_5}{\sigma_5} = \frac{1}{G}$$

$$C_{66} = \frac{\epsilon_6}{\sigma_6} = \frac{1}{G}$$

(3-13)

where G is the shearing modulus of elasticity for the isotropic material.

The three-dimensional stress-strain relations for an isotropic material that behaves linearly elastically now become

$$\epsilon_x = \frac{\sigma_x}{E} - \frac{v\sigma_y}{E} - \frac{v\sigma_z}{E}$$

$$\epsilon_y = -\frac{v\sigma_x}{E} + \frac{\sigma_y}{E} - \frac{v\sigma_z}{E}$$

$$\epsilon_z = -\frac{v\sigma_x}{E} - \frac{v\sigma_y}{E} + \frac{\sigma_z}{E}$$

$$\gamma_{xy} = \frac{\tau_{xy}}{G}$$

$$\gamma_{xz} = \frac{\tau_{xz}}{G}$$

$$\gamma_{yz} = \frac{\tau_{yz}}{G}$$

(3-14a)

Inverting Eqs. (3-14a), we can obtain the stress components expressed in terms of the strain components. These relations are given here for reference purposes.

$$\sigma_x = \frac{E}{(1 + v)(1 - 2v)} [(1 - v)\epsilon_x + v(\epsilon_y + \epsilon_z)]$$

$$\sigma_y = \frac{E}{(1 + v)(1 - 2v)} [(1 - v)\epsilon_y + v(\epsilon_z + \epsilon_x)]$$

$$\sigma_z = \frac{E}{(1 + v)(1 - 2v)} [(1 - v)\epsilon_z + v(\epsilon_x + \epsilon_y)]$$

$$\tau_{xy} = G\gamma_{xy}$$

$$\tau_{xz} = G\gamma_{xz}$$

$$\tau_{yz} = G\gamma_{yz}$$

(3-14b)

An important simplification of Eqs. (3-14a) occurs for so-called biaxial states of stress wherein all stress components lie in the same plane. Taking this common plane to be the xy plane, we have

$$\epsilon_x = \frac{\sigma_x}{E} - \frac{\nu\sigma_y}{E}$$

$$\epsilon_y = -\frac{\nu\sigma_x}{E} + \frac{\sigma_y}{E}$$ (3-15a)

$$\gamma_{xy} = \frac{\tau_{xy}}{G}$$

The inverted form of Eqs. (3-15a) is

$$\sigma_x = \frac{E}{1 - \nu^2}(\epsilon_x + \nu\epsilon_y)$$

$$\sigma_y = \frac{E}{1 - \nu^2}(\epsilon_y + \nu\epsilon_x)$$ (3-15b)

$$\tau_{xy} = G\gamma_{xy}$$

EXAMPLE 3-3

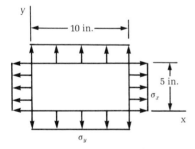

Figure 3-6

A thin steel rectangular plate with the dimensions shown in Figure 3-6 is deformed by uniform stresses σ_x and σ_y acting along its edges. The dimensions of the plate parallel to the x and y axes are observed to increase by 0.005 and 0.001 in., respectively. Determine the magnitudes of σ_x and σ_y that cause these deformations.

SOLUTION

The state of stress shown in Figure 3-6 is a plane state of stress; therefore, Eqs. (3-15b) apply. Thus

$$\left.\begin{array}{l} \sigma_x = \dfrac{E}{1 - \nu^2}(\epsilon_x + \nu\epsilon_y) \\[3mm] \sigma_y = \dfrac{E}{1 - \nu^2}(\epsilon_y + \nu\epsilon_x) \end{array}\right\}$$ (a)

The strains appearing in Eqs. (a) are calculated from the observed dimension changes. Thus

$$\left.\begin{array}{l} \epsilon_x = \dfrac{0.005}{10} = 0.0005 = 500\mu \text{ in./in.} \\[3mm] \epsilon_y = \dfrac{0.001}{5} = 0.0002 = 200\mu \text{ in./in.} \end{array}\right\}$$ (b)

Substituting the strains from Eqs. (b) into Eqs. (a) yields

$$\sigma_x = \frac{30 \times 10^6}{1 - 0.25^2} (500 + 0.25(200)) \times 10^{-6} = 17.6 \text{ ksi}$$

$$\sigma_y = \frac{30 \times 10^6}{1 - 0.25^2} (200 + 0.25(500)) \times 10^{-6} = 10.4 \text{ ksi}$$

EXAMPLE 3-4

Calculate the shearing stresses corresponding to the shearing strains calculated in Example 3-2 if the material is aluminum for which $G = 28$ GPa and linearly elastic behavior is assumed.

SOLUTION

The shearing stresses corresponding to γ_{xy} and γ_{xz} calculated in Example 3-2 are given by Eqs. (3-14a). Thus

$$\tau_{xy} = G\gamma_{xy} = 28 \times 10^9 (872.7 \times 10^{-6}) = 24.44 \text{ MPa}$$

and

$$\gamma_{xz} = G\gamma_{xz} = 28 \times 10^9 (349.1 \times 10^{-6}) = 9.77 \text{ MPa}$$

EXAMPLE 3-5

A rigid material has a smooth rectangular cavity of dimensions $a \times a \times h$ embedded in it. A cross section of the cavity is shown in Figure 3-7. The cavity is originally filled with a linearly elastic, isotropic material with modulus of elasticity E and Poisson's ratio v. The material in the cavity is compressed as shown in the figure by a rigid cap with a force P acting on it. Neglecting the weight of the rigid cap, determine the relationship between the force P and the decrease c in the height of the material.

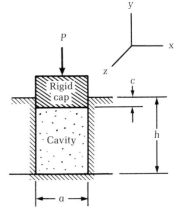

Figure 3-7

SOLUTION

This is a three-dimensional problem; however, no shearing strains occur in the material. Moreover, the only nonzero normal strain component is ϵ_y. The normal strain components in x and z directions are zero because the rigid medium surrounding the cavity does not allow the expansion of the material in these directions. Thus

$$\epsilon_y = -\frac{c}{h} \quad \text{and} \quad \epsilon_x = \epsilon_z = \gamma_{xy} = \gamma_{xz} = \gamma_{yz} = 0 \qquad \text{(a)}$$

Substituting these strain components into the constitutive relations

Eq. (3-14b), we obtain

$$\frac{1 - v}{v} \sigma_x = \frac{1 - v}{v} \sigma_z = \sigma_y = -\frac{E(1 - v)}{(1 + v)(1 - 2v)} \frac{c}{h} \qquad (b)$$

The shearing stress components are all zero.

The force P is obtained from a consideration of the free-body diagram of the rigid cap. Since the cap is weightless, we have

$$P = a^2 \sigma_y \qquad (c)$$

or

$$P = \frac{(1 - v)}{(1 + v)(1 - 2v)} \left(\frac{Ea^2}{h} \right) c \qquad (d)$$

The relationship between P and c is linear as expected because the material filling the cavity was assumed to be linearly elastic and because the displacement of the cap was restricted to small values.

Section 2-12 stated that, in any solid mechanics problem, there exist three basic sets of equations, and their simultaneous solution is essential for statically indeterminate problems. The examples in that section were for axially loaded members, resulting in a uniaxial stress state. The present example shows that the three-dimensionality of the problem does not relieve us of the need to consider these three basic ingredients. In the solution of this statically indeterminate problem, we have to use the equilibrium, compatibility, and constitutive equations simultaneously. Equations (a) are the compatibility equations since they were obtained strictly from the geometry of the problem. If compatibility equations contain strains, they are frequently called strain-displacement relations. Equations (b) were obtained from the stress-strain relations; that is, they describe material behavior. They correspond to the force-deformation relations of the previous uniaxial problems. Finally, Eq. (c) is the equilibrium equation obtained from the free-body diagram of the rigid cap. Simultaneous solution of Eqs. (a) through (c) yields the desired result (d). The conclusion is that simultaneous solution of the three sets of equations is inevitable for a statically indeterminate problem, whether it is a case of uniaxial stress state or general stress state.

PROBLEMS / Sections 3-3 and 3-4

3-4 The thin steel plate shown in Figure P3-4 is subjected to uniformly distributed normal forces along its edges. The dimensions parallel to the x and y axes are observed to increase 0.0020 in. and 0.0048 in., respectively. Determine the normal stresses σ_x and σ_y that correspond to these deformations ($E = 30 \times 10^6$ psi and $v = 0.25$).

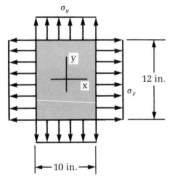

Figure P3-4

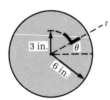

Figure P3-5

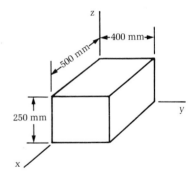

Figure P3-6

3-5 Electrical resistance strain gages are used to measure the radial and circumferential strains at a point 3 in. from the center of a thin 12-in. diameter aluminum plate as shown in Figure P3-5. These strains are $\epsilon_r = 300\mu$ and $\epsilon_\theta = -400\mu$. Calculate the normal stresses σ_r and σ_θ that are associated with these strains ($E = 10 \times 10^6$ psi and $v = 0.3$).

3-6 The x, y, and z dimensions of the aluminum block shown in Figure P3-6 are observed to change by the amounts 0.15 mm, -0.08 mm, and 0.10 mm, respectively, as the result of uniform stresses σ_x, σ_y, and σ_z being applied to its surfaces. Calculate the normal strains associated with these dimension changes. Calculate the corresponding normal stresses ($E = 70$ GPa and $v = 0.3$).

3-5 Strain Energy

An elastic spring subjected to an external force becomes deformed. The work done by the force in deforming the spring is said to be stored as elastic potential energy because the spring has the potential to give back an equal amount of work when it is allowed to return to its original configuration. In a similar manner, an arbitrary elastic body stores elastic energy when it is subjected to deformations. This elastic energy is commonly called **strain energy**. The strain energy associated with a unit volume of material at a point in the body is called the **strain energy density** of the material at that point, and it is denoted by U_0. Thus U_0 has units in.-lb/in^3 or N·m/m^3. A differential volume experiences an infinitesimal change in its strains corresponding to an infinitesimal change in the stresses, and, consequently, it experiences an infinitesimal change in its strain energy density. The infinitesimal change in U_0 is denoted by δU_0. To determine the total strain energy associated with a definite deformation process of a finite volume of material, we need two integrations: an integration relative to the deformation process (from which one obtains U_0) followed by an integration over the volume occupied by the body. Accordingly, the total strain energy associated with a finite volume of material is

$$U = \int_{\text{vol}} U_0 \, dv \qquad (3\text{-}16)$$

Consider an elastic spring of one of the types shown in Figure 3-8. Three types of springs are identified based on their load-deflection characteristics: nonlinear hardening springs, linear springs, and nonlinear softening springs. Regardless of the spring considered, the increment of strain energy associated with an increment of displacement δe can be obtained as the work done by the instantaneous force F on the infinitesimal displacement δe. Consequently,

$$\delta U = F \, \delta e \qquad (3\text{-}17)$$

It is not appropriate to think of strain energy density for the spring; consequently, we have considered the increment in total strain energy stored in the spring. For a continuous material, it is convenient to consider its strain energy density U_0 and from it compute the total strain energy associated with the finite volume it occupies.

If the force-displacement relationship for the spring can be expressed mathematically, Eq. (3-17) can be integrated. For example, the force-deflection relationship for a linear spring is

$$F = ke \qquad (3\text{-}18)$$

where k is the so-called spring constant and has units lb/in., N/m, or some similar convenient units. The strain energy stored in this linear spring at the end of a finite change e in the unstretched length of the spring is

$$U = \int_0^e ke \, de = \tfrac{1}{2}ke^2 \qquad (3\text{-}19)$$

or, in terms of the force F,

$$U = \int_0^F F \frac{dF}{k} = \frac{F^2}{2k} \qquad (3\text{-}20)$$

Expressions for the strain energy associated with nonlinear springs of either the hardening or of the softening kind can be developed in exactly the same manner. We must be able to express the force-displacement relationship as a mathematical equation.

Figure 3-9 depicts a planar torsional spring and its rotational characteristics. As with tension or compression springs, three types of torsional springs are identified: nonlinear hardening torsional springs, linear torsional springs, and nonlinear softening torsional springs.

The strain energy associated with an increment of rotation can be obtained as the work done by the instantaneous torque T on the

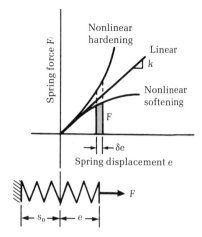

Figure 3-8

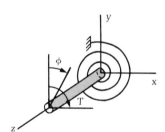

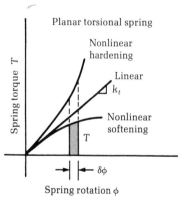

Figure 3-9

infinitesimal rotation $\delta\varphi$. Consequently,

$$\delta U = T\,\delta\varphi \qquad (3\text{-}21)$$

Again we note that it is not appropriate to think of strain energy density for the torsional spring; rather, we consider the increment of total strain energy stored in the spring.

If the torque-rotation relationship for the torsional spring can be expressed mathematically, Eq. (3-21) can be integrated. For example, the torque-rotation relationship for a linear torsional spring is

$$T = k_t\varphi \qquad (3\text{-}22)$$

where k_t is the torsional spring constant and has units in inch pounds per radian, Newton meters per radian, or some similar convenient units. The strain energy stored in the torsional spring at the end of a finite rotation φ from its unstrained configuration is

$$U = \int_0^\varphi k_t\varphi \; d\varphi = \tfrac{1}{2}k_t\varphi^2 \qquad (3\text{-}23)$$

or, in terms of the final torque T,

$$U = \int_0^T T\,\frac{dT}{k_t} = \frac{1}{2}\frac{T^2}{k_t} \qquad (3\text{-}24)$$

Figure 3-10a shows a differential volume of elastic material subjected to a normal stress acting along the x axis. The force acting on the plane area $(da_x = dy\,dz)$ perpendicular to the x axis that corresponds to a displacement of the area equal to e_x is $F_x = \sigma_x\,da_x$. The increment in strain energy stored in the differential volume dV as a result of an infinitesimal increment of displacement δe_x is

$$\delta(dU) = F_x\,\delta e_x = \sigma_x\,\delta\epsilon_x\,da_x\,dx = \sigma_x\,\delta\epsilon_x\,dV$$

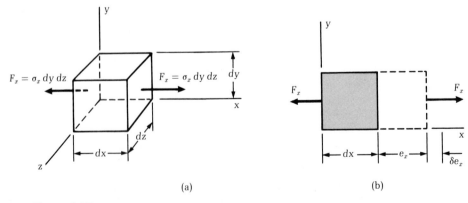

(a)   (b)

Figure 3-10

where use has been made of the relation $e_x = \epsilon_x \, dx$. From the last equation, we obtain

$$\delta\left(\frac{dU}{dV}\right) = \sigma_x \, \delta\epsilon_x \qquad (3\text{-}25)$$

We observe that dU/dV is the strain energy per unit volume of material, and it was previously designated by U_0. Accordingly, the increment in strain energy density associated with a uniaxial stress is

$$\delta U_0 = \sigma_x \, \delta\epsilon_x \qquad (3\text{-}26)$$

This is precisely the differential area of a stress-strain diagram as depicted in Figure 3-11.

For a linearly elastic material, $\sigma_x = E\epsilon_x$ so that

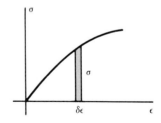

Figure 3-11

$$U_0 = \int_0^{\epsilon_x} E\epsilon_x \, \delta\epsilon_x = \tfrac{1}{2}E\epsilon_x^2 \qquad (3\text{-}27)$$

or, in terms of stresses,

$$U_0 = \int_0^{\sigma_x} \sigma_x \, \frac{\delta\sigma_x}{E} = \frac{1}{2}\frac{\sigma_x^2}{E} \qquad (3\text{-}28)$$

In a similar manner, we can obtain expressions for the increment in strain energy density for a normal stress acting along the y axis or along the z axis. Accordingly,

$$\delta U_0 = \sigma_y \, \delta\epsilon_y \qquad (3\text{-}29)$$

and

$$\delta U_0 = \sigma_z \, \delta\epsilon_z \qquad (3\text{-}30)$$

Formulas analogous to those given by Eqs. (3-27) and (3-28) are valid for stresses acting either along the y axis or along the z axis. We remark that the strain energy density resulting from the simultaneous action of three normal stress ($\sigma_x, \sigma_y, \sigma_z$) is *not* the algebraic sum of appropriate forms of Eq. (3-27) or (3-28). However, the superposition of increments of strain energy density is valid. Consequently, the increment in the strain energy density associated with a triaxial state of stress is

$$\delta U_0 = \sigma_x \, \delta\epsilon_x + \sigma_y \, \delta\epsilon_y + \sigma_z \, \delta\epsilon_z \qquad (3\text{-}31)$$

This relationship does *not* imply that the total energy of a triaxial state of stress can be obtained by superposition of the strain energy due to each stress acting separately.

Consider a three-dimensional differential volume of material such as that shown in Figure 3-10a. Instead of subjecting it to a normal stress

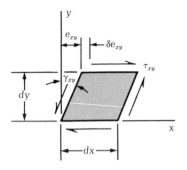

Figure 3-12

acting along the x axis, let a shearing stress τ_{xy} act on the area perpendicular to the x axis as shown by the two-dimensional view in Figure 3-12.

The increment in strain energy associated with the shearing stress can be determined as the work done by the shearing force $F_{xy} = \tau_{xy}\, da_y$ on the infinitesimal shearing displacement δe_{xy}. Consequently,

$$\delta(dU) = F_{xy}\,\delta e_{xy} = \tau_{xy}\,\delta\gamma_{xy}\,dV \tag{3-32}$$

since $\delta e_{xy} = \delta\gamma_{xy}\,dy$ and $da_y = dx\,dz$. Consequently, since $dU/dV \equiv U_0$, we write

$$\delta U_0 = \tau_{xy}\,\delta\gamma_{xy} \tag{3-33}$$

which is a differential element of area under a shearing stress-strain diagram similar to that of Figure 3-11.

For a linearly elastic material, $\tau_{xy} = G\gamma_{xy}$ so that

$$U_0 = \int_0^{\gamma_{xy}} G\gamma_{xy}\,\delta\gamma_{xy} = \tfrac{1}{2}G\gamma_{xy}^2 \tag{3-34}$$

or, in terms of stress,

$$U_0 = \int_0^{\tau_{xy}} \tau_{xy}\,\frac{\delta\tau_{xy}}{G} = \frac{\tau_{xy}^2}{2G} \tag{3-35}$$

Formulas for the increment in strain energy density for shearing stresses τ_{xz} or τ_{yz} can be developed in a manner analogous to the one used to develop Eq. (3-33). Accordingly,

$$\delta U_0 = \tau_{xz}\,\delta\gamma_{xz} \tag{3-36}$$

and

$$\delta U_0 = \tau_{yz}\,\delta\gamma_{yz} \tag{3-37}$$

As we shall see shortly, the strain energy density for shearing stresses τ_{xy}, τ_{xz}, and τ_{yz} acting simultaneously, in contrast to the conclusion for normal stresses, can be obtained by superposition of the strain energy densities associated with each shearing stress acting separately. This is a consequence of the fact that the relationship between the shearing stresses τ_{xy}, τ_{xz}, or τ_{yz} depends on only γ_{xy}, γ_{xz}, or γ_{yz}, respectively, whether these stresses act singly or simultaneously. The basic concept is that the increments in strain energy density can be superimposed. Consequently,

$$\delta U_0 = \tau_{xy}\,\delta\gamma_{xy} + \tau_{xz}\,\delta\gamma_{xz} + \tau_{yz}\,\delta\gamma_{yz} \tag{3-38}$$

which leads to

$$U_0 = \frac{\tau_{xy}^2}{2G} + \frac{\tau_{xz}^2}{2G} + \frac{\tau_{yz}^2}{2G} \tag{3-39}$$

for a linearly elastic material. Superposition of the individual strain energy densities would yield the same result.

Finally, the increment in the strain energy density of a material when all three normal stresses and all six shearing stresses act simultaneously is, by superposition,

$$\delta U_0 = \sigma_x \, \delta\epsilon_x + \sigma_y \, \delta\epsilon_y + \sigma_z \, \delta\epsilon_z + \tau_{xy} \, \delta\gamma_{xy} + \tau_{xz} \, \delta\gamma_{xz} + \tau_{yz} \, \delta\gamma_{yz}$$

(3-40)

Note that this formula is not restricted to a linear material. All that is required is that the relationship between stress and strain be expressible as a mathematical relation.

Axially loaded uniform rod. Consider a rod of uniform cross section A and length ℓ. Let the rod be loaded by an axial force P as shown in Figure 3-13a. The stress-strain diagram for the material from which the rod is made is shown in Figure 3-13b.

According to Eq. (3-28), the strain energy density for this rod is $U_0 = \sigma^2/2E$. The total energy associated with this rod is

$$U = \int_{\text{vol}} U_0 \, dV = \int_0^\ell \frac{\sigma^2}{2E} A \, dx = \int_0^\ell \frac{P^2}{2AE} \, dx \qquad (3\text{-}41)$$

or

$$U = \frac{P^2\ell}{2AE} \qquad (3\text{-}42)$$

Since $e = P\ell/AE$, Eq. (3-42) can be written as

$$U = \frac{EA}{2\ell} e^2 \qquad (3\text{-}43)$$

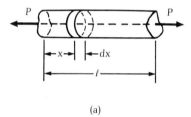

(a)

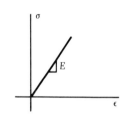

(b)

Figure 3-13

Biaxial states of stress. For a biaxial state of stress, the increment in strain energy density is

$$\delta U = \sigma_x \, \delta\epsilon_x + \sigma_y \, \delta\epsilon_y + \tau_{xy} \, \delta\gamma_{xy} \qquad (3\text{-}44)$$

Using the isotropic stress-strain relations given in Eqs. (3-15a) or (3-15b), we obtain

$$\delta U_0 = \left[\frac{E}{1 - v^2} \, (\epsilon_x + v\epsilon_y) \right] \delta\epsilon_x + \left[\frac{E}{1 - v^2} \, (\epsilon_y + v\epsilon_x) \right] \delta\epsilon_y$$

$$+ \frac{E}{2(1 + v)} \, \gamma_{xy} \, \delta\gamma_{xy}$$

(3-45)

or

$$\delta U_0 = \frac{E}{1 - v^2} \{\epsilon_x \, \delta\epsilon_x + \epsilon_y \, \delta\epsilon_y + v(\epsilon_x \, \delta\epsilon_y + \epsilon_y \, \delta\epsilon_x)\}$$

$$+ \frac{E}{2(1 + v)} \gamma_{xy} \, \delta\gamma_{xy}$$

(3-46)

Integration yields

$$U_0 = \frac{E}{2(1 - v^2)} \{\epsilon_x^2 + \epsilon_y^2 + 2v\epsilon_x\epsilon_y + \tfrac{1}{2}(1 - v)\gamma_{xy}^2\}$$

(3-47)

In terms of stress components,

$$\delta U_0 = \sigma_x \left[\frac{1}{E} \, (\delta\sigma_x - v \, \delta\sigma_y) \right] + \sigma_y \left[\frac{1}{E} \, (\delta\sigma_y - v \, \delta\sigma_x) \right]$$

$$+ \tau_{xy} \frac{2(1 + v)}{E} \, \delta\tau_{xy}$$

(3-48)

or, after integration,

$$U_0 = \frac{1}{2E} \{\sigma_x^2 + \sigma_y^2 - 2v\sigma_x\sigma_y + 2(1 + v)\tau_{xy}^2\}$$

(3-49)

Three-dimensional states of stress.   The strain energy density for three-dimensional states of stress can be derived in terms of either the stress components or the strain components in a manner analogous to that used to derive the strain energy density for a biaxial state of stress. The derivations of these formulas are left as exercises; however, the results are given here for reference purposes.

In terms of strain components,

$$U_0 = \frac{E}{2(1 + v)(1 - 2v)} \{(1 - v)(\epsilon_x^2 + \epsilon_y^2 + \epsilon_z^2)$$

$$+ 2v(\epsilon_x\epsilon_y + \epsilon_x\epsilon_z + \epsilon_y\epsilon_z)$$

$$+ \tfrac{1}{2}(1 - 2v)(\gamma_{xy}^2 + \gamma_{xz}^2 + \gamma_{yz}^2)\}$$

(3-50)

In terms of stress components,

$$U_0 = \frac{1}{2E} \{(\sigma_x^2 + \sigma_y^2 + \sigma_z^2) - 2v(\sigma_x\sigma_y + \sigma_x\sigma_z + \sigma_y\sigma_z)$$

$$+ 2(1 + v)(\tau_{xy}^2 + \tau_{xz}^2 + \tau_{yz}^2)\}$$

(3-51)

EXAMPLE 3-6

A thin aluminum plate is subjected to the edge stresses shown in Figure 3-14. The strain energy density for the plate under these stresses is 100 in.-lb/in^3. Determine the magnitude of the edge stresses. Assume a biaxial state of stress ($E = 10 \times 10^6$ psi and $v = 0.3$).

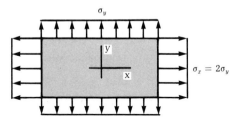

Figure 3-14

SOLUTION

The strain energy density for a biaxial state of stress is given by Eq. (3-49). Therefore,

$$100 = \frac{1}{2E} \{(2\sigma_y)^2 + \sigma_y^2 - 4(0.3)\sigma_y^2\} = \frac{1.9}{E} \sigma_y^2$$

or

$$\sigma_y = 22,940 \text{ psi}$$

and

$$\sigma_x = 2\sigma_y = 45,880 \text{ psi}$$

EXAMPLE 3-7

Assuming that the material contained in the cavity of Example 3-5 is steel ($E = 210$ GPa and $v = 0.25$), derive a formula for the strain energy density in terms of the load P. Let $a = 50$ mm and $h = 300$ mm.

SOLUTION

From Eq. (b) of Example 3-5, the normal stresses in the material are found to be

$$\left.\begin{array}{l} \sigma_x = \sigma_z = \dfrac{vE}{(1 + v)(1 - 2v)} \, \epsilon_y = \tfrac{2}{5}E\epsilon_y \\[2ex] \sigma_y = \dfrac{(1 - v)E}{(1 + v)(1 - 2v)} \, \epsilon_y = \tfrac{6}{5}E\epsilon_y \end{array}\right\} \qquad \text{(a)}$$

The strain energy density for an isotropic material under a three-dimensional state of stress is given by Eq. (3-51). Accordingly,

$$U_0 = \frac{1}{2E} \{(\tfrac{2}{5}E\epsilon_y)^2 + (\tfrac{6}{5}E\epsilon_y)^2 + (\tfrac{2}{5}E\epsilon_y)^2$$
$$- \tfrac{1}{2}[\tfrac{12}{25}E^2\epsilon_y^2 + \tfrac{4}{25}E^2\epsilon_y^2 + \tfrac{12}{25}E^2\epsilon_y^2]\}$$

or

$$U_0 = \tfrac{3}{5}E\epsilon_y^2 \qquad\qquad\qquad (b)$$

Now $\epsilon_y = -c/h$ from Eq. (a) of Example 3-5 so that

$$U_0 = \tfrac{3}{5}E\left(\frac{c}{h}\right)^2 \qquad\qquad\qquad (c)$$

From Eq. (d) of the same example,

$$P = \tfrac{6}{5}Ea^2\left(\frac{c}{h}\right)$$

so that Eq. (c) becomes

$$U_0 = \frac{5}{12}\frac{P^2}{Ea^4} \qquad (v = \tfrac{1}{4}) \qquad\qquad (d)$$

Substituting $a = 0.05$ m and $E = 210$ GPa,

$$U_0 = 0.3175P^2 \qquad\qquad\qquad (e)$$

If P is expressed in newtons, then U_0 has units N·m/m^3.

Volumetric and distortional strain energy.   The yielding of isotropic materials is closely related to the strain energy associated with the distortion of a volumetric element. Yielding is also nearly independent of volumetric strain energy. The strain energy associated with the distortion, or angle changes, of a differential volume element is called **strain energy of distortion**, while the strain energy associated with volumetric changes with no angle changes is called **volumetric strain energy**. Distortional strain energy plays an important role in theories of the failure of isotropic materials.

A general state of stress is usually decomposed into a spherical part and a deviator part because the volumetric strain energy depends only on the spherical part and the distortional energy depends only on the deviator part. Moreover, the derivations of formulas for the volumetric and distortional strain energies are simplified if the decomposition is accomplished relative to the principal axes of stress. Therefore, let the x, y, and z axes coincide with the principal axes of stress as shown in Figure 3-15a. The spherical state of stress is, by definition, a state of stress for which the stress in every direction has the same

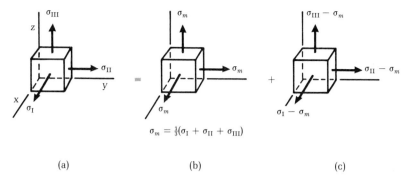

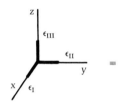

(a)   (b)   (c)

Figure 3-15

magnitude

$$\sigma_m = \tfrac{1}{3}(\sigma_I + \sigma_{II} + \sigma_{III}) \qquad (3\text{-}52)$$

as shown in Figure 3-15b. The difference between the original state of stress of Figure 3-15a and the spherical state of stress of Figure 3-15b is the deviator state of stress shown in Figure 3-15c.

The state of strain shown in Figure 3-16a corresponds to the stress state of Figure 3-15a. It is decomposed into the spherical state of strain shown in Figure 3-16b and the deviator state shown in Figure 3-16c in the manner just described.

The spherical strain is defined by the relation

$$e_m = \tfrac{1}{3}(\epsilon_I + \epsilon_{II} + \epsilon_{III}) \qquad (3\text{-}53)$$

Consequently, for an isotropic material, the stress-strain relation for the spherical stress and strain state is

$$e_m = \frac{1}{3}\left\{ \frac{\sigma_I - v(\sigma_{II} + \sigma_{III})}{E} + \frac{\sigma_{II} - v(\sigma_{III} + \sigma_I)}{E} + \frac{\sigma_{III} - v(\sigma_I + \sigma_{II})}{E} \right\}$$

or

$$e_m = \frac{1 - 2v}{E}\,\sigma_m = \frac{1 - 2v}{3E}(\sigma_I + \sigma_{II} + \sigma_{III}) \qquad (3\text{-}54)$$

A formula for the volumetric strain energy can be established by applying Eq. (3-31) to the spherical state of stress. Accordingly,

$$\delta U_0^v = 3\sigma_m\,\delta e_m \qquad (3\text{-}55)$$

Eqs. (3-54) and (3-55) yield, after integration,

$$U_0^v = \frac{3(1 - 2v)}{2E}\,\sigma_m^2 = \frac{1 - 2v}{6E}(\sigma_I + \sigma_{II} + \sigma_{III})^2 \qquad (3\text{-}56)$$

The distortional strain energy is the difference between the total strain energy associated with the state of stress shown in Figure 3-15a

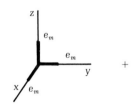

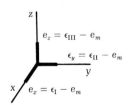

(a)

(b)

(c)

Figure 3-16

and the volumetric energy. The total strain energy is given by Eq. (3-51) with $\tau_{xy} = \tau_{xz} = \tau_{yz} = 0$ and $\sigma_x = \sigma_I$, $\sigma_y = \sigma_{II}$, and $\sigma_z = \sigma_{III}$. Consequently,

$$U_0 = \frac{1}{2E} \{\sigma_I^2 + \sigma_{II}^2 + \sigma_{III}^2 - 2v(\sigma_I\sigma_{II} + \sigma_I\sigma_{III} + \sigma_{II}\sigma_{III})\} \qquad (3\text{-}57)$$

Eqs. (3-56) and (3-57) yield, after some algebraic manipulation,

$$U_0^d = \frac{1+v}{6E} \{(\sigma_I - \sigma_{II})^2 + (\sigma_{II} - \sigma_{III})^2 + (\sigma_{III} - \sigma_I)^2\} \qquad (3\text{-}58)$$

EXAMPLE 3-8

Determine a formula for the strain energy of distortion per unit volume of material for a plane state of stress in terms of the general stress components σ_x, σ_y, and τ_{xy}.

SOLUTION

A formula for the distortional strain energy per unit volume for a plane state of stress is obtained from Eq. (3-58) by setting $\sigma_{III} = 0$. Consequently,

$$U_0^d = \frac{1+v}{3E} (\sigma_I^2 + \sigma_{II}^2 - \sigma_I\sigma_{II}) \qquad (a)$$

Now, for a plane state of stress, the principle stresses σ_I and σ_{II} are given by the formulas (see Chapter 7)

$$\left.\begin{array}{l} \sigma_I = \dfrac{\sigma_x + \sigma_y}{2} + \sqrt{\left(\dfrac{\sigma_x - \sigma_y}{2}\right)^2 + \tau_{xy}^2} \\[4mm] \sigma_{II} = \dfrac{\sigma_x + \sigma_y}{2} - \sqrt{\left(\dfrac{\sigma_x - \sigma_y}{2}\right)^2 + \tau_{xy}^2} \end{array}\right\} \qquad (b)$$

Direct substitution of Eqs. (b) into Eq. (a) yields

$$U_0^d = \frac{1+v}{3E} (\sigma_x^2 + \sigma_y^2 - \sigma_x\sigma_y + 3\tau_{xy}^2) \qquad (c)$$

EXAMPLE 3-9

The state of stress at a point in a material is $\sigma_x = 10$ ksi, $\sigma_y = 12$ ksi, $\tau_{xy} = 8$ ksi, and $\sigma_z = \tau_{zx} = \tau_{zy} = 0$. Determine the strain energy of distortion per unit volume of material. Assume $E = 30 \times 10^6$ psi and $v = 0.25$.

SOLUTION

From Eq. (c) of Example 3-8, the distortional strain energy density is

$$U_0^d = \frac{1 + 0.25}{3(30 \times 10^6)} \{10^2 + 12^2 - 10(12) + 3(8)^2\} \times 10^6$$

$$= \frac{1.25(316)}{3(30)} = 4.389 \text{ in.-lb/in}^3$$

EXAMPLE 3-10

The distortional strain energy per unit volume of an aluminum material is not to exceed 6 in.-lb/in^3. For a plane state of stress, $\sigma_x = 5$ ksi and $\sigma_y = 10$ ksi. Determine the magnitude of the largest permissible shearing stress that can accompany these normal stresses. Assume $E = 10 \times 10^6$ psi and $v = 0.3$.

SOLUTION

From Eq. (c) of Example 3-8 we find

$$6 = \frac{1 + 0.3}{3(10 \times 10^6)} \{5^2 + 10^2 - 5(10) + 3\tau_{xy}^2\} \times 10^6$$

or

$$\tau_{xy} = 4.6 \text{ ksi}$$

PROBLEMS / Section 3-5

3-7 Two rigid links are connected by smooth pins at their ends as shown in Figure P3-7a. Angular displacements at the pins are resisted by identical linear torsional springs. Characteristics for the springs are shown in Figure P3-7b. Determine the strain energy stored in each spring and in the system. Assume that the springs are torque-free when the links are aligned as indicated by the dotted lines.

3-8 A flexible beam rests on an elastic foundation as shown in Figure P3-8. The elastic foundation is represented by a continuously distributed set of nonlinear

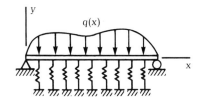

Figure P3-8

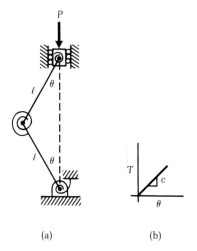

(a)   (b)

Figure P3-7

springs for which the force-displacement relationship is $F = k(y + \alpha y^3)$. Determine the strain energy associated with the foundation because of a small lateral displacement y of the beam. The force F is in.-lb/in.

3-9 Three rigid links are connected by smooth pins at their ends as shown in Figure P3-9a. Angular displacements at the pins are resisted by identical linear torsional springs. The torque-angular displacement relation for the springs is shown in Figure P3-8b. Determine a formula for the strain energy stored in each spring. Express the total energy of the system in terms of the angles θ and φ. The springs are torque-free when the links are aligned along the horizontal. Assume small angular displacements.

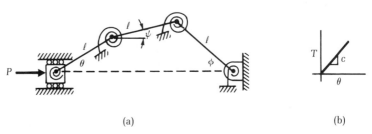

(a)   (b)

Figure P3-9

3-10 Two rigid links are connected at their ends by smooth pins. Lateral displacement of the central pin is resisted by a spring with the nonlinear characteristics shown in Figure P3-10. Determine a formula for the strain energy stored in the spring. Assume that the spring is force-free when the links are aligned as indicated by the dotted line.

3-11 A nonlinear spring has the force-displacement characteristics $F = kx^3$. Determine a formula for the strain energy associated with this spring.

3-12 Calculate the elastic strain energy stored in the steel rod shown in Figure P3-12. The forces $P_x = 300$ kN cause a uniform stress on the end sections of the rod ($E = 210$ GPa).

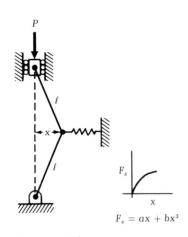

$F_s = ax + bx^3$

Figure P3-10

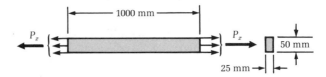

Figure P3-12

3-13 A straight rod with cross-sectional area A and length ℓ is subjected to the tensile forces P as shown in Figure P3-13a. The material of the rod has the stress-strain diagram shown in Figure P3-13b. Derive a formula for the strain energy stored in the rod. Express the result in terms P, ℓ, K, A, and n.

3-14 A thin aluminum plate 0.125 in. thick has the planar dimensions shown in Figure P3-14. The forces $P_x = 10$ k and $P_y = 25$ k cause the uniform stresses σ_x and σ_y acting on the edges of the plate. Calculate the strain energy density for this stress distribution and, subsequently, the strain energy stored in the plate ($E = 10 \times 10^6$ psi and $\nu = 0.3$).

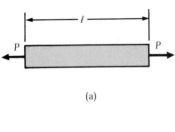

(a)

$\sigma = K\epsilon^n$

(b)

Figure P3-13

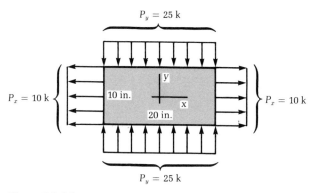

Figure P3-14

3-6 Dynamic Loading

Let us investigate the effect of applying a load on the deflection of an elastic system in a sudden or dynamic manner. The elastic system is perceived as a linear spring, and the dynamic load that is applied to the system as a weight W that falls freely through a prescribed height h is shown in Figure 3-17a. The force-deflection relation for the linear spring is depicted in Figure 3-17c.

We assume that the material behaves elastically with no energy being dissipated as a result of inelastic deformation at either the point of impact or at the system supports. Furthermore, we assume that the inertia of the system resisting the dynamic load is negligible. These assumptions imply that the deflections of the system are directly proportional to applied force regardless of whether the force is static or dynamic.

Since no energy is dissipated, mechanical energy is conserved. From the principle of conservation of mechanical energy,

$$W \cdot h + 0 = - W \delta_{DYN} + \tfrac{1}{2} k \delta_{DYN}^2 \tag{3-59}$$

where the reference for potential energy is as shown in Figure 3-17a and δ_{DYN} is the maximum dynamic deflection. Now $W = k \delta_{ST}$ where δ_{ST} is the static deflection of the spring—that is, the deflection that corresponds to the equilibrium position for the spring weight system. Equation (3-59) can now be written as

$$\delta_{DYN}^2 - 2 \delta_{DYN} \delta_{ST} - 2h \delta_{ST} = 0 \tag{3-60}$$

Solving this quadratic equation yields

$$\delta_{DYN} = \left(1 + \sqrt{1 + \frac{2h}{\delta_{ST}}} \right) \delta_{ST} \tag{3-61}$$

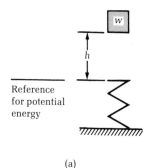

Reference
for potential
energy

(a)

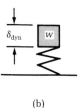

(b)

(c)

Figure 3-17

or

$$\delta_{DYN} = K\,\delta_{ST} \tag{3-62}$$

where

$$K = 1 + \sqrt{1 + \frac{2h}{\delta_{ST}}} \tag{3-63}$$

is referred to as an **impact factor**.

Evidently the dynamic deflection can be calculated from Eq. (3-62) after the impact factor has been calculated from Eq. (3-63). Observe that the impact factor depends only on the height h through which the weight falls freely and the deflection of the spring if the weight is applied statically.

Now, according to our third assumption,

$$P_{DYN} = k\,\delta_{DYN} = k(K\,\delta_{ST}) = KP_{ST} \tag{3-64}$$

This result is an important one because, as we shall see later, quantities such as shearing forces and bending moments in beams are related linearly to the applied force. Consequently, if the applied forces are of the dynamic type considered here, we can obtain the shearing forces and bending moments for beams loaded dynamically by multiplying the corresponding static quantities by the impact factor K.

EXAMPLE 3-11

A steel-step rod 60 in. long is shown in Figure 3-18. If a 6-lb weight is released from rest from the position shown, determine the resulting maximum deflection of the free end of the rod.

SOLUTION

The static deflection of the rod is

$$\delta_{ST} = \frac{6(30)}{(\pi/4)(\tfrac{1}{2})^2 30 \times 10^6} + \frac{6(30)}{(\pi/4)30 \times 10^6} = 38.2 \times 10^{-6} \text{ in.} \quad \text{(a)}$$

The impact factor for this system is

$$K = 1 + \sqrt{1 + 2\left(\frac{29}{38.2 \times 10^{-6}}\right)} = 1233 \quad \text{(b)}$$

so that the dynamic deflection based on the present theory is

$$\delta_{DYN} = 1233(38.2 \times 10^{-6}) = 0.047 \text{ in.} \quad \text{(c)}$$

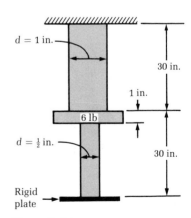

Figure 3-18

Note that the actual deflection would be less than th
because we have disregarded inertia effects, and there is l
some energy dissipated because of inelastic deformations at t
of impact. Also note that this dynamic deflection is not a perma
one since vibrations will begin immediately after the maximu
deflection occurs. In the presence of some damping in the system,
the free end of the rod will eventually come to rest at its static equilibrium position.

PROBLEMS / Section 3-6

3-15   The static deflection at the end of the cantilevered beam shown in Figure
P3-15 is given by the formula $\delta_{ST} = W\ell^3/3EI$, where ℓ is its length, I is the moment
of inertia of its cross-sectional area about the xx axis, and E is the modulus of
elasticity. Determine the maximum height through which the weight $w = 5$ lb
can be dropped if the maximum deflection at the end of the cantilever is not to
exceed 0.090 in. Use $E = 30 \times 10^6$ psi ($h = 0.789$ in.).

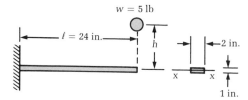

Figure P3-15

3-16   The mechanism shown in Figure P3-16 is a crude model of a torsional bar
for an automobile. The static angle of twist of the bar AB is given by the formula
$\varphi_{ST} = T\ell/JG$, where T is the torque applied to the bar, ℓ is the length, J is the polar
moment of inertia of the cross-sectional area of the bar relative to the axis of the
bar, and G is the shearing modulus of elasticity for the material from which the
rod is made. Determine the maximum angular rotation of the arm AC ($G = 84$ GPa)
($\varphi_{DYN} = 1.05$ degrees).

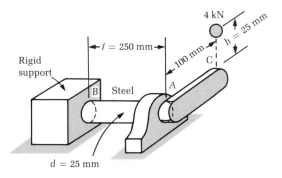

Figure P3-16

at calculated
kely to be
e point
ent

ns and shearing strains for a three-dimen-
e also established the following stress-strain
terials.

Strain Relations

$$\epsilon_x = \frac{\sigma_x}{E} - \frac{v\sigma_y}{E} - \frac{v\sigma_z}{E}$$

$$\sigma_x = \frac{E}{(1+v)(1-2v)}\left\{(1-v)\epsilon_x + v(\epsilon_y + \epsilon_z)\right\}$$

$$\epsilon_y = \frac{\sigma_y}{E} - \frac{v\sigma_z}{E} - \frac{v\sigma_x}{E}$$

$$\sigma_y = \frac{E}{(1+v)(1-2v)}\left\{(1-v)\epsilon_y + v(\epsilon_z + \epsilon_x)\right\}$$

$$\epsilon_z = \frac{\sigma_z}{E} - \frac{v\sigma_x}{E} - \frac{v\sigma_y}{E}$$

$$\sigma_z = \frac{E}{(1+v)(1-2v)}\left\{(1-v)\epsilon_z + v(\epsilon_x + \epsilon_y)\right\}$$

$$\gamma_{xy} = \frac{\tau_{xy}}{G}$$

$$\tau_{xy} = G\gamma_{xy}$$

$$\gamma_{xz} = \frac{\tau_{xz}}{G}$$

$$\tau_{xz} = G\gamma_{xz}$$

$$\gamma_{yz} = \frac{\tau_{yz}}{G}$$

$$\tau_{yz} = G\gamma_{yz}$$

Two-Dimensional Stress-Strain Relations

$$\epsilon_x = \frac{\sigma_x}{E} - \frac{v\sigma_y}{E}$$

$$\sigma_x = \frac{E}{1-v^2}(\epsilon_x + v\epsilon_y)$$

$$\epsilon_y = \frac{\sigma_y}{E} - \frac{\nu\sigma_x}{E}$$

$$\sigma_y = \frac{E}{1 - \nu^2}(\epsilon_y + \nu\epsilon_x)$$

$$\gamma_{xy} = \frac{\tau_{xy}}{G}$$

$$\tau_{xy} = G\gamma_{xy}$$

$$G = \frac{E}{2(1 + \nu)}$$

We also established basic formulas for the increment in strain energy density associated with uniaxial, biaxial, pure shear, and general states of stress. They are

$$\delta U_0 = \sigma_x \delta\epsilon_x \quad \text{(Uniaxial)}$$

$$\delta U_0 = \sigma_x \delta\epsilon_x + \sigma_y \delta\epsilon_y + \tau_{xy} \delta\gamma_{xy} \quad \text{(Biaxial)}$$

$$\delta U_0 = \tau_{xy} \delta\gamma_{xy} \quad \text{(Pure shear)}$$

$$\delta U_0 = \sigma_x \delta\epsilon_x + \sigma_y \delta\epsilon_y + \sigma_z \delta\epsilon_z + \tau_{xy} \delta\gamma_{xy} + \tau_{xz} \delta\gamma_{xz} + \tau_{yz} \delta\gamma_{yz}$$
$$\text{(General)}$$

Strain Energy Density Formulas for Isotropic Materials

$$U_0 = \frac{\sigma_x^2}{2E} = \frac{E\epsilon_x^2}{2} \quad \text{(Uniaxial)}$$

$$U_0 = \frac{\tau_{xy}^2}{2G} = \frac{G\gamma_{xy}^2}{2} \quad \text{(Pure shear)}$$

$$U_0 = \begin{cases} \dfrac{1}{2E}\{\sigma_x^2 + \sigma_y^2 - 2\nu\sigma_x\sigma_y + 2(1+\nu)\tau_{xy}^2\} \\[4mm] \dfrac{E}{2(1-\nu^2)}\{\epsilon_x^2 + \epsilon_y^2 + 2\nu\epsilon_x\epsilon_y + \tfrac{1}{2}(1-\nu)\gamma_{xy}^2\} \end{cases} \qquad \text{(Biaxial)}$$

$$U_0 = \begin{cases} \dfrac{1}{2E}\{\sigma_x^2 + \sigma_y^2 + \sigma_z^2) - 2\nu(\sigma_x\sigma_y + \sigma_x\sigma_z + \sigma_y\sigma_z) \\[2mm] \qquad + 2(1+\nu)(\tau_{xy}^2 + \tau_{xz}^2 + \tau_{yz}^2)\} \qquad \text{(General)} \\[4mm] \dfrac{E}{2(1+\nu)(1-2\nu)}\{(1-\nu)(\epsilon_x^2 + \epsilon_y^2 + \epsilon_z^2) + 2\nu(\epsilon_x\epsilon_y + \epsilon_x\epsilon_z + \epsilon_y\epsilon_z) \\[2mm] \qquad + \tfrac{1}{2}(1-2\nu)(\gamma_{xy}^2 + \gamma_{xz}^2 + \gamma_{yz}^2)\} \end{cases}$$

$$U_0^v = \dfrac{1-2\nu}{6E}(\sigma_I + \sigma_{II} + \sigma_{III})^2 \qquad \text{(Volumetric energy)}$$

$$U_0^d = \dfrac{1+\nu}{6E}\{\sigma_I - \sigma_{II})^2 + (\sigma_{II} - \sigma_{III})^2 + (\sigma_{III} - \sigma_I)^2\}$$
$$\text{(Deviatoric energy)}$$

The chapter closed with the development of a theory that can be used to calculate stresses and deflections due to dynamic loads. Pertinent formulas are

$$K = 1 + \sqrt{1 + \dfrac{2h}{\delta_{ST}}} \qquad \text{(Impact factor)}$$
$$P_{DYN} = KP_{ST} \qquad \text{and} \qquad \delta_{DYN} = K\delta_{ST}$$

PROBLEMS   CHAPTER 3

3-17 Calculate the dimensional changes for a unit cube of material under the stress system shown in Figure P3-17. The modulus of elasticity for the material is 30×10^6 psi, and Poisson's ratio is 0.25.

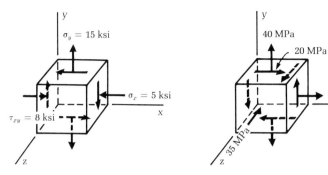

Figure P3-17   **Figure P3-19**

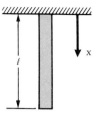

Figure P3-20

3-18   Determine the state of stress that corresponds to the following state of strain: $\epsilon_x = 0.001$, $\epsilon_y = -0.005$, $\epsilon_z = 0$, $\gamma_{xy} = -0.0025$, $\gamma_{yz} = 0.025$, $\gamma_{zx} = 0$ ($E = 210$ GPa and $\nu = 0.25$).

3-19   The stresses at a point in an aluminum member are shown in Figure P3-19. Calculate the normal and shearing strains associated with this element ($E = 70$ GPa and $\nu = \frac{1}{3}$).

3-20   The homogeneous rod with a uniform cross section A (in^2), weight density γ (lb/in^3), and length ℓ (in.) shown in Figure P3-20 is suspended from a rigid support. Derive a formula for the strain energy stored in the rod due to its own weight.

3-21   The weight density of the homogeneous conical frustrum shown in Figure P3-21 is $\gamma = 0.284$ lb/in^3. Determine the strain energy stored in the rod due to its own weight. Use $E = 30 \times 10^6$ psi as the elastic modulus for the material.

3-22   The ultimate strength of the 0.5-in. diameter steel rod shown in Figure P3-22 is 65 ksi. Determine the maximum height that the 100-lb weight can be dropped if the axial stress in the rod is not to exceed the ultimate strength reduced by a safety factor of 2. Neglect the weight of the bar and the flange ($h = 2.0$ in.).

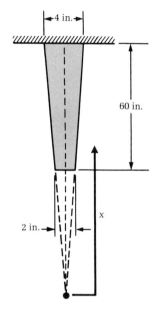

Figure P3-21

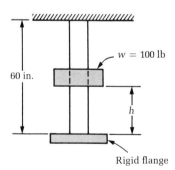

Figure P3-22

CHAPTER 4

Torsion

4-1 Introduction

This chapter establishes procedures required for the analyses of the strength and stiffness characteristics of straight rods that resist only torsional loads. Examples of this type of structural element can be readily observed in an industrial society. Thus examples of torsionally loaded shafts are shafts connecting motor-pump and motor-generator sets, propeller shafts in airplanes, helicopters, and ships, torsion bars in automobile suspension systems, and the drive shafts of automobiles. Many tool components can be viewed essentially as torsionally loaded shafts: screw drivers, drill bits, and router tools are examples. (These tool components also rely on an axial load component for their effectiveness.)

4-2 Power Transmission

The operating specifications for a motor frequently list the power it transmits in horsepower (HP) and the angular speed of the motor either in revolutions per minute (rpm) or in cycles per second (Hz). However, a designer needs to know the torque that the shaft will be required to transmit to select a shaft of appropriate dimensions. One of the most important structural components is the shaft used to transmit power between a power source and a machine of some variety. The shafts in motor-generator and motor-pump combinations, automobile drive shafts, propellor shafts in sea-going vessels, and various power train systems used in industry are examples of the widespread use of torsional shafts to transmit power.

Figure 4-1 shows a motor, a generator, or a pump and a segment of a shaft that connects either of them with another rotating machine. The angular velocity of the shaft is denoted by ω, and the torque transmitted by the shaft is denoted by T.

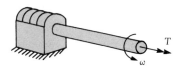

Figure 4-1

From physics, the power transmitted by a torque T is

$$P = T\omega \qquad (4\text{-}1)$$

In SI units, $\omega = 2\pi f$, where f is the frequency of the rotating shaft in Hertz. Accordingly, if T is expressed in newton meters (N·m), the power transmitted by the shaft is

$$P = 2\pi fT \text{ N·m/s} \qquad (4\text{-}2)$$

Now, 1 HP is 745.7 N·m/s so that the horse power transmitted by the shaft is

$$\text{HP} = \frac{2\pi fT}{745.7} = \frac{fT}{119} \qquad (4\text{-}3)$$

In English units, $\omega = 2\pi n$ where n is expressed in revolutions per minute. Accordingly, if T is expressed in inch pounds (in.·lb), the power transmitted is

$$P = 2\pi nT \text{ in. lb/min} \qquad (4\text{-}4)$$

One horsepower is 550 ft·lb/sec so that the horsepower transmitted is

$$\text{HP} = \frac{2\pi nT}{550(12)(60)} = \frac{nT}{63,000} \qquad (4\text{-}5)$$

4-3   Kinematics for Circular Shafts

Since shafts with solid or hollow circular cross sections are more widely used than shafts with noncircular cross sections, and since the theory of circular shafts is significantly simpler than theories for noncircular shafts, we shall focus our attention on shafts with circular cross sections first.

To take advantage of cylindrical symmetry, we use cylindrical coordinates shown in Figure 4-2a. We make three basic geometric assumptions regarding the deformational behavior of shafts with circular cross sections.

1. The circular cross sections of the shaft are rigid. To realize the implications of this assumption, consider the line elements shown in Figure 4-2b. One line element lies along a radial line, and the other lies along a circle of radius r. Since the cross section is rigid, $\epsilon_r = \epsilon_\theta = \gamma_{r\theta} = 0$. Here ϵ_r and ϵ_θ are the normal strains of the line elements in the radial and tangential directions, respectively, and $\gamma_{r\theta}$ is the shearing strain between these line elements.

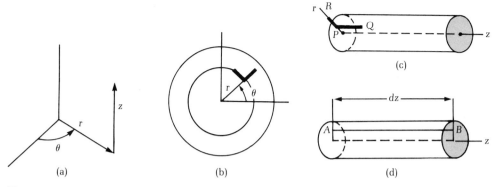

Figure 4-2

2. The shaft remains straight, and cross sections perpendicular to its geometric axis before deformation remain perpendicular to its geometric axis after deformation. Figure 4-2c depicts a small segment of a circular shaft. PQ and PR are line elements parallel to the geometric axis and a radial direction, respectively. The current assumption implies that the change in right angle between these two line elements is zero. Accordingly, there is no shearing in the rz plane either; that is, $\gamma_{rz} = 0$.

3. The distance between cross sections is left unchanged by torsional action. Clearly, a line element parallel to the axis of the shaft, such as line AB of Figure 4-2d, undergoes no strain; that is, the normal strain ϵ_z of a line element lying along the z axis is zero.

The assumptions listed here are geometric. They do not depend on material behavior and, therefore, they are valid for elastic and inelastic material behavior. These assumptions are limited to reasonably small deformations; otherwise they remain valid for various kinds of material behavior.

A general state of stress referred to cylindrical coordinates is shown in Figure 4-3a. Notice that only the stress components that act on the positive coordinate faces are shown.

Now, collectively, the foregoing geometric assumptions require that $\epsilon_r = \epsilon_\theta = \epsilon_z = 0$ and $\gamma_{r\theta} = \gamma_{rz} = 0$. Consequently, by Hooke's law (Chapter 3), we see that the corresponding stress components $\sigma_r = \sigma_\theta = \sigma_z = 0$ and $\tau_{r\theta} = \tau_{rz} = 0$. Thus the only nonzero stress component is $\tau_{\theta z}$. Figure 4-3b depicts this result. Notice that $\tau_{\theta z} = \tau_{z\theta}$ because shearing stresses must occur in pairs. From this point forward, the subscripts on $\tau_{\theta z}$ and $\gamma_{\theta z}$ will be dropped.

According to the three geometric assumptions listed here, the deformation of a circular shaft due to pure twisting is simply a rotation of a cross section about the geometric axis of the shaft. Figure 4-4a indicates two cross sections located at z and $z + \Delta z$, respectively. Consider two line elements on a cylindrical surface a distance r from the axis of the shaft. The line element AB lies parallel to the z axis

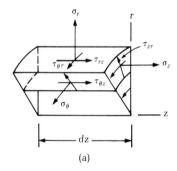

(a)

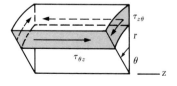

(b)

Figure 4-3

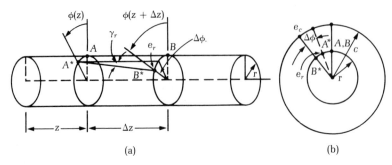

(a)   (b)

Figure 4-4

before twisting. The second line element lies along the circumference of the cylindrical surface and emanates from point A. Consequently, these axial and circumferential line elements are perpendicular before the twisting of the shaft occurs. After twisting, points A and B reside at points A^* and B^*, respectively. We note that points A^* and B^* still lie in the same circumference as they did prior to deformation. However, line A^*B^* is no longer parallel to the z axis. Accordingly, the right angle between the axial and circumferential line elements decreases by the angle γ_r shown in Figure 4-4a. This is the shearing strain induced by the applied torque. The subscript r indicates that γ_r is the shearing strain at a distance r from the axis of the shaft.

From Figure 4-4a, the shearing deformation at a distance r from the axis of the shaft can be expressed as

$$r\,\Delta\varphi = \gamma_r\,\Delta z \tag{4-6}$$

Here we have made use of the fact that the two triangles—the one on the cylindrical surface with angle γ_r and the one on the circular cross section with angle $\Delta\varphi$—have a common side. Moreover, the deformations are assumed to be small enough so that γ_r and $\Delta\varphi \ll 1$. In Eq. (4-6), upon letting the distance Δz become infinitesimally small, we obtain

$$\frac{d\varphi}{dz} = \frac{\gamma_r}{r} \tag{4-7}$$

Equation (4-7) expresses the relative rotation of the cross section at $z + dz$ with respect to the section at z in terms of the shearing strain at a distance r from the center of the shaft.

Now, from Figure 4-4b, we find that the shearing deformation e_r varies along a radius according to the relation

$$e_r = \frac{r}{c}\,e_c \tag{4-8}$$

where e_r and e_c are shearing deformations at distances r and c from the axis of the shaft. Clearly, the shearing deformations vary linearly with this distance. Note also that $e_r = r\,d\varphi$ and $e_c = c\,d\varphi$ as seen in Figure 4-4b.

Dividing Eq. (4-8) by dz and observing that $e_r/dz = \gamma_r$ and $e_c/dz = \gamma_c$, note that the shearing strains also vary linearly with the radial distance r. Hence

$$\gamma_r = \frac{r}{c}\,\gamma_c \tag{4-9}$$

Equations (4-7) through (4-9) have resulted from purely geometric considerations. Therefore, the formulas remain valid for any material behavior.

4-4 Equilibrium Considerations

An end view of the infinitesimal element shown in Figure 4-3b is shown in its position in the circular cross section in Figure 4-5. A concentrated torque T_R that is statically equivalent to the torque produced by the shearing stress τ is

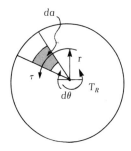

$$T_R = \int_{\text{area}} (\tau\,da)r \tag{4-10}$$

Figure 4-5

This relation depends on only statical considerations, and it is independent of material behavior. It is, therefore, valid for elastic and inelastic material behavior.

T_R is the resisting torque at the section, and it can be expressed in terms of the externally applied torques by summing moments about the axis of the shaft. Accordingly,

$$T_R = \Sigma M_z \tag{4-11}$$

EXAMPLE 4-1

Figure 4-6a shows a circular shaft rigidly clamped at its left end and free to rotate in a frictionless bearing at its right end. A 16-in. diameter pulley and a 12-in. diameter pulley are keyed to the shaft as shown. Cables attached to the pulleys exert the forces shown. The internal or resisting torque T_R is required for the intervals $0 \leq z \leq 2$ and $2 \leq z \leq 6$.

SOLUTION

To obtain these internal reactions, construct the free-body diagrams shown in Figures 4-6b and 4-6c. Accordingly, from Figure 4-6b,

$$\Sigma M_z = 0: \quad T_R - 1000(16) - 1500(12) = 0$$

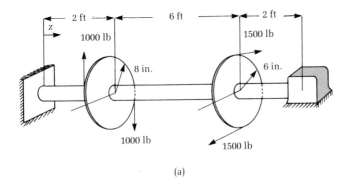

(a)

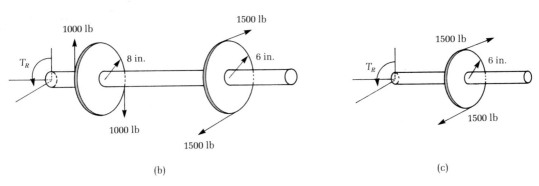

(b)   (c)

Figure 4-6

or

$$T_R = 34,000 \text{ in.-lb} \qquad (a)$$

and, from Figure 4-6c,

$$\sum M_z = 0: \quad T_R - 1500(12) = 0$$

or

$$T_R = 18,000 \text{ in.-lb} \qquad (b)$$

This procedure can always be used to express the internal torque T_R in terms of the applied torques and, in the case of statically indeterminate shafts, torques that occur at rigid supports such as at the left end of the shaft of Figure 4-6a. However, for statically indeterminate problems, static equilibrium considerations alone are not sufficient to provide a solution for the unknown torques occurring at rigid supports. In these cases, it is necessary to make use of the remaining ingredients of mechanics: geometric compatibility and material behavior. Statically indeterminant torsion problems are considered later.

4-5 Material Behavior

Explicit formulas for shearing stress and angle of twist can be developed for shafts with circular cross sections provided that stress-strain relations are available for the material from which they are made. We consider several types of material behavior here.

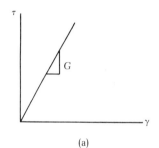

(a)

Linear elastic behavior. The stress-strain diagram for an isotropic material that behaves in a linearly elastic manner is shown in Figure 4-7a. Accordingly,

$$\tau = G\gamma \tag{4-12}$$

at any point r on the cross section.

The distribution of the stress on the cross section is

$$\tau = G\gamma_c \frac{r}{c} = \tau_{\max} \frac{r}{c} \tag{4-13}$$

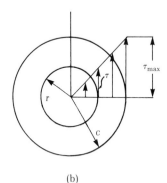

(b)

Figure 4-7

Here $\tau_{\max}$ is the shearing stress at a distance c from the axis of the shaft. Equation (4-13) shows that the shearing stress varies linearly with radial distance r. This stress distribution is shown in Figure 4-7b.

Elastic, perfectly plastic behavior. The stress-strain diagram for a material that behaves in an elastic-plastic manner is shown in Figure 4-8a. Two separate mathematical relations are required to represent the elastic and plastic regions. Accordingly

$$\tau = \begin{cases} G\gamma & 0 \le \gamma \le \gamma_{YP} \\ \tau_{YP} & \gamma_{YP} \le \gamma \le \gamma_L \end{cases} \tag{4-14}$$

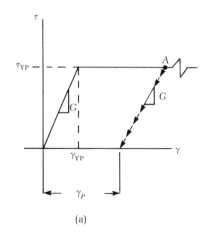

(a)

where γ_L represents a strain beyond which the flat-top portion of the stress-strain diagram no longer holds.

Since the strain varies linearly over the entire cross section, even in the presence of inelastic material behavior, the stress distribution on the cross section can be written as

$$\tau = \begin{cases} \tau_{YP} \dfrac{r}{a} & 0 \le r \le a \\ \tau_{YP} & a \le r \le c \end{cases} \tag{4-15}$$

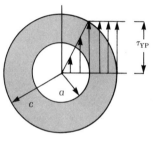

(b)

Figure 4-8

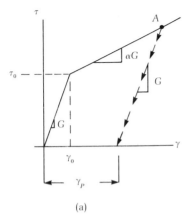

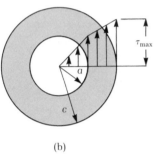

Figure 4-9

The radius of the elastic portion of the cross section is denoted by the letter a. Therefore, yielding has penetrated to a depth of $c - a$.

Elastic-linear hardening material. The stress-strain diagram for several materials can be closely represented by two sloping straight lines as shown in Figure 4-9a. The stress and strain corresponding to the intersection of these straight lines are denoted by τ_0 and γ_0. Accordingly, the stress-strain relation for a material of this kind is

$$\tau = \begin{cases} G\gamma & 0 \leq \gamma \leq \gamma_0 \\ (1 - \alpha)\tau_0 + \alpha G\gamma & \gamma_0 \leq \gamma \leq \gamma_{\mathrm{L}} \end{cases} \tag{4-16}$$

Here α is the ratio between the slopes of the two straight line segments. Again the strain varies linearly on the cross section so that

$$\tau = \begin{cases} \dfrac{r}{a}\tau_0 & 0 \leq r \leq a \\ (1 - \alpha)\tau_0 + \alpha\dfrac{r}{a}\tau_0 & a \leq r \leq c \end{cases} \tag{4-17}$$

The stress distribution corresponding to the elastic linear hardening material is shown in Figure 4-9b.

Nonlinear elastic hardening material. The stress-strain diagram for a nonlinear elastic hardening material is shown in Figure 4-10a. It can be represented mathematically by the relation

$$\tau = K\gamma^n \tag{4-18}$$

where K and n are parameters adjusted so that Eq. (4-18) approximates the stress-strain data as closely as possible. Since $\gamma = (r/c)\gamma_c$, it follows from Eq. (4-18) that

$$\tau = (r/c)^n\tau_c \tag{4-19}$$

where $\tau_c = K\gamma_c^n$. This stress distribution is shown in Figure 4-10b.

4-6 Elastic Twisting of Circular Shafts

Explicit formulas for the angle of twist per unit length and the shearing stress at any point in an elastic shaft with a circular cross section are easily derived from Eqs. (4-7), (4-10), and (4-12). These equations embody the concepts of geometrically compatible deforma-

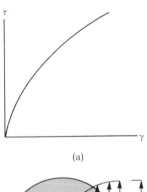

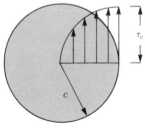

Figure 4-10

tions, equilibrium, and material behavior, respectively. Accordingly,

$$T_R = \underbrace{\int_{area} \tau r \, da}_{\text{Equilibrium}} = \underbrace{\int_{area} G\gamma_r r \, da}_{\text{Material behavior}}$$

$$= \underbrace{\int_{area} \frac{G\gamma_r}{r} r^2 \, da}_{\substack{\text{Rearrangement} \\ \text{of integrand}}} = \underbrace{\int_{area} G \frac{d\varphi}{dz} r^2 \, da}_{\substack{\text{Geometrically} \\ \text{compatible} \\ \text{deformations}}} \qquad (4\text{-}20)$$

In Eq. (4-20), the first integral is the result of equilibrium considerations only. The assumed material behavior has been incorporated in the second integral. The third integral is obtained by simply rearranging the terms in the integrand. In the last integral, the geometric compatibility condition has been used. In this manner, it is possible to establish a relationship between the internal resisting torque T_R and the unit angle of twist, $d\varphi/dz$. This method is a general one in mechanics; internal forces are related to geometrically compatible deformations via the use of constitutive equations describing material behavior.

Since the shearing modulus of elasticity G and the unit angle of twist $d\varphi/dz$ do not depend upon the radial coordinate r, they may be factored from the last integral in Eq. (4-20) so that the resisting torque becomes

$$T_R = G \frac{d\varphi}{dz} \int_{area} r^2 \, da \qquad (4\text{-}21)$$

The integral appearing in Eq. (4-21) is the polar moment of inertia of the circular cross-sectional area with respect to the geometric axis of shaft. This integral is denoted by J. Consequently, Eq. (4-21) gives

$$\frac{d\varphi}{dz} = \frac{T_R}{JG} \qquad (4\text{-}22)$$

Note that T_R has units inch pounds (in.-lb) or newton meters (N·m), J has units inches to the fourth power (in^4) or meters to the fourth power (m^4), and G has units pounds per square inch (psi) or newtons per square meter (N/m^2). Accordingly, $d\varphi/dz$ has units radians per inch (rad/in.) or radians per meter (rad/m).

A formula from which the shearing stress at any radial point r can be calculated is obtained from the stress-strain relation Eq. (4-12), the geometric relation Eq. (4-7), and the elastic unit angle of twist Eq. (4-22).

$$\underbrace{\tau = G\gamma_r}_{\substack{\text{Stress-strain} \\ \text{relation}}} \quad \underbrace{= G\,\frac{\gamma_r}{r}\,r}_{\text{Rearrangement}} \quad \underbrace{= G\,\frac{d\varphi}{dz}\,r}_{\substack{\text{From geometric} \\ \text{compatibility}}} \quad \underbrace{= \frac{T_R r}{J}}_{\text{From Eq. (4-22)}}$$

(4-23)

Equations (4-22) and (4-23) provide a means for analyzing the stiffness and the strength of shafts with circular cross sections. These formulas are also valid for hollow shafts when the hollow and solid portions are concentric. Formulas for polar moments of inertia of solid and annular circular cross sections are

$$J = \begin{cases} \dfrac{\pi}{32}\,d^4 & \text{(Solid)} \\[2ex] \dfrac{\pi}{32}\,(d_0^4 - d_i^4) & \text{(Hollow)} \end{cases}$$

(4-24)

where d_0 denotes the outer diameter and d_i represents the inner diameter of the cross-sectional area of the hollow shaft.

The relative rotation of two cross sections separated by a finite distance is obtained by integrating Eq. (4-22). Therefore, if z_{j+1} and z_j denote two distinct cross sections, then the relative angle of twist associated with them is given by the formula

$$\varphi_{j+1/j} = \varphi_{j+1} - \varphi_j = \int_{z_j}^{z_{j+1}} \frac{T_R}{JG}\,dz$$

(4-25)

Uniform shaft. A particularly important formula that can be obtained from Eq. (4-25) is the one that expresses the relative angle of twist of the ends of a uniform shaft. A uniform shaft is one for which neither the cross section nor the shearing modulus varies along its axis. Moreover, torques are applied only at its ends. Figure 4-11 depicts a uniform shaft of length ℓ subjected to equilibrating torques at its ends. Because T_R/JG is independent of z, Eq. (4-22) yields

Figure 4-11

$$\varphi = \frac{T\ell}{JG}$$

(4-26)

as the relative rotation of the end cross sections.

Nonuniform shafts. Equation (4-22) must be thought of as the sum of several integrals when the integrand T_R/JG contains discontinuities. Figure 4-12 depicts the T_R/JG diagram for a shaft for which T_R/JG is discontinuous at the points z_A and z_B. Discontinuities in the T_R/JG diagram are caused by abrupt changes in applied torques,

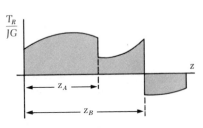

Figure 4-12

abrupt changes in cross section, and/or abrupt changes in material. The integral of Eq. (4-25) becomes a sum of several integrals—one for each region of the T_R/JG diagram for which T_R/JG is smooth and continuous.

Three important procedures are used to analyze elastic circular shafts for which the T_R/JG contains discontinuities. They are (a) the direct integration procedure, (b) the discrete element procedure, and (c) the superposition procedure.

Direct integration procedure.   The direct integration procedure consists of the direct integration of Eq. (4-22). Because the integrand of this equation is discontinuous at cross sections where T_R, J, or G change abruptly, the integral must be divided into intervals for which T_R/JG is smooth and continuous. Consequently, for most torsion problems, the integration of Eq. (4-22) implies a sum of several integrals—one for each interval for which T_R/JG is smooth and continuous.

Discrete element procedure.   The discrete element procedure requires the shaft to be divided into a finite number of segments for each of which T_R/JG is constant. Consequently, we can preceive the complete shaft as a series of connected uniform shafts. The relative rotation of the ends of each segment can be calculated from Eq. (4-26). The relative angle of twist of a series of connected segments is given by

$$\varphi = \Sigma \varphi_i \qquad (4-27)$$

Superposition procedure.   The superposition precedure asserts that the relative rotation of two cross sections due to several external torques acting simultaneously is equal to the algebraic sum of the relative angles of twists of the same cross sections due to each external torque applied separately. The superposition principle is valid in this case because angles of twist are linearly related to the external torques.

The computational procedures described here are illustrated in the examples.

EXAMPLE 4-2

A step-shaft constructed from aluminum ($G = 4 \times 10^6$ psi) and steel ($G = 12 \times 10^6$ psi) bar stock carries the torsional loads shown in Figure 4-13a. Determine the angle of twist of the section at A relative to the section at D. Assume that all materials behave in a linearly elastic manner. Solve this problem using (a) the direct integration procedure, (b) the discrete element procedure, and (c) the superposition procedure.

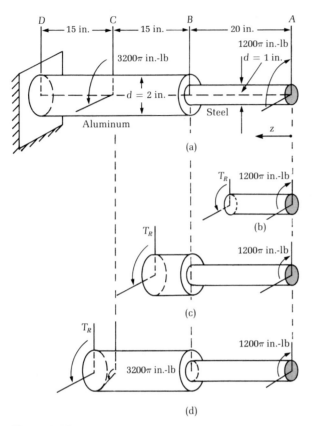

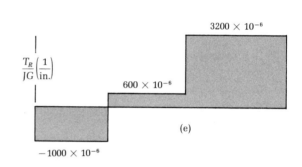

Figure 4-13

SOLUTION

Direct integration procedure. The internal resisting torques for the segments AB, BC, and CD are obtained from the free-body diagrams exhibited in Figures 4-13b through 4-13d. Accordingly, by summing moments about the axis of the shaft,

$$
\left.
\begin{aligned}
T_R &= 1200\pi \text{ in.-lb} & 0 \leq z \leq 20\\
T_R &= 1200\pi \text{ in.-lb} & 20 \leq z \leq 35\\
T_R &= -2000\pi \text{ in.-lb} & 35 \leq z \leq 50
\end{aligned}
\right\} \tag{a}
$$

Clearly, the resisting torque T_R is discontinuous at the section $z = 35$ in. Discontinuities in T_R/JG that are attributable to T_R are characterized by the appearance of new external torques.

The polar moments of inertia for the two sections of shaft are, from Eq. (4-24),

$$
\left.
\begin{aligned}
J_{AB} &= \frac{\pi}{32}\,(1)^4 = \frac{\pi}{32}\,\text{in}^4 & 0 \leq z \leq 20\\[2mm]
J_{BD} &= \frac{\pi}{32}\,(2)^4 = \frac{\pi}{2}\,\text{in}^4 & 20 \leq z \leq 50
\end{aligned}
\right\} \tag{b}
$$

Notice that T_R/JG is also discontinuous at the section where J changes abruptly. Discontinuities due to abrupt changes in J are characterized by sudden changes in the cross section of the shaft.

Finally, the material changes from steel ($G = 12 \times 10^6$ psi) to aluminum (4×10^6 psi) at $z = 20$ in. Consequently, T_R/JG has a discontinuity due to the abrupt change in shearing modulus.

The numerical values for T_R/JG for the three intervals over which T_R/JG is smooth and continuous are

$$\left.\begin{array}{ll}
\dfrac{T_R}{JG} = \dfrac{1200\pi}{\dfrac{\pi}{32}(12 \times 10^6)} = 3200 \times 10^{-6} \text{ in}^{-1} & 0 \le z \le 20 \\[3em]
\dfrac{T_R}{JG} = \dfrac{1200\pi}{\dfrac{\pi}{2}(4 \times 10^6)} = 600 \times 10^{-6} \text{ in}^{-1} & 20 \le z \le 35 \\[3em]
\dfrac{T_R}{JG} = \dfrac{-2000\pi}{\dfrac{\pi}{2}(4 \times 10^6)} = -1000 \times 10^{-6} \text{ in}^{-1} & 35 \le z \le 50
\end{array}\right\} \quad \text{(c)}$$

The graph of T_R/JG is shown in Figure 4-13e.

The relative angle of twist of a section at A with respect to a section at D is obtained by integrating Eq. (4-22) between A and D. The result is

$$\begin{aligned}
\varphi_{A/D} &= \int_0^{20} (3200 \times 10^{-6})\, dz + \int_{20}^{35} (600 \times 10^{-6})\, dz \\
&\quad + \int_{35}^{50} (-1000 \times 10^{-6})\, dz \\
&= 0.0640 + 0.0090 - 0.0150 \\
&= 0.058 \text{ rad}
\end{aligned}$$

or

$$\varphi_{A/D} = 3.323 \text{ degrees} \quad \text{(d)}$$

Discrete element procedure. Figure 4-13e indicates that three distinct segments are required; that is, there are three segments for which T_R/JG is constant. Consequently,

$$\begin{aligned}
\varphi_{A/D} &= \varphi_{A/B} + \varphi_{B/C} + \varphi_{C/D} \\
&= \dfrac{(1200\pi)20}{\dfrac{\pi}{32}(12 \times 10^6)} + \dfrac{(1200\pi)15}{\dfrac{\pi}{2}(4 \times 10^6)} + \dfrac{(-2000\pi)15}{\dfrac{\pi}{2}(4 \times 10^6)} \\
&= 0.0640 + 0.0090 - 0.0150 \\
&= 0.058 \text{ rad} \\
&= 3.323 \text{ degrees} \quad \text{(e)}
\end{aligned}$$

This result is precisely the same one that was obtained by the integration method. Indeed, we should expect this result since both methods determine the area under the T_R/JG diagram. We remark that the rotation of any section P relative to another section Q is equal to the area under the T_R/JG diagram between points P and Q.

Superposition procedure.   From Figures 4-14a and 4-14b, we calculate the relative rotation of section A with respect to section D for each load acting separately. Accordingly,

$$\varphi'_{A/D} = \frac{(1200\pi)20}{\frac{\pi}{32}(12 \times 10^6)} + \frac{(1200\pi)(30)}{\frac{\pi}{2}(4 \times 10^6)}$$

$$= 0.0640 + 0.0180$$

$$= 0.082 \text{ rad} \qquad\qquad\qquad (f)$$

and

$$\varphi''_{A/D} = 0 + 0 + \frac{(-3200\pi)(15)}{\frac{\pi}{2}(4 \times 10^6)}$$

$$= -0.0240 \text{ rad} \qquad\qquad\qquad (g)$$

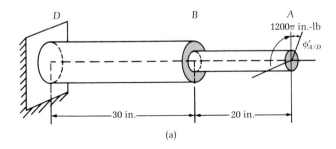

(a)

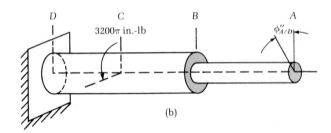

(b)

Figure 4-14

In accordance with the superposition principle,

$$\varphi_{A/D} = \varphi'_{A/D} + \varphi''_{A/D}$$
$$= 0.082 - 0.024$$
$$= 0.058 \text{ rad}$$
$$= 3.323 \text{ degrees}$$

(h)

EXAMPLE 4-3

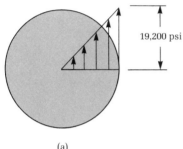

(a)

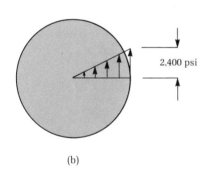

(b)

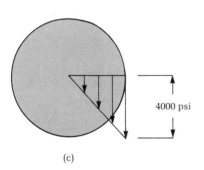

(c)

Figure 4-15

Calculate the maximum shearing stress that occurs in each of the three segments AB, BC, and CD of the shaft shown in Figure 4-13a of Example 4-2.

SOLUTION

Elastic behavior is assumed, so the shearing stress at any point on any cross section is given by the torsion formula

$$\tau = \frac{T_R r}{J}$$

(a)

The internal resisting torques for the three segments of the shaft were determined in Example 4-2 as Eqs. (a). Accordingly, the maximum shearing stresses are

$$\tau_{max} = \frac{1200\pi(0.5)}{\dfrac{\pi}{32}} = 19{,}200 \text{ psi} \qquad 0 \leq z \leq 20$$

$$\tau_{max} = \frac{1200\pi(1.0)}{\dfrac{\pi}{2}} = 2400 \text{ psi} \qquad 20 \leq z \leq 35$$

$$\tau_{max} = \frac{-2000\pi(1.0)}{\dfrac{\pi}{2}} = -4000 \text{ psi} \qquad 35 \leq z \leq 50$$

(b)

The stress distributions for the intervals AB, BC, and CD are shown in Figures 4-15a, 4-15b, and 4-15c, respectively.

Note that elastic behavior of the material of the shaft was assumed from the outset. Consequently, the stresses computed in Eqs. (b) should be compared with the shearing proportional limit of the appropriate material to determine if the assumption is justified. If the stress in the aluminum or the stress in the steel does not exceed the shearing proportional limit of the aluminum or of the steel, respectively, the elastic analysis is justified. If either shearing stress exceeds the shearing proportional limit of the corresponding material, the elastic analysis is invalid.

EXAMPLE 4-4

A pump is connected to an electric motor through steel shafting as shown in Figure 4-16a. An offset necessary to connect the pump to its power source is provided by the gear arrangement as shown. If the motor delivers 100π HP at 330 rpm at its shaft, determine (a) the maximum shearing stress in the shaft and (b) the relative rotation of sections A and C ($G = 12 \times 10^6$ psi for steel).

SOLUTION

(a) The torque delivered to shaft AB by the motor is calculated from Eq. (4-5). Thus

$$T_{AB} = \frac{550(12)60}{2\pi(330)} 100\pi = 60,000 \text{ in.-lb} \tag{a}$$

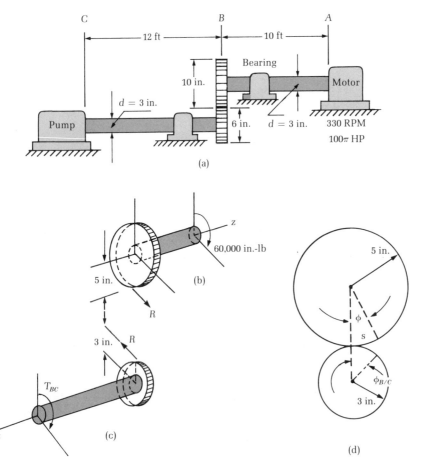

Figure 4-16

Accordingly, the maximum shearing stress in shaft AB is

$$\tau_{max} = \frac{Tc}{J} = \frac{60,000(1.5)}{7.952} = 11,318 \text{ psi} \tag{b}$$

To calculate the maximum stress in shaft BC, we need to know the torque in shaft BC. From the free-body diagram of shaft AB shown in Figure 4-16b, we calculate the reaction R of the 6-in. gear on the 10-in. gear. Thus

$$\Sigma M_z = 0: \quad 5R = 60,000$$

or

$$R = 12,000 \text{ lb} \tag{c}$$

From the free-body diagram of shaft BC shown in Figure 4-16c,

$$\Sigma M_z = 0: \quad T_{BC} = 12,000(3) = 36,000 \text{ in.-lb} \tag{d}$$

The maximum shearing stress in shaft BC is

$$\tau_{max} = \frac{36,000(1.5)}{7.952} = 6791 \text{ psi} \tag{e}$$

(b) The relative rotation of the sections at A and C is

$$\varphi_{A/C} = \varphi_{A/B} + \varphi \tag{f}$$

where $\varphi_{A/B}$ is the rotation of a diameter at section A relative to a parallel diameter at B in shaft AB, and φ is the rotation of the 10-in. gear that results from the elastic deformations in shaft BC (see Figure 4-16d). Now

$$\varphi_{A/B} = \frac{T_{AB} \ell_{AB}}{JG} = \frac{60,000(120)}{7.952(12 \times 10^6)} = 0.07545 \text{ rad} \tag{g}$$

and $5\varphi = 3\varphi_{B/C}$, where $\varphi_{B/C} = T_{BC}\ell_{BC}/JG$, so that $\tag{h}$

$$\varphi = \frac{3}{5} \frac{36,000(144)}{7.952(12 \times 10^6)} = 0.03259 \text{ rad} \tag{i}$$

Substituting the results shown in Eqs. (g) and (i) into Eq. (f) yields

$$\varphi_{A/C} = 0.07545 + 0.03259 = 0.108 \text{ rad} = 6.19 \text{ degrees.} \tag{j}$$

EXAMPLE 4-5

Figure 4-17a shows a composite step-shaft consisting of an aluminum section 50 mm in diameter and a steel section 25 mm in diameter. The ends of the shaft are fixed so that rotation cannot occur there. Determine

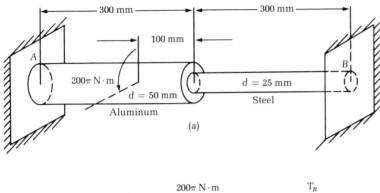

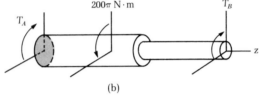

(b)

Figure 4-17

(a) the resisting torques exerted by the supports on the shaft and (b) the maximum stress in the aluminum and the maximum stress in the steel.

SOLUTION

(a) The shaft is statically indeterminate. Therefore, the equilibrium equation (moment equation about the z axis) must be supplemented with an equation expressing the geometric compatibility of the angular deformations.

Equilibrium.   First write the equilibrium equation for moments about the axis of the shaft. Thus, from Figure 4-17b,

$$\Sigma M_z = 0: \quad T_A + T_B = 200\pi \tag{a}$$

This single equilibrium equation is not sufficient, in itself, to solve for the two unknown torques.

Geometric compatibility.   The required compatibility equation is

$$\varphi_{B/A} = 0 \tag{b}$$

which expresses the condition that the ends of the shaft cannot rotate. Tentatively, consider a shaft fixed at the left end and subjected to the torques 200π N·m and T_B. At this point, T_B is *not* regarded as the torque exerted by the support at B. It is merely a torque that is applied at the right end of the shaft. The rotation of a diameter at B relative to a parallel diameter at A is, by the superposition method,

$$\varphi_{B/A} = \frac{T_B(0.3)}{\dfrac{\pi}{32}(0.025)^4(84 \times 10^9)} + \frac{T_B(0.3)}{\dfrac{\pi}{32}(0.05)^4(28 \times 10^9)}$$
$$- \frac{200\pi(0.2)}{\dfrac{\pi}{32}(0.05)^4(28 \times 10^9)} \tag{c}$$

Since the compatibility Eq. (b) requires that $\varphi_{B/A}$ be zero, it follows that

$$T_B = 21.05\pi \text{ N·m} \tag{d}$$

and Eq. (a) yields

$$T_A = 178.95\pi \text{ N·m} \tag{e}$$

(b) Since we know the internal torques that act in the aluminum and in the steel portions of the shaft, we can calculate the maximum shearing stresses in each material. Accordingly,

$$(\tau_{AL})_{max} = \frac{(178.95\pi)(0.025)}{\dfrac{\pi}{32}(0.05)^4} = 22.90 \text{ MPa} \tag{f}$$

and

$$(\tau_{ST})_{max} = \frac{(21.05\pi)(0.0125)}{\dfrac{\pi}{32}(0.025)^4} = 21.55 \text{ MPa} \tag{g}$$

EXAMPLE 4-6

The steel shaft shown in Figure 4-18 is required to transmit 20π HP at 5.5 Hz. If the allowable angle of twist per meter of shaft is not to exceed 4.5 degrees, and if the allowable shearing stress is not to exceed 84 MPa, calculate the minimum permissible diameter d.

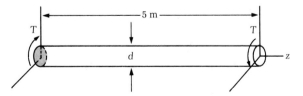

Figure 4-18

SOLUTION

The solution of the problem requires that the minimum permissible diameter satisfy the inequalities $\varphi_{A/B} \leq 4.5$ degrees/m and $\tau_{max} \leq$ 84 MPa. There are two approaches to a solution of this problem.

First, calculate a minimum diameter based on the strength requirement, and then calculate a minimum diameter based on stiffness requirement. The minimum permissible diameter is the larger of the two diameters.

The torque that the shaft must transmit is

$$T = \frac{745.7(20\pi)}{2\pi(5.5)} = 1355.8 \text{ N·m} \tag{a}$$

Strength requirement. Calculate the minimum diameter required if the allowable shearing stress is not exceeded. From Eq. (4-23),

$$84 \times 10^6 = \frac{1355.8(d/2)}{\dfrac{\pi}{32}(d)^4}$$

or

$$d = 43.5 \text{ mm} \tag{b}$$

Stiffness requirement. Now calculate the minimum diameter required if the allowable angle of twist is not to be exceeded. Since the allowable angle of twist per unit of length is 4.5 degrees or 0.07854 rad, the allowable angle of twist between the ends of the shaft is

$$\varphi_{\text{ALL}} = 0.07854(5) \text{ rad} \tag{c}$$

The required diameter is calculated from the equation $\varphi = T\ell/JG$, which yields

$$0.07854(5) = \frac{1355.8(5)}{\dfrac{\pi}{32}d^4(84 \times 10^9)}$$

or

$$d = 38 \text{ mm} \tag{d}$$

The minimum permissible diameter is 43.5 mm.

Alternate solution. As an alternate approach to a solution for this type of problem, assume that shearing stress governs so that, according to Eq. (b), $d = 43.5$ mm. Now check the angle of twist that results from the use of a shaft with this diameter.

The angle of twist associated with $d = 43.5$ mm is

$$\varphi = \frac{1355.8(5)}{\dfrac{\pi}{32}(0.0435)^4(84 \times 10^9)} = 0.23 \text{ rad} = 13.15 \text{ degrees} \tag{e}$$

The angle of twist per unit of length is, then, $13.15/5 = 2.63$ degrees. Since this angle of twist does not exceed the allowable angle of twist, we conclude that $d = 43.5$ mm is the minimum permissible diameter.

EXAMPLE 4-7

The solid, tapered steel shaft shown in Figure 4-19a is fastened to a fixed support at its left end. The shaft has a uniform torque distribution applied to its surface. Assume that the distributed torque has a uniform intensity $\overline{T}$ N·m/m. Determine the angle of twist at the free end of the shaft. The dimensions of the shaft and the torque distribution are shown in Figures 4-19b and 4-19c. Consider the origin of the coordinate axes at the vertex of the extended shaft as shown in Figure 4-19d. Assume that the geometric assumptions regarding strain in prismatic shafts with circular cross sections apply.

SOLUTION

Equation (4-22), which gives the angle of twist per unit length, is applicable here also. However, the resisting torque T_R and the polar moment of inertia are both functions of z. Therefore,

$$\frac{d\varphi}{dz} = \frac{T_R(z)}{GJ(z)} \tag{a}$$

From Figure 4-19d one obtains

$$h = \frac{\ell\, d_1}{d_2 - d_1} \tag{b}$$

and, in a similar manner,

$$d(z) = \frac{d_1}{h}\, z \tag{c}$$

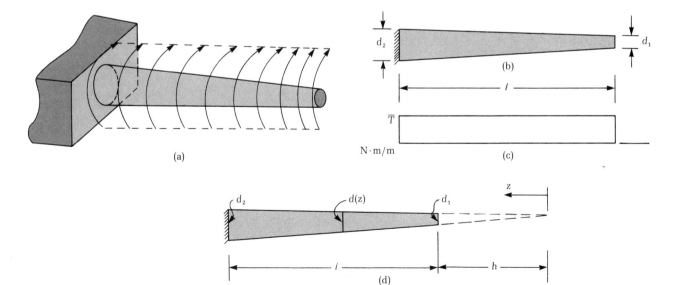

Figure 4-19

Accordingly,

$$J(z) = \frac{\pi}{32} d^4 = \frac{\pi}{32} \frac{d_1^4}{h^4} z^4 \tag{d}$$

The resisting torque $T_R(z)$ at any section can be expressed in terms of the externally applied, uniformly distributed torque $\bar{T}$ as

$$T_R(z) = \begin{cases} 0 & 0 \le z \le h \\ \bar{T}(z - h) & h \le z \le \ell \end{cases} \tag{e}$$

The angle of twist of the free end is then obtained by integrating Eq. (a) from h to $h + \ell$:

$$\varphi = \int_h^{h+\ell} \frac{\bar{T}(z - h)}{G \dfrac{\pi d_1^4}{32 h^4} z^4} \, dz \tag{f}$$

or

$$\varphi = \frac{32 h^4 \bar{T}}{\pi G \, d_1^4} \left\{ \int_h^{h+\ell} \frac{dz}{z^3} - h \int_h^{h+\ell} \frac{dz}{z^4} \right\} \tag{g}$$

The final result is

$$\varphi = \frac{32 h^4 \bar{T}}{\pi G \, d_1^4} \frac{(3h + \ell)\ell^2}{6 h^2 (h + \ell)^3} \tag{h}$$

where h is to be substituted from Eq. (b).

PROBLEMS / Section 4-6

4-1 Determine the maximum shearing stress in the 50-mm diameter solid circular shaft shown in Figure P4-1. Sketch the shearing stress distribution for a cross section between points B and C.

4-2 Determine the maximum shearing stress in the torsion member shown in Figure P4-2. The external diameter of the shaft is 2 in. and the diameter of the concentric hole is 1 in. Sketch the shearing stress distribution for a cross section in the portion of the shaft containing the hole.

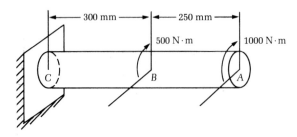

Figure P4-1

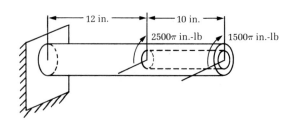

Figure P4-2

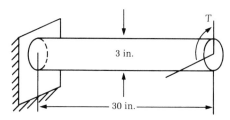

Figure P4-4

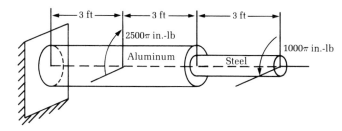

Figure P4-5

4-3 A solid, circular steel ($G = 84$ GPa) shaft is to be used to couple a motor and a pump. The motor delivers 100π HP to the shaft at a shaft speed of 5 Hz. Determine the minimum permissible diameter for the shaft if the shearing stress is not to exceed 56 MPa, and the angle of twist per unit length is not to exceed 1.5 degrees per meter (87.9 mm).

4-4 Determine the maximum torque T that can be applied to the solid aluminum ($G = 4 \times 10^6$ psi) shaft shown in Figure P4-4 if the angle of twist for the 30-in. length is not to exceed 0.5 degrees, and the shearing stress is not to exceed the shearing ultimate stress (33 ksi) reduced by a safety factor of 3.

4-5 Use the superposition method to determine the angular displacement of the free end of the composite steel and aluminum step-shaft shown in Figure P4-5. The diameters of the steel ($G = 12 \times 10^6$ psi) and the aluminum ($G = 4 \times 10^6$ psi) sections are 1 in. and 2 in., respectively. Calculate the magnitude of the maximum shearing stress in the aluminum and in the steel.

4-6 The solid, circular step-shaft shown in Figure P4-6 consists of an aluminum segment ($G = 4 \times 10^6$ psi) and a bronze segment ($G = 5 \times 10^6$ psi) whose diameters are 2 in. and 1 in., respectively. (a) Construct the T/JG diagram for the loads shown in the figure. (b) Use the discrete element method to determine the angular displacement of the free end. (c) Calculate the maximum shearing stress in the aluminum and in the brass. (d) Determine the angle of twist of section BB relative to section AA.

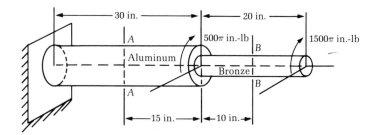

Figure P4-6

4-7 Two machines are driven by belts through pulleys whose diameters are 300 mm and 400 mm, respectively, as shown in Figure P4-7. The 50-mm diameter steel shaft to which the pulleys are connected is driven by a 75-kW electric motor at 660 rpm.

Determine the maximum torsional shearing stress in the shaft and the relative angle of twist of one pulley relative to the other.

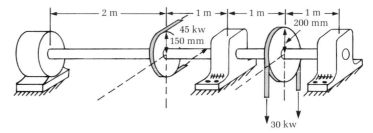

Figure P4-7

4-8 A 100-kW electric motor drives a set of gears at 990 rpm as shown in Figure P4-8. The gear on the left drives a mixer that requires 70 kW, and the gear on the right drives a second mixer that requires 30 kW. Determine the minimum permissible diameter of a steel ($G = 84$ GPa) shaft to connect these gears if the shearing stress is not to exceed 60 MPa.

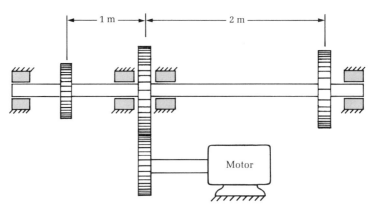

Figure P4-8

4-9 Determine the reactions at the rigid supports for the statically indeterminate shaft shown in Figure P4-9. Calculate the maximum shearing stress in the aluminum ($G = 4 \times 10^6$ psi) and in the steel ($G = 12 \times 10^6$ psi).

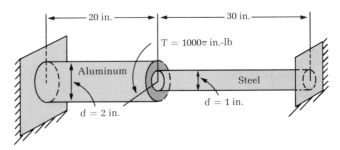

Figure P4-9

4-10 Establish formulas for the reactions of the rigid supports at A and B on the shaft shown in Figure P4-10. Establish a formula for the maximum shearing stress. Express your results in terms of T_0, a, b, d, and ℓ.

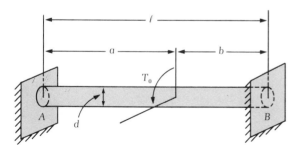

Figure P4-10

4-11 A circular shaft of length ℓ and diameter d is rigidly supported at the right end as shown in Figure P4-11. The shaft is resisted by a linear torsional spring at its left end. The torque-angular displacement diagram for the spring is shown in the figure. Express the reactions at the spring and the rigid support in terms of T_0, a, b, ℓ, J, G, and k.

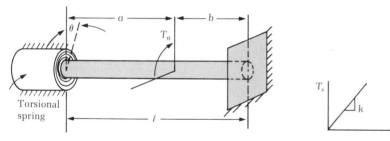

Figure P4-11

4-7 Elastic Twisting of Composite Shafts

Circular shafts whose cross sections deform elastically and are composed of two or more isotropic materials placed symmetrically about the shaft axis can be treated in a manner similar to a one-material elastic shaft. The key assumptions are the three geometric assumptions enumerated in Section 4-3. Under these assumptions, the only nonzero strain is, as before, the shearing strain component $\gamma_{z\theta}$ and, hence, the only nonzero stress is the shearing stress component $\tau_{z\theta}$. These hypotheses result in a linear variation of the shearing strain as shown in Figure 4-20a. In other words, nonuniformity of the materials does not change the linearity of strain distribution on the cross section.

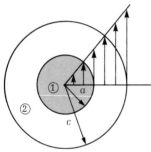

(a) Strain distribution

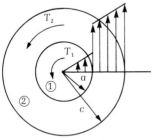

(b) Stress distribution

Figure 4-20

The material in each portion of the cross section is assumed to behave in a linearly elastic manner. Therefore, the stress-strain relation for each material is

$$\tau = \begin{cases} G_1\gamma_r & 0 \le \gamma_r \le \gamma_a \\ G_2\gamma_r & \gamma_a \le \gamma_r \le \gamma_c \end{cases} \tag{4-28}$$

Here it is assumed that the cross section is composed of two different materials. (Generalization to more materials can be made without difficulty.) Shear moduli of the materials of the inner circular cylinder and of the outer circular annulus are denoted by G_1 and G_2, respectively. The outer radii of these two regions are a and c. The materials are bonded securely at the interface a, and the areas of the inner and outer regions are denoted by A_1 and A_2, respectively. Subscripts 1 and 2 used with other pertinent parameters signify the region to which they belong.

The statically equivalent torque associated with each material of the shaft is

$$\left. \begin{aligned} T_1 &= \int_{A_1} \tau r\, da = \int_{A_1} G_1 \frac{\gamma_r}{r} r^2\, da = G_1 J_1 \frac{d\varphi}{dz} \\[2ex] T_2 &= \int_{A_2} \tau r\, da = \int_{A_2} G_2 \frac{\gamma_r}{r} r^2\, da = G_2 J_2 \frac{d\varphi}{dz} \end{aligned} \right\} \tag{4-29}$$

Clearly, the angle of twist can be calculated from either of the formulas

$$\left. \begin{aligned} \frac{d\varphi}{dz} &= \frac{T_1}{J_1 G_1} \\[2ex] \frac{d\varphi}{dz} &= \frac{T_2}{J_2 G_2} \end{aligned} \right\} \tag{4-30}$$

since it is assumed that no slippage occurs at the interface.

The stress distribution associated with each material is calculated from Eqs. (4-28) with the aid of Eqs. (4-30). Consequently,

$$\left. \begin{aligned} \tau_1 &= G_1\left(\frac{\gamma_r}{r}\right)r = \frac{T_1}{J_1}r \\[2ex] \tau_2 &= G_2\left(\frac{\gamma_r}{r}\right)r = \frac{T_2}{J_2}r \end{aligned} \right\} \tag{4-31}$$

This stress distribution is shown in Figure 4-20b.

It is not possible to determine the internal torques T_1 and T_2 associated with different portions of the shaft through the equilibrium equations alone. Consequently, shafts made of two or more materials are inherently statically indeterminate. The example that follows illustrates a procedure that can be used to solve such statically indeterminate problems.

EXAMPLE 4-8

A composite aluminum and steel shaft is used to transmit a torque $T = 6000\pi$ in.-lb as shown in Figure 4-21a. The two materials are assumed to act as a unit; that is, no relative motion occurs between the aluminum and the steel portions at their common interface. Determine (a) the resisting torque T_{AL} in the aluminum and the torque T_{ST} in the steel, (b) the angle of twist of the free end relative to the fixed end, and (c) the maximum stress in the aluminum and in the steel. Assume linearly elastic behavior for both materials.

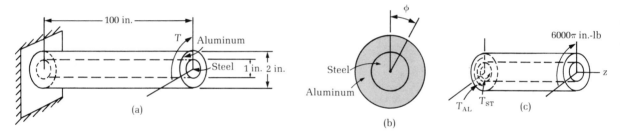

Figure 4-21

SOLUTION

(a) We note that since we have assumed that cross sections are rigid, it follows that diameters remain straight during twisting as shown in Figure 4-21b. Thus the angle of twist in the aluminum and in the steel must be the same.

Geometric compatibility. The appropriate compatibility equation is therefore

$$\varphi_{ST} = \varphi_{AL} \tag{a}$$

or

$$\frac{T_{ST}\ell_{ST}}{J_{ST}G_{ST}} = \frac{T_{AL}\ell_{AL}}{J_{AL}G_{AL}} \tag{b}$$

From Eq. (b), determine

$$T_{ST} = \frac{J_{ST}}{J_{AL}} \frac{G_{ST}}{G_{AL}} T_{AL} \tag{c}$$

Now

$$J_{ST} = \frac{\pi}{32} \text{ in}^4 \quad \text{and} \quad J_{AL} = \frac{15\pi}{32} \text{ in}^4$$

so that $J_{ST}/J_{AL} = \frac{1}{15}$. Also,

$$\frac{G_{ST}}{G_{AL}} = \frac{12 \times 10^6}{4 \times 10^6} = 3$$

Consequently,

$$T_{ST} = \tfrac{1}{5}T_{AL} \tag{d}$$

Equilibrium.   Equilibrium of moments with respect to the z axis of the free-body diagram shown in Figure 4-21c yields the equation

$$\Sigma M_z = 0: \quad T_{AL} + T_{ST} = 6000\pi \tag{e}$$

Simultaneous solution of Eqs. (d) and (e) gives

$$\left.\begin{array}{l} T_{AL} = 5000\pi \text{ in.-lb} \\ T_{ST} = 1000\pi \text{ in.-lb} \end{array}\right\} \tag{f}$$

(b) The angle of twist of the free end relative to the fixed end can be calculated from either of the following formulas.

$$\left.\begin{array}{l} \varphi = \dfrac{T_{AL}\ell}{J_{AL}G_{AL}} \\[2em] \varphi = \dfrac{T_{ST}\ell}{J_{ST}G_{ST}} \end{array}\right\} \tag{g}$$

Accordingly,

$$\varphi = \frac{5000\pi(100)}{\dfrac{15\pi}{32}(4 \times 10^6)} = 0.267 \text{ rad} \tag{h}$$

(c) The maximum shearing stresses in the aluminum and in the steel are

$$\tau_{max} = \frac{5000\pi(1)}{\dfrac{15\pi}{32}} = 10{,}667 \text{ psi (aluminum)}$$

$$\tau_{max} = \frac{1000\pi(0.5)}{\dfrac{\pi}{32}} = 16{,}000 \text{ psi (steel)} \tag{i}$$

Note that elastic stress-strain relations have been used in arriving at Eq. (c), and that an elastic formula has been used to calculate the stresses in Eqs. (i). Accordingly, the computed stresses in Eqs. (i) should be compared with the shearing proportional limits for aluminum and steel, respectively. If neither of the computed stresses exceeds the shearing proportional limit for the corresponding material, the analysis based on elastic formulas is valid. If either stress exceeds the associated shearing proportional limit, the analysis is invalid and another approach must be taken.

PROBLEMS / Section 4-7

4-12 A steel $(G = 84 \text{ GPa})$ sleeve is shrunk on an aluminum $(G = 28 \text{ GPa})$ shaft to form the composite torsion member shown in Figure P4-12. If the allowable stresses for the steel and aluminum are 84 MPa and 70 MPa, respectively, and if the composite shaft is not to twist more than 0.025 rad, determine the maximum torque T that can be applied to the member.

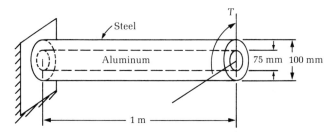

Figure P4-12

4-13 A tubular aluminum rod is fit over a solid steel rod as shown in Figure P4-13. The left end of the composite shaft is rigidly attached to a support at A, and a torque is applied at its right end through a bar that is rigidly attached to both materials. Calculate the maximum shearing stress in each material and the relative rotation of its ends.

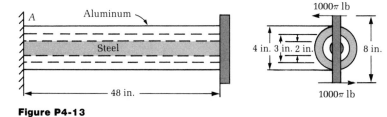

Figure P4-13

4-8 Inelastic Twisting of Shafts with Circular Cross Sections

When the shearing strain at any point $r = a$ in the cross section of a circular shaft exceeds the shearing yield strain γ_{YP}, yielding of the material has occurred in the region $a \leq r \leq c$. The material in the inner region $0 \leq r \leq a$ remains elastic. The stress distribution for an elastic, perfectly plastic material is shown in Figure 4-8b, and the stress distribution for an elastic, linear hardening material is shown in Figure 4-9b.

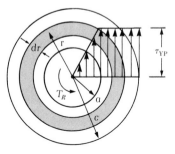

Figure 4-22

Elastic, perfectly plastic material.   The stress distribution for an elastic, perfectly plastic material is repeated in Figure 4-22. Recall that a is the radius of the inner core of the shaft that behaves elastically.

The resisting torque T_R that is statically equivalent to the moment of the shearing stress distribution about the geometric axis of the shaft is given by Eq. (4-10). The shearing stress is expressed as a function of the radial coordinate r in Eq. (4-15). Using a thin ring of thickness dr as the differential area for which $da = 2\pi r\, dr$, we can perform the integration in Eq. (4-10) as follows:

$$T_R = \int_{\text{area}} \tau (da) r = \int_0^a \left(\tau_{\text{YP}} \frac{r}{a} \right)(2\pi r\, dr)r + \int_a^c \tau_{\text{YP}}(2\pi r\, dr)r \quad (4\text{-}32)$$

Integration of these integrals and evaluation at their limits give the formula

$$T_R = \frac{\pi \tau_{\text{YP}}}{6}(4c^3 - a^3) \qquad (4\text{-}33)$$

The angle of twist per unit length is obtained through the use of Eq. (4-7). Since the shearing strain at the interface between the elastic and the plastic regions of the shaft is known to be γ_{YP}, Eq. (4-7) gives

$$\frac{d\varphi}{dz} = \frac{\gamma_{\text{YP}}}{a} = \frac{\tau_{\text{YP}}}{aG} \qquad (4\text{-}34)$$

Replacing τ_{YP} in Eq. (4-34) with its equivalent obtained from Eq. (4-33) yields

$$\frac{d\varphi}{dz} = \frac{6\, T_R}{\pi G a(4c^3 - a^3)} \qquad (4\text{-}35)$$

Equations (4-33) and (4-35) are strength and stiffness formulas for inelastic twisting; they are analogous to the strength and stiffness formulas for elastic twisting of circular shafts.

Elastic, linear hardening material. The stress distribution for an elastic, linear hardening material is repeated in Figure 4-23 for convenience. Again, the material in the region $0 \leq r \leq a$ behaves elastically, while the material in the region $a \leq r \leq c$ experience strains that exceed γ_0. Note that γ_0 and τ_0 were defined in Figure 4-9a.

The resisting torque T_R that is statically equivalent to the moment of this stress distribution about the geometric axis of the shaft is given, again, by Eq. (4-10). The shearing stress is expressed as a function of the radial coordinate r in Eq. (4-17).

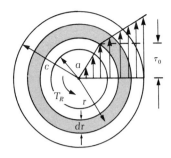

Figure 4-23

Accordingly,

$$T_R = \int_{area} \tau (da)r = \int_0^a \left(\tau_0 \frac{r}{a} \right)(2\pi r \, dr)r$$

$$+ \int_a^c \left\{ (1 - \alpha)\tau_0 + \alpha\tau_0 \frac{r}{a} \right\}(2\pi r \, dr)r \qquad (4\text{-}36)$$

Direct integration of the integrals in Eq. (4-35) and evaluation at their limits yields the formula

$$T_R = 2\pi\tau_0 \left\{ \frac{a^3}{4} + \frac{1-\alpha}{3}(c^3 - a^3) + \frac{\alpha}{4a}(c^4 - a^4) \right\} \qquad (4\text{-}37)$$

This formula reduces to Eq. (4-33) when $\alpha = 0$; that is, for an elastic, perfectly plastic material. Clearly, the resisting torque T_R can be calculated when the material properties τ_0 and α and the radius a of the elastic core are known. Conversely, if the resisting torque T_R and the material properties τ_0 and α are known, the depth of yielding, $c - a$, can be computed.

The angle-of-twist formula is obtained through the use of Eq. (4-7). Again, the shearing strain at the interface between the elastic and inelastic regions is known to be γ_0. Consequently, by Eq. (4-7),

$$\frac{d\varphi}{dz} = \frac{\gamma_0}{a} = \frac{\tau_0}{Ga} \qquad (4\text{-}38)$$

Replacing τ_0 in Eq. (4-38) with its equivalent from Eq. (4-37) yields

$$\frac{d\varphi}{dz} = \frac{T_R}{2\pi aG \left\{ \dfrac{a^3}{4} + \dfrac{1-\alpha}{3}(c^3 - a^3) + \dfrac{\alpha}{4a}(c^4 - a^4) \right\}} \qquad (4\text{-}39)$$

Equations (4-37) and (4-39) are strength and stiffness formulas for inelastic twisting of an elastic, linear hardening material; they are analogous to the strength and stiffness formulas for elastic twisting of circular shafts.

Residual stress. If the shearing strain at a point in the cross section of a shaft has exceeded the shearing yield strain, subsequent reduction of the external torque causes the shearing stress at this point to decrease along a straight line parallel to the initial elastic region of its stress-strain diagram. Accordingly, reduction of the shearing stress at a point for which the shearing strain has exceeded the shearing yield strain follows a linear elastic law. This linear elastic unloading law is shown in Figure 4-8a for an elastic, perfectly plastic material, and in Figure 4-9a for an elastic, linear hardening material, by the dashed lines that emanate from point A in the inelastic region of the respective stress-strain diagrams. Note that a permanent strain γ_P exists at a point where the material has been stressed beyond the elastic region and subsequently unloaded. The interaction of the unloading strains in the elastic region of the cross section with the unloading strains in the inelastic region results in a complicated residual strain distribution that, in turn, results in a complicated shearing stress distribution.

Residual stress distributions can be derived quite easily for torsional shafts with circular cross sections. Residual shearing stress distributions for noncircular torsion members are difficult to determine.

EXAMPLE 4-9

Determine the residual shearing stress distribution for a shaft with a circular cross section made from an elastic, perfectly plastic material.

SOLUTION

From Eqs. (4-15), the shearing stress distribution associated with a loading torque T_R that causes strains in the outer region of the cross section to exceed the shearing yield strain is

$$\tau = \begin{cases} \tau_{YP} \dfrac{r}{a} & 0 \leq r \leq a \\[2mm] \tau_{YP} & a \leq r \leq c \end{cases} \tag{a}$$

This stress distribution is shown in Figure 4-24a.

Since the stresses decrease in a linear manner when the torque T_R is gradually released, the shearing stress during the unloading process is

$$\tau = \begin{cases} \tau_{YP} \dfrac{r}{a} + Gr \dfrac{d\varphi'}{dz} & 0 \leq r \leq a \\[3mm] \tau_{YP} + Gr \dfrac{d\varphi'}{dz} & a \leq r \leq c \end{cases} \tag{b}$$

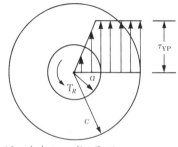

(a) Loaded stress distribution

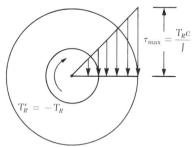

(b) Unloading stress distribution

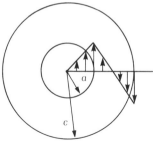

(c) Residual stress distribution

Figure 4-24

We note that $Gr(d\varphi'/dz) = \tau'$ is the stress due to an angle of twist $d\varphi'/dz$ applied with a sense opposite to the angle of twist that occurs during the loading process.

Now the resisting torque at the cross section at any stage of the unloading process is, from Eq. (4-10),

$$
T'_R = \int_0^a \left(\tau_{YP} \frac{r}{a} + Gr \frac{d\varphi'}{dz} \right) 2\pi r^2 \, dr
$$
$$
+ \int_a^c \left(\tau_{YP} + Gr \frac{d\varphi'}{dz} \right) 2\pi r^2 \, dr \tag{c}
$$

Integration of Eq. (c) yields

$$
T'_R = \frac{\pi \tau_{YP}}{6} (4c^3 - a^3) + \frac{\pi Gc^4}{2} \frac{d\varphi'}{dz} \tag{d}
$$

Once the shaft has been completely unloaded, the torque $T'_R = 0$, and the angle of twist that occurs during the unloading process is

$$
\frac{d\varphi'}{dz} = -\frac{\tau_{YP}(4c^3 - a^3)}{3Gc^4} = -\frac{\pi \tau_{YP}(4c^3 - a^3)}{6GJ} = -\frac{T_R}{JG} \tag{e}
$$

Note that this angle of twist is the one that results from the use of the elastic formula, Eq. (4-22), for a torque equal to the loading torque T_R.

Substituting Eq. (e) into Eq. (b) yields the residual stress distribution for a circular cross section made from an elastic, perfectly plastic material. Accordingly,

$$
\tau = \begin{cases} \tau_{YP} \dfrac{r}{a} - \dfrac{T_R r}{J} & 0 \le r \le a \\[2ex] \tau_{YP} - \dfrac{T_R r}{J} & a \le r \le c \end{cases} \tag{f}
$$

The structure of the residual stress distribution suggests that the superposition principle applies. Indeed, the most convenient way to obtain a residual stress distribution for numerical problems is to calculate the stress distribution resulting from the loading torque T_R as shown in Figure 4-24a, then independently calculate the elastic stress distribution for a torque $T'_R = -T_R$ as shown in Figure 4-24b, and add the two distributions algebraically. The resultant residual stress distribution is shown in Figure 4-24c. Note that τ_{YP} in terms of T_R in the loading phase is obtained from Eq. (4-33).

EXAMPLE 4-10

Determine the resisting torque T_R associated with a stress distribution for which yielding has occurred at every point in the cross section. Calculate the residual stress distribution that results when the torque T_R is reduced to zero.

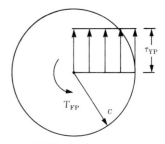

(a) Fully plastic stress distribution

SOLUTION

The stress distribution corresponding to a so-called fully plastic condition is shown in Figure 4-25a. The fully plastic torque, designated by T_{FP}, is calculated from Eq. (4-33) by setting $a = 0$. Accordingly,

$$T_{FP} = \frac{2\pi\tau_{YP}}{3}c^3 = \frac{4}{3}\frac{\tau_{YP}}{c}\frac{\pi c^4}{2} = \frac{4}{3}\frac{\tau_{YP}J}{c}$$

or

$$T_{FP} = \tfrac{4}{3}T_{FE} \tag{a}$$

where T_{FE} is the fully elastic torque—the torque that causes the maximum stress on the cross section to reach τ_{YP}.

   The unloading stress distribution due to a torque $T'_R = -T_{FP} = -\tfrac{4}{3}T_{FE}$ is

$$\tau = \frac{-\tfrac{4}{3}T_{FE}}{J}r \tag{b}$$

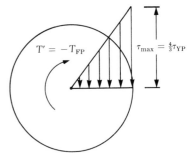

(b) Unloading stress distribution

When $r = c$, the maximum stress for the unloading process is

$$\tau_{max} = -\frac{4}{3}\frac{T_{FE}c}{J} = -\frac{4}{3}\tau_{YP}. \tag{c}$$

The unloading stress distribution is shown in Figure 4-25b.

   The residual stress distribution is obtained by adding the stress distributions due to loading and unloading algebraically. This distribution is shown in Figure 4-25c.

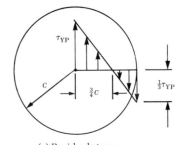

(c) Residual stress distribution

Figure 4-25

EXAMPLE 4-11

A 150-mm diameter shaft has yielded to a depth equal to two-thirds of the radius of the cross section. If the shaft is made from the material whose stress-strain diagram is shown in Figure 4-26a, determine (a) the resisting torque T_R, (b) the stress distribution, (c) the residual stress distribution, and (d) the residual angle of twist (G = 28 GPa).

SOLUTION

(a) The partially plastic torque T_{PP} that causes the shaft to yield as stipulated is calculated from Eq. (4-33). Accordingly, with $a = 25$ mm,

$$T_{PP} = \frac{140 \times 10^6 \pi}{6}[4(0.075)^3 - (0.025)^3] = 122.6 \text{ kN} \cdot \text{m} \tag{a}$$

(b) The stress distribution associated with the prescribed depth of yielding is shown in Figure 4-26b.

(c) The stress distribution due to unloading elastically from a torque level $T_{PP} = 122.6$ kN·m is

$$\tau = \frac{122.6 \times 10^3}{\dfrac{\pi}{32}(0.15)^4} r = \frac{122.6 \times 10^3}{49 \times 10^{-6}} r = 2.467 \times 10^9 r \text{ N/m}^2 \tag{b}$$

This distribution of shearing stress is shown in Figure 4-26c.

The residual stress distribution is obtained by superposition of the stress distributions of Figures 4-26b (loading) and 4-26c (unloading). The result of this superposition is shown in Figure 4-26d.

(d) The residual angle of twist is the difference between the angle of twist that corresponds to the loaded state and the angle of twist that occurs for the unloading process. Thus

$$\left(\frac{d\varphi}{dz}\right)_{\text{residual}} = \left(\frac{d\varphi}{dz}\right)_{\text{loading}} - \left(\frac{d\varphi}{dz}\right)_{\text{unloading}} \tag{c}$$

Now,

$$\left(\frac{d\varphi}{dz}\right)_{\text{loading}} = \frac{\gamma_a}{a} = \frac{0.005}{0.025} = 0.20 \text{ rad/m} \tag{d}$$

and

$$\left(\frac{d\varphi}{dz}\right)_{\text{unloading}} = \frac{T}{JG} = \frac{122.6 \times 10^3}{49.7 \times 10^{-6}(28 \times 10^9)}$$
$$= 0.0881 \text{ rad/m} \tag{e}$$

Consequently,

$$\left(\frac{d\varphi}{dz}\right)_{\text{residual}} = 0.20 - 0.0881 = 0.1119 \text{ rad/m} \tag{f}$$
$$= 6.41 \text{ degrees per meter}$$

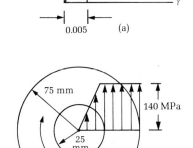

(a)

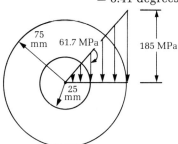

(b) Stress distribution for
the loaded shaft

(c) Stress distribution for
elastic unloading

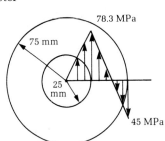

(d) Residual stress
distribution

Figure 4-26

PROBLEMS / Section 4-8

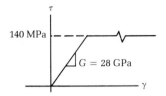

Figure P4-15

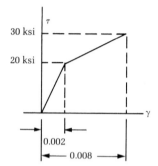

Figure P4-16

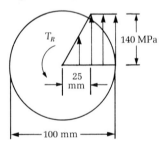

Figure P4-17

4-14   An annular shaft with the cross section shown in Figure P4-14a has the stress-strain diagram shown in Figure P4-14b. Calculate the ratio of the fully plastic torque to the fully elastic torque ($T_{FP}/T_{FE} = 1.245$).

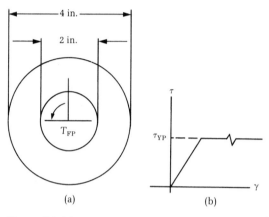

Figure P4-14

4-15   The stress-strain diagram for a 100-mm diameter shaft that is subjected to a torque of 35.5 kN·m at its ends is shown in Figure P4-15. Determine (a) the depth to which the yielding has penetrated, (b) the stress distribution on the cross section, and (c) the unit angle of twist ($a = 25$ mm).

4-16   A 4-in. diameter shaft is made from a material with the stress-strain diagram shown in Figure P4-16. Determine the torque required to cause yielding to penetrate to a depth of 1 in., and determine the corresponding unit angle of twist ($T = 267,035$ in.-lb and $d\varphi/dz = 0.002$ rad/in.).

4-17   Determine the residual stress distribution that results when the applied torque shown in Figure P4-17 is reduced to zero.

4-9   Elastic Twisting of Thin-Walled, Closed Sections

An important structural component that falls within the scope of the three basic geometric assumptions enumerated in Section 4-3 is the thin-walled, closed cross section, torsional member. Because of these geometric assumptions, the only nonzero strain the tube experiences is the shearing strain shown in Figure 4-27a.

A free-body diagram of a portion of the tube is shown in Figure 4-27b. Note that the thickness of the wall may vary on a cross section, but it is assumed to be constant along any line parallel to the axis of the tube. Moreover, since the walls are thin, it is assumed that the shearing strain, and hence the shearing stress, is distributed uniformly over the thickness, either in a longitudinal plane or in a circumferential plane. We let z be a coordinate along the length of the tube and s be a coordinate along the mean perimeter of the cross section as depicted in Figure 4-27b. We note that the length of the free-body diagram parallel to the axis of the tube is infinitesimal (dz), while the length along a cross-sectional arc is finite (s). This situation means that the stresses τ_1 and τ_2 are the same at any point on the longitudinal planes on which they act, and the stresses τ_3 and τ_4 can be different. These stresses are shown in Figure 4-27b.

The stress distributions τ_1 and τ_2 can be replaced by the statically equivalent shearing forces

$$\left.\begin{array}{l} F_1 = \tau_1 t_1 \, dz \\ F_2 = \tau_2 t_2 \, dz \end{array}\right\} \tag{4-40}$$

shown in Figure 4-27c. Likewise, the stress distributions acting in the cross-sectional planes can be replaced by the statically equivalent forces F_3 and F_4.

Equilibrium of forces along the axis of the element yields

$$F_1 = F_2 \tag{4-41}$$

or

$$\tau_1 t_1 = \tau_2 t_2 = \text{constant} \tag{4-42}$$

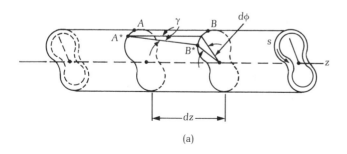

(a)

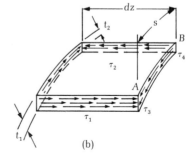

(b)

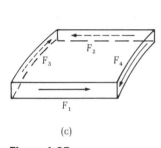

(c)

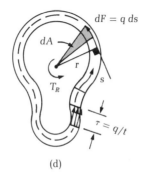

(d)

Figure 4-27

This relation implies that the product of the shearing stress acting in a longitudinal plane and the thickness of the wall in this plane is the same for any longitudinal plane.

Now the shearing stresses τ_1 and τ_3 at the corner A of Figure 4-27b must be equal since shearing stresses always occur in pairs that are at right angles to each other. Likewise, the shearing stresses τ_2 and τ_4 at corner B must be equal. Replacing τ_1 and τ_2 by their equivalents in Eq. (4-42) yields

$$\tau_3 t_1 = \tau_4 t_2 \tag{4-43}$$

Physically, Eq. (4-43) expresses the condition that the product of the shearing stress at any point s on the cross section and the thickness at this point is the same for any point on the cross section. Accordingly, we write

$$\tau t = q \tag{4-44}$$

where q is called the **shear flow** for the cross section and has units in pounds per inch or newtons per meter depending on the system of units employed. Observe that the shearing stress can be calculated for any point on a cross section provided that the shear flow q and the thickness t at the point are known.

A formula for the shear flow q can be established as follows. The resisting torque at a cross section of the tube is

$$T_R = \int dF\, r = \int_{\substack{\text{mean} \\ \text{perimeter}}} q\, ds\, r \tag{4-45}$$

where $dF = q\, ds$, and r is the perpendicular distance from the action line of dF to the center of twist of the tube; that is, the point in the cross section about which the tube tends to twist (see Figure 4-27d). Its exact location is irrelevant in the present context.

We observe that the shaded area in Figure 4-27d is

$$dA = \tfrac{1}{2} r\, ds \tag{4-46}$$

since it is a triangle with base ds and height r. Now, noting that q is a constant, Eq. (4-45) becomes

$$T_R = 2q \int_{A_m} dA = 2qA_m \tag{4-47}$$

A_m is the area contained within the mean perimeter of the cross section of the tube. Consequently, the shear flow may be calculated from the formula

$$q = \frac{T_R}{2A_m} \tag{4-48}$$

and the shearing stress from the formula

$$\tau = \frac{q}{t} \tag{4-49}$$

Equations (4-48) and (4-49) constitute formulas that are analogous to the strength formula for elastic twisting of circular shafts.

To derive a formula for the unit angle of twist for the elastic twisting of tubular members, use the principle of conservation of energy, the physical law that asserts that, for a conservative system, the work done by the internal forces must be equal to the work done by the external forces.

The work done by the internal forces acting in the tube is equal to the strain energy stored in it. Consequently,

$$dU = \frac{\tau^2}{2G} (t\,ds\,dz) \tag{4-50}$$

where $\tau^2/2G$ is the strain energy density for a state of pure shear and $t\,ds\,dz$ is the volume of a differential element of the tube. Now Eqs. (4-48) through (4-50) yield

$$dU = \frac{T_R^2}{8GA_m^2} \frac{ds}{t}\,dz \tag{4-51}$$

so that the strain energy associated with a length dz of the tube is

$$U = \frac{T_R^2}{8GA_m^2} \left(\int_{s_m} \frac{ds}{t} \right) dz \tag{4-52}$$

Here s_m denotes the mean perimeter of the cross section.

The work done by the external torque on a length dz of the tube is

$$W_e = \left(\frac{d\varphi}{dz}\,dz \right) \frac{T_R}{2} \tag{4-53}$$

Since the tube deforms elastically, the system is conservative and, therefore, Eqs. (4-52) and (4-53) yield

$$\frac{d\varphi}{dz} = \frac{T_R}{4A_m^2 G} \int_{s_m} \frac{ds}{t} \tag{4-54}$$

Notice that, for cross sections made of a series of rectangular parts, for each of which the thickness is constant, the integral in Eq. (4-54) can be replaced by the sum $\sum_{i=1}^{n} (s_i/t_i)$ and, thus, the unit angle of

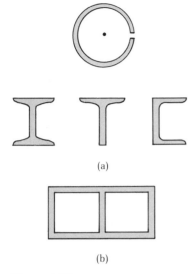

(a)

(b)

Figure 4-28

twist is obtained from

$$\frac{d\varphi}{dz} = \frac{T_R}{4GA_m^2} \sum_{i=1}^{n} \frac{s_i}{t_i}$$

(4-55)

It should be emphasized that the formulas developed in this section are valid only for thin-walled cross sections that are closed. They are not valid for so-called open cross sections such as those typified by Figure 4-28a.

Also excluded from the class of problems for which the formulas developed in this section are valid are torsion members whose cross sections are composed of two or more closed cells such as the one shown in Figure 4-28b. Treatments of torsional members of this type are beyond the scope of this text. They can be found in advanced books on the topic. For an introductory treatment, refer to *Advanced Mechanics of Materials* by Seely, Smith, Sidebottom, and Boresi.

EXAMPLE 4-12

The box beam shown in Figure 4-29a is to be used as a torsion member. The beam is 10 ft long and must resist a maximum torque of 1000 in.-lb. Determine the maximum shearing stress on a cross section of the torsional member and the relative angle of twist of its ends. Assume the beam is aluminum.

SOLUTION

The shear flow for a cross section is given by Eq. (4-48). The area enclosed by the mean perimeter of the cross section is

$$A_m = 3.9(1.925) = 7.51 \text{ in}^2$$ (a)

so that the shear flow is

$$q = \frac{1000}{2(7.51)} = 66.6 \text{ lb/in.}$$ (b)

Consequently, the shearing stresses on the three legs whose thickness is 0.10 in. are obtained from Eq. (4-49). They are

$$\tau = \frac{66.6}{0.10} = 666 \text{ psi}$$ (c)

and the shearing stress on the leg whose thickness is 0.05 in. is equal to

$$\tau = \frac{66.6}{0.05} = 1332 \text{ psi}$$ (d)

The maximum shearing stress occurs on the thinnest leg of the cross section. The shear flow q is shown in Figure 4-29b acting along the center line of the cross section. The stress distributions for the 0.10-in.

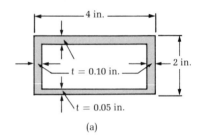

(a)

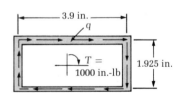

(b)

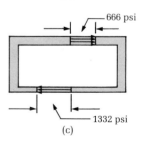

(c)

Figure 4-29

and the 0.05-in. thick legs are shown in Figure 4-29c. The stress distributions for the vertical legs are identical to that of the top, horizontal leg.

The relative unit angle of twist of the ends of the box beam is given by Eq. (4-55). Thus the angle of twist for the entire beam is

$$\varphi = \frac{(1000)(10 \times 12)}{4(7.51)^2(4 \times 10^6)}\left\{\frac{3.9}{0.10} + 2\left(\frac{1.925}{0.10}\right) + \frac{3.9}{0.05}\right\} = 0.021 \text{ rad}$$

or

$$\varphi = 1.18 \text{ degrees} \tag{e}$$

EXAMPLE 4-13

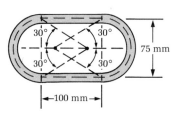

Figure 4-30

The aluminum $(G = 28 \text{ GPa})$ beam shown in Figure 4-30 resists a torque of 1000 N·m. If the beam is 4 m long, determine both the maximum shearing stress acting on a cross section and the angle of twist of its ends. The ends of the cross section are circular arcs and the thickness is 2 mm everywhere.

SOLUTION

The shear flow for a cross section is given by Eq. (4-48). The area enclosed by the mean perimeter of the cross section is

$$A_m = 0.1(0.075) + 2\left[0.5(0.075)^2\frac{\pi}{3} - 0.5(0.075)^2(0.866)\right]$$

$$= 0.0085 \text{ m}^2 \tag{a}$$

so that the shear flow is

$$q = \frac{1000}{2(0.0085)} = 58.82 \text{ kN/m} \tag{b}$$

Since the thickness of the cross section is the same at any point, the shearing stress must be constant on the cross section and equal to

$$\tau_{max} = \frac{58,820}{0.002} = 29.4 \text{ MPa} \tag{c}$$

The relative angle of twist of the free ends of the beam is obtained from Eq. (4-55). Accordingly, the angle of twist for the entire beam is

$$\varphi = \frac{(1000)(4)}{4(0.0085)^2(28 \times 10^9)}\left\{\frac{2\left[0.10 + \frac{\pi}{3}(0.075)\right]}{0.002}\right\} = 0.088 \text{ rad}$$

or

$$\varphi = 5.04 \text{ degrees} \tag{d}$$

PROBLEMS / Section 4-9

4-18 Calculate the torsional capacity for the thin-walled, circular tube shown in Figure P4-18 if the shearing stress is not to exceed 8 ksi, and the angle of twist is not to exceed 1.5 degrees per foot. Use (a) the basic torsion formulas and (b) the thin-wall approximate formulas. Compare the results ($G = 12 \times 10^6$ psi).

4-19 Compute the torsional capacity for the thin-walled tube shown in Figure P4-19 if the shearing stress is not to exceed 4 ksi. Determine the unit angle of twist for this limiting torque ($G = 4 \times 10^6$ psi).

4-20 A straight, thin-walled aluminum ($G = 28$ GPa) tube with the cross section shown in Figure P4-20 is subjected to a torque of 140 N·m. Determine (a) the shear flow q, (b) the shearing stress in each segment, and (c) the angle of twist per unit length. Neglect stress concentrations at the apexes of the triangle.

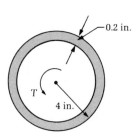

Figure P4-18

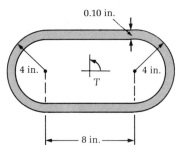

Figure P4-19

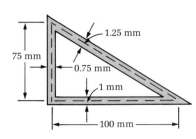

Figure P4-20

4-10   Elastic Twisting of Solid Noncircular Sections

The basic kinematic assumptions enumerated in Section 4-3 for the twisting of shafts with circular cross sections are not valid for the twisting of shafts with noncircular cross sections. In particular, transverse cross sections warp during twisting of shafts with solid, noncircular cross sections. Warping means that points that lie in a plane perpendicular to the axis of the shaft before twisting do not lie in a plane after twisting. In other words, points in a transverse plane may experience displacements parallel to the axis of the shaft. This behavior has a profound effect on the stress distributions for noncircular cross sections.

This section presents some formulas that are useful in the calculation of shearing stresses for solid, noncircular cross sections. The rigorous development of the formulas presented here is beyond the pedagogical scope of this text. The interested reader is referred to an advanced text such as *Applied Elasticity* by Chi-Teh Wang for further details.

Elliptical cross section.   The components of shearing stress for an elliptical cross section whose semi-major and semi-minor axes are a and b, respectively, are

$$\left.\begin{aligned}\tau_{zx} &= -\frac{2Ty}{\pi ab^3}\\[2mm]\tau_{zy} &= \frac{2Tx}{\pi a^3 b}\end{aligned}\right\} \tag{4-56}$$

The unit angle of twist is given by the formula

$$\frac{d\varphi}{dz} = \frac{T_R}{JG} \tag{4-57}$$

where

$$J = \frac{\pi a^3 b^3}{a^2 + b^2} \tag{4-58}$$

Figure 4-31 depicts an elliptical cross section and the stress components τ_{zx} and τ_{zy}. The maximum stress occurs at the ends of the minor axis of the ellipse. Its magnitude, from Eq. (4-56), is

$$\tau_{max} = \frac{2Tb}{\pi ab^3} = \frac{2T}{\pi ab^2} \tag{4-59}$$

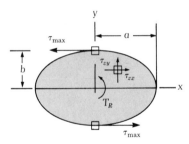

Figure 4-31

Rectangular cross section.   Analytical solutions have been obtained for torsion of rectangular, elastic members of sides $2a$ and $2b$ as shown in Figure 4-32. The components of shearing stress are given by formulas in terms of infinite series. The mathematical treatment is beyond the scope of this book. The analysis shows that shearing stresses at the corners are zero. Note that the corners of a rectangular cross section are its most remote points. Accordingly, the stress is zero at the most remote point of the rectangular cross section. Recall that the stress was a maximum at the most remote point for circular cross sections. This observation emphasizes how different the stress distributions are for these two cross sections. The maximum shearing stress for a rectangular cross section occurs at the midpoint of the long side. It can be calculated analytically and the result expressed in terms of an infinite series.

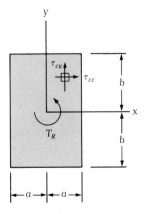

Figure 4-32

Table 4-1

$\dfrac{b}{a}$	Table of coefficients for rectangular sections								
	1.00	1.50	2.00	2.50	3.00	4.00	6.00	10.0	∞
α	0.208	0.231	0.246	0.256	0.267	0.282	0.299	0.312	0.333
β	0.141	0.196	0.229	0.249	0.263	0.281	0.299	0.312	0.333

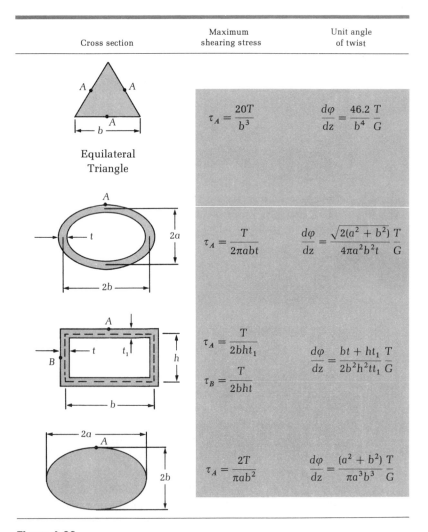

Cross section	Maximum shearing stress	Unit angle of twist
Equilateral Triangle	$\tau_A = \dfrac{20T}{b^3}$	$\dfrac{d\varphi}{dz} = \dfrac{46.2}{b^4}\dfrac{T}{G}$
	$\tau_A = \dfrac{T}{2\pi a b t}$	$\dfrac{d\varphi}{dz} = \dfrac{\sqrt{2(a^2+b^2)}}{4\pi a^2 b^2 t}\dfrac{T}{G}$
	$\tau_A = \dfrac{T}{2bht_1}$ $\tau_B = \dfrac{T}{2bht}$	$\dfrac{d\varphi}{dz} = \dfrac{bt + ht_1}{2b^2h^2 t t_1}\dfrac{T}{G}$
	$\tau_A = \dfrac{2T}{\pi a b^2}$	$\dfrac{d\varphi}{dz} = \dfrac{(a^2+b^2)}{\pi a^3 b^3}\dfrac{T}{G}$

Figure 4-33

Frequently, the maximum shearing stress and the unit angle of twist are expressed in the forms

$$
\left.\begin{aligned}
\tau_{max} &= \frac{T_R}{\alpha(2b)(2a)^2} \\
\frac{d\varphi}{dz} &= \frac{T_R}{\beta(2b)(2a)^3 G}
\end{aligned}\right\}
\tag{4-60}
$$

where $2a$ is the length of the short side of the rectangle, and $2b$ is the length of the long side. The coefficients α and β are determined so that, when they are substituted into Eq. (4-60), τ_{max} and $d\varphi/dz$ coincide with values of τ_{max} and $d\varphi/dz$ calculated from the more advanced analyses. Numerical values for α and β that correspond to several values of b/a are listed in Table 4-1. For thin rectangles—that is, for $2b$ much greater than $2a$—$\alpha = \beta = \frac{1}{3}$.

Approximate formulas for the torsional shearing stresses and angles of twist for several noncircular cross sections are listed in Figure 4-33.

PROBLEMS / Section 4-10

4-21   A steel ($G = 84$ GPa) shaft with the elliptical cross section shown in Figure P4-21 resists an applied torque $T = 500$ N·m. Determine both the magnitude and location of the maximum shearing stress on a cross section and the unit angle of twist.

4-22   An aluminum ($G = 28$ GPa) shaft with the square cross section shown in Figure P4-22 is subjected to a torque $T = 1000$ in.-lb. Determine both the magnitude and location of the maximum shearing stress on a cross section and the unit angle of twist.

4-23   A steel ($G = 84$ GPa) shaft with the rectangular cross section shown in Figure P4-23 is subjected to a torque $T = 500$ N·m. Determine both the magnitude and location of the maximum shearing stress on a cross section and the unit angle of twist.

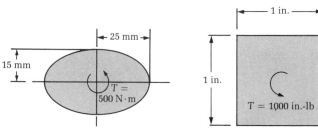

Figure P4-21   **Figure P4-22**   **Figure P4-23**

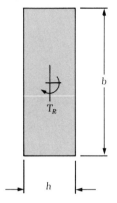

Figure 4-34

4-11 Sections Composed of Narrow Rectangles

For the long, narrow rectangular section of height b and base h shown in Figure 4-34, the following approximations are obtained from the exact analysis

$$J = \tfrac{1}{3}bh^3 \atop \tau_{\text{max}} = \dfrac{T_R h}{J}\Bigg\} \tag{4-61}$$

Here b is the long side of the narrow rectangle, and h is the short side. The maximum shearing stress occurs at the midpoint of the long side of the rectangle.

The value of J for sections composed of several long, narrow rectangles is approximately

$$J = \sum_{i=1}^{n} J_i, \tag{4-62}$$

where J_i is the value of J for ith rectangular segment. This formula can be used to calculate the value of J for angles, channels, and wide-flange and other sections that are made up of narrow rectangles.

The corresponding approximate angle of twist is

$$\frac{d\varphi}{dz} = \frac{T_R}{JG} \tag{4-63}$$

EXAMPLE 4-14

Determine the magnitude and location of the maximum shearing stress in each leg of the aluminum angle section shown in Figure 4-35 when a 5-ft length is subjected to twisting couples of 3000 in.-lb at its ends. Determine the relative rotation of the ends of this torsion member.

SOLUTION

This cross section consists of two narrow rectangles. An approximate value of polar moment of inertia is obtained from Eqs. (4-61) and (4-62). Accordingly.

$$J = J_1 + J_2 = \tfrac{1}{3}(4)(0.5)^3 + \tfrac{1}{3}(6)(0.25)^3 = 0.1979 \text{ in}^4 \tag{a}$$

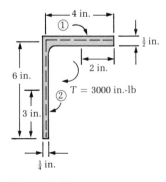

Figure 4-35

The magnitude of the maximum shearing stress in each leg is given by the second formula in Eqs. (4-61). Thus

$$\left.\begin{array}{l} \tau_{\max}^{(1)} = \dfrac{3000}{0.1979}(0.5) = 7580 \text{ psi} \\[4mm] \tau_{\max}^{(2)} = \dfrac{3000}{0.1979}(0.25) = 3790 \text{ psi} \end{array}\right\} \qquad \text{(b)}$$

These stresses occur at the midlengths of the horizontal and vertical legs of the section.

The relative rotation of the ends of the bar is obtained through the use of Eq. (4-63). Accordingly,

$$\varphi = \frac{3000(60)}{0.1979(4 \times 10^6)} = 0.227 \text{ rad} \qquad \text{(c)}$$

PROBLEMS / Section 4-11

4-24   Consider steel ($G = 12 \times 10^6$ psi) shafts with the cross sections shown in Figures P4-24a, b, and c. Calculate both the maximum shearing stress in each section and the unit angle of twist when $T = 1000$ in.-lb.

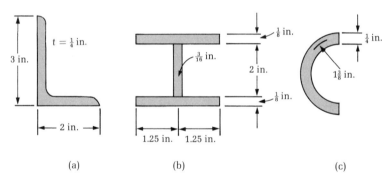

(a)   (b)   (c)

Figure P4-24

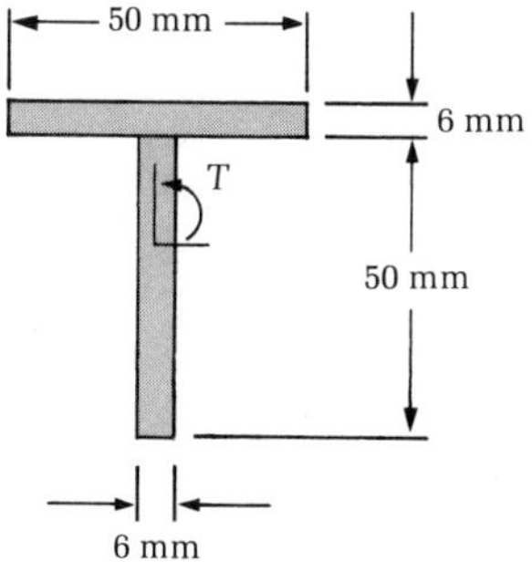

Figure P4-25

4-25   Calculate the maximum torque T that the aluminum shaft shown in Figure P4-25 can carry if the shearing stress is not to exceed 20 MPa, and the angle of twist is not to exceed 30 degrees per meter.

4-12   SUMMARY

In this chapter, we developed formulas that are used to calculate shearing stress and angular displacements for pure twisting of straight shafts. We developed these formulas for the elastic twisting of straight s with cross sections that are: circular (solid or annular), closed thin-walled, solid rectangular or elliptical, fabricated from a ⸺es of interconnected thin rectangles, and made from more than one ⸺ ⸺terial. These formulas are summarized in Figure 4-36.

We developed analogous formulas for the twisting of shafts with circular (solid or annular) cross sections that undergo inelastic strains.

Cross section	Stress formula	Angle of twist formula
Circular or Annular	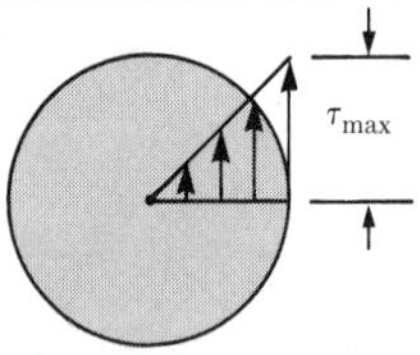 $\tau = \dfrac{Tr}{J}$	$\dfrac{d\varphi}{dz} = \dfrac{T}{JG}$
Closed Thin-walled	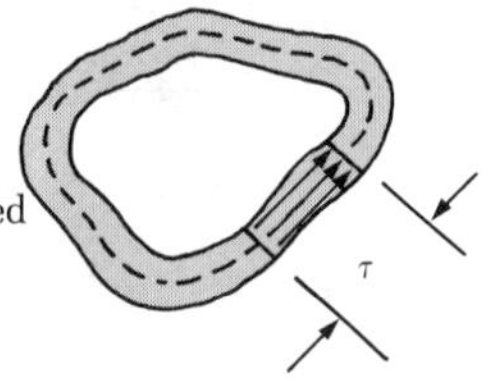 $\tau = q/t$ $q = \dfrac{T}{2A_m}$	$\dfrac{d\varphi}{dz} = \dfrac{T}{4GA_m^2}\int\dfrac{ds}{t}$
Rectangle	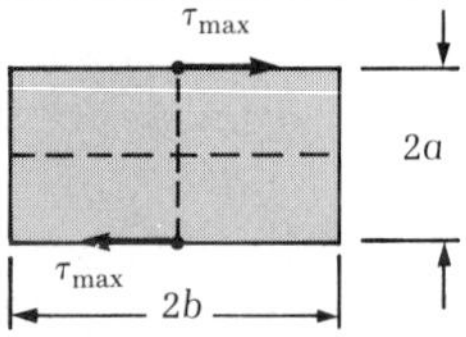 $\tau_{max} = \dfrac{T}{\alpha(2b)(2a)^2}$	$\dfrac{d\varphi}{dz} = \dfrac{T}{\beta(2b)(2a)^3 G}$
Elliptical	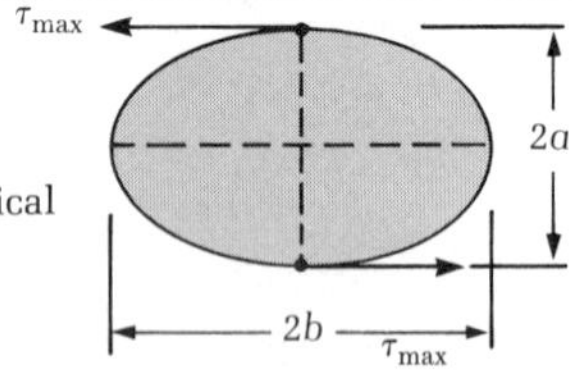 $\tau_{max} = \dfrac{2T}{\pi ab^2}$	$\dfrac{d\varphi}{dz} = \dfrac{(a^2+b^2)T}{\pi Ga^3b^3}$
Two-material Circular	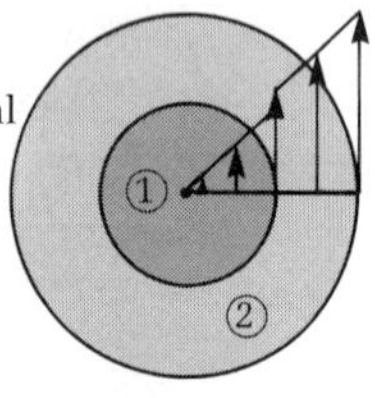 $\tau_1 = \dfrac{T_1 r}{J_1}$ $\tau_2 = \dfrac{T_2 r}{J_2}$	$\dfrac{d\varphi}{dz} = \dfrac{T_1}{J_1 G_1}$ or $\dfrac{d\varphi}{dz} = \dfrac{T_2}{J_2 G_2}$

Figure 4-36

Elastic, perfectly plastic:

$$T_{PP} = \frac{\pi \tau_{YP}}{6} (4c^3 - a^3)$$

$$\frac{d\varphi}{dz} = \frac{6T}{\pi Ga(4c^3 - a^3)}$$

Elastic, linear hardening:

$$\begin{cases} T_{PP} = 2\pi\tau_0 \left\{ \frac{a^3}{4} + \frac{1-\alpha}{3} (c^3 - a^3) + \frac{\alpha}{4a} (c^4 - a^4) \right\} \\ \dfrac{d\varphi}{dz} = \dfrac{T}{2\pi aG \left\{ \frac{a^3}{4} + \frac{1-\alpha}{3} (c^3 - a^3) + \frac{\alpha}{4a} (c^4 - a^4) \right\}} \end{cases}$$

In addition, we demonstrated a procedure that can be used to determine residual stress distributions due to unloading of **circular** shafts from stress levels that exceed the yield point. We developed formulas that connect horsepower, the angular velocity of a shaft, and the torque transmitted through a rotating shaft in English and SI units.

$$HP = \begin{cases} \dfrac{2\pi NT}{33,000(12)} & \text{English units} \\[2ex] \dfrac{2\pi fT}{745.7} & \text{SI units} \end{cases}$$

The methods of direction integration, discrete element, and super-position were used to determine relative angles of twist for circular elastic shafts for statically determinate torsional problems. These methods were extended to statically indeterminate elastic shafts where we emphasized the three basic ingredients required for a solution: equilibrium of torques, compatibility of angular displacements, and material behavior.

PROBLEMS / CHAPTER 4

4-26 The diameters of the three segments of the steel ($G = 84$ GPa) step-shaft shown in Figure P4-26 are 150 mm, 100 mm, and 50 mm. (a) Calculate the maximum shearing stress in the shaft. (b) Use the discrete element method to determine the angular displacement of the section 2 m from the left end.

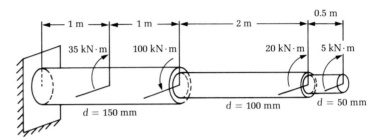

Figure P4-26

4-27 The aluminum ($G = 28$ GPa) step-shaft shown in Figure P4-27 has a 25-mm diameter concentric hole as shown. Use the superposition method to determine the angular displacement of the free end. Calculate the maximum shearing stress in the shaft.

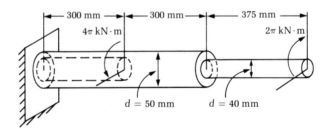

Figure P4-27

4-28 An external torque t (in.-lb) is distributed continuously along the surface of a circular rod of length ℓ and diameter d as shown in Figure P4-28. If $t(z) = 200 - 100\ (z/\ell)$, determine the angular displacement of the free end.

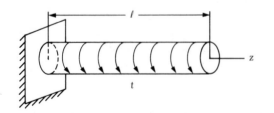

Figure P4-28

4-29 A solid circular shaft tapers uniformly from a diameter d_s to a diameter d_l as shown in Figure P4-29. If the taper angle α is small, establish a formula for the angular displacement of one end relative to the other.

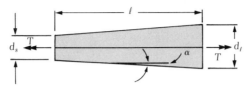

Figure P4-29

4-30 A simple torsion bar spring is shown in Figure P4-30. If the shearing stress in the steel shaft is not to exceed 100 MPa, and if the vertical deflection of point A is not to exceed 10 mm, determine the required diameter of the shaft. Neglect the bending of the shaft.

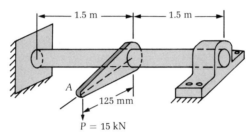

Figure P4-30

4-31 Determine the minimum permissible diameter for the aluminum segment of the composite step-shaft shown in Figure P4-31 if the maximum permissible angle of twist of the free end is 10 degrees and the allowable shearing stress in the aluminum is 50 MPa (G_{ST} = 84 GPa and G_{AL} = 28 GPa).

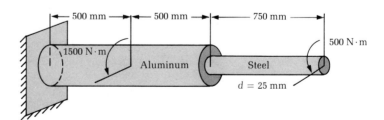

Figure P4-31

4-32 A system of gears transmits power via two solid steel ($G = 12 \times 10^6$ psi) shafts as shown in Figure P4-32. Determine the horsepower that can be safely transmitted from gear A to gear D if the shearing stress is not to exceed 12 ksi in either shaft and if shaft AB rotates at 330 rpm.

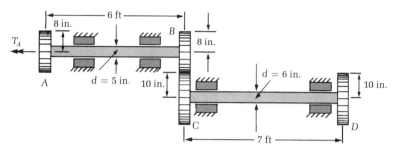

Figure P4-32

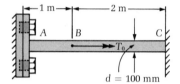

Figure P4-33

4-33   An aluminum ($G = 28$ GPa) shaft is rigidly supported at its right end as shown in Figure P4-33. The left end of the shaft is connected to a rigid support via a flange plate. The flange plate allows the shaft to rotate 0.010 rad before the bolts provide rigid support. Determine the maximum torque that can be applied at point B if the shearing stress is not to exceed 50 MPa.

4-34   A composite aluminum ($G = 4 \times 10^6$ psi) and steel ($G = 12 \times 10^6$ psi) shaft is used to transmit a torque of 6000π in.-lb as shown in Figure P4-34. The two materials are assumed to behave as a unit. Compute the maximum stress in the aluminum and in the steel and the rotation of the free end.

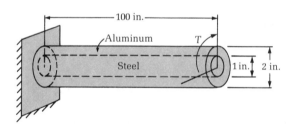

Figure P4-34

4-35   The steel step-shaft shown in Figure P4-35 is required to transmit 20 HP at 330 rpm. If the angle of twist per foot of length is not to exceed 5 degrees, and if the shearing stress is not to exceed 8 ksi, determine the minimum permissible diameter d of the smaller segment.

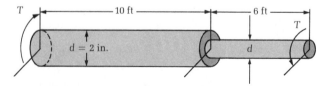

Figure P4-35

4-36 Determine the maximum shearing stress in the shafting used in the motor-pump combination shown in Figure P4-36. The motor delivers 75π kw at 330 rpm. Assume there is no power loss due to friction in the bearings or in the gears. Shafts AB and CD are made of steel ($G = 84$ GPa), and each has a diameter of 75 mm.

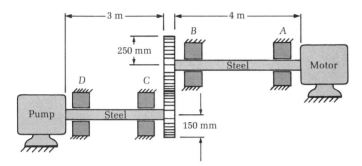

Figure P4-36

4-37 A diesel engine for a commercial fishing boat delivers 1000 HP at 200 rpm. The engine is connected by solid steel shafting through a gear box with a 1:4 (the angular velocity of the propeller shaft is four times that of the main shaft) ratio to the boat's propeller. Determine the minimum permissible diameters for the two shafts if the shearing stress is not to exceed 18 ksi and the angle of twist is not to exceed 5 degrees per 10-ft length of the propeller shaft. Neglect power loss in the gear box and assume that the propeller shaft is subjected to pure torsion. (See Figure P4-37.)

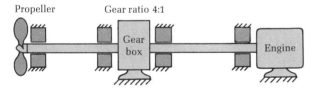

Figure P4-37

4-38 A composite aluminum ($G = 4 \times 10^6$ psi) and steel ($G = 12 \times 10^6$ psi) shaft is rigidly supported at each end as shown in Figure P4-38. Determine the maximum shearing stress in the aluminum and in the steel.

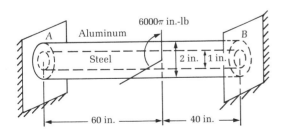

Figure P4-38

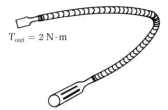

Figure P4-39

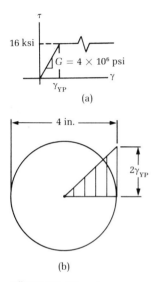

(a)

(b)

Figure P4-40

4-39 The flexible shaft shown in Figure P4-39 consists of a 5 mm diameter steel wire in a flexible hollow tube that imposes a frictional torque of 0.45 N·m/m. The lengths of the steel wire and flexible tube are 1 m. The unit is to be used to tighten screws that are otherwise inaccessible. Determine the maximum shearing stress in the steel wire if an output torque of 2 N·m is required. Determine the relative angle of twist of the handle relative to the blade.

4-40 A 4-in. diameter shaft is made from a material whose stress-strain diagram is shown in Figure P4-40a. The shaft is subjected to a torque T that causes the shearing strain distribution shown in Figure P4-40b. (a) Determine the radius of the portion of the cross section that exhibits elastic behavior. (b) Construct the stress distribution on a cross section. (c) Calculate the torque T that is associated with this strain distribution. (d) Determine the residual stress distribution and the residual unit angle of twist.

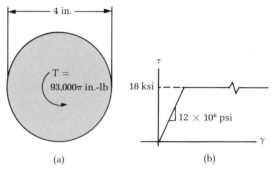

(a)   (b)

Figure P4-41

4-41 A 4-in. diameter steel shaft 10 ft long is subjected to a torque of $93,000\pi$ in.-lb as shown in Figure P4-41a. Determine (a) the shearing stress distribution on a cross section, (b) the angle of twist of one end relative to the other, (c) the residual stress distribution, and (d) the residual angle of twist.

4-42 A circular shaft of radius c is subjected to a torque T that causes the stress distribution shown in Figure P4-42. Establish a formula for the torque T in terms of τ_{max} and c.

4-43 The shearing strain distribution varies linearly on a cross section as shown in Figure P4-43a. The stress-strain relation for the material from which

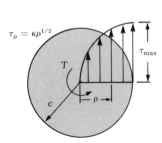

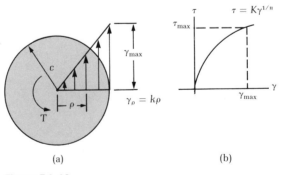

(a)   (b)

Figure P4-42   **Figure P4-43**

the shaft is made is given in Figure P4-43b. Determine a formula for the torque T associated with this strain distribution in terms of τ_{max}, n, K, and c. Derive a formula for the unit angle of twist in terms of τ_{max}, n, K, and c.

4-44 The stress-strain diagram for a certain material is shown in Figure P4-44. (a) Establish the equations that express the shearing stress in terms of shearing strain. (b) Sketch the stress distribution corresponding to a 1-in. depth of yielding in a 4-in. diameter shaft made of this material. (c) Calculate the torque that will cause a 1-in. depth of yielding. (d) Calculate the unit angle of twist that accompanies this torque. (e) Determine the residual stress distribution that results when the torque is removed. (f) Calculate the residual angle of twist.

4-45 A hollow steel shaft is made from an ideally elastic, plastic material with a shearing yield point of 140 MPa. The outside and inside diameters of the shaft's cross section are 100 mm and 50 mm, respectively. (a) Determine the maximum elastic torque that the shaft can resist. (b) Calculate the torque corresponding to a fully plastic condition. (c) Compute the percentage increase in the load carrying capacity of the fully plastic condition over the fully elastic condition.

4-46 A 4-in. diameter solid circular steel shaft made of mild steel ($\tau_{YP} = 18$ ksi) is twisted by a torque that causes yielding of a cross section to penetrate to a depth equal to 40% of the radius. Calculate the torque.

4-47 Calculate the torsional capacity of the thin-walled aluminum tube with the elliptical cross section shown in Figure P4-47. The shearing stress is not to exceed 28 MPa, and the angle of twist is not to exceed 1.5 degrees per meter of tube.

4-48 The thin-walled tube with the square cross section shown in Figure P4-48 is made from an ideally elastic plastic material with a shearing yield point of 20 ksi. (a) Determine the fully plastic torque for the cross section. (b) Compare it with the fully elastic torque for the same section.

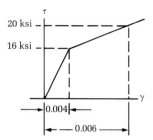

Figure P4-44

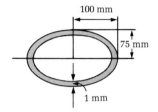

Figure P4-47

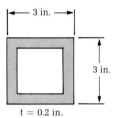

Figure P4-48

CHAPTER 5

Bending of Straight Beams

5-1 Introduction

A beam is a structural member for which one dimension is large compared to the other two dimensions, and it carries loads perpendicular to the large dimension. The beam is, perhaps, the most widely used structural element. Its uses are evident in all sorts of structures, including buildings, bridges, and machines. Beams can be straight or curved, statically determinate or statically indeterminate, and symmetric or unsymmetric in cross section. In this chapter, we develop in detail the theory of straight beams with cross sections that have at least one axis of symmetry. Chapter 12 extends this theory to curved beams. This chapter is also restricted to the analysis of beams that bend in a single plane—the plane containing the axes of symmetry of cross sections along the beam axis. Chapter 12 extends the theory to beams with unsymmetric cross sections and to beams that bend simultaneously in two perpendicular planes.

5-2 Classifications of Straight Beams

Straight beams form an important class of structural members. Their use is frequent and varied. The following definitions help to identify different types of beams.

Loads. There are essentially three types of loads to deal with in the analysis of straight beams: concentrated forces, concentrated couples, and distributed forces. Figure 5-1a shows these three load classifications.

Supports. A beam is ordinarily supported at one or more points along its axis. There are essentially three different support mechanisms.

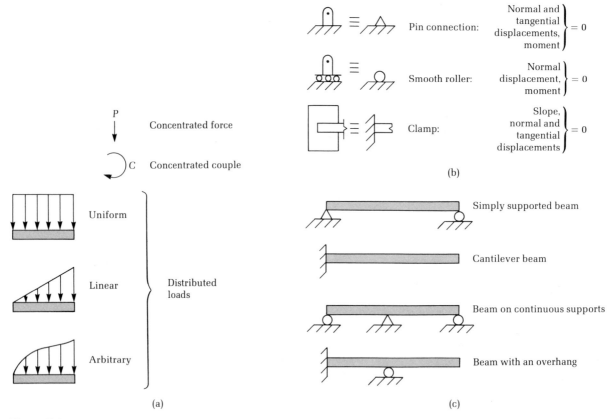

Figure 5-1

The representations of the support mechanisms used in this text are shown in Figure 5-1b. They are the smooth pin connection, the smooth roller, and the clamp.

The support mechanisms shown in Figure 5-1b are idealizations (of the actual beam supports) that simplify the analyses of straight beams. Actual beam supports may differ sharply from the three idealizations in Figure 5-1b. Nevertheless, these idealizations are of significant practical value.

We learned in statics that a rigid body can undergo at most three plane motions: two mutually perpendicular translations and a rotation. The purpose of a beam support is to prevent one or more of these motions from occurring. The purpose of a combination of supports is to prevent all three rigid-body motions. Consequently, when the supports are sufficient in number and type to prevent the rigid-body motions of a beam, the externally applied forces can only cause the beam to deform.

A support prevents a so-called rigid body motion by providing appropriate reactions (forces and/or couples). Various properties of the three support mechanisms are shown in Figure 5-1b. For example, a reaction provided by a roller prevents movement of a point on a beam in the direction parallel to the roller reaction. The point can, however, move in a direction perpendicular to the reaction. Moreover, the roller offers no resistance to a tendency of the beam to rotate about the contact point as a pivot. The rotation to which we refer is due to bending action caused by loads acting on the beam. The pin connection prevents both the component of displacement perpendicular to the axis of the beam and the component parallel to it. Finally, the clamp mechanism prevents both components of displacement and the tendency of the beam to rotate about the clamp as a pivot.

We note that a support mechanism provides reactions that act on the beam in accordance with the displacement and/or rotation that it prevents. Therefore, a roller exerts a single force on the beam whose direction is perpendicular to the plane on which the roller rests, a pin connection exerts a single force of unknown magnitude and direction (usually perceived as two mutually perpendicular components), and a clamp exerts two rectangular force components and a couple.

Beams. There is no standard method of naming beams. However, certain beams, because of their frequent use, have labels that are used to identify them. For example, a **simply supported beam** is one that is supported at two points by smooth pin connectors, rollers, or a combination of the two. A **cantilever beam** is supported by a clamp at one end and is unsupported elsewhere. A beam on continuous supports is supported at three or more points by rollers, pin connectors, or a combination of the two. These beams are depicted in Figure 5-1c.

The theory of straight beams is usually developed under the assumption that all the loads are applied perpendicular to the axis of the beam. As a consequence, the normal force does not appear; that is, it is identically zero. Thus only the transverse shearing force V and the bending moment M appear on any free-body diagram. It is not difficult to introduce the effect of the normal force after thorough analyses of the effects of the shearing force and the bending moment have been made. In the following examples, the normal force is not explicitly shown on a free-body diagram when equilibrium in the direction of the normal force is zero.

5-3 Internal Reactions

Consider a straight beam that is loaded and supported as shown in Figure 5-2a. The loads are assumed to lie in the plane of the paper and to act in such a way that the beam is deformed in the plane containing them.

An imaginary section is established at a distance x from the left end of the beam. A free-body diagram of the segment of the beam to the left of the imaginary section is shown in Figure 5-2b, and a free-body diagram of the segment to the right of the imaginary section is

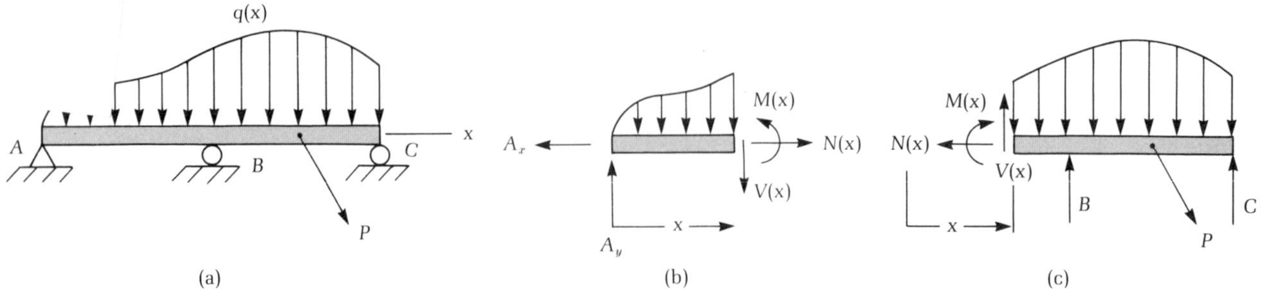

Figure 5-2

shown in Figure 5-2c. The components of the force-couple system that is equivalent to the distributed force system acting on the section are designated by $V(x)$, $N(x)$, and $M(x)$. $V(x)$ is the transverse shearing force at the section, $N(x)$ is the normal force, and $M(x)$ is the bending moment.

It is common practice (although not universal) for the directions shown in Figures 5-2b and 5-2c to represent a positive internal shearing force $V(x)$, a positive internal normal force $N(x)$, and a positive internal bending moment $M(x)$. Note that Newton's third law connects V, N, and M for the two free-body diagrams. That is, the directions for V, N, and M shown in Figure 5-2c follow directly from Newton's third law if positive directions for V, N, and M are assigned via the free-body diagram of Figure 5-2b.

It is appealing and desirable to maintain sign conventions for the shearing force, the normal force, and the bending moment (in a straight beam) that are consistent with the definitions established in Eq. (1-4). The sign conventions established in the previous paragraph for the normal force and the bending moment in a straight beam are consistent with those given by the first of Eq. (1-3) and the third of Eq. (1-4). However, the sign convention for the shearing force in a straight beam is contrary to the definition given by the second of Eq. (1-4).

Practical considerations make it desirable to adopt this anomaly. Experience shows that maintaining the sign convention established in Eq. (1-3) between the positive direction for shearing stress and the corresponding shearing force resultant results in an annoying negative sign that must be accounted for when constructing shear and moment diagrams by the so-called summation technique. The presence of this negative sign is a nuisance that can be avoided by adopting the sign convention just described.

EXAMPLE 5-1

Determine the internal shearing force and the internal bending moment acting on a section 5 ft from point A on the straight beam shown in Figure 5-3a.

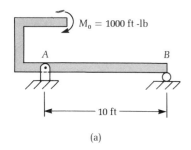

(a)

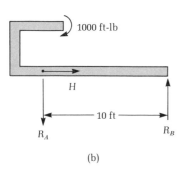

(b)

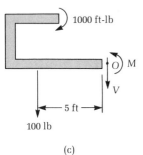

(c)

Figure 5-3

SOLUTION

The external reactions at the beam supports must be determined before the internal shearing force and the internal bending moment can be established.

Consider the free-body diagram of the entire system shown in Figure 5-3b. The forces that the supports provide are shown in the figure. Clearly,

$$\left(+\ \Sigma M_A = 0: \quad 10R_B - 1000 = 0, \qquad R_B = 100 \text{ lb} \right. \tag{a}$$

and

$$\left(+\ \Sigma M_B = 0: \quad 10R_A - 1000 = 0, \qquad R_A = 100 \text{ lb} \right. \tag{b}$$

Consideration of equilibrium in the horizontal direction yields $H = 0$.

As a check that no arithmetical errors have been made in calculating the external reactions, observe that the vertical forces acting on the beam must be in equilibrium. Thus, if

$$\uparrow + \Sigma F_y = 0: \quad -100 + 100 = 0 \tag{c}$$

we can conclude that reactions have been calculated correctly.

The internal reactions at a section 5 ft from point A can be calculated from the free-body diagram shown in Figure 5-3c. Equilibrium of the portion of beam shown in Figure 5-3c yields

$$+\uparrow \Sigma F_y = 0: \quad -V - 100 = 0 \qquad \text{or} \qquad V = -100 \text{ lb}$$
$$\left(+\ \Sigma M_O = 0: \quad M - 1000 + 5(100) = 0 \right. \tag{d}$$
$$\text{or} \qquad M = 500 \text{ ft-lb}$$

The negative sign on the shearing force indicates that V acts in the direction opposite the one shown in Figure 5-3c.

EXAMPLE 5-2

Determine the internal shearing force, the internal normal force, and the internal bending moment acting on a cross section 7 ft from point A of the inclined beam shown in Figure 5-4a.

SOLUTION

The external reactions can be determined with the aid of the free-body diagram of the entire system shown in Figure 5-4b. Taking the sign convention shown in Figure 5-4b, we obtain

$$\Sigma M_A = 0: \quad 10R_B - 600(5) = 0 \qquad \text{or} \qquad R_B = 300 \text{ lb} \tag{a}$$
$$\Sigma M_B = 0: \quad -10A_y + 600(5) = 0 \qquad \text{or} \qquad A_y = 300 \text{ lb} \tag{b}$$
$$\Sigma F_x = 0: \quad A_x - 800 = 0 \qquad \text{or} \qquad A_x = 800 \text{ lb} \tag{c}$$

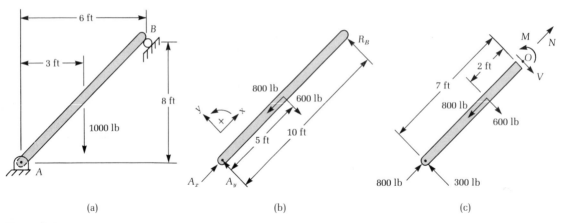

(a)　　　　　　　　　(b)　　　　　　　　　(c)

Figure 5-4

Having determined the external reactions, we can calculate the internal reactions at the specified section. Since the free-body diagram shown in Figure 5-4c must be in equilibrium, it follows that

$$\Sigma F_y = 0: \quad -V - 600 + 300 = 0 \qquad \text{or} \quad V = -300 \text{ lb}$$

$$\Sigma M_0 = 0: \quad M + 600(2) - 300(7) = 0 \quad \text{or} \quad M = 900 \text{ ft-lb} \quad \text{(d)}$$

$$\Sigma F_x = 0: \quad N + 800 - 800 = 0 \qquad \text{or} \quad N = 0$$

The assumed direction of M is correct, but V acts in the direction opposite the one shown in Figure 5-4c.

EXAMPLE 5-3

Determine the internal shearing force and the internal bending moment 3 m from the left end of the straight beam shown in Figure 5-5a.

SOLUTION

The reactions at the supports can be calculated in the usual manner. They are found to be

$$\left. \begin{array}{l} R_A = 25 \text{ kN} \\ R_B = 35 \text{ kN} \end{array} \right\} \qquad \text{(a)}$$

The internal shearing force and the internal bending moment 3 m from the left end of the beam are obtained through force and moment equilibrium conditions for the free-body diagram shown in Figure 5-5b. Taking the sign convention shown in Figure 5-5c, equilibrium conditions yield

$$\left. \begin{array}{lll} \Sigma F_y = 0: & -V + 25 - 15 = 0 & \text{or} \quad V = 10 \text{ kN} \\ \Sigma M_0 = 0: & M + 15(1.5) - 25(3) = 0 & \text{or} \quad M = 52.5 \text{ kN·m} \end{array} \right\} \quad \text{(b)}$$

The assumed directions of V and M are correct.

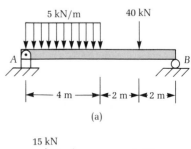

(a)

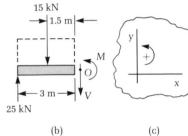

(b)　　　　　　(c)

Figure 5-5

PROBLEMS / Section 5-3

5-1 Determine the reactions at the supports for the straight beam shown in Figure P5-1. Also determine the internal reactions at section *aa*.

5-2 Determine the external reactions at the supports for the beam shown in Figure P5-2. Determine the internal reactions at section *aa* 0.5 m to the left of the right support.

5-3 Determine the external reactions at the supports and the internal reactions at sections *aa* and *bb* for the beam shown in Figure P5-3.

5-4 Determine the external reactions at the supports for the beam in Figure P5-4. Determine the internal reactions at section *aa*.

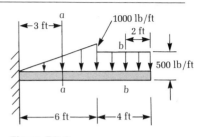

Figure P5-3

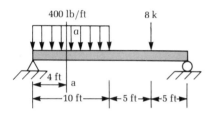

Figure P5-1

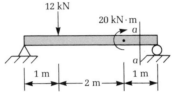

Figure P5-2

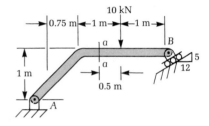

Figure P5-4

5-4   Shear and Moment Equations and Diagrams

The preceding examples show the methods required to determine the magnitudes and senses of the internal normal force, the shearing force, and the bending moment at a specified section of a beam. A natural extension of these methods is to write expressions for these internal reactions as functions of distance along the beam axis. The analytical expressions for $N(x)$, $V(x)$, and $M(x)$ can be plotted to obtain a graphic representation of their variation along the beam axis. This method is illustrated in the following examples.

EXAMPLE 5-4

For the straight beam loaded and supported as shown in Figure 5-6a, determine algebraic expressions for the internal shearing force and the internal bending moment at an arbitrary section along the centroidal line of the beam. Plot these expressions.

SOLUTION

First determine the external reactions at the supports at points A and B. Equilibrium of the entire beam shown in Figure 5-6a yields

$$\Sigma M_B = 0: \quad -12R_A + 45(4) + 48 + 36(10) = 0$$

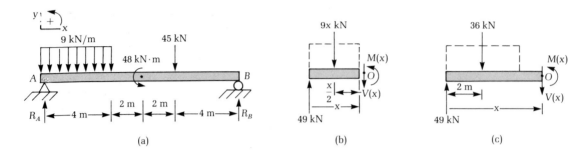

(a)   (b)   (c)

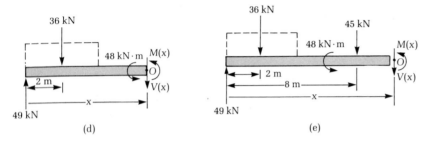

(d)   (e)

Figure 5-6

or

$$R_A = 49 \text{ kN} \qquad\qquad\qquad (a)$$

and

$$\Sigma M_A = 0: \quad 12R_B - 45(8) + 48 - 36(2) = 0$$

or

$$R_B = 32 \text{ kN} \qquad\qquad\qquad (b)$$

Note that the 36-kN force used in the preceding equations is the resultant of the distributed load.

As a check on the correctness of the reactions calculated in
Eqs. (a) and (b), observe that the transverse forces must satisfy the
vertical equilibrium equation. Hence, if

$$\Sigma F_y = 0: \quad 49 + 32 - 36 - 45 = 0$$

then the reactions have been calculated correctly.

Several algebraic expressions are required to represent the internal
shearing force and the internal bending moment at an arbitrary section
along the beam axis.

Since the internal shearing force and the internal bending moment
are established through the equilibrium of transverse forces and moment equilibrium, it should be clear that a new set of equations for V
and M is required whenever an abrupt change in the force system acting
on a free-body diagram of any segment of the beam occurs. Thus the

free-body diagram depicted in Figure 5-6b remains valid in the interval $0 \leq x \leq 4$ only; a second free-body diagram is required for the interval $4 \leq x \leq 6$ as shown in Figure 5-6c; a third free-body diagram is required for the interval $6 \leq x \leq 8$ as shown in Figure 5-6d; and a fourth is required for the interval $8 \leq x \leq 12$ as shown in Figure 5-6e.

Using the sign convention shown in Figure 5-6a, we determine the expressions for the internal shearing force and the internal bending moment in each interval as follows.

From the free-body diagram of Figure 5-6b,

$$\left.\begin{array}{ll} \Sigma F_y = 0: & -V - 9x + 49 = 0 \\ \Sigma M_0 = 0: & M + 9x(x/2) - 49x = 0 \end{array}\right\} \qquad \text{(c)}$$

which yield

$$\left.\begin{array}{l} V = 49 - 9x \text{ kN} \\ M = 49x - 4.5x^2 \text{ kN·m} \end{array}\right\} \text{for } 0 \leq x \leq 4 \qquad \text{(d)}$$

Likewise, from Figure 5-6c,

$$\left.\begin{array}{ll} \Sigma F_y = 0: & -V - 36 + 49 = 0 \\ \Sigma M_0 = 0: & M + 36(x - 2) - 49x = 0 \end{array}\right\} \qquad \text{(e)}$$

which yield

$$\left.\begin{array}{l} V = 13 \text{ kN,} \\ M = 49x - 36(x - 2) = 13x + 72 \text{ kN·m} \end{array}\right\} \text{for } 4 \leq x \leq 6 \qquad \text{(f)}$$

Similarly, from Figure 5-6d,

$$\left.\begin{array}{ll} \Sigma F_y = 0: & -V - 36 + 49 = 0 \\ \Sigma M_0 = 0: & M + 48 + 36(x - 2) - 49x = 0 \end{array}\right\} \qquad \text{(g)}$$

so that

$$\left.\begin{array}{l} V = 13 \text{ kN} \\ M = 13x + 24 \text{ kN·m} \end{array}\right\} \text{for } 6 \leq x \leq 8 \qquad \text{(h)}$$

Finally, from Figure 5-6e,

$$\left.\begin{array}{ll} \Sigma F_y = 0: & -V - 45 - 36 + 49 = 0 \\ \Sigma M_0 = 0: & M + 45(x - 8) + 48 + 36(x - 2) - 49x = 0 \end{array}\right\} \qquad \text{(i)}$$

so that

$$\left.\begin{array}{l} V = -32 \text{ kN} \\ M = 384 - 32x \text{ kN·m} \end{array}\right\} \text{for } 8 \leq x \leq 12 \qquad \text{(j)}$$

Equations (d), (f), (h), and (j) are mathematical expressions for the internal shearing force and internal bending moment for the beam loaded and supported as shown in Figure 5-7a. They are plotted in Figures 5-7b and 5-7c, respectively.

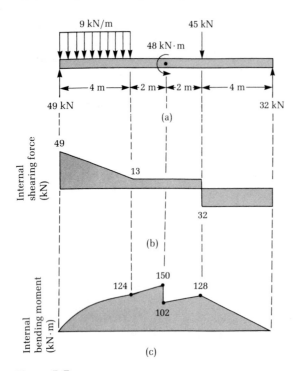

Figure 5-7

Figure 5-7b is called the **shear force diagram**. It gives a visual picture of the way the internal shearing force varies along the length of the beam. This is a valuable aid in beam design.

Figure 5-7c is called the **moment diagram** for the beam, and it affords a visual picture of the way the internal bending moment varies along the axis of the beam. The moment diagram is also a valuable aid in beam design.

Most generally, shear and moment diagrams for a beam are constructed directly beneath the beam as has been done in Figure 5-7. Abscissas usually have the same units as the length of the beam. That is, if distances along the beam axes are given in meters, the abscissas of the shear and moment diagrams are expressed in meters.

PROBLEMS / Section 5-4

5-5–5-8 Using an origin of coordinates at the left end of each the beams shown in Figure P5-1 through P5-4, determine shear and moment equations that are valid for every segment of the beams. Construct the shear and moment diagrams by plotting these equations.

5-5 Shear and Moment Diagrams by Summation

It is not necessary to construct shear and moment diagrams by first writing shear and moment equations and subsequently plotting these equations as was done in Example 5-4. Shear and moment diagrams can also be constructed via the so-called summation procedure. The bases for the summation procedure are incremental relations that express the *change* in internal shearing force in terms of the area under the load intensity diagram and the *change* in internal moment in terms of the area under the shear diagram.

Distributed loads. First consider a beam that is loaded by a continuously distributed load as shown in Figure 5-8a. A free-body diagram of a small but finite segment of the beam is shown in Figure 5-8b. The distributed load is replaced by a statically equivalent load $q(x + \lambda \cdot \Delta x) \Delta x$, where $0 \le \lambda \le 1$.

According to a mean value theorem of calculus, an ordinate of the load intensity curve can be determined between x and $x + \Delta x$ such that, when it is multiplied by Δx, the result will be the area under the load intensity diagram for the interval Δx. Moreover, from statical considerations, the resultant force must act at the centroid of the area shown in Figure 5-8b—that is, at $\mu \cdot \Delta x$ where $0 \le \mu \le 1$.

Transverse force equilibrium and moment equilibrium about the point O of Figure 5-8b gives

$$\Delta F_y = 0: \quad -V(x + \Delta x) + V(x) + q(x + \lambda \cdot \Delta x)\Delta x = 0 \quad (5\text{-}1)$$

$$\Sigma M_0 = 0: \quad M(x + \Delta x) - M(x) - V(x)\Delta x$$
$$- [q(x + \lambda \cdot \Delta x)\Delta x]\mu \cdot \Delta x = 0 \quad (5\text{-}2)$$

Dividing Eqs. (5-1) and (5-2) by Δx and letting $\Delta x \to 0$ yield

$$\lim_{\Delta x \to 0} \frac{V(x + \Delta x) - V(x)}{\Delta x} = \lim_{\Delta x \to 0} q(x + \lambda \cdot \Delta x) \quad (5\text{-}3)$$

and

$$\lim_{\Delta x \to 0} \frac{M(x + \Delta x) - M(x)}{\Delta x} = \lim_{\Delta x \to 0} V(x)$$
$$+ \lim_{\Delta x \to 0} [q(x + \lambda \cdot \Delta x)\mu \cdot \Delta x] \quad (5\text{-}4)$$

According to differential calculus, the ratios on the left-hand sides of Eqs. (5-3) and (5-4) are the derivatives dV/dx and dM/dx. Moreover, since $0 \le \lambda \le 1$ and $0 \le \mu \le 1$, it follows from Eqs. (5-3) and (5-4) that

$$\frac{dV}{dx} = q(x) \quad (5\text{-}5)$$

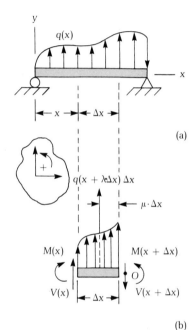

Figure 5-8

and

$$\frac{dM}{dx} = V(x) \tag{5-6}$$

Geometrically, Eq. (5-5) asserts that the slope of the shear diagram at any point x is equal to the load intensity at the same point. Similarly, Eq. (5-6) asserts that the slope of the moment diagram at any point x is equal to the shearing force at the same point. Eqs. (5-5) and (5-6) show that the load intensity, the shearing force, and the bending moment are all interrelated.

To establish incremental relations analogous to the differential relations given in Eqs. (5-5) and (5-6), integrate each equation between two sections that are separated by a finite distance. Denote these two sections by x_j and x_{j+1} and signify the shear force and the bending moment at these sections by V_j, M_j, and V_{j+1}, M_{j+1}, respectively, as shown in Figure 5-9. Accordingly,

$$\int_{V_j}^{V_{j+1}} dV = \int_{x_j}^{x_{j+1}} q(x)\,dx \tag{5-7}$$

and

$$\int_{M_j}^{M_{j+1}} dM = \int_{x_j}^{x_{j+1}} V(x)\,dx \tag{5-8}$$

The left-hand sides of Eqs. (5-7) and (5-8) are the difference between the shearing forces and the bending moments at the sections x_j and x_{j+1}, respectively. The right-hand sides are the areas under the load intensity diagram and the shear force diagram between sections x_j and x_{j+1}, respectively.

Equations (5-7) and (5-8) yield the useful incremental relations

$$\Delta V = V_{j+1} - V_j = \begin{array}{l}\text{the area under the load intensity} \\ \text{diagram between sections } x_j \text{ and} \\ x_{j+1}\end{array} \tag{5-9}$$

and

$$\Delta M = M_{j+1} - M_j = \begin{array}{l}\text{the area under the shear force} \\ \text{diagram between sections } x_j \text{ and} \\ x_{j+1}\end{array} \tag{5-10}$$

The incremental relations given in Eqs. (5-9) and (5-10) are valid for any segment of a beam over which the loads acting on it vary in a

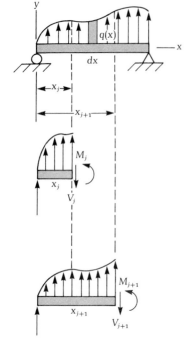

Figure 5-9

continuous manner. If several distributed loads of different intensities act on different segments of a beam, Eqs. (5-9) and (5-10) can be used for each segment.

Concentrated loads. Consider a beam that is loaded by a concentrated force at a point x as shown in Figure 5-10a.

A free-body diagram of a small but finite segment of the beam surrounding the load P is shown in Figure 5-10b. The shear and moment at section $x + \epsilon$ are denoted by $V(x + \epsilon)$ and $M(x + \epsilon)$, respectively, and the shear and moment at section $x - \epsilon$ are denoted by $V(x - \epsilon)$ and $M(x - \epsilon)$, respectively.

Transverse force equilibrium and moment equilibrium about point O of Figure 5-10b yields

$$\Sigma F_y = 0: \quad -V(x + \epsilon) + V(x - \epsilon) + P = 0$$

and

$$\Sigma M_0 = 0: \quad M(x + \epsilon) - M(x - \epsilon) - V(x - \epsilon)2\epsilon - P\epsilon = 0$$

Rearranging these relations and letting $\epsilon \to 0$ leads to

$$\lim_{\epsilon \to 0} \{V(x + \epsilon) - V(x - \epsilon)\} = P \tag{5-11}$$

and

$$\lim_{\epsilon \to 0} \{M(x + \epsilon) - M(x - \epsilon)\} = \lim_{\epsilon \to 0} \{V(x - \epsilon)2\epsilon - P\epsilon\} \tag{5-12}$$

Let $\lim_{\epsilon \to 0} V(x + \epsilon) \equiv V(x+)$ and $\lim_{\epsilon \to 0} V(x - \epsilon) \equiv V(x-)$. Similarly, $\lim_{\epsilon \to 0} M(x + \epsilon) \equiv M(x+)$ and $\lim_{\epsilon \to 0} M(x - \epsilon) \equiv M(x-)$. Physically, $V(x+)$ and $V(x-)$ are the shearing forces at sections located infinitesimal distances to the positive and negative x sides of the concentrated load P. Likewise, $M(x+)$ and $M(x-)$ are the moments at these sections. From Eqs. (5-11) and (5-12),

$$V(x+) - V(x-) = P \tag{5-13}$$

and

$$M(x+) - M(x-) = 0 \tag{5-14}$$

Equations (5-13) and (5-14) are useful relations that control the change in shearing force and the change in bending moment that occur across a concentrated load. Accordingly,

$$\Delta V = V(x+) - V(x-) = P \tag{5-15}$$

and

$$\Delta M = M(x+) - M(x-) = 0 \tag{5-16}$$

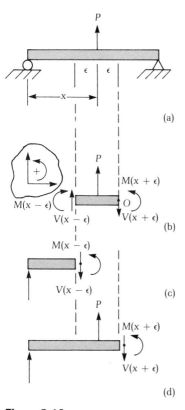

Figure 5-10

Clearly, the change in shearing force across a concentrated load is equal to the magnitude of the concentrated force. Moreover, the sense of the change in shear coincides with the sense of the concentrated force. The change in moment that occurs across a concentrated force is zero.

Equations (5-15) and (5-16) serve the same purposes for concentrated forces as Eqs. (5-9) and (5-10) serve for distributed loads.

Concentrated couples. Consider a beam loaded by a concentrated couple of magnitude C at a point x as shown in Figure 5-11a. A free-body diagram of a small but finite segment of the beam surrounding the couple C is shown in Figure 5-11b.

Transverse force equilibrium and moment equilibrium about the point O of Figure 5-11b yield

$$\Sigma F_y = 0: \quad -V(x + \epsilon) + V(x - \epsilon) = 0 \tag{5-17}$$

and

$$\Sigma M_0 = 0: \quad M(x + \epsilon) - M(x - \epsilon) - V(x - \epsilon)2\epsilon - C = 0 \tag{5-18}$$

Rearranging these equations and letting $\epsilon \to 0$ leads to

$$\Delta V = V(x+) - V(x-) = 0 \tag{5-19}$$

and

$$\Delta M = M(x+) - M(x-) = C \tag{5-20}$$

Thus the change in the shearing force across a concentrated couple is equal to zero. The change in the bending moment across a concentrated couple is equal to the magnitude of the couple. Moreover, the sense of the change in moment coincides with the sense of the concentrated couple.

As we stated previously, there are three essentially different types of loads that straight beams are expected to carry: distributed loads, concentrated forces, and concentrated couples. Consequently, when the internal shearing force is known at one point along the length of a beam, the internal shearing force can be calculated for any other point along the beam axis from Eqs. (5-9), (5-15), or (5-19). Similarly, when the internal bending moment is known at one point along the length of the beam, the internal bending moment can be calculated for any other point from Eqs. (5-10), (5-16), or (5-20). The use of these equations is illustrated in the examples that follow.

Shear and moment curves for distributed loads. The shape of a curve connecting adjacent points on a shear diagram, or on a moment diagram, under a distributed load can be established from the integral

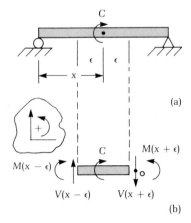

(a)

(b)

Figure 5-11

relations

$$V(x) = \int q(x)\,dx + C_1$$
$$M(x) = \int V(x)\,dx + C_2$$

(5-21)

which have been obtained simply by integrating Eqs. (5-5) and (5-6), respectively.

Figure 5-12 illustrates the mathematical nature of the curves representing shear and moment under a distributed load $q(x)$.

In general, when $q(x) \neq 0$, the successive representations follow the pattern

if $q(x) \sim x^n$

then $V(x) \sim x^{n+1}$

and $M(x) \sim x^{n+2}$

(5-22)

The most commonly occurring patterns are shown in Figures 5-12a, 5-12b, and 5-12c.

The following geometric interpretations of the differential relations given in Eqs. (5-5) and (5-6) are useful.

$$\frac{dV}{dx} = q(x)$$

{ The magnitude and sign of the slope at point x of the shear diagram are equal to the magnitude and sign of the load intensity at point x.

(5-23)

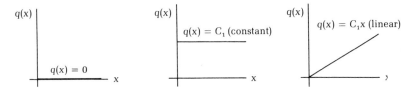

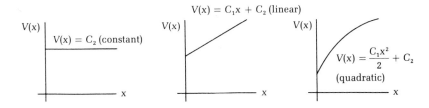

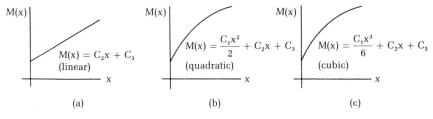

(a)   (b)   (c)

Figure 5-12

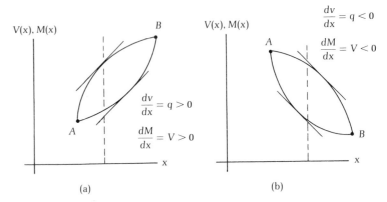

Figure 5-13

and

$$\frac{dM}{dx} = V(x)$$

$\begin{cases} \text{The magnitude and sign of the slope} \\ \text{at point x of the moment diagram are} \\ \text{equal to the magnitude and sign of} \\ \text{the shearing force at point x.} \end{cases}$ (5-24)

Equations (5-23) and (5-24) establish one of two possible classes of curves into which the shear or moment equations may fall. These two classes of curves are shown in Figures 5-13a and 5-13b.

To establish which of the two curves from a particular set is the correct curve for a given distributed load, we need to make use of a second set of differential relations. Accordingly, by differentiation of Eqs. (5-23) and (5-24),

$$\frac{d}{dx}\left(\frac{dV}{dx}\right) = \frac{dq}{dx}$$

$\begin{cases} \text{The tangent to the shear diagram} \\ \text{rotates counterclockwise or} \\ \text{clockwise according to whether} \\ \text{the rate of change of the load} \\ \text{intensity is positive or negative.} \end{cases}$ (5-25)

and

$$\frac{d}{dx}\left(\frac{dM}{dx}\right) = \frac{dV}{dx}$$

$\begin{cases} \text{The tangent to the moment} \\ \text{diagram rotates counterclockwise} \\ \text{or clockwise according to} \\ \text{whether the slope of the shear} \\ \text{diagram is positive or negative.} \end{cases}$ (5-26)

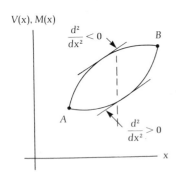

Figure 5-14

The use of these relations is illustrated in Figure 5-14.

EXAMPLE 5-5

Determine the shear and moment diagrams for the simply supported beam shown in Figure 5-15a.

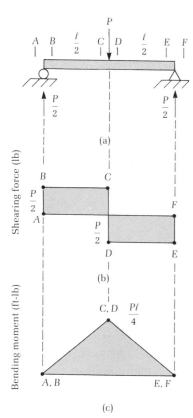

Figure 5-15

Shearing force (lb)

Bending moment (ft-lb)

SOLUTION

The reactions at the supports are determined through the symm requirements, and they are shown in Figure 5-15a.

The shear diagram is constructed as follows. The shearing force ω section A of the beam just to the left of the left support reaction is zero. Therefore, begin at point A on the shear diagram. Now the change in shear across the left reaction is $P/2$, and it has the same sense as the reaction. If B is a section infinitesimally to the right of the left support, from Eq. (5-13)

$$V_B - V_A = \frac{P}{2}$$

or

$$V_B = \frac{P}{2} \tag{a}$$

since V_A is known to be zero.

This value is shown at point B of the shear diagram. Now consider sections B and C, which are infinitesimally close to the left support reaction and the applied load, respectively. Consequently, from Eq. (5-9),

$$V_C = V_B + \text{area under the load intensity diagram}$$
$$\text{between } x_B \text{ and } x_C$$
$$= \frac{P}{2} + 0 = \frac{P}{2} \tag{b}$$

This value of the shearing force is shown at point C on the shear diagram. To determine how to connect points B and C, refer to Figure 5-12. Since the shear force is constant for an unloaded region of a beam as shown in Figure 5-12a, connect points B and C by a straight line.

Across the concentrated load P, again from Eq. (5-13),

$$V_D - V_C = -P \tag{c}$$

so that point D of the shear diagram is obtained.

If D and E are sections that are infinitesimally close to the applied load and the right reaction, respectively, then, from Eq. (5-9),

$$V_E = V_D + \text{area under the load intensity diagram between}$$
$$x_D \text{ and } x_E$$
$$= -\frac{P}{2} + 0 = -\frac{P}{2} \tag{d}$$

This locates point E on the shear diagram. Point D and E are connected by a straight line for the reason stated earlier.

Finally, taking sections E and F infinitesimally close to the right support reaction, we see that

$$V_F = V_E + \frac{P}{2} = -\frac{P}{2} + \frac{P}{2} = 0 \tag{e}$$

This value of the shearing force is shown as point F on the shear diagram. Note that points A and B and points E and F are actually infinitesimally close to one another. Consequently, lines AB and EF are simply vertical, straight lines.

The moment diagram is constructed in the same manner as the shear diagram. Beginning at section A, the bending moment is zero. From Eq. (5-16), the change in moment across a concentrated force is zero, so that

$$M_B = M_A + 0 = 0 \tag{f}$$

To determine the moment M_C at section C, simply evaluate the area under the shear force diagram between sections B and C and add it to M_B. Accordingly, for the interval $0 \le x \le \ell/2$,

$$M_C = M_B + \frac{P\ell}{4} = \frac{P\ell}{4} \tag{g}$$

Moreover, the change in moment across the concentrated force P is zero, so that

$$M_D = M_C + 0 = \frac{P\ell}{4} \tag{h}$$

Note that the area under the shear force diagram between sections D and E is negative. Therefore, for the interval $\ell/2 \le x \le \ell$,

$$M_E = M_D - \frac{P\ell}{4} = 0 \tag{i}$$

Finally, across the right support reaction, the change in moment is zero, so that

$$M_F = M_E + 0 = 0 \tag{j}$$

After a little practice, we need only to visualize the sections that have been designated by the letters A, B, C, D, E, and F.

EXAMPLE 5-6

Using the summation technique, construct the shear and moment diagrams for the simply supported beam shown in Figure 5-16a.

SOLUTION

Prior to constructing the shear and moment diagrams, calculate the reactions at the supports and the reactions of the extension at C.

From the free-body diagram shown in Figure 5-16e, it is seen that

$$\left. \begin{array}{l} N_0 = 1500 \text{ lb} \\ M_0 = 1500 \text{ ft-lb} \end{array} \right\} \tag{a}$$

N_0 and M_0 are the reactions that the beam exerts on the extension. Accordingly, the reactions of the extension on the beam have equal

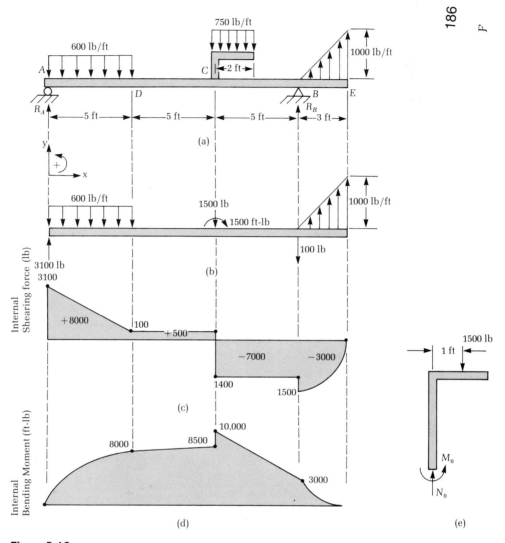

Figure 5-16

magnitudes and opposite senses. These reactions are shown in Figure 5-16b.

The reactions at the supports are obtained from moment equilibrium about points A and B, respectively. Equilibrium of transverse forces is used as a check on the computations. Therefore,

$$\stackrel{+}{\curvearrowright}\Sigma M_A = 0: \quad 15R_B + 1500(17) - 1500(10) - 1500$$
$$- 3000(2.5) = 0 \tag{b}$$

$$\text{or} \quad R_B = -100 \text{ lb}$$

$$\stackrel{+}{\curvearrowright}\Sigma M_B = 0: \quad -15R_A + 3000(12.5) + 1500(5) - 1500$$
$$+ 1500(2) = 0 \tag{c}$$

$$\text{or} \quad R_A = 3100 \text{ lb}$$

$$\uparrow + \Sigma F_y \stackrel{?}{=} 0: \quad 3100 - 3000 - 1500 - 100 + 1500 = 0 \tag{d}$$

quation (d) shows that the reactions at the supports have been calculated correctly.

The shear and moment diagrams are shown in Figures 5-16c and 5-16d, respectively. You should use the summation procedure to verify the various points on these diagrams. Observe that the equations of the lines connecting the points are obtained from the relationships between the load intensity, the shearing force, and the bending moment as shown in Figure 5-12. Thus, if the load intensity is represented by a curve of degree n, the curve representing the shearing force is of degree $n + 1$, and the curve representing the bending moment is of degree $n + 2$. Since the load in the interval AD of Figure 5-16a is uniformly distributed, the shear diagram in the same interval is an inclined straight line, and the corresponding moment diagram is described by a quadratic equation. The load in the interval BE is triangular; that is, its equation is linear and, thus, the shear and moment diagrams in that interval are quadratic and cubic, respectively. In the intervals DC and CB, no external load is acting, and the shear diagrams in these two intervals are horizontal lines and the moment diagrams are both inclined straight lines. Note that the uniformly distributed load on the extension at C affects the main beam indirectly via the reactions it causes at C, *not* as a distributed load acting on part of the interval CB.

PROBLEMS / Section 5-5

5-9–5-12 Use the summation procedure to construct the shear and moment diagrams for the beams shown in Figures P5-9–P5-12.

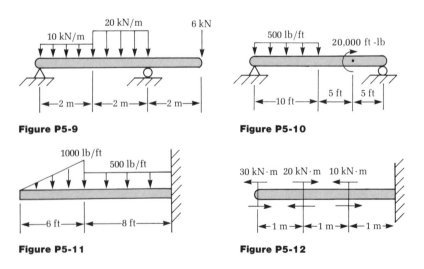

Figure P5-9

Figure P5-10

Figure P5-11

Figure P5-12

5-6 Bending Stresses in Straight Beams

The bending moment at a cross section of a straight beam is the couple of a force-couple system that is statically equivalent to the distribution of stress that acts perpendicular to the cross section. To establish an algebraic relationship between this bending moment and its associated stress distribution, we employ the three basic ingredients of mechanics. In this section, we develop kinematic and equilibrium relations. The behaviors of several materials are examined in the next section.

Kinematic considerations.   We need to make two basic geometric assumptions regarding the deformational behavior of straight beams.

Assumption 1: Transverse cross sections perpendicular to the centroidal line of the beam before bending remain plane and perpendicular to the centroidal line in its deformed configuration as shown in Figures 5-17a and 5-17b. This is the famous **Bernoulli hypothesis** for straight beams.

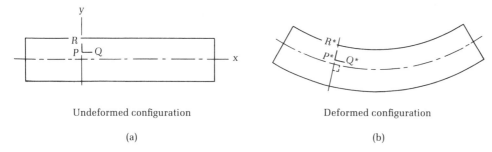

Undeformed configuration   Deformed configuration

(a)   (b)

Figure 5-17

In the undeformed configuration, line elements PQ and PR are perpendicular. In the deformed configuration, PQ and PR become P^*Q^* and P^*R^*. Because transverse cross sections remain plane and perpendicular to the deformed centroidal line, the angle between P^*Q^* and P^*R^* remains 90 degrees. Accordingly, γ_{xy}, which is the shearing strain in the plane containing the line elements PQ and PR, is identically zero.

Consider a third line element PS that is perpendicular to the line elements PQ and PR of Figure 5-17a. Because bending is assumed to occur in a single plane (the plane containing the loads) for elementary beam theory, there is no deformation in the plane containing PQ and PS, and thus $\gamma_{zx} = 0$. For beams that bend in two perpendicular planes, the shearing strain γ_{zx} is still zero because the angle between PQ and PS remains unchanged according to the plane section hypothesis.

Assumption 2: Unlike the elementary theory of circular rods, transverse cross sections of a straight beam are not assumed to be

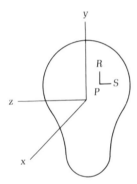

Undeformed
shape of the
cross section

(a)

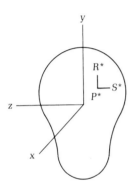

Deformed
shape of the
cross section

(b)

Figure 5-18

rigid. Figures 5-18a and 5-18b show a typical cross section in its undeformed and deformed configurations, respectively. The line elements PR and PS are parallel to the centroidal x and y axes in the undeformed configuration. PR and PS become P*R* and P*S* in the deformed configuration as shown in Figure 5-18b. The lengths of the new line elements, in general, differ from those of their original counterparts.

Elementary beam theory assumes that, since the normal stresses σ_y and σ_z are zero on the longitudinal surfaces of the beam, $\sigma_y = \sigma_z = 0$ throughout the cross section. This is a reasonable assumption because of the slenderness of the beam. This assumption leads, via the three-dimensional stress-strain relations, to the kinematic consequences $\epsilon_y = \epsilon_z = -\nu\epsilon_x$. Moreover, we assume that the angle between PR and PS does not change due to bending of the beam. Thus the shearing strain γ_{yz} in the yz plane is zero.

Notice that, collectively, these hypotheses require that $\epsilon_y = \epsilon_z = -\nu\epsilon_x$ (Poisson strains) and $\gamma_{xy} = \gamma_{xz} = \gamma_{yz} = 0$. We remark that the assumptions listed here are geometric. They do no depend on material behavior and, therefore, they are valid for elastic and inelastic material behavior. These hypotheses are limited to reasonably small deformations; otherwise they remain valid for various kinds of material behavior.

The variation of the bending strain through the depth of a beam can be obtained via the definition of engineering normal strain. Two cross sections separated by an infinitesimal distance ds in the undeformed configuration of a beam are shown in Figure 5-19a. Observe that, when a beam bends as in Figure 5-19b, line elements along its upper surface are compressed while line elements along its lower surface are stretched. Since plane sections remain plane, there exists a line element RS that undergoes no extension or contraction. This line element lies in a surface called the **neutral surface**. The edge of the neutral surface seen by an observer looking along the axis of the beam is called the **neutral axis** for the cross section viewed.

Let ds denote the length of the line element RS in the undeformed configuration of the beam and, let ds* be its length in the deformed configuration. Because of the special character of line element RS, $ds^* = ds$.

Now let dS denote the length of a line element PQ that is located at an arbitrary distance y from the neutral surface as depicted in Figure 5-19a. PQ becomes P*Q* in the deformed configuration, and its new length is denoted by dS* as shown in Figure 5-19b.

The engineering strain of the line element PQ is

$$\epsilon_x = \frac{dS^* - dS}{dS} = \frac{dS^* - ds}{ds} \tag{5-27}$$

because $dS = ds$.

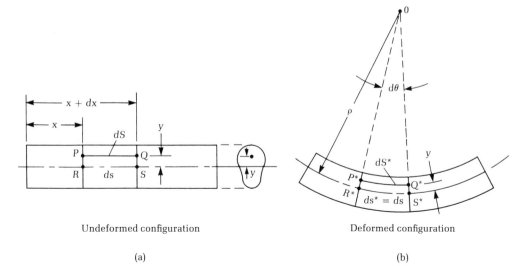

Undeformed configuration

Deformed configuration

(a)

(b)

Figure 5-19

From Figure 5-19b, by similar triangles,

$$\frac{dS^*}{\rho - y} = \frac{ds^*}{\rho} = \frac{ds}{\rho} \tag{5-28}$$

so that

$$\frac{dS^*}{ds} = 1 - \frac{y}{\rho} \tag{5-29}$$

Substituting Eq. (5-29) into Eq. (5-27) yields

$$\epsilon_x = -\frac{y}{\rho} \tag{5-30}$$

where ρ is called the radius of curvature of the deformed centroidal line at section x. Equation (5-30) indicates that the bending strain varies linearly through the depth of the beam. Observe that the strain distribution expressed by Eq. (5-30) has been developed via geometric assumptions and is, therefore, independent of material behavior. It remains valid as long as the two basic geometric assumptions remain valid.

Equilibrium considerations. According to the elastic stress-strain relations for isotropic materials, the kinematic assumptions cited in the previous section imply that $\sigma_y = \sigma_z = \tau_{xy} = \tau_{zx} = \tau_{zy} = 0$. Consequently, the only nonzero stress is the normal stress σ_x acting normal to a transverse cross section. The normal stress that acts on an infinitesimal element of area dA at an arbitrary cross section of a straight beam is shown in Figure 5-20.

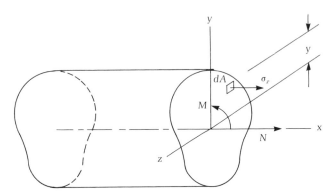

Figure 5-20

The force-couple system that is statically equivalent to the distribution of stress σ_x over the cross section consists of a normal force N and a couple M. For statical equivalency,

$$N = \int_{\text{area}} \sigma_x \, dA \tag{5-31}$$

and

$$M = -\int_{\text{area}} \sigma_x y \, dA \tag{5-32}$$

Notice that Eq. (5-32) is consistent with the sign convention adopted previously for a positive bending moment.

Equations (5-31) and (5-32) depend on only statical considerations and are independent of material behavior. They are therefore valid for elastic and inelastic material behavior. Also, note that the neutral axis for pure bending, elastic or inelastic, can be determined via Eq. (5-31).

5-7 Material Behavior

An explicit formula for the normal stress in a straight beam due to bending can be developed provided that the stress-strain relation for the material from which the beam is made is available. The behavior of several types of material is considered in this section.

Linear elastic behavior. Figure 5-21a shows the stress-strain diagram for a material that behaves elastically. The mathematical

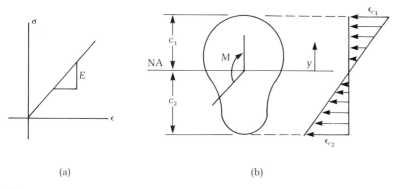

(a)   (b)

Figure 5-21

expression for the linear region of the diagram is

$$\sigma_x = E\epsilon_x \tag{5-33}$$

and it is valid for any point y on a cross section.

The strain distribution, according to Eq. (5-30), varies linearly through the depth of the beam as shown in Figure 5-21b. Consequently,

$$\sigma_x = -\frac{Ey}{\rho} \tag{5-34}$$

Elastic-plastic behavior. The stress-strain diagram for a material that behaves in an ideally elastic, plastic manner is shown in Figure 5-22a. Two separate mathematical relations are required to represent the stresses in the elastic and plastic regions. The shaded portions of the cross section shown in Figure 5-22b represent plastic

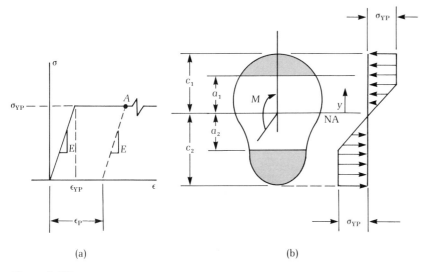

(a)   (b)

Figure 5-22

regions. Appropriate mathematical representations for the stress-strain relation are

$$\sigma_x = \begin{cases} E\epsilon_x & 0 \leq \epsilon_x \leq \epsilon_{YP} \\ \sigma_{YP} & \epsilon_{YP} \leq \epsilon_x \leq \epsilon_\ell \end{cases} \tag{5-35}$$

where ϵ_ℓ represents a strain beyond which the flat top portion of the stress-strain diagram ceases to be valid.

Since the strain varies linearly on the cross section even in the presence of inelastic material behavior, we can write the stress distributed on the cross section as

$$|\sigma_x| = \begin{cases} -\dfrac{Ey}{\rho} & -a_2 \leq y \leq a_1 \\ \sigma_{YP} & -c_2 \leq y \leq -a_2 \quad \text{and} \quad a_1 \leq y \leq c_1 \end{cases} \tag{5-36}$$

When a cross section possesses an axis of symmetry parallel to the bending axis, $c_1 = c_2$ and $a_1 = a_2$, and the intervals for y in Eq. (5-36) simplify accordingly.

Elastic, linear hardening material. The stress-strain diagram for several materials can be represented closely by two sloping straight lines as shown in Figure 5-23a. We denote the stress and strain corresponding to the intersection of these straight lines by σ_0 and ϵ_0. The stress-strain relations for this material are

$$\sigma_x = \begin{cases} E\epsilon_x & 0 \leq \epsilon_x \leq \epsilon_0 \\ (1-\alpha)\sigma_0 + \alpha E\epsilon_x & \epsilon_0 \leq \epsilon_x \leq \epsilon_\ell \end{cases} \tag{5-37}$$

Again the strain varies linearly on the cross section ($\epsilon_x = -y/\rho$) so that

$$\sigma_x = \begin{cases} -\dfrac{Ey}{\rho} & -a_2 \leq y \leq a_1 \\ (1-\alpha)\sigma_0 - \dfrac{\alpha Ey}{\rho} & -c_2 \leq y \leq -a_2, \quad a_1 \leq y \leq c_1 \end{cases} \tag{5-38}$$

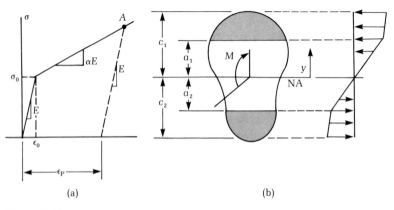

(a)   (b)

Figure 5-23

The stress distribution corresponding to the elastic, linear hardening material is shown in Figure 5-23b.

Nonlinear elastic material. The stress-strain diagram for a nonlinear elastic material is shown in Figure 5-24. It can be represented mathematically by the relation

$$\sigma_x = K\epsilon_x^n \tag{5-39}$$

where K and n are parameters that are adjusted to make Eq. (5-39) approximate the stress-strain data as closely as possible. Since $\epsilon_x = -y/\rho$, it follows from Eq. (5-39) that

$$\sigma_x = K(-y/\rho)^n = K(-1)^n(y/\rho)^n \tag{5-40}$$

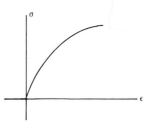

Figure 5-24

5-8 Elastic Bending of Straight Beams

For linear, elastic behavior, we can derive an explicit formula for the normal stress on a cross section due to bending from Eqs. (5-31), (5-32), and (5-34). These equations embody the concepts of geometrically compatible deformations, equilibrium, and material behavior. For beams of uniform material,

$$M = -\underbrace{\int_{area} \sigma_x y \, dA}_{\text{Equilibrium}} = -\underbrace{\int_{area} E\epsilon_x y \, dA}_{\text{Material behavior}}$$

$$= -\underbrace{\int_{area} \frac{E\epsilon_x}{y} y^2 \, dA}_{\text{Rearrangement}} = +\frac{E}{\rho} \underbrace{\int_{area} y^2 \, dA}_{\substack{\text{Geometric} \\ \text{compatibility of} \\ \text{deformations}}} \tag{5-41}$$

The last integral appearing in Eq. (5-41) is the moment of inertia of the cross-sectional area with respect to the axis from which y is measured. Denote the integral by I. Observe that y is measured from the neutral axis of the cross section. Its location is undetermined at this point in the analysis.

Consequently, from Eq. (5-41),

$$\frac{1}{\rho} = \frac{M}{EI} \tag{5-42}$$

This is the so-called moment-curvature relationship for the elastic bending of straight beams.

A formula from which the normal stress at any distance y from the neutral surface can be calculated is obtained from Eqs. (5-34) and

(5-42). Accordingly,

$$\sigma_x = -\frac{Ey}{\rho} = -\frac{My}{I} \qquad (5\text{-}43)$$

Equation (5-43) is the well-known **flexure formula**. M is the resisting moment at the cross section where the normal stress is to be calculated, I is the moment of inertia of the cross section relative to its neutral axis, and y is the distance of a point in the cross section from the neutral axis. Appendix B lists important formulas that can be used to calculate the areas, centroidal coordinates, and moments of inertia for several plane areas. Appendix C reviews the procedures used to calculate the centroidal coordinates and moments of inertia of composite areas. Finally, Appendix D contains properties for structural steel shapes such as the *wideflange section* (W), the *I section* (S), the *channel section* (C), *equal* and *unequal* leg *angle sections* (L), and standard steel pipe. The usual units for M and I are inch pounds and inches to the fourth power in the English system and newton meters and meters to the fourth power in the SI system. The distance y is usually either in inches or meters depending on the system of units being used.

Before I can be calculated, it is necessary to locate the neutral axis of the cross section. The neutral axis can be located by observing that the normal force N acting on a cross section is zero when no external axial loads are applied to the beam. Therefore,

$$N = \int_{\text{area}} \sigma_x \, dA = -\frac{E}{\rho} \int_{\text{area}} y \, dA = -\frac{E}{\rho} A\bar{y} = 0$$

where A and $\bar{y}$ denote the area and y coordinate of the centroid of the cross section which is, as before, measured from the neutral axis. Note that $E \neq 0$ and $A \neq 0$. Moreover, a cross section can always be selected for which the radius of curvature $\rho \neq \infty$. These observations lead to the conclusion that $\bar{y} = 0$. Since y is measured from the neutral axis of the cross section, we conclude that the neutral axis must coincide with the centroidal axis of the cross section. The location of the neutral axis has now been established for elastic bending.

EXAMPLE 5-7

Determine the normal stress distribution on a rectangular cross section for which the bending moment is 21 100 N·m. This cross section is shown in Figure 5-25.

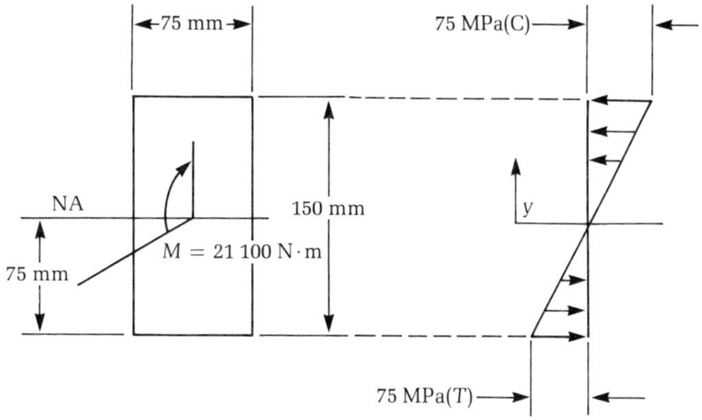

Figure 5-25

SOLUTION

The centroidal axis and therefore the neutral axis for the rectangular section is determined by symmetry. The moment of inertia of the cross-sectional area relative to the neutral axis is

$$I = \tfrac{1}{12}bh^3 = \tfrac{1}{12}(0.075)(0.150)^3$$
$$= 21.1 \times 10^{-6} \text{ m}^4 \tag{a}$$

The stress distribution is given by the equation

$$\sigma_x = -\frac{My}{I} = -\frac{(21\ 100)y}{21.1 \times 10^{-6}} = -1000y \text{ MPa} \tag{b}$$

This stress distribution is shown in Figure 5-25. Notice that the maximum tensile stress occurs along the lower edge of the cross section, and the maximum compressive stress occurs along the upper edge of the cross section. Also observe that the maximum tensile and compressive stresses are numerically equal. This situation will always be true for a cross section that is being bent about a horizontal axis of symmetry.

EXAMPLE 5-8

Determine the normal stress distribution on the T section shown in Figure 5-26a if the bending moment $M = 136,000$ in.-lb.

SOLUTION

We first locate the neutral axis. Since it coincides with the centroidal axis, we have

$$A\bar{y} = A_1\bar{y}_1 + A_2\bar{y}_2 = 12(7) + 12(3) = 120 \text{ in}^3$$

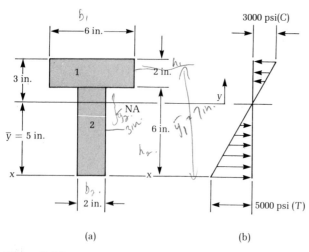

Figure 5-26

so that

$$\bar{y} = \frac{120}{24} = 5 \text{ in.} \tag{a}$$

The moment of inertia with respect to the centroidal axis is determined using the parallel axis theorem. Therefore,

$$I = \sum_{i=1}^{2} (\bar{I} + A\bar{y}^2)_i = [\tfrac{1}{12}(6)2^3 + 12(2)^2] + [\tfrac{1}{12}(2)6^3 + 12(2)^2]$$

$$= 136 \text{ in}^4 \tag{b}$$

Accordingly, the stress-distribution for the T section is

$$\sigma_x = -\frac{136,000}{136} y = -1000y \text{ psi} \tag{c}$$

The stress distribution is shown in Figure 5-26b. Notice that the stresses at the extreme edges are different for cross sections that do not possess an axis of symmetry parallel to the axis about which the bending takes place.

EXAMPLE 5-9

Determine the maximum tensile flexural stress and the maximum compressive flexural stress in the beam shown in Figure 5-27a. The cross section of the beam is shown in Figure 5-27b. Assume that the stresses do not exceed the tensile and compressive proportional limits of the material from which the beam is made.

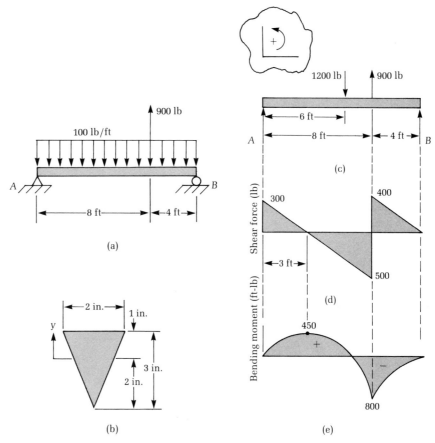

Figure 5-27

SOLUTION

First calculate the section properties. The location of the centroidal axis is shown in Figure 5-27b. The centroidal moment of inertia is

$$I = \tfrac{1}{36}bh^3 = \tfrac{1}{36}(2)3^3 = 1.5 \text{ in}^4 \tag{a}$$

The support reactions can be determined from the free-body diagram shown in Figure 5-27c, where the uniformly distributed load has been replaced by its equivalent resultant force. It should be emphasized again that this replacement is allowed only for the purpose of determining external reactions. The distributed load, rather than its resultant shown in Figure 5-27c, must be used to determine the internal reactions and, hence, to draw shear and moment diagrams. Referring to Figure 5-27c and using the sign convention indicated in the figure,

$$\Sigma M_A = 0: \quad 12B + 900(8) - 1200(6) = 0 \quad \text{or} \quad B = 0 \tag{b}$$

and

$$\Sigma M_B = 0: \quad -12A - 900(4) + 1200(6) = 0 \quad \text{or} \quad A = 300 \text{ lb} \tag{c}$$

The horizontal component of the reaction at A is identically zero. To check the correctness of these reactions, note that the transverse forces, including the support reactions, must satisfy force equilibrium. Consequently,

$$\Sigma F_y \overset{?}{=} 0: \quad 300 - 1200 + 900 + 0 = 0 \tag{d}$$

Shear and moment diagrams for the beam are shown in Figures 5-27d and 5-27e.

To determine the magnitude of the maximum tensile flexural stress, we must calculate the tensile stresses at the sections where the maximum positive and negative moments occur. The bending moment diagram shows that the maximum positive moment is 450 ft-lb, and the maximum negative moment is 800 ft-lb. The tensile flexural stresses at the sections 3 ft and 8 ft from the left end are, respectively,

$$\left. \begin{aligned} \sigma_{3\,ft} &= -\frac{(450 \times 12)(-2)}{1.5} = 7200 \text{ psi } (T) \\[2mm] \sigma_{8\,ft} &= -\frac{(-800 \times 12)(1)}{1.5} = 6400 \text{ psi } (T) \end{aligned} \right\} \tag{e}$$

To determine the maximum compressive flexural stress, calculate in a similar manner, the compressive stress at the sections where the maximum positive and negative moments occur. Thus

$$\left. \begin{aligned} \sigma_{3\,ft} &= -\frac{(450 \times 12)(1)}{1.5} = 3600 \text{ psi } (C) \\[2mm] \sigma_{8\,ft} &= -\frac{(-800 \times 12)(-2)}{1.5} = 12,800 \text{ psi } (C) \end{aligned} \right\} \tag{f}$$

This example illustrates that maximum flexural stress (tensile or compressive) does not necessarily occur at the section that resists the largest moment. To calculate either the maximum tensile stress or compressive flexural stress, we must maximize the product of the moment and the distance from the neutral axis to the corresponding most remote point on the cross section.

Observe that it is generally more convenient to determine the tensile or compressive character of the flexural stress by inspection rather than from Eq. (5-43). Thus a positive moment causes a tensile stress on the bottom edge of the cross section and a compressive stress on the top edge. A negative moment causes tensile stresses on the top edge and compressive stresses on the bottom edge. Consequently, the stresses at the top and bottom edges of a cross section can be calculated using absolute values for M and y. The tensile or compressive character is then assigned in accordance with the foregoing discussion.

PROBLEMS / Section 5-8

5-13 For each cross section shown in Figure P5-13, determine (a) the location of the neutral axis, (b) the moment of inertia of the cross section relative to the neutral axis, and (c) the maximum tensile and compressive stresses if each section resists a bending moment of $+136,000$ in.-lb ($I_{NA} = 136$ in^4, 428 in^4).

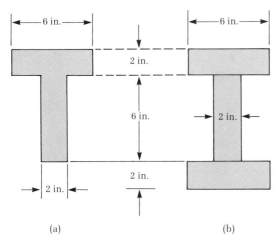

(a)   (b)

Figure P5-13

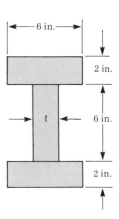

Figure P5-14

5-14 Calculate the section modulus $S = I_{NA}/c$ for the I section shown in Figure P5-14 when $t = 2$ in., 1 in., $\frac{1}{2}$ in., and 0.0 in. Show that the flexural stiffness is diminished by only 8.4% for the limiting case $t = 0$, and that the weight of a beam with this cross section is reduced by $\frac{1}{3}$ compared to the case when $t = 2$ in.

5-15 Two L $2 \times 2 \times \frac{3}{8}$ are riveted to a 6-in. $\times \frac{1}{2}$-in. steel plate to form the cross section shown in Figure P5-15. (a) Verify the location of the neutral axis shown in the figure, (b) calculate the moment of inertia of the composite cross section relative to the neutral axis, and (c) determine the maximum tensile and compressive stresses on the cross section due to a bending moment of 50,000 in.-lb ($I_{NA} = 10.86$ in^4, $\sigma = 20,810$ psi (C), 6814 psi (T)).

5-16 Select the channel section of least weight to act as a simply supported beam with a concentrated load at its center as shown in Figure P5-16. The bending stress is not to exceed 18 ksi.

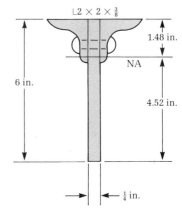

Figure P5-15

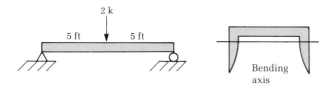

Figure P5-16

5-17 A cantilever beam with the triangular cross section shown in Figure P5-17 carries a uniformly distributed load of intensity 2 kN/m over its entire length. If the depth h of the cross section is twice its base, and if the allowable tensile or compressive stress is 24 MPa, determine the minimum dimensions the cross section should have.

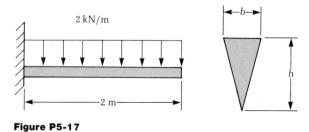

Figure P5-17

5-9 Inelastic Bending of Straight Beams

If the moment acting at a cross section of a straight beam causes strains which, at some fibers, exceed the yield point strain ϵ_{YP} shown in Figure 5-22a, the material of the cross section at these fibers is said to have yielded. In this case, the analysis of bending stresses is significantly different from that of the previous section.

Ideally elastic, plastic bending. Consider a beam with a cross section that possesses an axis of symmetry perpendicular to the plane containing the applied forces. Three common cross sections of this type are shown in Figures 5-28a, 5-29a, and 5-30a.

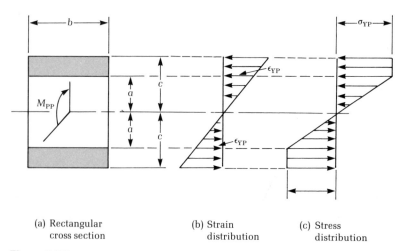

(a) Rectangular   (b) Strain   (c) Stress
   cross section   distribution   distribution

Figure 5-28

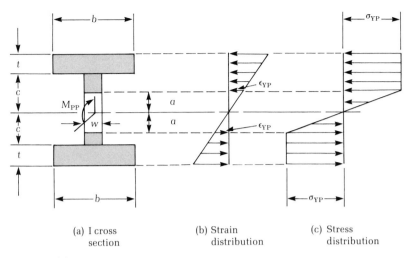

(a) I cross section

(b) Strain distribution

(c) Stress distribution

Figure 5-29

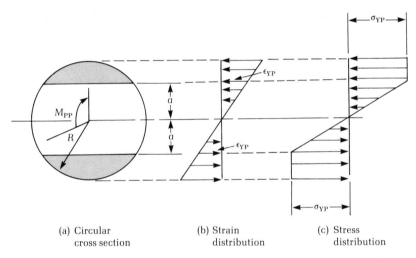

(a) Circular cross section

(b) Strain distribution

(c) Stress distribution

Figure 5-30

Let M_{PP} represent the bending moment that causes the material of the cross section to yield in the regions that are shaded in Figures 5-28a, 5-29a, and 5-30a. M_{PP} is referred to as a **partially plastic** bending moment.

Notice that because of the assumed symmetry of the cross section, the magnitudes of the strains at the extreme edges of the cross section are equal for elastic behavior. Thus yielding will be initiated simultaneously in the material at these edges. As the magnitude of the bending moment is increased, the yielding action penetrates deeper into the cross section at both locations at the same rate.

To establish the distribution of normal stress over a cross section, observe that, because plane sections remain plane even during inelastic material behavior, the normal strain varies linearly over the depth of the cross section. The strain distribution corresponding to a prescribed partially plastic bending moment M_{PP} is shown in Figures 5-28b, 5-29b, and 5-30b for each of the three cross sections considered. Notice that the magnitude of the strain at the interface between the elastic and the plastic regions is ϵ_{YP}, the strain at the yield point. Consequently, the magnitudes of the strains associated with the shaded regions of a cross section exceed ϵ_{YP} and, therefore, the stresses in these regions must be σ_{YP}, as can be determined from the stress-strain diagram of Figure 5-22a. Since the strains in the inner region of a cross section $(-a \leq y \leq a)$ do not exceed ϵ_{YP}, the stresses vary linearly in this region. The stress distribution corresponding to the strain distribution caused by the bending moment M_{PP} is shown for each cross section in Figures 5-28c, 5-29c, and 5-30c. Observe that the stress distributions have precisely the same shape for all three cross sections. This observation remains valid for ideally elastic, plastic bending of beams with cross sections that possess an axis of symmetry perpendicular to the plane containing the applied forces.

The neutral axis for inelastic bending coincides with the centroidal axis for the symmetric cross sections considered in the preceding discussion. This assertion can be verified by noting that the normal force N on a cross section must vanish for pure bending. This condition means that the magnitudes of the net tensile force and the net compressive force on the section must be equal. This condition requires the neutral axis to coincide with the centroidal axis of the cross section, as was the case for elastic bending.

We stated previously that the yielded regions of a cross section enlarge at the same rate as the partially plastic bending moment increases. The bending moment associated with yielding that has penetrated the entire depth of the cross section is called the **fully plastic** bending moment M_{FP}. (This bending moment is frequently referred to as the **ultimate** bending moment.) The stress distribution corresponding to the so-called fully plastic condition is shown in Figure 5-31. Note that this stress distribution is an idealization of a limiting condition. The stress at the centroidal axis cannot reach σ_{YP} because reaching σ_{YP} would require that the strain at the centroidal axis reach a value equal to ϵ_{YP} which, in turn, would require that the strains at the most remote fibers become infinite. Nevertheless, the fully plastic moment M_{FP} provides an important measure of the load carrying capacity of a cross section.

Let us denote the bending moment that causes the maximum elastic stress on a cross section to reach the yield stress for the material by M_{FE}: the fully elastic bending moment. An important measure of the reserve strength of a beam designed to deform elastically is the ratio M_{FP}/M_{FE}. This ratio is the ratio of the moment at which the entire

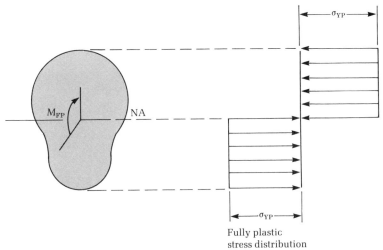

Figure 5-31

cross section has yielded to the moment at which the most remote fibers of the cross section have reached the yield stress. Computation of this ratio for several cross-sectional shapes is demonstrated in the examples that follow.

EXAMPLE 5-10

For the rectangular cross section shown in Figure 5-32a, determine (a) a formula connecting the partially plastic bending moment and the depth of yielding a, (b) the fully plastic moment, and (c) the ratio of the fully plastic to the fully elastic bending moments.

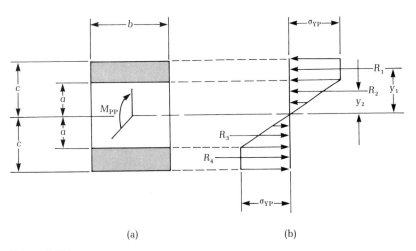

(a)

(b)

Figure 5-32

SOLUTION

(a) From Figure 5-32b, noting that $R_1 = R_4$ and $R_2 = R_3$ in magnitude, the partially plastic bending moment is

$$M_{PP} = 2(R_1 y_1 + R_2 y_2) = 2\left\{[\sigma_{YP}b(c - a)]\frac{c + a}{2} + \left[\frac{\sigma_{YP}}{2}ba\right]\frac{2a}{3}\right\}$$

or

$$M_{PP} = \frac{\sigma_{YP}b}{3}(3c^2 - a^2) \tag{a}$$

The fully elastic moment is obtained by setting $a = c$ in Eq. (a). Thus

$$M_{FE} = \tfrac{2}{3}\sigma_{YP}bc^2 = \frac{\sigma_{YP}I}{c} \tag{b}$$

as expected.

(b) The stress distribution corresponding to a fully plastic stress distribution is shown in Figure 5-33. The fully plastic moment is determined to be

$$M_{FP} = R_1 y_1 + R_2 y_2 \tag{c}$$

Now

$$R_1 = R_2 = \sigma_{YP}bc \tag{d}$$

and

$$y_1 = y_2 = \frac{c}{2} \tag{e}$$

so that

$$M_{FP} = 2(\sigma_{YP}bc)\frac{c}{2} = \sigma_{YP}bc^2 \tag{f}$$

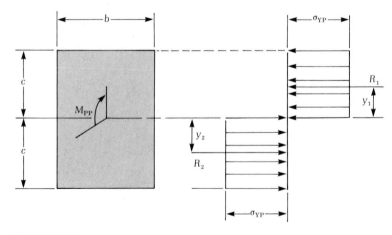

Figure 5-33

This result also follows from Eq. (a) by setting $a = 0$.

(c) The ratio of the fully plastic to the fully elastic bending moments is

$$\frac{M_{FP}}{M_{FE}} = \frac{\sigma_{YP}bc^2}{\dfrac{\sigma_{YP}I}{c}} = \frac{bc^3}{\frac{2}{3}bc^3} = 1.5 \tag{g}$$

Equation (g) indicates that a beam with a rectangular cross section designed to behave elastically has an ultimate strength nearly 50% greater than the elastic strength of the beam.

We note that the fully plastic condition is not a desirable design objective because the section or sections at which this condition occurs are incapable of resisting moments slightly in excess of M_{FP}, and the beam tends to rotate about this section in much the same manner as a door rotates on its hinges. Consequently, the fully plastic condition is frequently referred to as a **plastic hinge**.

EXAMPLE 5-11

For the I section shown in Figure 5-34a, determine (a) a formula connecting the partially plastic bending moment and the depth of yielding $(c - a)$, (b) the fully plastic bending moment, and (c) the ratio of the fully plastic to the fully elastic bending moments if $b = 6$ in., $c = 3$ in., and $w = t = 2$ in.

SOLUTION

(a) To establish a formula connecting M_{PP} and the depth of yielding, first replace the stress distribution with a statically equivalent one

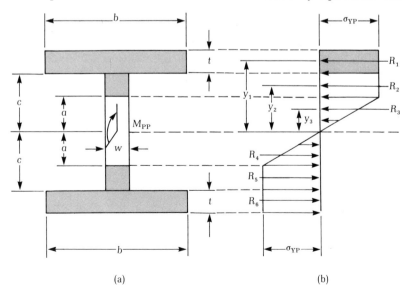

(a)   (b)

Figure 5-34

consisting of concentrated forces as shown in Figure 5-34b. Assume that yielding has penetrated into the web of the cross section. Forces R_1 and R_6 are the compressive and tensile forces in the top and bottom flanges, respectively; R_2 and R_5 are compressive and tensile forces associated with the yielded regions of the web; and R_3 and R_4 are similarly associated with the elastic region of the cross section. Note that $R_1 = R_6$, $R_2 = R_5$, and $R_3 = R_4$ in magnitude. Consequently,

$$M_{PP} = 2(y_1 R_1 + y_2 R_2 + y_3 R_3)$$

$$= 2\left\{ \left(c + \frac{t}{2}\right)\sigma_{YP}bt + \frac{(c + a)}{2}\sigma_{YP}(c - a)w + (\tfrac{2}{3}a)\frac{\sigma_{YP}}{2}aw \right\}$$

or, after simplification,

$$M_{PP} = \sigma_{YP}\{bt(2c + t) + \tfrac{1}{3}w(3c^2 - a^2)\} \tag{a}$$

This equation is valid only if $a \le c$. If yielding has not crossed through the flange—that is, if $a > c$—Eq. (a) needs to be modified.

(b) A formula for the fully plastic moment is obtained from Eq. (a) by setting $a = 0$. Consequently,

$$M_{FP} = \sigma_{YP}\{bt(2c + t) + wc^2\} \tag{b}$$

(c) For the specified dimensions,

$$M_{FP} = \sigma_{YP}\{6(2)(6 + 2) + 2(3)^2\} = 114\sigma_{YP} \tag{c}$$

and

$$M_{FE} = \frac{\sigma_{YP}I}{c + t} = \frac{\sigma_{YP}(428)}{5} = 85.6\sigma_{YP} \tag{d}$$

It should be emphasized that Eq. (d) cannot be obtained from Eq. (a) simply by setting $a = c + t$ in the latter. Equations (c) and (d) yield the ratio

$$\frac{M_{FP}}{M_{FE}} = \frac{114}{85.6} = 1.33 \tag{e}$$

Notice that when yielding has just penetrated the flange, the partially plastic bending moment is

$$M_{PP} = 108\sigma_{YP} \tag{f}$$

or 94.7% of the ultimate moment for the section. Equation (f) is obtained from Eq. (a) by setting $a = c = 3$ in. This calculation illustrates the following useful observation: it is not necessary for a plastic hinge to form to realize the nearly full plastic strength of the material. The interior region of a cross section is frequently allowed to remain elastic, thus limiting the deflections of the beam to magnitudes of the order of elastic deflections.

Now consider a beam with a cross section that does not possess an axis of symmetry perpendicular to the plane containing the applied forces. Two common sections of this type are shown in Figures 5-35a and 5-36a. The neutral axis for pure bending is not located so easily for this type of cross section. The neutral axis for inelastic bending does not coincide with the centroidal axis for cross sections that are not symmetric about the bending axis. Nevertheless, the neutral axis is still located through the requirement that the net normal force N acting on a cross section is zero for pure bending.

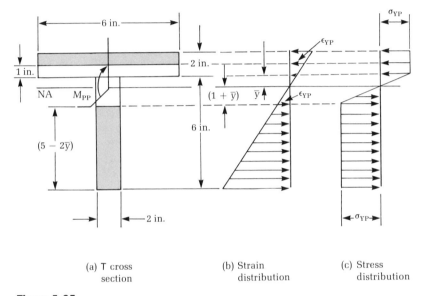

(a) T cross section

(b) Strain distribution

(c) Stress distribution

Figure 5-35

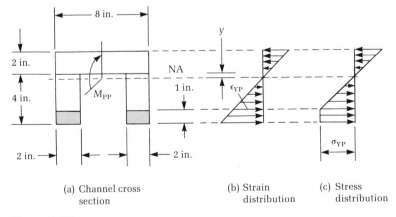

(a) Channel cross section

(b) Strain distribution

(c) Stress distribution

Figure 5-36

To establish the stress distribution for the cross sections considered, observe that, even for the present inelastic and unsymmetric problem, the strain distribution still varies linearly over the cross section in accordance with the assumption that plane sections remain plane. The strain distributions are shown to the right of each cross section. It is assumed that the strain distribution for the T section is such that yielding has occurred in the material at both the top and bottom portions of the cross section. The strain distribution for the channel section is assumed to be such that yielding has occurred in only the lower portion of the cross section. The stress distribution corresponding to the given strain distribution is shown in Figures 5-35c and 5-36c for the T section and channel section, respectively.

For the dimensions of the sections shown, it is easy to verify that the centroidal axis for the T section lies 5 in. above its lower edge, and the centroidal axis for the channel section lies 3.5 in. above its lower edge. Since points on the lower edge of either cross section are the most remote from the centroidal axis (neutral axis for elastic bending), the strains along this edge for either section will reach the yield strain first. Accordingly, first yielding occurs at points most remote from the centroidal axis of the cross section. As M_{PP} is increased, the yielded region enlarges until, eventually, the strain at points along the upper edge reaches the yield strain. Further increases in M_{PP} cause both yielded regions to enlarge until $M_{PP} = M_{FP}$, in which case a plastic hinge is formed.

Details of the computations required to locate the neutral axis for inelastic bending are demonstrated in the examples that follow.

EXAMPLE 5-12

For the T section shown in Figure 5-35a and the indicated depth of yielding, determine (a) the location of the neutral axis, (b) the partially plastic moment, and (c) the ratio of the fully plastic moment to fully elastic moment.

SOLUTION

(a) The strain distribution is established by the specified amount of yielding. Since yielding penetrates to a depth of 1 in. in the flange, the strain at this level is known to be ϵ_{YP} as shown in Figure 5-35b. Now the location of the neutral axis is designated by the variable distance $\bar{y}$ measured from the bottom of the flange. Finally, the strain varies linearly with distance from the neutral axis so that the extent of yielding in the web is determined as a function of $\bar{y}$.

The stress distribution is determined from the strain distribution and the stress-strain diagram for the material from which the beam is made. The stress distribution is shown in Figure 5-35c.

The net force acting on the cross section is

$$N = -\sigma_{YP}6 - \frac{1}{2}\left(\sigma_{YP} + \frac{\overline{y}}{\overline{y}+1}\sigma_{YP}\right)6 - \left(\frac{\overline{y}}{\overline{y}+1}\frac{\sigma_{YP}}{2}\right)2\overline{y}$$

$$+ \frac{\sigma_{YP}}{2}2(\overline{y}+1) + \sigma_{YP}(5-2\overline{y})2 \tag{a}$$

where $\frac{1}{2}\{\sigma_{YP} + [\overline{y}/(\overline{y}+1)]\sigma_{YP}\}$ is the average stress acting on the elastic portion of the flange, and $\frac{1}{2}[\overline{y}/(y+1)]\sigma_{YP}$ is the average stress acting on the elastic portion of the web that lies above the neutral axis.

Setting $N = 0$ yields the quadratic equation

$$\overline{y}^2 + \overline{y} - \tfrac{1}{2} = 0 \tag{b}$$

with roots

$$\overline{y} = -\frac{1}{2} \pm \frac{\sqrt{3}}{2} = 0.366 \text{ in.}, -1.367 \text{ in.} \tag{c}$$

The last root is discarded because it is inconsistent with the specified depth of yielding. Thus the neutral axis for the specified amount of yielding is situated 0.366 in. below the lower edge of the flange as shown in Figure 5-37a.

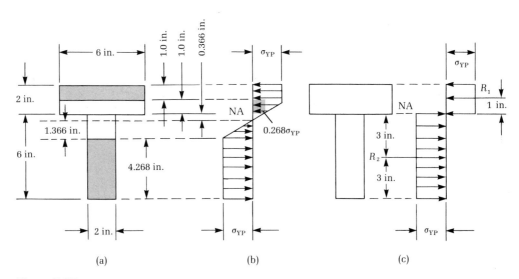

(a)   (b)   (c)

Figure 5-37

(b) The partially plastic bending moment is determined with the aid of Figure 5-37b. A convenient way to accomplish this is to divide the stress diagram into several separate parts: the two regions of the stress diagram associated with the yielded portions of the cross section, two additional regions associated with the elastic portions of the web, and a region associated with the elastic portion of the flange. Consequently,

$$\overbrace{M_{PP} = \underset{\text{flange}}{6\sigma_{YP}(1.866)} + \underset{\text{web}}{[\sigma_{YP}(4.268)2]3.5}}^{\text{yielded regions}}$$

$$+ \overbrace{\left[\underset{\text{above neutral axis}}{\left(\frac{0.268\sigma_{YP}}{2}\right)(2 \times 0.366)\right](\tfrac{2}{3} \times 0.366)} + \underset{\text{below neutral axis}}{\left[\frac{\sigma_{YP}}{2}(2 \times 1.366)(\tfrac{2}{3} \times 1.366)\right]}}^{\text{elastic region of the web}}$$

$$+ \overbrace{\underset{\substack{\text{shaded part} \\ \text{of stress diagram}}}{[0.268\sigma_{YP}(6)]0.866} + \left[\underset{\substack{\text{unshaded part of} \\ \text{stress diagram}}}{\left(\frac{1.0 - 0.268}{2}\sigma_{YP}\right)(6)(0.366 + \tfrac{2}{3} \times 1)\right]}}^{\text{elastic region of the flange}} = 46\sigma_{YP}$$

$$(d)$$

(c) The neutral axis corresponding to the fully plastic condition for any cross section is located by observing that the area above the neutral axis must be equal to the area below the axis. Thus, for the dimensions specified in the present example, the neutral axis must coincide with the lower edge of the flange. The stress distribution corresponding to the fully plastic condition is depicted in Figure 5-37c. The fully plastic moment is

$$M_{FP} = 1 \times R_1 + 3R_2 = \sigma_{YP}(12) + 3(\sigma_{YP}\,12) = 48\sigma_{YP} \qquad (e)$$

The fully elastic moment is

$$M_{FE} = \frac{\sigma_{YP}I}{c} = \frac{136\sigma_{YP}}{5} = 27.2\sigma_{YP} \qquad (f)$$

so that the ratio

$$\frac{M_{FP}}{M_{FE}} = \frac{48}{27.2} = 1.76 \qquad (g)$$

EXAMPLE 5-13

Determine the location of the neutral axis for the channel cross section shown in Figure 5-36a when the amount of yielding is as shown. Also determine the ratio M_{FP}/M_{FE}.

SOLUTION

The stresses σ_1 and σ_2 shown in Figure 5-38b are related to σ_{YP} by similar triangles. Accordingly,

$$\sigma_1 = \frac{2 + \bar{y}}{3 - \bar{y}} \sigma_{YP} \tag{a}$$

and

$$\sigma_2 = \frac{\bar{y}}{3 - \bar{y}} \sigma_{YP} \tag{b}$$

The net normal force acting on the cross section is

$$N = R_1 + R_2 - R_3 - R_4 \tag{c}$$

where

$$
\left.
\begin{aligned}
R_1 &= 4\sigma_{YP} \\[6pt]
R_2 &= \frac{\sigma_{YP}}{2}(3 - \bar{y})4 \\[6pt]
R_3 &= \left(\frac{\bar{y}}{3 - \bar{y}}\frac{\sigma_{YP}}{2}\right)4\bar{y} \\[6pt]
R_4 &= \tfrac{1}{2}(\sigma_1 + \sigma_2)16 = \frac{16(1 + \bar{y})}{3 - \bar{y}}\sigma_{YP}
\end{aligned}
\right\} \tag{d}
$$

Substituting Eqs. (d) into Eq. (c) leads to the equation

$$14 - 32\bar{y} = 0 \tag{e}$$

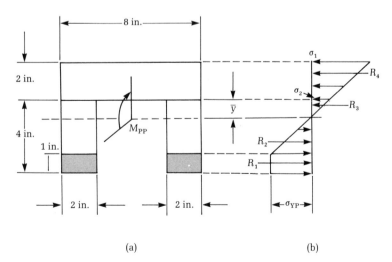

(a)   (b)

Figure 5-38

or

$$\bar{y} = 0.4375 \text{ in.} \tag{f}$$

Consequently, the neutral axis for the yield condition described in Figure 5-38a lies 0.4375 in. below the lower edge of the flange.

(b) The neutral axis for the fully plastic condition coincides with the lower edge of the flange. This situation is verified by observing that the area above this line is equal to the area below it. The fully plastic bending moment is

$$M_{\text{FP}} = 16\sigma_{\text{YP}}(1) + 16\sigma_{\text{YP}}(2) = 48\sigma_{\text{YP}} \tag{g}$$

and the fully elastic bending moment is

$$M_{\text{FE}} = \frac{\sigma_{\text{YP}}I}{c} \tag{h}$$

Now the centroidal axis is located 3.5 in. from the bottom of the section so that the centroidal moment of inertia is

$$I = \tfrac{1}{12}(8)2^3 + 16(1.5)^2 + 2[\tfrac{1}{12}(2)4^3 + 8(1.5)^2] = 98.67 \text{ in}^4 \tag{i}$$

Thus the ratio of the fully plastic bending moment to the fully elastic bending moment is

$$\frac{M_{\text{FP}}}{M_{\text{FE}}} = \frac{48\sigma_{\text{YP}}}{\dfrac{98.67\sigma_{\text{YP}}}{3.5}} = 1.70 \tag{j}$$

PROBLEMS / Section 5-9

5-18 Determine the partially plastic moment that causes yielding to penetrate a rectangular cross section to the extent shown in Figure P5-18.

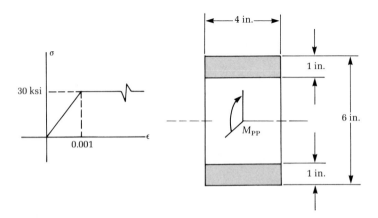

Figure P5-18

5-19 Calculate the dimension of the elastic core for the rectangular cross section shown in Figure P5-19 when $M_{PP} = 24$ kN·m.

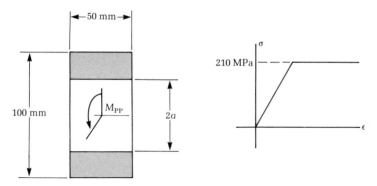

Figure P5-19

5-20 Determine the fully plastic moment for the cross section shown in Figure P5-20. The tensile yield stress for the ideally elastic, ideally plastic material is 30 ksi. Calculate the ratio M_{FP}/M_{FE}.

5-21 Determine the fully plastic moment M_{FP} for the T section shown in Figure P5-21a. The material properties are shown in Figure P5-21b. Calculate the ratio M_{FP}/M_{FE}.

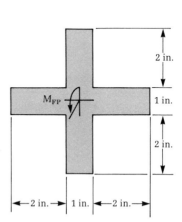

Figure P5-20

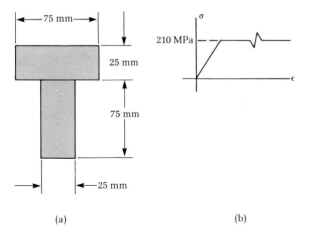

(a)   (b)

Figure P5-21

5-10 Residual Stresses

If the normal strain at a point on a cross section of a beam has exceeded the yield strain, subsequent reduction of the bending moment causes the normal stress at this point to decrease along a straight line parallel to the initial elastic region of the stress-strain diagram. Thus reduction of the normal stress at the point for which the normal strain has exceeded the yield strain follows a linear elastic law. This linear elastic unloading is shown in Figure 5-22a for an ideally elastic, plastic material and in Figure 5-23a for an ideally elastic, linear hardening material by the dashed lines that emanate from point A.

The most convenient way to obtain a residual stress distribution is to calculate (a) the stress distribution that results from the loading moment M_{PP}, (b) the elastic stress distribution corresponding to a moment equal to M_{PP} but which is applied with the opposite sense, and (c) the algebraic sum of these two stress distributions. This procedure is demonstrated in Example 5-14.

EXAMPLE 5-14

A beam with a rectangular cross section is made from an ideally elastic, plastic material for which $\sigma_{YP} = 210$ MPa and $E = 210$ GPa. At a specific cross section along the axis of the beam, a bending moment that causes yielding to penetrate to a depth of 25 mm on both the upper and lower edges of the section as shown in Figure 5-39a occurs. Determine (a) the partially plastic moment M_{PP} and (b) the residual stress distribution that results when the section is unloaded.

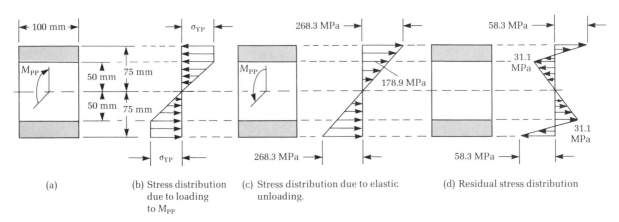

(a)

(b) Stress distribution due to loading to M_{PP}

(c) Stress distribution due to elastic unloading.

(d) Residual stress distribution

Figure 5-39

SOLUTION

(a) From Figure 5-39b,

$$M_{PP} = 2\left\{0.1(0.025\sigma_{YP})(0.0625) + 0.1\left(0.050\,\frac{\sigma_{YP}}{2}\right)(\tfrac{2}{3} \times 0.050)\right\}$$

$$= 0.0004792\sigma_{YP} = 100.63 \text{ kN·m} \tag{a}$$

(b) The elastic stress distribution due to a bending moment of magnitude 100.63 kN·m applied with a sense opposite that of M_{PP} is calculated from the elastic formula

$$\sigma = -\frac{My}{I} \tag{b}$$

or

$$\sigma = \frac{-100.63y}{28.3 \times 10^{-6}} = -3.578 \times 10^6 y \text{ kPa} \tag{c}$$

This stress distribution is shown in Figure 5-39c.

The residual stress distribution is obtained by the superposition of the stress distributions due to loading and unloading. In this way we obtain the stress distribution shown in Figure 5-39d.

PROBLEMS / Section 5-10

5-22–5-24 Determine the residual stress distributions that result when the applied moments in Problems 5-19, 5-20, and 5-21 are released.

5-11 Shearing Stresses in Beams

The kinematic assumptions adopted in Section 5-5 preclude the existence of transverse shearing stresses in a straight beam. Recall that Bernoulli's hypothesis leads to $\gamma_{xy} \equiv 0$. Since $\tau_{xy} = G\gamma_{xy}$ and $V = \int_{\text{area}} \tau_{xy}\, dA$, it follows that $V = 0$ as a consequence of the hypothesis. Recall that the assumptions made were for beams subjected to pure bending. Consider the cantilever beam shown in Figure 5-40a. The free-body diagram of a segment of the cantilever beam shown in Figure 5-40b indicates that transverse equilibrium of forces cannot be satisfied under these circumstances. This observation is true, generally, for a segment of a beam that is loaded and supported in a general manner so that pure bending conditions are no longer valid. (We remark that the flexure formula remains approximately valid when the bending moment varies along the length of the beam.)

By adopting the Bernoulli hypothesis as the kinematic basis for the flexure formula, we have adopted an hypothesis that unwittingly prevents an essential ingredient of any sound structural theory. This paradox is resolved by conceding that plane sections actually warp, but that the shearing strain that accompanies the warping has little effect on the normal strain and, thus, on the normal stress. Consequently, it is permissible to establish the flexure formula on the kinematic hypothesis that sections that are plane before deformation remain plane after deformation. Having conceded that cross sections actually warp, we can admit the existence of the transverse shearing force V that is required if transverse equilibrium is to be satisfied.

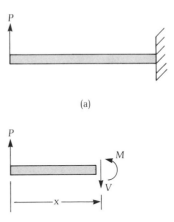

(a)

(b)

Figure 5-40

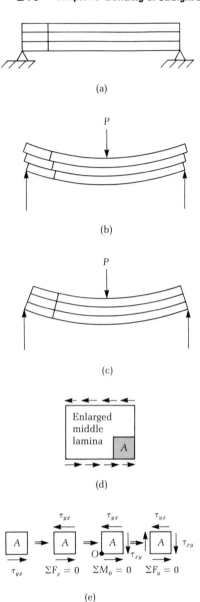

(a)

(b)

(c)

Enlarged
middle
lamina

(d)

τ_{yx} τ_{yx} τ_{yx}

A $\Rightarrow$ A $\Rightarrow$ A $\Rightarrow$ A τ_{xy}

τ_{yx} $\Sigma F_x = 0$ $\Sigma M_0 = 0$ $\Sigma F_y = 0$

(e)

Figure 5-41

It is possible to determine τ_{xy} and hence V directly by integration of the strain-displacement relations; however, the strength-of-materials approach abandons the kinematic assumptions that serve as the basis of the flexure formula and determines τ_{xy} directly from equilibrium considerations.

Before developing a formula from which the transverse shearing stresses can be calculated, we should consider the physical aspects that give rise to these stresses. In this regard, consider a straight beam composed of three smooth lamina as shown in Figure 5-41a. These lamina may be wooden planks or, for visual purposes, a deck of cards. The essential feature is that the contacting surfaces are frictionless.

Suppose, further, that the laminated beam is loaded at midspan. Figure 5-41b shows the beam in its deformed configuration. Since the surfaces of the lamina are smooth, relative motion at their interfaces can occur freely. Accordingly, cross sections do not remain plane for the total beam thickness. Instead, the cross section of each lamina remains plane as shown in Figure 5-41b.

Now suppose that the lamina are glued securely at their interfaces so that relative motion is resisted by the glue. The same type of resistance can be realized by nails, bolts, rivets, or welds (for metals). The deformed configuration of this laminated beam is shown in Figure 5-41c. The kinematic behavior has been altered drastically. Because relative motion between the lamina is prevented, cross sections remain plane. Shearing stress must therefore exist between the lamina as shown in Figure 5-41d. This shearing stress is denoted by τ_{yx} in accordance with the subscript convention adopted earlier in this book. It is frequently referred to as the horizontal shearing stress in a straight beam.

If a differential volume element is taken from the middle lamina as shown in Figure 5-41d, $\tau_{xy} = \tau_{yx}$ from equilibrium considerations. Figure 5-41e indicates the reasoning used to validate this conclusion. τ_{xy} is referred to as the transverse shearing stress in a straight beam.

The strength-of-materials approach to establishing a formula for the transverse shearing stress in a straight beam is to determine a formula for the horizontal shearing stress directly and, hence, a formula for the transverse shearing stress indirectly.

5-12   Shear Flow and Shearing Stress

To develop a formula from which the shearing stresses in a straight beam can be calculated, we isolate a portion of the beam that is contained between the cross sections at x and $x + \Delta x$ and a longitudinal plane a distance y from the neutral surface of the beam as shown in Figure 5-42a. Denote the normal stresses acting on the cross sections at x and $x + \Delta x$ by $\sigma(x, y)$ and $\sigma(x + \Delta x, y)$, respectively.

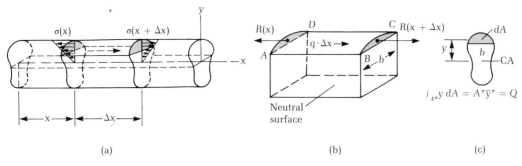

(a)   (b)   (c)

Figure 5-42

According to the flexure formula, these stress distributions are

$$\sigma(x, y) = -\frac{M(x)y}{I}$$

$$\sigma(x + \Delta x, y) = -\frac{M(x + \Delta x)}{I}y \Bigg\}$$

(5-44)

Observe that, in general, $M(x) \neq M(x + \Delta x)$ so that $\sigma(x, y) \neq \sigma(x + \Delta x, y)$. The normal stress distributions given by Eqs. (5-44) are depicted in Figure 5-42a.

Figure 5-42b shows concentrated forces $R(x)$ and $R(x + \Delta x)$ that are statically equivalent to the normal stress distributions that act on the shaded areas shown in Figures 5-42a. They are

$$R(x) = \int_{A^*} \sigma(x, y)\, dA = -\frac{M(x)}{I} \int_{A^*} y\, dA$$

$$R(x + \Delta x) = \int_{A^*} \sigma(x + \Delta x, y)\, dA = -\frac{M(x + \Delta x)}{I} \int_{A^*} y\, dA \Bigg\}$$

(5-45)

Here A^* is the shaded area shown in Figure 5-42. We assume that the shaded areas at x and $x + \Delta x$ are equal. The integral appearing in Eqs. (5-45) is the first moment of the shaded area of Figure 5-42b with respect to the centroidal axis of the cross section. This quantity is customarily signified by the letter Q Accordingly,

$$Q = \int_{A^*} y\, dA = A^* \overline{y}^*$$

(5-46)

With this designation, the statically equivalent forces of Eqs. (5-45) become

$$R(x) = -\frac{M(x)}{I} Q$$

$$R(x + \Delta x) = -\frac{M(x + \Delta x)}{I} Q \Bigg\}$$

(5-47)

From elementary differential calculus,

$$\frac{dR}{dx} = \lim_{\Delta x \to 0} \frac{R(x + \Delta x) - R(x)}{\Delta x} = -\lim_{\Delta x \to 0} \frac{M(x + \Delta x) - M(x)}{\Delta x} \frac{Q}{I}$$

so that

$$\frac{dR}{dx} = -\frac{dM}{dx} \frac{Q}{I} \tag{5-48}$$

Now, from Eq. (5-6), we have $dM/dx = V(x)$ and, hence,

$$\frac{dR}{dx} = -\frac{VQ}{I} \tag{5-49}$$

Equation (5-49) allows us to calculate the change in the normal force per unit of length for a portion A^* of the cross-sectional area.

Since $\sigma(x, y) \neq \sigma(x + \Delta x, y)$ in general, $R(x) \neq R(x + \Delta x)$, in general. Examining the free-body diagram in Figure 5-42b, we observe that if equilibrium of forces parallel to the axis of the beam prevails, a force parallel to these forces must exist in the cutting plane $ABCD$ in the same figure. This force is denoted by $q\,\Delta x$, where q is a force per unit length and is called shear flow. Equilibrium of the element of Figure 5-42b gives

$$R(x + \Delta x) - R(x) + q\,\Delta x = 0 \tag{5-50}$$

or, after division by Δx and passing to the limit ($\Delta x \to 0$),

$$\frac{dR}{dx} = -q \tag{5-51}$$

Equating Eqs. (5-49) and (5-51) leads to the famous shear flow formula for straight beams,

$$q = \frac{VQ}{I} \tag{5-52}$$

We remark that, consistent with our sign convention for positive shearing forces V, q acts as shown in Figure 5-42a for points that lie above the centroidal axis in the cross section; q acts in the opposite direction for points below the centroidal axis. For negative shearing forces, the directions of q for points above and below the centroidal axis are the reverse of those for positive V.

The shearing stress acting on the plane $ABCD$ of Figure 5-42b has a complicated distribution over the width of the cutting plane. A useful approximation of this stress distribution is obtained by assuming that it is constant over the width of the cutting plane.

Consequently,

$$\tau_{ave} = \frac{q}{b} = \frac{VQ}{Ib} \tag{5-53}$$

This procedure is analogous to the one used to calculate the shearing stress in thin-walled closed tubes.

The following examples illustrate the use of the shear flow formula and the shearing stress formula given by Eqs. (5-52) and (5-53), respectively.

EXAMPLE 5-15

A beam is fabricated from wooden planks to obtain the cross section shown in Figure 5-43a. Flange plank A is nailed to web plank B, and flange plank C is glued to web plank B. Side boards are attached to plank A by means of nails whose resistance in shear is 98 lb. The

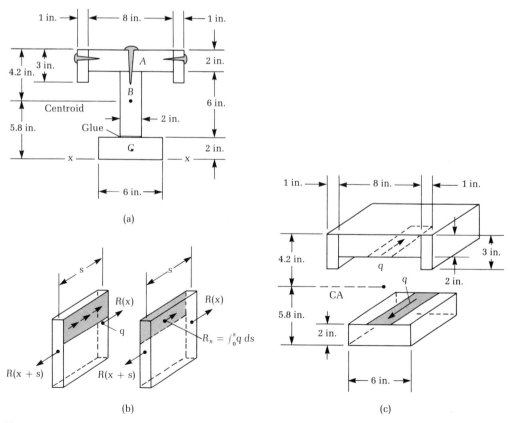

Figure 5-43

internal shearing force for a certain length of the beam is 1000 lb. Determine (a) the minimum spacing for the nails connecting the side boards to plank A, (b) the minimum shearing strength for nails connecting plank A to plank B if the spacing is 2 in., and (c) the minimum strength for the glue at the interface between planks B and C.

SOLUTION

The properties of cross-sectional area are calculated first. The area of the section is $A = 46$ in^2, and the centroidal axis is located at

$$A\bar{y} = (12)(1) + 12(5) + 16(9) + 2(3)(8.5) = 267 \text{ in}^3$$

so that

$$\bar{y} = \frac{267}{46} = 5.80 \text{ in.} \tag{a}$$

The moment of inertia of the section relative to the reference axis XX is

$$\begin{aligned}
I_{XX} &= \tfrac{1}{3}(6)2^3 + \{\tfrac{1}{12}(2)6^3 + 12(5)^2\} \\
&\quad + \{\tfrac{1}{12}(8)2^3 + 16(9)^2\} \\
&\quad + 2\{\tfrac{1}{12}(1)3^3 + 3(8.5)^2\} \\
&= 2091.3 \text{ in}^4 \tag{b}
\end{aligned}$$

By the parallel axis theorem, the centroidal moment of inertia is

$$\bar{I}_{XX} = 2091.3 - 46(5.8)^2 = 543.9 \text{ in}^4$$

(a) Figure 5-43b shows the process used to determine the shearing force each nail must carry. First, imagine the side boards to be an integral part of plank A; that is, imagine them to be continuously attached to plank A as would be the case if the configuration were cut directly from stock material. The shear flow q would be distributed continuously along the length of the beam as shown in the first part of Figure 5-43b. Since the side boards are not integral with plank A, each nail must resist a shearing force equal to

$$R_n = \int_0^s q \, ds \tag{c}$$

where s is the spacing between nails. This notion is shown in the second part of Figure 5-43b. For a region of a beam for which the shearing force V is constant, the shear flow will be constant. Thus Eq. (c) is replaced by the relation

$$R_n = qs \tag{d}$$

To calculate the required spacing for the nails in the side boards, calculate the shear flow q and then use Eq. (d). Accordingly,

$$q = \frac{VQ}{I} = \frac{1000}{543.9}[3(4.2 - 1.5)] = 14.9 \text{ lb/in.} \tag{e}$$

and, consequently,

$$s = \frac{98}{14.9} = 6.58 \text{ in.} \tag{f}$$

(b) To calculate the minimum shearing strength for the nails that connect plank A to plank B, calculate q for the interface between the two planks. Figure 5-43c shows the area whose moment is required in the calculation of q. Therefore,

$$Q = 16(3.2) + 6(2.7) = 67.4 \text{ in}^3 \tag{g}$$

and the shear flow is

$$q = \frac{1000(67.4)}{543.9} = 123.9 \text{ lb/in.} \tag{h}$$

The minimum strength of the nails is calculated from Eq. (d).

$$R_n = 123.9(2) = 247.8 \text{ lb} \tag{i}$$

(c) The strength of the glue used to attach plank C to plank B must be at least equal to the shearing stress induced in the beam at the interface between these planks. Accordingly, the shear flow at this interface is (see Figure 5-43c)

$$q = \frac{1000}{543.9} [12(5.8 - 1.0)] = 105.9 \text{ lb/in.} \tag{j}$$

and the shearing stress is

$$\tau = \frac{q}{t} = \frac{105.9}{2} = 52.95 \text{ psi} \tag{k}$$

An appropriate glue must have a shearing strength of at least 52.95 psi.

EXAMPLE 5-16

Determine the shear flow distribution for the thin box beam of thickness t shown in Figure 5-44a.

SOLUTION

The shear flow in the vertical legs of the box beam can be obtained directly from the free-body diagram of Figure 5-44b. Assuming that $R(x + \Delta x) > R(x)$,

$$\frac{dR}{dx} = 2q \tag{a}$$

so that

$$q = \frac{VQ}{2I} \tag{b}$$

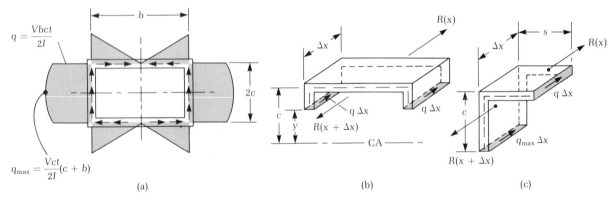

Figure 5-44

Now

$$Q = (bt)c + \left[(c - y)t\left(\frac{c + y}{2}\right)\right]2$$

or

$$Q = bct + (c^2 - y^2)t \qquad \text{(c)}$$

Consequently, the shear flow at any level y in a vertical leg is

$$q = \frac{V}{2I}[bc + (c^2 - y^2)]t \qquad \text{(d)}$$

The largest shear flow in a vertical leg occurs at the centroidal axis as can be seen from Eq. (d). Thus

$$q_{max} = \frac{Vct}{2I}(b + c) \qquad \text{(e)}$$

The shear flow for the vertical legs is shown in Figure 5-44a.

The shear flow distribution for the upper horizontal leg can be obtained from the equilibrium of the element shown in Figure 5-44c. Since

$$\frac{dR}{dx} = q_{max} + q \qquad \text{(f)}$$

we have

$$q = \frac{dR}{dx} - q_{max} \qquad \text{(g)}$$

where

$$\frac{dR}{dx} = \frac{VQ}{I} = \frac{V}{I}\left\{\frac{c^2 t}{2} + stc\right\} = \frac{Vct}{2I}(c + 2s) \qquad \text{(h)}$$

and q_{max} is given in Eq. (e).

Substituting Eqs. (h) and (e) into Eq. (g) yields an expression for the shear flow distribution in the horizontal leg; thus

$$q = \frac{Vct}{2I}(2s - b) \tag{i}$$

This relation is shown in Figure 5-44a. The direction of the shear flow for the portion of the beam in the second quadrant is determined from the free-body diagrams of Figures 5-44b and 5-44c. For $R(x + \Delta x) > R(x)$, the shear flow in the legs must be upward. From Eq. (i) it is seen that q is negative for the horizontal leg lying in the second quadrant. Figure 5-44c then shows that the shear flow must proceed from left to right. The sense of the shear flow q for the remaining quadrants can be obtained from similar free-body diagrams.

EXAMPLE 5-17

Determine a formula for the shear flow and the shearing stress at any point y in the cross section of the thin circular cross section shown in Figure 5-45a.

SOLUTION

To determine the shear flow for the tubular beam, use the free-body diagram of material shown in Figure 5-45b. Assuming that $R(x + \Delta x) > R(x)$, we find from equilibrium that

$$\frac{dR}{dx} = 2q = \frac{VQ}{I} \tag{a}$$

The statical moment of the shaded area of Figure 5-45a is

$$Q = \int_{A^*} y \, dA = \int_{\theta_1}^{\pi - \theta_1} (R \sin \theta)(R \, d\theta \, t)$$

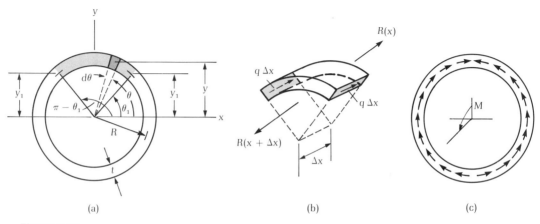

(a)   (b)   (c)

Figure 5-45

or

$$Q = 2R^2 t \cos \theta_1 \qquad\qquad\qquad \text{(b)}$$

Now $\cos \theta_1 = \sqrt{1 - (y_1/R)^2}$ so that Eq. (b) becomes

$$Q = 2R^2 t \sqrt{1 - (y_1/R)^2} \qquad\qquad\qquad \text{(c)}$$

Since the cross section is thin, $I = \pi R^3 t$ and $A = 2\pi R t$. Consequently, the shear flow at a point y_1 on the cross section is

$$q = \frac{VQ}{2I} = \frac{VR^2 t \sqrt{1 - (y_1/R)^2}}{\pi R^3 t} = \frac{V}{\pi R} \sqrt{1 - (y_1/R)^2} \qquad\qquad \text{(d)}$$

The direction of the shear flow on the section is shown in Figure 5-45c. Directions for the upper half of the cross section are obtained from the free-body diagram shown in Figure 5-45b.

The shear stress at any level y_1 is calculated by dividing the shear flow given in Eq. (d) by the wall thickness t. Therefore,

$$\tau = \frac{V}{\pi R t} \sqrt{1 - (y_1/R)^2} \qquad\qquad\qquad \text{(e)}$$

The maximum shearing stress occurs at $y_1 = 0$ and is equal to

$$\tau_{\max} = \frac{2V}{2\pi R t} = 2 \frac{V}{A} = 2\tau_{\text{ave}} \qquad\qquad\qquad \text{(f)}$$

Equation (f) shows that maximum shearing stress in a thin tubular beam is twice the average shearing stress.

EXAMPLE 5-18

Determine the shear flow q and the shearing stress distribution for the rectangular cross section shown in Figure 5-46a. Sketch these distributions.

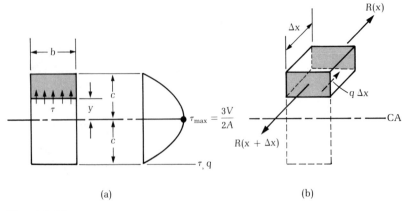

(a)   (b)

Figure 5-46

SOLUTION

Figure 5-46b shows a free-body diagram of a portion of a rectangular beam that is appropriate for the determination of the shear flow q and, hence, the shearing stress τ.

From equilibrium considerations,

$$q = \frac{dR}{dx} = \frac{VQ}{I} \tag{a}$$

where Q is the first moment of the shaded area with respect to the centroidal axis of the rectangular section. Now, from Figure 5-46b,

$$Q = (c - y)b\left(\frac{c + y}{2}\right) = \frac{b}{2}(c^2 - y^2) \tag{b}$$

Consequently,

$$q = \frac{Vb}{2I}(c^2 - y^2) \tag{c}$$

and

$$\tau = \frac{q}{b} = \frac{V}{2I}(c^2 - y^2) \tag{d}$$

These parabolic distributions are depicted in Figure 5-46a.

Note that the maximum stress occurs at the centroidal axis of the cross section where $y = 0$ and has the value

$$\tau_{max} = \frac{Vc^2}{2I} \tag{e}$$

Now $I = \frac{2}{3}bc^3$ so that

$$\tau_{max} = \frac{3}{2}\frac{V}{2bc} = \frac{3}{2}\frac{V}{A} = \frac{3}{2}\tau_{ave} \tag{f}$$

Equation (f) shows that the maximum shearing stress in a beam with a rectangular cross section is 1.5 times the average shearing stress for the section.

EXAMPLE 5-19

A channel-type beam is constructed from thin sheets of stainless steel and four equal leg angles as shown in Figure 5-47a. If the area of the thin sheets is small compared to the area of the equal leg angles, and if we can therefore neglect the normal stresses in the thin sheets due to bending, determine the shearing stress in the thin sheets caused by a vertical shear force V.

SOLUTION

Beams of this type are frequently idealized like the one shown in Figure 5-47b. Since the area of the sheets is small compared to the area

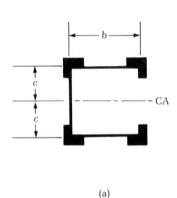

(a)

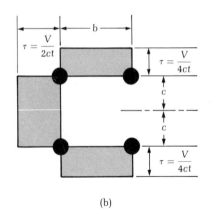

(b)

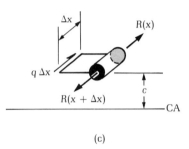

(c)

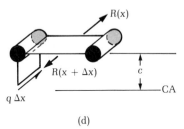

(d)

Figure 5-47

of the angles, the sheets are shown as lines without any thickness. The angles or stiffeners, whatever shape they may be, are shown as concentrated areas.

From equilibrium of forces in Figure 5-47c we see that

$$q = \frac{dR}{dx} = \frac{VQ}{I} = \frac{V}{I} (Ac) \tag{a}$$

where A is the cross-sectional area of an angle. Similarly, equilibrium of forces in Figure 5-47d gives

$$q = \frac{dR}{dx} = \frac{VQ}{I} = \frac{V}{I} (Ac)2 \tag{b}$$

Now $I = 4Ac^2$. (Moments of inertia of the angle sections about their centroidal axis are considered to be small relative to the transfer moment of inertia.) Consequently, the shearing stresses for the flanges and the web are

$$\tau_f = \frac{V}{4ct}$$

and

$$\tau_w = \frac{V}{2ct}$$

respectively. These stresses are shown in Figure 5-47b.

5-13   Limitations on Shear Flow and Shear Stress Formulas

The shear flow formula $[q = (VQ/I)]$ is based on equilibrium considerations. It does not result from initial kinematic hypotheses as the flexure formula does. Indeed, it is one of the few formulas that falls within the scope of the strength of materials that is not the result of initial kinematic assumptions. It should be recognized that the flexure formula was used to calculate the force $R(x)$ and $R(x + \Delta x)$ and, consequently, the kinematic assumptions pertaining to it have indirectly played a role in the development of the shear flow formula. Moreover, since the flexure formula assumes linear elastic material behavior, the shear flow formula is restricted to this type of behavior.

The shearing stress is calculated as an average over the width of the cutting plane ($\tau = q/b$). Therefore, we expect this average shearing stress to be a reasonably accurate approximation of the maximum shearing stress if the width of the beam is small compared to the depth of the beam. The accuracy of this approximation diminishes as the ratio of the thickness of the cutting plane to the depth of the section increases. In fact, the shearing stress calculated via $\tau = q/t$ may be worthless for the flanges of an S beam or a W beam. The shear flow formula is not afflicted with a loss of accuracy because of this averaging effect.

The shear flow calculated from $q = VQ/I$ is not always a realistic quantity. For example, at the free edge A of the T section shown in Figure 5-48a, the shearing stress as calculated from $\tau = VQ/Ib$ is not zero. However, the shearing stress is obviously zero because surface A is a free surface. As is seen from the differential element of Figure 5-48b, if $\tau \neq 0$, then, for equilibrium, surface A must have applied to it a shearing stress of magnitude τ. But surface A is a free surface so that $\tau = 0$ there. Consequently, the shearing stress formula leads to erroneous results there.

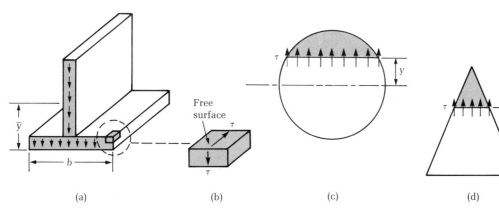

(a)   (b)   (c)   (d)

Figure 5-48

Erroneous results are also obtained for the solid circular and triangular cross sections shown in Figure 5-48c and 5-48d. At level y, for either section, the vertical shearing stresses that result from the formula $\tau = VQ/Ib$ have a component that is perpendicular to the

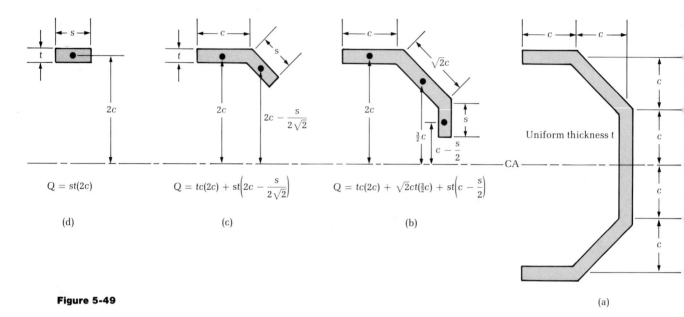

$$Q = st(2c)$$

(d)

$$Q = tc(2c) + st\left(2c - \frac{s}{2\sqrt{2}}\right)$$

(c)

$$Q = tc(2c) + \sqrt{2}ct(\tfrac{3}{2}c) + st\left(c - \frac{s}{2}\right)$$

(b)

Figure 5-49

(a)

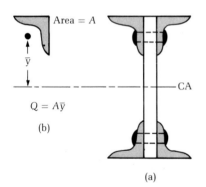

$$Q = A\bar{y}$$

(b)

(a)

Figure 5-50

free surface. This is obviously a contradiction with a free surface. The formula, however, predicts an accurate value for the shearing stress at the centroidal axis for the circular cross section. Note that this level is the only one in the cross section for which the shearing stress as calculated from $\tau = VQ/Ib$ does not have a component perpendicular to the free surface.

Consider the thin cross section shown in Figure 5-49a. Figures 5-49b through 5-49d show proper selections of free-body diagrams for calculating the shear flow at various points on the section.

Figure 5-50a is a composite section consisting of a plate and four angle sections. Figure 5-50b shows the typical free-body diagram required to calculate the force on the rivet shown in Figure 5-50a.

PROBLEMS / Sections 5-11–5-13

5-25 The maximum shearing force in a beam with the cross section shown in Figure P5-25 is 4450 N. Calculate the shearing stress at the planes aa and bb. The thickness of each segment of the cross section is 2.5 mm.

5-26 A beam is to be fabricated from two sections of 2-in. standard steel pipe and a $\frac{1}{4}$-in. thick steel plate by four fillet welds as depicted in Figure P5-26. If the maximum transverse shearing force that the beam must resist is 2000 lb, determine whether $\frac{1}{8}$-in. fillet welds with a capacity of 1000 lb/in. each will be sufficient.

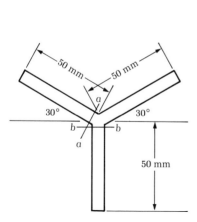

Figure P5-25

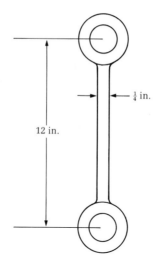

Figure P5-26

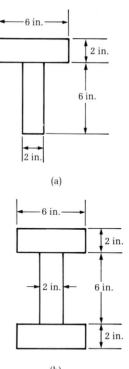

(a)

(b)

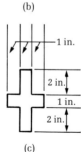

(c)

Figure P5-27

5-27 Plot the shearing stress distributions for the cross sections shown in Figures P5-27a through P5-27c.

5-28 A C 12 × 20.7 is bolted to a W 18 × 50 to increase the lateral stiffness of the beam as shown in Figure P5-28. If the shearing resistance of the bolts is 3000 lb, and if the maximum transverse shearing force that the beam must resist is 15,000 lb, calculate the maximum permissible spacing for the bolts.

5-29 Two 100-mm × 25-mm boards and two 150-mm × 25-mm boards are to be used in the fabrication of a box beam as shown in Figures P5-29a or P5-29b. If the shearing resistance of each nail is 1000 N and if the maximum shearing force in the beam is 1 kN, determine the maximum permissible spacing of the nails for each scheme.

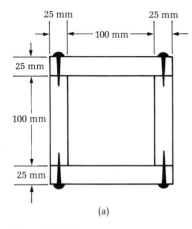

(a)

(b)

Figure P5-29

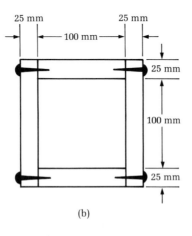

Figure P5-28

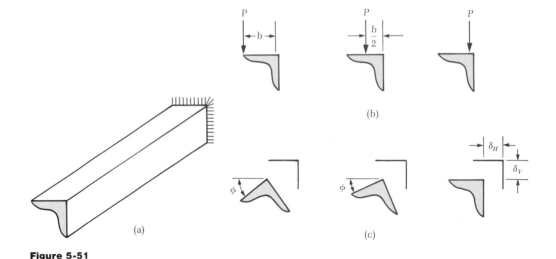

Figure 5-51

5-14 Shear Center

Let us introduce the concept of a shear center in a purely physical manner first. Consider a structural steel angle section that is used as the cantilever beam shown in Figure 5-51a.

Imagine that a concentrated force P is applied at its end. Figure 5-51b shows the application point of P at three different locations along the horizontal leg of the angle. Intuitively, the cantilever is expected to bend downward and to twist simultaneously. This tendency is shown in Figure 5-51c for each of the three positions of P shown in Figure 5-51b. We note that when P is applied at the outer tip of the horizontal leg, the end section translates sidewise, as well as downward, while it rotates through the angle φ. We sense that this same action occurs when P is applied midway along the horizontal leg, except that the rotation diminishes as shown in the second part of Figure 5-51c. Now, when P is applied along the center line of the vertical leg, it becomes uncertain as to whether the sidewise translation and the rotation will still occur. Indeed, our intuition tells us that since the rotation tends to decrease continuously as the location of P approaches the center line of the vertical leg, there must be some location for the action line of P such that no twisting of the cross section will occur. For the angle considered, no twisting occurs when the action line coincides with the center line of the vertical leg.

Similarly, if the action line of a horizontal force Q coincides with the center line of the horizontal leg as shown in Figure 5-52, the section will displace laterally and vertically, but without rotating.

Intuitively, we suspect that the end cross section will simply displace horizontally and vertically, without rotation, for any force P that passes through the intersection of the center lines of the two legs as shown in Figure 5-53. This is, indeed, what occurs. Note that the action line of P must pass through this point when the cross section

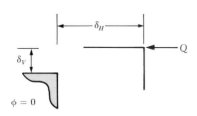

Figure 5-52

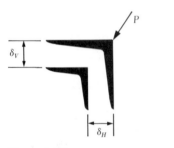

Figure 5-53

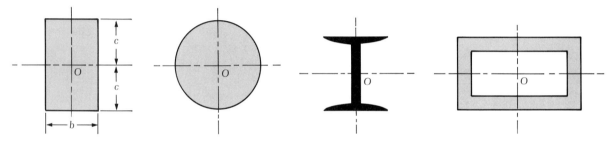

Figure 5-54

does not twist. This point is called the **shear center** for the angle cross section. The point is frequently called the **center of twist** for the cross section for obvious reasons.

Each cross-sectional shape has its own shear center or center of twist. Shear centers for cross sections composed of thin members can be determined rather easily. Determination of the shear centers for solid sections is more difficult.

An extremely useful observation is that the shear center for any cross section always lies in an axis of symmetry of the section. For cross sections with two axes of symmetry, the shear center lies at their intersection. The shear centers for several common cross sections that possess two axes of symmetry are shown in Figure 5-54. Examples of common cross sections that possess one plane of symmetry are shown in Figure 5-55.

The locus of the shear centers of the cross sections along the length of a beam form a straight line called the **axis of twist**. For pure torsional loads, points on this axis remain stationary while cross sections rotate. This axis displaces for bending loads whose action lines pass through it, but cross sections do not rotate.

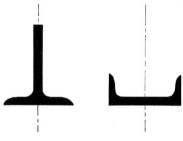

Figure 5-55

The shear center for a cross section will now be defined in more concrete terms. The definition of the shear center for a cross section is closely connected with the concept of statically equivalent force systems. Figure 5-56a shows the cross section of a beam at which the bending moment and the shearing force are denoted by M and V, respectively. In the previous section, we showed that the transverse shearing force V gives rise to a shear flow distribution as shown in Figure 5-56b. Signify the shear flow in the various rectangular segments of the cross section by q_1, q_2, and q_3. These shear flows can be calculated readily from the shear flow formula in the previous section.

Figure 5-56c shows a system of concentrated forces that are statically equivalent to the distributed force system in Figure 5-56b. The forces of this system are given by the formula

$$V_i = \int q_i \, ds \tag{5-54}$$

where the integral extends over the length of the center line of the ith rectangular segment. Note that the action line of V_i coincides with the center line of the ith rectangular segment.

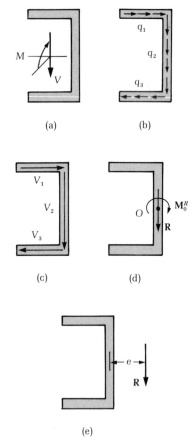

(a)

(b)

(c)

(d)

(e)

Figure 5-56

Since the force system of Figure 5-56c is coplanar and nonconcurrent, it can be reduced to an equivalent force system consisting of a single force. This reduction is done in two steps to emphasize the importance of the concept of equivalent force systems in locating the shear center of a cross section.

Figure 5-56d shows a force-couple system that is statically equivalent to the system of forces of Figure 5-56c provided that

$$\left.\begin{array}{l} \mathbf{R} = \Sigma\mathbf{V}_i \\ \mathbf{M}_0^R = \Sigma\mathbf{r}_i \times \mathbf{V}_i \end{array}\right\} \tag{5-55}$$

where $\mathbf{V}_i$ signifies the forces of the system shown in Figure 5-56c, and $\mathbf{r}_i$ is a position vector that emanates from a common point O and terminates at a convenient point on the action line of $\mathbf{V}_i$.

Finally, the force-couple system of Figure 5-56d is reduced to a single force whose action line satisfies the relation

$$eR = M_0^R \tag{5-56}$$

where e is shown in Figure 5-56e. Observe that $\mathbf{R}$ is the resultant of the internal shearing forces $\mathbf{V}_i$ that result from pure bending of a beam with this cross section. Now, if the action line of the external force at the section coincides with action line of $\mathbf{R}$, and if it has a sense opposite that of $\mathbf{R}$, the twisting stresses will be negated and the cross section will not rotate. Consequently, the shear center must lie on the action line of the resultant of the shearing forces acting on the cross section. Moreover, the shear center always lies along an axis of symmetry. Thus the shear center for this channel section lies at the intersection of its axis of symmetry and the action line of the resultant of the shearing forces.

EXAMPLE 5-20

Determine the shear center for the cross section shown in Figure 5-57a. The thicknesses of the flanges and the web are equal to t.

SOLUTION

First we note that the section possesses an axis of symmetry. The shear center must lie along this symmetry axis. Consequently, only the location along this axis remains to be determined.

To establish consistent directions for the shear flow in the three rectangular segments of the section, assume that $R(x + \Delta x) > R(x)$. Then the proper senses for the q's are as shown in Figure 5-57a.

Following the procedures established in Section 5-12, derive formulas for the shear flow in each rectangular segment.

$$q_1 = \frac{V}{I}\,(st)c \qquad 0 \leq s \leq b \tag{a}$$
$$\text{(top flange)}$$

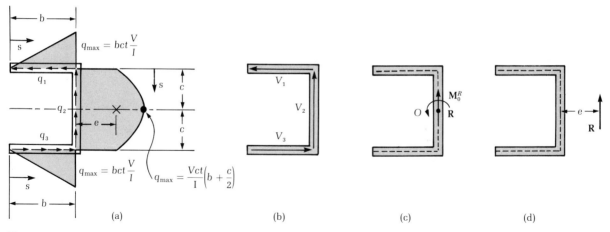

Figure 5-57

$$q_2 = \frac{V}{I}\left\{(bt)c + st\left(c - \frac{s}{2}\right)\right\} \qquad 0 \le s \le 2c \qquad \text{(b)}$$
$$\text{web}$$

and

$$q_3 = -\frac{V}{I}(st)c \qquad 0 \le s \le b \qquad \text{(c)}$$
$$\text{(bottom flange)}$$

The distributions of q_1, q_2, and q_3 are shaded in Figure 5-57a. We note that the shear flow distributions in the flanges vary linearly along the center lines of the flanges, while the shear flow distribution in the web varies parabolically along its center line.

The statically equivalent concentrated shearing forces V_1, V_2, and V_3 are obtained as the area under the corresponding shear flow diagrams or by integration of the shear flow over the length of the center line. In either case,

$$\left.\begin{aligned}
V_1 &= \int_0^b q_1\, ds = \tfrac{1}{2}bq_{max} = \frac{b^2ctV}{2I}\\[4pt]
V_2 &= 2\int_0^c q_2\, ds = \frac{Vt}{I}(2c^2b + \tfrac{2}{3}c^3)\\[4pt]
V_3 &= \int_0^b q_3\, ds = \tfrac{1}{2}bq_{max} = \frac{b^2ctV}{2I}
\end{aligned}\right\} \qquad \text{(d)}$$

These forces are shown in Figure 5-57b.

The resultant **R** of the system of forces shown in Figure 5-57b has the following components.

$$\left.\begin{aligned}
R_x &= -V_1 + V_3 = 0\\
R_y &= V_2
\end{aligned}\right\} \qquad \text{(e)}$$

Consequently, the resultant of the forces acting on the section is a vertical force of magnitude

$$R = V_2 = \frac{Vt}{I}(2c^2b + \tfrac{2}{3}c^3) \tag{f}$$

We digress briefly to calculate the moment of inertia for the section. It is

$$I = \tfrac{1}{12}t(2c)^3 + 2\,[bt(c)^2]$$
$$= (\tfrac{2}{3}c^3 + 2bc^2)t \tag{g}$$

where the moments of inertia of the flanges with respect to their centroidal axes have been neglected in comparison to the transfer portions. Equations (g) and (f) show that

$$R = V \tag{h}$$

The magnitude of the couple that accompanies **R** in its position as shown in Figure 5-56c is

$$M_0^R = 2cV_1 = \frac{b^2c^2tV}{I} \tag{i}$$

Finally, determine the action line of **R** so that its moment with respect to point O is equal to M_0^R. Hence

$$eR = M_0^R \tag{j}$$

or, using Eqs. (h) and (i),

$$eV = \frac{b^2c^2tV}{I} \tag{k}$$

so that

$$e = \frac{b^2c^2t}{I} \tag{l}$$

The shear center for the channel section lies e units to the right of the web center line along the axis of symmetry of the cross section as shown by the star in Figure 5-57a.

EXAMPLE 5-21

Determine the shear center for the cross section shown in Figure 5-58a. The rectangular segments each have a thickness t.

SOLUTION

By symmetry, the shear center must lie along the xx axis of Figure 5-58a. The intersection of this axis of symmetry with the action line of the resultant of the shearing forces acting on the cross section is the shear center.

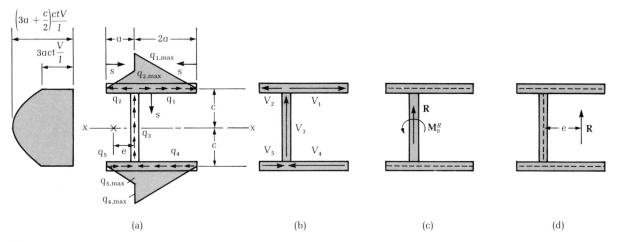

Figure 5-58

The shear flows q_i for the various rectangular segments of the cross section are

$$q_1 = \frac{V}{I}(st)c \qquad\qquad 0 \leq s \leq 2a$$

$$q_2 = \frac{V}{I}(st)c \qquad\qquad 0 \leq s \leq a \qquad\qquad (a)$$

$$q_3 = \frac{V}{I}\left\{(3at)c + st\left(c - \frac{s}{2}\right)\right\} \qquad 0 \leq s \leq c$$

The shear flows q_4 and q_5 have magnitudes equal to q_1 and q_2, respectively. These distributions are shown in Figure 5-58a by the shaded areas.

The statically equivalent system of concentrated forces shown in Figure 5-58b are

$$V_1 = \tfrac{1}{2}(2a)q_{1,\,max} = 2a^2ct\,\frac{V}{I}$$

$$V_2 = \tfrac{1}{2}(a)q_{2,\,max} = \tfrac{1}{2}a^2ct\,\frac{V}{I}$$

$$V_3 = 2\int_0^c \frac{Vt}{I}\left\{3ac + cs - \frac{s^2}{2}\right\}ds \qquad\qquad (b)$$

$$= \frac{2Vt}{I}\left\{3acs + \frac{cs^2}{2} - \frac{s^3}{6}\right\}_0^c$$

$$= \frac{Vt}{I}(6ac^2 + \tfrac{2}{3}c^3)$$

The shearing forces V_4 and V_5 have magnitudes equal to those of V_1 and V_2, respectively.

The statically equivalent force-couple system is shown in Figure 5-58c. The components of the resultant force are

$$R_x = V_1 - V_2 + V_5 - V_4 = 0 \atop R_y = V_3 \Bigg\}$$ (c)

The moment of inertia of the cross section is

$$I = (6ac^2 + \tfrac{2}{3}c^3)t$$ (d)

where the moments of inertia of the flanges relative to their centroidal axes are neglected.

We find from the last of Eqs. (b), by means of Eq. (d), that $V_3 = V$. Consequently, the resultant force is

$$R = V$$ (e)

The magnitude of the accompanying couple is

$$M_0^R = 2cV_2 - 2cV_1 = -3a^2c^2t\,\frac{V}{I}$$ (f)

Now determine the location of the action line of **R** so that its moment about point O is the same as the moment of $\mathbf{M}_0^R$ about the same point. Accordingly,

$$eR = M_0^R$$ (g)

or

$$eV = -3a^2c^2t\,\frac{V}{I}$$ (h)

so that

$$e = \frac{-3a^2c^2t}{I}$$ (i)

The shear center of the cross section lies at the intersection of the axis of symmetry and the vertical line $e = -3a^2c^2t/I$. A star in Figure 5-58a marks the location of the shear center for the section considered in this example.

EXAMPLE 5-22

Determine the shear center for the beam shown in Figure 5-59a. The dark areas represent the cross-sectional areas of longitudinal stiffeners. These areas are considered to be large in comparison with the area of the thin plates that connect them. The area of a stiffener is denoted by A.

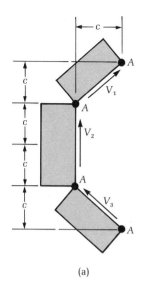

(a)

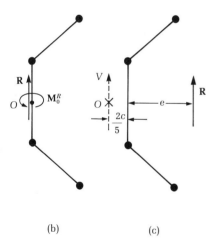

(b)   (c)

Figure 5-59

SOLUTION

The moment of inertia of the cross section with respect to the centroidal axis of the composite section is

$$I = 2A(2c)^2 + 2Ac^2 = 10Ac^2 \tag{a}$$

Expressions for the shear flow in the three straight sections are

$$\left. \begin{array}{l} q_1 = \dfrac{V}{I} A(2c) = \dfrac{V}{5c} \\[2mm] q_2 = \dfrac{V}{I} [A(2c) + Ac] = \dfrac{3V}{10c} \\[2mm] q_3 = \dfrac{V}{I} A(2c) = \dfrac{V}{5c} \end{array} \right\} \tag{b}$$

These distributions are shown by the shaded areas in Figure 5-59a.

The shearing forces that correspond to the shear flow distributions given in Eq. (b) are

$$\left. \begin{array}{l} V_1 = \dfrac{V}{5c} (\sqrt{2}c) = \dfrac{\sqrt{2}}{5} V \\[2mm] V_2 = \dfrac{3}{10} \dfrac{V}{c} (2c) = \dfrac{3}{5} V \\[2mm] V_3 = \dfrac{V}{5c} (\sqrt{2}c) = \dfrac{\sqrt{2}}{5} V \end{array} \right\} \tag{c}$$

The force-couple system that is statically equivalent to the force system of Eqs. (b) is

$$\left. \begin{array}{l} R_x = \dfrac{V_1}{\sqrt{2}} - \dfrac{V_3}{\sqrt{2}} = 0 \\[2mm] R_y = \dfrac{V_1}{\sqrt{2}} + V_2 + \dfrac{V_3}{\sqrt{2}} = V \\[2mm] M_0^R = -\dfrac{V_1}{\sqrt{2}} (2c) = -\tfrac{2}{5}Vc \end{array} \right\} \tag{d}$$

Now locate the action line of **R** so that the moment of **R** about point O is equal to the moment of $\mathbf{M}_0^R$ about point O. From Figure 5-59c,

$$eR = M_0^R$$

or

$$eV = -\tfrac{2}{3}Vc, \qquad e = -\tfrac{2}{5}c \tag{e}$$

The shear center is marked by the star in Figure 5-59c.

PROBLEMS / Section 5-14

5-30 Locate the shear center for the section shown in Figure P5-30. The thickness is uniform throughout the cross section.

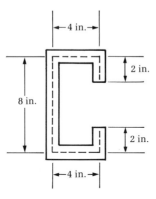

Figure P5-30

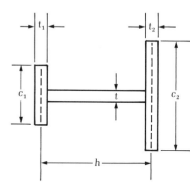

Figure P5-31

5-31 Locate the shear center for the *H* section shown in Figure P5-31.

5-32 Determine the shear centers for the cross sections shown in Figure P5-32.

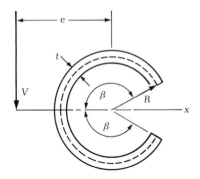

Figure P5-33

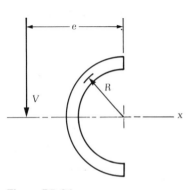

Figure P5-34

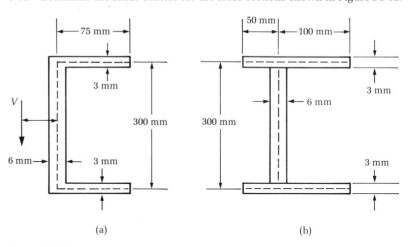

(a)   (b)

Figure P5-32

5-33 Derive a formula for the *x* coordinate of the shear center for the cross section shown in Figure P5-33. Consider the ring to be thin.

$$\left(e = 2R \left(\frac{\sin \beta - \beta \cos \beta}{\beta - \sin \beta \cos \beta} \right) \right)$$

5-34 Show that the shear center for the thin semicircular tube shown in Figure P5-34 is $e = 4R/\pi$.

5-15 Composite Beams

Beams fabricated from two or more materials are called composite beams. The advantage of this type of construction is that large quantities of a low-modulus material can be used in regions of low stress, and small quantities of high-modulus material can be used in regions of high stress. Two common applications of this concept are reinforced concrete beams and wooden beams whose bending strength is bolstered by steel strips, either along its sides or along its top and/or bottom.

The customary method of treating composite beams is to devise a cross section of a single material whose deformational response is equivalent to that of the actual composite section. An equivalent cross section for a composite beam is defined to consist of a single material (usually one of the materials from which the composite section is made) for which:

1. The normal strain ϵ_y^* at any level y is equal to the normal strain ϵ_y at the same level in the composite section.
2. The net normal force R^* is equal to the net normal force R acting on the composite section.
3. The net bending moment M^* is equal to the net bending moment M acting on the composite section.

The following definitions facilitate the development of the theory of bending of composite beams. Let M_1 and M_2 represent the materials that compose the composite section. Let A_1 and E_1 denote the area and modulus of elasticity of material M_1; A_2 and E_2 denote the area and modulus of elasticity of material M_2. Moreover, let R^* and M^* represent the net normal force and the net bending moment acting on the equivalent cross section. R and M denote their counterparts for the composite section. These quantities are shown in Figures 5-60a and 5-60e.

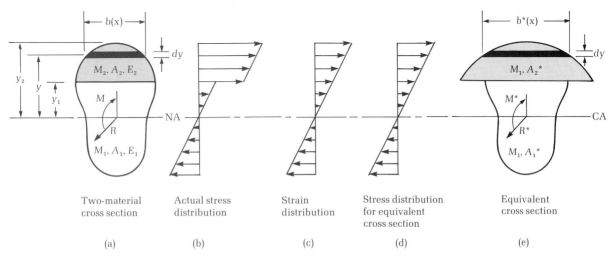

Two-material cross section	Actual stress distribution	Strain distribution	Stress distribution for equivalent cross section	Equivalent cross section
(a)	(b)	(c)	(d)	(e)

Figure 5-60

In the development of an equivalent cross section, we shall assume that the material M_2 is transformed to an equivalent amount of material M_1. This choice is arbitrary and does not cause any loss of generality.

The kinematic assumptions that prevail for homogeneous beams are assumed to apply to composite beams. Accordingly, the strain distribution for the composite section is linear as shown in Figure 5-60c. Multiplying the strain at any level y in the composite section by the appropriate modulus of elasticity yields the stress distribution shown in Figure 5-60b. Notice the discontinuity in the stress distribution at the interface between the two materials. It is assumed that $E_2 > E_1$.

The stress distribution for the **equivalent section** is obtained by multiplying the strain at any level y by the modulus of elasticity of the material M_1. (Recall that material M_2 is being transformed into an equivalent amount of material M_1.) It is shown in Figure 5-60d.

The first requirement for the transformed and composite cross sections to be equivalent is that the normal strains be equal at identical distances from the centroidal axis of the composite section. Thus

$$\epsilon_y^* = \epsilon_y \tag{5-57}$$

where ϵ_y^* is the strain at level y for the transformed section, and ϵ_y is the strain at the same level for the composite section. If both materials behave elastically, Eq. (5-57) can be written as

$$\frac{\sigma_y^*}{E^*} = \frac{\sigma_y}{E} \tag{5-58}$$

where σ_y^* and σ_y are stresses at level y in the transformed and composite sections, respectively; E^* is the modulus of elasticity of the material of the transformed section (E_1 under our agreement); and E is either E_1 or E_2 according to whether y defines a point in material M_1 or M_2. Therefore, the formula that connects stresses associated with the transformed section to the actual stresses of the composite section is

$$\sigma_y = \frac{E}{E^*} \sigma_y^* \tag{5-59}$$

Thus, if a procedure can be established to determine the stresses due to bending of the transformed beam, the stresses in the composite beam can be calculated from Eq. (5-59).

Since the transformed section is a homogeneous section for which the kinematic hypotheses for homogeneous sections are assumed to prevail,

$$\sigma_y^* = -\frac{My}{I^*} \tag{5-60}$$

where I^* is the moment of inertia of the area of the transformed section with respect to its centroidal axis, and y is the distance to any point in the section measured from the centroidal axis.

The second and third requirements for a transformed cross section are expressed by the integrals,

$$\int_{A_1^*} \sigma_y^* \, dA + \int_{A_2^*} \sigma_y^* \, dA = \int_{A_1} \sigma_y \, dA + \int_{A_2} \sigma_y \, dA \qquad (5\text{-}61)$$

and

$$\int_{A_1^*} \sigma_y^* y \, da + \int_{A_2^*} \sigma_y^* y \, da = \int_{A_1} \sigma_y y \, dA + \int_{A_2} \sigma_y y \, dA \qquad (5\text{-}62)$$

Equation (5-61) expresses the equivalence of the normal forces $(R^* = R)$, and Eq. (5-62) expresses the equivalence of the bending moments $(M^* = M)$.

Now $A_1^* = A_1$ and $\sigma_y^* = \sigma_y$ in the region occupied by material M_1. Therefore, Eqs. (5-61) and (5-62) reduce to

$$\int_{A_2^*} \sigma_y^* \, dA = \int_{A_2} \sigma_y \, dA \qquad (5\text{-}63)$$

and

$$\int_{A_2^*} \sigma_y^* y \, dA = \int_{A_2} \sigma_y y \, dA \qquad (5\text{-}64)$$

Multiply both sides of Eqs. (5-63) and (5-64) by y/y, and replace σ_y by Eq. (5-59) to obtain

$$\int_{A_2^*} \frac{\sigma_y^*}{y} y \, dA = \int_{A_2} \frac{E}{E^*} \frac{\sigma_y^*}{y} y \, dA \qquad (5\text{-}65)$$

and

$$\int_{A_2^*} \frac{\sigma_y^*}{y} y^2 \, dA = \int_{A_2} \frac{E}{E^*} \frac{\sigma_y^*}{y} y^2 \, dA \qquad (5\text{-}66)$$

Notice that σ_y^*/y is the gradient of the stress distribution for the transformed cross section, and it is independent of y. This observation leads to the relations

$$\int_{A_2^*} y \, dA = \int_{A_2} \frac{E}{E^*} y \, dA \qquad \text{or} \qquad A_2^* \bar{y}_2^* = \frac{E}{E^*} A_2 \bar{y}_2 \qquad (5\text{-}67)$$

and

$$\int_{A_2^*} y^2 \, dA = \int_{A_2} \frac{E}{E^*} y^2 \, dA \qquad \text{or} \qquad I_2^* = \frac{E}{E^*} I_2 \qquad (5\text{-}68)$$

Equation (5-67) asserts that the first moment of the area A_2^* of the transformed section must be equal to E/E^* times the moment of this area prior to the transformation. Equation (5-68) asserts that the moment of inertia of the area A_2 of the transformed section must be equal to E/E^* times the moment of inertia of this area prior to the transformation.

Equations (5-67) and (5-68) will be satisfied if the width $b^*(x)$ at any level y of the transformed section is related to the width $b(x)$ at the same level in the composite section by the equation

$$b^*(x) = \frac{E}{E^*} b(x) \qquad (5\text{-}69)$$

To verify the truth of this assertion, it is only necessary to write Eqs. (5-67) and (5-68) as follows:

$$A_2^* \bar{y}_2^* = \int_{y_1}^{y_2} y \left[\frac{E}{E^*} b(x) \right] dy = \int_{y_1}^{y_2} y b^*(x)\, dy \qquad (5\text{-}70)$$

and

$$I_2^* = \int_{y_1}^{y_2} y^2 \left[\frac{E}{E^*} b(x) \right] dy = \int_{y_1}^{y_2} y^2 b^*(x)\, dy \qquad (5\text{-}71)$$

The analysis just presented is illustrated in the following examples.

EXAMPLE 5-23

The composite section shown in Figure 5-61a is used as a simply supported beam. If the maximum bending moment in the beam is 216,000 in.-lb, determine the stress distribution in the aluminum and in the wood. The elastic moduli for aluminum and wood are 10×10^6 psi and 1.25×10^6 psi, respectively.

SOLUTION

The equivalent wooden section is shown in Figure 5-61b. Notice that the width of the aluminum portion has been increased by a multiplicative factor $E/E^* = 8$.

The centroidal axis for the equivalent section is obtained as follows.

$$A\bar{y} = A_1 y_1 + A_2 y_2$$
$$100\bar{y} = 96(3) + 4(0.5) = 290 \text{ in}^3$$

or

$$\bar{y} = 2.9 \text{ in.} \qquad (a)$$

The moment of inertia of the equivalent section relative to the x axis as a reference is

$$I_{xx} = \tfrac{1}{3}(4)(1)^3 + \tfrac{1}{36}(32)(6)^3 + 96(3)^2 = 1057 \text{ in}^4 \qquad (b)$$

By the parallel axis theorem, the centroidal moment of inertia is found to be

$$\bar{I}_{xx} = 1057 - 100(2.9)^2 = 216 \text{ in}^4 \qquad (c)$$

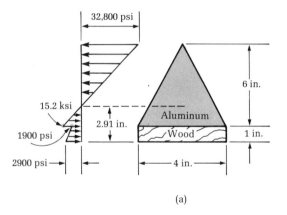

(a)

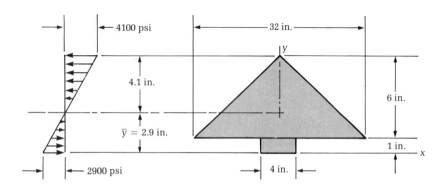

(b) Equivalent wooden section

Figure 5-61

Accordingly, the stress distribution for the equivalent section is

$$\sigma_y = -\frac{216,000}{216}\, y = -1000y \qquad \text{(for the wood)} \qquad \text{(d)}$$

This stress distribution is shown in Figure 5-61b. Equation (d) is directly valid for the wooden part of the composite beam; however, according to Eq. (5-59), the stress in the steel is obtained by multiplying σ_y by $E/E^* = 8$. Therefore, the stress in the steel is

$$\sigma_y = -8000y \text{ psi} \qquad \text{(for the steel)} \qquad \text{(e)}$$

The stress distribution for the composite section is shown in Figure 5-61a.

EXAMPLE 5-24

Determine the stress distribution for the composite section shown in Figure 5-62a if the moment acting on it is 80 000 N·m. The moduli of elasticity for steel and aluminum are 204 GPa and 68 GPa, respectively.

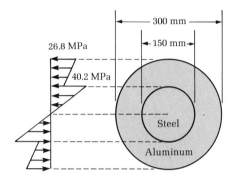

(a) Composite section

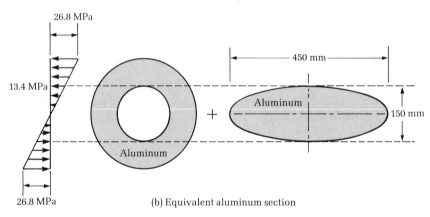

(b) Equivalent aluminum section

Figure 5-62

SOLUTION

The equivalent aluminum cross section is shown in Figure 5-62b. Note that the shape of the equivalent amount of aluminum needed to replace the steel core is elliptical. The moment of inertia of the equivalent section is

$$I = \frac{\pi}{64}\{(0.3)^4 - (0.15)^4\} + \frac{\pi}{4}(0.225)(0.075)^3 \tag{a}$$

$$= 4.473 \times 10^{-4} \text{ m}^4$$

Consequently, the stress distribution for the equivalent section is

$$\sigma_y^* = -\frac{80{,}000y}{4.47 \times 10^{-4}} \tag{b}$$

$$= -178.9 \times 10^6 y \text{ Pa} \qquad \text{for} \qquad 0.075 \le y \le 0.150$$

This stress distribution is shown in Figure 5-62b. Note that the stress predicted by Eq. (b) is valid in the region occupied by the aluminum.

The stress distribution in the steel is given by the formula

$$\sigma_y = \frac{E}{E^*}\sigma_y^* = 3(-178.9 \times 10^6)y$$

$$= -536.7y \text{ MPa} \qquad 0 \leq y \leq 0.075$$

(c)

This stress distribution is shown in Figure 5-62a.

EXAMPLE 5-25

A concrete beam is reinforced with steel rods as shown in Figure 5-63a. Establish (a) a procedure for the determination of the normal stresses in the concrete and in the steel and (b) a procedure for a balanced design.

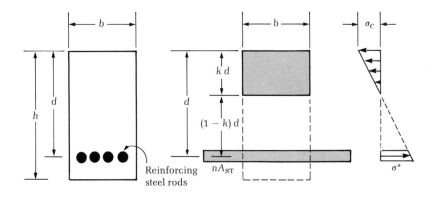

(a) Reinforced concrete beam   (b) Equivalent concrete section

Figure 5-63

SOLUTION

(a) The concrete portion of the cross section of a reinforced concrete beam is assumed to be homogeneous. Moreover, since concrete is weak in tension, concrete that lies below the neutral axis for bending is not likely to contribute much to the bending strength of the composite beam. Therefore, this material is neglected when devising an equivalent cross section.

The shaded area at the top of Figure 5-63b is the concrete area that contributes to the bending strength of the beam. This material experiences only compressive stress. The shaded area near the bottom of this figure is the amount of concrete equivalent to the area of the steel reinforcing rods. This area experiences tension only.

Let d represent the distance from the top of the beam to the centroid of the transformed area of steel. Then kd locates the neutral axis relative to the top of the beam.

The concrete that has been neglected serves two useful purposes. It provides the necessary bond between the steel and the concrete when the beam cross section must transmit a shearing force as well as a bending moment, and it provides fireproofing that minimizes the possibility of the steel softening and thereby losing enough strength to collapse.

Since the moment (with respect to the centroidal axis) of the area above the centroidal axis of the equivalent concrete section must be equal to the moment with respect to the same axis of the area below it,

$$(kd)b\left(\frac{kd}{2}\right) = (1 - k)d(nA_{ST}) \tag{a}$$

where A_{ST} is the area of the steel reinforcing rods and $n = E/E^* = E_{ST}/E_c$. This equation simplifies to

$$k^2 + 2npk - 2np = 0 \tag{b}$$

where

$$p = \frac{A_{ST}}{bd} \tag{c}$$

is called the **steel ratio**.

Equation (b) is a quadratic equation for k; consequently,

$$k = -np + \sqrt{(np)^2 + 2np} \tag{d}$$

If the steel ratio $p = A_{ST}/bd$ is known, the multiplier k can be determined from Eq. (d) and, hence, the location of the neutral axis for the equivalent section is established. A stress analysis proceeds in a manner precisely identical to a stress analysis for any composite section. We note that the moment of inertia for the transformed section is

$$I = \tfrac{1}{12}b(kd)^3 + b(kd)\left(\frac{kd}{2}\right)^2 + (nA_{ST})\,[(1 - k)d\,]^2 \tag{e}$$

(b) The proper design of a reinforced concrete beam requires that the beam not be overreinforced or underreinforced. To satisfy a balanced reinforcement, the neutral axis for the equivalent concrete section should be located according to the relation

$$\frac{\sigma_c}{kd} = \frac{\sigma^*}{(1 - k)d} \tag{f}$$

where σ^* and σ_c are shown in Figure 5-63b. Equation (f) imposes the condition that the allowable stress in the concrete and the allowable stress in the steel occur simultaneously. Now $\sigma_{ST} = (E/E^*)\sigma^* = n\sigma^*$ so that Eq. (f) can be written as

$$\frac{\sigma_c}{kd} = \frac{\sigma_{ST}}{n(1-k)d} \tag{g}$$

Solving for k gives

$$k = \frac{1}{1 + \dfrac{\sigma_{ST}}{n\sigma_c}} \tag{h}$$

Observe that, when the allowable stresses for the concrete and the steel are prescribed, k can be calculated from Eq. (h). Subsequently, the required steel ratio can be determined from Eq. (b). The solution of Eq. (b) for np is

$$np = \frac{k^2}{2(1-k)} \tag{i}$$

EXAMPLE 5-26

A concrete beam contains three number 9 steel reinforcing rods as shown in Figure 5-64a. The cross-sectional area of each rod is 1 in^2. Determine the maximum stress in the concrete and in the steel if the beam is to resist a maximum bending moment of 100,000 ft-lb. The moduli of elasticity for the concrete and the steel are 1.5×10^6 psi and 30×10^6 psi, respectively.

SOLUTION

The modulus ratio is

$$n = \frac{E}{E^*} = 20 \tag{a}$$

and the steel ratio is

$$p = \frac{A_{ST}}{bd} = \frac{3}{12(20)} = \frac{1}{80} \tag{b}$$

Consequently, $np = \frac{1}{4}$ so that Eq. (b) of Example 5-25 becomes

$$k^2 + \tfrac{1}{2}k - \tfrac{1}{2} = 0 \tag{c}$$

The roots of Eq. (c) are

$$k = \tfrac{1}{2}, \ -1 \tag{d}$$

Since $k = -1$ has no physical significance, the centroidal axis for the equivalent concrete section is

$$kd = 0.5(20) = 10 \text{ in.} \tag{e}$$

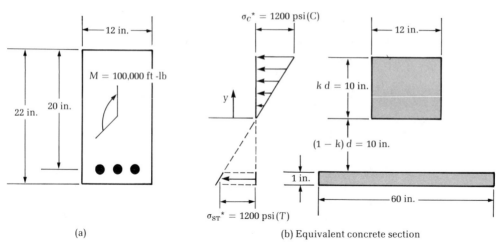

(a)

(b) Equivalent concrete section

Figure 5-64

from the top of the beam as shown in Figure 5-64b. The centroid of the equivalent steel area is located

$$(1 - k)d = 0.5(20) = 10 \text{ in.} \tag{f}$$

below the centroidal axis as shown in Figure 5-64b.

The moment of inertia of the equivalent concrete section is

$$I = \tfrac{1}{12}(12)10^3 + 120(5)^2 + 60(10)^2 = 10,000 \text{ in}^4 \tag{g}$$

Consequently, the stress distribution for the equivalent concrete section is

$$\sigma^* = \frac{-100,000(12)}{10,000} y = -120y \text{ psi} \tag{h}$$

This stress distribution is shown in Figure 5-64b.

Recall that Eq. (h) gives the actual stress in the concrete portion of the composite section ($0 \leq y \leq 10$); however, the stress in the steel portion of the composite section is given by the formula

$$\sigma_{ST} = n\sigma^* = 20(1200) = 24,000 \text{ psi}(T) \tag{i}$$

PROBLEMS / Section 5-15

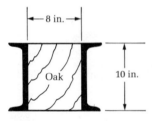

Figure P5-35

5-35 A beam is fabricated from two C 10 × 30 steel sections and a 8-in. × 10-in. oak timber as shown in Figure P5-35. If the fiber stresses in the steel and in the oak are not to exceed 18 ksi and 1.8 ksi, respectively, determine the maximum moment the section can resist ($E_{ST} = 30 \times 10^6$ psi and $E_{OAK} = 1.5 \times 10^6$ psi).

5-36 The reinforced concrete beam shown in Figure P5-36 resists a bending moment of 40,000 ft-lb. Calculate the maximum stress in the steel and in the concrete. The area of each of the four reinforcing rods is 1 in². Assume $E_{ST} = 30 \times 10^6$ psi and $E_C = 2 \times 10^6$ psi.

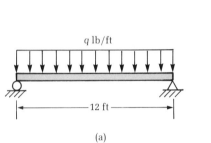

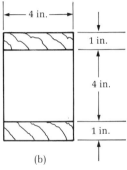

(a)

(b)

Figure P5-36

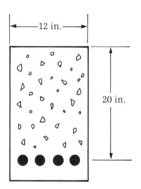

Figure P5-37

5-37 A simply supported beam carries a uniformly distributed load over its entire span as shown in Figure P5-37a. The beam is fabricated by gluing three timbers together to form the cross section shown in Figure P5-37b. Determine the maximum permissible load intensity (q lb/ft) that the beam can carry if the shearing stresses in the glue and in the wood are not to exceed 100 psi and 120 psi, respectively, and the bending stress is not to exceed 1800 psi.

5-16 SUMMARY

In this chapter, we described procedures used to establish shear and moment equations for straight beams. We obtained shear and moment diagrams by plotting these shear and moment equations and by the summation procedure. Formulas pertinent to the construction of shear and moment diagrams by the summation procedure are

$$\left.\begin{array}{l} \Delta V = P \\ \Delta M = 0 \end{array}\right\} \quad \text{across a concentrated force}$$

$$\left.\begin{array}{l} \Delta V = 0 \\ \Delta M = C \end{array}\right\} \quad \text{across a couple}$$

$$\left.\begin{array}{l} \Delta V = \text{the area under the load} \\ \qquad \text{intensity diagram between} \\ \qquad \text{appropriate sections} \\ \Delta M = \text{the area under the shear} \\ \qquad \text{diagram between} \\ \qquad \text{appropriate sections} \end{array}\right\} \quad \begin{array}{l}\text{under distributed} \\ \text{loads}\end{array}$$

Formulas required to calculate stresses in elastic beams are

One Material:

$$\begin{cases} \sigma = -\dfrac{My}{I} & \text{(normal stress)} \\[2mm] \tau = \dfrac{q}{b} & \text{with } q = \dfrac{VQ}{I} \quad \text{(shearing stress)} \end{cases}$$

Two Materials:

$$\sigma_y = \frac{E}{E^*}\,\sigma_y^* \qquad \text{where } \sigma_y^* = -\frac{My}{I^*}$$

I^* = centroidal moment of inertia of the transformed section

y = distance measure from the centroidal axis of the transformed section

E^* = modulus of elasticity of the material of the transformed section

E = modulus of elasticity of the material corresponding to the value of y in the original section

We developed procedures for the determination of partially plastic and fully plastic bending moments for elastic, perfectly plastic materials and elastic, linear hardening materials. Finally, we developed procedures for the determination of the shear center for a thin-wall cross section.

PROBLEMS / CHAPTER 5

5-38–5-41 Calculate the reactions at the supports for the beams shown in Figures P5-38–P5-41. Also calculate the internal reactions at the sections designated as aa or bb.

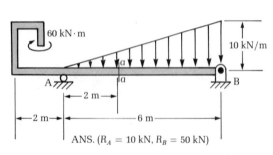

ANS. $(R_A = 10 \text{ kN}, R_B = 50 \text{ kN})$

Figure P5-38

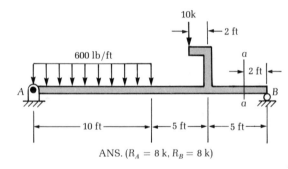

ANS. $(R_A = 8 \text{ k}, R_B = 8 \text{ k})$

Figure P5-39

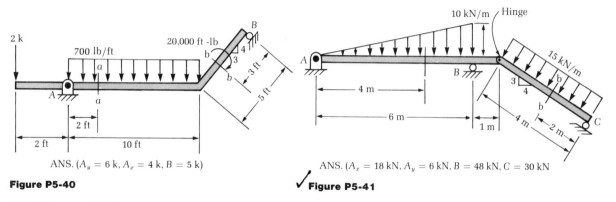

ANS. $(A_y = 6$ k, $A_x = 4$ k, $B = 5$ k)

Figure P5-40

ANS. $(A_x = 18$ kN, $A_y = 6$ kN, $B = 48$ kN, $C = 30$ kN$)$

Figure P5-41

5-42–5-49 Establish general expressions for the transverse shearing force V and the bending moment M for the beams shown in Figures P5-42–P5-49. Use a coordinate system with the origin located at the left end of each beam. Construct shear and moment diagrams by plotting the shear and moment equations.

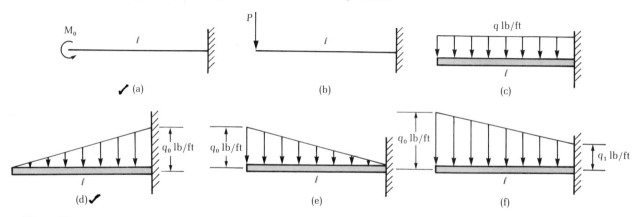

Figure P5-42

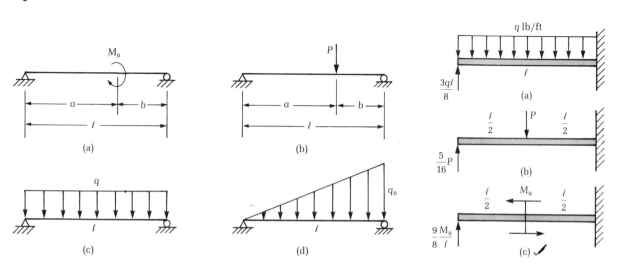

Figure P5-43

Figure P5-44

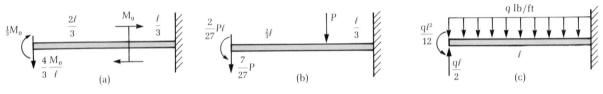

Figure P5-45

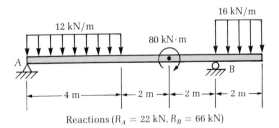

Figure P5-46

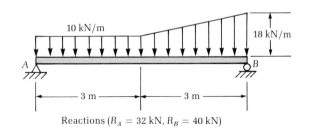

Figure P5-47

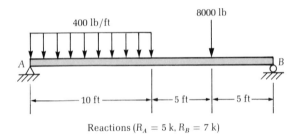

Figure P5-48

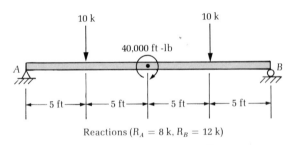

Figure P5-49

5-50–5-57 Construct shear and moment diagrams for the beams shown in Figures P5-50–P5-57 using the summation process. Indicate the mathematical nature of the various segments in each diagram and, where appropriate, verify the concavity of each segment.

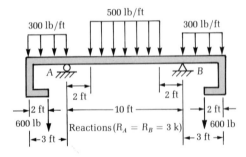

Figure P5-50

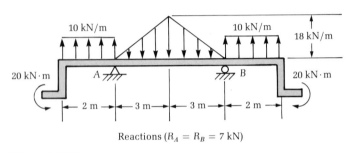

Figure P5-51

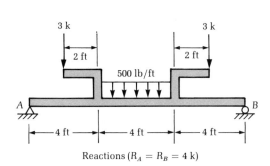

Reactions ($R_A = R_B = 4$ k)

Figure P5-52

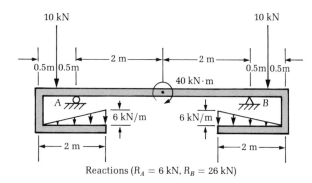

Reactions ($R_A = 6$ kN, $R_B = 26$ kN)

Figure P5-53

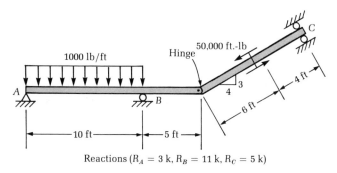

Reactions ($R_A = 3$ k, $R_B = 11$ k, $R_C = 5$ k)

Figure P5-54

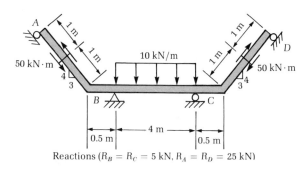

Reactions ($R_B = R_C = 5$ kN, $R_A = R_D = 25$ kN)

Figure P5-55

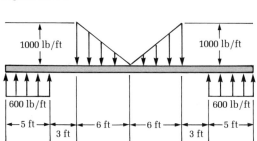

Figure P5-56

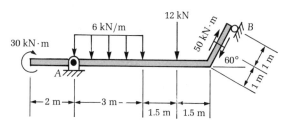

Figure P5-57

5-58 The lateral stiffness of a W 10×26 section is enhanced by riveting a C 12×25 to one of its flanges as shown in Figure P5-58. Determine (a) the moment of inertia of the composite section with respect to the centroidal axis, (b) the section modulus ($S = I_{NA}/c$), and (c) the maximum tensile and compressive stresses if the bending moment resisted by the section is $-50{,}000$ ft-lb ($I_{NA} = 224$ in^4, 8330 psi(T), 20 ksi(C)).

5-59 Two sections of 2-in. standard steel pipe are connected by a $\frac{1}{4}$-in. steel plate by $\frac{1}{8}$-in. fillet welds as shown in Figure P5-59. Determine the moment of inertia of the cross section relative to its centroidal axis and, subsequently, calculate the maximum tensile and compressive stresses due to a bending moment of 160,000 in.-lb.

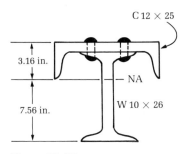

Figure P5-58

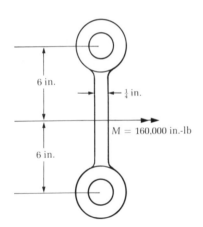

Figure P5-59

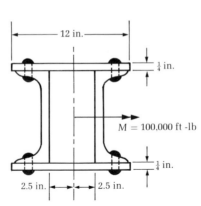

Figure P5-60

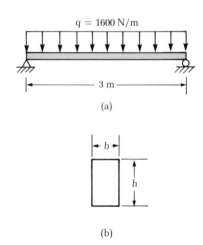

Figure P5-61

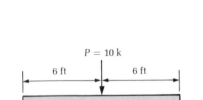

Figure P5-62

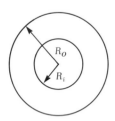

Figure P5-63

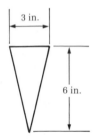

Figure P5-64

5-60 The composite steel section shown in Figure P5-60 consists of two C 12 × 25 and two 12-in. × $\frac{1}{4}$-in. plates. Determine (a) the centroidal moment of inertia of the section, (b) the section modulus, and (c) the maximum tensile and compressive stresses if the section resists a moment of 100,000 ft-lb.

5-61 A simply supported beam with a rectangular cross section carries a uniformly distributed load of 1600 N/m on a span of 3 m as shown in Figure P5-61a. The depth h of the cross section is to be twice the width b. If the fiber stress is not to exceed 21 MPa at any point in the beam, determine the minimum cross-sectional dimensions.

5-62 Select a W section of least weight to act as a simply supported beam with a concentrated load at its center as shown in Figure P5-62. The bending stress is not to exceed 18 ksi.

5-63 A beam with the cross-sectional dimensions shown in Figure P5-63 is subjected to a bending moment that causes a plastic hinge to form. Show that the ratio of fully plastic moment to the fully elastic moment is

$$M_{FP}/M_{FE} = \frac{16}{3\pi} \frac{R_O(R_O^3 - R_i^3)}{(R_O^4 - R_i^4)}$$

5-64 The cross section of a beam is a triangle with the dimensions shown in Figure P5-64. Determine the ratio of the fully plastic moment to the fully elastic moment.

5-65 The cross section of a beam has the shape and dimensions shown in Figure P5-65a. The stress-strain diagram for the material of the beam is shown in Figure P5-65b. Determine (a) the bending moment that will cause yielding to penetrate throughout the narrow portion, (b) the stress distribution corresponding to this moment, and (c) the residual stress distribution that results when this moment is released.

5-66 Determine the ratio M_{FP}/M_{FE} for the cross section shown in Figure P5-66. Determine the residual stress distribution that results when the fully plastic moment is released.

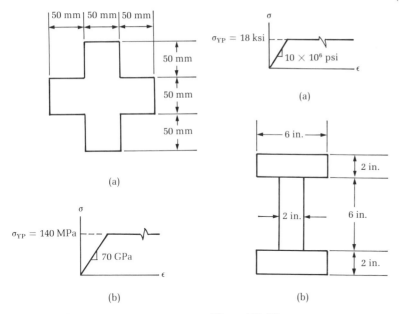

Figure P5-65

Figure P5-66

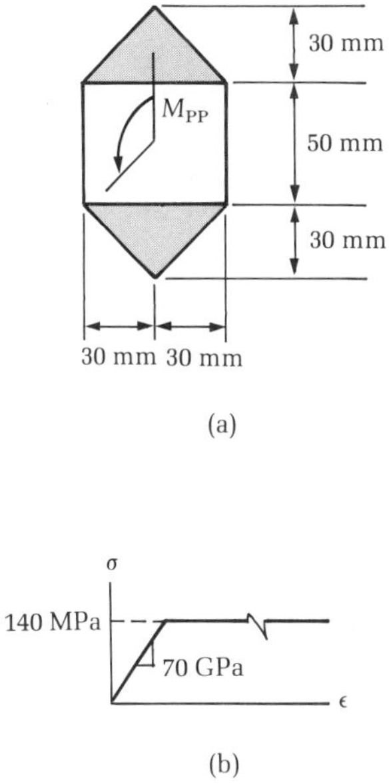

Figure P5-67

5-67 A partially plastic bending moment causes yielding of the cross section shown in Figure P5-67 to penetrate to a depth of 30 mm. Determine the magnitude of M_{PP}. Determine the stress distribution corresponding to M_{PP} and, subsequently, plot the residual stress distribution.

5-68 A beam with a rectangular cross section is subjected to a moment M_{PP} that causes inelastic strains to the extent shown by the shaded areas in Figure P5-68a. The material from which the beam is made has the stress-strain diagram shown in Figure P5-68b. Determine the partially plastic moment that corresponds to the indicated inelastic strains. Determine the residual stress distribution that results when this moment is released.

5-69 A beam is made from a nonlinear elastic material with a stress-strain diagram like the one shown in Figure P5-69b. For a rectangular cross section with the dimensions shown in Figure P5-69, show that $M = [2bc^2/(n + 2)]\sigma_{max}$, where σ_{max} is the maximum stress that acts on the cross section.

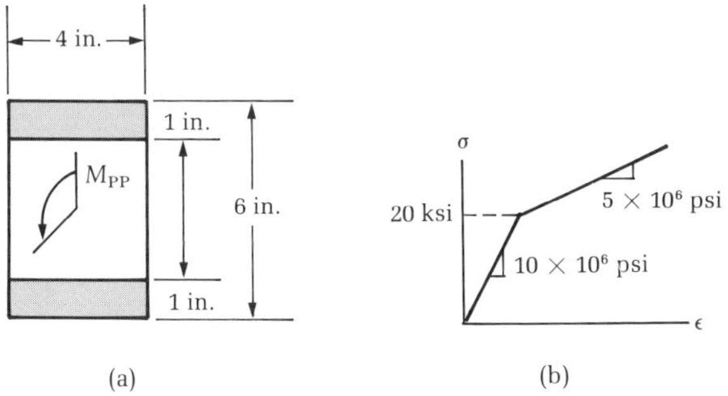

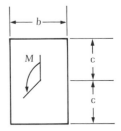

(a)

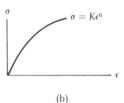

(b)

Figure P5-69

Figure P5-68

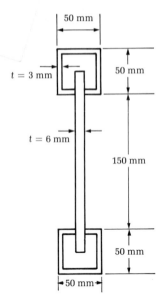

Figure P5-70

5-70 A beam is fabricated from two sections of 50-mm × 50-mm × 3-mm steel pipe and a 200-mm × 6-mm steel plate as shown in Figure P5-70. Calculate the maximum permissible transverse shearing force that the beam can resist if 3 mm fillet welds with a capacity of 200 kN/m are used. Calculate the shearing stress at the centroidal axis due to this shear force.

5-71 The wooden cantilever beam shown in Figure P5-71a is fabricated by gluing three timbers together to form the cross section shown in Figure P5-71b. Determine the maximum permissible load P that the beam can resist if the shearing stresses in the glue and in the wood are not to exceed 120 psi and 165 psi, respectively, and the bending stress is not to exceed 1800 psi.

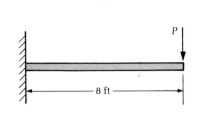

(a)

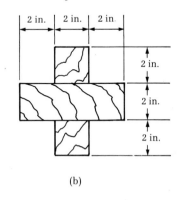

(b)

Figure P5-71

5-72 The simply supported beam shown in Figure P5-72a is fabricated from four 4-in. × 1-in. planks to form the cross section shown in Figure P5-72b. Determine the maximum permissible force P acting at midspan that the beam can resist if the shearing stress in the wood is not to exceed 165 psi, the bending stress is not to exceed 2000 psi, and the nails are rated at 225 lb in direct shear. The nails are spaced uniformly on 5-in. centers.

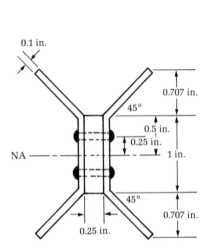

Figure P5-73

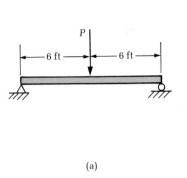

(a)

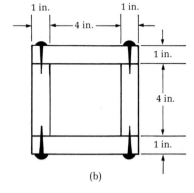

(b)

Figure P5-72

5-73 A 1-in. × ¼-in. strip of sheet aluminum is used as a spacer for two sheet aluminum channels to form the cross section of a beam as shown in Figure P5-73. If the beam resists a maximum transverse shearing force of 8000 lb, determine the maximum shearing stress in the beam. What shearing force must each rivet carry if the spacing is 12 in. on center?

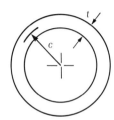

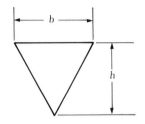

Figure P5-74   **Figure P5-75**   **Figure P5-76**

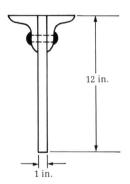

Figure P5-77

5-74 A beam has a circular cross section as shown in Figure P5-74. Show that the maximum shearing stress on the cross section is $\tau_{max} = (4/3)(V/A)$, where V is the transverse shearing force that acts on the cross section and $A = \pi c^2$.

5-75 The thin-wall circular tube shown in Figure P5-75 is to be used as a straight beam. Show that the maximum shearing stress that acts on the cross section is $\tau_{max} = 2(V/A)$, where V is the transverse shearing force that acts on the cross section and $A = 2\pi ct$. Note that $I \approx \pi c^3 t$.

5-76 Show that the maximum shearing stress in the triangular cross section shown in Figure P5-76 occurs at the level $h/2$. If $h = 100$ mm and $b = 75$ mm, determine the maximum shearing stress on the cross section when $V = 2000$ N.

5-77 A beam is fabricated by riveting two L 4 × 4 × $\frac{1}{2}$ angles to a 12-in. × 1-in. steel plate as shown in Figure P5-77. If $\frac{1}{2}$-in. diameter rivets with a transverse shearing capacity of 3000 lb are used, calculate the maximum permissible spacing of the rivets. The maximum transverse shearing force that the beam must resist is 6200 lb.

5-78 Determine the x coordinate of the shear center for the thin section of Figure P5-78. The thickness is uniform throughout.

5-79 Figure P5-79 depicts a model of a beam cross section that consists of a very thin semicircular tube and two relative stiff edge stiffeners, such as channels or bars. Assume that the thin curved portion does not carry any of the bending stress; that is, it resists a shear flow of constant magnitude. Show that the x coordinate of the shear center is located at $e = \pi R/2$.

5-80 A simply supported beam with the composite cross section shown in Figure P5-80 supports a concentrated force of 10k at the center of a 10-ft span. The cross section consists of a 9-in. × 18-in. oak timber with two 18-in. × $\frac{1}{4}$-in.

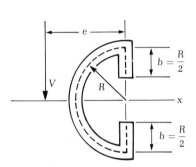

Figure P5-78

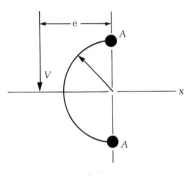

Figure P5-79

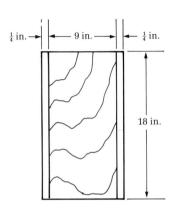

Figure P5-80

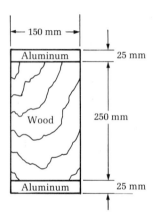

Figure P5-81

aluminum side plates. The plates are rigidly attached to the timber so that the three structural elements act as a unit. Calculate the maximum tensile and compressive stresses in the aluminum and in the wood ($E_{AL} = 10 \times 10^6$ psi and $E_{WOOD} = 2 \times 10^6$ psi; $\sigma_{AL} = 2416$ psi, $\sigma_{OAK} = 433$ psi).

5-81 A beam is fabricated from a 150-mm × 250-mm oak timber and two 150-mm × 25-mm aluminum face plates as shown in Figure P5-81. If the fiber stresses in the aluminum and in the wood are not to exceed 210 MPa and 42 MPa, respectively, calculate the maximum moment that the section can resist ($E_{AL} = 70$ GPa and $E_{OAK} = 14$ GPa).

5-82 A beam is fabricated from two 150-mm × 50-mm oak timbers and two 350-mm × 10-mm aluminum plates as shown in Figure P5-82. The screws used to attach the plates to the oak timbers are spaced uniformly at 150 mm and can resist 6000 N in direct shear. Determine the maximum transverse shearing force that the beam can support ($E_{AL} = 70$ GPa and $E_{OAK} = 14$ GPa).

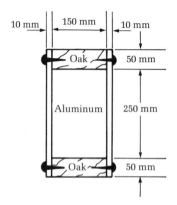

Figure P5-82

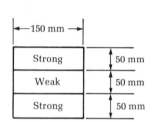

Figure P5-83

5-83 A laminated beam has the cross section shown in Figure P5-83. The direction of the grain for the top and bottom layers is parallel to the axis of the beam ($E = 14$ GPa) and the direction of the grain of the intermediate layer is perpendicular to the beam axis ($E = 3.5$ GPa). The stresses at the proportional limit parallel and perpendicular to the grain are 8.4 MPa and 2.1 MPa, respectively. Determine the maximum permissible uniform load q (N/m) that a simply supported beam of span 8 m can safely carry over its entire length.

Compound Stress

6-1 Introduction

Figure 6-1a shows the components of a force-couple system that is statically equivalent to a distribution of stress at an arbitrary section of a three-dimensional solid. The origin of the xyz coordinate system is usually taken at the centroid because this choice results in the least complicated formulas connecting the stress components with the components of the statically equivalent force and couple.

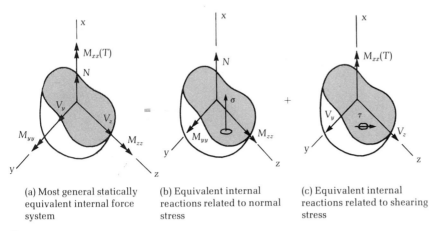

(a) Most general statically equivalent internal force system

(b) Equivalent internal reactions related to normal stress

(c) Equivalent internal reactions related to shearing stress

Figure 6-1

Fundamentally, there are two types of stress: normal stress and shearing stress. Figure 6-1b depicts the internal reactions associated with normal stress, and Figure 6-1c depicts the internal reactions associated with shearing stress. In general, the normal force N and the bending moments M_{yy} and M_{zz} are associated with a normal stress.

The direct shearing forces V_y and V_z and the twisting moment $M_{xx}(T)$ are associated with a shearing stress.

In the preceding chapters, we established formulas from which we can calculate a stress distribution due to a single internal reaction. In practice, it frequently happens that these reactions occur in various combinations. This situation gives rise to an important question: How can the normal stress and/or the shearing stress at a point on a section be calculated when the internal reactions occur in combination? The question is usually resolved by invoking the **superposition principle**. Application of the superposition principle is justified through the observation that each stress formula developed in this text shows a linear relationship between the stress (normal or shearing) and the internal reaction that corresponds to it. This condition is precisely the one required for the superposition principle to be applicable to compound stress calculations.

Recall that in the preceding chapters we presented the derivation for, and the application of, several fundamental stress formulas such as the normal stress due to axial force ($\sigma = N/A$), the normal stress due to bending ($\sigma = (My/I)$), the shearing stress due to twisting of circular shafts ($\tau = T\rho/J$), and the shearing stress due to a transverse shearing force in a beam ($\tau = (VQ/Ib)$). In this chapter, we are concerned with the stress formulas just summarized.

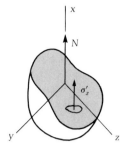

(a) Normal stress due to a normal force

6-2 Normal Stress

The superposition principle applied to the normal stress produced by N, M_{yy}, and M_{zz} asserts that the normal stress acting on an infinitesimal element of area due to N, M_{yy}, and M_{zz} acting simultaneously is equal to the **algebraic sum** of the normal stresses acting on the same element of area due to N, M_{yy}, and M_{zz} acting separately.

Figures 6-2a through 6-2c depict the normal stresses σ_x', σ_x'', and σ_x''' due to N, M_{yy}, and M_{zz}, respectively. According to the superposition

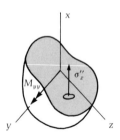

(b) Normal stress due to a bending moment about the y axis

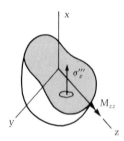

(c) Normal stress due to a bending moment about the z axis

Figure 6-2

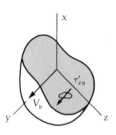

(a) Shearing stress due to a direct shearing force V_y

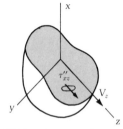

(b) Shearing stress due to a direct shearing force V_z

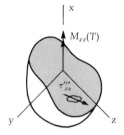

(c) Shearing stress due to a twisting moment $M_{xx}(T)$

Figure 6-3

principle, the normal stress acting on an infinitesimal element of area is

$$\sigma_x = \sigma_x' + \sigma_x'' + \sigma_x'''$$ (6-1)

where σ_x', σ_x'', and σ_x''' are calculated from appropriate formulas established for separate internal reactions. Typical computations of compounded normal stresses are illustrated in the following examples. Superposition of shearing stresses is explained in Section 6-3.

EXAMPLE 6-1

Determine the maximum tensile and compressive stresses acting on the section AA of the structural member shown in Figure 6-4a. The moment of inertia for the cross section is $I = 10.38$ in.^{4}.

SOLUTION

Figure 6-4c depicts a free-body diagram that isolates the portion of the member to the left of plane AA. Equilibrium requires that

$$\Sigma F_x = 0: \quad N = 3000 \text{ lb} \tag{a}$$

$$\Sigma F_y = 0: \quad V = 4000 \text{ lb} \tag{b}$$

$$\Sigma M_0 = 0: \quad M + 3000(36 - 2) - 4000(48) = 0,$$

$$M = 90,000 \text{ in.-lb} \tag{c}$$

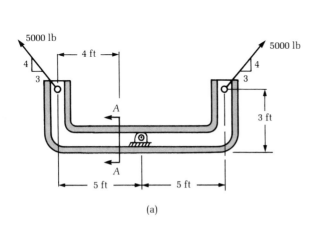

(a)

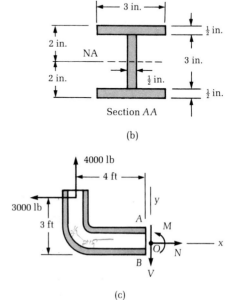

Section AA

(b)

(c)

Figure 6-4

The maximum tensile stress occurs at point B, and the maximum compressive stress occurs at point A. They are, respectively,

$$\sigma_B = \frac{MC}{I} + \frac{N}{A} = \frac{90{,}000(2)}{10.38} + \frac{3000}{4.5} = 17{,}341 + 667 \qquad \text{(d)}$$

$$= 18{,}008 \text{ psi}$$

$$\sigma_A = -\frac{MC}{I} + \frac{N}{A} = -\frac{90{,}000(2)}{10.38} + \frac{3000}{4.5} \qquad \text{(e)}$$

$$= -17{,}341 + 667 = -16{,}674 \text{ psi}$$

The stress distribution on the cross section AA due to the bending moment $M = 90{,}000$ in.-lb is shown in Figure 6-5a, the stress distribution due to the axial load $N = 3000$ lb is shown in Figure 6-5b, and the stress distribution due to the combined effects of M and N is shown in Figure 6-5c.

The neutral axis no longer coincides with the centroidal axis of the cross section as it does for pure bending. The neutral axis for the combined loading is located by making use of the linearity of the stress distribution. Therefore,

$$\frac{a}{16{,}674} = \frac{4 - a}{18{,}008} \qquad \text{or} \qquad a = 1.923 \text{ in.} \qquad \text{(f)}$$

measured from the top of the cross section.

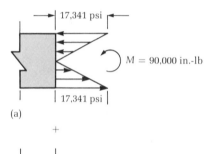

(a)

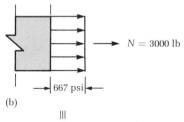

(b)

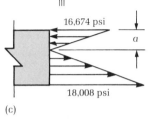

(c)

Figure 6-5

EXAMPLE 6-2

A short compression block is loaded eccentrically as shown in Figure 6-6a. Write an algebraic equation expressing the normal stress at an arbitrary point on a cross section perpendicular to its axis in terms of the load P, the eccentricities e_y and e_z, and the dimensions of the cross section. Determine the region of the cross section within which P must be applied if no tensile stresses are to develop on the cross section.

SOLUTION

The normal stress at a point (y, z) is obtained by superposition.

$$\sigma(y, z) = -\frac{M_{yy}}{I_{yy}} z + \frac{M_{zz}}{I_{zz}} y + \frac{N}{A} \qquad \text{(a)}$$

Here I_{yy} and I_{zz} are the moments of inertia of the cross section about the y and z axes, respectively. N, M_{yy}, and M_{zz} are shown in Figure 6-6b, which depicts a free-body diagram that isolates the required cross section. Note that the directions of N, M_{yy}, and M_{zz} are simply assumed. For these assumed directions, the normal stress at any point (y, z) of the cross section is given by Eq. (a). The sign of the first term on the right-hand side of Eq. (a) is opposite that of the remaining two terms

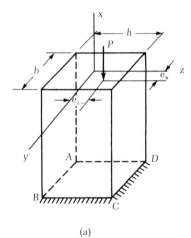

(a)

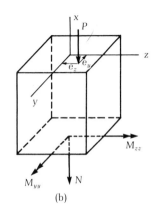

(b)

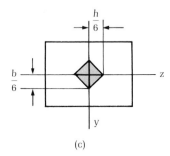

(c)

Figure 6-6

because, at a point (y, z) in the first quadrant of the yz plane, M_{yy} causes compressive stresses while M_{zz} and N cause tensile stresses.

Recall that the stresses are to be superposed algebraically. Equilibrium of the free-body diagram of Figure 6-6b requires that

$$\begin{array}{lll} \Sigma F_x = 0: & -N - P = 0, & N = -P \\ \Sigma M_{yy} = 0: & M_{yy} - Pe_z = 0, & M_{yy} = Pe_z \\ \Sigma M_{zz} = 0: & M_{zz} + Pe_y = 0, & M_{zz} = -Pe_y \end{array} \right\} \qquad (b)$$

These results indicate that the assumed directions shown in Figure 6-6b for N and M_{zz} are *not* correct for the loading considered. Either the directions for N and M_{zz} can be changed on the free-body diagram and Eq. (a) modified appropriately, or the negative signs can be retained and Eq. (a) used directly. The latter course is followed here.

The rectangular moments of inertia and cross-sectional area are

$$\begin{aligned} I_{yy} &= \tfrac{1}{12}bh^3 \\ I_{zz} &= \tfrac{1}{12}hb^3 \\ A &= bh \end{aligned} \right\} \qquad (c)$$

Consequently, Eq. (a) assumes the form

$$\sigma(y, z) = -\frac{P}{bh}\left\{\frac{12e_y}{b^2}y + \frac{12e_z}{h^2}z + 1\right\} \qquad (d)$$

The normal stress at any point on the cross section can be calculated from Eq. (d). Observe that y and z appearing in Eq. (d) must be given algebraic values. For example, to calculate the maximum tensile stress on the cross section, observe that it occurs at point A with coordinates $y = -b/2$ and $z = -h/2$. Therefore, the maximum tensile stress is given by the formula

$$(\sigma_{max})_{tension} = \frac{P}{bh}\left(\frac{6e_y}{b} + \frac{6e_z}{h} - 1\right) \qquad (e)$$

This value is the maximum positive value that σ can acquire. Clearly, $(6e_y/b) + (6e_z/h) > 1$ if tensile stresses occur on the cross section. If the eccentricities e_y and e_z are sufficiently small, Eq. (e) may result in a compressive stress at point A. Accordingly, no tensile stresses will develop on the cross section when

$$\frac{6}{b}e_y + \frac{6}{h}e_z \leq 1 \qquad (f)$$

For given b and h, the equality sign in Eq. (f) represents a straight line with e_y and e_z as independent variables. This line is shown in the first quadrant of Figure 6-6c.

If the force P is applied in the second quadrant, the maximum tensile stress occurs at point B. If no tensile stresses are to develop on the cross section,

$$-\frac{6}{b}e_y + \frac{6}{h}e_z \le 1 \tag{g}$$

The equality sign in Eq. (g) represents the straight line shown in the second quadrant of Figure 6-6c.

Similar expressions can be obtained for the third and fourth quadrants. In this way, we can identify the region of the cross section within which P must be applied so that no tensile stresses develop. This region is the shaded area shown in Figure 6-6c, and it is called the **kern** of the section.

EXAMPLE 6-3

A short rectangular block of material is loaded eccentrically as shown in Figure 6-7a. Write an algebraic equation expressing the normal stress on any section perpendicular to its axis. Determine the normal stresses at the corners A, B, C, and D and locate the intersection of the neutral axis with the edges of the cross section.

SOLUTION

Figure 6-7b depicts a free-body diagram that isolates the required cross section. Equilibrium requires that

$$\left.\begin{array}{lll} \Sigma F_x = 0: & -N - 240 = 0, & N = -240 \text{ kN} \\ \Sigma M_{yy} = 0: & M_{yy} - 240(0.05) = 0, & M_{yy} = 12 \text{ kN·m} \\ \Sigma M_{zz} = 0: & M_{zz} + 240(0.025) = 0, & M_{zz} = -6 \text{ kN·m} \end{array}\right\} \tag{a}$$

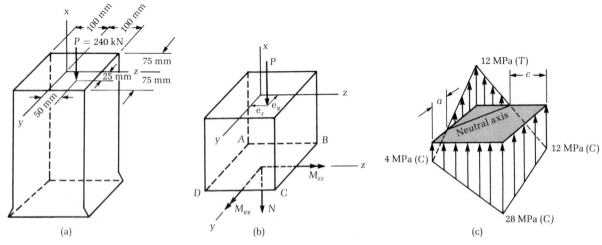

(a)   (b)   (c)

Figure 6-7

The normal stress at an arbitrary point (y, z) on the c

$$\sigma = -\frac{M_{yy}}{I_{yy}}z + \frac{M_{zz}}{I_{zz}}y + \frac{N}{A}$$

The moments of inertia are $I_{yy} = 0.10 \times 10^{-3}$ m^4 and 10^{-4} m^4, and the cross-sectional area is $A = 0.03$ m^2. Consequ....,, the normal stress on the cross section varies according to the formula

$$\sigma(y, z) = \frac{-12{,}000}{0.10 \times 10^{-3}}z + \frac{(-6000)}{0.5625 \times 10^{-4}}y - \frac{240{,}000}{0.03}$$

$$= (-120z - 106.7y - 8) \times 10^6 \text{ N/m}^2 \qquad (c)$$

The normal stresses at the corners are

$$
\left.
\begin{aligned}
\sigma_A &= -120(-0.1) - 106.7(-0.075) - 8.0 = 12.0 \text{ MPa} \\
\sigma_B &= -120(0.1) - 106.7(-0.075) - 8.0 = -12.0 \text{ MPa} \\
\sigma_C &= -120(0.1) - 106.7(0.075) - 8.0 = -28.0 \text{ MPa} \\
\sigma_D &= -120(-0.1) - 106.7(0.075) - 8.0 = -4.0 \text{ MPa}
\end{aligned}
\right\} \qquad (d)
$$

The intersections of the neutral axis with the edges of the cross section are found by making use of similar triangles; accordingly.

$$
\left.
\begin{aligned}
\frac{a}{4} &= \frac{150 - a}{12} \qquad \text{so that } a = 37.5 \text{ mm} \\[2mm]
\frac{c}{12} &= \frac{200 - c}{12} \qquad \text{so that } c = 100 \text{ mm}
\end{aligned}
\right\} \qquad (e)
$$

PROBLEMS / Section 6-2

6-1 Determine the maximum tensile and compressive stresses that act on the cross sections aa and bb of the beam shown in Figure P6-1.

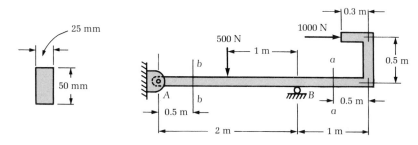

Figure P6-1

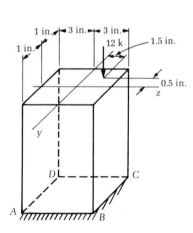

Figure P6-2

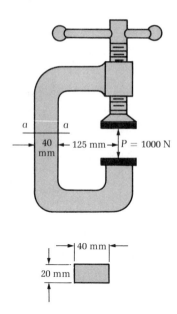

Figure P6-3

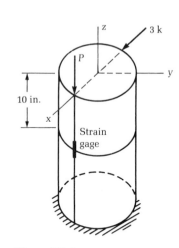

Figure P6-4

6-2 Calculate the normal stress at the points A, B, C, and D of the compression block shown in Figure P6-2. Sketch the profile of the stress distribution and locate the neutral axis.

6-3 Determine the maximum tensile stress that acts on section aa of the C clamp shown in Figure P6-3.

6-4 The axial strain at a point 10 in. below the upper surface of a 4-in. diameter aluminum compression block is measured by an electrical resistance strain gage to be -800μ in./in. Determine the magnitude of the force P, shown in Figure P6-4, that causes this strain.

6-3 Shearing Stress

The superposition principle applied to the shearing stress produced by V_y, V_z, and M_{xx} of Figure 6-1 asserts that the shearing stress acting on an infinitesimal element of area due to V_y, V_z, and M_{xx} acting simultaneously is equal to the **vector sum** of the shearing stresses acting on the same element of area due to V_y, V_z, and M_{xx} acting separately.

Figures 6-3a through 6-3c depict the shearing stresses τ'_{xy}, τ''_{xz}, and τ'''_{xs} due to V_y, V_z, and M_{xx}, respectively. According to the superposition principle, the shearing stress acting on an infinitesimal element of area is

$$\tau = \tau'_{xy} + \tau''_{xz} + \tau'''_{xs} \tag{6-2}$$

where τ'_{xy}, τ''_{xz}, and τ'''_{xs} are calculated from appropriate formulas that have been established for separate internal reactions. Note that the summation given in Eq. (6-2) is **vectorial**. This application of the super-position principle is different from the superposition of the normal stresses discussed in the previous section because the normal stresses all occur in the direction normal to the cross section. Thus an algebraic sum for that case is sufficient. The shearing stresses to be superposed, on the other hand, may occur in different directions. This fact is taken into account by summing the shear stresses vectorially. Technically, the separate shearing forces should be added vectorially. However, because the shearing stress components act on a common area, the various components of shearing stress can be added vectorially.

EXAMPLE 6-4

An angle bracket is riveted to a rigid support as shown in Figure 6-8a. The horizontal and vertical segments of the angle bracket sustain axial loads of 7850 lb and 15,700 lb, respectively, that pass through the centroid of the rivet areas. The diameter of each rivet is 1 in. Determine the shearing stress to which each rivet is subjected as a result of these loads.

SOLUTION

Figures 6-8b and 6-8c show the forces each rivet sustains due to Q and P, respectively. Notice that each shearing area carries a proportionate amount of the horizontal load Q and of the vertical load P. That is,

$$V_x = \frac{7850}{4} = 1963 \text{ lb} \quad \text{and} \quad V_y = \frac{15,700}{4} = 3925 \text{ lb}$$

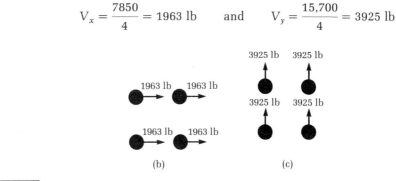

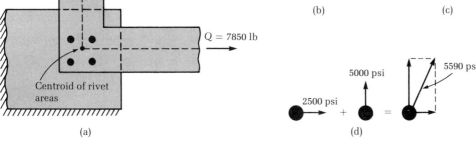

Figure 6-8

The cross-sectional area of one rivet is 0.785 in^2. Thus the components of shearing stress acting on a rivet due to Q and P are

$$\tau_x = \frac{1963}{0.785} = 2500 \text{ psi} \tag{a}$$

and

$$\tau_y = \frac{3925}{0.785} = 5000 \text{ psi} \tag{b}$$

The resultant shearing stress acting on a rivet is

$$\tau = \sqrt{(2{,}500)^2 + (5{,}000)^2} = 5{,}590 \text{ psi} \tag{c}$$

The stresses given in Eqs. (a), (b), and (c) are shown in Figure 6-8d.

PROBLEMS / Section 6-3

6-5 Calculate the shearing stress that acts on element A at the end of a horizontal diameter of the 2-in. diameter shaft shown in Figure P6-5.

6-6 Determine the shearing stress that acts on each of the $\frac{1}{2}$-in. diameter bolts of the structural connection shown in Figure P6-6. Assume that each load acts through the centroid of the bolt areas and that the loads divide equally among the bolts.

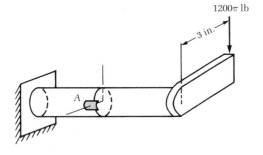

Figure P6-5

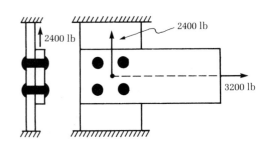

Figure P6-6

6-4 General States of Stress

The examples in Section 6-2 deal only with normal stresses, while the examples in Section 6-3 deal only with shearing stresses. General loading situations usually result in normal *and* shearing stresses acting on the same element of area. In such cases, the normal stress is calculated as described in Section 6-2, and the shearing stress is calculated as described in Section 6-3. The following examples illustrate the procedure.

EXAMPLE 6-5

A cantilever of length ℓ is loaded with a concentrated force P at its free end as shown in Figure 6-9a. The cross section of the beam and the point of application of P are shown in Figure 6-9b. Determine the maximum shearing stress at the fixed end of the cantilever. Assume that the ratio $h/b = 2$.

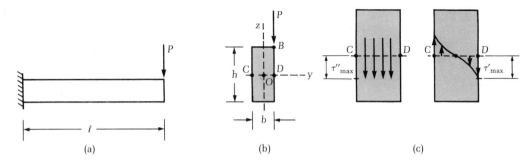

(a)   (b)   (c)

Figure 6-9

SOLUTION

The shear center of the rectangular cross section is at point O. Since the external load P does not pass through the shear center, it causes a twisting moment $T = Pb/2$. This moment causes a shearing stress in the cross section. It was shown in Chapter 4 that the maximum shearing stresses τ'_{max} due to the twisting moment occur at the midpoints C and D of the long sides of the cross section as shown in Figure 6-9c. Their common magnitude is

$$\tau'_{max} = \frac{T}{\alpha h b^2} \tag{a}$$

where $\alpha = 0.246$ as given in Table 4-1. Note that the shearing stresses at points C and D have opposite directions.

In addition to the shearing stress caused by the twisting moment, there is a shearing stress caused by the direct shear $V = P$. The maximum value of this second shearing stress τ''_{max} occurs at the level CD as shown in Figure 6-9c. Its magnitude is

$$\tau''_{max} = \frac{VQ_{max}}{Ib} \tag{b}$$

where I is the moment of inertia of the entire cross section about CD, and Q_{max} is the moment of the area above CD about the axis CD.

Accordingly,

$$I = \tfrac{1}{12}bh^3 \Bigg\}$$
$$Q_{max} = \tfrac{1}{8}bh^2 \Bigg\} \qquad \text{(c)}$$

Substituting Eq. (c) into Eq. (b) yields

$$\tau''_{max} = \frac{3}{2}\frac{P}{bh} \qquad \text{(d)}$$

Similarly, Eq. (a) yields

$$\tau'_{max} = \frac{1}{0.492}\frac{P}{bh} \qquad \text{(e)}$$

Since the two shearing stress components are arithmetically additive at point D, the maximum shearing stress is

$$\tau_{max} = \tau'_{max} + \tau''_{max} = 3.53\,\frac{P}{bh} \qquad \text{(f)}$$

EXAMPLE 6-6

A steel shaft 100 mm in diameter is supported in flexible bearings at the ends. Two pulleys, each 0.5 m in diameter, are keyed to the shaft. The pulleys carry belts whose tensions are shown in Figure 6-10a. Determine the state of stress at the point C on the surface of the shaft.

SOLUTION

The reactions of the flexible bearings on the shaft at points A and B are obtained as follows:

$$\Sigma M_{Ax} = 0: \quad 1(1.5) + 5(1.5) - 2B_y = 0 \quad \text{or} \quad B_y = 4.5 \text{ kN} \qquad \text{(a)}$$
$$\Sigma M_{Ay} = 0: \quad 5(0.5) + 1(0.5) - 2B_x = 0 \quad \text{or} \quad B_x = 1.5 \text{ kN} \qquad \text{(b)}$$
$$\Sigma M_{Bx} = 0: \quad 5(0.5) + 1(0.5) - 2A_y = 0 \quad \text{or} \quad A_y = 1.5 \text{ kN} \qquad \text{(c)}$$
$$\Sigma M_{By} = 0: \quad 5(1.5) + 1(1.5) - 2A_x = 0 \quad \text{or} \quad A_x = 4.5 \text{ kN} \qquad \text{(d)}$$

The internal reactions acting on a cross section through point C can be calculated from the free-body diagram shown in Figure 6-10b. Accordingly,

$$\begin{aligned} T &= -1000 \text{ N} \cdot \text{m} \\ M_x &= -1500 \text{ N} \cdot \text{m} \\ M_y &= 1500 \text{ N} \cdot \text{m} \\ V_x &= 1500 \text{ N} \\ V_y &= -1500 \text{ N} \end{aligned} \Bigg\} \qquad \text{(e)}$$

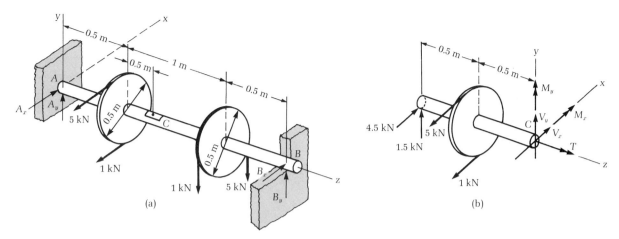

(a)

(b)

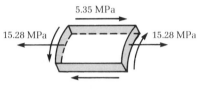

(c)

Figure 6-10

Note that, for the 100-mm diameter shaft, $I = (\pi/64)(0.10)^4 = 4.909 \times 10^{-6}$ m^4 and $J = 9.817 \times 10^{-6}$ m^4. Also note that, for this shaft,

$$Q = \frac{\pi}{2}(0.05)^2 \frac{4(0.05)}{3\pi} = 83.33 \times 10^{-6} \text{ m}^3$$

with respect to the diameter.

The normal stress at point C due to each reaction is:

Direct Axial Stress:

$$\sigma = \frac{N}{A} = 0$$

Bending About x Axis:

$$\sigma = \frac{M_x y_C}{I} = \frac{1500(0.05)}{4.909 \times 10^{-6}} = 15.28 \text{ MPa(C)}$$

Bending About y Axis:

$$\sigma = \frac{M_y x_C}{I} = 0, \ (x_C = 0)$$

(f)

The shearing stress at point C due to each reaction is:

Direct Shear Due to V_x:

$$\tau = \frac{V_x Q}{Ib} = \frac{1500(83.33 \times 10^{-6})}{4.909 \times 10^{-6}(0.10)} = 0.255 \text{ MPa}$$

Direct Shear Due to V_y:

$$\tau = \frac{V_y Q}{Ib} = 0, \ (Q = 0)$$

(g)

Shear Due to Twisting:

$$\tau = \frac{T(d/2)}{J} = \frac{1000(0.05)}{9.817 \times 10^{-6}} = 5.09 \text{ MPa}$$

The resultant state of stress at point A of the shaft is shown on the differential element in Figure 6-10c.

EXAMPLE 6-7

A 100-mm $\times$ 200-mm (full size) timber is used as a cantilever beam to carry the loads shown in Figure 6-11a. Determine (a) the normal stress acting at corner A of the section at the support and (b) the shearing stress at point B of the same section.

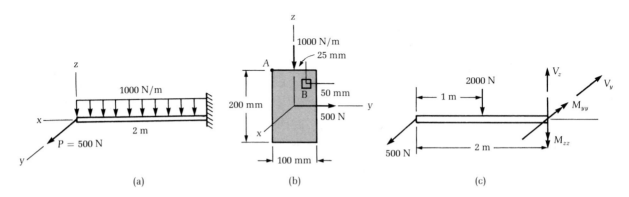

(a)   (b)   (c)

SOLUTION

The internal reactions at the support are obtained from the free-body diagram shown in Figure 6-11c. Accordingly,

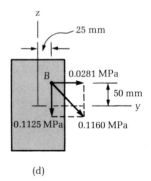

$$V_y = 500 \text{ N} \tag{a}$$

$$V_z = 2000 \text{ N} \tag{b}$$

$$M_{yy} = 2000 \text{ N} \cdot \text{m} \tag{c}$$

$$M_{zz} = 1000 \text{ N} \cdot \text{m} \tag{d}$$

(d)

Figure 6-11

The principal moments of inertia of the cross section are

$$I_{yy} = \tfrac{1}{12}(0.10)(0.20)^3 = 66.67 \times 10^{-6} \text{ m}^4 \tag{e}$$

and

$$I_{zz} = \tfrac{1}{12}(0.20)(0.10)^3 = 16.67 \times 10^{-6} \text{ m}^4 \tag{f}$$

(a) The normal stress at point A of the cross section at the support is

$$\sigma = \frac{M_{yy}C_z}{I_{yy}} + \frac{M_{zz}C_y}{I_{zz}} = \frac{2000(0.10)}{66.67 \times 10^{-6}} + \frac{1000(0.05)}{16.67 \times 10^{-6}}$$

$$= 6.0 \text{ MPa} \tag{g}$$

(b) The components of shearing stress at point B of the cross section at the support are

$$\tau_y = \frac{V_y Q_z}{I_{zz} b_z} = \frac{500(0.025 \times 0.2 \times 0.0375)}{16.67 \times 10^{-6}(0.2)} = 0.0281 \text{ MPa} \qquad \text{(h)}$$

and

$$\tau_z = \frac{V_z Q_y}{I_{yy} b_y} = \frac{2000(0.10 \times 0.05 \times 0.075)}{66.67 \times 10^{-6}(0.1)} = 0.1125 \text{ MPa} \qquad \text{(i)}$$

These stress components and their vector sum are shown in Figure 6-11d.

EXAMPLE 6-8

Determine the normal and shearing stresses at point A of the prismatic member shown in Figure 6-12a.

SOLUTION

To calculate the internal reactions at the appropriate section, a free-body diagram is constructed as shown in Figure 6-11b. Force and moment equilibrium requirements yield:

$$\Sigma F_x = 0: \quad N = 1000 \text{ lb} \qquad \text{(a)}$$

$$\Sigma F_y = 0: \quad V_y = 2000 \text{ lb} \qquad \text{(b)}$$

$$\Sigma F_z = 0: \quad V_z = 3000 \text{ lb} \qquad \text{(c)}$$

$$\Sigma M_{xx} = 0: \quad M_{xx} + 2000(9) - 3000(17) = 0$$
$$\text{or} \quad M_{xx} = 33{,}000 \text{ in.-lb} \qquad \text{(d)}$$

$$\Sigma M_{yy} = 0: \quad M_{yy} + 3000(8) - 1000(9) = 0$$
$$\text{or} \quad M_{yy} = -15{,}000 \text{ in.-lb} \qquad \text{(e)}$$

$$\Sigma M_{zz} = 0: \quad M_{zz} + 2000(8) - 1000(17) = 0$$
$$\text{or} \quad M_{yy} = 1000 \text{ in.-lb} \qquad \text{(f)}$$

The normal stress at point A is determined according to the superposition principle. Thus the normal stress at point A due to all internal reactions acting simultaneously is equal to the algebraic sum of the normal stresses at point A due to each reaction acting separately. The stress distribution for each reaction is shown in Figures 6-12c through 6-12h. Figure 6-12c shows only the stress at point A due to the twisting moment.

The normal stress at point A is

$$\sigma_A = -5625 + 0 + 125 = 5500 \text{ psi}(C) \qquad \text{(g)}$$

The shearing stress at point A possesses two components,

$$\tau_{xy} = \frac{M_{xx}}{\alpha b c^2} = \frac{33{,}000}{0.246(4)(2)^2} = 8384 \text{ psi}$$

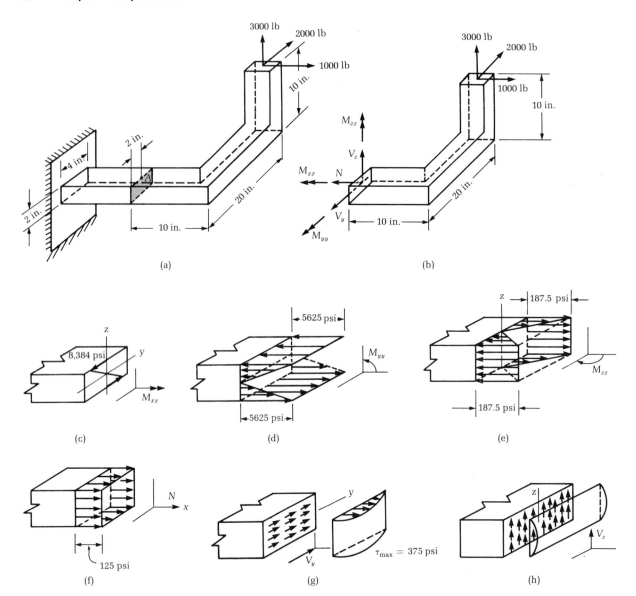

Figure 6-12

along the negative y axis as shown in Figure 6-12c and

$$\tau_{xy} = \frac{V_y Q}{Ib} = \frac{3}{2}\frac{V_y}{A} = \frac{3}{2}\frac{2000}{8} = 375 \text{ psi}$$

along the positive y axis as shown in Figure 6-12g. The shearing stress τ_{xz} due to the direct shearing force V_z is zero at point A as shown in Figure 6-12h. Therefore,

$$\tau_A = -8384 + 375 = -8009 \text{ psi} \qquad\qquad (h)$$

A differential element at point A of the member is shown pictorially in Figure 6-13a. Figure 6-13b illustrates a differential volume element surrounding point A and the stresses that act at that point. Figure 6-13c depicts the customary two-dimensional representation of the three-dimensional volume element when stresses are confined to a single plane. A state of stress of this type is called a **biaxial state of stress**.

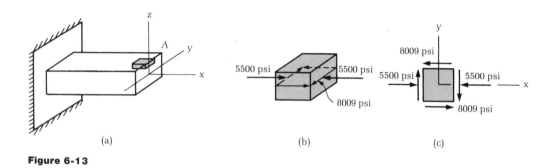

(a) (b) (c)

Figure 6-13

PROBLEMS / Section 6-4

6-7 Determine the components of stress that act on an element of material located at the end of a horizontal diameter (point A) at the fixed end of the member shown in Figure P6-7.

6-8 Determine the magnitudes of the stress components that act on an element of material at point A of the 2-in. diameter bar shown in Figure P6-8.

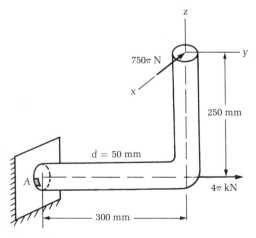

Figure P6-7

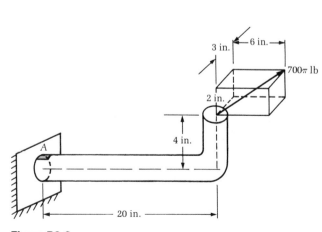

Figure P6-8

6-9 Determine the components of stress on an element of material of the 75-mm diameter shaft located midway between the pulleys in Figure P6-9.

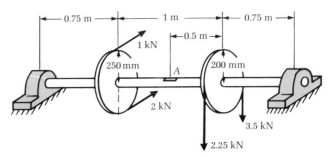

Figure P6-9

6-5   Thin-Walled Pressure Vessels

A pressure vessel can be thought of as being a closed surface in three-dimensional space with a finite thickness t. Customarily, the closed surface is a surface of revolution; that is, it is a surface obtained by rotating a plane curve, called the **generating curve**, about a fixed axis called the **symmetry axis**. For example, a right circular cylinder is obtained by rotating the straight line $x = a$ of Figure 6-14a about the y axis. A spherical surface is obtained by rotating the circular arc of Figure 6-14b about the y axis.

When the ratio of the wall thickness to the radius of a cylindrical or spherical pressure vessel is less than about $\frac{1}{10}$, the pressure vessel is said to be thin. The stress distribution over the thickness of a thin-walled pressure vessel is essentially constant. Consequently, a pressure vessel behaves like a thin membrane with a small thickness; that is, no bending of the walls occurs.

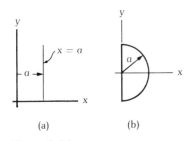

Figure 6-14

Thin-walled cylinders.   Figure 6-15a shows a right circular cylinder of thickness t and internal radius R subjected to an internal pressure p in excess of the external pressure.

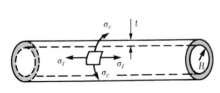

(a) Right circular cylinder

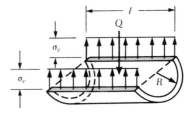

(b) Free-body diagram required to determine the circumferential normal stress

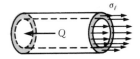

(c) Free-body diagram required to determine the longitudinal normal stress

Figure 6-15

Figure 6-15b shows a free-body diagram of a finite length of the cylinder. From equilibrium of forces,

$$Q = 2\sigma_c lt$$

where

$$Q = p2Rl$$

Thus

$$\sigma_c = \frac{pR}{t} \tag{6-3}$$

Note that the force Q developed by the internal pressure is simply the pressure times the projected area of the cylindrical segment onto the diametric plane. Equation (6-3) permits the calculation of the **circumferential** or the so-called **hoop** stress in a thin-walled cylinder.

Figure 6-15c shows a free-body diagram that can be used to calculate the longitudinal normal stress in a thin-walled cylinder. Axial force equilibrium gives

$$(2\pi Rt)\sigma_l = Q$$

where

$$Q = \pi R^2 p$$

Thus

$$\sigma_l = \frac{pR}{2t} \tag{6-4}$$

Observe that, for cylindrical pressure vessels,

$$\sigma_c = 2\sigma_l \tag{6-5}$$

The circumferential and longitudinal stresses are shown on a differential element on the surface of the cylinder in Figure 6-15a. Note that, because of the symmetry of the pressure distribution, there is no angular distortion of the element, and consequently the shearing stresses on this element are zero. Consequently, σ_c and σ_l are **principal stresses**, as we shall see in the next chapter.

Thin-walled spheres. Thin-walled spherical pressure vessels can be analyzed in a manner analogous to that used to analyze thin-walled cylindrical pressure vessels.

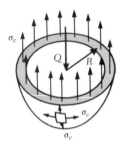

Figure 6-16

Figure 6-16 shows a portion of a spherical pressure vessel that has been obtained by cutting the sphere along a great circle. Denote the radius and thickness of the sphere by R and t, respectively.

Equilibrium of forces yields

$$(2\pi Rt)\sigma_c = Q$$

where

$$Q = \pi R^2 p$$

Thus

$$\sigma_c = \frac{pR}{2t} \tag{6-6}$$

Here again, the concept of projected area has been used. Now, cutting the spherical surface along any other great circle leads to the same free-body diagram which, in turn, leads to Eq. (6-6). We concluded that the normal stress in a spherical pressure vessel is the same in all directions. This situation is shown on a differential element of material at the surface of the spherical vessel shown in Figure 6-16.

The analysis given here shows that a sphere is an optimum shape for an internally pressurized closed vessel. The maximum normal stress in a cylindrical vessel is twice that of a spherical vessel for the same internal pressure and the same R/t ratio.

EXAMPLE 6-9

A cylindrical tank 5 ft in diameter is made from steel plate $\frac{3}{4}$ in. thick and is used to store a certain gas under pressure. Determine the maximum pressure the tank can resist if the allowable stress is 20,000 psi in tension.

SOLUTION

The maximum normal stress in the cylindrical pressure vessel is given by the formula

$$\sigma_c = \frac{pR}{t} \tag{a}$$

Since σ_c is the maximum normal stress in the cylinder, it cannot exceed 20,000 psi; consequently,

$$20,000 = \frac{p(30)}{\frac{3}{4}}$$

or

$$p_{max} = 500 \text{ psi} \tag{b}$$

Of course, the cylindrical tank has ends. They can be flat, hemi-spherical, or some other shape. Flat ends are apparently the least desirable because incompatible geometric deformations at the juncture are most pronounced in this case. This incompatibility is present for other ends also. Thus there will always be additional local stresses developed at the juncture. In the simple example given here, these local stresses were not taken into consideration.

EXAMPLE 6-10

Thin-walled cylindrical pressure vessels have been used by many investigators to study the accuracy with which various failure theories predict the onset of inelastic material behavior. Suppose that the cylinder is fixed at its left end and has applied to its free end twisting torque $T = 31,400$ in.-lb and a direct force $P = 3140$ lb as shown in Figure 6-17a. Calculate the state of stress at point A on the surface of the surface of the cylinder if the internal pressure is 50 psi.

SOLUTION

A free-body diagram of the portion of the cylinder to the right of a transverse plane through point A is shown in Figure 6-17b. Equilibrium

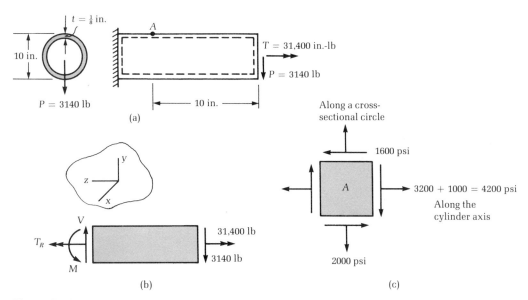

Figure 6-17

requirements are

$$\Sigma F_y = 0: \quad V = 3140 \text{ lb} \tag{a}$$

$$\Sigma M_{xx} = 0: \quad M = 31{,}400 \text{ in.-lb} \tag{b}$$

$$\Sigma M_{zz} = 0: \quad T_R = 31{,}400 \text{ in.-lb} \tag{c}$$

The moment of inertia of the annular cross section with respect to the x axis is

$$I = \frac{\pi}{4}(R_o^4 - R_i^4) = \frac{\pi}{4}(R_o + R_i)(R_o^2 + R_i^2)(R_o - R_i) \tag{d}$$

Now, since $R_i \approx R_o$, $R_i + R_o \approx 2R_o$. Also, $R_o - R_i = t$. Consequently, if the thickness to radius ratio is small (say about 1/10), then an acceptable approximation for rectangular moment of inertia is

$$I = \pi R^3 t \tag{e}$$

and since

$$J = 2I$$

$$J = 2\pi R^3 t \tag{f}$$

At point A on the cylinder,

Stress Due to Bending:

$$\sigma_B = \frac{31{,}400(5)}{\pi(5)^3(\frac{1}{8})} = 3200 \text{ psi}$$

Stress Due to Torsion:

$$\tau_B = \frac{31{,}400(5)}{2\pi(5)^3(\frac{1}{8})} = 1600 \text{ psi}$$

Stress Due to Pressure:

$$\begin{cases} \sigma_c = \dfrac{pR}{t} = \dfrac{50(5)}{\frac{1}{8}} = 2000 \text{ psi} \\[3mm] \sigma_l = \dfrac{pR}{2t} = \tfrac{1}{2}\sigma_c = 1000 \text{ psi} \end{cases}$$

Direct Shearing Stress:

$$\tau = \frac{VQ}{Ib} = 0 \text{ since } Q \equiv 0 \text{ for point } A$$

The state of stress at point A due to the three reactions and the internal pressure acting simultaneously is obtained via the super-position principle. This state of stress is shown in Figure 6-17c.

Note that the torsion formula, flexure formula, direct shearing stress formula, and the thin-walled pressure formulas are not valid in the vicinity of the ends of the cylinder. The end plates are rigid in their planes and prevent the cylinder from expanding uniformly along its length. As a consequence, bending stresses of considerable magnitude occur near the junctions of the cylindrical surface with the end plates. These bending stresses dissipate rapidly, and the membrane stresses calculated from the basic strength-of-materials formulas are accurate predictions at small distances from the ends of the cylinder. Localized bending stresses of the kind described are often referred to as edge effects or boundary layer effects.

PROBLEMS / Section 6-5

6-10 A cylindrical pressure vessel has a diameter of 20 in. and a wall thickness of 0.125 in. If the internal pressure is 100 psi, calculate the longitudinal and circumferential normal stresses in the cylinder.

6-11 A 10-m diameter spherical pressure vessel has a wall thickness $t = 10$ mm and is subjected to an internal pressure of 0.3 MPa. Calculate the maximum normal stress in the vessel.

6-12 A 6-m diameter cylindrical pressure vessel is to be made from a material whose ultimate tensile strength is 1200 MPa. If the cylinder is to be capable of containing a gas at a pressure of 0.35 MPa, determine the minimum thickness for the tank. A safety factor of 4 is required.

6-13 A water pipe operates under an internal pressure of 500 kN/m². It is constructed of wooden staves wrapped with steel hoops of mean diameter 1.2 m as shown in Figure P6-13. If the cross-sectional area of the steel hoops is 300 mm² and the allowable tensile strength for the steel is 140 MPa, determine the required spacing of the hoops. Assume that the steel governs the design.

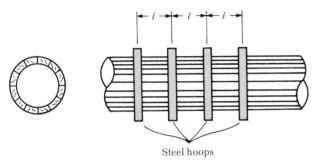

Steel hoops

Figure P6-13

6-6   Close-Coiled Helical Springs

An interesting and useful example of combined torsional and direct shearing stresses is a close-coiled helical spring. The spring is assumed to be constructed by winding a wire of circular cross section on a cylindrical mandrel. Consequently, we consider only helical springs whose coils are circular in cross section.

Let R denote the mean radius, or core radius, of the spring, and let d denote the diameter of the wire used to construct the spring. Let n represent the number of coils in the spring.

A free-body diagram of a portion of a close-coiled helical spring is shown in Figure 6-18a. The reactions at a cross section of a coil are shown in Figure 6-18a and 6-18b.

The direct shearing force P leads to the average shearing stress $\tau'_{ave} = P/A$, and the torque T leads to the torsional shearing stress $\tau'' = 16PR/\pi d^3$. These stresses augment one another at the innermost point on the coil cross section as shown in Figure 6-18c. Thus the maximum shearing stress in a close-coiled helical spring is

$$\tau_{max} = \frac{P}{A} + \frac{16PR}{\pi d^3} = \frac{16PR}{\pi d^3}\left(1 + \frac{d}{4R}\right) \tag{6-7}$$

The torsion formula $(\tau = TC/J)$ applies strictly to straight shafts. In using it to calculate the shearing stress due to twisting of a curved rod, an error is introduced. The error is small as long as the ratio d/R is small. However, as $d/2 \to R$, the term $d/4R$ becomes important. Equation 6-7 is replaced by the more accurate expression derived by

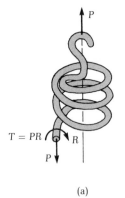

(a)

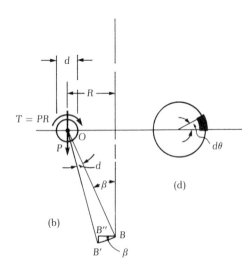

(b)

(d)

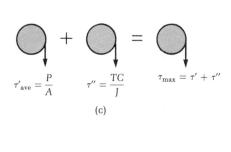

$\tau'_{ave} = \dfrac{P}{A}$ $\tau'' = \dfrac{TC}{J}$ $\tau_{max} = \tau' + \tau''$

(c)

Figure 6-18

A. M. Wahl[*],

$$\tau_{max} = \frac{16PR}{\pi d^3}\left(\frac{4m-1}{4m-4} + \frac{0.615}{m}\right) \tag{6-8}$$

where $m = 2R/d$.

The axial deflection of a close-coiled helical spring can be obtained in the following manner. Tentatively consider the spring to be rigid except for a differential element of length $R\,d\theta$ as shown in Figure 6-18d. Let B denote the bottom end of the spring in its undeformed configuration. Because of the deformation of the differential element of coil, point B will rotate about the center O of the coil cross section to point B' as depicted in Figure 6-18b. The vertical deflection of the spring due to the deformability of the differential coil element is

$$d\delta = B'B'' = BB'\sin\beta = BB'\frac{R}{a} = R\,d\varphi \tag{6-9}$$

where $d\varphi$ is the rotation of one end of the differential coil element with respect to its other end.

The total vertical deflection is the sum of the vertical deflections due to all differential coil elements. Consequently,

$$\delta = \int_0^{2\pi n} R\,\frac{T}{JG}\,R\,d\theta = \int_0^{2\pi n}\frac{PR^3}{JG}\,d\theta = \frac{PR^3}{JG}\,2\pi n$$

or

$$\delta = \frac{64nPR^3}{G\,d^4} \tag{6-10}$$

where n is the total number of coils. This equation gives the force-deflection relation for helical springs. It corresponds to the stress-strain relations (constitutive equations) of deformable bodies.

EXAMPLE 6-11

A close-coiled helical spring has the following properties: $R = 1.5$ in., $d = \frac{1}{8}$ in., $n = 20$ coils, and $G = 12 \times 10^6$ psi. If the shearing stress in the spring is not to exceed 12,000 psi, determine the maximum allowable load P for the spring and the corresponding maximum deflection.

* Wahl, A. M., 1944. *Mechanical Springs*, Penton Publishing Co., Cleveland, Ohio.

SOLUTION

The maximum shearing stress in the spring is not to exceed 12,000 psi; thus the maximum allowable load P is given by Eq. (6-7). Consequently,

$$12{,}000 = \frac{16P(1.5)}{\pi(0.125)^3}\left(1 + \frac{0.125}{4(1.5)}\right) \tag{a}$$

or

$$P = 3.0 \text{ lb}$$

The corresponding maximum deflection is given by Eq. (6-10), or

$$\delta_{max} = \frac{64(20)3(1.5)^3}{(0.125)^4 12 \times 10^6} = 4.43 \text{ in.} \tag{b}$$

EXAMPLE 6-12

Two close-coiled helical springs are identical except in the number coils. Spring 1 has 20 coils and spring 2 has 30 coils. These springs are used in the manner shown in Figure 6-19a. Determine the number of coils of a single spring of like construction equivalent to the two springs operating in parallel.

SOLUTION

The spring constants for each of the original springs are obtained from the force-deflection relation Eq. (6-10). They are

$$\left. \begin{aligned} k_1 &= \frac{P_1}{\delta_1} = \frac{d^4 G}{64 n_1 R^3} \\ k_2 &= \frac{P_2}{\delta_2} = \frac{d^4 G}{64 n_2 R^3} \end{aligned} \right\} \tag{a}$$

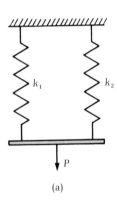

From the free-body diagram shown in Figure 6-19b,

$$P = P_1 + P_2 \tag{b}$$

This equilibrium equation is not sufficient to determine the two unknowns P_1 and P_2. Therefore, the problem is statically indeterminate, and we must use a compatibility condition. Assume that the plank shown in Figure 6-19a remains horizontal after load P is applied. Geometric compatibility requires that the deflections δ_1 and δ_2 of the springs must be equal. Let

$$\delta_1 = \delta_2 = \delta$$

Now, $P_1 = k_1 \delta$ and $P_2 = k_2 \delta$ so that

$$k_{eq}\delta = k_1 \delta + k_2 \delta$$

or

$$k_{eq} = k_1 + k_2 \tag{c}$$

Here k_{eq} is the spring constant for the equivalent spring.

(a)

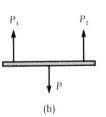

(b)

Figure 6-19

Substituting from Eqs. (a),

$$k_{eq} = \frac{d^4G}{64R^3}\left(\frac{1}{n_1} + \frac{1}{n_2}\right) = \frac{d^4G}{64R^3}\frac{n_1 + n_2}{n_1 n_2} = \frac{d^4G}{64R^3}\frac{1}{n_{eq}} \tag{d}$$

Thus a formula for the number of coils in the equivalent spring is

$$n_{eq} = \frac{n_1 + n_2}{n_1 n_2} \tag{e}$$

Consequently, for the two springs considered,

$$n_{eq} = \frac{20(30)}{20 + 30} = \frac{600}{50} = 12 \text{ coils} \tag{f}$$

EXAMPLE 6-13

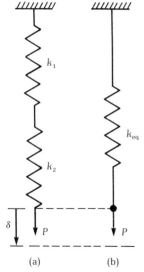

(a)   (b)

Figure 6-20

Suppose the two springs described in Example 6-12 are to be used in series as shown in Figure 6-20. Determine the number of coils required of a single equivalent spring.

SOLUTION

The single spring will be equivalent to the two springs in series provided that the same force P produces the same deflection δ. Thus the compatibility condition becomes

$$\delta = \delta_1 + \delta_2 \tag{a}$$

Now $\delta = P/k_{eq}$, $\delta_1 = P/k_1$, and $\delta_2 = P/k_2$ so that Eq. (a) yields

$$\frac{1}{k_{eq}} = \frac{1}{k_1} + \frac{1}{k_2} \tag{b}$$

Substituting from Eqs. (a) of Example 6-12 yields

$$\frac{1}{k_{eq}} = \frac{64R^3}{d^4G}(n_1 + n_2) \tag{c}$$

Accordingly,

$$k_{eq} = \frac{d^4G}{64R^3}\frac{1}{n_1 + n_2} \tag{d}$$

so that

$$n_{eq} = n_1 + n_2 \tag{e}$$

Consequently, for the two springs considered,

$$n_{eq} = 20 + 30 = 50 \text{ coils} \tag{f}$$

PROBLEMS / Section 6-6

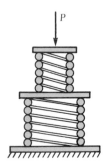

Figure P6-14

6-14 Two springs are compressed between two parallel plates as shown in Figure P6-14. Each spring is made from steel ($G = 84$ GPa) wire 25 mm in diameter and has 75 coils. Determine the maximum compressive force that can be applied to the parallel plates if the shearing stress in the springs is not to exceed 140 MPa. The mean diameters of the springs are 140 mm and 200 mm, respectively.

6-15 A close-coiled helical tension spring is made from 1-mm diameter steel wire having a torsional strength of 700 MPa. The spring has a mean diameter of 10 mm, 40 active coils, and hooked ends. A prestress of 70 MPa during winding causes the coils to be closed under no-load conditions. Calculate the no-load force in the spring, the maximum load that can be applied to the spring, and the maximum extension of the spring ($G = 84$ GPa).

6-16 A close-coiled helical spring is made from $\frac{3}{8}$-in. diameter steel ($G = 12 \times 10^6$ psi) wire. The spring has a mean diameter of 3 in. and 15 active coils. If the maximum shearing stress is not to exceed 18 ksi, determine the maximum permissible force that can be applied to the spring. Calculate the maximum deflection the spring can undergo before the shearing stress exceeds the stipulated limit.

6-7 SUMMARY

We computed normal stress that results from the combined action of axial forces and bending moments by invoking the superposition principle for stresses. According to this principle, the normal stress due to several sources is equal to the **algebraic sum** of the normal stresses due to each source acting separately.

Basic Formulas for Normal Stress

Axial:

$$\sigma = \frac{N}{A}$$

Bending:

$$\sigma = -\frac{My}{I}$$

Pressure:

$$\left.\begin{array}{l} \sigma_c = \dfrac{pr}{t} \\[2mm] \sigma_\ell = \dfrac{pr}{2t} \end{array}\right\} \text{thin cylinders} \qquad \sigma = \dfrac{pr}{2t} \text{ thin spheres}$$

We computed shearing stress that results from the combined action of direct shear forces in beams or rivets and twisting by invoking the superposition principle for shearing stresses. This principle asserts that the shearing stress due to several sources is equal to the **vectorial sum** of the shearing stresses due each source acting separately.

Basic Formulas for Shearing Stress

Bolts, Rivets, and Welds:

$$\tau = \frac{V}{A}$$

Straight Beams:

$$\tau = \frac{VQ}{Ib}$$

Circular Shafts:

$$\tau = \frac{Tr}{J}$$

Thin-Walled Closed Shafts:

$$\tau = \frac{q}{t} \text{ with } q = \frac{T}{2A_m}$$

See Sections 4-10 and 4-11 for shearing-stress formulas for noncircular solid cross sections and for sections made of a series of interconnected thin rectangles.

PROBLEMS / CHAPTER 6

6-17   Determine the maximum tensile and compressive stresses acting on section *aa* of the structural member shown in Figure P6-17.

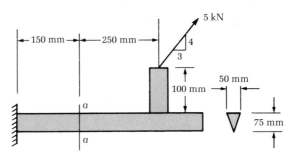

Figure P6-17

6-18   The frame structure shown in Figure P6-18 is fabricated from 3-in. standard steel pipe. Determine the maximum tensile and compressive stresses on section *aa*.

6-19   The 1-in. × 1-in. steel bar shown in Figure P6-19 supports a 200 lb force. Calculate the maximum tensile and compressive stresses on section *aa*.

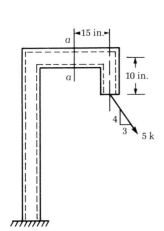

Figure P6-18

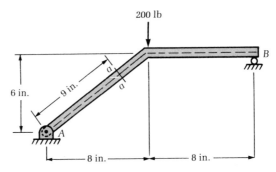

Figure P6-19

6-20   Determine the normal stress distribution on a cross section 250 mm from the pin connection of the eccentrically loaded member of Figure P6-20. Locate the neutral axis.

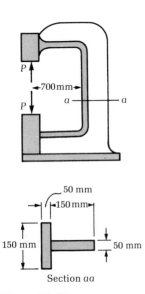

Section *aa*

Figure P6-21

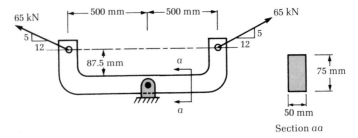

Figure P6-20

6-21 The cast iron frame of a small press is shaped as shown in Figure P6-21. Determine the maximum permissible load P that the press can resist if the compressive stress is not to exceed 140 MPa and the tensile stress is not to exceed 28 MPa.

6-22 The vertical support for a signboard shown in Figure P6-22 is a 3-in. standard steel pipe. The pipe is embedded in concrete, and the center of gravity of the 200 lb sign is located 3 ft from the centerline of the pipe. Calculate the maximum tensile and compressive stresses in the pipe.

6-23 Determine the maximum tensile and compressive stresses in the carver's clamp shown in Figure P6-23 when the clamping force is 250 N.

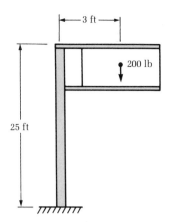

Figure P6-22

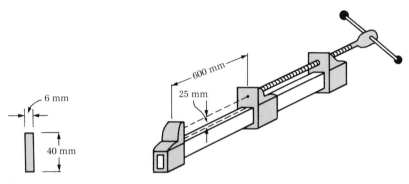

Figure P6-23

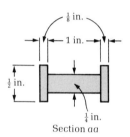

Section aa

6-24 A steel C clamp has the dimensions shown in Figure P6-24. Determine the maximum permissible clamping force if the normal stress is not to exceed 12 ksi.

6-25 Calculate the maximum tensile and compressive stresses in the frame of the coping saw shown in Figure P6-25 when the tension in the blade is 50 N.

6-26 The beam AB shown in Figure P6-26 is a C 4×7.25 channel with its flanges directed away from the 900-lb ball. Calculate the maximum compressive stress in the channel.

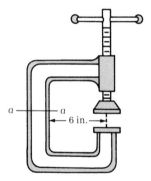

Figure P6-24

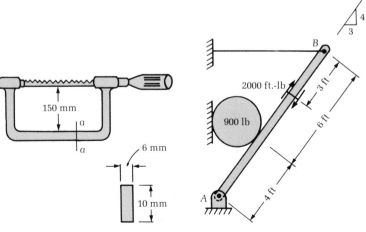

Figure P6-25   **Figure P6-26**

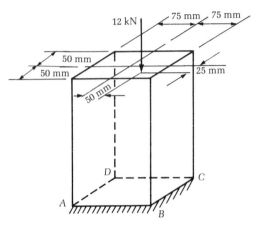

Figure P6-27

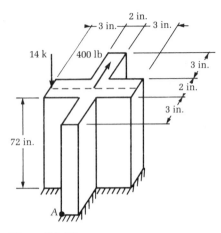

Figure P6-28

6-27 Determine the normal stresses at points A, B, C, and D of the compression block shown in Figure P6-27. Sketch the stress surface and locate the neutral axis.

6-28 Calculate the magnitude of the normal stress at point A of the engineering member shown in Figure P6-28.

6-29 An elliptical cylinder is used as a compression block as shown in Figure P6-29. Write an expression for the normal stress that acts on the boundary of the elliptical cross section that lies in the first quadrant. Determine the maximum compressive stress acting on the section.

6-30 A W 8 × 18 section is used as a short compression block as shown in Figure P6-30. If the normal stress on any cross section is not to exceed 18 ksi, determine the maximum permissible force P that can be applied in the location shown.

6-31 The concrete (23.6 kN/m³) dam shown in Figure P6-31 is used to maintain the water supply of a municipality. If water has a density of 9.81 kN/m³, determine the maximum depth of water that can be tolerated so that no tensile stresses develop on the base section of the dam.

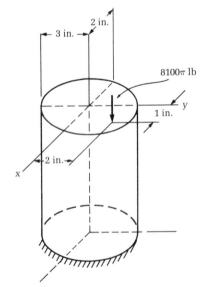

Figure P6-29

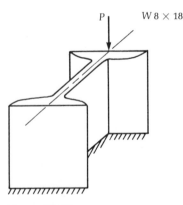

Figure P6-30

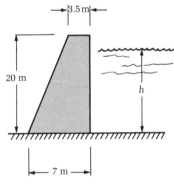

Figure P6-31

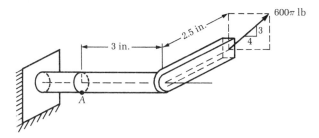

600π lb

2.5 in.

3 in.

3

4

A

Figure P6-32

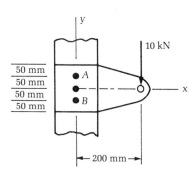

y

10 kN

50 mm

50 mm

50 mm

50 mm

● A

● B

x

← 200 mm →

Figure P6-33

6-32 Determine the components of stress that act on an element of material located at point A of the 2-in. diameter shaft shown in Figure P6-32.

6-33 A steel bracket is connected to a support by means of three 12-mm diameter rivets as shown in Figure P6-33. Determine the maximum shearing stress in the rivets. Hint: Replace the 10-kN force with a force applied at the centroid of the rivet areas and a couple. Assume that the centroidal force divides equally among the rivets and that the force on a rivet due to the couple varies linearly with distance from the centroid of the rivet areas.

6-34 An 8-in. × 10-in. timber is used as a cantilever beam required to support a uniformly distributed load as shown in Figure P6-34a. The orientation of the cross section of the timber relative to direction of the applied load is shown in Figure P6-34b. Determine the maximum permissible load intensity (q lb/ft) the beam can resist if the bending stress is not to exceed 1925 psi in either tension or compression.

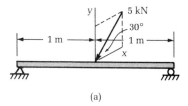

y

5 kN

30°

1 m

1 m

x

(a)

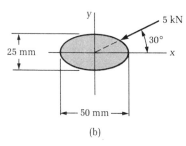

y

5 kN

30°

25 mm

x

← 50 mm →

(b)

Figure P6-35

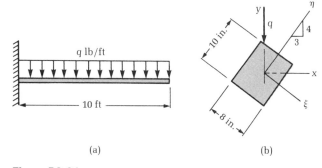

y

q

η

4

3

10 in.

x

ξ

8 in.

q lb/ft

10 ft

(a)

(b)

Figure P6-34

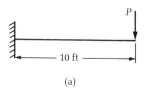

P

10 ft

(a)

6-35 A simply supported beam with an elliptical cross section carries a 5 kN force at midspan as shown in Figure P6-35a. The 5 kN force makes an angle of 60 degrees with the y axis as shown. Write the expression for the normal stress at the boundary of the elliptical cross section in the first quadrant at the center of the beam. Determine the maximum compressive stress on the cross section.

6-36 A W 10 × 39 steel section is used as a cantilever beam to carry a concentrated force at its end. The action line of the applied force passes through the shear center of the W section but makes an angle of 36.8 degrees with its web as shown in Figure P6-36b. Determine the maximum tensile and compressive stresses that occur in the wide flange section.

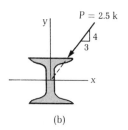

y

P = 2.5 k

4

3

x

(b)

Figure P6-36

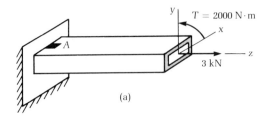

Figure P6-37

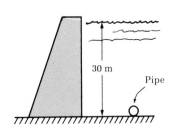

30 m

Pipe

Figure P6-38

6-37 A thin aluminum box beam is required to resist the axial force and twisting moment shown in Figure P6-37a. If the dimensions of the cross section are those given in Figure P6-37b, determine the state of stress at point A near the wall.

6-38 A 1-m diameter steel pipe ($E = 210$ GPa) transports water under a pressure of 1 MPa and rests on the bottom of a reservoir under 30 m of water (9.81 kN/m³). If the pipe thickness is 10 mm and its ends are prevented from displacing axially, determine (a) the circumferential and longitudinal stresses in the pipe and (b) the increase in diameter of the pipe.

6-39 The 5-ft diameter cylindrical tank shown in Figure P6-39 has a wall thickness of $\frac{3}{4}$ in. and is being considered as a means of storing a gas. If the gas must be stored at a pressure of 400 psi, and if the tensile stresses in the tank are not to exceed 20 ksi, determine the adequacy of the tank.

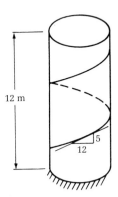

12 m

5

12

Figure P6-40

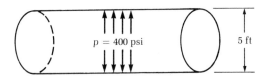

$p = 400$ psi

5 ft

Figure P6-39

6-40 The 4-m diameter penstock shown in Figure P6-40 is made from 12-mm thick steel plate by welding along a spiral seam with pitch of $\frac{5}{12}$. Determine the normal and shearing forces (per meter) that the weld near the base must resist when the tank is full of water (9.81 kN/m³).

6-41 Determine the maximum gas pressure and the corresponding radial displacement that a 10-ft diameter spherical tank of thickness $\frac{1}{4}$ in. can sustain if the tensile stress is not to exceed 20 ksi.

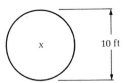

x

10 ft

Figure P6-41

6-42 Consider the two thin-walled, open ended cylinders whose cross sections are shown in Figures P6-42a and P6-42b. The outer radius of cylinder 1 is denoted by r, and the inner radius of cylinder 2 is denoted by R. The thicknesses and moduli of elasticity for cylinders 1 and 2 are signified by t_1 and t_2, and E_1 and E_2, respectively. Suppose that r is slightly larger than R, and that it is desirable to insert cylinder 1 into cylinder 2 by heating the latter appropriately. Show that the pressure at the common surface between the two cylinders after cooling is

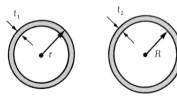

t_1 t_2

r R

(a) Cylinder 1   (b) Cylinder 2

Figure P6-42

$$p = (r - R) \frac{E_1 t_1}{r^2 \left(1 + \dfrac{E_1}{E_2} \dfrac{t_1}{t_2}\right)}$$

and, consequently, that the circumferential stress in each cylinder has the approximate magnitude

$$\sigma = -(r - R) \frac{E_1}{r\left(1 + \dfrac{E_1}{E_2}\dfrac{t_1}{t_2}\right)} \text{ (inner cylinder)}$$

or

$$\sigma = (r - R) \frac{E_2}{r\left(1 + \dfrac{E_2}{E_1}\dfrac{t_2}{t_1}\right)} \text{ (outer cylinder)}$$

6-43 Two concentric close-coiled helical springs are compressed between two parallel plates as shown in Figure P6-43. Each spring is made from steel (G = 84 GPa) wire 25 mm in diameter and has 75 active coils. If the maximum shearing stresses in either spring are not to exceed 140 MPa, determine the maximum permissible concentric force P that can be applied to the springs. Calculate the deflection of the spring due to P. What is the effective spring constant for the system?

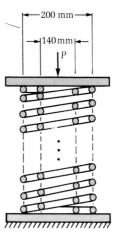

Figure P6-43

CHAPTER 7

Transformations in Mechanics

It is frequently necessary in engineering analysis to determine the maximum normal stress and/or the maximum shearing stress that exist at a point in an engineering member due to prescribed loads. Conversely, formulas that relate these quantities to the prescribed loads are required to determine the size of the members of the structure so that the maximum normal stress and/or the maximum shearing stress do not exceed the strength of the material in tension or in shear. Similar comments can be made regarding principal normal and principal shearing strains.

In this chapter, we develop transformation equations for stress, strain, and area moments of inertia. Furthermore, we establish formulas and procedures for calculating the principal values of normal stress, normal strain, and moments of inertia. We also establish formulas and procedures for calculating the principal values of shearing stress and shearing strain.

7-1 Biaxial States of Stress

Consider an infinitesimal volume of material at a point P in an engineering member. The most general state of stress at point P is shown in Figure 7-1. Stress components acting on the negative faces of the element are not included in the figure.

The sign convention for normal stresses is that tensile stresses are positive and compressive stresses are negative. A shear stress component is positive if it acts in the positive coordinate direction when it is associated with a positive coordinate area, or if it acts in a negative coordinate direction when it is associated with a negative coordinate area. Recall that a positive coordinate area (positive face) was defined as the face of the rectangular volume element whose normal vector is in the positive coordinate direction. For example, the face of the volume element marked A in Figure 7-1 is a positive coordinate face. Thus the

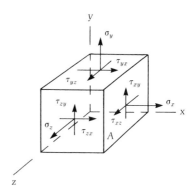

Figure 7-1

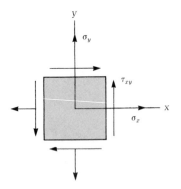

Figure 7-2

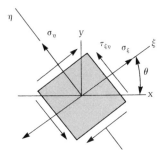

Figure 7-3

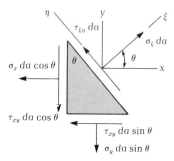

Figure 7-4

shearing stresses on this face are positive when they act in the positive coordinate directions. Otherwise, they are negative.

A **biaxial state of stress** is a special case of a three-dimensional state of stress that is realized when all the stress components act in the same plane. Let this plane be the xy plane, and represent this two-dimensional state of stress by a diagram such as the one shown in Figure 7-2. Note that the relation $\tau_{xy} = \tau_{yx}$ has been taken into account in this figure.

Figure 7-3 depicts a generic orientation of the element at point P. The element in this orientation is referred to the $\xi\eta$ coordinate system whose angular orientation relative to the xy coordinate system is signified by the angle θ.

Functional relationship are sought between the normal stress σ_ξ (or σ_η) and the coordinate stresses σ_x, σ_y, and τ_{xy}, and between the shearing stress $\tau_{\xi\eta}$ and the coordinate stresses. To obtain these relationships, the forces acting on the prismatic element shown in Figure 7-4—which is part of the element in Figure 7-2—must satisfy Newton's first law. For equilibrium of the prismatic element in Figure 7-4, the sum of the **forces** acting in two perpendicular directions (in the ξ and η directions or in the x and y directions) must be zero. Note that it is the forces that correspond to the stresses that are to be added vectorially, not the stresses themselves. The stresses must be first converted into forces. Denote the area of the inclined surface by da; then the area of the surface perpendicular to the x axis is $da \cos\theta$ and the area perpendicular to the y axis is $da \sin\theta$. The forces acting on the exposed faces of the element are shown in Figure 7-4.

Equilibrium of forces in the ξ and η directions leads to

$$\sigma_\xi = \sigma_x \cos^2\theta + \sigma_y \sin^2\theta + 2\tau_{xy} \sin\theta \cos\theta \tag{7-1}$$

and

$$\tau_{\xi\eta} = -(\sigma_x - \sigma_y)\sin\theta\cos\theta + \tau_{xy}(\cos^2\theta - \sin^2\theta) \tag{7-2}$$

We list some useful trigonometric identities here for convenience.

$$\left.\begin{array}{l} \cos^2\theta = \dfrac{1 + \cos 2\theta}{2} \\[2mm] \sin^2\theta = \dfrac{1 - \cos 2\theta}{2} \\[2mm] \sin\theta\cos\theta = \tfrac{1}{2}\sin 2\theta \\[2mm] \cos^2\theta - \sin^2\theta = \cos 2\theta \end{array}\right\} \tag{7-3}$$

Using these trigonometric identities, we can express Eqs. (7-1) and (7-2) in the more common form

$$\sigma_\xi = \frac{\sigma_x + \sigma_y}{2} + \frac{\sigma_x - \sigma_y}{2}\cos 2\theta + \tau_{xy}\sin 2\theta \tag{7-4}$$

and

$$\tau_{\xi\eta} = -\frac{\sigma_x - \sigma_y}{2} \sin 2\theta + \tau_{xy} \cos 2\theta \qquad (7\text{-}5)$$

Equations (7-4) and (7-5) determine the components of stress that act on a plane perpendicular to the ξ axis as shown in either Figure 7-3 or Figure 7-4.

To complete the description of the state of stress at point P relative to the $\xi\eta$ coordinate system, we require the normal stress σ_η. This stress can be calculated from Eq. (7-4) by substituting $\theta = \theta + (\pi/2)$. Consequently,

$$\sigma_\eta = \frac{\sigma_x + \sigma_y}{2} - \frac{\sigma_x - \sigma_y}{2} \cos 2\theta - \tau_{xy} \sin 2\theta \qquad (7\text{-}6)$$

The shearing stress acting on the plane normal to the η axis is, of course, equal to the shearing stress acting on the plane normal to the ξ axis.

We remark that σ_η is most easily calculated from the relation

$$\sigma_x + \sigma_y = \sigma_\xi + \sigma_\eta \qquad (7\text{-}7)$$

which is obtained by adding Eqs. (7-4) and (7-6). The foregoing analysis shows that, if the stress components on any two orthogonal faces of an element are known, the stress components on all faces with normal vectors lying in the plane of the element can be calculated using Eqs. (7-4), (7-5), and (7-7).

7-2 Principal Normal Stresses

Equation (7-4) shows that σ_ξ depends continuously on the angle θ. Consequently, it is natural to inquire whether there are certain values of θ for which the normal stress attains extreme values. According to the calculus, extreme values of $\sigma_\xi(\theta)$ will occur for values of θ that satisfy the relation $d\sigma_\xi/d\theta = 0$. Thus the extreme values of σ_ξ correspond to the values of θ that satisfy the transcendental equation

$$\tan 2\theta_p = \frac{\tau_{xy}}{\left(\dfrac{\sigma_x - \sigma_y}{2}\right)} \qquad (7\text{-}8)$$

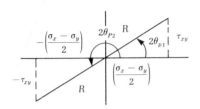

Figure 7-5

There are two values of θ_p that satisfy Eq. (7-8). These values of θ_p are denoted by θ_{p1} and θ_{p2} and are shown in Figure 7-5. Let

$$R = \sqrt{\left(\frac{\sigma_x - \sigma_y}{2}\right)^2 + \tau_{xy}^2} \tag{7-9}$$

From Figure 7-5,

$$\sin 2\theta_{p1} = \frac{\tau_{xy}}{R}, \quad \cos 2\theta_{p1} = \frac{\dfrac{\sigma_x - \sigma_y}{2}}{R} \tag{7-10a}$$

and

$$\sin 2\theta_{p2} = -\frac{\tau_{xy}}{R}, \quad \cos 2\theta_{p2} = -\frac{\left(\dfrac{\sigma_x - \sigma_y}{2}\right)}{R} \tag{7-10b}$$

Substituting Eqs. (7-10a) and Eqs. (7-10b) into Eq. (7-4), in succession, yields

$$\left\{\begin{matrix} \sigma_\xi(2\theta_{p1}) \\ \sigma_\xi(2\theta_{p2}) \end{matrix}\right\} = \frac{\sigma_x + \sigma_y}{2} + \frac{\sigma_x - \sigma_y}{2}\left(\pm\frac{\sigma_x - \sigma_y}{2}\right) + \tau_{xy}\left(\pm\frac{\tau_{xy}}{R}\right) \tag{7-11}$$

where the upper sign (positive) is associated with $2\theta_{p1}$ and the lower sign (negative) is associated with $2\theta_{p2}$. Taking account of Eq. (7-9), we determine that

$$\left\{\begin{matrix} \sigma_{max} \\ \sigma_{min} \end{matrix}\right\} = \frac{\sigma_x + \sigma_y}{2} \pm \sqrt{\left(\frac{\sigma_x - \sigma_y}{2}\right)^2 + \tau_{xy}^2} \tag{7-12}$$

These are the magnitudes of the **principal normal stresses**. Their orientations are given by Eq. (7-8). Observe that two magnitudes are calculated from Eq. (7-12), and orientations for two **principal planes** on which the principal normal stresses act are calculated from Eq. (7-8). A further calculation is required to assign the proper stress with the proper principal plane. This correlation calculation is demonstrated in the examples.

It is important to note that if the trigonometric values for $2\theta_{p1}$ and $2\theta_{p2}$ from Eqs. (7-10) are substituted into the transformation equation for shearing stress, Eq. (7-5), it turns out that $\tau_{\xi\eta} \equiv 0$. Thus the planes on which the principal normal stresses act are always free of shearing stresses.

Finally, observe from Figure 7-5 that

$$2\theta_{p2} = 2\theta_{p1} + \pi \tag{7-13}$$

from which it is concluded that the planes on which the principal normal stresses act are perpendicular to each other.

7-3 Principal Shearing Stresses

Equation (7-5) shows that $\tau_{\xi\eta}$ also depends continuously on the angle θ. Therefore, it is again natural to inquire whether there are certain values of θ for which the shearing stress attains extreme values. Extreme values of $\tau_{\xi\eta}$ occur for values of θ that satisfy the relation $d\tau_{\xi\eta}/d\theta = 0$. Consequently, the extreme values of $\tau_{\xi\eta}$ correspond to values of θ that satisfy the transcendental equation

$$\tan 2\theta_s = -\frac{\dfrac{\sigma_x - \sigma_y}{2}}{\tau_{xy}} \tag{7-14}$$

There are two values of θ_s that satisfy Eq. (7-14). These values of θ_s are denoted by $2\theta_{s1}$ and $2\theta_{s2}$ and are shown in Figure 7-6. Thus

$$\sin 2\theta_{s1} = \frac{\sigma_x - \sigma_y}{2}, \quad \cos 2\theta_{s1} = -\frac{\tau_{xy}}{R} \tag{7-15a}$$

and

$$\sin 2\theta_{s2} = -\frac{\dfrac{\sigma_x - \sigma_y}{2}}{R}, \quad \cos 2\theta_{s2} = \frac{\tau_{xy}}{R} \tag{7-15b}$$

where R is given by Eq. (7-9).

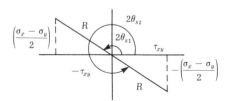

Figure 7-6

Substituting Eqs. (7-15a) and (7-15b) into Eq. (7-5), in succession, yields

$$\begin{Bmatrix} \tau_{\xi\eta}(2\theta_{s1}) \\ \tau_{\xi\eta}(2\theta_{s2}) \end{Bmatrix} = -\frac{\sigma_x - \sigma_y}{2}\left(\pm\frac{\sigma_x - \sigma_y}{2}\right) + \tau_{xy}\left(\mp\frac{\tau_{xy}}{R}\right) \tag{7-16}$$

where the upper sign is associated with $2\theta_{s1}$ and the lower sign with $2\theta_{s2}$. Taking account of Eq. (7-9), we determine that

$$\begin{Bmatrix} \tau_{max} \\ \tau_{min} \end{Bmatrix} = \pm\sqrt{\left(\frac{\sigma_x - \sigma_y}{2}\right)^2 + \tau_{xy}^2} \tag{7-17}$$

These magnitudes are the magnitudes of the **principal shearing stresses**. The orientations of the planes on which they act are calculated from Eq. (7-14). A further calculation is required to correlate the proper shearing stress with the proper plane. This correlation calculation is demonstrated in the examples.

It is important to observe that, when the trigonometric values for $2\theta_{s1}$ and $2\theta_{s2}$ are substituted into the transformation equation for normal stresses, Eqs. (7-4) and (7-6), we find that

$$\sigma_\xi = \sigma_\eta = \frac{\sigma_x + \sigma_y}{2} \qquad (7\text{-}18)$$

That is, the normal stresses that act on the **principal shear planes** are each equal to the average of the coordinate normal stresses (σ_x, σ_y).

Furthermore, observe from Figure 7-6 that

$$2\theta_{s2} = 2\theta_{s1} + \pi \qquad (7\text{-}19)$$

from which we conclude that the planes on which the principal shearing stresses act are perpendicular to each other.

Finally, note that Eq. (7-8) is the negative reciprocal of Eq. (7-14). Thus

$$\tan 2\theta_p = -\frac{1}{\tan 2\theta_s}$$

or

$$2\theta_s = 2\theta_p + \frac{\pi}{2} \qquad (7\text{-}20)$$

from which we conclude that the element associated with the principal shearing stresses is oriented 45 degrees from the element on which the principal normal stresses act.

Summary of stress transformation equations. The stress transformation equations established in the preceding section are summarized here for convenience.

General Stress Transformation Equations

$$\left.\begin{aligned}
\sigma_\xi &= \frac{\sigma_x + \sigma_y}{2} + \frac{\sigma_x - \sigma_y}{2}\cos 2\theta + \tau_{xy}\sin 2\theta \\[2mm]
\tau_{\xi\eta} &= -\frac{\sigma_x - \sigma_y}{2}\sin 2\theta + \tau_{xy}\cos 2\theta \\[2mm]
\sigma_\xi + \sigma_\eta &= \sigma_x + \sigma_y
\end{aligned}\right\} \qquad (7\text{-}21)$$

Principal Normal Stress Transformation Equations

Magnitudes:

$$\begin{Bmatrix} \sigma_{max} \\ \sigma_{min} \end{Bmatrix} = \frac{\sigma_x + \sigma_y}{2} \pm \sqrt{\left(\frac{\sigma_x - \sigma_y}{2}\right)^2 + \tau_{xy}^2} \text{ and } \tau_{\xi\eta} = \tau_{\eta\xi} \equiv 0$$

Orientations:

$$\tan 2\theta_p = \frac{\tau_{xy}}{\dfrac{\sigma_x - \sigma_y}{2}}, \; \theta_{p1} \perp \theta_{p2}$$

(7-22)

Principal Shearing Stress Transformation Equations

Magnitudes:

$$\begin{Bmatrix} \tau_{max} \\ \tau_{min} \end{Bmatrix} = \pm \sqrt{\left(\frac{\sigma_x - \sigma_y}{2}\right)^2 + \tau_{xy}^2}$$

and $\quad \sigma_\xi = \sigma_\eta = \dfrac{\sigma_x + \sigma_y}{2}$

(7-23)

Orientations:

$$\tan 2\theta_s = -\frac{\dfrac{\sigma_x - \sigma_y}{2}}{\tau_{xy}}, \qquad \theta_{s1} \perp \theta_{s2}$$

EXAMPLE 7-1

The state of stress at a point in an engineering member has been determined to be that shown in Figure 7-7a. (a) Determine the components of stress associated with an element orientation $\theta = 15$ degrees and show them on a properly oriented element. (b) Determine the principal normal stresses and show them on a properly oriented element. (c) Determine the principal shearing stresses and show them on a properly oriented element.

SOLUTION

According to the stated sign convention for stresses, the stress components shown in Figure 7-7a are all positive.

(a) The orientation of the element for which the components of stress are required is shown in Figure 7-7b. The magnitudes of these

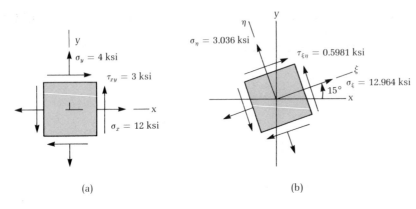

(a)   (b)

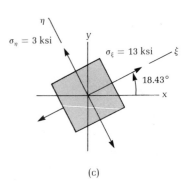

(c)

components of stress are

$$\sigma_\xi(15°) = \frac{12+4}{2} + \frac{12-4}{2} \cos 30° + 3 \sin 30° = 12.964 \text{ ksi}$$

$$\tau_{\xi\eta}(15°) = -\frac{12-4}{2} \sin 30° + 3 \cos 30° = 0.5981 \text{ ksi}$$

$$\sigma_\eta = 12 + 4 - 12.964 = 3.036 \text{ ksi}$$

(b) The magnitudes of the principal normal stresses are

$$\begin{Bmatrix} \sigma_{max} \\ \sigma_{min} \end{Bmatrix} = \frac{12+4}{2} \pm \sqrt{\left(\frac{12-4}{2}\right)^2 + 3^2} = 8 \pm 5 = \begin{cases} 13 \text{ ksi,} \\ 3 \text{ ksi} \end{cases}$$

and the orientations of the planes on which these principal normal stresses act are given by

$$\tan 2\theta_p = \frac{3}{\dfrac{12-4}{2}} = \frac{3}{4}$$

Hence

$$2\theta_p = 36.87°, 216.87°$$

or

$$\theta_p = 18.43°, 108.43°$$

To correlate the stresses and the planes on which they act, substitute one of these angles, preferably the acute angle, into the transformation equation for normal stress. Consequently, from Eqs. (7-21),

$$\sigma_\xi(18.43°) = 8 + 4(\tfrac{4}{5}) + 3(\tfrac{3}{5}) = 13 \text{ ksi}$$

This result shows that $\sigma_{max} = 13$ ksi acts on the plane perpendicular to the ξ axis whose direction is defined by the angle $\theta_p = 18.43$ degrees. Since the principal planes are mutually perpendicular, $\sigma_{min} = 3$ ksi must act on a plane perpendicular to the η axis. Moreover, the shearing stresses are always zero on the principal normal stress planes. The

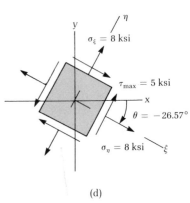

(d)

Figure 7-7

principal normal stresses are shown on a properly oriented element in Figure 7-7c.

(c) The magnitudes of the principal shearing stresses are

$$\begin{Bmatrix} \tau_{max} \\ \tau_{min} \end{Bmatrix} = \pm \sqrt{\left(\frac{12-4}{2}\right)^2 + 3^2} = \pm 5 \text{ ksi}$$

and the orientations of the planes on which they act are given by

$$\tan 2\theta_s = -\frac{\dfrac{12-4}{2}}{3} = -\frac{4}{3}$$

Hence

$$2\theta_s = -53.13°, \ -233.13°$$

or

$$\theta_s = -26.57°, \ -116.57°$$

To correlate the shearing stresses and the planes on which they act, substitute one of these angles, preferably the acute angle, into the transformation equation for shearing stress. Thus, from Eqs. (7-21),

$$\tau_{\xi\eta}(-26.57°) = -4(-\tfrac{4}{5}) + 3(\tfrac{3}{5}) = +5 \text{ ksi}$$

This result shows that $\tau_{max} = +5$ ksi acts on the plane perpendicular to the ξ axis whose direction is defined by the angle $\theta = -26.57$ degrees.

The normal stresses that act on the principal shearing planes are

$$\sigma_\xi = \sigma_\eta = \frac{12+4}{2} = 8 \text{ ksi}$$

The principal shearing stresses are shown on a properly oriented element in Figure 7-7d.

EXAMPLE 7-2

The state of stress at a point in a material is shown in Figure 7-8a. (a) Determine the principal normal stresses and show them on a properly oriented element. (b) Determine the principal shearing stresse' and show them on a properly oriented element.

SOLUTION

(a) The normal stress component in the y direction is negative, and the other two stress components are positive. The magnitudes of the

principal normal stresses are

$$\begin{Bmatrix} \sigma_{max} \\ \sigma_{min} \end{Bmatrix} = \frac{50-30}{2} + \sqrt{\left(\frac{50+30}{2}\right)^2 + 30^2}$$

$$= 10 \pm 50 = 60 \text{ MPa}, -40 \text{ MPa}$$

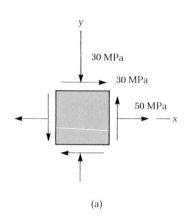

(a)

and the orientations of the planes on which they act are given by

$$\tan 2\theta_p = \tfrac{30}{40}$$

so that

$$\theta_p = 18.43°, 108.43°$$

The correlation calculation shows that

$$\sigma_\xi(18.43°) = 10 + 40(\tfrac{4}{5}) - 30(\tfrac{3}{5}) = +60 \text{ MPa}$$

Consequently, the maximum principal normal stress acts on the plane perpendicular to the ξ axis whose direction is defined by the angle $\theta_p = 18.43$ degrees.

The principal normal stresses are shown on a properly oriented element in Figure 7-8b.

(b) The magnitudes of the principal shearing stresses are

$$\begin{Bmatrix} \tau_{max} \\ \tau_{min} \end{Bmatrix} = \pm\sqrt{\left(\frac{50+30}{2}\right)^2 + 30^2} = \pm 50 \text{ MPa}$$

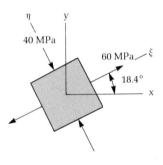

(b)

and the orientation of the planes on which they act are given by

$$\tan 2\theta_s = -\tfrac{40}{30}$$

so that

$$\theta_s = -26.57°, -116.57°$$

The correlation calculation shows that

$$\tau_{\xi\eta}(-26.57°) = -40(-\tfrac{4}{5}) + 30(\tfrac{3}{5}) = +50 \text{ MPa}$$

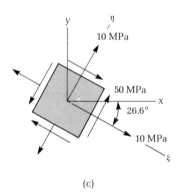

(c)

Thus $\tau_{max} = +50$ MPa acts on the plane perpendicular to the ξ axis whose direction is defined by the angle $\theta = -26.57$ degrees.

The normal stresses that act on the principal shear planes are

$$\sigma_\xi = \sigma_\eta = \frac{50 + (-30)}{2} = 10 \text{ MPa}$$

The principal shearing stresses are shown on a properly oriented element in Figure 7-8c.

Figure 7-8

PROBLEMS / Section 7-3

7-1–7-4   Figures P7-1 through P7-4 are associated with Problems 7-1 through 7-4. For the state of stress shown in a figure, (a) Determine analytically the magnitudes and directions of the principal normal stresses and show these stresses on a properly oriented element. (b) Determine analytically the magnitudes and directions of the principal shearing stresses and show these stresses on a properly oriented element. (c) Determine analytically the magnitudes of the stress components associated with an element orientation of 15 degrees counterclockwise and show these stresses on a properly oriented element.

7-5   Calculate the components of stress at point A of the bracket shown in Figure P7-5. Determine the principal normal stresses and the principal shearing stresses, and show them on properly oriented elements.

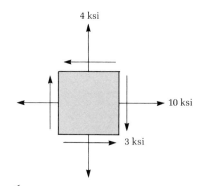

✓**Figure P7-1**

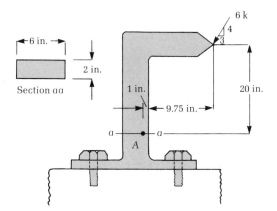

Figure P7-5

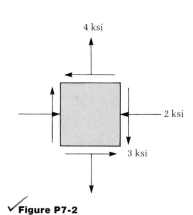

✓**Figure P7-2**

7-6   Determine the principal normal stresses and the principal shearing stresses at point A of the machine element shown in Figure P7-6. Show these stresses on properly oriented elements.

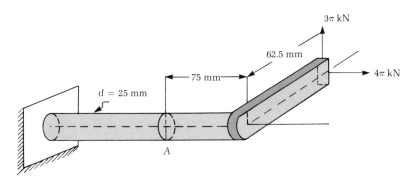

Figure P7-6

Figure P7-3

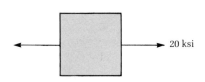

Figure P7-4

✓ **7-7**   The allowable normal and shearing stresses for the shaft of Figure P7-7 are 130 MPa and 70 MPa, respectively. Determine the maximum permissible value of P.

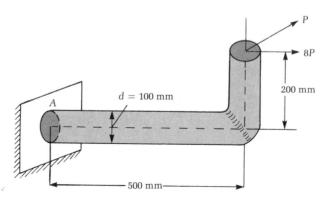

Figure P7-7

7-4   Biaxial States of Strain

The general transformation equations for biaxial states of strain are obtained from geometric considerations as demonstrated in the following developments.

In Figures 7-9a, 7-9b, and 7-9c, line segments $\overline{OP}$ and $\overline{OQ}$ of lengths ds_1 and ds_2 represent fibers of a material that are mutually perpendicular in an unstressed state. The coordinates of the points P and Q in the xy system are (dx_1, dy_1) and (dx_2, dy_2), respectively. For convenience, orient a $\xi\eta$ system of axes so that the ξ axis coincides with segment $\overline{OP}$ and the η axis coincides with segment $\overline{OQ}$. The angle θ locates the $\xi\eta$ coordinate system relative the xy coordinate system.

Line segments $\overline{OP}^*$ and $\overline{OQ}^*$ denote the positions of the fibers $\overline{OP}$ and $\overline{OQ}$ respectively, in the deformed state. Since the strains that are tolerable in most engineering applications are small (the proportional limit strain for structural steel is about 0.001), the angle θ^* that the segment $\overline{OP}^*$ makes with the x axis is essentially equal to the angle θ.

For small strains, the superposition principle is applicable. That is, the normal strain of the line element along the ξ axis due to simultaneous strains ϵ_x, ϵ_y, and γ_{xy} can be obtained as the algebraic sum of the normal strains of this line element due to ϵ_x, ϵ_y, and γ_{xy} acting individually.

Figure 7-9a indicates the positions of segments $\overline{OP}^*$ and $\overline{OQ}^*$ due to a strain ϵ_x only. Similarly, Figures 7-9b and 7-9c indicate the positions of segments $\overline{OP}^*$ and $\overline{OQ}^*$ due to strains ϵ_y and γ_{xy}, respectively.

Suppose that the strains ϵ_x, ϵ_y, and γ_{xy} are known, and that we must determine expressions for the normal strains ϵ_ξ and ϵ_η and the shearing strain $\gamma_{\xi\eta}$ in terms of the coordinate strains ϵ_x, ϵ_y, and γ_{xy}.

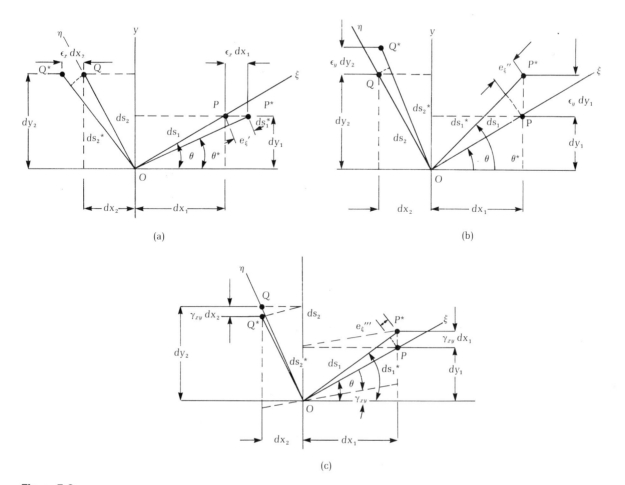

Figure 7-9

Engineering normal strain was defined as the change in length of a line element divided by its original length. Thus the normal strain associated with the line element $\overline{OP}$ is

$$\epsilon_\xi = \epsilon'_\xi + \epsilon''_\xi + \epsilon'''_\xi \qquad (7\text{-}24)$$

where

$$\epsilon'_\xi = \frac{e'_\xi}{ds_1} = \frac{\epsilon_x \, dx_1 \cos \theta}{ds_1} = \epsilon_x \cos^2 \theta \qquad (7\text{-}25a)$$

$$\epsilon''_\xi = \frac{e''_\xi}{ds_1} = \frac{\epsilon_y \, dy_1 \sin \theta}{ds_1} = \epsilon_y \sin^2 \theta \qquad (7\text{-}25b)$$

$$\epsilon'''_\xi = \frac{e'''_\xi}{ds_1} = \frac{\gamma_{xy} \, dx_1 \sin \theta}{ds_1} = \gamma_{xy} \sin \theta \cos \theta \qquad (7\text{-}25c)$$

By the superposition principle, the normal strain of a line element situated along the ξ axis becomes

$$\epsilon_\xi = \epsilon_x \cos^2 \theta + \epsilon_y \sin^2 \theta + \gamma_{xy} \sin \theta \cos \theta \qquad (7\text{-}26)$$

The engineering shearing strain associated with line elements that are situated along the ξ and η axes prior to the deformation is defined as the change in the right angle between these elements as they pass into the deformed state. Consequently, by the superposition principle,

$$\gamma_{\xi\eta} = \gamma'_{\xi\eta} + \gamma''_{\xi\eta} + \gamma'''_{\xi\eta} \tag{7-27}$$

where $\gamma'_{\xi\eta}$, $\gamma''_{\xi\eta}$, and $\gamma'''_{\xi\eta}$ are the changes in right angle that occur when ϵ_x, ϵ_y, and γ_{xy} are applied separately. From Figures 7-9a, 7-9b, and 7-9c,

$$\gamma'_{\xi\eta} = -\frac{\epsilon_x \, dx_1 \sin\theta}{ds_1} - \frac{\epsilon_x \, dx_2 \cos\theta}{ds_2} = -2\epsilon_x \sin\theta \cos\theta \tag{7-28a}$$

$$\gamma''_{\xi\eta} = \frac{\epsilon_y \, dy_1 \cos\theta}{ds_1} + \frac{\epsilon_y \, dy_2 \sin\theta}{ds_2} = 2\epsilon_y \sin\theta \cos\theta \tag{7-28b}$$

$$\gamma'''_{\xi\eta} = \frac{\gamma_{xy} \, dx_1 \cos\theta}{ds_1} - \frac{\gamma_{xy} \, dx_2 \sin\theta}{ds_2} = \gamma_{xy}(\cos^2\theta - \sin^2\theta) \tag{7-28c}$$

By the superposition principle, the shearing strain associated with the line elements situated along the ξ and η axes becomes

$$\gamma_{\xi\eta} = -2(\epsilon_x - \epsilon_y)\sin\theta\cos\theta + \gamma_{xy}(\cos^2\theta - \sin^2\theta)$$

Making use of the trigonometric identities listed in Eqs. (7-3), we can express the transformation equations for normal and shearing strains as

$$\epsilon_\xi = \frac{\epsilon_x + \epsilon_y}{2} + \frac{\epsilon_x - \epsilon_y}{2}\cos 2\theta + \frac{\gamma_{xy}}{2}\sin 2\theta \tag{7-29}$$

and

$$\frac{\gamma_{\xi\eta}}{2} = -\frac{\epsilon_x - \epsilon_y}{2}\sin 2\theta + \frac{\gamma_{xy}}{2}\cos 2\theta \tag{7-30}$$

Note the similarities between the stress transformation Eqs. (7-4) and (7-5) and the strain transformation Eqs. (7-29) and (7-30). Indeed, the strains ϵ_x, ϵ_y, and $\gamma_{xy}/2$ transform in precisely the same manner as the stresses σ_x, σ_y, and τ_{xy}. Quantities that transform in this manner under a rotation of coordinates are called **tensor quantities of the second order** or simply a **second-order tensor**. Quantities of this kind do not obey the laws of vector addition and subtraction.

To complete the description of the state of strain at point P, we need the strain associated with a line element situated along the η axis prior to deformation. This strain can be obtained by substituting

$\theta = \theta + \pi/2$ in Eq. (7-29). Thus

$$\epsilon_\eta = \frac{\epsilon_x + \epsilon_y}{2} - \frac{\epsilon_x - \epsilon_y}{2} \cos 2\theta - \frac{\gamma_{xy}}{2} \sin 2\theta \qquad (7\text{-}31)$$

Observe that ϵ_η is most easily calculated from the relation

$$\epsilon_\xi + \epsilon_\eta = \epsilon_x + \epsilon_y \qquad (7\text{-}32)$$

which is obtained by adding Eqs. (7-29) and (7-31).

We should emphasize again that the strain transformation equations and the stress transformation equations are identical in structure. Indeed, σ_x and ϵ_x, σ_y and ϵ_y, and τ_{xy} and $\gamma_{xy}/2$ play identical roles as coefficients in these equations. Consequently, we must expect that analyses for principal normal strains and principal shearing strains will be similar to those for principal normal stresses and principal shearing stresses.

Summary of strain transformation equations. The strain transformation equations established in this section are listed here for convenience.

General Strain Transformation Equations

$$\left.\begin{aligned}
\epsilon_\xi &= \frac{\epsilon_x + \epsilon_y}{2} + \frac{\epsilon_x + \epsilon_y}{2} \cos 2\theta + \frac{\gamma_{xy}}{2} \sin 2\theta \\
\frac{\gamma_{\xi\eta}}{2} &= -\left(\frac{\epsilon_x - \epsilon_y}{2}\right) \sin 2\theta + \frac{\gamma_{xy}}{2} \cos 2\theta \\
\epsilon_\xi + \epsilon_\eta &= \epsilon_x + \epsilon_y
\end{aligned}\right\} \qquad (7\text{-}33)$$

Principal Normal Strain Transformation Equations

$$\left.\begin{aligned}
\textit{Magnitudes:} \quad & \begin{Bmatrix} \epsilon_{\max} \\ \epsilon_{\min} \end{Bmatrix} = \frac{\epsilon_x + \epsilon_y}{2} \pm \sqrt{\left(\frac{\epsilon_x - \epsilon_y}{2}\right)^2 + \left(\frac{\gamma_{xy}}{2}\right)^2} \\
\text{and} \quad & \gamma_{\xi\eta} = \gamma_{\eta\xi} \equiv 0 \\
\textit{Orientations:} \quad & \tan 2\theta_p = \frac{\dfrac{\gamma_{xy}}{2}}{\dfrac{\epsilon_x - \epsilon_y}{2}}, \qquad \theta_{p1} \perp \theta_{p2}
\end{aligned}\right\} \qquad (7\text{-}34)$$

Principal Shearing Strain Transformation Equations

$$\left.\begin{array}{l} \left\{\begin{array}{l} \left(\dfrac{\gamma_{\xi\eta}}{2}\right)_{max} \\[2ex] \left(\dfrac{\gamma_{\xi\eta}}{2}\right)_{min} \end{array}\right\} = \pm\sqrt{\left(\dfrac{\epsilon_x - \epsilon_y}{2}\right)^2 + \left(\dfrac{\gamma_{xy}}{2}\right)^2} \\[6ex] \text{and} \quad \epsilon_\xi = \epsilon_\eta = \dfrac{\epsilon_x + \epsilon_y}{2} \\[4ex] \textit{Orientations:} \\[2ex] \tan 2\theta_s = -\dfrac{\dfrac{\epsilon_x - \epsilon_y}{2}}{\dfrac{\gamma_{xy}}{2}}, \quad \theta_{s1} \perp \theta_{s2} \end{array}\right\} \qquad (7\text{-}35)$$

Magnitudes:

EXAMPLE 7-3

The components of strain associated with fibers situated along the x and y axes of Figure 7-10a are $\epsilon_x = 1200\mu$ in./in., $\epsilon_y = 400\mu$ in./in., and $\gamma_{xy} = 600\mu$ in./in. (a) Determine the components of strain associated with a pair of line elements that are situated along the ξ and η axes when $\theta = 15$ degrees. Show the fibers that undergo these strains on a diagram. (b) Determine the principal normal strains and show the fibers that experience these strains on a diagram. (c) Determine the principal shearing strains and indicate the fibers that experience them on a diagram.

SOLUTION

(a) The strain components are all positive. The orientations of the fibers for which the components of strain are required are shown in Figure 7-10b. The magnitudes of these strains are

$$\epsilon_\xi(15°) = \frac{1200 + 400}{2} + \frac{1200 - 400}{2}\cos 30° + \frac{600}{2}\sin 30°$$

$$= 1296.4\mu \text{ in./in.}$$

$$\frac{\gamma_{\xi\eta}}{2}(15°) = -\left(\frac{1200 - 400}{2}\right)\sin 30° + \frac{600}{2}\cos 30°$$

$$= 59.81\mu \text{ in./in.}$$

$$\epsilon_\eta(15°) = 1200 + 400 - 1296.4 = 303.6\mu \text{ in./in.}$$

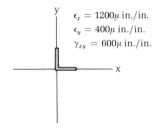

$\epsilon_x = 1200\mu$ in./in.
$\epsilon_y = 400\mu$ in./in.
$\gamma_{xy} = 600\mu$ in./in.

(a)

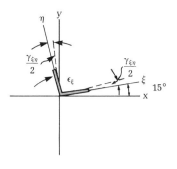

(b)

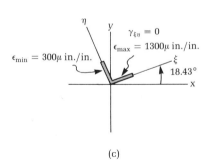

$\epsilon_{min} = 300\mu$ in./in.
$\gamma_{\xi\eta} \equiv 0$
$\epsilon_{max} = 1300\mu$ in./in.
$18.43°$

(c)

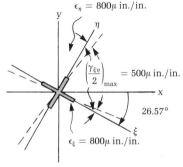

$\epsilon_\eta = 800\mu$ in./in.
$\left(\dfrac{\gamma_{\xi\eta}}{2}\right)_{max} = 500\mu$ in./in.
$26.57°$
$\epsilon_\xi = 800\mu$ in./in.

(d)

Figure 7-10

Recall that a positive shearing strain corresponds to a decrease in the original right angle between the fibers along the ξ and η axes as shown by the dotted line in Figure 7-10b.

(b) The magnitudes of the principal normal strains are

$$\begin{Bmatrix} \epsilon_{max} \\ \epsilon_{min} \end{Bmatrix} = \frac{1200 + 400}{2} \pm \sqrt{\left(\frac{1200 - 400}{2}\right)^2 + \left(\frac{600}{2}\right)^2}$$

$$= 800 \pm 500 = 1300\mu \text{ in./in.}, 300\mu \text{ in./in.}$$

and the orientations of the fibers that experience these strains are given by

$$\tan 2\theta_p = \frac{\dfrac{600}{2}}{\dfrac{1200 - 400}{2}} = \frac{3}{4}$$

Hence

$$2\theta_p = 36.87°, 216.87°$$

or

$$\theta_p = 18.43°, 108.43°$$

To correlate the principal normal strains and the fibers that experience them, substitute one of these angles—preferably the acute angle—into the transformation equation for normal strains. Thus, from Eqs. (7-33),

$$\epsilon_\xi(18.43°) = 800 + 400(\tfrac{4}{5}) + 300(\tfrac{3}{5}) = 1300\mu \text{ in./in.}$$

This calculation shows that the fiber situated along the ξ axis whose direction is $\theta = 18.43$ degrees experiences the maximum normal strain. The fiber that experiences the minimum normal strain is perpendicular to the ξ axis and is, therefore, situated along the η axis. The shearing strain $\gamma_{\xi\eta} \equiv 0$.

The principal normal strains and the fibers that experience them are shown in Figure 7-10c.

(c) The magnitudes of the principal shearing strains are

$$\begin{Bmatrix} \left(\dfrac{\gamma_{\xi\eta}}{2}\right)_{max} \\ \left(\dfrac{\gamma_{\xi\eta}}{2}\right)_{min} \end{Bmatrix} = \pm \sqrt{\left(\frac{1200 - 400}{2}\right)^2 + \left(\frac{600}{2}\right)^2} = \pm 500\mu \text{ in./in.}$$

and the orientations of fibers that experience these strains are given by

$$\tan 2\theta_s = -\frac{\dfrac{1200 - 400}{2}}{\dfrac{600}{2}} = -\frac{4}{3}$$

Hence

$$2\theta_s = -53.13°, \; -233.13°$$

or

$$\theta_s = -26.57°, \; -116.57°$$

To correlate the principal shearing strains and the fibers that experience them, substitute one of the angles—preferably the acute angle—into the transformation equation for shearing strain. Thus, from Eqs. (7-33),

$$\frac{\gamma_{\xi\eta}}{2}(-26.57°) = -400(-\tfrac{4}{5}) + 300(\tfrac{3}{5}) = +500\mu \; \text{in./in.}$$

This calculation shows that the maximum shearing strain $(+1000\mu$ in./in.) occurs between the fibers that are situated along the ξ and η axes when $\theta = -26.57$ degrees.

The normal strains associated with these fibers are

$$\epsilon_\xi = \epsilon_\eta = \frac{1200 + 400}{2} = 800\mu \; \text{in./in.}$$

The fibers that experience the principal shearing strains are shown in Figure 7-10d. Note that the positive shearing strain implies a decrease in the angle between the positive ξ and the positive η axes. The negative shearing strain $(\gamma_{min} = -1000\mu$ in./in.) is associated with the increase in the angle between the positive ξ and the negative η axes. These relationships are indicated by the dotted lines in Figure 7-10d.

Observe that strains are dimensionless quantities so the results obtained here are valid in the SI system of units as well.

PROBLEMS / Section 7-4

For each state of strain listed in Problems 7-8, 7-9, and 7-10, determine: (a) The magnitudes and directions of the principal normal strains. Show the orientation of the line elements that experience these strains. (b) The magnitudes and directions of the principal shearing strains. Show the orientation of the line elements that experience these strains. (c) The components of strain associated with a set of line elements oriented 15 degrees counterclockwise with respect to the x axis.

✓**7-8** $\epsilon_x = 400\mu$ in./in., $\epsilon_y = -600\mu$ in./in., $\gamma_{xy} = -2400\mu$ in./in.

7-9 $\epsilon_x = -300\mu$ mm/mm, $\epsilon_y = -400\mu$ mm/mm, $\gamma_{xy} = -600\mu$ mm/mm

7-10 $\epsilon_x = 100\mu$ mm/mm, $\epsilon_y = -300\mu$ mm/mm, $\gamma_{xy} = 400\mu$ mm/mm

7-11 Three strain gages are mounted on a machine part as shown in Figure P7-11. The recorded strains for these gages are $\epsilon_1 = 200\mu$ in./in., $\epsilon_2 = -100\mu$ in./in., and $\epsilon_3 = -600\mu$ in./in. Determine the shearing strain between the line

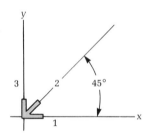

Figure P7-11

elements that lie along the x and y axes. Determine the principal normal strains and the principal shearing strains. For each case, show the set of line elements that experience these strains. If the material of the machine part is aluminum ($E = 10 \times 10^6$ psi, $G = 4 \times 10^6$ psi), determine the principal normal and shearing stresses.

7-5 Moments of Inertia of Plane Areas

Principal moments of inertia of plane areas play important roles in the unsymmetric bending of straight beams and in the buckling of concentrically and eccentrically loaded columns. In this section, we establish transformation equations for plane moments of inertia. We also establish formulas for the principal moments of inertia.

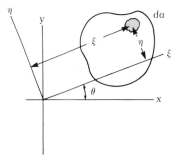

Figure 7-11

Consider a plane area lying in the xy plane and denote by da an infinitesimal element of area. Let ξ and η be a rectangular coordinate system whose origin coincides with the origin of the xy coordinate system. Let θ be the angular orientation of the ξ axis relative to the x axis, as shown in Figure 7-11.

The coordinate transformation that relates the coordinates of a point in the two systems is

$$\left.\begin{array}{l} \xi = x \cos \theta + y \sin \theta \\ \eta = -x \sin \theta + y \cos \theta \end{array}\right\} \tag{7-36}$$

From the definition of the moment of inertia relative to the ξ axis,

$$I_{\xi\xi} = \int_a \eta^2 \, da = \int_a \{x^2 \sin^2 \theta - 2xy \sin \theta \cos \theta + y^2 \cos^2 \theta\} \, da$$

or

$$I_{\xi\xi} = I_{xx} \cos^2 \theta + 2I_{xy}^* \sin \theta \cos \theta + I_{yy} \sin^2 \theta \tag{7-37}$$

where

$$I_{xx} = \int_a y^2 \, da$$

$$I_{yy} = \int_a x^2 \, da$$

$$I_{xy} = \int_a xy \, da$$

$$I_{xy}^* = -I_{xy}$$

are rectangular moments of inertia with respect to the x and y axes and the product of inertia with respect to these axes, respectively. Using the trigonometric identities in Eq. (7-3),

$$I_{\xi\xi} = \frac{I_{xx} + I_{yy}}{2} + \frac{I_{xx} - I_{yy}}{2} \cos 2\theta + I_{xy}^* \sin 2\theta \tag{7-38}$$

and

$$I_{\eta\eta} = \frac{I_{xx} + I_{yy}}{2} - \frac{I_{xx} - I_{yy}}{2} \cos 2\theta - I_{xy}^* \sin 2\theta \qquad (7\text{-}39)$$

Equation (7-39) has been obtained from Eq. (7-38) by setting $\theta = \theta + \pi/2$. Actually, $I_{\eta\eta}$ is more easily calculated from the relation

$$I_{\xi\xi} + I_{\eta\eta} = I_{xx} + I_{yy} \qquad (7\text{-}40)$$

which is obtained by adding Eqs. (7-38) and (7-39).

The transformation for products of inertia is obtained from its definition,

$$I_{\xi\eta}^* = -\int_a \xi\eta \, da = -\int_a \{y^2 \sin\theta \cos\theta + xy(\cos^2\theta - \sin^2\theta) \\ - x^2 \sin\theta \cos\theta\} \, da$$

or

$$I_{\xi\eta}^* = -\left(\frac{I_{xx} - I_{yy}}{2}\right) \sin 2\theta + I_{xy}^* \cos 2\theta \qquad (7\text{-}41)$$

The transformation equation for products of inertia is not ordinarily required in engineering analysis. It is included here to show that moments of inertia of plane areas transform according to the same rules as stresses and strains. Observe that Eqs. (7-38) and (7-41) have precisely the same form as Eqs. (7-21) for stresses and the same form as Eq. (7-33) for strains. Moreover, we observe that σ_x, ϵ_x, and I_{xx}; σ_y, ϵ_y, and I_{yy}; and τ_{xy}, $\gamma_{xy}/2$, and I_{xy}^* play identical roles as coefficients in these equations. Consequently, analyses for principal moments of inertia will be identical to those for principal normal stresses or principal normal strains.

Summary of moments of inertia transformation equations.
The transformation equations pertinent to moments of inertia of plane areas are listed here for convenience.

General Transformation Equations for Moments of Inertia

$$\left.\begin{aligned} I_{\xi\xi} &= \frac{I_{xx} + I_{yy}}{2} + \frac{I_{xx} - I_{yy}}{2} \cos 2\theta + I_{xy}^* \sin 2\theta \\[2mm] I_{\xi\eta}^* &= -\left(\frac{I_{xx} - I_{yy}}{2}\right) \sin 2\theta + I_{xy}^* \cos 2\theta \\[2mm] I_{\xi\xi} + I_{\eta\eta} &= I_{xx} + I_{yy} \end{aligned}\right\} \qquad (7\text{-}42)$$

Principal Moments of Inertia Transformation Equations

Magnitudes:
$$\begin{Bmatrix} I_{\max} \\ I_{\min} \end{Bmatrix} = \frac{I_{xx} + I_{yy}}{2} \pm \sqrt{\left(\frac{I_{xx} - I_{yy}}{2}\right)^2 + I_{xy}^{*2}} \quad \text{and} \quad I_{\xi\eta} = 0$$

Orientations:
$$\tan 2\theta_p = \frac{I_{xy}^*}{\dfrac{I_{xx} - I_{yy}}{2}}, \quad \theta_{p1} \perp \theta_{p2}$$

(7-43)

To correlate the principal moments of inertia with the axes about which they occur, substitute one of these angles—preferably the acute angle—into the transformation equation for moments of inertia.

Equations for principal products of inertia are not recorded because principal products of inertia play no role in engineering analysis.

EXAMPLE 7-4

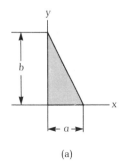

(a)

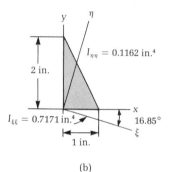

(b)

Figure 7-12

Determine the magnitudes of the principal moments of inertia and the orientation of the axes with which they are associated for the triangular cross section shown in Figure 7-12a. Let $a = 1$ in. and $b = 2$ in.

SOLUTION

The moments of inertia of the triangular area of Figure 7-12a are

$$I_{xx} = \tfrac{1}{12}ab^3 = \tfrac{4}{6} \text{ in.}^4$$
$$I_{yy} = \tfrac{1}{12}ba^3 = \tfrac{1}{6} \text{ in.}^4$$
$$I_{xy}^* = -\tfrac{1}{24}a^2b^2 = -\tfrac{1}{6} \text{ in.}^4$$

The magnitudes of the principal moments of inertia are

$$\begin{Bmatrix} I_{\max} \\ I_{\min} \end{Bmatrix} = \frac{\tfrac{4}{6} + \tfrac{1}{6}}{2} + \sqrt{\left(\frac{\tfrac{4}{6} - \tfrac{1}{6}}{2}\right)^2 + \left(-\tfrac{1}{6}\right)^2}$$
$$= \tfrac{1}{12}(5 \pm \sqrt{13}) = 0.7171 \text{ in.}^4, \ 0.1162 \text{ in.}^4$$

and the orientation of the axes associated with these principal moments of inertia are given by the relation

$$\tan 2\theta_p = \frac{-\tfrac{1}{6}}{\dfrac{\tfrac{4}{6} - \tfrac{1}{6}}{2}} = -\frac{2}{3}$$

Hence

$$2\theta_p = -33.69°, \ 146.31°$$

or

$$\theta_p = -16.85°, 73.15°$$

To correlate the magnitudes of the principal moments of inertia with these two axes, substitute one angle—preferably the acute angle—into the transformation for moments of inertia. Thus from Eq. (7-42),

$$I_{\xi\xi}(-16.85°) = 0.4167 + 0.25 \cos(-33.69°) - 0.1667 \sin(-33.69°)$$
$$= 0.7171 \text{ in}^4$$

This calculation shows that $I_{max} = 0.7171 \text{ in}^4$ is associated with the ξ axis when $\theta = -16.85$ degrees. $I_{min} = 0.1162 \text{ in}^4$ must be associated with the η axis of the $\xi\eta$ system. The principal axes of inertia for the triangular area are shown in Figure 7-12b.

EXAMPLE 7-5

Determine the magnitudes of the principal centroidal moments of inertia, and the orientation of the axes with which they are associated, for the plane area shown in Figure 7-13.

SOLUTION

The centroid of the area has the coordinates

$$\bar{x} = \bar{y} = \frac{150(50)75 + 100(50)25}{12,500} = 55 \text{ mm}$$

Moreover, the coordinate centroidal moments of inertia are

$$\bar{I}_{xx} = \tfrac{1}{12}(150)50^3 + (150 \times 50)30^2$$
$$+ \tfrac{1}{12}(50)100^3 + (100 \times 50)45^2$$
$$= 22.6 \times 10^6 \text{ mm}^4 = 22.6 \times 10^{-6} \text{ m}^4$$
$$\bar{I}_{xy} = 150(50)(-30)20 + 100(50)(-30)45$$
$$= -11.25 \times 10^6 \text{ mm}^4 = -11.25 \times 10^{-6} \text{ m}^4$$

and $\bar{I}_{yy} = \bar{I}_{xx}$ by symmetry.

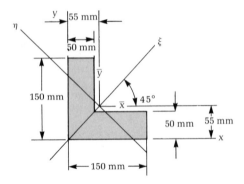

Figure 7-13

The magnitudes of the principal moments of inertia are

$$\begin{Bmatrix} I_{max} \\ I_{min} \end{Bmatrix} = \left\{ \frac{22.6 + 22.6}{2} \pm \sqrt{0 + (+11.25)^2} \right\} \times 10^{-6}$$

$$= 33.85 \times 10^{-6} \text{ m}^4, 11.35 \times 10^{-6} \text{ m}^4$$

and the orientation of the axes associated with these principal moments of inertia are given by the relation

$$\tan 2\theta_p = \frac{+11.25}{\dfrac{22.6 - 22.6}{2}} = +\infty$$

so that $\theta_p = +45°, -45°$.

The correlation calculation shows that

$$I_{\xi\xi}(+45°) = 22.6 \times 10^{-6} + (+11.25 \times 10^{-6}) = 33.85 \times 10^{-6} \text{ m}^4$$

Thus $I_{max} = 33.85 \times 10^{-6}$ m^4 is associated with the ξ axis oriented at $+45$ degrees as shown in Figure 7-13.

PROBLEMS / Section 7-5

7-12 Determine the principal centroidal moments of inertia for the angle section of Figure P7-12 and show the orientation of the axes with which they are associated.

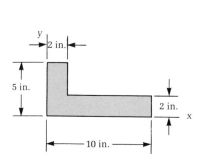

Figure P7-12

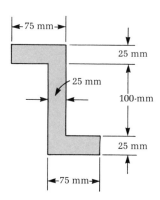

Figure P7-13

7-13 Calculate the principal centroidal moments of inertia for the z section shown in Figure P7-13. Sketch the orientation of the axes with which they are associated.

7-14 Calculate the principal centroidal moments of inertia for the triangular area shown in Figure P7-14. Sketch the orientation of the principal axes on the figure and indicate the axis about which the moment of inertia has its least value.

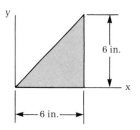

Figure P7-14

7-6 Mohr's Circle

A simple graphic representation called Mohr's circle can be used to facilitate the application of the transformation equations. Mohr's circle can be used as a purely graphic method for solving biaxial stress or biaxial strain or moment of inertia problems. However, it is more frequently used as a semi-graphic method of solving these types of problems. Before establishing Mohr's circle, it is instructive to bring together the transformation equations associated with stress, strain, and moments of inertia so that the similarity in their structures can be examined.

Stress Transformations

$$\left. \begin{aligned} \sigma_\xi &= \frac{\sigma_x + \sigma_y}{2} + \frac{\sigma_x - \sigma_y}{2} \cos 2\theta + \tau_{xy} \sin 2\theta \\ \tau_{\xi\eta} &= -\frac{\sigma_x - \sigma_y}{2} \sin 2\theta + \tau_{xy} \cos 2\theta \\ \sigma_\xi + \sigma_\eta &= \sigma_x + \sigma_y \end{aligned} \right\} \qquad (7\text{-}44)$$

Strain Transformations

$$\left. \begin{aligned} \epsilon_\xi &= \frac{\epsilon_x + \epsilon_y}{2} + \frac{\epsilon_x - \epsilon_y}{2} \cos 2\theta + \frac{\gamma_{xy}}{2} \sin 2\theta \\ \frac{\gamma_{\xi\eta}}{2} &= -\frac{\epsilon_x - \epsilon_y}{2} \sin 2\theta + \frac{\gamma_{xy}}{2} \cos 2\theta \\ \epsilon_\xi + \epsilon_\eta &= \epsilon_x + \epsilon_y \end{aligned} \right\} \qquad (7\text{-}45)$$

Inertia Transformations

$$\left. \begin{aligned} I_{\xi\xi} &= \frac{I_{xx} + I_{yy}}{2} + \frac{I_{xx} - I_{yy}}{2} \cos 2\theta + I_{xy}^* \sin 2\theta \\ I_{\xi\eta}^* &= -\frac{I_{xx} - I_{yy}}{2} \sin 2\theta + I_{xy}^* \cos 2\theta \\ I_{\xi\xi} + I_{\eta\eta} &= I_{xx} + I_{yy} \end{aligned} \right\} \qquad (7\text{-}46)$$

The general law of transformation that characterizes the transformation of stress, strain and moments of inertia is

$$P = A + B \cos 2\theta + C \sin 2\theta \qquad (7\text{-}47)$$

and

$$Q = -B \sin 2\theta + C \cos 2\theta \qquad (7\text{-}48)$$

Now transpose A to left-hand side of Eq. (7-47) and then square both sides. Also, square both sides of Eq. (7-48) and add the squares to the preceding result. The result is

$$(P - A)^2 + Q^2 = B^2 + C^2 = R^2 \qquad (7\text{-}49)$$

This equation is of a circle of radius $R = \sqrt{B^2 + C^2}$ whose center lies at the point $P = A$, $Q = 0$. This circle is called **Mohr's circle** and is a graphic representation of the transformation equations exhibited in Eqs. (7-44) through (7-46). Mohr's circle is depicted in Figure 7-14a.

There are several features of Mohr's circle that should be observed.

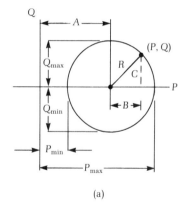

(a)

* The center of the circle is always situated on the P axis a distance $P = A$ from the origin of the P, Q space.
* The radius of the circle is always equal to $R = \sqrt{B^2 + C^2}$.
* The coordinates of a point on Mohr's circle are P, Q. In terms of the physical problems considered, the coordinates of a point on the circle are either:
  (a) σ_ξ, $\tau_{\xi\eta}$, the components of stress that act on an area perpendicular to the ξ axis.
  (b) ϵ_ξ, $\gamma_{\xi\eta}/2$, the normal strain of a fiber situated along the ξ axis and $\frac{1}{2}$ of the shearing strain between fibers lying along the ξ and η axes.
  (c) $I_{\xi\xi}$, $I_{\xi\eta}^*$, the moment of inertia with respect to the ξ axis and the product of inertia relative to the ξ and η axes.
* The principal values of P for any of the physical problems considered are obtained as the abscissas of the extreme points on the circle. Clearly the magnitudes of the principal values of P are

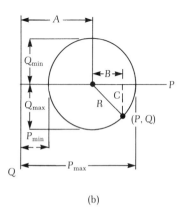

$$\begin{Bmatrix} P_{max} \\ P_{min} \end{Bmatrix} = A \pm R$$

* The principal values of Q for any of the physical problems considered are obtained as the ordinates of the most extreme points in the Q direction on the circle. The magnitudes of the principal values of Q are

$$\begin{Bmatrix} Q_{max} \\ Q_{min} \end{Bmatrix} = \pm R$$

Mohr's circle referred to a left-hand coordinate system.

Figure 7-14

* The orientation of the $\xi\eta$ coordinate system with respect to which the principal values of P and Q occur are determined by the following rules:
  Rule 1. Angular displacement on the circle is opposite angular rotation of the element.
  Rule 2. Angular displacement on the circle is equal to twice the angular rotation of the element.

The use of Mohr's circle and its relevant features are best demonstrated via examples. Examples 7-1, 7-3, and 7-4 are solved via Mohr's circle to illustrate important interpretations of its use for solving stress, strain, and moment of inertia problems.

Alternate Mohr's circle.   An alternate Mohr's circle that is sometimes found to be appealing is shown in Figure 7-14b. The difference between Mohr's circle of Figure 7-14a and the one shown in Figure 7-14b is that the P,Q-axis of the former circle forms a *right-hand* coordinate system, while the P,Q-axis of the latter circle forms a *left-hand* coordinate system. Right-hand coordinate systems are usually preferred in the formulation of mathematical problems, and our experiences have been restricted principally to such coordinate systems. Consequently, right-hand systems are chosen almost instinctively.

The justification for referring Mohr's circle to a left-hand coordinate system is that Rule 1 is reversed. That is, the sense of angular displacement on the circle is the *same* as the rotation of the element. The user is freed from the need to be cognizant of Rule 1, but must now remember to plot Q downward. The procedure used to construct a Mohr's circle is precisely the same for either a right-hand or a left-hand coordinate system.

Mohr's circle referred to a right-hand coordinate system is used in this book. You should rework examples 7-6, 7-7, and 7-8 using Mohr's circles referred to a left-hand coordinate system to demonstrate that the correct magnitudes for the principal values for P and Q, as well as the correct orientation for the principal axes for P and Q, are obtained.

EXAMPLE 7-6

Use Mohr's circle for stress to obtain the solution for Example 7-1. The state of stress given in Example 7-1 is repeated in Figure 7-15a for convenience.

SOLUTION

Two circumferential points are required to establish Mohr's circle for stress. The coordinates of a point on Mohr's circle for stress correspond to the components of stress on a plane through a point in the material. Such a plane is shown in Figure 7-15b. Thus one point on the circle can be established by recognizing that

$$\sigma_\xi = \sigma_x = 12 \text{ ksi} \qquad \text{and} \qquad \tau_{\xi\eta} = \tau_{xy} = 3 \text{ ksi}$$

when $\theta = 0$. This is point X in Figure 7-15c. A second point is established by letting $\theta = 90°$ and observing that $\sigma_\xi = \sigma_y = 4$ ksi and $\tau_{\xi\eta} = -\tau_{xy} = -3$ ksi. This is point Y in Figure 7-15c. The line XY connecting these two points is a diameter of the circle, and the intersection of this diameter with the σ_ξ axis is the center of the circle. This construction is shown in Figure 7-15c. Note that the coordinates of points X and Y in Mohr's circle correspond to the stress components acting on the x and y faces of the element in Figure 7-15a, respectively. (a) The magnitudes of the principal normal stresses are obtained from

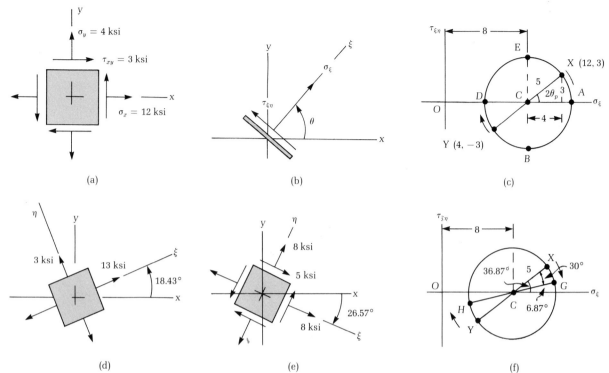

(a)   (b)   (c)

(d)   (e)   (f)

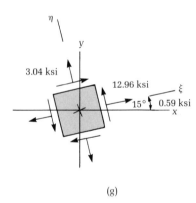

(g)

Figure 7-15

the circle

$$\begin{cases} \sigma_{max} \\ \sigma_{min} \end{cases} = 8 \pm 5 = 13 \text{ ksi, } 3 \text{ ksi}$$

Observe that, geometrically, σ_{max} is equal to the abscissa of the center of the circle plus the radius of the circle. It is the distance OA in Figure 7-15c. Likewise, σ_{min} is equal to the center of the circle minus the radius. It is the distance OD in Figure 7-15c.

The orientations of the planes on which these principal stresses act are obtained from the circle by observing that point X corresponds to an orientation of the $\xi\eta$ coordinate system such that the ξ and η axes coincide with x and y axes, respectively. Consequently, point A on the circle is located in a clockwise direction relative to point X. Moreover, the radii to points X and A subtend an angle of $2\theta_p = 36.87$ degrees, which can be obtained easily from the associated triangle. Rules 1 and 2 require that the ξ axis be located at $36.87/2 = 18.43$ degrees in a counterclockwise direction from the x axis. The element on which the principal normal stresses act is shown in Figure 7-15d.

(b) The magnitudes of the principal shearing stresses are

$$\begin{cases} \tau_{max} \\ \tau_{min} \end{cases} = \pm 5 \text{ ksi}$$

Note that these magnitudes are equal, geometrically, to the radius of the circle, and they are the ordinates of points E and B in Figure 7-15c. Moreover, the magnitudes of the normal stresses associated with the planes on which these principal shearing stresses act are, geometrically, equal to the abscissas of points E and B. Thus

$$\sigma_\xi = \sigma_\eta = 8 \text{ ksi}$$

The orientations of the planes on which the principal shearing stresses act are obtained from the circle by observing that point E on the circle is located in a counterclockwise direction relative to point X. Moreover, the radii to points X and E subtend an angle of 53.13 degrees. Rules 1 and 2 require that the ξ axis be located at $53.13/2 = 26.57$ degrees in a clockwise direction relative to the x axis. The element on which the principal shearing stresses act is shown in Figure 7-15e.

The proper senses for the shearing stresses are determined by noting that the coordinates of point E on the circle correspond to the components of stress on the plane perpendicular to the ξ axis. Since the shearing stress is positive on this plane, its sense must be the same as the positive η axis. The senses of the other three shearing stresses are then fixed by equilibrium requirements.

(c) The components of stress associated with a counterclockwise element orientation of $\theta = 15$ degrees are equal, geometrically, to the abscissas and ordinates of points G and H on the circle. A 15 degree counterclockwise orientation of the element corresponds to a 30 degree clockwise rotation of the diameter XY of the circle. In other words, the angle subtended by the radii CX and CG is 30 degrees. The angle that the diameter GH makes with the horizontal axis is, therefore, $36.87 - 30.00$ or 6.87 degrees. This construction is shown in Figure 7-15f where Mohr's circle of Figure 7-15c has been repeated for clarity. Thus, considering point G,

$$\sigma_\xi = 8 + 5 \cos 6.87° = 12.96 \text{ ksi}$$
$$\tau_{\xi\eta} = 5 \sin 6.87° = 0.59 \text{ ksi}$$

and, considering point H,

$$\sigma_\eta = 8 - 5 \cos 6.87° = 3.04 \text{ ksi}$$
$$\tau_{\xi\eta} = -5 \sin 6.87° = -0.59 \text{ ksi}$$

These stresses are shown on a properly oriented element in Figure 7-15g. The results presented in this example are the same as those of Example 7-1 as expected.

EXAMPLE 7-7

Use Mohr's circle for strain to obtain the solution for Example 7-3. The state of strain given in Example 7-3 is repeated in Figure 7-16a for convenience.

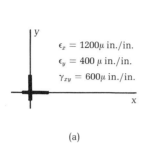

(a)

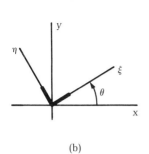

(b)

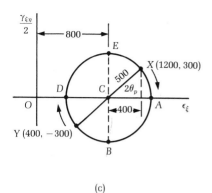

(c)

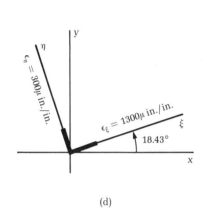

(d)

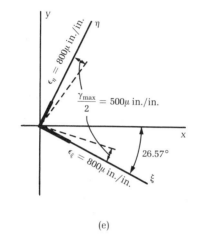

(e)

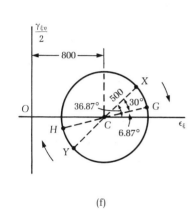

(f)

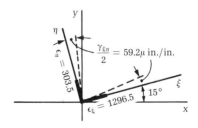

(g)

Figure 7-16

SOLUTION

Two circumferential points are required to establish Mohr's circle for strain. The coordinates of a point on Mohr's circle for strain correspond to the normal strain of a line element situated along the ξ axis and one-half of the shearing strain between line elements situated along the ξ and η axes. The $\xi\eta$ coordinate system is shown in Figure 7-16b. Consequently, point X on the circle can be established by recognizing that $\epsilon_\xi = \epsilon_x = 1200\mu$ in./in. and $\gamma_{\xi\eta}/2 = \gamma_{xy}/2 = 300\mu$ in./in. when $\theta = 0$. A second point, point Y, is established by letting $\theta = 90$ degrees and observing that $\epsilon_\xi = \epsilon_y = 400\mu$ in./in. and $\gamma_{\xi\eta}/2 = -\gamma_{xy}/2 = -300\mu$ in./in. The line XY connecting these two points is a diameter of the circle, and the intersection of this diameter with the ξ axis is the center C of the circle. This construction is shown in Figure 7-16c.

(a) The magnitudes of the principal normal strains are obtained from the circle

$$\left\{\begin{array}{c}\epsilon_{\max}\\\epsilon_{\min}\end{array}\right\} = 800 \pm 500 = 1300\mu \text{ in./in., } 300\mu \text{ in./in.}$$

Geometrically, ϵ_{max} is equal to the abscissa of the center of the circle plus the radius. It is the distance OA in Figure 7-16c. Similarly, ϵ_{min} is equal to the abscissa of the center of the circle minus the radius. It is the distance OD in the same figure.

The orientations of the line elements that experience these principal normal strains are obtained from the circle by observing that point X corresponds to an orientation on the $\xi\eta$ coordinate system such that the ξ and η axes coincide, respectively, with the x and y axes. Thus point A on the circle is located in a clockwise direction relative to point X. Moreover, the radii to points X and A subtend an angle of $2\theta_p = 36.87$ degrees. Rules 1 and 2 require that the ξ axis be located at 18.43 degrees in a counterclockwise direction from the x axis. The elements that experience the principal normal strains are shown in Figure 7-16d.

(b) The magnitudes of the principal shearing strains are

$$\left\{ \begin{array}{c} \left(\dfrac{\gamma}{2}\right)_{max} \\[2mm] \left(\dfrac{\gamma}{2}\right)_{min} \end{array} \right\} = \pm 500\mu \text{ in./in.}$$

Geometrically, these magnitudes are equal to the radius of the circle, and they are the ordinates of points E and B on the circle. Moreover, the magnitudes of the normal strains associated with the line elements that experience these shearing strains are equal to the abscissas of points E and B. Accordingly,

$$\epsilon_\xi = \epsilon_\eta = 800\mu \text{ in./in.}$$

The orientations of the line elements that experience these strains are obtained from the circle by observing that point E on the circle is located in a counterclockwise direction relative to point X. Moreover, the radii to points X and E subtend an angle of 53.13 degrees. Rules 1 and 2 require that the ξ axis be located at 26.57 degrees in a clockwise direction relative to the x axis. The line elements that experience the principal shearing strains are shown in Figure 7-16e.

Observe that the shearing strain corresponding to point E on the circle is positive. In addition, this strain occurs between elements that are situated along the positive ξ and positive η axes. By definition, a positive shearing strain represents a decrease in the right angle between these fibers. This result is shown by the dotted lines in Figure 7-16e.

(c) The components of strain associated with line element situated along the coordinate axes of the $\xi\eta$ system for $\theta = 15$ degrees are geometrically, equal to, the abscissas and ordinates of points G and H on the circle. Points G and H are obtained by rotating the XY diameter through 30 degrees in a clockwise direction. This construction is shown in Figure 7-16f where Mohr's circle of Figure 7-16c has been repeated for clarity. Considering point G,

$$\epsilon_\xi = 800 + 500 \cos 6.87° = 1296.5\mu \text{ in./in.}$$

$$\frac{\gamma_{\xi\eta}}{2} = 500 \sin 6.87° = 59.5\mu \text{ in./in.}$$

and, considering point H,

$$\epsilon_\eta = 800 - 500 \cos 6.87° = 303.5\mu \text{ in./in.}$$

$$\frac{\gamma_{\xi\eta}}{2} = -500 \sin 6.87° = -59.5\mu \text{ in./in.}$$

The line elements that experience these strains are shown in Figure 7-16g. The present results agree with those of Example 7-3 as expected.

EXAMPLE 7-8

Use Mohr's circle for moments of inertia to obtain the solution for Example 7-4. The moments of inertia for Example 7-4 are repeated in Figure 7-17a for convenience.

SOLUTION

Two circumferential points are required to establish Mohr's circle for moments of inertia. The coordinates of a point on Mohr's circle for moments of inertia correspond to the moment of inertia with respect to the ζ axis, $I_{\xi\xi}$, and the product of inertia with respect to the $\xi\eta$ axes, $I_{\xi\eta}^*$.

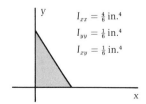

(a)

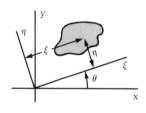

(b)

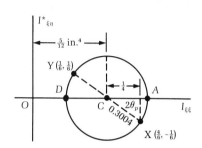

(c)

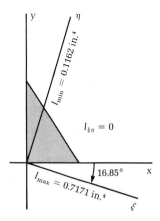

(d)

Figure 7-17

Thus point X on the circle can be established by recognizing that $I_{\xi\xi} = I_{xx} = \frac{4}{6}$ in^4 and $I_{\xi\eta}^* = I_{xy}^* = -\frac{1}{6}$ in^4 when $\theta = 0$. This can be seen from Figure 7-17b by imagining $\theta = 0$. A second point, point Y, on the circle can be established by letting $\theta = 90$ degrees and observing that $I_{\xi\xi} = I_{yy} = \frac{1}{6}$ in^4 and $I_{\xi\eta}^* = -I_{xy}^* = \frac{1}{6}$ in^4. The line XY connecting these two points is a diameter of the circle, and the intersection of this diameter with the $I_{\xi\xi}$ axis is the center of the circle. This construction is shown in Figure 7-17c.

The magnitudes of the principal moments of inertia are obtained from the circle,

$$\left\{ \begin{matrix} I_{max} \\ I_{min} \end{matrix} \right\} = \frac{5}{12} \pm 0.3004 = 0.7171 \text{ in}^4, 0.1162 \text{ in}^4$$

Geometrically, I_{max} is equal to the abscissa of the center of the circle plus the radius of the circle. It is the distance OA in Figure 7-17c. Likewise, I_{min} is equal to the center of the circle minus the radius. It is the distance OD in the figure.

The orientation of the principal axes of inertia are obtained from the circle by observing that point X corresponds to an orientation of the $\xi\eta$ coordinate system when $\theta = 0$. Thus point A on the circle is located in a counterclockwise direction relative to point X. Moreover, the radii to points X and A subtend an angle of $2\theta_p = 33.69$ degrees. Rules 1 and 2 require that the ξ axis be located at 16.85 degrees in a clockwise direction relative to the x axis. The principal axes of inertia are shown in Figure 7-17d. The results are the same as those of Example 7-4 as expected.

EXAMPLE 7-9

Two separate uniaxial states of stress are shown in Figure 7-18a. Determine (a) the state of stress, referred to an element whose sides are parallel to the xy axes, that results from a superposition of these two stress states and (b) the magnitudes and directions of the principal normal stresses associated with the combined stress state.

SOLUTION

(a) Recall that stress is not a vector. Consequently, the two stress states shown in Figure 7-18a cannot simply be added vectorially.

To superimpose the given states of stress, it is necessary to calculate the state of stress associated with the xy axes that is equivalent to the uniaxial state of stress II. This is done using Mohr's circle.

Mohr's circle for state of stress II is shown in Figure 7-18b. Note that points A and D on the circle correspond to the planes that are normal to the ξ and η axes, respectively.

The stress components of an equivalent state of stress referred to the xy axes are the coordinates of the points X and Y. Therefore,

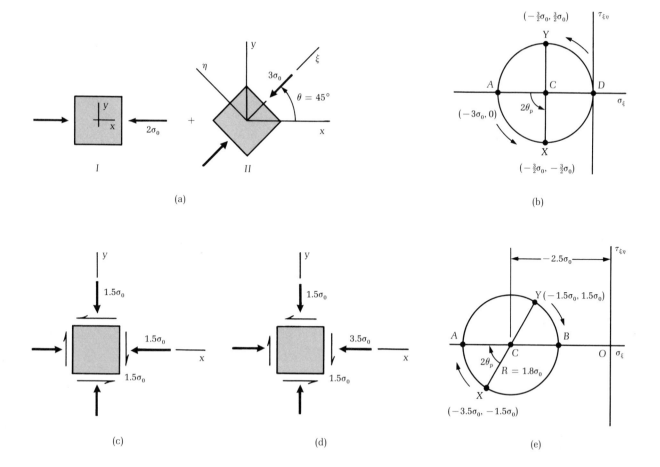

(a)

(b)

(c)

(d)

(e)

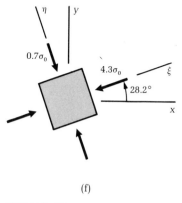

(f)

Figure 7-18

the coordinates of point X are the components of stress that act on the plane perpendicular to the x axis, and the coordinates of point Y are the components of stress that act on the plane perpendicular to the y axis.

The state of stress, referred to the xy axes, that is equivalent to the state of stress II is shown in Figure 7-18c.

The superposition of states of stress I and II can now be obtained by adding, algebraically, corresponding components of stress of state I with those of state II from Figures 7-18a and 7-18c. The result is shown in Figure 7-18d.

(b) The magnitudes and directions of the principal normal stresses corresponding to the superposed state of stress is now a straightforward application of the Mohr's circle. Point X on the circle of Figure 7-18e is established by recognizing that $\sigma_\xi = \sigma_x = -3.5\sigma_0$ and $\tau_{\xi\eta} = \tau_{xy} = -1.5\sigma_0$ when $\theta = 0$. Point Y on this circle is established by observing that $\sigma_\xi = \sigma_y = -1.5\sigma_0$ and $\tau_{\xi\eta} = -\tau_{xy} = 1.5\sigma_0$ when $\theta = 90$ degrees.

The magnitudes of the principal normal stresses correspond to the abscissas of points A and B in Figure 7-18e. Thus

$$\begin{Bmatrix} \sigma_{max} \\ \sigma_{min} \end{Bmatrix} = -2.5\sigma_0 \pm \sqrt{3.25}\sigma_0 = -0.7\sigma_0, \ -4.3\sigma_0$$

The orientations of the planes on which the principal normal stresses act are also obtained from the circle. The radii to points X and A subtend an angle $2\theta_p = 56.3$ degrees. Therefore, the ξ axis must be located 28.2 degrees in a counterclockwise direction from the x axis. The element on which the principal normal stresses act are shown in Figure 7-18f.

7-7   Relationship between E, v, and G for Isotropic Materials

The relationship between E, v, and G for an isotropic material can be established as follows. First, note that the biaxial stress-strain relations for an isotropic material are

$$\sigma_x = \frac{E}{1 - v^2}(\epsilon_x + v\epsilon_y) \tag{7-49a}$$

$$\sigma_y = \frac{E}{1 - v^2}(\epsilon_y + v\epsilon_x) \tag{7-49b}$$

$$\tau_{xy} = G\gamma_{xy} \tag{7-49c}$$

Since the material considered is isotropic, these same relations are valid regardless of the orientation of the xy system of axes.

Second, recall that the planes on which the principal normal stresses act are characterized by zero shearing stresses. Moreover, the shearing strain between the two line elements that experience the principal normal strains is also zero. From Eq. (7-49c), if $\gamma_{xy} = 0$, then $\tau_{xy} = 0$ so that the principal stress directions coincide with the principal strain directions.

Now consider the state of pure shearing stress shown in Figure 7-19a. Mohr's circle for this state of stress is shown in Figure 7-19b, and the principal normal stresses are shown on a properly oriented element in Figure 7-19c. Clearly,

$$\begin{Bmatrix} \sigma_{max} = \tau \\ \sigma_{min} = -\tau \end{Bmatrix} \tag{7-50}$$

The state of strain corresponding to the state of pure shearing stress is shown in Figure 7-20a, and the corresponding Mohr's circle is shown in Figure 7-20b. The line elements that experience the principal

(a)

(b)

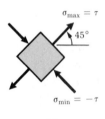

(c)

Figure 7-19

normal strains are shown in Figure 7-20c. Clearly,

$$\left. \begin{aligned} \epsilon_{max} &= \frac{\gamma}{2} \\[2em] \epsilon_{min} &= -\frac{\gamma}{2} \end{aligned} \right\} \qquad (7\text{-}51)$$

Since Eqs. (7-49) apply to any set of axes, it follows that

$$\sigma_{max} = \frac{E}{1 - v^2} (\epsilon_{max} + v\epsilon_{min}) \qquad (7\text{-}52)$$

or

$$\tau = \frac{E}{1 - v^2} \left\{ \frac{\gamma}{2} + v \left(-\frac{\gamma}{2} \right) \right\} \qquad (7\text{-}53)$$

Now $\tau = G\gamma$ so that Eq. (7-53) yields

$$G = \frac{E}{2(1 + v)} \qquad (7\text{-}54)$$

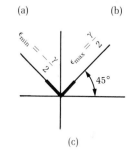

Figure 7-20

PROBLEMS / Section 7-6

7-15 Use Mohr's circle for stress to solve Problem 7-1.

7-16 Use Mohr's circle for stress to solve Problem 7-2.

7-17 Use Mohr's circle for stress to solve Problem 7-3.

7-18 Use Mohr's circle for strain to solve Problem 7-8.

7-19 Use Mohr's circle for strain to solve Problem 7-9.

7-20 Use Mohr's circle for strain to solve problem 7-10.

7-21 Use Mohr's circle for moments of inertia to solve Problem 7-12.

7-22 Use Mohr's circle for moments of inertia to solve Problem 7-13.

7-23 Use Mohr's circle for moments of inertia to solve Problem 7-14.

7-8 SUMMARY

This chapter developed transformation equations associated with plane stress, plane strain, and moments of inertia of plane areas. We also developed formulas and procedures required to determine the magnitudes and orientations of the principal values of normal and shearing stress, normal and shearing strain, and moments of inertia. We demonstrated the construction and use of Mohr's circle for the solution of plane stress, plane strain, and moments of inertia problems.

General Two-Dimensional Transformation Equations

$$P = A + B \cos 2\theta + C \sin 2\theta$$
$$Q = -B \sin 2\theta + C \cos 2\theta$$

where

$$P = \sigma_\xi, \ \epsilon_\xi, \ \text{or} \ I_{\xi\xi}$$

$$Q = \tau_{\xi\eta}, \ \frac{\epsilon_{\xi\eta}}{2}, \ \text{or} \ I^*_{\xi\eta}$$

$$A = \frac{\sigma_x + \sigma_y}{2}, \frac{\epsilon_x + \epsilon_y}{2}, \ \text{or} \ \frac{I_{xx} + I_{yy}}{2}$$

$$B = \frac{\sigma_x - \sigma_y}{2}, \frac{\epsilon_x - \epsilon_y}{2}, \ \text{or} \ \frac{I_{xx} - I_{yy}}{2}$$

$$C = \tau_{xy}, \frac{\epsilon_{xy}}{2}, \ \text{or} \ I^*_{xy} = -I_{xy}$$

Principal Values and Orientations of P

$$\begin{Bmatrix} P_{max} \\ P_{min} \end{Bmatrix} = A \pm \sqrt{B^2 + C^2} \qquad \tan 2\theta_p = \frac{C}{B}$$

Principal Values and Orientations of Q

$$\begin{Bmatrix} Q_{max} \\ Q_{min} \end{Bmatrix} = \pm \sqrt{B^2 + C^2} \qquad \tan 2\theta_s = -\frac{B}{C}$$

PROBLEMS / CHAPTER 7

Figures P7-24 through P7-35 correspond to Problems 7-24 through 7-35. For the state of stress shown in a figure, (a) Determine analytically the magnitudes and directions of the principal normal stresses. Show these stresses on a properly oriented element. (b) Determine analytically the magnitudes and directions of the principal shearing stresses. Show these stresses on a properly oriented element. (c) Determine analytically the magnitudes of the stress components associated with an element orientation of 15 degrees counterclockwise. Show these stresses on a properly oriented element.

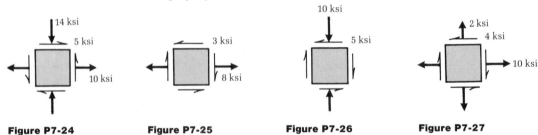

Figure P7-24   Figure P7-25   Figure P7-26   Figure P7-27

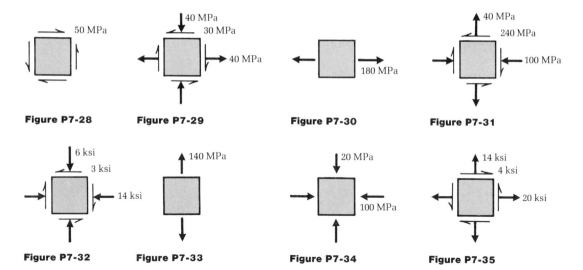

Figure P7-28   Figure P7-29   Figure P7-30   Figure P7-31

Figure P7-32   Figure P7-33   Figure P7-34   Figure P7-35

Problems 7-36 through 7-47 pertain, successively, to the states of stress shown in Figures P7-24 through P7-35. Construct Mohr's circle for the state of stress shown in the figure. (a) Determine the magnitudes and directions of the principal normal stresses. Show them on a properly oriented element. (b) Determine the magnitudes and directions of the principal shearing stresses. Show them on a properly oriented element. (c) Determine the magnitudes of the stress components associated with an element orientation of 15 degrees counterclockwise. Show these stresses on a properly oriented element.

7-48   For the 2-in. diameter shaft shown in Figure P7-48, determine the components of stress acting on an element of material at point A. Determine the principal normal stresses and the principal shearing stresses and show them on a properly oriented element.

7-49   For the 50-mm diameter shaft shown in Figure P7-49, determine the components of stress acting on an element of material at point A. Determine the principal normal stresses and the principal shearing stresses and show them on properly oriented elements.

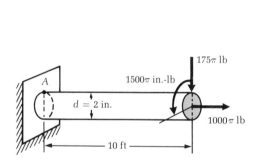

Figure P7-48

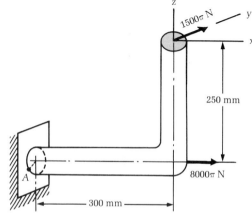

Figure P7-49

7-50 Determine the components of stress at point A of the 2-in. diameter shaft of Figure P7-50. Determine the principal normal stresses and the principal shearing stresses and show them on properly oriented elements.

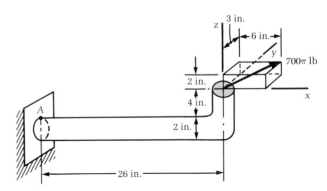

Figure P7-50

Problems 7-51 through 7-61 pertain to prescribed states of strain. For the state of strain listed, (a) Determine analytically the magnitudes and directions of the principal normal strains. Show the orientation of the line elements that experience these strains. (b) Determine analytically the magnitudes and directions of the principal shearing strains. Show the orientation of the line elements that experience these strains. (c) Determine analytically the components of strain associated with a set of line elements oriented 15 degrees counterclockwise with respect to the x axis.

7-51 $\epsilon_x = 65\mu$ in./in., $\epsilon_y = -35\mu$ in./in., $\gamma_{xy} = 240\mu$ in./in.

7-52 $\epsilon_x = 1500\mu$ in./in., $\epsilon_y = 500\mu$ in./in., $\gamma_{xy} = 2400\mu$ in./in.

7-53 $\epsilon_x = 60\mu$ in./in., $\epsilon_y = -80\mu$ in./in., $\gamma_{xy} = 0$

7-54 $\epsilon_x = 16\mu$ in./in., $\epsilon_y = 8\mu$ in./in., $\gamma_{xy} = 6\mu$ in./in.

7-55 $\epsilon_x = -1800\mu$ in./in., $\epsilon_y = 600\mu$ in./in., $\gamma_{xy} = 1000\mu$ in./in.

7-56 $\epsilon_x = 0$, $\epsilon_y = 0$, $\gamma_{xy} = 1000\mu$ in./in.

7-57 $\epsilon_x = -1500\mu$ mm/mm, $\epsilon_y = -500\mu$ mm/mm, $\gamma_{xy} = 2400\mu$ in./in.

7-58 $\epsilon_x = 0$, $\epsilon_y = 0$, $\gamma_{xy} = -800\mu$ mm/mm

7-59 $\epsilon_x = -400\mu$ mm/mm, $\epsilon_y = +200\mu$ mm/mm, $\gamma_{xy} = 600\mu$ mm/mm

7-60 $\epsilon_x = -140\mu$ mm/mm, $\epsilon_y = -300\mu$ mm/mm, $\gamma_{xy} = -120\mu$ mm/mm

7-61 $\epsilon_x = -40\sqrt{3}\mu$ mm/mm, $\epsilon_y = 20\sqrt{3}\mu$ mm/mm, $\gamma_{xy} = 60\mu$ mm/mm

7-62 A thin aluminum plate is subjected to a uniform stress of 10 ksi as shown in Figure P7-62. If Poisson's ratio is 0.25 for the material, calculate the normal strains ϵ_ξ, ϵ_η, and $\gamma_{\xi\eta}$. Assume Hooke's law applies.

Problems 7-63 through 7-73. Construct Mohr's circle for the states of strain given in Problems 7-51 through 7-61. (a) Determine the magnitudes and directions of the principal normal strains. Show the orientation of the line elements that experience these strains. (b) Determine the magnitudes and directions of the principal shearing strains. Show the orientations of the line elements that

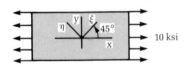

Figure P7-62

experience these strains. (c) Determine the magnitudes of the strain components associated with an orientation of line elements of 15 degrees counterclockwise. Show the orientations of the line elements that experience these strains.

7-74 Three strain gages are mounted on a machine part as shown in Figure P7-74. Show that the strain components associated with line elements situated along the x and y axes are

$$\epsilon_x = \epsilon_1, \epsilon_y = \epsilon_3, \text{ and } \gamma_{xy} = (2\epsilon_2 - \epsilon_1 - \epsilon_3)$$

where ϵ_1, ϵ_2, and ϵ_3 are the strains indicated by gages 1, 2, and 3, respectively.

If $\epsilon_1 = 300\mu$ in./in., $\epsilon_2 = 200\mu$ in./in., and $\epsilon_3 = -500\mu$ in./in., determine the magnitudes and directions of the principal normal strains and of the principal shearing strains.

7-75 Three strain gages are mounted on a structure in the configuration shown in Figure P7-75. Show that the strain components associated with line elements situated along the x and y axes are

$$\epsilon_x = \epsilon_1, \epsilon_y = \tfrac{1}{3}(2\epsilon_2 + 2\epsilon_3 - \epsilon_1), \gamma_{xy} = \frac{2}{\sqrt{3}}(\epsilon_2 - \epsilon_3)$$

where ϵ_1, ϵ_2, and ϵ_3 are the strains indicated by gages 1, 2, and 3, respectively.

If $\epsilon_1 = 500\mu$ mm/mm, $\epsilon_2 = 300\mu$ mm/mm, and $\epsilon_3 = -500\mu$ mm/mm, determine the magnitudes and directions of the principal normal strains and of the principal shearing strains.

7-76 Calculate the principal centroidal moments of inertia for the isosceles triangle shown in Figure P7-76. ($I_{xx} = 54$ in^4, $I_{yy} = 3.375$ in^4, and $I_{xy} = 0$.) Sketch the orientation of the axes with respect to which these principal centroidal moments of inertia occur.

7-77 Determine the principal centroidal moments of inertia for the right triangle shown in Figure P7-77. Sketch the orientation of the principal axes.

7-78 Determine the principal centroidal moments of inertia for the hollow rectangular cross section shown in Figure P7-78. Show the orientation of the principal axes.

7-79 Calculate the principal centroidal moments of inertia for the cross section shown in Figure P7-79. Show the orientation of the axes with which they are associated.

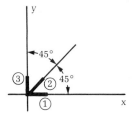

Figure P7-74

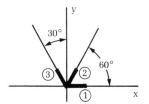

Figure P7-75

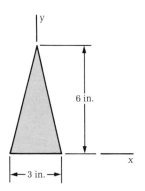

Figure P7-76

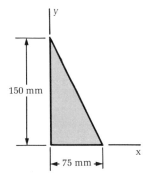

Figure P7-77

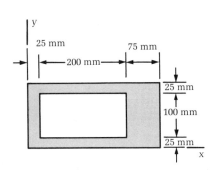

Figure P7-78

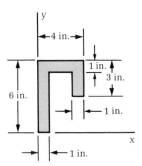

Figure P7-79

Figure P7-80

7-80   A C 10 × 20 section and a S 10 × 25.4 section are used to form the cross section shown in Figure P7-80. Calculate the principal centroidal moments of inertia. Sketch the directions of the principal centroidal axes.

7-81 and **7-82**   Use Mohr's circle to determine the magnitude of the principal centroidal moments of inertia for the plane areas shown in Figures P7-77 and P7-79. Sketch the orientation of the axes with which they are associated.

CHAPTER 8

Deflections of Elastic Beams

8-1 Introduction

The design or analysis of straight beams (either statically determinate or statically indeterminate) for strength presumes that means are available whereby we can determine the stresses in an engineering member resulting from the application of external forces. It also presumes that the strength of the material from which the member is made is readily available. In Chapter 5, we established formulas that permit the determination of normal stresses and shearing stresses on a cross section. In Chapter 2, we discussed methods for obtaining various mechanical properties of engineering materials, particularly strength parameters such as proportional limit, yield stress or yield strength, and ultimate strength. These formulas and strength parameters make it possible to ensure that the loads applied to a beam do not result in stresses that exceed an allowable or design stress. This aspect of the design of beams is called **design for strength**.

It is important to observe that a beam that is properly designed for strength is not necessarily properly designed for stiffness; that is, the lateral displacements of the centroidal line of the beam may be intolerable for the intended purpose of the beam. The design aspect that aims at proportioning the beam so that excessive deformations do not occur is called **design for stiffness**.

That a beam properly designed for strength may not be properly designed for stiffness is easily seen by considering a long slender cantilever beam such as a hacksaw blade. A force that causes extremely large deflections can be applied at one end of the blade; however, when the force is removed, the blade will return to its original straight configuration. The fact that large deflections have occurred in the blade for a force that results in stresses that do not exceed the proportional limit of the material suggests that a separate design for stiffness is required.

The purpose of this chapter is to develop procedures through which we can establish formulas required to facilitate the design of beams for stiffness. A further purpose is to develop some of the more basic beam deflection formulas.

Essentially four methods are used to develop formulas for the slope and deflection of straight beams: double integration, moment-area, superposition, and successive integrations. The physical and mathematical foundation for each of these methods and effective procedures to facilitate their use are developed in this chapter.

The deflection of a beam needs to be defined more precisely. Consider the locus of the centroids of the cross sections along the axis of a beam. This locus is a straight line that is referred to as the **centroidal line** of the beam. Now, as in Chapter 5, the cross section of the beam is assumed to be rigid so that every point in a cross section displaces the same amount laterally. Consequently, as far as the lateral deflections of the beam are concerned, the beam can be perceived as a line of particles along its centroidal line. The object of all of the four methods enumerated here is to determine the deviation (displacement) of a point on the centroidal line in its deformed configuration from its position in its undeformed configuration.

8-2 Double Integration Procedure

The method of double integration permits an algebraic expression for the slope and an algebraic expression for the deflected centroidal line of the beam to be obtained by direct integration.

Figure 8-1a shows a segment of a straight beam in its position before deformation. Two vertical sections are identified by the dotted lines. The length of a fiber situated along the centroidal line is denoted by ds, while the length of a fiber parallel to the centroidal line, but at a distance η from it, is denoted by dS. In the undeformed state, $dS \equiv ds$ for the straight beam.

Figure 8-1b shows the beam segment in its deformed configuration. The length of a fiber situated along the centroidal line in the

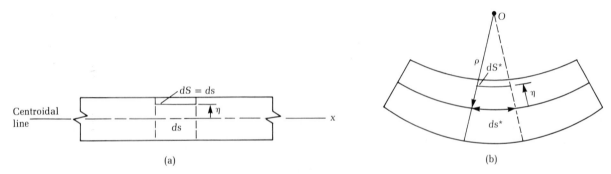

(a)

(b)

Figure 8-1

deformed configuration is ds^*, and the fiber whose length was dS before deformation acquires the length dS^* after deformation. Denote the **radius of curvature** at a section by ρ. Referring to Figure 8-1b, from geometrical considerations,

$$\frac{ds^*}{\rho} = -\frac{dS^* - ds^*}{\eta} \tag{8-1}$$

The strain at any point on a cross section is denoted by $\epsilon(\eta)$ so that

$$\left. \begin{aligned} dS^* &= \{1 + \epsilon(\eta)\}\, ds \\ ds^* &= \{1 + \epsilon(0)\}\, ds \end{aligned} \right\} \tag{8-2}$$

Observe that $\epsilon(0)$ is the strain associated with a line element lying along the **centroidal line** of the beam. Under general loading conditions, this strain is not zero because the centroidal line may not lie in the neutral surface of the beam.

Thus

$$\frac{dS^* - ds^*}{\eta} = \left\{ \frac{\epsilon(\eta) - \epsilon(0)}{\eta} \right\} ds \tag{8-3}$$

Now, if the beam behaves in an elastic manner,

$$\epsilon(\eta) = \frac{\sigma(\eta)}{E} = -\frac{M\eta}{EI} \tag{8-4}$$

so that Eq. (8-3) can be expressed as

$$\frac{dS^* - ds^*}{\eta} = -\frac{M}{EI}\, ds \tag{8-5}$$

Equations (8-1) and (8-5) combine to yield

$$\frac{1}{\rho} = \frac{M}{EI} \frac{ds}{ds^*} \tag{8-6a}$$

From Eq. (8-2), $ds/ds^* = 1/[1 + \epsilon(0)] = 1$ for elastic bending of beams by transverse loads only. Therefore, the **moment-curvature** relation of an elastic beam is

$$\frac{1}{\rho} = \frac{M}{EI} \tag{8-6b}$$

Observe that the assumption $\epsilon(0) \ll 1$ does not preclude large displacements of the centroidal line of the beam. This assertion is substantiated by considering a slender beam, such as a hacksaw blade, that is clamped at one end and loaded at its free end by a concentrated load. The blade can undergo extremely large deflections while the strains remain within the proportional limit.

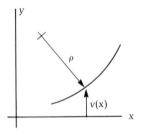

Figure 8-2

The curvature of the plane curve shown in Figure 8-2 is

$$\frac{1}{\rho} = \frac{\dfrac{d^2v}{dx^2}}{\left\{1 + \left(\dfrac{dv}{dx}\right)^2\right\}^{3/2}} \tag{8-7}$$

If we restrict our attention to the class of plane curves for which $(dv/dx)^2 \ll 1$, then an approximate expression for the curvature of this class of curves is

$$\frac{1}{\rho} \approx \frac{d^2v}{dx^2} \tag{8-8}$$

This approximation for the curvature is accurate for so-called shallow curves.

Equating Eqs. (8-6b) and (8-8) leads to the differential equation governing the deflection of points on the centroidal line of the beam,

$$EIv''(x) = M(x) \tag{8-9}$$

The centroidal line of the deformed beam is frequently referred to as the elastic curve of the beam. Note that the approximation $(dv/dx)^2 \ll 1$ restricts the use of this equation to relatively small displacements. This restriction does not negate the practical importance of this formula, however. Most engineering beam problems fall well within the limits imposed by this restriction.

Boundary and continuity conditions.   Equation (8-9) applies to any segment of a straight beam for which the moment equation is continuous. Consequently, if n separate moment equations are required to describe the bending moments along the axis of the beam, n separate differential equations of the type shown in Eq. (8-9) are required to describe the deflected centroidal line of the beam—one for each segment for which a moment equation has been written. Two constants arise from the integration of each of these differential equations. Consequently, $2n$ constants of integration result.

Beam with a single moment equation.   For beams for which a single moment equation and, hence, a single differential equation, suffices, the constants of integration are determined from algebraic equations that account for the manner in which the beam is supported. For example, at a roller or a pin connection, the transverse deflection v is prevented. At a clamp support, both the transverse deflection v and the slope v' are prevented. These support conditions are more commonly

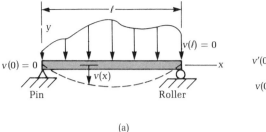

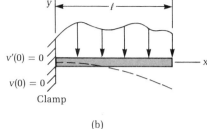

(a)   (b)

Figure 8-3

referred to as **boundary conditions** and are expressed symbolically as

$$v(a) = 0 \quad \text{(pin connection or roller)} \tag{8-10}$$

or

$$\left.\begin{array}{l} v'(a) = 0 \\ v(a) = 0 \end{array}\right\} \text{(clamp)} \tag{8-11}$$

Here a signifies the value of x at which the support condition occurs. In Figures 8-3a and 8-3b, the pin connector and the clamp occur at $x = 0$, and the roller occurs at $x = \ell$ in the coordinate systems shown in the figures.

Beams with _n_ moment equations. For beams that require n moment equations, $2n$ undetermined constants arise because of the integration of the n differential equations associated with the n moment equations. The support conditions discussed in the previous paragraphs are not sufficient to permit the determination of all of these integration constants. Thus supplementary conditions called **continuity conditions** are required. Continuity conditions are frequently called matching conditions because these conditions arise from the process that matches the slopes and deflections between adjacent segments of the beam.

Physically, the slope is required to be continuous between adjacent intervals because no corners are permitted in the deflection curve. Similarly, the deflection is required to be continuous between adjacent intervals because no breaks in the deflection curve are permitted. Figure 8-4a represents a deflection curve for which the slope and deflection are continuous at $x = a$. Figure 8-4b shows a sharp corner at $x = a$ that is indicative of a discontinuity in slope. Figure 8-4c shows

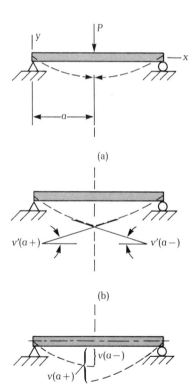

Figure 8-4

a break in the deflection curve that is indicative of a discontinuity in the deflection curve.

Mathematically, the continuity conditions can be expressed as

$$v'(a-) = v'(a+) \tag{8-12}$$

and

$$v(a-) = v(a+) \tag{8-13}$$

Here $a-$ and $a+$ are used to indicate values of x infinitesimally to the left and to the right of $x = a$, respectively. Thus Eqs. (8-12) and (8-13) state that the slope and deflection of the segment to the left of $x = a$ must be equal, respectively, to the slope and deflection of the segment to the right of $x = a$ as the point $x = a$ is approached.

For statically determinate beams, there will always exist just two boundary conditions. For statically indeterminate beams, the number of boundary conditions must exceed two by the degree of statical indeterminacy of the beam. Thus, if a beam is clamped at both ends, it is statically indeterminate to the second degree and requires four boundary conditions for a solution. The additional boundary conditions lead to algebraic equations that allow the determination of the support reactions that cannot be determined from statical considerations alone.

Often the important deflection of a beam is the maximum deflection it undergoes. The location of the maximum deflection can frequently be determined by inspection. For example, the maximum deflection of a cantilever with a concentrated force at its free end occurs under the load. When we cannot determine the location of the maximum deflection by inspection, we must make use of analytical means.

In general, the maximum deflection of a beam occurs at a point along the beam for which the tangent to the deflection curve (slope $v'(x)$) is zero or at an overhang. To locate the point along the beam that undergoes the maximum deflection, we must locate all points for which a horizontal tangent occurs. We can construct a sequence whose members are the values of deflection at these points of horizontal tangency. By inspection, the largest numerical member of the sequence is the maximum of the deflections associated with horizontal tangents of the deflection curve. If there is an overhang, the deflection there should be calculated and compared with the maximum value of the deflections associated with the horizontal tangents.

EXAMPLE 8-1

Determine equations for the slope and deflection of the cantilever beam shown in Figure 8-5. The flexural rigidity EI is constant over the entire span ℓ.

SOLUTION

The moment equation for the interval $0 \leq x \leq \ell$ is

$$M(x) = -Px \tag{a}$$

so that the differential equation that governs the deflections of the centroidal line of the beam is

$$EIv''(x) = -Px \tag{b}$$

Since the beam is clamped at the end $x = \ell$, the boundary conditions associated with differential Eq. (b) are

$$v'(\ell) = 0 \tag{c}$$

and

$$v(\ell) = 0 \tag{d}$$

Equations (b), (c), and (d) constitute a so-called **boundary value problem**. To obtain a solution, Eq. (b) is integrated directly.

$$EIv'(x) = -\frac{Px^2}{2} + C_1 \tag{e}$$

$$EIv(x) = -\frac{Px^3}{6} + C_1x + C_2 \tag{f}$$

Boundary conditions (c) and (d) are used to obtain two algebraic equations connecting the constants of integration C_1 and C_2. Consequently,

$$v'(\ell) = 0: \quad 0 = -\frac{P\ell^2}{2} + C_1$$

$$v(\ell) = 0: \quad 0 = -\frac{P\ell^3}{6} + C_1\ell + C_2$$

or, expressed in standard form,

$$\left. \begin{array}{l} C_1 = \dfrac{P\ell^2}{2} \\[2mm] C_1\ell + C_2 = \dfrac{P\ell^3}{6} \end{array} \right\} \tag{g}$$

This is a particularly simply set of simultaneous equations that can be solved successively. That is, C_1 can be calculated directly from the first equation and, subsequently, C_2 can be calculated from the second equation. Solution of algebraic equations that arise through the application of boundary conditions can frequently be obtained in this

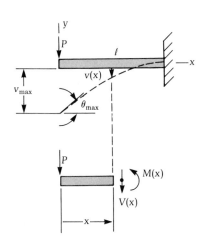

Figure 8-5

manner for more complicated problems. Equations (g) yield

$$C_1 = \frac{P\ell^2}{2} \\ C_2 = -\frac{P\ell^3}{3}$$

(h)

Accordingly, the equations representing the slope and deflection of the centroidal line of the beam at an arbitrary point are

$$v'(x) = \frac{1}{EI}\left(-\frac{Px^2}{2} + \frac{P\ell^2}{2}\right) = \frac{P}{2EI}(\ell^2 - x^2)$$

(i)

and

$$v(x) = \frac{1}{EI}\left(-\frac{Px^3}{6} + \frac{P\ell^2 x}{2} - \frac{P\ell^3}{3}\right) = -\frac{P}{EI}\left(\frac{x^3}{6} - \frac{\ell^2 x}{2} + \frac{\ell^3}{3}\right)$$

(j)

Frequently, formulas are required for the maximum slope and deflection. These formulas can be obtained by setting $x = 0$ in Eqs. (i) and (j), respectively. Thus

$$v'(0) \equiv \theta_{max} = \frac{P\ell^2}{2EI}$$

(k)

and

$$v(0) \equiv v_{max} = -\frac{P\ell^3}{3EI}$$

(l)

EXAMPLE 8-2

Determine equations for the slope and deflection of the cantilever beam shown in Figure 8-6. The flexural rigidity EI is constant over the entire span ℓ.

SOLUTION

The moment equation for the interval $0 \le x \le \ell$ is

$$M(x) = -\frac{qx^2}{2}$$

(a)

so that the differential equation that governs the deflections of the centroidal line of this beam is

$$EIv''(x) = -\frac{qx^2}{2}$$

(b)

Again, the clamped support at $x = \ell$ leads to the boundary conditions

$$v'(\ell) = 0$$

(c)

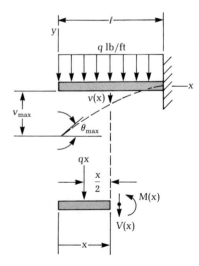

Figure 8-6

and

$$v(\ell) = 0 \qquad\qquad\qquad\text{(d)}$$

Direct integration of Eq. (b) leads to

$$EIv'(x) = -\frac{qx^3}{6} + C_1 \qquad\qquad\qquad\text{(e)}$$

and

$$EIv(x) = -\frac{qx^4}{24} + C_1 x + C_2 \qquad\qquad\qquad\text{(f)}$$

Application of the boundary conditions (c) and (d) lead to the two algebraic equations

$$\left.\begin{aligned} C_1 &= \frac{q\ell^3}{6} \\ C_1\ell + C_2 &= \frac{q\ell^4}{24} \end{aligned}\right\} \qquad\qquad\qquad\text{(g)}$$

which yield

$$\left.\begin{aligned} C_1 &= \frac{q\ell^3}{6} \\ C_2 &= -\frac{q\ell^4}{8} \end{aligned}\right\} \qquad\qquad\qquad\text{(h)}$$

Consequently, the equations representing the slope and deflection of the centroidal line of the beam at an arbitrary point are

$$v'(x) = \frac{1}{EI}\left(-\frac{qx^3}{6} + \frac{q\ell^3}{6}\right) = \frac{q}{6EI}(\ell^3 - x^3) \qquad\qquad\text{(i)}$$

and

$$v(x) = \frac{1}{EI}\left(-\frac{qx^4}{24} + \frac{q\ell^3 x}{6} - \frac{q\ell^4}{8}\right) = \frac{-q}{24EI}(x^4 - 4\ell^3 x + 3\ell^4) \quad\text{(j)}$$

Formulas for the maximum slope and deflection are obtained by setting $x = 0$ in Eqs. (i) and (j), respectively. Thus

$$v'(0) \equiv \theta_{max} = \frac{q\ell^3}{6EI} \qquad\qquad\qquad\text{(k)}$$

and

$$v(0) \equiv v_{max} = -\frac{q\ell^4}{8EI} \qquad\qquad\qquad\text{(l)}$$

EXAMPLE 8-3

Determine equations for the slope and deflection of the simply supported beam shown in Figure 8-7. The flexural rigidity EI is constant over the entire span.

SOLUTION

A separate moment equation is applicable for the segments of the beam to the left and to the right of the applied force P. Thus

$$\underline{0 \le x \le a} \qquad\qquad \underline{a \le x \le \ell}$$

$$M(x) = \frac{bP}{\ell}\,x \qquad M(x) = \frac{bP}{\ell}\,x - P(x-a) \qquad\text{(a)}$$

A separate differential equation must be satisfied in each interval for which a separate moment equation is required. Thus

$$EIv''(x) = \frac{Pb}{\ell}\,x \quad\text{and}\quad EIv''(x) = \frac{Pb}{\ell}\,x - P(x-a) \qquad\text{(b)}$$

Integration of the differential equations applicable to the intervals $0 \le x \le a$ and $a \le x \le \ell$ gives

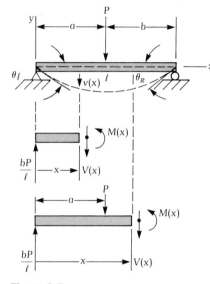

Figure 8-7

$$\underline{0 \le x \le a} \qquad\qquad \underline{a \le x \le \ell}$$

$$EIv'(x) = \frac{Pb}{\ell}\frac{x^2}{2} + C_1 \qquad EIv'(x) = \frac{Pb}{\ell}\frac{x^2}{2} - \frac{P(x-a)^2}{2} + C_3 \qquad\text{(c)}$$

$$EIv(x) = \frac{Pb}{\ell}\frac{x^3}{6} \qquad\qquad EIv(x) = \frac{Pb}{\ell}\frac{x^3}{6} - \frac{P(x-a)^3}{6}$$

$$+ C_1 x + C_2 \qquad\qquad\qquad + C_3 x + C_4 \qquad\text{(d)}$$

Notice that each integration results in a new constant of integration so that two integration constants arise for each segment of the beam that requires an independent moment equation. In the present problem, four constants of integration arise. For beams where n moment equations are required, $2n$ constants of integration arise.

Since the beam is simply supported at its ends, appropriate boundary conditions are

$$v(0) = 0 \qquad\qquad\qquad\qquad\qquad\qquad \text{(e)}$$

and

$$v(\ell) = 0 \qquad\qquad\qquad\qquad\qquad\qquad \text{(f)}$$

These conditions will lead to two algebraic equations between the four integration constants. Two additional algebraic equations are clearly required if unique solutions for C_1, C_2, C_3, and C_4 are to be obtained. The additional equations are obtained from the **requirements**

of continuity of slope and deflection at the point of load application. Consequently, at $x = a$,

$$v'(a-) = v'(a+) \tag{g}$$

and

$$v(a-) = v(a+) \tag{h}$$

The boundary conditions and continuity relations lead to the set of algebraic equations

$$\left.\begin{array}{l} C_2 = 0 \\ C_1 = C_3 \\ C_2 = C_4 = 0 \\ \dfrac{bP\ell^2}{6} - \dfrac{P(\ell - a)^3}{6} + C_3\ell = 0 \end{array}\right\} \tag{i}$$

Accordingly,

$$C_1 = C_3 = -\frac{Pb}{6\ell}(\ell^2 - b^2) \tag{j}$$

The required equations for the slope and deflection of the centroidal line of the beam are obtained by substituting Eqs. (i) and (j) into Eqs. (c) and (d). This substitution is not done here since the results of these substitutions are obvious.

Suppose that $a = b = \ell/2$. Then, from symmetry considerations, the slope at the left end (θ_l) and the slope at the right end (θ_r) are numerically equal. From Eq. (c), for the interval $0 \le x \le a$,

$$v'_{max}(0) = -\frac{P\ell^2}{16EI} \tag{k}$$

and then

$$v'_{max}(\ell) = +\frac{P\ell^2}{16EI} \tag{l}$$

Moreover, by observation, the maximum deflection occurs at the center of the span under the load P. It is given by the formula

$$v_{max}\left(\frac{\ell}{2}\right) = -\frac{P\ell^3}{48EI} \tag{m}$$

EXAMPLE 8-4

For the straight beam shown in Figure 8-8a, determine (a) equations for the slope and deflection at an arbitrary point along the beam, (b) the location of the maximum deflection, and (c) the magnitude of the maximum deflection.

SOLUTION

(a) Appropriate moment equations for the intervals $0 \leq x \leq \ell$ and $\ell \leq x \leq 2\ell$ are, from Figures 8-8b and 8-8c,

$$M(x) = -\frac{M_0}{2\ell} x \qquad \text{and} \qquad M(x) = -\frac{M_0}{2\ell} x + M_0 \qquad \text{(a)}$$

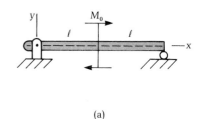

(a)

respectively. Thus differential equations that describe the deflected centroidal line of the beam in the two intervals are

$$\underline{0 \leq x \leq \ell} \qquad\qquad\qquad \underline{\ell \leq x \leq 2\ell}$$

$$EIv''(x) = -\frac{M_0}{2\ell} x \qquad \text{and} \qquad EIv''(x) = -\frac{M_0}{2\ell} x + M_0 \qquad \text{(b)}$$

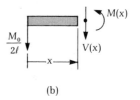

(b)

The boundary conditions and continuity requirements are

$$\left.\begin{aligned} v(0) &= 0 \\ v(2\ell) &= 0 \\ v'(\ell-) &= v'(\ell+) \\ v(\ell-) &= v(\ell+) \end{aligned}\right\} \qquad \text{(c)}$$

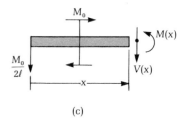

(c)

Figure 8-8

Direct integration of Eqs. (b) yields

$$\underline{0 \leq x \leq \ell} \qquad\qquad\qquad \underline{\ell \leq x \leq 2\ell}$$

$$EIv'(x) = -\frac{M_0}{2\ell} \frac{x^2}{2} + C_1 \qquad EIv'(x) = -\frac{M_0}{2\ell} \frac{x^2}{2} + M_0 x + C_3 \qquad \text{(d)}$$

$$EIv(x) = -\frac{M_0}{2\ell} \frac{x^3}{6} \qquad\quad EIv(x) = -\frac{M_0}{2\ell} \frac{x^3}{6} + \frac{M_0 x^2}{2}$$

$$+ C_1 x + C_2 \qquad\qquad\quad + C_3 x + C_4 \qquad\qquad \text{(e)}$$

Conditions (c) lead to the set of algebraic equations

$$\left.\begin{aligned} C_2 &= 0 \\ 2C_3\ell + C_4 &= -\tfrac{4}{3}M_0\ell^2 \\ C_1 - C_3 &= M_0\ell \\ C_1\ell - C_3\ell - C_4 &= \tfrac{1}{2}M_0\ell^2 \end{aligned}\right\} \qquad \text{(f)}$$

which have the solution

$$\left.\begin{aligned} C_1 &= \tfrac{1}{12}M_0\ell \\ C_2 &= 0 \\ C_3 &= -\tfrac{11}{12}M_0\ell \\ C_4 &= \tfrac{1}{2}M_0\ell^2 \end{aligned}\right\} \qquad \text{(g)}$$

The required equations for slope and deflection of the centroidal line of the beam are

$$\underline{0 \leq x \leq \ell} \qquad\qquad\qquad \underline{\ell \leq x \leq 2\ell}$$

$$EIv'(x) = -\frac{M_0}{2\ell}\frac{x^2}{2} + \tfrac{1}{12}M_0\ell \qquad EIv'(x) = -\frac{M_0}{2\ell}\frac{x^2}{2} + M_0 x \tag{h}$$
$$- \tfrac{11}{12}M_0\ell$$

$$EIv(x) = -\frac{M_0}{2\ell}\frac{x^3}{6} + \tfrac{1}{12}M_0\ell x \qquad EIv(x) = -\frac{M_0}{2\ell}\frac{x^3}{6} + \frac{M_0 x^2}{2} \tag{i}$$
$$- \tfrac{11}{12}M_0\ell x + \tfrac{1}{2}M_0\ell^2$$

(b) The locations along the centroidal line at which the maximum deflections occur are obtained by determining points of the deflected centroidal line for which its tangent is horizontal. The procedure consists of setting the slope of the deflected centroidal line in each interval equal to zero. Real roots of these polynomials correspond to points of horizontal tangency for the beam, provided that they do not fall outside the interval of validity of the appropriate slope equation. For the present problem, we need to set the slope equations in the intervals $0 \leq x \leq \ell$ and $\ell \leq x \leq 2\ell$ equal to zero. Equations (h) lead to the algebraic equations for points of horizontal tangency,

$$-x^2 + \frac{\ell^2}{3} = 0 \qquad \text{for} \qquad 0 \leq x \leq \ell \tag{j}$$

and

$$-x^2 + 4\ell x - \tfrac{11}{3}\ell^2 = 0 \qquad \text{for} \qquad \ell \leq x \leq 2\ell \tag{k}$$

The roots of Eqs. (j) and (k) that are physically significant are

$$x = +\frac{\ell}{\sqrt{3}} \tag{l}$$

and

$$x = \left(2 - \frac{1}{\sqrt{3}}\right)\ell \tag{m}$$

(c) The deflections occurring at the points of horizontal tangency are obtained by substituting the roots $x = \ell/\sqrt{3}$ and $x = [2 - (1/\sqrt{3})]\ell$ into first and second of Eqs. (i), respectively. The results are

$$v\left(\frac{\ell}{\sqrt{3}}\right) = \frac{M_0\ell^2}{18\sqrt{3}EI} \tag{n}$$

and

$$v\left[\left(2 - \frac{1}{\sqrt{3}}\right)\ell\right] = -\frac{M_0\ell^2}{18\sqrt{3}EI} \tag{o}$$

EXAMPLE 8-5

For the statically indeterminate beam of Figure 8-9a, determine (a) the reaction at the roller, (b) equations for the slope and deflection of the centroidal line of the beam, (c) the location and magnitude of maximum deflection, and (d) the shear and moment diagrams.

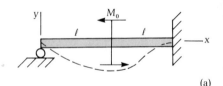

(a)

SOLUTION

(a) and (b) Moment equations for the intervals $0 \leq x \leq \ell$ and $\ell \leq x \leq 2\ell$ are, respectively,

$$M(x) = Rx \quad \text{and} \quad M(x) = Rx - M_0 \qquad \text{(a)}$$

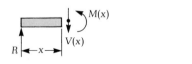

(b)

where R is the unknown roller reaction. Consequently, differential equations that describe the deflected centroidal line of the beam in the two separate segments are

$$\underline{0 \leq x \leq \ell} \qquad \underline{\ell \leq x \leq 2\ell}$$

$$EIv''(x) = Rx \qquad EIv''(x) = Rx - M_0 \qquad \text{(b)}$$

Boundary conditions and continuity relations are, respectively,

$$\left.\begin{aligned} v(0) &= 0 \\ v'(2\ell) &= 0 \\ v(2\ell) &= 0 \end{aligned}\right\} \qquad \text{(c)}$$

and

$$\left.\begin{aligned} v'(\ell-) &= v'(\ell+) \\ v(\ell-) &= v(\ell+) \end{aligned}\right\} \qquad \text{(d)}$$

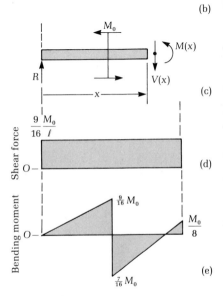

Figure 8-9

Direct integration of Eqs. (b) leads to

$$\underline{0 \leq x \leq \ell} \qquad\qquad \underline{\ell \leq x \leq 2\ell}$$

$$EIv'(x) = \frac{Rx^2}{2} + C_1 \qquad EIv'(x) = \frac{Rx^2}{2} - M_0 x + C_3 \qquad \text{(e)}$$

$$EIv(x) = \frac{Rx^3}{6} + C_1 x + C_2 \qquad EIv(x) = \frac{Rx^3}{6} - \frac{M_0 x^2}{2} \qquad \text{(f)}$$
$$+ C_3 x + C_4$$

Observe that these equations involve five unknown quantities: R, C_1, C_2, C_3, and C_4. We also observe that there are three boundary conditions and two continuity relations that will lead to five algebraic equations in these unknowns. Application of Eqs. (c) and (d) leads to the system of algebraic equations,

$$\left.\begin{array}{r} C_2 = 0 \\ C_3 + 2R\ell^2 = 2M_0\ell \\ 2C_3\ell + C_4 + \tfrac{4}{3}R\ell^3 = 2M_0\ell^2 \\ C_1 - C_3 = -M_0\ell \\ C_1\ell - C_3\ell - C_4 = -\tfrac{1}{2}M_0\ell^2 \end{array}\right\} \tag{g}$$

The solution of Eqs. (g) is

$$\left.\begin{array}{l} C_1 = -\tfrac{1}{8}M_0\ell \\ C_2 = 0 \\ C_3 = \tfrac{7}{8}M_0\ell \\ C_4 = -\tfrac{1}{2}M_0\ell^2 \\ R = \dfrac{9}{16}\dfrac{M_0}{\ell} \end{array}\right\} \tag{h}$$

The required equations for the slope and deflection of the centroidal line of the beam are,

| $0 \le x \le \ell$ | $\ell \le x \le 2\ell$ |

$$EIv'(x) = \frac{9}{32}\frac{M_0}{\ell}x^2 - \tfrac{1}{8}M_0\ell \qquad EIv'(x) = \frac{9}{32}\frac{M_0}{\ell}x^2 - M_0x$$
$$+ \tfrac{7}{8}M_0\ell \tag{i}$$

$$EIv(x) = \frac{3}{32}\frac{M_0}{\ell}x^3 - \tfrac{1}{8}M_0\ell x \qquad EIv(x) = \frac{3}{32}\frac{M_0}{\ell}x^3 - \tfrac{1}{2}M_0x^2$$
$$+ \tfrac{7}{8}M_0\ell x - \tfrac{1}{2}M_0\ell^2 \tag{j}$$

(c) The locations of points of the centroidal line that undergo maximum displacements are determined by setting the slope equations for each interval equal to zero. Thus Eqs. (i) lead to the algebraic equations

$$x^2 - \tfrac{4}{9}\ell^2 = 0 \qquad \text{for} \qquad 0 \le x \le \ell \tag{k}$$

and

$$9x^2 - 32\ell x + 28\ell^2 = 0 \qquad \text{for} \qquad \ell \le x \le 2\ell \tag{l}$$

Roots of the first equation are $x = \pm\tfrac{2}{3}\ell$. Since $x = -\tfrac{2}{3}\ell$ does not lie in the interval of validity of Eqs. (i), it is discarded. The deflection corresponding to $x = +\frac{2}{3}\ell$ is obtained from the first of Eqs. (j),

$$v\left(\frac{2\ell}{3}\right) = -\frac{M_0\ell^2}{18EI} \tag{m}$$

Roots of Eq. (l) are $x = 2\ell$, $\frac{14}{9}\ell$. The second root leads to the deflection

$$v\left(\frac{14}{9}\ell\right) = +0.0041 \frac{M_0\ell^2}{EI} \tag{n}$$

The root $x = 2\ell$ clearly verifies that the clamped boundary condition has been applied correctly since it leads to zero slope and deflection at that boundary.

(d) The shear and moment diagrams for the beam of Figure 8-9a are obtained in the usual manner using the summation procedure. They are shown directly beneath the beam in Figures 8-9d and 8-9e, respectively.

EXAMPLE 8-6

For the statically indeterminate beam shown in Figure 8-10a, determine (a) the equations for the slope and deflection of the centroidal line of the beam, (b) the location and magnitude of the maximum deflection, and (c) the shear and moment diagrams.

SOLUTION

(a) Moment equations for the intervals $0 \le x \le 2\ell$ and $2\ell \le x \le 3\ell$ are, respectively,

$$M(x) = M_\ell - V_\ell x \quad \text{and} \quad M(x) = M_\ell - V_\ell x + M_0 \tag{a}$$

where M_ℓ and V_ℓ are unknown support reactions. Consequently, differential equations that describe the deflected centroidal line of the beam are

$$\underline{0 \le x \le 2\ell} \qquad\qquad \underline{2\ell \le x \le 3\ell}$$

$$EIv''(x) = M_\ell - V_\ell x \qquad EIv''(x) = M_\ell - V_\ell x + M_0 \tag{b}$$

Boundary conditions and continuity relations are, respectively,

$$\left.\begin{array}{l} v'(0) = 0 \\ v(0) = 0 \\ v'(3\ell) = 0 \\ v(3\ell) = 0 \end{array}\right\} \tag{c}$$

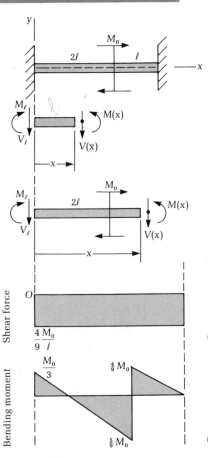

Figure 8-10

and

$$\left.\begin{array}{l} v'(2\ell-) = v'(2\ell+) \\ v(2\ell-) = v(2\ell+) \end{array}\right\} \tag{d}$$

Direct integration of Eqs. (b) leads to

$$\underline{0 \leq x \leq 2\ell} \qquad\qquad \underline{2\ell \leq x \leq 3\ell}$$

$$EIv'(x) = M_\ell x - V_\ell \frac{x^2}{2} + C_1 \qquad EIv'(x) = M_\ell x - \frac{V_\ell x^2}{2}$$
$$+ M_0 x + C_3 \tag{e}$$

$$EIv(x) = \frac{M_\ell x^2}{2} - \frac{V_\ell x^3}{6} \qquad EIv(x) = \frac{M_\ell x^2}{2} - \frac{V_\ell x^3}{6}$$
$$+ C_1 x + C_2 \qquad\qquad + \frac{M_0 x^2}{2} + C_3 x + C_4 \tag{f}$$

Observe that these equations involve six unknown quantities: M_ℓ, V_ℓ, C_1, C_2, C_3, and C_4. Also note that there are four boundary conditions and two continuity relations that will lead to six algebraic equations in these unknowns. Application of Eqs. (c) and (d) leads to the system of algebraic equations.

$$\left.\begin{array}{r} C_1 = 0 \\ C_2 = 0 \\ 3M_\ell\ell - \frac{9}{2}V_\ell\ell^2 = -M_0\ell \\ \frac{9}{2}M_\ell\ell^2 - \frac{27}{6}V_\ell\ell^3 = -\frac{1}{2}M_0\ell^2 \\ C_3 = -2M_0\ell \\ 2C_3\ell + C_4 = -2M_0\ell^2 \end{array}\right\} \tag{g}$$

The solutions of Eqs. (g) are

$$\left.\begin{array}{l} C_1 = 0 \\ C_2 = 0 \\ C_3 = -2M_0\ell \\ C_4 = 2M_0\ell^2 \\ V_\ell = \dfrac{4}{9}\dfrac{M_0}{\ell} \\ M_\ell = \frac{1}{3}M_0 \end{array}\right\} \tag{h}$$

The required equations for the slope and deflection of the centroidal line of the beam are obtained by substituting Eqs. (h) into Eqs. (e) and (f). They are

$$0 \le x \le 2\ell \qquad\qquad\qquad 2\ell \le x \le 3\ell$$

$$EIv'(x) = \tfrac{1}{3}M_0 x - \tfrac{2}{9}M_0 \frac{x^2}{\ell} \qquad EIv'(x) = \tfrac{1}{3}M_0 x - \frac{2}{9}\frac{M_0}{\ell}x^2$$

$$+ M_0 x - 2M_0\ell \qquad\qquad (i)$$

$$EIv(x) = \tfrac{1}{6}M_0 x^2 - \frac{2}{27}\frac{M_0}{\ell}x^3 \qquad EIv(x) = \tfrac{1}{6}M_0 x^2 - \frac{2}{27}\frac{M_0}{\ell}x^3$$

$$+ \frac{M_0 x^2}{2} - 2M_0\ell x \qquad (j)$$

$$+ 2M_0\ell^2$$

(b) The location of points on the centroidal line that undergo the maximum displacements are determined by setting the slope equation for each interval equal to zero. Thus Eqs. (i) lead to the algebraic equations

$$x(x - \tfrac{3}{2}\ell) = 0 \qquad \text{for} \qquad 0 \le x \le 2\ell \qquad (k)$$

and

$$x^2 - 6x\ell + 9\ell^2 = 0 \qquad \text{for} \qquad 2\ell \le x \le 3\ell \qquad (l)$$

The roots of Eq. (k) are $x = 0$, 1.5ℓ. The first root affirms the clamped-end boundary condition at $x = 0$. The second root gives the deflection

$$v_{max}(\tfrac{3}{2}\ell) = \frac{M_0\ell^2}{8EI} \qquad (m)$$

The roots of Eq. (l) are the repeated roots $x = +3\ell$. These roots verify that the beam is clamped at its right end. Equation (m) is therefore the maximum deflection that the centroidal line experiences.

(c) The shear and moment diagrams for the beam shown in Figure 8-10a are shown in Figures 8-10b and 8-10c.

EXAMPLE 8-7

For the statically indeterminate beam shown in Figure 8-11a, determine (a) the reactions at the supports and (b) the shear and moment diagrams.

SOLUTION

(a) Let the unknown support reactions be R_1, R_2, and R_3. Then the moment equations for the intervals $0 \le x \le \ell$ and $\ell \le x \le 2\ell$ are, respectively,

$$M(x) = R_1 x - \tfrac{1}{2}qx^2 \quad \text{and} \quad M(x) = R_1 x - \tfrac{1}{2}qx^2 + R_2(x - \ell)$$

$$(a)$$

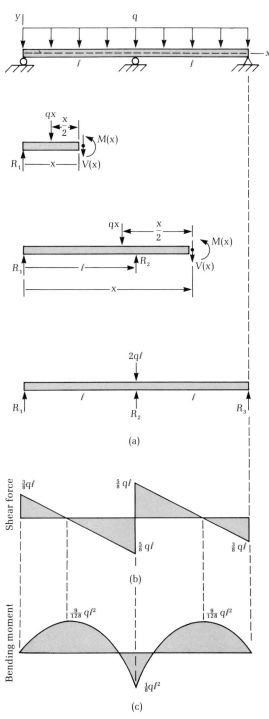

Figure 8-11

Consequently, appropriate differential equations describing the centroidal line of the beam are

$$\underline{0 \leq x \leq \ell}$$

$$EIv''(x) = R_1 x - \tfrac{1}{2}qx^2 \tag{b}$$

and

$$\underline{\ell \leq x \leq 2\ell}$$

$$EIv''(x) = R_1 x - \tfrac{1}{2}qx^2 + R_2(x - \ell) \tag{c}$$

Boundary conditions and continuity relations are, respectively,

$$\left.\begin{array}{c} v(0) = 0 \\ v(\ell) = 0 \\ v(2\ell) = 0 \end{array}\right\} \tag{d}$$

and

$$\left.\begin{array}{c} v'(\ell-) = v'(\ell+) \\ v(\ell-) = v(\ell+) \end{array}\right\} \tag{e}$$

Direct integration of Eqs. (b) and (c) leads to

$$\underline{0 \leq x \leq \ell}$$

$$\left.\begin{array}{l} EIv'(x) = R_1 \dfrac{x^2}{2} - \dfrac{qx^3}{6} + C_1 \\[2mm] EIv(x) = R_1 \dfrac{x^3}{6} - \dfrac{qx^4}{24} + C_1 x + C_2 \end{array}\right\} \tag{f}$$

and

$$\underline{\ell \leq x \leq 2\ell}$$

$$\left.\begin{array}{l} EIv'(x) = R_1 \dfrac{x^2}{2} - \dfrac{qx^3}{6} + \dfrac{R_2(x - \ell)^2}{2} + C_3 \\[2mm] EIv(x) = R_1 \dfrac{x^3}{6} - \dfrac{qx^4}{24} + \dfrac{R_2(x - \ell)^3}{6} + C_3 x + C_4 \end{array}\right\} \tag{g}$$

These equations contain six unknown quantities: R_1, R_2, C_1, C_2, C_3, and C_4. Note that only three boundary conditions and two continuity relations apply. However, an additional equation relating R_1 and R_2 can be obtained from moment equilibrium of the entire beam. Consequently, from Figure 8-11a,

$$\Sigma M_{R_3} = 0: \quad 2\ell R_1 + \ell R_2 = 2q\ell^2 \tag{h}$$

Application of Eqs. (d) and (e) leads to the system of algebraic equations,

$$\left.\begin{array}{c} C_2 = 0 \\[4pt] \frac{1}{6}R_1\ell^3 + C_1\ell = \frac{q\ell^4}{24} \\[4pt] \frac{4}{3}R_1\ell^3 + \frac{1}{6}R_2\ell^3 + 2\ell C_3 + C_4 = \frac{2}{3}q\ell^4 \\[4pt] C_1 = C_3 \\[4pt] C_2 = C_4 \end{array}\right\} \tag{i}$$

The solutions of Eqs. (h) and (i) are

$$\left.\begin{array}{l} C_1 = -\dfrac{q\ell^3}{48} \\[8pt] C_2 = 0 \\[8pt] C_3 = -\dfrac{q\ell^3}{48} \\[8pt] C_4 = 0 \\[4pt] R_1 = \frac{3}{8}q\ell \\[4pt] R_2 = \frac{5}{4}q\ell \end{array}\right\} \tag{j}$$

Force equilibrium perpendicular to the beam axis gives

$$R_1 + R_2 + R_3 = 2q\ell \tag{k}$$

so that

$$R_3 = \tfrac{3}{8}q\ell$$

Observe that this problem involves seven unknown quantities: R_1, R_2, R_3 (the reactions at the three simple supports) and C_1, C_2, C_3, and C_4 (constants of integration). Boundary conditions provide three algebraic equations among these unknowns, continuity relations provide two additional algebraic equations, and equilibrium provides two further equations. Consequently, the number of independent algebraic equations is equal to the number of unknown quantities. This analysis is typical of the thought process that is involved in the solution of statically indeterminate problems.

(b) The shear and moment diagrams for this beam are shown Figures 8-11b and 8-11c, respectively.

PROBLEMS / Section 8-2

8-1 An initially straight thin strip of aluminum alloy is wrapped around a 40-in. diameter cylinder as shown in Figure P8-1. Determine the bending stress in the aluminum. Assume that the material remains elastic and use $E = 10 \times 10^6$ psi for the modulus of elasticity for the aluminum.

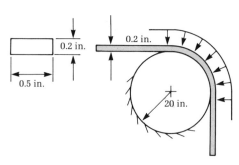

Figure P8-1

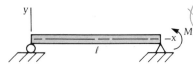

Figure P8-2

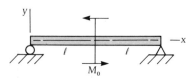

Figure P8-3

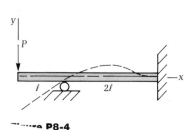

Figure P8-4

8-2 The flexural rigidity EI for the simply supported beam shown in Figure P8-2 is uniform along its length. Determine algebraic equations for the slope and deflection of the deflected centroidal axis of the beam. Use the coordinate system shown in the figure.

8-3 Determine algebraic equations for the slope and deflection of the centroidal line of the simply supported beam shown in Figure P8-3. Assume that the flexural rigidity EI is constant for the beam.

8-4 For the statically indeterminate beam shown in Figure P8-4, determine (a) algebraic equations for the slope and deflection of the centroidal line of the beam, (b) formulas for the slope and deflection at the end of the overhang, and (c) the location and magnitude of the maximum deflection between the supports.

8-5 A schematic drawing of a torque wrench is shown in Figure P8-5. It consists of a pointer that is rigidly attached to a socket at A, a thin steel beam that is also rigidly attached to the socket, and a scale that is attached to the free end of the handle of the wrench. Determine suitable graduations for the scale.

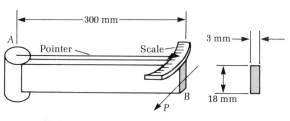

8-5

Moment Diagrams by Parts

In Chapter 5, we developed the summation procedure to obtain a graphic representation of the bending moment at any section along centroidal axis of a straight beam. A moment diagram constructed this procedure is called a **composite moment diagram** because an ordinate to this diagram represents the net moment due to all forces and couples that act on one side of the section. That is, an ordinate to this moment diagram of this type is the algebraic sum of the moments of forces and couples that act on one side of the section.

Another graphic representation of the bending moment that is more convenient for use with the moment area method is the so-called **moment diagram by parts**. The foundation for this representation is the superposition principle applied to moments.

Suppose that the contribution of each concentrated force, distributed load, or couple to the net moment at a section is denoted by M', M'', M''', ... The superposition principle asserts that the net moment at the section is equal to the algebraic sum of the contributions of each concentrated force, distributed force, or couple that acts on an appropriate beam segment. Symbolically,

$$M(x) = M'(x) + M''(x) + M'''(x) + \cdots + \tag{8-14}$$

In constructing a moment diagram by parts, we elect a specific section to serve as a reference. The section is called a **reference section** and can be any section along the axis of the beam. An effective procedure that can be used to construct the moment diagram by parts is one where segments of the beam to the left and right of the reference section are cantilevers with the reference section serving as a clamp. The moment diagram for *each* load that acts to the left of the reference section is constructed directly beneath that segment of the beam. The moment diagram for *each* load that acts to the right of the reference section is constructed directly beneath the right segment of the beam.

For statically determinant beams, it is usually convenient to take a reference section at one end of a distributed load because this minimizes the complexity of the moment diagram for the distributed load. For statically indeterminate beams, it is convenient to take a reference section that will eliminate as many unknown reactions as possible. This choice reduces the order of the system of algebraic equations that must be solved simultaneously.

EXAMPLE 8-8

Construct the moment diagram by parts for the statically indeterminate beam shown in Figure 8-12a.

SOLUTION

The moment equation at an arbitrary section x is

$$M(x) = Rx - \tfrac{1}{2}qx^2 \tag{a}$$

Note that the net moment at section x depends linearly on the loads R and q. Equation (a) can be rewritten in the symbolic form

$$M(x) = M'(x) + M''(x) \tag{b}$$

where $M'(x)$ and $M''(x)$ are the contributions to the net moment due to R and q acting separately as indicated in Figure 8-12e. The moment diagram by parts is obtained by determining the ordinary moment diagrams for R and q acting separately. This determination amounts

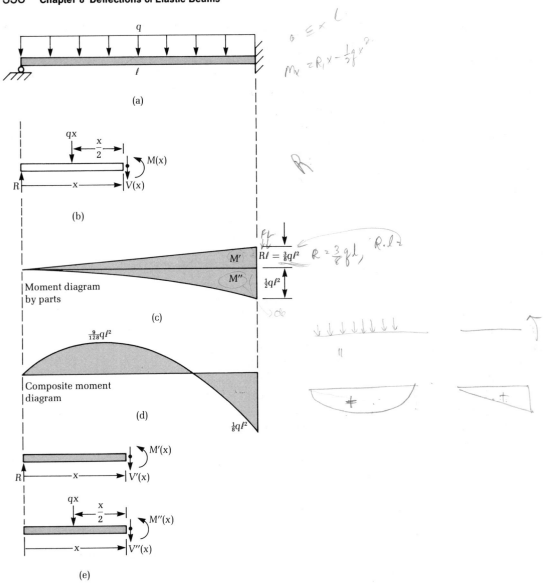

Figure 8-12

to plotting $M'(x)$ and $M''(x)$ separately as shown in Figure 8-12c. The algebraic sum of $M'(x)$ and $M''(x)$ gives the composite moment diagram shown in Figure 8-12d.

The importance of this representation of the moment diagram is that it results in geometric shapes for which formulas for areas and centroids are readily available. For the present example, the representation leads to a triangle and a parabola with the apex at the left end of the beam.

EXAMPLE 8-9

Determine the moment diagram by parts for the beam shown in Figure 8-13a. Use a reference section at the right end of the distributed load.

SOLUTION

Using a reference section at the right end of the distributed load, draw free-body diagrams for the segments of the beam to the left and to the right of the reference section for each load or reaction acting on the left segment and the right segment, respectively. These free-body diagrams are shown in Figure 8-13c. From the free-body diagrams of

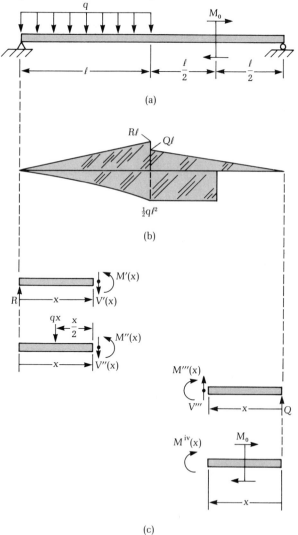

(a)

(b)

(c)

Figure 8-13

the Figure 8-13c,

$$M'(x) = Rx \qquad 0 \le x \le \ell$$

$$M''(x) = -\frac{qx^2}{2} \qquad 0 \le x \le \ell$$

$$M'''(x) = Qx \qquad 0 \le x \le \ell$$

$$M^{iv}(x) = \begin{cases} 0 & 0 \le x \le \dfrac{\ell}{2} \\ -M_0 & \dfrac{\ell}{2} \le x \le \ell \end{cases}$$

(a)

Each of these relations is plotted in Figure 8-13b. This representation results in two triangles, a rectangle, and a parabolic spandrel.

EXAMPLE 8-10

Determine the moment diagram by parts for the beam of Example 8-9 using a reference section at the right end of the beam. The beam is reproduced in Figure 8-14a for convenience.

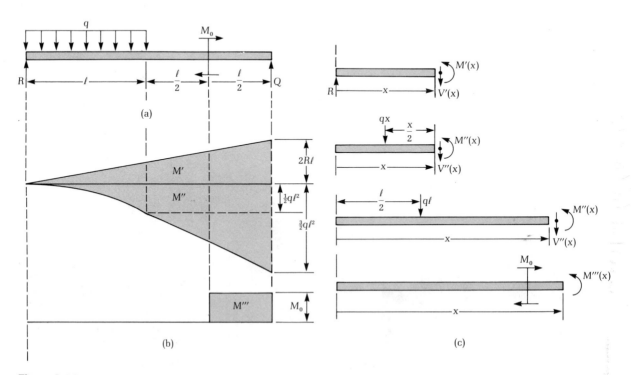

(a)

(b)

(c)

Figure 8-14

SOLUTION

Appropriate free-body diagrams for each load or reaction acting to the left of the reference section are shown in Figure 8-14c. From these free-body diagrams,

$$M'(x) = Rx \qquad\qquad\qquad 0 \le x \le 2\ell$$

$$M''(x) = \begin{cases} -\dfrac{qx^2}{2} & 0 \le x \le \ell \\[3mm] -q\ell\left(x - \dfrac{\ell}{2}\right) & \ell \le x \le 2\ell \end{cases}$$

$$M'''(x) = \begin{cases} 0 & 0 \le x \le \dfrac{3\ell}{2} \\[3mm] M_0 & \dfrac{3\ell}{2} \le x \le 2\ell \end{cases}$$

(a)

Each of these relations is plotted in Figure 8-14b. Observe that, by taking a reference section at the right end of the beam, the moment diagram by parts consists of two rectangles, two triangles, and one parabolic spandrel. This representation is a bit more complicated than the representation obtained in Example 8-9 for a reference section at the end of the distributed load.

For statically determinate beams, it is usually convenient to take a reference section at one end of a distributed load. For statically indeterminate beams, it may be convenient to elect a reference that will eliminate as many unknown reactions as possible.

PROBLEMS / Section 8-3

8-6 Construct the moment diagram by parts for the beam shown in Figure P8-6 using a reference section (a) at the left end of the distributed load and (b) at the right end of the distributed load.

8-7 Construct the moment diagram by parts for the beam shown in Figure P8-7 using a reference section (a) at the left support and (b) at the right support.

8-8 Construct the moment diagram by parts for the beam shown in Figure P8-8. Select a reference section at the left end of the distributed load.

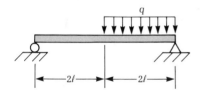

Figure P8-6

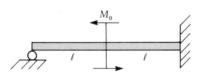

Figure P8-7

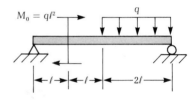

Figure P8-8

8-4 Moment-Area Method

The slope and/or deflection at a specific point on the deformed centroidal line of a beam is frequently all that is required to complete an analysis. Consequently, the double integration procedure, which leads to algebraic expressions for slope and deflection along the beam, may not be the most feasible way to obtain the required quantities.

The moment-area method allows us to determine formulas for the slope and/or the deflection at prescribed points without establishing algebraic expressions for the slope and deflection along the beam. Moreover, its semi-graphic approach to beam deflection analyses has certain attractive features.

The curvature for an elastic beam was found to be

$$\frac{1}{\rho} = \frac{M}{EI} \tag{8-15}$$

Recall that the curvature of a plane curve is

$$\frac{1}{\rho} = \frac{d\theta}{ds} \tag{8-16}$$

where θ is the angle that the tangent at a point on the curve makes with the x axis, and s is the arc length along the curve. These parameters are shown in Figure 8-15. Equations (8-15) and (8-16) give

$$d\theta = \frac{M}{EI} ds \tag{8-17}$$

where $d\theta$ is the angle between the tangent lines at the left and right ends of a differential element ds as shown in Figure 8-16a. Now

$$ds^2 = dx^2 + dy^2 = \left\{ 1 + \left(\frac{dy}{dx} \right)^2 \right\} dx^2 \tag{8-18}$$

and, if we restrict our attention to the class of curves for which $(dy/dx)^2 \ll 1$,

$$ds \approx dx \tag{8-19}$$

Consequently, the angle between the tangent lines at the left and right ends of the differential element ds becomes

$$d\theta = \frac{M}{EI} dx \tag{8-20}$$

Integrate this relation between two sections of the beam that are a finite distance apart. To be specific, integrate between the section at x_A and the section at x_B shown in Figure 8-16a. The result is

$$\theta_B - \theta_A = \int_{x_A}^{x_B} \frac{M}{EI} dx \tag{8-21}$$

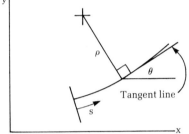

Figure 8-15

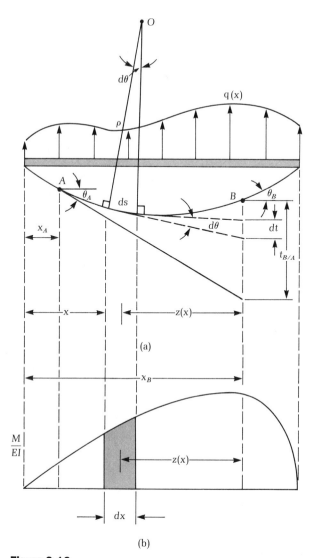

Figure 8-16

The integrand in Eq. (8-21) is a differential element of area under M/EI diagram as shown in Figure (8-16b). Consequently, the **integral** represents the area under the M/EI diagram between the sections at x_A and x_B. This physical interpretation of Eq. (8-21) leads to the first moment-area proposition.

First moment-area proposition. The difference between the angles that lines tangent to the deformed centroidal line of a beam at x_A and x_B make with the undeformed centroidal line of the beam is equal to the area under the M/EI diagram between the sections at x_A

and x_B. A positive change in moment causes a positive difference between these angles. It is convenient to speak of the difference $\theta_B - \theta_A$ as an angle change.

To establish the second moment-area proposition, consider the distance subtended by the intersections of the tangent lines to the deformed centroidal line at the ends of the differential element ds with a vertical line through point B on the deformed centroidal line of the beam. This construction is shown in Figure 8-16a where the subtended distance is denoted by dt. Mathematically, this distance is approximately

$$dt = z \, d\theta = z \, \frac{M}{EI} \, dx \tag{8-22}$$

Now suppose this expression is integrated between the limits corresponding to sections at x_A and x_B. The result is

$$\int dt = \int_{x_A}^{x_B} z \, \frac{M}{EI} \, dx \tag{8-23}$$

The integral on the left of the equality sign is the distance from point B on the deformed centroidal line to the intersection of the tangent line at A with a vertical line through point B. This distance is signified by $t_{B/A}$ in Figure 8-16a, and it is called the **tangential deviation** of point B with respect to a tangent line at A. Observe that tangential deviations are not deflections, but they can often be related to deflections.

The integral on the right of the equality sign in Eq. (8-23) is the first moment of the area under the M/EI diagram between sections at x_A and x_B with respect to a vertical line through point B. This is shown in Figure 8-16b.

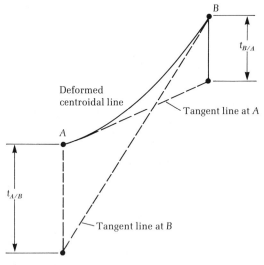

Figure 8-17

These physical interpretations lead to the statement of the second moment-area proposition.

Second moment-area proposition.   The tangential deviation of a point B on the deformed centroidal line of an initially straight beam with respect to a tangent line at point A on the deformed centroidal line is equal to the first moment of the area under the M/EI diagram between points A and B with respect to a vertical line through the point whose deviation is being sought (point B). Since a positive change in moment causes the beam to bend so that the tangent line at the left end of an element slants below the point at the right end, it follows that a positive tangential deviation implies that the point whose deviation is sought lies above the tangent line.

Notice that $t_{B/A} \neq t_{A/B}$ generally. This inequality of $t_{B/A}$ and $t_{A/B}$ is depicted in Figure 8-17.

EXAMPLE 8-11

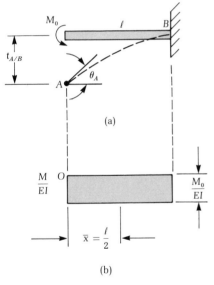

(a)

(b)

Figure 8-18

For the straight beam shown in Figure 8-18a, use the moment-area method to derive formulas for the slope and deflection at the free end of the beam. The flexural rigidity EI is constant.

SOLUTION

The M/EI diagram is shown in Figure 8-18b. The deflection at the free end of the cantilever is seen to be equal to the absolute value of the tangential deviation of point A on the centroidal line with respect to a tangent line at B. Thus

$$t_{A/B} = \left(-\frac{M_0 \ell}{EI} \right) \frac{\ell}{2} = -\frac{M_0 \ell^2}{2EI} \tag{a}$$

The minus sign signifies that point A lies below the tangent line at B. This position signifies that the deflection of point A lies below the undeformed centroidal line. Consequently,

$$v_{max} = \frac{M_0 \ell^2}{2EI} \downarrow \tag{b}$$

The slope at the free end of the cantilever is obtained by use of the first moment-area proposition. Therefore,

$$\theta_B - \theta_A = -\frac{M_0 \ell}{EI}$$

or, since $\theta_B = 0$,

$$\theta_A = \frac{M_0 \ell}{EI} \tag{c}$$

EXAMPLE 8-12

For the statically determinate beam shown in Figure 8-19a, use the moment-area method to determine (a) the slope at points A and B and (b) the location and magnitude of the maximum deflection. The flexural rigidity EI is constant and the beam is simply supported.

SOLUTION

(a) Figure 8-19b depicts the free-body diagram of the entire beam, and the M/EI diagram for the beam is shown in Figure 8-19c.

The magnitude of the slope at point A can be determined from the relations (for small angles)

$$\left|\tan \theta_A\right| \approx \left|\theta_A\right| = \frac{\left|t_{B/A}\right|}{\ell} \tag{a}$$

and its sense can be determined by examining the sign of $t_{B/A}$. Now

$$t_{B/A} = \left(\frac{M_0\ell}{EI}\right)\frac{\ell}{2} - \left(\frac{1}{2}\frac{M_0\ell}{EI}\right)\frac{\ell}{3} = \frac{M_0\ell^2}{3EI} \tag{b}$$

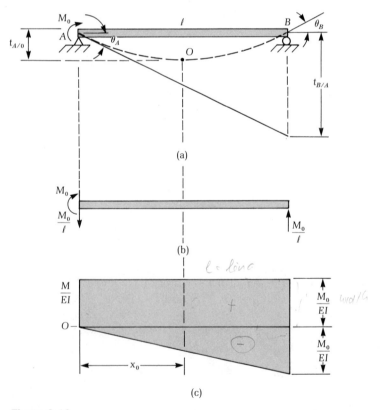

(a)

(b)

(c)

Figure 8-19

and, from Eq. (a), it is seen that

$$|\theta_A| = \frac{M_0 \ell}{3EI} \tag{c}$$

To establish the sense of θ_A, observe that $t_{B/A}$ is positive, which implies that point B lies above the tangent line at A. This allows the the tangent at A to be drawn in the proper direction as shown in Figure 8-19a. Then, from the geometry of the figure, θ_A is observed to be negative. Consequently,

$$\theta_A = \frac{-M_0 \ell}{3EI} \tag{d}$$

The slope at point B can be established using the first moment-area proposition,

$$\theta_B - \theta_A = \frac{M_0 \ell}{EI} - \frac{1}{2} \frac{M_0 \ell}{EI} = \frac{M_0 \ell}{2EI}$$

or, upon substituting θ_A from Eq. (d), it is seen that

$$\theta_B = + \frac{M_0 \ell}{6EI} \tag{e}$$

(b) The location of the point on the centroidal line that undergoes the maximum deflection is determined through the use of the first moment-area proposition. This point is denoted by O in Figure 8-19a.

By the first moment-area proposition, we determine from Figure 8-19c that

$$\theta_0 - \theta_A = \frac{M_0}{EI} x_0 - \frac{1}{2} \frac{M_0}{EI\ell} x_0^2 \tag{f}$$

Substituting from Eq. (d) and observing that $\theta_0 = 0$ yields the equation

$$x_0^2 - 2x_0\ell + \tfrac{2}{3}\ell^2 = 0 \tag{g}$$

which determines the point or points on the centroidal line at which the tangent is horizontal. The roots of this equation are

$$x_0 = \left(1 \pm \frac{1}{\sqrt{3}}\right)\ell \tag{h}$$

Since the root $x_0 = [1 + 1/\sqrt{3}]\ell$ does not lie in the interval containing the beam, it is discarded. Consequently, the maximum deflection occurs at

$$x_0 = \left(1 - \frac{1}{\sqrt{3}}\right)\ell \tag{i}$$

from the left end of the beam.

The magnitude of the deflection at point O on the centroidal line is obtained by observing in Figure 8-19a that $|t_{A/O}| = |v_0|$. Accordingly,

$$t_{A/O} = \left(\frac{M_0 x_0}{EI}\right)\frac{x_0}{2} - \left(\frac{1}{2}\frac{M_0}{EI\ell}x_0^2\right)\frac{2x_0}{3} \qquad (j)$$

Substitution of Eq. (i) into Eq. (j) leads to the formula

$$t_{A/O} = \frac{M_0 \ell^2}{9\sqrt{3}EI} \qquad (k)$$

Since this tangential deviation is positive, point A must lie above the tangent line at O. Consequently,

$$v_0 \equiv v_{\max} = \frac{M_0 \ell^2}{9\sqrt{3}EI} \quad \downarrow \qquad (l)$$

EXAMPLE 8-13

Determine formulas for the slope at points A and B of the uniformly loaded simply supported beam of Figure 8-20a. Determine the magnitude of the maximum deflection.

SOLUTION

Appropriate M/EI diagrams are constructed in Figure 8-20c.

The slope at point A is obtained from the relation

$$|\theta_A| = \frac{|t_{B/A}|}{\ell} \qquad (a)$$

Thus,

$$t_{B/A} = \left(\frac{1}{2}\frac{q\ell^2}{2EI}\ell\right)\frac{\ell}{3} - \left(\frac{1}{3}\frac{q\ell^2}{2EI}\ell\right)\frac{\ell}{4} = \frac{q\ell^4}{24EI} \qquad (b)$$

$$\underbrace{\phantom{\left(\frac{1}{2}\frac{q\ell^2}{2EI}\ell\right)\frac{\ell}{3}}}_{\text{triangle}} \quad \underbrace{\phantom{\left(\frac{1}{3}\frac{q\ell^2}{2EI}\ell\right)\frac{\ell}{4}}}_{\text{spandrel}}$$

so that

$$|\theta_A| = \frac{q\ell^3}{24EI} \qquad (c)$$

Since the sign of $t_{B/A}$ is positive, point B must lie above the tangent line at A. This information enables us to establish the sign of θ_A. Accordingly,

$$\theta_A = -\frac{q\ell^3}{24EI} \qquad (d)$$

The sense of θ_A could be obtained in this case by observation; however, a determination of the sense of θ will not always be possible by inspection.

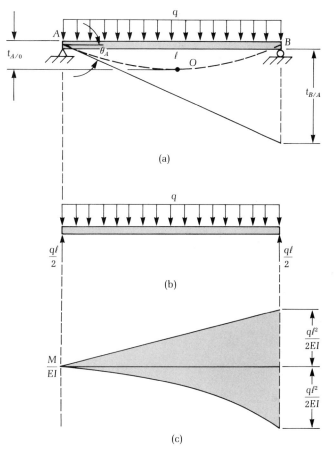

Figure 8-20

To obtain a formula for the maximum deflection, observe from the geometry of Figure 8-20a that the absolute value of this deflection coincides with the absolute value of the tangential deviation $t_{A/0}$. Consequently,

$$t_{A/0} = \left(\frac{1}{2}\frac{q\ell^2}{4EI}\frac{\ell}{2}\right)\left[\frac{2}{3}\left(\frac{\ell}{2}\right)\right] - \left(\frac{1}{3}\frac{q\ell^2}{8EI}\frac{\ell}{2}\right)\left[\frac{3}{4}\left(\frac{\ell}{2}\right)\right]$$

$$= \frac{5}{384}\frac{q\ell^4}{EI}$$

(e)

Since $t_{A/0}$ is positive, point A lies above the tangent line at O. Therefore, the relative vertical position of points A and O in Figure 8-20a is correct and, hence, the maximum deflection is seen to be

$$v_{\max} = \frac{5}{384}\frac{q\ell^4}{EI}\downarrow$$

(f)

EXAMPLE 8-14

The flexural rigidities for the simply supported step-shaft shown in Figure 8-21a are indicated in the figure. Use the moment-area method to determine the location and magnitude of the maximum deflection.

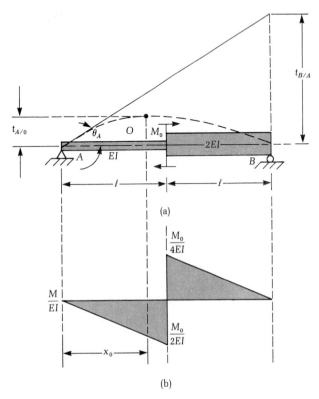

(a)

(b)

Figure 8-21

SOLUTION

Appropriate M/EI diagrams are shown in Figure 8-21b. The central section of the beam is used as a reference section.

The tangential deviation of point B with respect to point A is

$$t_{B/A} = \left(-\frac{1}{2}\frac{M_0}{2EI}\ell\right)\frac{4\ell}{3} + \left(\frac{1}{2}\frac{M_0}{4EI}\ell\right)\frac{2\ell}{3} = -\frac{M_0\ell^2}{4EI} \qquad \text{(a)}$$

Accordingly, the magnitude of the rotation at point A is

$$|\theta_A| = \frac{|t_{B/A}|}{2\ell} = \frac{M_0\ell}{8EI} \qquad \text{(b)}$$

Interpreting the sign of $t_{B/A}$, we see that point B must lie below the tangent line so that θ_A must be positive as shown in Figure 8-21a.

Denote the point of horizontal tangency by O and invoke the first moment-area proposition to obtain

$$\theta_0 - \theta_A = -\frac{1}{2} \frac{M_0}{2EI\ell} x_0^2 \tag{c}$$

Observing that $\theta_0 = 0$ and substituting $\theta_A = M_0\ell/8EI$, we obtain

$$x_0 = +\frac{\ell}{\sqrt{2}} \tag{d}$$

as the location of the point that undergoes the maximum deflection.

The magnitude of the maximum deflection is established by observing that the deflection at point O coincides with the tangential deviation of A relative to a tangent at O. Thus

$$t_{A/O} = \left(-\frac{1}{2}\frac{M_0}{2EI\ell}x_0^2\right)\frac{2x_0}{3} = -\frac{M_0 x_0^3}{6EI\ell} \tag{e}$$

Consequently, substituting Eq. (d) into Eq. (e) we obtain

$$v_0 \equiv v_{max} = \frac{M_0\ell^2}{12\sqrt{2}EI} \uparrow \tag{f}$$

EXAMPLE 8-15

A simply supported beam with an overhang is subjected to a concentrated moment at the end of the overhang. If the flexural rigidity of the beam is constant, determine (a) the deflection of the overhang and (b) the location and magnitude of the maximum deflection between supports. Refer to Figure 8-22a.

SOLUTION

(a) The M/EI diagram for the beam is shown in Figure 8-22b. The reference section has been taken at point B.

From the figure, we observe geometrically that

$$|v_C| = |t_{C/A}| - z \tag{a}$$

Now calculate

$$t_{C/A} = \left(-\frac{1}{2}\frac{M_0}{EI}2\ell\right)\frac{5\ell}{3} - \left(\frac{M_0\ell}{EI}\right)\frac{\ell}{2} = -\frac{13}{6}\frac{M_0\ell^2}{EI} \tag{b}$$

and

$$t_{B/A} = \left(-\frac{1}{2}\frac{M_0}{EI}2\ell\right)\frac{2\ell}{3} = -\frac{2}{3}\frac{M_0\ell^2}{EI} \tag{c}$$

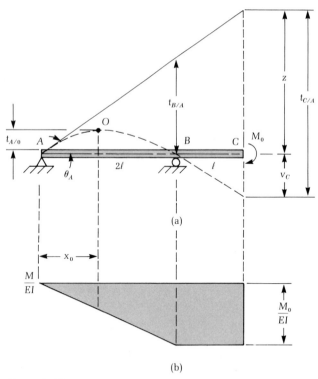

(a)

(b)

Figure 8-22

By similar triangles,

$$z = \frac{3}{2}\left|t_{B/A}\right| = \frac{M_0\ell^2}{EI} \tag{d}$$

Consequently, Eqs. (a), (b), and (d) yield

$$\left|v_C\right| = \frac{13}{6}\frac{M_0\ell^2}{EI} - \frac{M_0\ell^2}{EI} = \frac{7}{6}\frac{M_0\ell^2}{EI} \tag{e}$$

Since $t_{C/A}$ is negative, point C must lie below the tangent line at point A. This information establishes the relative position of the tangent line at A and point C as shown on the diagram. From the geometry of the figure, the deflection of point C must then be downward.

(b) To determine the location of the point of maximum deflection between supports, we first calculate θ_A. Clearly,

$$\theta_A = \frac{\left|t_{B/A}\right|}{2\ell} = \frac{M_0\ell}{3EI} \tag{f}$$

By the first moment-area proposition,

$$\theta_0 - \theta_A = -\frac{1}{2}\frac{M_0}{2\ell EI}x_0^2 \tag{g}$$

and, since $\theta_0 = 0$, we obtain from Eqs. (f) and (g)

$$x_0 = +\frac{2}{\sqrt{3}}\ell \qquad\qquad \text{(h)}$$

Finally, the magnitude of the deflection at x_0 is seen to coincide with $t_{A/0}$. Accordingly, since

$$t_{A/0} = \left(-\frac{1}{2}\frac{M_0}{2EI\ell}x_0^2\right)\frac{2x_0}{3} = -\frac{M_0 x_0^3}{6EI\ell} = -\frac{4}{9\sqrt{3}}\frac{M_0\ell^2}{EI} \qquad\qquad \text{(i)}$$

it follows that

$$v_0 = \frac{4}{9\sqrt{3}}\frac{M_0\ell^2}{EI}\uparrow \qquad\qquad \text{(j)}$$

The sense of the deflection v_0 is determined by interpreting the sign of $t_{A/0}$ and sketching the deflection curve as shown in Figure 8-22a.

EXAMPLE 8-16

For the statically indeterminate beam of Figure 8-23a, determine (a) the reaction at the roller, (b) the slope at the roller, and (c) the location and magnitude of the maximum deflection.

SOLUTION

(a) The M/EI diagram is shown in Figure 8-23c.

The reaction at the roller is obtained by observing that the moment-area equivalent of the condition that the deflection at point A is zero is

$$t_{A/B} = 0 \qquad\qquad \text{(a)}$$

Therefore, calculate

$$t_{A/B} = \left(\frac{1}{2}\frac{2R\ell}{EI}2\ell\right)\frac{2}{3}(2\ell) - \left(\frac{M_0\ell}{EI}\right)\frac{3\ell}{2} = \frac{8}{3}\frac{R\ell^3}{EI} - \frac{3}{2}\frac{M_0\ell^2}{EI} \qquad\qquad \text{(b)}$$

Applying the constraint relation of Eq. (a) yields

$$R = \frac{9}{16}\frac{M_0}{\ell} \qquad\qquad \text{(c)}$$

(b) The magnitude of the slope at the roller is given by the geometric relation

$$|\theta_A| = \frac{|t_{B/A}|}{2\ell} \qquad\qquad \text{(d)}$$

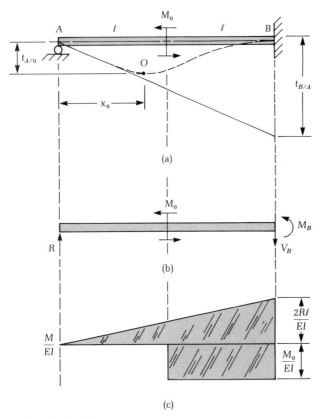

Figure 8-23

We determine

$$t_{B/A} = \left(\frac{1}{2} \frac{2R\ell}{EI} 2\ell \right) \frac{2\ell}{3} - \left(\frac{M_0\ell}{EI} \right) \frac{\ell}{2} = \frac{M_0\ell^2}{4EI} \qquad (e)$$

and, consequently, that

$$\theta_A = -\frac{M_0\ell}{8EI} \qquad (f)$$

(c) The maximum deflection occurs at the point x_0 where the tangent to the deformed centroidal line is horizontal. Thus, by the first moment-area proposition and Figure 8-23c,

$$\theta_0 - \theta_A = \frac{1}{2} \frac{Rx_0}{EI} x_0 \qquad (g)$$

Now $\theta_0 = 0$ by definition, and θ_A is given by Eq. (f) so that

$$x_0 = \pm \tfrac{2}{3}\ell \qquad (h)$$

We exclude the negative root on physical grounds.

The magnitude of the deflection at $x_0 = 2\ell/3$ coincides with the magnitude of $t_{A/0}$. Therefore,

$$t_{A/0} = \left(\frac{1}{2}\frac{9}{16}\frac{M_0}{\ell EI}x_0^2\right)\frac{2}{3}x_0 = \frac{M_0\ell^2}{18EI} \tag{i}$$

Consequently,

$$v_0 = v_{max} = \frac{M_0\ell^2}{18EI} \quad \downarrow \tag{j}$$

The sign of the deflection is again determined by noting that $t_{A/0}$ is positive, which means that point A must lie above the tangent line at point O.

EXAMPLE 8-17

For the statically indeterminate beam shown in Figure 8-24a, determine (a) the reactions at point A and (b) the location and magnitude of the maximum deflection. The flexural rigidity is constant.

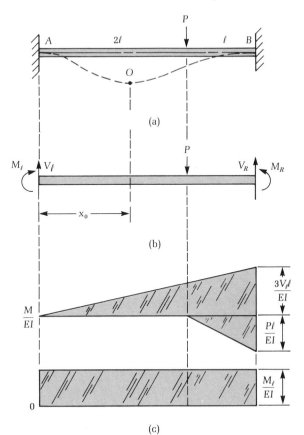

Figure 8-24

SOLUTION

(a) A reference section is established at the right end of the beam, and M/EI diagrams are shown in Figure 8-24c.

The reactions at the left end of the beam can be obtained by noting that the moment-area equivalents of the conditions that both the slope and deflection at point A are zero are

$$\theta_B - \theta_A = 0 \tag{a}$$

and

$$t_{A/B} = 0 \tag{b}$$

Accordingly, Eq. (a) leads to the algebraic equation

$$\frac{1}{2} \frac{3V_\ell \ell}{EI} 3\ell - \frac{1}{2} \frac{P\ell}{EI} \ell + \frac{M_\ell}{EI} 3\ell = 0$$

or, after rearranging,

$$\tfrac{9}{2}V_\ell \ell + 3M_\ell = \tfrac{1}{2}P\ell \tag{c}$$

Likewise, Eq. (b) leads to the algebraic equation

$$\left(\frac{1}{2} \frac{3V_\ell \ell}{EI} 3\ell \right) 2\ell + \left(\frac{M_\ell}{EI} 3\ell \right) \frac{3\ell}{2} - \left(\frac{1}{2} \frac{P\ell}{EI} \ell \right) \frac{8\ell}{3} = 0$$

or, after rearranging,

$$9V_\ell \ell + \tfrac{9}{2}M_\ell = \tfrac{4}{3}P\ell \tag{d}$$

Equations (c) and (d) can be solved simultaneously for the reactions V_ℓ and M_ℓ. Thus

$$V_\ell = \tfrac{7}{27}P \tag{e}$$

and

$$M_\ell = -\tfrac{2}{9}P\ell \tag{f}$$

Reactions V_R and M_R at the right can now be calculated from a free-body diagram of the entire beam. The calculations are not shown here. However, the reader may easily verify that $V_R = 20/27P$ and $M_R = -4/9P\ell$.

(b) The location of the maximum deflection is given by the equation

$$\theta_0 - \theta_A = \frac{1}{2} \frac{V_\ell x_0^2}{EI} + \frac{M_\ell x_0}{EI} = 0 \tag{g}$$

or, after substituting Eqs. (e) and (f) into Eq. (g),

$$x_0(x_0 - \tfrac{12}{7}\ell) = 0 \tag{h}$$

The root $x_0 = 0$ clearly corresponds to the boundary condition at the left end of the beam. Drawing a tangent at $x_0 = 12/7\ell$, we determine

that

$$t_{A/0} = \left(\frac{1}{2}\frac{V_\ell x_0^2}{EI}\right)\frac{2}{3}x_0 + \left(\frac{M_\ell x_0}{EI}\right)\frac{x_0}{2}$$

$$= 0.1088\frac{P\ell^3}{EI} \tag{i}$$

Now $t_{A/0} > 0$, which indicates that point A lies above the tangent at point O and, hence, that the deflection at point O is downward. Consequently,

$$v_0 = v_{\max} = 0.1088\frac{P\ell^3}{EI} \quad\downarrow \tag{j}$$

PROBLEMS / Section 8-4

8-9 Determine formulas for the slope and deflection at the free end of the cantilever beam shown in Figure P8-9. Calculate the numerical value of the deflection when $P = 500$ lb, $\ell = 60$ in., $E = 30 \times 10^6$ psi, and the cross section is a S 5×10 with $I = 12.3$ in.^{4}

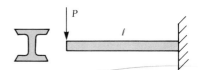

Figure P8-9

8-10 Determine formulas for the slopes at the ends of the simply supported beam shown in Figure P8-10. Also establish a formula for the deflection under the load.

8-11 For the statically indeterminate beam shown in Figure P8-11, determine (a) the reaction at the roller, (b) the slope at the roller, and (c) the location and magnitude of the maximum deflection.

8-12 For the statically indeterminate beam shown in Figure P8-12, determine (a) the reactions at the supports and (b) the location and magnitude of the maximum deflection.

8-13 Determine slope at the left end of the beam shown in Figure P8-13. Also derive a formula for the deflection at the end of the overhang.

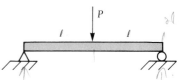

Figure P8-10

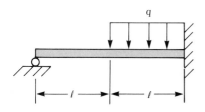

Figure P8-11

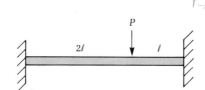

Figure P8-12

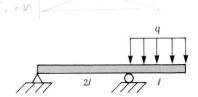

Figure P8-13

8-5 Method of Superposition

The superposition principle applied to beam deflections or beam slopes asserts that the deflection or slope at a point on the deformed centroidal line of a beam due to the simultaneous action of several loads (concentrated forces, couples, and/or distributed loads) is equal to the algebraic sum of the deflections or slopes at the same point due to each load acting separately.

The superposition principle is valid for slopes and deflections because they are linearly related to applied loads. To verify this statement, examine the deflection equation $EIv''(x) = M(x)$, and observe that the moment $M(x)$ is always linearly related to loads. Mathematically, this observation is expressed as

$$M(x) = M_1(x) + M_2(x) + \cdots + M_n(x) \tag{8-24}$$

where $M_i(x)$ $(i = 1, 2, \ldots, n)$ is the moment at the same section due to the ith load acting to one side of the section. We write

$$v(x) = v_1(x) + v_2(x) + \cdots + v_n(x) \tag{8-25}$$

so that, if the $v_i(x)$ satisfy the differential equations

$$\left. \begin{array}{l} EIv_1''(x) = M_1(x) \\ EIv_2''(x) = M_2(x) \\ \quad \vdots \\ EIv_n''(x) = M_n(x) \end{array} \right\} \tag{8-26}$$

the sum will satisfy $EIv''(x) = M(x)$. It is assumed that each $y_i(x)$ satisfies the same boundary conditions and continuity relations as $y(x)$.

The superposition method of determining beam slopes and deflections is usually used to determine a slope or a deflection at a prescribed point. A reasonably complete table of formulas for slope and deflection of several elementary beams is required to make the method effective. Such a table is provided in Appendix E.

Another attractive feature of the method is the relative ease with which it can be used to translate geometric support constraints into algebraic equations that are required to augment equilibrium equations in the solution of statically indeterminate problems.

Several uses of the superposition method for slopes and deflections are illustrated in the following example.

EXAMPLE 8-18

Use the method of superposition to determine formulas for the slope and deflection at the free end of the cantilever beam shown in Figure 8-25a.

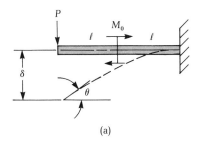

(a)

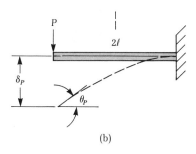

(b)

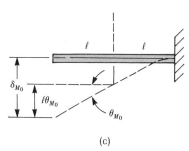

(c)

Figure 8-25

SOLUTION

According to the superposition principle as it applies to beam slopes and deflections, the slope or deflection at the free end of the cantilever due to P and M_0 acting simultaneously is equal to the algebraic sum of the slope or deflection at the same point due to P and M_0 acting separately. Thus

$$\left.\begin{aligned} \delta &= \delta_P + \delta_{M_0} \\ \theta &= \theta_P + \theta_{M_0} \end{aligned}\right\} \tag{a}$$

Now

$$\delta_P = \frac{P(2\ell)^3}{3EI} = \frac{8}{3}\frac{P\ell^3}{EI} \tag{b}$$

and

$$\delta_{M_0} = \frac{M_0\ell^2}{2EI} + \ell\frac{M_0\ell}{EI} \tag{c}$$

Consequently, from Eq. (a),

$$\delta = \frac{8}{3}\frac{P\ell^3}{EI} + \frac{3}{2}\frac{M_0\ell^2}{EI} \tag{d}$$

The slopes due to P and M_0 acting separately are

$$\theta_P = \frac{P(2\ell)^2}{2EI} = \frac{2P\ell^2}{EI} \tag{e}$$

and

$$\theta_{M_0} = \frac{M_0\ell}{EI} \tag{f}$$

Consequently,

$$\theta = \frac{2P\ell^2}{EI} + \frac{M_0\ell}{EI} \tag{g}$$

EXAMPLE 8-19

Use the superposition principle to establish formulas for the slope and deflection for the cantilever beam of Figure 8-26a.

SOLUTION

The partial load q can be replaced with a load of equal intensity q acting over the entire span of the beam and a distributed load of equal magnitude but opposite sense acting over the half beam nearest the support as shown in Figures 8-26b and 8-26c. According to the superposition principle, the slope and deflection are given by the relations

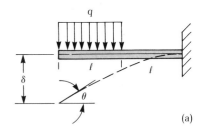

$$\left.\begin{array}{l} \theta = \theta' + \theta'' \\ \delta = \delta' + \delta'' \end{array}\right\} \qquad \text{(a)}$$

(a)

Now

$$\theta' = \frac{q(2\ell)^3}{6EI} = \frac{8}{6}\frac{q\ell^3}{EI} \qquad \text{and} \qquad \theta'' = -\frac{q\ell^3}{6EI} \qquad \text{(b)}$$

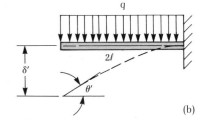

so that

$$\theta = \frac{8}{6}\frac{q\ell^3}{EI} - \frac{q\ell^3}{6EI} = \frac{7}{6}\frac{q\ell^3}{EI} \qquad \text{(c)}$$

(b)

Similarly,

$$\delta' = \frac{q(2\ell)^4}{8EI} = 2\frac{q\ell^4}{EI} \qquad \text{and} \qquad \delta'' = \frac{q\ell^4}{8EI} + \frac{q\ell^3}{6EI}\ell = \frac{7}{24}\frac{q\ell^4}{EI} \qquad \text{(d)}$$

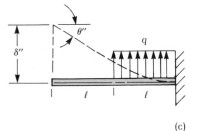

so that

$$\delta = -2\frac{q\ell^4}{EI} + \frac{7}{24}\frac{q\ell^4}{EI} = -\frac{41}{24}\frac{q\ell^4}{EI} \qquad \text{(e)}$$

(c)

Figure 8-26

EXAMPLE 8-20

For the statically indeterminate beam of Figure 8-27a, determine the reactions at the roller and at the fixed end.

SOLUTION

To determine the roller reaction, consider the reaction R at the roller to be an arbitrary force of unknown magnitude applied to the end of the cantilever. The deflection at the end of the cantilever due to P and R acting simultaneously can be established by the method of super-

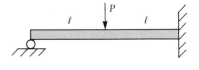

(a)

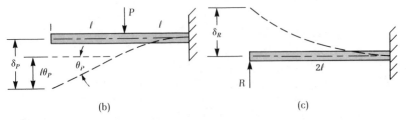

(b)   (c)   (d)

Figure 8-27

position. Thus

$$\delta = \delta_P + \delta_R = -\left\{ \frac{P\ell^3}{3EI} + \frac{P\ell^2}{2EI}\ell \right\} + \frac{R(2\ell)^3}{3EI}$$

$$= -\frac{5}{6}\frac{P\ell^3}{EI} + \frac{8}{3}\frac{R\ell^3}{EI} \tag{a}$$

Now the roller prevents the end of the beam from deflecting. Consequently, because of this geometric constraint ($\delta = 0$),

$$R = \tfrac{5}{16}P \tag{b}$$

Then, from the free-body diagram of Figure 8-27d

$$V_R = \tfrac{11}{16}P \tag{c}$$

and

$$M_R = -\tfrac{3}{8}P\ell \tag{d}$$

EXAMPLE 8-21

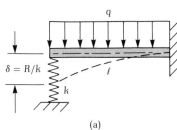

(a)

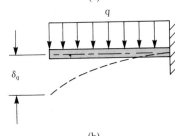

(b)

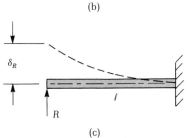

(c)

Figure 8-28

Determine the reaction that the linear spring of Figure 8-28a exerts on the uniformly loaded cantilever beam.

SOLUTION

Figures 8-28b and 8-28c depict the deflections that occur at the end of the cantilever due to q and R acting independently. Formulas for these deflections are

$$\delta_q = -\frac{q\ell^4}{8EI} \tag{a}$$

and

$$\delta_R = \frac{R\ell^3}{3EI} \tag{b}$$

From the superposition principle,

$$\delta = -\frac{q\ell^4}{8EI} + \frac{R\ell^3}{3EI} \tag{c}$$

and, since the load-deflection response of the spring is linear,

$$\delta = -\frac{R}{k} \tag{d}$$

Consequently, Eqs. (c) and (d) lead to

$$R = \frac{\frac{3}{8}q\ell}{1 + \dfrac{3EI}{k\ell^3}} \tag{e}$$

Note that, for a rigid spring (effectively a roller), $k \to \infty$ and $R \to \frac{3}{8}q\ell$. For a free end, $k \to 0$, $R \to 0$, and $\delta = -q\ell^4/8EI$.

PROBLEMS / Section 8-5

8-14   Use the superposition method to establish formulas for the slope and deflection of the free end of the cantilever beam shown in Figure P8-14.

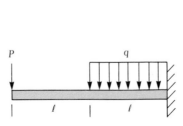

Figure P8-14

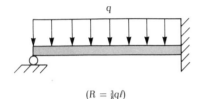

$(R = \frac{3}{8}q\ell)$

(a)

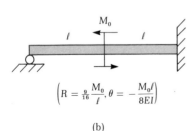

$\left(R = \frac{9}{16}\frac{M_0}{\ell}, \theta = -\frac{M_0\ell}{8EI}\right)$

(b)

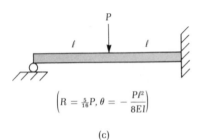

$\left(R = \frac{5}{16}P, \theta = -\frac{P\ell^2}{8EI}\right)$

(c)

Figure P8-16

Figure P8-15

8-15   Establish formulas for the slope at the left end and the deflection at the center of the simply supported beam shown in Figure P8-15. Use the superposition method.

8-16   Use the superposition method to establish a formula for the roller reaction on each of the statically indeterminate beams shown in Figure P8-16. Construct composite shear and moment diagrams. Determine a formula for the slope of the centroidal line at the left end of each beam.

8-17   The midpoint of a simply supported beam rests against a linear spring as shown in Figure P8-17. Prior to application of the uniformly distributed load, the deflection in the spring is zero. Use the superposition method to determine a formula for the force exerted by the spring on the beam after the distributed load is applied. The load-deflection relation for the spring is shown in the figure.

8-18   A gap Δ exists between the midpoint of the beam shown in Figure P8-18 and a linear spring prior to the application of the distributed load. Assuming that the magnitude of the distributed load is sufficient to close the gap, determine a formula for the force exerted by the spring on the beam. The spring constant for the linear spring is k.

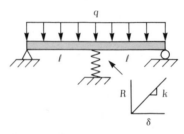

Figure P8-17

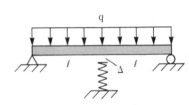

Figure P8-18

8-6 Method of Successive Integration

The method of double integration for determining algebraic expressions for the slope and deflection of the centroidal line of a beam began with the differential equation

$$EIv''(x) = M(x) \qquad (8\text{-}27)$$

By differentiation of Eq. (8-27),

$$EIv'''(x) = \frac{dM}{dx} = V(x) \qquad (8\text{-}28)$$

and

$$EIv^{iv}(x) = \frac{dV}{dx} = q(x) \qquad (8\text{-}29)$$

The method of determining algebraic expressions for slope and deflection of the centroidal line of a beam that begins with Eq. (8-29) is called the **method of successive integrations**. This method is helpful when the loading on the beam is a complicated distribution for which the expression for the bending moment is difficult to obtain. Otherwise, the double integration procedure is preferred.

The following example illustrates the mechanics involved in the use of Eq. (8-29).

EXAMPLE 8-22

Use the method of successive integration to establish algebraic expressions for the slope and deflection of the centroidal line of the beam shown in Figure 8-29a. Assume that the flexural rigidity is constant. Determine the slope at point A and the location and magnitude of the maximum deflection.

SOLUTION

The load intensity at an arbitrary section x is

$$q(x) = -\frac{q_0}{\ell} x \qquad (a)$$

so that the differential equation that must be satisfied at each point of the centroidal line is

$$EIv^{iv}(x) = -\frac{q_0}{\ell} x \qquad (b)$$

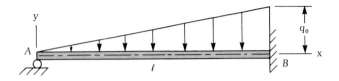

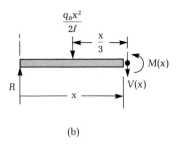

(a)

(b)

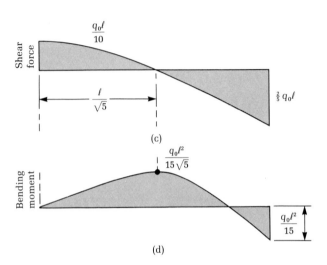

(c)

(d)

Figure 8-29

This differential equation must be augmented by the following boundary conditions to complete the boundary value problem for the deflected centroidal line.

$$\left.\begin{array}{l} v(0) = 0 \\ M(0) = EIv''(0) = 0 \\ v'(\ell) = 0 \\ v(\ell) = 0 \end{array}\right\} \qquad (c)$$

Direct integration of Eq. (b) gives

$$\left.\begin{array}{l} EIv'''(x) = -\dfrac{q_0}{\ell}\dfrac{x^2}{2} + C_1 \\[2mm] EIv''(x) = -\dfrac{q_0}{\ell}\dfrac{x^3}{6} + C_1 x + C_2 \\[2mm] EIv'(x) = -\dfrac{q_0}{\ell}\dfrac{x^4}{24} + C_1 \dfrac{x^2}{2} + C_2 x + C_3 \\[2mm] EIv(x) = -\dfrac{q_0}{\ell}\dfrac{x^5}{120} + C_1 \dfrac{x^3}{6} + C_2 \dfrac{x^2}{2} + C_3 x + C_4 \end{array}\right\} \qquad (d)$$

The first two boundary conditions clearly yield $C_4 = 0$ and $C_2 = 0$, respectively. The last two boundary conditions lead easily to the algebraic equations

$$\left. \begin{array}{l} \frac{1}{2}C_1\ell^2 + C_3 = \dfrac{q_0\ell^3}{24} \\[4mm] \frac{1}{6}C_1\ell^2 + C_3 = \dfrac{q_0\ell^3}{120} \end{array} \right\} \tag{e}$$

which have the solution

$$\left. \begin{array}{l} C_1 = \dfrac{q_0\ell}{10} \\[4mm] C_3 = -\dfrac{q_0\ell^3}{120} \end{array} \right\} \tag{f}$$

Consequently, Eqs. (d) yield

$$EIv'''(x) = -\frac{q_0}{\ell}\frac{x^2}{2} + \frac{q_0\ell}{10} \tag{g}$$

$$EIv''(x) = -\frac{q_0}{\ell}\frac{x^3}{6} + \frac{q_0\ell}{10}x \tag{h}$$

$$EIv'(x) = -\frac{q_0}{\ell}\frac{x^4}{24} + \frac{q_0\ell}{20}x^2 - \frac{1}{120}q_0\ell^3 \tag{i}$$

$$EIv(x) = -\frac{q_0}{\ell}\frac{x^5}{120} + \frac{q_0\ell}{60}x^3 - \frac{1}{120}q_0\ell^3 x \tag{j}$$

To calculate the reaction at the roller, observe from the free-body diagram of Figure 8-29b that $R = \lim_{x \to 0} V(x) = V(0)$. Accordingly, from Eq. (g),

$$EIv'''(0) \equiv V(0) = \frac{q_0\ell}{10} \tag{k}$$

Therefore,

$$R = \frac{q_0\ell}{10} \tag{l}$$

The shearing force and the bending moment at the fixed support are determined from Eqs. (g) and (h) in the same fashion. Thus

$$V(\ell) = EIv'''(\ell) = -\frac{q_0\ell}{2} + \frac{q_0\ell}{10} = -\tfrac{2}{5}q_0\ell \tag{m}$$

and

$$M(\ell) = EIv''(\ell) = -\frac{q_0\ell^2}{6} + \frac{q_0\ell^2}{10} = -\tfrac{1}{15}q_0\ell^2 \tag{n}$$

The shear and moment diagrams for the beam under consideration are shown in Figures 8-29c and 8-29d. The point of zero shear can be obtained by setting Eq. (g) equal to zero, and the peak positive moment can be calculated from Eq. (h).

The slope at point A is obtained from Eq. (i) by setting $x = 0$. Therefore,

$$\theta_A \equiv v'(0) = -\frac{1}{120}\frac{q_0\ell^3}{EI} \tag{o}$$

The maximum deflection occurs at the value of x for which the slope is zero. Let x_0 represent the location of the point of zero slope. Then, from Eq. (i),

$$-\frac{q_0}{\ell}\frac{x_0^4}{24} + \frac{q_0\ell}{20}x_0^2 - \frac{1}{120}q_0\ell^2 = 0$$

or

$$5x_0^4 - 6\ell^2 x_0^2 + \ell^4 = 0 \tag{p}$$

Consequently, the maximum deflection occurs at

$$x_0 = \frac{\ell}{\sqrt{5}} \tag{q}$$

from the left end of the beam. The magnitude of the deflection at $x = x_0$ is obtained from Eq. (j). Direct substitution of Eq. (q) into Eq. (j) yields

$$v_{max} = -\frac{2}{375\sqrt{5}}\frac{q_0\ell^4}{EI} \tag{r}$$

PROBLEMS / Section 8-6

8-19 Derive expressions for the slope and deflection at any point along the beam shown in Figure P8-19. Note that $q(x) = -(q_0 + q_0 x/\ell)$.

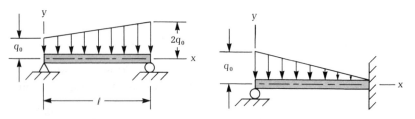

Figure P8-19   **Figure P8-20**

8-20 For the statically indeterminate beam shown in Figure P8-20, use the the method of successive integration to determine formulas for the transverse shearing force and the bending moment as functions of the beam coordinate x. Determine the reaction at the roller.

8-7 Singularity Function Method

The method of double integration discussed in Section 8-2 requires a distinct differential equation for each segment of a beam for which a distinct moment equation is required. The solutions to these differential equations are required to satisfy continuity of slope and deflection at common junctions; this situation can lead to unwieldy algebraic equations as the number of required moment equations increases. To avoid the need to carry a number of differential equations, and to eliminate the need to enforce the continuity requirements and thereby reduce the order of the algebraic system that needs to be solved, A. Clebsch and later W. H. Macaulay introduced the so-called Macaulay functions.

The motivation for the development of the singularity functions resides in the differential relations

$$
\left.
\begin{aligned}
\frac{dV}{dx} &= q(x) \\[6pt]
\frac{dM}{dx} &= V(x) \\[6pt]
EI\,\frac{d^2v}{dx^2} &= M(x)
\end{aligned}
\right\}
\qquad (8\text{-}30)
$$

Observe that, when the transverse load $q(x)$ varies in a continuous manner along the axis of the beam, the shearing force, bending moment, slope, and deflection equations can be obtained from Eqs. (8-30) by integration. The nature of the singularity functions that are about to be introduced permits the expression of *any* system of loads (distributed loads, concentrated forces, couples, and combinations of them) as an equivalent continuously distributed load. With this observation, Eqs. (8-30) become valid for *any* system of loads. This value is the principal one of the singularity function approach to beam deflections.

Macaulay functions.   The Macaulay functions are defined by the relations

$$
\langle x - a \rangle^n =
\begin{cases}
0 & x < a \\
(x-a)^n & a \le x < \infty
\end{cases}
\quad \text{for } n = 0, 1, 2, 3, \ldots,
$$

$$(8\text{-}31)$$

Note that the Macaulay functions are zero for values of x less than a, and they become the regular functions $(x - a)^n$ for values of x equal to or greater than a.

It is easy to show that the Macaulay functions have the following integration property

$$\int_0^x \langle \xi - a \rangle^n d\xi = \frac{\langle x - a \rangle^{n+1}}{n+1} \tag{8-32}$$

To show this, it is necessary to write only

$$\int_0^x \langle \xi - a \rangle^n d\xi = \int_a^x (\xi - a)^n d\xi = \frac{(\xi - a)^{n+1}}{n+1}\Big|_a^x$$

$$= \frac{(x - a)^{n+1}}{n+1} \equiv \frac{\langle x - a \rangle^{n+1}}{n+1}$$

It is by means of these Macaulay functions that *any system of distributed loads can be expressed as an equivalent continuously distributed load.*

EXAMPLE 8-23

Determine the equivalent continuously distributed load associated with the distributed load shown in Figure 8-30.

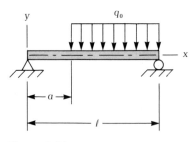

Figure 8-30

SOLUTION

Using the Macaulay functions, write

$$q_e = q_0 \langle x - a \rangle^0 \tag{a}$$

According to the definition of the Macaulay functions, $\langle x - a \rangle^0 = 0$ for $x < a$, and $\langle x - a \rangle^0 = 1$ for $x \geq a$. Thus Eq. (a) is the equivalent continuously distributed load associated with the partial distribution shown in Figure 8-30.

EXAMPLE 8-24

Determine the equivalent continuously distributed load associated with the partial load shown in Figure 8-31a.

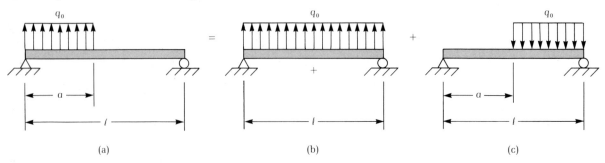

(a)   (b)   (c)

Figure 8-31

SOLUTION

The superposition principle permits us to perceive the given load to be the sum of the distributions shown in Figures 8-31b and 8-31c.

The equivalent continuously distributed loads for these cases are

$$\left.\begin{array}{l} q'_e = q_0\langle x - 0 \rangle^0 \\ q''_e = -q_0\langle x - a \rangle^0 \end{array}\right\} \tag{a}$$

Consequently,

$$q_e = q'_e + q''_e = q_0\langle x - 0 \rangle^0 - q_0\langle x - a \rangle^0 \tag{b}$$

EXAMPLE 8-25

Determine the equivalent continuously distributed load for the linear load distribution shown in Figure 8-32a.

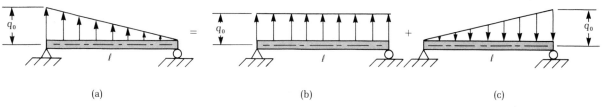

(a)   (b)   (c)

Figure 8-32

SOLUTION

The prescribed loading can be perceived as the sum of the distributions shown in Figures 8-32b and 8-32c. The equivalent continuously distributed loads for these cases are

$$\left.\begin{array}{l} q'_e = q_0\langle x - 0 \rangle^0 \\ q''_e = -\dfrac{q_0}{\ell}\langle x - 0 \rangle^1 \end{array}\right\} \tag{a}$$

By the superposition principle,

$$q_e = q_0\langle x - 0 \rangle^0 - \frac{q_0}{\ell}\langle x - 0 \rangle^1 \tag{b}$$

EXAMPLE 8-26

Derive an integration formula for the integral $\int_0^x x^n\langle x - a \rangle^m \, dx$.

SOLUTION

First write

$$\int_0^x \xi^n\langle \xi - a \rangle^m \, d\xi = \int_0^x [(\xi - a) + a]^n\langle \xi - a \rangle^m \, d\xi \tag{a}$$

Now expand the expression in brackets using the binomial expansion formula,

$$\int_0^x \xi^n \langle \xi - a \rangle^m \, d\xi$$

$$= \int_0^x \left\{ (\xi - a)^n + \frac{na(\xi - a)^{n-1}}{1!} + \frac{n(n-1)a^2(\xi - a)^{n-2}}{2!} \right.$$

$$\left. + \frac{n(n-1)(n-2)a^3(\xi - a)^{n-3}}{3!} + \cdots \right\} \langle \xi - a \rangle^m \, d\xi \qquad \text{(b)}$$

Using the definition (8-31) of the Macaulay functions, Eq. (b) can be expressed as

$$\int_0^x \xi^n (\xi - a)^m \, d\xi = \int_0^x \left\{ \langle \xi - a \rangle^{n+m} + \frac{na\langle \xi - a \rangle^{n+m-1}}{1!} \right.$$

$$+ \frac{n(n-1)a^2\langle \xi - a \rangle^{n+m-2}}{2!}$$

$$+ \frac{n(n-1)(n-2)a^3\langle \xi - a \rangle^{n+m-3}}{3!}$$

$$\left. + \cdots \right\} d\xi \qquad \text{(c)}$$

Using the integration formula (8-32), Eq. (c) becomes

$$\int_0^x \xi^n (\xi - a)^m d\xi = \frac{\langle x - a \rangle^{n+m+1}}{m+n+1} + \frac{na\langle x - a \rangle^{n+m}}{m+n}$$

$$+ \frac{n(n-1)a^2\langle x - a \rangle^{n+m-1}}{(m+n-1)2!} + \cdots \qquad \text{(d)}$$

In particular, if $n = m = 1$,

$$\int_0^x \xi \langle \xi - a \rangle^1 \, d\xi = \frac{\langle x - a \rangle^3}{3} + \frac{a\langle x - a \rangle^2}{2} \qquad \text{(e)}$$

and if $n = 2$ and $m = 1$,

$$\int_0^x \xi^2 \langle \xi - a \rangle^1 \, d\xi = \frac{\langle x - a \rangle^4}{4} + \frac{2a\langle x - a \rangle^3}{3} + \frac{a^2\langle x - a \rangle^2}{4} \qquad \text{(f)}$$

Singularity functions. To represent a **concentrated force** as an equivalent distributed load, it is necessary to introduce a new function that has the properties

$$\langle x - a \rangle^{-1} = \begin{cases} 0 & x \neq a \\ \infty & x = a \end{cases} \qquad \text{(8-33a)}$$

and

$$\int_0^x \langle \xi - a \rangle^{-1} d\xi = \langle x - a \rangle^0 \equiv 1 \qquad (8\text{-}33\text{b})$$

This function is zero everywhere except at $x = a$ where it becomes infinite. Notice that even though the new function itself becomes infinite at $x = a$, its integral remains finite. This new function is variously called the **Dirac delta** or the unit impulse function.

Physically, a concentrated force is perceived as a distributed force of high intensity that is distributed over a vanishingly small length of beam. Then, using the singularity function (8-33), the equivalent continuously distributed load corresponding to the concentrated load P at $x = a$ shown in Figure 8-33 can be represented as

$$q_e = P\langle x - a \rangle^{-1} \qquad (8\text{-}34)$$

Observe that the area under this load distribution is

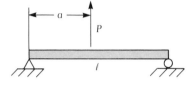

Figure 8-33

$$\int_0^x q_e \, d\xi = \int_0^x P\langle x - a \rangle^{-1} d\xi = P\langle x - a \rangle^0 \equiv P$$

To represent a **concentrated couple** by an equivalent continuously distributed load, we introduce a second singular function that has the properties

$$\langle x - a \rangle^{-2} = \begin{cases} 0 & x \neq 0 \\ \infty & x = 0 \end{cases} \qquad (8\text{-}35\text{a})$$

and

$$\int_0^x \langle \xi - a \rangle^{-2} d\xi = \langle x - a \rangle^{-1} \qquad (8\text{-}35\text{b})$$

The equivalent continuously distributed load corresponding to a concentrated couple, such as that shown in Figure 8-34, is

$$q_e = C\langle x - a \rangle^{-2} \qquad (8\text{-}36)$$

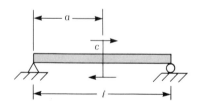

Figure 8-34

EXAMPLE 8-27

Determine the equivalent continuously distributed load associated with the beam shown in Figure 8-35. Determine the shear and moment equations, and the slope and deflection equations, using the Macaulay functions and the singularity functions.

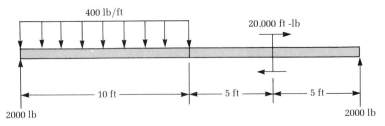

Figure 8-35

SOLUTION

The equivalent continuously distributed load corresponding to all applied forces and reactions is

$$q_e = 2\langle x - 0 \rangle^{-1} - [0.4\langle x - 0 \rangle^0 - 0.4\langle x - 10 \rangle^0]$$
$$+ 20\langle x - 15 \rangle^{-2} + 2\langle x - 20 \rangle^{-1} \tag{a}$$

The shearing force equation is obtained by integrating Eq. (a); consequently,

$$V(x) = 2\langle x - 0 \rangle^0 - [0.4\langle x - 0 \rangle^1 - 0.4\langle x - 10 \rangle^1]$$
$$+ 20\langle x - 15 \rangle^{-1} + 2\langle x - 20 \rangle^0 \tag{b}$$

The moment equation is obtained by integrating Eq. (b); thus

$$M(x) = 2\langle x - 0 \rangle^1 - [0.2\langle x - 0 \rangle^2 - 0.2\langle x - 10 \rangle^2]$$
$$+ 20\langle x - 15 \rangle^0 + 2\langle x - 20 \rangle^1 \tag{c}$$

Notice that neither equation requires a constant of integration because we included the reactions in the expression for the equivalent continuously distributed load. If the reactions had not been included in q_e, a constant of integration would be required for each integration.

The equations for slope and deflection follow from Eq. (c):

$$EIv'(x) = \langle x - 0 \rangle^2 - \left[\frac{0.2}{3} \langle x - 0 \rangle^3 - \frac{0.2}{3} \langle x - 10 \rangle^3 \right]$$
$$+ 20\langle x - 15 \rangle^1 + \langle x - 20 \rangle^2 + C_1 \tag{d}$$

and

$$EIv(x) = \tfrac{1}{3}\langle x - 0 \rangle^3 - \left[\frac{0.2}{12} \langle x - 0 \rangle^4 - \frac{0.2}{12} \langle x - 10 \rangle^4 \right]$$
$$+ 10\langle x - 15 \rangle^2 + \tfrac{1}{3}\langle x - 20 \rangle^3 + C_1 x + C_2 \tag{e}$$

A constant of integration has been included for each integration that leads to the last two equations. These constants are required so that boundary conditions appropriate to the problem can be satisfied. In the present case, the boundary conditions yield

$$v(0) = 0: \quad C_2 = 0 \tag{f}$$

$$v(20) = 0: \quad \frac{20^3}{3} - \left[\frac{0.2}{12}(20)^4 - \frac{0.2}{12}(10)^4 \right]$$
$$+ 10(5)^2 + 20C_1 = 0 \tag{g}$$

Accordingly,

$$C_1 = -\frac{500}{24} \text{k} \tag{h}$$

Let us write the shear and moment equations for the intervals $0 \leq x \leq 10$ and $10 \leq x \leq 15$. From Eqs. (b) and (c), we determine that

$$\left. \begin{array}{ll} 0 \leq x \leq 10 & 10 \leq x \leq 15 \\ V(x) = 2 - 0.4x & \text{and} \quad V(x) = 2 - 0.4x + 0.4(x-10) \\ M(x) = 2x - 0.2x^2 & M(x) = 2x - 0.2x^2 + 0.2(x-10)^2 \end{array} \right\} \tag{i}$$

Verify that these equations are correct by drawing appropriate free-body diagrams and invoking force and moment equilibrium as was done in Chapter 5.

EXAMPLE 8-28

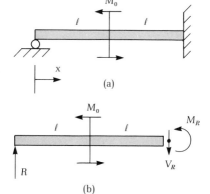

(a)

(b)

Figure 8-36

Determine the reactions at the roller and the wall for the statically indeterminate beam shown in Figure 8-36a.

SOLUTION

The equivalent continuously distributed load for the beam shown in Figure 8-36b is

$$q_e = R\langle x - 0 \rangle^{-1} - M_0 \langle x - \ell \rangle^{-2}$$
$$- V_R \langle x - 2\ell \rangle^{-1} - M_R \langle x - 2\ell \rangle^{-2} \tag{a}$$

Then, by integration,

$$V(x) = R\langle x - 0 \rangle^0 - M_0 \langle x - \ell \rangle^{-1}$$
$$- V_R \langle x - 2\ell \rangle^0 - M_R \langle x - 2\ell \rangle^{-1} \tag{b}$$

$$M(x) = R\langle x - 0 \rangle^1 - M_0 \langle x - \ell \rangle^0$$
$$- V_R \langle x - 2\ell \rangle^1 - M_R \langle x - 2\ell \rangle^0 \tag{c}$$

$$EIv'(x) = \frac{R\langle x - 0 \rangle^2}{2} - M_0 \langle x - \ell \rangle^1$$
$$- \frac{V_R \langle x - 2\ell \rangle^2}{2} - M_R \langle x - 2\ell \rangle^1 + C_1 \tag{d}$$

$$EIv(x) = \frac{R\langle x - 0 \rangle^3}{6} - \frac{M_0 \langle x - \ell \rangle^2}{2}$$
$$- \frac{V_R \langle x - 2\ell \rangle^3}{6} - \frac{M_R \langle x - 2\ell \rangle^2}{2} + C_1 x + C_2 \tag{e}$$

The boundary conditions yield

$$v(0) = 0: \quad C_2 = 0 \tag{f}$$

$$v'(2\ell) = 0: \quad \frac{R(2\ell)^2}{2} - M_0\ell + C_1 = 0 \tag{g}$$

$$v(2\ell) = 0: \quad \frac{R(2\ell)^3}{6} - \frac{M_0\ell^2}{2} + 2C_1\ell = 0 \tag{h}$$

Eqs. (g) and (h) give

$$C_1 = -\tfrac{1}{2}M_0\ell \tag{i}$$

and

$$R = \frac{9}{16}\frac{M_0}{\ell} \tag{j}$$

The reactions at the clamped end are obtained by substituting $x = 2\ell$ into Eqs. (b) and (c). Thus we obtain

$$R - \underline{M_0\langle x - \ell\rangle^{-1}} - V_R = 0 \tag{k}$$

and

$$2R\ell - M_0 - M_R = 0 \tag{l}$$

Notice that the underlined term in Eq. (k) has no physical significance and must be discarded. From Eqs. (k) and (l), we calculate

$$V_R = R = \frac{9}{16}\frac{M_0}{\ell} \tag{m}$$

and

$$M_R = \frac{M_0}{8} \tag{n}$$

PROBLEMS / Section 8-7

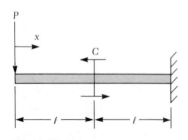

Figure P8-21

Use singularity functions in the solution of each of the following problems.

8-21   Determine formulas for the slope and deflection of the free end of the cantilever beam shown in Figure P8-21.

8-22   Determine the equations for the slope and deflection of the simply supported beam shown in Figure P8-22.

8-23   Determine formulas for the slope and deflections at the end of the overhang of the beam shown in Figure P8-23.

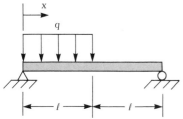

Figure P8-22

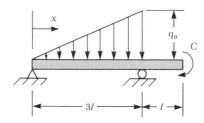

Figure P8-23

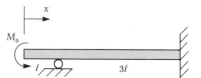

Figure P8-24

8-24 For the statically indeterminate beam shown in Figure P8-24, determine both the roller reaction and formulas for the slope and deflection at the end of the overhang.

8-25 For the statically indeterminate beam shown in Figure P8-25, determine the reactions at the left support.

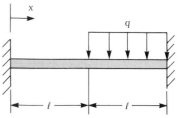

Figure P8-25

8-8 SUMMARY

We developed the methods of double integration, moment-area, super-position, successive integration, and singularity functions for determining the slope and deflection of straight elastic beams in this chapter. Each of these methods was applied to the solution of statically determinate and statically indeterminate beams.

Double Integration:

$$EIv''(x) = M(x)$$ with suitable boundary conditions and continuity conditions whenever more than one moment equation is required.

Moment-Area:

Two-moment-area propositions are required: one pertains to change in slope along the beam axis, and the other pertains to a tangential deviation.

First Moment-Area Proposition:

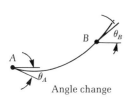

Angle change

$$\theta_B - \theta_A = \text{area under the } M/EI \text{ diagram between points } A \text{ and } B$$

Second Moment-Area Proposition:

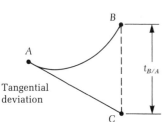

Tangential deviation

$$t_{B/A} = \text{first moment of the area under the } M/EI \text{ diagram between points } A \text{ and } B \text{ taken with respect to the line } BC$$

Superposition:

> The deflection (slope) at a point on the deformed centroidal line of a beam due to several loads acting simultaneously is equal to the algebraic sum of the deflections (slopes) at the same point due to each load acting separately. A table of beam deflections (slopes) is required.

Successive Integration:

> The double integration method begins with the moment expression; the method of successive integration begins with the load intensity acting on the beam.

$$EIv^{iv}(x) = q(x) \text{ (load intensity)}$$
$$EIv'''(x) = V(x) \text{ (shear force)}$$
$$EIv''(x) = M(x) \text{ (moment)}$$
$$v'(x) = \text{slope}$$
$$v(x) = \text{deflection}$$

Singularity Functions:

> Macaulay and singularity functions were introduced to express concentrated forces, couples, and general distributed loads as equivalent continuously distributed loads.

PROBLEMS / Chapter 8

8-26 Using an origin of coordinates at the left end of each of the cantilever beams shown in Figure P8-26, establish algebraic equations for the slope and deflection of the centroidal line of each beam. Establish formulas for the slope and deflection at the free end of each beam.

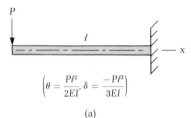

$$\left(\theta = \frac{P\ell^2}{2EI}, \delta = \frac{-P\ell^3}{3EI}\right)$$

(a)

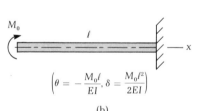

$$\left(\theta = -\frac{M_0\ell}{EI}, \delta = \frac{M_0\ell^2}{2EI}\right)$$

(b)

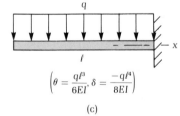

$$\left(\theta = \frac{q\ell^3}{6EI}, \delta = \frac{-q\ell^4}{8EI}\right)$$

(c)

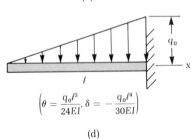

$$\left(\theta = \frac{q_0\ell^3}{24EI}, \delta = -\frac{q_0\ell^4}{30EI}\right)$$

(d)

Figure P8-26

8-27 Using an origin of coordinates at the left end of each of the simply supported beams shown in Figure P8-27, establish algebraic equations for the slope and deflection of the centroidal line of each beam. Establish formulas for the slopes at the ends of each beam.

8-28 Using an origin of coordinates at the left end of the simply supported beam of Figure P8-28, determine algebraic expressions for the slope and deflection of an arbitrary point on the centroidal line. Determine formulas for the location and magnitude of the maximum deflection of the beam.

$$x_0 = \sqrt{2}\ell, \delta_{max} = \frac{2\sqrt{2}}{9}\frac{M_0\ell^2}{EI}$$

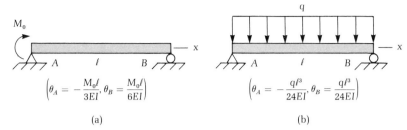

$$\left(\theta_A = -\frac{M_0\ell}{3EI}, \theta_B = \frac{M_0\ell}{6EI}\right)$$

(a)

$$\left(\theta_A = -\frac{q\ell^3}{24EI}, \theta_B = \frac{q\ell^3}{24EI}\right)$$

(b)

Figure P8-27

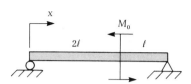

Figure P8-28

8-29 Use the method of double integration to determine the reaction at the roller for the statically indeterminate beam shown in Figure P8-29. Determine the reactions of the fixed support on the beam. Construct shear and moment diagrams.

$$R = \frac{3q\ell}{8}$$

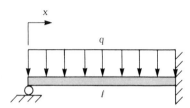

Figure P8-29

8-30 For the statically indeterminate beam shown in Figure P8-30, determine the roller reaction and the reactions of the fixed support on the beam. Construct shear and moment diagrams. Establish formulas for the location and magnitude of the maximum deflection of the beam.

$$R = \frac{9}{16}\frac{M_0}{\ell}, M_R = \frac{M_0}{8}, V_R = \frac{9}{16}\frac{M_0}{\ell}, x_0 = \frac{2\ell}{3}, \delta_{max} = -\frac{M_0\ell^2}{18EI}$$

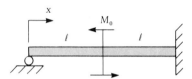

Figure P8-30

8-31 Use the method of double integration to determine the roller reaction for the statically indeterminate beam shown in Figure P8-31. Establish formulas for the location and magnitude of the maximum deflection of the beam.

$$x_0 = \frac{\ell}{3}, \delta_{max} = \frac{1}{27}\frac{M_0\ell^2}{EI}$$

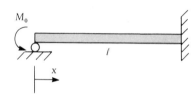

Figure P8-31

8-32 Use the method of double integration to determine the roller reaction for the statically indeterminate beam shown in Figure P8-32. Construct shear and moment diagrams. Establish formulas for the location and magnitude of the maximum deflection of the beam.

$$R = \tfrac{5}{16}P, V_R = \tfrac{11}{16}P, M_R = -\tfrac{3}{8}P\ell$$

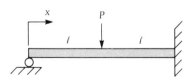

Figure P8-32

8-33 Use the method of double integration to determine the reactions at the left end of the statically indeterminate beam shown in Figure P8-33. Construct shear and moment diagrams. Establish formulas for the location and magnitude of the maximum deflection of the beam.

$$V_R = -V_L = \frac{4}{9}\frac{M_0}{\ell}, M_L = \frac{M_0}{3}, M_R = 0, x_0 = \frac{3}{2}\ell, \delta_{max} = \frac{M_0\ell^2}{8EI}$$

8-34 A 150-mm × 300-mm simply supported oak ($E = 8.4$ GPa) beam supports a 10-kN force at the midpoint of a 6-m span. A pointer is rigidly attached to its

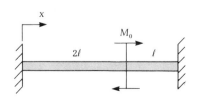

Figure P8-33

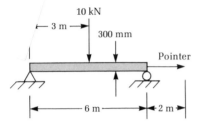

Figure P8-34

right end as shown in Figure P8-34. Determine the vertical movement of the end of the pointer and the maximum deflection of the beam.

8-35 Figure P8-35 shows a device used to calibrate certain electronic strain measuring equipment. It consists of an essentially rigid base, a cantilever beam (shaded in the figure), and a micrometer. Electrical resistance strain gages are attached to the beam as shown. If the output of the top gage is 50μ in./in., determine the deflection at the micrometer.

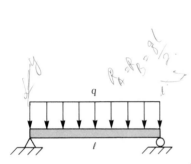

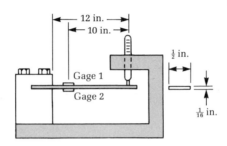

Figure P8-35

Figure P8-36

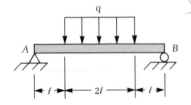

Figure P8-37

8-36 For the statically determinate, simply supported beam shown in Figure P8-36, (a) construct the moment diagram by parts using a reference section at the right end and (b) use the moment-area to establish formulas for the slopes at its ends and the deflection at its center.

8-37 Construct the moment diagram by parts for the beam shown in Figure P8-37. Use a reference section at the center of the span. Derive formulas for the slopes at points A and B and the deflection at the center using the moment-area method.

$$\theta_B = -\theta_A = \frac{11}{6}\frac{q\ell^3}{EI}, \, \delta_C = -\frac{57}{24}\frac{q\ell^4}{EI}$$

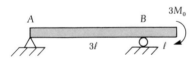

Figure P8-38

8-38 For the simply supported beam shown in Figure P8-38, determine (a) formulas for the location and magnitude of the maximum deflection between the supports and (b) the deflection of the end of the overhang.

$$x_0 = \sqrt{3\ell}, \, \delta_{\max} = \frac{\sqrt{3}M_0\ell^2}{EI}$$

Figure P8-39

8-39 Show that the slope at point A and the deflection of point C of the simply supported beam shown in Figure P8-39 are given by the formulas

$$\theta_A = \frac{M_0\ell}{3EI} \quad \text{and} \quad \delta_C = \frac{7}{6}\frac{M_0\ell^2}{EI}$$

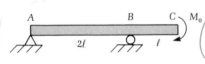

Figure P8-40

8-40 Use the moment-area method to show, for the beam depicted in Figure P8-40, that: (a) The slope at point A is

$$\theta_A = \frac{P\ell^2}{3EI}$$

(b) The deflection of the end of the overhang is

$$\delta_C = \frac{P\ell^3}{EI}$$

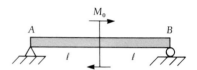

Figure P8-41

(c) The location and magnitude of the maximum deflection between supports are given by

$$x_0 = \frac{2}{\sqrt{3}}\ell \quad \text{and} \quad \delta_{max} = \frac{4}{9\sqrt{3}}\frac{P\ell^3}{EI}$$

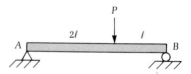

Figure P8-42

8-41   For the statically determinate beam shown in Figure P8-41, determine, using the moment-area method, (a) a formula for the slope of the centroidal line at point A and (b) the location and magnitude of the maximum positive deflection.

$$\theta_A = \frac{M_0\ell}{12EI}, \; x_0 = \frac{\ell}{\sqrt{3}} \text{ from point } A, \; v_{max} = \frac{M_0\ell^2}{18\sqrt{3}EI}$$

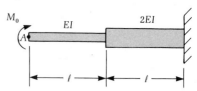

Figure P8-43

8-42   For the statically determinate, simply supported beam of Figure P8-42, determine, using the moment-area method, (a) a formula for the slope of the centroidal line at point A and (b) a formula for the deflection of a point on the centroidal line at the load P.

$$\theta_A = -\frac{11}{27}\frac{P\ell^2}{EI}, \; \delta_P = \frac{10}{27}\frac{P\ell^3}{EI}$$

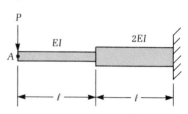

Figure P8-44

8-43   The flexural rigidities for the step beam shown in Figure P8-43 are EI and 2EI as shown. Use the moment-area method to determine formulas for the slope and deflection of point A on its centroidal line.

$$\theta_A = -\frac{3}{2}\frac{M_0\ell}{EI}, \; \delta_A = \frac{5}{4}\frac{M_0\ell^2}{EI}$$

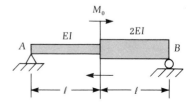

Figure P8-45

8-44   The flexural rigidities for the step beam shown in Figure P8-44 are EI and 2EI. Construct the moment diagram by parts using a reference section at the clamped end. Use the moment-area method to establish formulas for the slope and deflection of the free end.

$$\theta = \frac{5}{4}\frac{P\ell^2}{EI}, \; \delta = \frac{3}{2}\frac{P\ell^3}{EI}$$

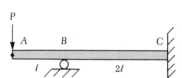

Figure P8-46

8-45   For the simply supported step beam shown in Figure P8-45, determine formulas for the slope at point A and the deflection at the center.

8-46   Construct the moment diagram by parts for the statically indeterminate beam shown in Figure P8-46. Use a reference section at the fixed support. Via the moment-area method, determine the reaction at the roller and the slope and deflection of the end of the overhang.

8-47   Use the moment-area method to determine the reaction of the roller on the beam shown in Figure P8-47. Determine formulas for the slope and deflection of point A.

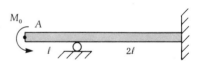

Figure P8-47

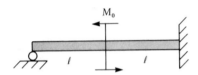

Figure P8-48

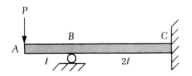

Figure P8-49

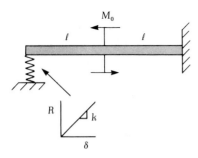

Figure P8-50

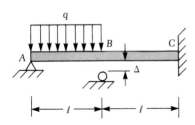

Figure P8-51

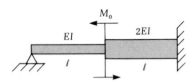

Figure P8-52

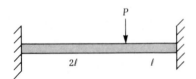

Figure P8-53

8-48 Use the moment-area method to determine the reaction at the roller for the beam shown in Figure P8-48. Establish formulas for the shearing force and bending moment at the fixed support. Construct composite shear and moment diagrams.

$$R = \frac{9}{16}\frac{M_0}{\ell}, \; M_R = \frac{M_0}{8}, \; V_R = \frac{9}{16}\frac{M_0}{\ell}$$

8-49 Construct a moment diagram by parts for the beam shown in Figure P8-49 using a reference section at the right end. Subsequently, use the moment-area method to determine the reaction at the roller and the slope and deflection at point A.

8-50 Show that the reaction of the linear spring on the beam shown in Figure P8-50 is

$$R = \frac{\dfrac{9}{16}\dfrac{M_0}{\ell}}{1 + \dfrac{3}{8}\dfrac{EI}{k\ell^3}}$$

and the moment exerted by the clamp on the beam is

$$M_C = \frac{M_0}{8} \frac{1 - \dfrac{3EI}{k\ell^3}}{1 + \dfrac{3}{8}\dfrac{EI}{k\ell^3}}$$

Use the moment-area method.

8-51 Use the moment-area method to determine the reactions at supports A, B, and C for the beam shown in Figure P8-51. Assume that q is large enough to cause a deflection at B that exceeds the gap Δ.

8-52 Use the moment-area method to determine the reaction of the roller on the beam shown in Figure P8-52. Determine the reactions of the fixed support on the beam. Construct composite shear and moment diagrams.

$$R = \frac{M_0}{2\ell}$$

8-53 Use the moment-area method to calculate the reactions of the left support on the statically indeterminate beam shown in Figure P8-53. Determine the reactions of the clamp support at the right end of the beam. Construct composite shear and moment diagrams. Also locate the section that undergoes the largest deflection. Determine a formula for this maximum deflection.

$$V_L = \tfrac{7}{27}P, \; M_L = -\tfrac{2}{9}P\ell, \; V_R = \tfrac{20}{27}P, \; M_R = -\tfrac{4}{9}P\ell$$

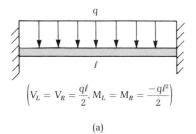
$$\left(V_L = V_R = \frac{q\ell}{2}, M_L = M_R = \frac{-q\ell^2}{2}\right)$$
(a)

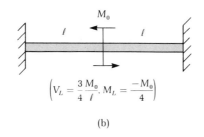
$$\left(V_L = \frac{3}{4}\frac{M_0}{\ell}, M_L = \frac{-M_0}{4}\right)$$
(b)

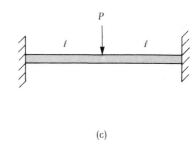
(c)

Figure P8-54

8-54 Use the superposition method to establish formulas for the reactions exerted by the left support on each of the beams shown in Figure P8-54. Construct composite shear and moment diagrams for each beam.

8-55 One end of a cantilever beam rests against a linear spring as shown in Figure P8-55. Before the distributed load q is applied, the deflection in the spring is zero. The load-deflection relationship for the spring is shown in the figure. Use the superposition method to show that the roller reaction is given by

$$R = \frac{\frac{3}{8}q\ell}{\left(1 + \frac{3EI}{k\ell^3}\right)}$$

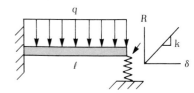

Figure P8-55

8-56 Prior to applying the distributed load to the cantilever beam shown in Figure P8-56, a gap Δ exists between its free end and a roller. Assuming that q is large enough to close the gap, show that the roller reaction is given by the formula

$$R = \frac{3}{8}q\ell - \frac{3EI}{\ell^3}\Delta$$

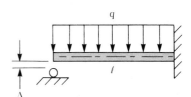

Figure P8-56

8-57 A cantilever beam carrying a uniformly distributed load is supported by an elastic rod at one end as shown in Figure P8-57. Designate the extensional stiffness of the rod by $E_R A$ and the flexural stiffness of the cantilever by $E_B I$. Using the superposition method show that the force exerted by the rod on the cantilever is

$$R = \frac{\dfrac{q\ell^4}{8E_B I}}{\left(\dfrac{a}{E_R I} + \dfrac{\ell^3}{3E_B I}\right)}$$

8-58 The right end of the cantilever beam AB shown in Figure P8-58 rests against the midpoint of the simply supported beam CBD. The deflection of the midpoint of the simply supported beam is zero prior to the application of the distributed load q. If the flexural rigidity EI is the same for each beam, show that the reaction between the beams and the deflection of their common point are

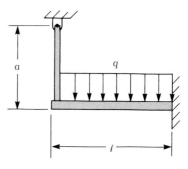

Figure P8-57

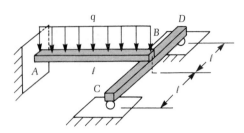

Figure P8-58

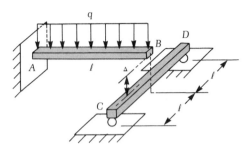

Figure P8-59

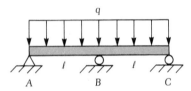

Figure P8-60

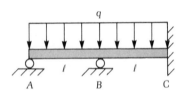

Figure P8-61

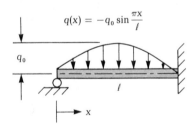

Figure P8-62

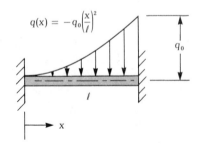

Figure P8-63

given by

$$R = \frac{q\ell}{4} \quad \text{and} \quad \delta = \frac{q\ell^4}{24EI}$$

8-59   A small gap Δ exists between the right end of the cantilever beam AB shown in Figure P8-59 and the midpoint of the simply supported beam CBD. The flexural rigidities EI of the two beams are equal. Use the superposition method to show that the reaction between the two beams and the deflection of the cantilever are given by the formulas

$$R = \frac{q\ell}{4} - \frac{2EI}{\ell^3}\Delta \quad \text{and} \quad \delta = \frac{q\ell^4}{24EI} + \frac{2}{3}\Delta$$

8-60   Use the superposition method to establish formulas for the reactions at the three supports of uniformly loaded beam shown in Figure P8-60. Construct composite shear and moment diagrams.

$$R_A = R_C = \tfrac{3}{8}q\ell, \; R_B = \tfrac{5}{4}q\ell$$

8-61   Use the superposition method to establish formulas for the reactions at the supports for the beam shown in Figure P8-61. Construct shear and moment diagrams.

8-62   Use the method of successive integrations to obtain a formula for the roller reaction for the statically indeterminate beam shown in Figure P8-62. The beam carries a distributed load that varies according to the equation $q = -q_0 \sin (\pi x/\ell)$. Develop expressions for the shearing force and bending moment as a function of x.

8-63   The statically indeterminate beam shown in Figure P8-63 carries a distributed load that varies according to the equation $q(x) = -q_0(x/\ell)^2$. Use the method of successive integration to determine formulas for the transverse shearing force and the bending moment as functions of the beam coordinate x. Determine the reactions at the left support.

$$V(x) = \frac{q_0\ell}{15}\left[1 - 5\left(\frac{x}{\ell}\right)^2\right], \, M(x) = -\frac{q_0\ell^2}{60}\left[1 + 5\left(\frac{x}{\ell}\right)^4 - 4\left(\frac{x}{\ell}\right)\right]$$

8-64   Use the singularity function method to obtain formulas for the slope and deflection of the free end of the cantilever shown in Figure P8-64.

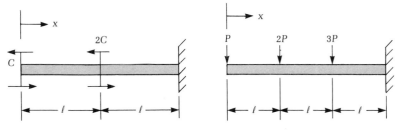

Figure P8-64   **Figure P8-65**

8-65 Use the singularity function method to obtain the deflection under each of the forces shown in Figure P8-65.

8-66 Use the singularity function method to determine the reaction at the roller of the statically indeterminate beam shown in Figure P8-66.

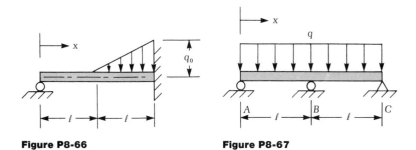

Figure P8-66   **Figure P8-67**

8-67 Use the singularity function method to determine the roller reactions for the statically indeterminate beam shown in Figure P8-67.

CHAPTER 9

Stability of Columns

9-1 Introduction

In the previous chapters of this book, we determined the load carrying capacity of an engineering member on the basis of strength *or* stiffness considerations. An engineering member is perceived to have failed when it ceases to be capable of performing its intended purpose. Consequently, a failure occurs whenever the magnitude of the stress or the displacement at any point in a member exceeds the allowable values for these quantities. The manner in which a failure occurs is called a **mode of failure**. Two easily identifiable modes of failure are: failure because a stress exceeds tolerable magnitudes or failure because a displacement exceeds tolerable magnitudes. The more descriptive terms **failure by yielding** and **failure by fracture** are frequently used.

A third mode of failure is associated with a loss of load carrying capacity because of a radical change in the geometric shape of an engineering member. This mode of failure is a geometric instability called **buckling**. Generally, any mechanical system that experiences compressive stress is subject to a loss of load carrying capacity by buckling. This chapter is concerned principally with the buckling behavior of straight columns subjected to axial compressive forces. However, the three criteria currently used to predict the load carrying capacity of columns are stated in general terms so that they apply to static mechanical systems for which energy is conserved.

A **perfect mechanical system** is one for which there are neither initial deviations from an ideal geometric shape nor eccentricities of loading. A deviation from an ideal geometric shape or an eccentricity of loading is usually referred to as an imperfection. Consequently, a mechanical system that possesses either imperfection is said to be an **imperfect mechanical system**. A column is said to be a perfect column if it is straight and if the compressive force is concentric; that is, the action line of the compressive force coincides with centroidal line of the column. A column is said to be an imperfect column if it is not initially straight and/or if the applied force is not concentric. A circular cylindrical pressure vessel is said to be a perfect cylinder if there is no deviation from the mathematical definition of a circular cylinder.

A mechanical system is said to be **imperfection sensitive** if the load carrying capacity of the imperfect system is substantially less than the load carrying capacity of the perfect system. Conversely, a mechanical system is said to be **imperfection insensitive** if there is no loss of load carrying capacity because of imperfections. It has been shown experimentally and analytically* that columns are imperfection insensitive. Flat plates are also known to be imperfection insensitive. However, cylinders and spheres under external pressure are examples of structural shapes that are imperfection sensitive.

9-2 Stability Criteria

Three stability criteria are commonly used to determine the static load carrying capacity of a mechanical system:

> 1. The adjacent equilibrium criterion.
> 2. The imperfection criterion.
> 3. The energy criterion.

Adjacent equilibrium criterion.   The adjacent equilibrium criterion asserts that an equilibrium configuration of a perfect mechanical system ceases to be stable whenever infinitesimally close equilibrium configurations exist for the same load. For the column, this means that at the stability limit, bent configurations of the column—as well as the straight configuration—can be sustained by the same axial load.

Imperfection criterion.   The imperfection criterion asserts that an equilibrium configuration of an imperfect mechanical system ceases to be stable whenever its deflections become unbounded or excessive under an applied load. For the imperfect column, the compressive force causes lateral displacements that become unbounded or excessive as the magnitude of the force is increased.

Energy criterion.   The energy criterion asserts that an equilibrium configuration of a perfect system is stable as long as the second-order variation of its total potential energy remains positive. The

---

* Timoshenko, S. P., and Gere, J. M., 1961. *Theory of Elastic Stability*. McGraw-Hill Book Company, Inc., New York.

equilibrium configuration becomes unstable when the second-order variation of the total potential energy vanishes.

The following examples examine a simple two-link mechanism to sharpen the basic ideas in the first two stability criteria. The same mechanism is also examined in the next section using the energy criterion.

EXAMPLE 9-1

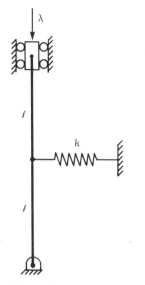

Figure 9-1

Investigate the stability of the straight form of the two-link mechanism shown in Figure 9-1 using the adjacent equilibrium criterion. The links of the mechanism are rigid and the elastic resisting forces of the system are represented by a linear spring with spring constant k. A compressive force λ is applied along the axis of the links when they are perfectly aligned.

SOLUTION

According to the adjacent equilibrium criterion, the smallest value of λ, for which equilibrium configurations other than the straight configuration exist, is sought. The value of λ that corresponds to a slightly displaced equilibrium configuration can be obtained by expressing equilibrium of forces and moments acting on the system in its displaced configuration. Thus we need to examine the free-body diagram shown in Figure 9-2.

Considering the force equilibrium of the complete, displaced system of Figure 9-2a, we see that the lateral reactions have the values indicated on the diagram. Moment equilibrium for the lower link (Figure 9-2b) leads to

$$(\lambda - \tfrac{1}{2}k\ell \cos \vartheta)x = 0 \tag{a}$$

Now $x = 0$ corresponds to the straight equilibrium configuration. Consequently, equilibrium configurations are sought for which $x \neq 0$. Such configurations are possible only if

$$\lambda = \tfrac{1}{2}k\ell \cos \vartheta \tag{b}$$

which gives the load-displacement relation for equilibrium configurations for which $x \neq 0$. For equilibrium configurations that are infinitesimally close to the straight equilibrium configuration, $\cos \vartheta \approx 1$ so that Eq. (b) yields

$$\lambda_{CR} = \tfrac{1}{2}k\ell, \tag{c}$$

where λ_{CR} is called the critical load for the two-link mechanism.

The equilibrium path defined by Eq. (b) is plotted in Figure 9-3 and is labeled **secondary equilibrium path**. The equilibrium path (load-deflection relationship) for the straight equilibrium configuration is indicated by the heavy dots and is labeled **primary equilibrium path**.

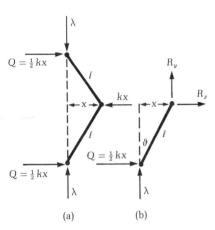

Figure 9-2

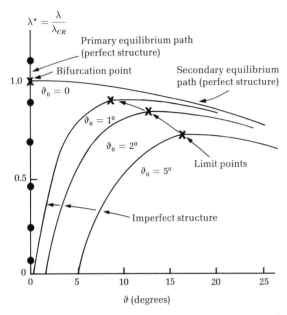

Figure 9-3

The intersection of the primary and secondary equilibrium paths is called a **bifurcation point**, and the load corresponding to the bifurcation point is called a **bifurcation load**. Note that the primary path (straight equilibrium configuration) remains **stable** for $\lambda < \lambda_{CR} = \frac{1}{2}k\ell$, but it becomes **unstable** for $\lambda \geq \lambda_{CR}$.

EXAMPLE 9-2

Investigate the stability of the two-link mechanism of Example 9-1 when a small initial geometric imperfection is present. The imperfect system is shown in Figure 9-4a.

SOLUTION

According to the imperfection criterion, the smallest value of the load λ for which the displacements of the imperfect system cease to be bounded is sought. A *small* geometric imperfection of the two-link system is represented by $\bar{x}$ in Figure 9-4a. The linear spring is assumed to be stress free in this initial configuration. The displacement of the system, in which the spring is compressed by an amount x, is shown in Figure 9-4b. The total angle that the lower link makes with the vertical is denoted by ϑ. Observe that prior to loading this angle is ϑ_0.

Moment equilibrium of the lower link shown in the free-body diagram of Figure 9-4c gives

$$(\lambda - \tfrac{1}{2}k\ell \cos \vartheta)x = -\lambda\bar{x} \tag{a}$$

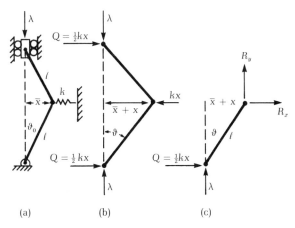

Figure 9-4

Let $\lambda^* = \lambda/\tfrac{1}{2}k\ell$. Then Eq. (a) can be solved explicitly for x.

$$x = \frac{-\bar{x}}{1 - \dfrac{\cos \vartheta}{\lambda^*}} \qquad (b)$$

The lateral displacement becomes infinite when

$$\lambda^* = \cos \vartheta \qquad (c)$$

which is the same load-displacement relation obtained via the adjacent equilibrium method. For infinitesimal excursions from the initial equilibrium configuration, $\cos \vartheta \approx 1$ so that

$$\lambda_{CR} = \tfrac{1}{2}k\ell \qquad (d)$$

The load-displacement relation for the imperfect system ($\vartheta_0 \neq 0$) is obtained from Eq. (b). Expressed in terms of the angle ϑ, it is

$$\lambda^* = \frac{\sin \vartheta - \sin \vartheta_0}{\tan \vartheta} \qquad (e)$$

where the relations $\bar{x} = \ell \sin \vartheta_0$ and $\bar{x} + x = \ell \sin \vartheta$ have been used. Plots of Eq. (e) for several magnitudes of the initial geometric parameter ϑ_0 are shown in Figure 9-3. The load-displacement curves for the imperfect system possess local maximums. The smallest value of the load parameter λ^* corresponding to a local maximum is the critical load for the mechanism with a given initial imperfection. Note the progressive deterioration in the buckling resistance of the mechanism as the magnitude of the initial imperfection increases. Notice also that as the initial imperfection approaches zero, the corresponding load-displacement curve approaches the load-displacement curve for the perfect system. Finally, observe that for sufficiently large values of displacement, the load-displacement curves of the imperfect mechanism approach the load-displacement curve for the perfect mechanism.

A real relationship between the local maximum λ_{CR}^* and the associated geometric imperfection ϑ_0 is obtained by differentiation of Eq. (e). Thus

$$\lambda_{CR}^* = \{1 - (\sin \vartheta_0)^{2/3}\}^{3/2} \tag{f}$$

Equation (f) is plotted in Figure 9-5. Recall that λ_{CR}^* corresponds to the local maximum of the equilibrium path of the imperfect mechanism. Figure 9-5 clearly exhibits the sensitivity of this mechanism to initial geometric imperfections.

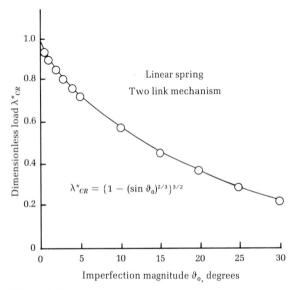

Figure 9-5

EXAMPLE 9-3

Investigate the stability of the imperfect two-link mechanism of Example 9-2 when the linear spring is replaced by a nonlinear hardening spring for which the spring force is $F_s = ax + bx^3$.

SOLUTION

The load-displacement relation for this problem is obtained in precisely the same manner as the load-displacement relation for the linear system established in Example 9-2. Consequently,

$$\lambda^* = \frac{(\sin \vartheta - \sin \vartheta_0) + \alpha(\sin \vartheta - \sin \vartheta_0)^3}{\tan \vartheta} \tag{a}$$

where

$$\lambda^* = \frac{2\lambda}{a\ell} \quad \text{and} \quad \alpha = \frac{b}{a}\ell^2 \tag{b}$$

Equation (a) is plotted in Figure 9-6 for $\alpha = 1$ for several values of initial imperfection ϑ_0. Note that the system is insensitive to small geometric imperfections because the load increases monotonically with the displacement ϑ for equilibrium configurations that are "near" the straight configuration.

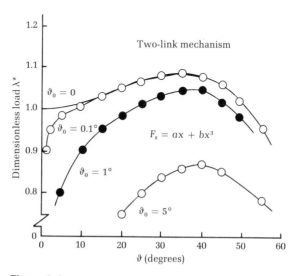

Figure 9-6

PROBLEMS / Section 9-2

9-1 A rigid link of length ℓ is pinned at its lower end and supports an eccentric load P as shown in Figure P9-1. Angular displacement of the link is resisted by a linear torsional spring (spring constant c, in.-lb/rad) at the pinned end. Determine the critical value of P using the imperfection criterion as a stability criterion. Construct a plot with the dimensionless load parameter $P\ell/c$ as ordinate and angular displacement θ as abscissa for values of dimensionless eccentricity parameter $e/\ell = 0.0$, 0.01, and 0.10. What conclusions can be reached regarding the sensitivity of the mechanism to small initial imperfections?

9-2 A rigid, weightless link is connected to a rigid support by a smooth pin as shown in Figure P9-2. Angular displacement of the link is resisted by a linear torsional spring with modulus c. A cord connected to the free end of the link passes over a smooth peg at point A. Use the adjacent equilibrium criterion as a basis for determining the magnitude of the smallest force P for which the vertical configuration of the link becomes unstable. Construct a graph of the load parameter $P\ell/4c$ versus angular displacement θ. Comment on the sensitivity of the mechanism to initial geometric imperfections. $(P\ell/c = 2)$

Figure P9-1

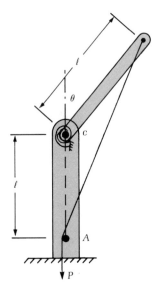

Figure P9-2

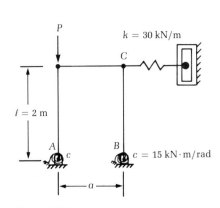

Figure P9-3

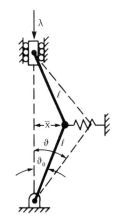

Figure P9-4

9-3　Use the adjacent equilibrium criterion as a basis for establishing the largest value of P for which the mechanism shown in Figure P9-3 remains stable. Plot the load-deflection curve corresponding to the secondary equilibrium path for the mechanism. The three links are weightless and rigid and are interconnected by smooth pins. Angular displacement of the vertical links is resisted by linear torsional springs at A and B and a linear spring at point C. [$P\ell/c = 2 + k\ell^2/c$, 75 kN]

9-4　The links of the mechanism shown in Figure P9-4 are rigid, and the elastic restoring forces of the system are represented by a nonlinear spring whose force-displacement relation is $F_s = ax - bx^3$. The spring is assumed to be in a stress-free state when the mechanism has the form indicated by the imperfection $\bar{x}$. Derive the force-deflection relation. Construct a plot of $\lambda^* = 2\lambda/a\ell$ versus ϑ for $\vartheta_0 = 0.1$, 1.0, and 2.0 degrees. Let $\alpha = (b/a)\ell^2 = 1$. Discuss the imperfection sensitivity of the mechanism.

9-3　Energy Criterion

　　　　To establish a connection between the incipient loss of stability of a conservative mechanical system and its potential energy, consider the law of conservation of mechanical energy applied to the bead shown in Figure 9-7. The small bead of mass m is constrained to slide along a smooth wire that lies in the x, y plane. The potential energy of the bead is $V = mgy$ and, hence, the potential energy function for the bead has the same shape as its path.

　　　　Figure 9-7 indicates four different equilibrium configurations for the bead. Positions 1 and 2 are maximum and minimum points on the potential energy curve, position 3 is an inflection point, and a position 4 lies in a locally flat region.

　　　　The bead in position 1 is given an instantaneous disturbance that causes it to deviate infinitesimally from its equilibrium position. This

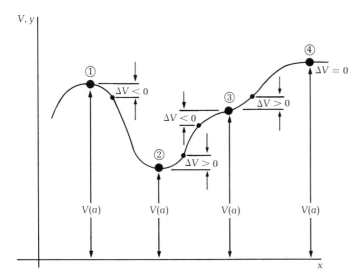

Figure 9-7

disturbance can be either an initial velocity and/or an initial displacement. It is convenient to think of it as an initial velocity. Mechanical energy is conserved as the bead departs from its equilibrium configuration. Consequently, if T denotes its kinetic energy and V its potential energy,

$$T + V = \text{constant} \qquad (9\text{-}1)$$

At a local maximum, the potential energy V decreases as a result of *any* admissible displacement of the bead. Therefore, the kinetic energy must increase, carrying the bead away from its equilibrium configuration. An equilibrium configuration for which infinitesimal deviations cause the bead to depart permanently from that configuration is said to be **unstable**. Thus an equilibrium configuration is unstable if the change in potential energy is negative as shown in Figure 9-7; that is, if

$$\Delta V < 0 \qquad (9\text{-}2a)$$

At equilibrium position 2, an instantaneous velocity causes the bead to depart from its equilibrium position at a local minimum on the potential energy curve. The potential energy increases, but the kinetic energy decreases. The kinetic energy that the bead acquires because of the initial instantaneous velocity is converted eventually to potential energy, and the bead reverses its motion. The bead oscillates about the equilibrium configuration at the minimum on the potential energy curve. An equilibrium configuration for which infinitesimal disturbances cause the bead to undergo bounded oscillations about this configuration is said to be **stable**. Consequently, an equilibrium configuration is stable if the change in potential energy is positive as shown in Figure 9-7; that is, if

$$\Delta V > 0 \tag{9-2b}$$

At equilibrium configuration 3, an instantaneous velocity can cause the potential energy of the bead to increase or decrease depending on the direction of the velocity. Accordingly, at an inflection point on the potential energy curve, $\Delta V > 0$ for some disturbances and $\Delta V < 0$ for other disturbances. Consequently, an equilibrium configuration for which $\Delta V > 0$ for some disturbances and $\Delta V < 0$ for other disturbances is said to be **unstable**.

At equilibrium position 4, the initial disturbance is considered to be a displacement. Since the region of the potential energy curve is locally flat, the derivatives $dV/dx = d^2V/dx^2 = \cdots = d^nV/dx^n = 0$ in the neighborhood of position 4. This means that infinitesimal disturbances about position 4 leave the potential energy unchanged. Thus the bead remains in the position to which it is displaced. An equilibrium configuration characterized by this behavior is called a **neutral equilibrium configuration**. Therefore, a neutral equilibrium position is characterized by

$$\Delta V = 0 \tag{9-2c}$$

Mathematical criterion. A mathematical criterion that establishes the loss of stability of a mechanical system can be obtained by expanding the potential energy function of the bead in a Taylor series about an equilibrium position. Let a signify the equilibrium positions of the bead, and let h denote an infinitesimal displacement of the bead from any of its equilibrium positions. Then

$$V(a + h) = V(a) + \frac{dV}{dx}\bigg|_{x=a} h + \frac{1}{2!}\frac{d^2V}{dx^2}\bigg|_{x=a} h^2 + \frac{1}{3!}\frac{d^3V}{dx^3}\bigg|_{x=a} h^3$$

$$+ \cdots + \frac{1}{n!}\frac{d^nV}{dx^n}\bigg|_{x=a} h^n \tag{9-3}$$

Because $x = a$ is an equilibrium position, $(dV/dx)\big|_{x=a} = 0$. Therefore, the change in potential energy is

$$\Delta V = \frac{1}{2!}\frac{d^2V}{dx^2}\bigg|_{x=a} h^2 + \frac{1}{3!}\frac{d^3V}{dx^3}\bigg|_{x=a} h^3 + \cdots + \frac{1}{n!}\frac{d^nV}{dx^n}\bigg|_{x=a} h^n \tag{9-4}$$

Now if h is small enough, the sign of ΔV is controlled by the sign of the first nonzero term in Eq. (9-4). Accordingly,

1. If $\Delta V > 0$ for any disturbance, then

$$\frac{d^2V}{dx^2}\bigg|_{x=a} > 0$$

and the system is stable.

2. If $\Delta V < 0$ for any disturbance, then

$$\left. \frac{d^2V}{dx^2} \right|_{x=a} < 0$$

and the the system is unstable.

The neutral equilibrium configuration characterizes the transition of the system from a stable to an unstable equilibrium position. Consequently, the critical load for a mechanical system is characterized by the relation

$$\left. \frac{d^2V}{dx^2} \right|_{x=a} = 0 \tag{9-5}$$

EXAMPLE 9-4

Use the energy stability criterion to investigate the stability of the perfect two-link system of Example 9-1.

SOLUTION

The potential energy for the two-link system consists of the energy stored in the spring and the negative of the work done by the axial force λ. Thus

$$V(x) = \tfrac{1}{2}kx^2 - 2\lambda\ell(1 - \cos \vartheta) \tag{a}$$

and, hence,

$$\frac{d^2V}{dx^2} = k - \frac{\dfrac{2\lambda}{\ell}}{\left[1 - \left(\dfrac{x}{\ell}\right)^2\right]^{3/2}} \tag{b}$$

The system remains stable as long as $(d^2V/dx^2)|_{x=0} > 0$. It ceases to be stable when $(d^2V/dx^2)|_{x=0} = 0$. From Eq. (b), the critical load for the two-link mechanism is

$$\left. \frac{d^2V}{dx^2} \right|_{x=0} = k - \frac{2\lambda}{\ell} = 0$$

or

$$\lambda_{CR} = \tfrac{1}{2}k\ell \tag{c}$$

This result is the same one that was obtained by both the adjacent equilibrium criterion and the imperfection criterion.

EXAMPLE 9-5

Determine the critical load P for the two-link mechanism shown in Figure 9-8. The rigid links are connected by smooth pins at their ends. Angular displacements at the pins are resisted by linear torsional springs with spring constant c. Plot the primary and secondary equilibrium paths for the mechanism.

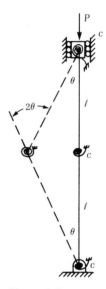

Figure 9-8

SOLUTION

The potential energy for the mechanism is

$$V(\theta) = \tfrac{1}{2}c\theta^2 + \tfrac{1}{2}c\theta^2 + \tfrac{1}{2}c(2\theta)^2 + 2P\ell \cos \theta + \text{constant}$$

or

$$V(\theta) = 3c\theta^2 + 2P\ell \cos \theta + \text{constant} \tag{a}$$

The first and second derivatives are

$$\frac{dV}{d\theta} = 6c\theta - 2P\ell \sin \theta \tag{b}$$

and

$$\frac{d^2V}{d\theta^2} = 6c - 2P\ell \cos \theta \tag{c}$$

The configuration corresponding to $\theta = 0$ is clearly an equilibrium configuration since $dV/d\theta|_{\theta=0} = 0$.

According to the energy criterion, the system remains stable as long as

$$\frac{d^2V}{d\theta^2}\bigg|_{\theta=0} = 6c - 2P\ell > 0 \tag{d}$$

or as long as

$$P < \frac{3c}{\ell} \tag{e}$$

The load-angular displacement relation for the secondary equilibrium path is

$$\sigma^* = \frac{\theta}{\sin \theta} \tag{f}$$

where

$$\sigma^* = \frac{P\ell}{3c} \tag{g}$$

The secondary path is shown by the open circles in Figure 9-9, and the primary equilibrium path is indicated by the crosses in Figure 9-9.

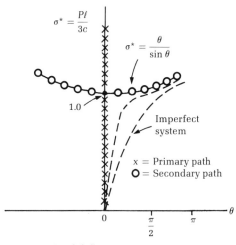

Load-deflection curves

Figure 9-9

The bifurcation point is obtained from Eq. (f). In this case,

$$\sigma_{CR}^* = 1 \tag{h}$$

or

$$\frac{P\ell}{c} = 3 \tag{i}$$

PROBLEMS / Section 9-3

9-5 Use the energy criterion as a basis for determining the largest value of P for which the mechanism shown in Figure P9-5 remains stable. The mechanism consists of three rigid links, each of length ℓ, that are connected at their ends by smooth pins as shown. Angular deformation of the mechanism is resisted by linear torsional springs at the pins A and B and by a linear spring at C. The spring constants for the torsional and linear springs are denoted by c and k, respectively. The mechanism has a slight side sway characterized by the angle θ_0 prior to the application of the force P. Derive the exact load-deflection relationship for the mechanism and plot the $P\ell/c$ versus θ curves for $\theta_0 = 1$ degree and 5 degrees for $k\ell^2/c = 8$.

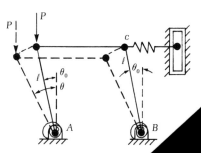

Figure P9-5

9-6 The rods shown in Figure P9-6 are rigid compared to the linear springs. The legs of the mechanism are each of length ℓ prior to load application and are inclined at α degrees with the horizontal as shown. Show that the equilibrium equation for the system is

$$\lambda = \sin\theta\left(1 - \frac{\cos\alpha}{\cos\theta}\right)$$

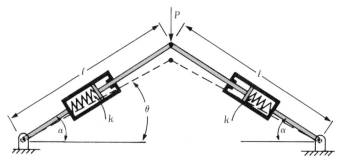

Figure P9-6

where $\lambda = P/2k\ell$ is a dimensionless load parameter. Construct the equilibrium path (λ versus θ) for the system when $\alpha = 30°$. Let λ be the ordinate and θ the abscissa. The linear springs have identical spring constants k.

9-4 Geometric Considerations

To use the adjacent equilibrium and imperfection criteria to establish stability limits for concentrically loaded columns, we must develop a moment-curvature relation when the bending of a straight beam is accomplished by a uniform axial compression.

Consider a straight beam that is subjected to bending and axial loads. The deformed configuration of this beam is shown in Figure 9-10a.

The length of a line element situated along the centroidal axis of the beam in its undeformed configuration is ds. Its length in the deformed configuration is ds^*. Moreover, a line element located at a distance η from the centroidal line in the deformed configuration is dS^*. Its length in the undeformed configuration is $dS \equiv ds$.

Assume that plane sections before deformation remain plane sections after deformation, and that plane sections perpendicular to

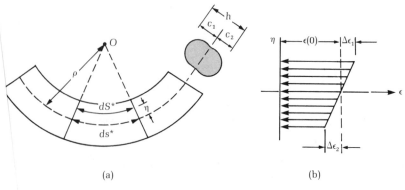

(a)   (b)

Figure 9-10

the centroidal line before deformation remain perpendicular to the centroidal line in the deformed configuration.

From Figure 9-10a,

$$\left.\begin{array}{l} \dfrac{ds^*}{\rho} = \dfrac{dS^*}{\rho - \eta} \\[2mm] -\dfrac{\eta}{\rho} = \dfrac{dS^* - ds^*}{ds^*} \end{array}\right\} \tag{9-6}$$

Now

$$\left.\begin{array}{l} dS^* = [1 + \epsilon(\eta)]\, ds \\[1mm] ds^* = [1 + \epsilon(0)]\, ds \end{array}\right\} \tag{9-7}$$

so that

$$-\frac{\eta}{\rho} = \frac{\epsilon(\eta) - \epsilon(0)}{1 + \epsilon(0)} \tag{9-8}$$

Since the strain at the centroidal line $\epsilon(0)$ is small compared to unity, Eq. (9-8) becomes

$$-\frac{\eta}{\rho} = \epsilon(\eta) - \epsilon(0) \tag{9-9}$$

Also, since plane sections remain plane, the strain must vary linearly over a cross section. Accordingly,

$$\epsilon(\eta) = -\frac{\Delta\epsilon}{h}\,\eta + \epsilon(0) \tag{9-10}$$

where

$$\Delta\epsilon = \Delta\epsilon_1 + \Delta\epsilon_2 \tag{9-11}$$

This strain distribution is shown in Figure 9-10b.

Substituting Eq. (9-10) into Eq. (9-9) yields a formula for the curvature of the beam in terms of the bending strain $\Delta\epsilon$.

$$\frac{1}{\rho} = \frac{\Delta\epsilon}{h} \tag{9-12}$$

Notice that this formula is independent of the material behavior; that is, it is valid for both elastic and inelastic material behavior.

9-5 The Euler Column Theory

Leonhard Euler, a famous eighteenth-century Swiss mathematician, assumed that an axially loaded rod would buckle elastically. The stress distribution on a cross section of the column in a slightly bent configuration is shown in Figure 9-11a. The stresses at points A, B,

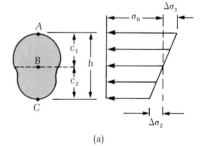

(a)

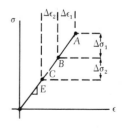

(b)

Figure 9-11

and C of a cross section are shown on a stress-strain diagram in Figure 9-11b. The bending stresses and bending strains are seen to be related by the equations

$$\left.\begin{aligned} \Delta\epsilon_1 &= \frac{\Delta\sigma_1}{E} \\[2mm] \Delta\epsilon_2 &= \frac{\Delta\sigma_2}{E} \end{aligned}\right\} \tag{9-13}$$

so that

$$\Delta\epsilon = \Delta\epsilon_1 + \Delta\epsilon_2 = \frac{1}{E}(\Delta\sigma_1 + \Delta\sigma_2) \tag{9-14}$$

Since the column buckles elastically, the bending stresses can be calculated from the flexure formula. Accordingly,

$$\Delta\epsilon = \frac{1}{E}\left\{\frac{MC_1}{I} + \frac{MC_2}{I}\right\} = \frac{Mh}{EI} \tag{9-15}$$

Consequently, the moment-curvature relation for **elastic** buckling is

$$\frac{1}{\rho} = \frac{M}{EI} \tag{9-16}$$

Consider a straight rod of length ℓ whose ends are simply supported as shown in Figure 9-12a. Let the rod be loaded by a concentric load P.

According to the adjacent equilibrium criterion, the critical load for a column is the smallest value of P for which a slightly bent equilibrium configuration is possible. Since the stability criterion stipulates a *slightly* bent configuration, only bent configurations that lie near the straight form of the column are considered. From the calculus, the curvature of a plane curve (see Figure 8-2)

$$\frac{1}{\rho} = \frac{v''(x)}{[1 + (v'(x))^2]^{3/2}} \tag{9-17}$$

where primes denote differentiation with respect to x. Then, for equilibrium configurations that deviate only slightly from the straight configurations, $[v'(x)]^2 \ll 1$ so that Eq. (9-16) becomes

$$EIv''(x) = M(x) \tag{9-18}$$

which is the differential equation for the bending of straight beams by lateral loads derived in Chapter 8.

From the free-body diagram shown in Figure 9-12b, the bending moment due to P is seen to be

$$M(x) = -Pv(x) \tag{9-19}$$

Thus, the differential equation of the bent configuration of the column is

$$v''(x) + \lambda^2 v(x) = 0 \tag{9-20}$$

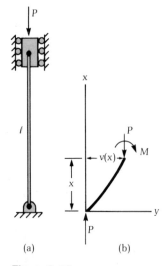

(a)   (b)

Figure 9-12

where

$$\lambda^2 = \frac{P}{EI} \tag{9-21}$$

The boundary conditions

$$\left.\begin{array}{l} v(0) = 0 \\ v(\ell) = 0 \end{array}\right\} \tag{9-22}$$

are added to this differential equation.

Differential equation (9-20) and boundary conditions (9-22) constitute a **linear eigenvalue problem**. Note that $v(x) \equiv 0$ is a solution. However, only solutions for which $v(x) \not\equiv 0$ are of interest.

The general solution of Eq. (9-20) is

$$v(x) = C_1 \sin \lambda x + C_2 \cos \lambda x \tag{9-23}$$

which can be checked readily by direct substitution in Eq. (9-20).

The first boundary condition in Eqs. (9-22) requires that $C_2 = 0$, and the second boundary condition requires that

$$C_1 \sin \lambda\ell = 0 \tag{9-24}$$

Since solutions for which $v(x) \not\equiv 0$ are sought, $C_1 \neq 0$. Accordingly, nonzero solutions exist for values of $\lambda\ell$ that satisfy the relation

$$\sin \lambda\ell = \sin n\pi \tag{9-25}$$

or

$$\lambda\ell = \pm n\pi \qquad (n = 1, 2, \ldots) \tag{9-26}$$

Equations (9-21) and (9-26) yield the **Euler loads** for the simply supported, concentrically loaded column as

$$P_n = \frac{n^2\pi^2 EI}{\ell^2} \qquad (n = 1, 2, \ldots) \tag{9-27}$$

and the corresponding buckled shapes are

$$v_n(x) = C_1 \sin \frac{n\pi x}{\ell} \tag{9-28}$$

The critical load for the column is the smallest of the Euler loads

$$P_{CR} = \frac{\pi^2 EI}{\ell^2} \tag{9-29}$$

Figure 9-13 shows the buckled shapes for the first two Euler loads.

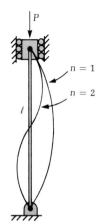

Figure 9-13

Recall that $I = Ar^2$ where r is the radius of gyration of the cross-sectional area A. Equation (9-29) can be written in the form

$$\sigma_{CR} = \frac{P_{CR}}{A} = \frac{\pi^2 E}{\left(\dfrac{\ell}{r}\right)^2} \tag{9-30}$$

The ratio ℓ/r is called the **slenderness ratio** for the column, and it plays an extremely important role in column analysis. Observe that the only material property that enters into this formula is the modulus of elasticity. No strength parameter such as proportional limit stress, yield stress or yield strength, or ultimate strength enters into the formula. Moreover, the modulus of elasticity for most alloys of metals is practically the same so that no advantage is gained by selecting a material of high strength. (This conclusion does not hold true for inelastic buckling.)

Other end conditions.   Consider next the elastic buckling of a column that is clamped at one end and simply supported at the other as shown in Figure 9-14a.

The differential equation of the slightly bent configuration is

$$v''(x) + \lambda^2 v(x) = \frac{Q}{EI} x, \tag{9-31}$$

and the pertinent boundary conditions are

$$\left.\begin{array}{l} v(0) = 0 \\ v'(\ell) = 0 \\ v(\ell) = 0 \end{array}\right\} \tag{9-32}$$

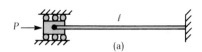

(a)

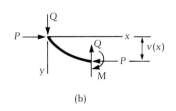

(b)

Figure 9-14

Note that three boundary conditions are required.

The general solution of Eq. (9-31) is

$$v(x) = C_1 \sin \lambda x + C_2 \cos \lambda x + \frac{Q}{EI\lambda^2} x \tag{9-33}$$

Now boundary conditions (9-32), when applied to Eq. (9-33), lead to the following system of homogeneous algebraic equations.

$$\left.\begin{array}{l} v(0) = 0: \quad C_2 = 0 \\[2mm] v'(\ell) = 0: \quad C_1\lambda \cos \lambda\ell + \dfrac{Q}{EI\lambda^2} = 0 \\[3mm] v(\ell) = 0: \quad C_1 \sin \lambda\ell + \dfrac{Q}{EI\lambda^2}\ell = 0 \end{array}\right\} \tag{9-34}$$

From the theory of homogeneous algebraic equations, we know that nonzero values for the quantities C_1 and Q exist if and only if the determinant of the coefficients is zero. Bent equilibrium configurations of the column exist for values of λ that satisfy the determinant equation

$$\begin{vmatrix} \lambda \cos \lambda \ell & \dfrac{1}{EI\lambda^2} \\[4mm] \sin \lambda \ell & \dfrac{\ell}{EI\lambda^2} \end{vmatrix} = 0 \qquad (9\text{-}35)$$

Evaluation of this determinant leads to the so-called **characteristic equation** for the load parameter λ.

$$\tan \lambda \ell = \lambda \ell \qquad (9\text{-}36)$$

This equation can easily be solved by trial and error or by the Newton method. The smallest root is $\lambda \ell = 4.4934$, and it occurs at the intersection of the curve $u = \lambda \ell$ and the second branch of the tangent function $u = \tan \lambda \ell$ as shown in Figure 9-15. Since

$$\lambda^2 = \frac{P}{EI} = \left(\frac{4.4934}{\ell}\right)^2 \qquad (9\text{-}37)$$

the critical load for the column is found to be

$$P_{CR} = 2.05 \frac{\pi^2 EI}{\ell^2} \qquad (9\text{-}38)$$

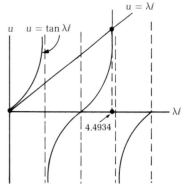

Figure 9-15

or

$$\sigma_{CR} = \frac{\pi^2 E}{\left(0.7 \dfrac{\ell}{r}\right)^2} \qquad (9\text{-}39)$$

Suppose we must determine the value of x for which the bending moment is zero for the bent equilibrium configuration. From Figure 9-14b,

$$M(x) = -Pv(x) + Qx \qquad (9\text{-}40)$$

Substituting from Eq. (9-33) with $C_2 = 0$, we find that the moment is zero when $\sin \lambda x_0 = 0$ or when $\lambda x_0 = \pi$. Consequently, the section at which $M(x) = 0$ is

$$x_0 = \frac{\pi}{\lambda} = \frac{\dfrac{\pi}{4.4934}}{\ell} = 0.7\ell \qquad (9\text{-}41)$$

In Figure 9-16, notice that the column between $0 \le x \le 0.7\ell$ behaves like a simply supported column of length 0.7ℓ. Thus, using the Euler formula for a simply supported column and an effective slenderness

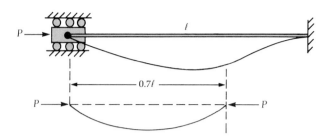

Figure 9-16

ratio of $0.7\ell/r$ we find that

$$\sigma_{CR} = \frac{\pi^2 E}{\left(0.7\,\dfrac{\ell}{r}\right)^2} \tag{9-42}$$

which is precisely the same result that was obtained from the rigorous solution. Thus, if two points along the length of a column for which the moment is zero are known, the critical stress can be calculated using the Euler formula for simply supported ends.

Figures 9-17a and 9-17b indicate effective lengths for two additional important end conditions. Euler's formula can be written to account for each of the four end conditions. Thus

$$\sigma_{CR} = \frac{\pi^2 E}{\left(\dfrac{k\ell}{r}\right)^2} \tag{9-43}$$

where

$$k = \begin{cases} 1.0 & \text{pinned-pinned} \\ 0.7 & \text{pinned-clamped} \\ 0.5 & \text{clamped-clamped} \\ 2.0 & \text{clamped-free} \end{cases} \tag{9-44}$$

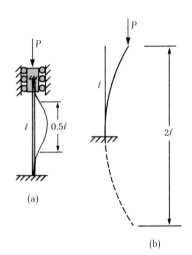

(a)

(b)

Figure 9-17

The ratio $k\ell/r$ is called the **effective slenderness ratio**. Of course, the k factor for the columns shown in Figure 9-17a and 9-17b can be derived directly as was done for the simply supported and the pinned-clamped end conditions.

Limitation of the Euler column formula. The Euler formula is based on the assumption that the column buckles elastically. Consequently, it remains valid for effective slenderness ratios that result in axial stresses that do not exceed the proportional limit of the material; that is, as long as $\sigma_{CR} \leq \sigma_{P\ell}$. The effective slenderness ratio that determines the limit on the use of Euler's formula is calculated from

Eq. (9-43) with $\sigma_{CR} = \sigma_{P\ell}$. Thus

$$\left(\frac{k\ell}{r}\right)_0 = \sqrt{\frac{\pi^2 E}{\sigma_{P\ell}}} \tag{9-45}$$

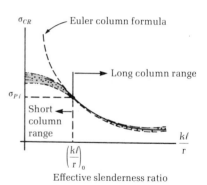

Figure 9-18

Consequently, Euler's column formula is valid for $k\ell/r$ ratios greater than the limiting effective slenderness ratio given by Eq. (9-45).

The Euler column formula is a hyperbola as shown in Figure 9-18. The Euler formula is shown dashed for $k\ell/r < (k\ell/r)_0$ to emphasize that it is no longer valid in this range of slenderness ratios. It agrees remarkably well with experimental results for long slender columns. For this reason, the Euler formula is used in almost every column design procedure for the design of long slender columns. Each column code will carefully stipulate the range of $k\ell/r$ for which the code permits use of the Euler column formula.

The shaded area in Figure 9-18 indicates a general trend for buckling experiments on columns. Clearly, a rational design formula for columns whose $k\ell/r < (k\ell/r)_0$ is required. This range of slenderness ratios is usually referred to as the short column range.

EXAMPLE 9-6

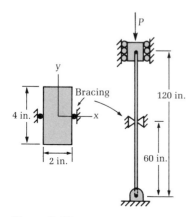

Figure 9-19

A 2-in. × 4-in. × 120-in. red oak timber is to be used in making a painter's scaffold. The ends are fitted with ball-and-socket joints so that the end conditions can be considered to be the pinned-end conditions. Moreover, bracing is provided as shown in Figure 9-19. Determine the largest axial load P that can be applied to this timber ($E = 1.8 \times 10^6$ psi and $\sigma_{P\ell} = 4.6$ ksi).

SOLUTION

The column lengths for buckling about the x and y axes are 120 in. and 60 in., respectively. Thus the effective slenderness ratios for the oak column are

$$\left.\begin{array}{l} \left(\dfrac{k\ell}{r}\right)_x = \dfrac{120}{\dfrac{4}{2\sqrt{3}}} = 60\sqrt{3} = 104 \\[4ex] \left(\dfrac{k\ell}{r}\right)_y = \dfrac{60}{\dfrac{2}{2\sqrt{3}}} = 60\sqrt{3} = 104 \end{array}\right\} \tag{a}$$

The least value of $k\ell/r$ for which Euler's formula remains valid is

$$\left(\frac{k\ell}{r}\right)_0 = \sqrt{\frac{\pi^2(1.8 \times 10^6)}{4.6 \times 10^3}} = 62.2 \tag{b}$$

Thus Euler's formula is applicable and yields as the critical load for the column

$$P_{CR} = \frac{\pi^2 EA}{\left(\dfrac{k\ell}{r}\right)^2} = \frac{\pi^2(1.8 \times 10^6)8}{(104)^2} = 13{,}140 \text{ lb} \qquad\qquad (c)$$

EXAMPLE 9-7

The crank arm and the steel connecting link shown in Figure 9-20a are connected at points A, B, and C by smooth pins, and the piston translates between smooth surfaces. Consider the reactions at pins B and C to be applied at the centroid of the cross section of the connecting link, and neglect stress concentrations. Determine the maximum magnitude of a moment M that may be applied to the crank arm AB if pins B and C are assumed to provide complete fixing with regard to bending in a plane perpendicular to the plane of the link and no fixing with regard to bending in the plane of the link. The crank arm is assumed to be designed to carry this maximum moment, and the system is assumed to be in equilibrium in the configuration shown. Take $E = 30 \times 10^6$ psi and $\sigma_{P\ell} = 30$ ksi for steel.

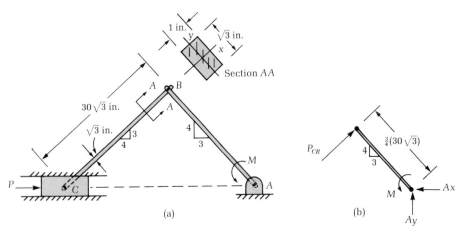

(a)   (b)

Figure 9-20

SOLUTION

The radius of gyration for a rectangular cross section of height h and base b is

$$r = \frac{h}{2\sqrt{3}} \qquad\qquad (a)$$

with respect to the centroidal axis parallel with the base. Consequently, the effective slenderness ratios for buckling about the x or y axes are

$$\left. \begin{array}{l} \left(\dfrac{k\ell}{r}\right)_x = 1.0 \, \dfrac{30\sqrt{3}}{\frac{1}{2}} = 60\sqrt{3} = 104 \\[3ex] \left(\dfrac{k\ell}{r}\right)_y = 0.5 \, \dfrac{30\sqrt{3}}{\dfrac{1}{2\sqrt{3}}} = 90 \end{array} \right\} \tag{b}$$

The critical load for the link corresponds to the maximum effective slenderness ratio. Thus buckling will occur in the plane of the link about the x axis. The least value of $k\ell/r$ for which Euler's column formula is applicable is

$$\left(\frac{k\ell}{r}\right)_0 = \sqrt{\frac{\pi^2\, 30 \times 10^6}{30 \times 10^3}} = 99.3 \tag{c}$$

Since $k\ell/r$ is greater than $(k\ell/r)_0$, the column under consideration falls in the long column range. The critical load for the link is therefore

$$P_{\mathrm{CR}} = \frac{\pi^2(30 \times 10^6)\sqrt{3}}{(104)^2} = 47{,}415 \text{ lb} \tag{d}$$

and the magnitude of M is, from Figure 9-20b,

$$M = \tfrac{3}{4}(30\sqrt{3})P_{\mathrm{CR}} = 1.848 \times 10^6 \text{ in.-lb} \tag{e}$$

Observe that a factor of safety would ordinarily be used to offset the uncertainties that arise because end conditions may not be ideal and the link may not be perfectly straight.

PROBLEMS / Section 9-5

9-7   Figure P9-7 shows the cross sections of four columns. (a) Derive formulas for the centroidal radii of gyration with respect to the x and y axes for each cross section. (b) If $b = 2$ in., $h = 4$ in., $d_o = 4$ in., and $d_i = 3.5$ in., and if each column is 120 in. long, compute the slenderness ratio for each column. Assume pinned ends.

9-8   Figure P9-8 shows the cross section of a column that has been fabricated from four L $2 \times 2 \times \frac{1}{4}$ steel sections by thin lattice bars. If the area of the lattice bars can be neglected in the computations for the moments of inertia of the

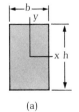
(a)

(b)

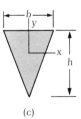
(c)

(d)

Figure P9-7

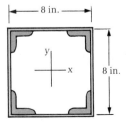

Figure P9-8

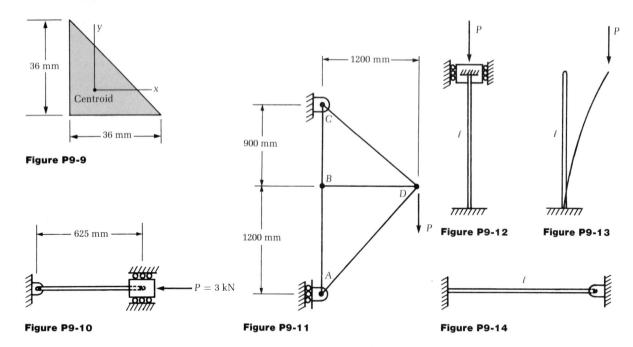

Figure P9-9

Figure P9-10

Figure P9-11

P

Figure P9-12

P

Figure P9-13

Figure P9-14

section, determine the radius of gyration for the composite section. If the length of the column is 15 ft and if both ends can be considered to be pin connected, compute its effective slenderness ratio.

9-9 Compute the minimum radius of gyration associated with the cross section shown in Figure P9-9.

9-10 Determine the minimum permissible diameter of the steel rod shown in Figure P9-10. The rod is concentrically loaded, and the ends of the rod can be considered to be pinned. A factor of safety of 2 is required. Verify the use of Euler's formula: $\sigma_{YP} = 252$ MPa and $E = 210$ GPa.

9-11 Calculate the maximum safe load P that can be applied to the pin-connected truss of Figure P9-11 if it is assumed that failure occurs by buckling. Assume that Euler's formula is applicable and that the diameter of member AD is 20 mm. $E = 210$ GPa and $\sigma_{YP} = 252$ MPa. Substantiate the validity of the use of Euler's formula.

9-12 Use the adjacent equilibrium criterion to establish the end-fixing factor $k = 0.5$ for a Euler column with both ends fixed as shown in Figure P9-12.

9-13 Use the adjacent equilibrium criterion to establish the end-fixing factor $k = 2.0$ for a Euler column with one end fixed and the other free as shown in Figure P9-13.

9-14 A straight aluminum rod fits stress free between two supports at a certain reference temperature. One end of the rod is rigidly clamped, while the other end is pinned as shown in Figure P9-14. Determine the maximum value of slenderness ratio that the rod may possess before a Euler instability occurs for a temperature increase of $120°F$ ($\alpha = 13 \times 10^{-6}$ F^{-1}, $\ell/r = 113.6$).

9-6   The Engesser Inelastic Column Theory

The Euler column theory applies to columns that buckle elastically; that is, for columns whose slenderness ratios exceed the limiting effective slenderness ratio $(k\ell/r)_0$. F. R. Engesser proposed a column theory similar to the Euler theory for columns that buckle inelastically. The adjacent equilibrium stability criterion is used to predict critical loads in the inelastic column range.

Engesser's hypothesis.   Consider a straight column for which the effective slenderness ratio is less than $(k\ell/r)_0$. Load the column with a concentric load to a load level that is slightly less than its critical load as stipulated by the adjacent equilibrium stability criterion. The stress distribution corresponding to this load level is shown in Figure 9-21a. Now apply the small increment necessary to bring the load to its critical value P_T, and simultaneously apply an infinitesimal lateral displacement. The resulting stress distribution is shown in Figure 9-21b.

Engesser hypothesized that the stress at every point of the cross section increases over its previous values even though bending stresses

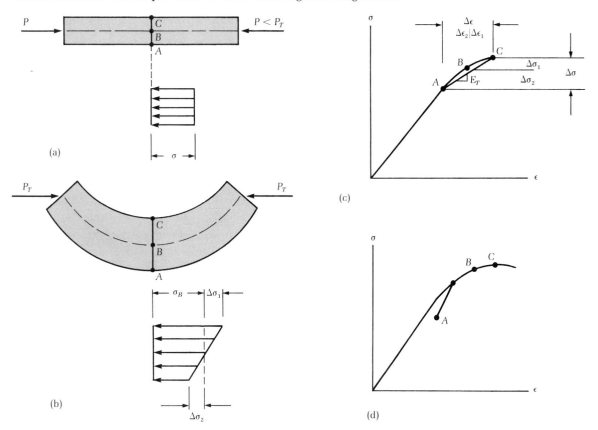

Figure 9-21

appear for this final loading path. This situation means that the stress on the side of the column where the bending stress is tensile does not decrease from its value before the last increment of load is applied. Thus the stresses at three points A, B, and C of a cross section remain on the stress-strain curve as shown in Figure 9-21c. If the stress at point A on the cross section had decreased from its value before the last increment of load was applied, the point would have ended up on a line parallel to the initial straight portion of the curve as shown in Figure 9-21d.

Moreover, Engesser reasoned that the adjacent equilibrium stability criterion requires that only equilibrium configurations infinitesimally close to the straight configuration exist. Consequently, points A, B, and C are infinitesimally close to each other on the stress-strain diagram. Consequently, Engesser reasoned that

$$\Delta\sigma = E_T\,\Delta\epsilon \qquad (9\text{-}46)$$

where E_T is called the **tangent modulus**.

Observe from Figure 9-21c that

$$\Delta\sigma = \Delta\sigma_1 + \Delta\sigma_2 \qquad (9\text{-}47)$$

so that

$$\Delta\sigma = \frac{Mc_1}{I} + \frac{Mc_2}{I} = \frac{Mh}{I} \qquad (9\text{-}48)$$

Substituting Eq. (9-48) into Eq. (9-46) results in

$$\Delta\epsilon = \frac{Mh}{E_T I} \qquad (9\text{-}49)$$

and, consequently, Eq. (9-12) yields the moment-curvature relation for inelastic buckling of a column. Thus

$$\frac{1}{\rho} = \frac{M}{E_T I} \qquad (9\text{-}50)$$

Using the approximation $v''(x) = 1/\rho$, we find the differential equation that determines axial loads that will just maintain bent equilibrium configurations is

$$E_T I v''(x) = M(x) \qquad (9\text{-}51)$$

This equation and appropriate boundary conditions constitute a boundary value problem that is completely analogous to the Euler boundary value problem. Therefore, the critical loads and critical stresses corresponding to inelastic buckling are

$$P_{CR} = \frac{\pi^2 E_T I}{\ell^2} \qquad (9\text{-}52)$$

and

$$\sigma_{CR} = \frac{\pi^2 E_T}{\left(\dfrac{k\ell}{r}\right)^2}$$ (9-53)

These formulas are referred to as the **Engesser formulas** or the **modified Euler formulas** in light of their resemblance to the Euler formulas.

Design of columns by the modified Euler formulas is a trial-and-error procedure illustrated in Example 9-8. The solution is greatly facilitated if a graph of σ versus E_T is constructed prior to beginning computations. Such a graph is shown in Figure 9-22 for the material whose stress-strain diagram also appears in the figure. The design procedure can be further facilitated by constructing so-called **column curves** such as those shown in Figure 9-23. These column curves have been constructed using tangent modulus data obtained from Figure 9-22.

To facilitate the trial-and-error procedure that is illustrated in the following example, it is necessary to establish a **stability test**. The stress that is calculated from the modified Euler formula will be called the **calculated allowable stress**. Note that a calculated allowable stress corresponds to a specific value of tangent modulus. The stress that is obtained from the stress-strain diagram at the same value of tangent

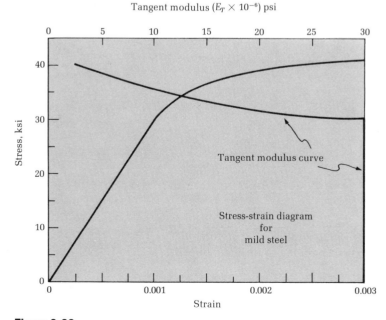

Figure 9-22

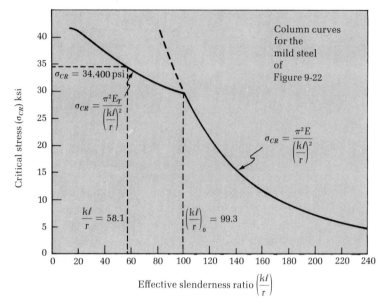

Figure 9-23

modulus will be called the **actual stress**. Thus a column will remain stable as long as the calculated allowable stress does not exceed the actual stress.

EXAMPLE 9-8

Calculate the largest concentric compressive load that a W 12 × 53 column 12 ft. long can support if its ends are assumed to be simply supported.

SOLUTION

From Appendix D, the area and minimum radius of gyration are 15.60 in^2 and 2.48 in., respectively. Therefore, the largest slenderness ratio is

$$\left(\frac{k\ell}{r}\right) = \frac{12(12)}{2.01} = 58.1 \tag{a}$$

The least value of $k\ell/r$ for which the Euler formula remains valid is

$$\left(\frac{k\ell}{r}\right)_0 = \sqrt{\frac{\pi^2 30 \times 10^6}{30 \times 10^3}} = 99.3 \tag{b}$$

Comparing the actual $k\ell/r$ for the column with this limiting value we see that the column buckles inelastically. Consequently, we must use the modified Euler column formula.

The stress-strain diagram for the material from which this section is assumed to be made is shown in Figure 9-22. Also shown in this figure is a plot of stress versus tangent modulus.

The calculated allowable stress for the column is calculated from the modified Euler buckling formula. Thus,

$$\sigma_{ALL} = \frac{\pi^2 E_T}{\left(\dfrac{k\ell}{r}\right)^2} = \frac{\pi^2}{(58.1)^2} E_T = 2927 \times 10^{-6} E_T \tag{c}$$

Table 9-1

σ (psi)	E_T (psi)	σ_{ALL} (psi)
35,000	11×10^6	32,197
34,000	13.3×10^6	38,929
34,250	12.37×10^6	36,207
34,500	11.75×10^6	34,392

The determination of the critical load requires a trial-and-error procedure. The results of the following calculations are listed in Table 9-1.

Select a stress above the proportional limit and find the tangent modulus that corresponds to it from the stress-strain diagram. Try as a first trial $\sigma = 35,000$ psi for which $E_T = 11 \times 10^6$ psi. Now use Eq. (c) to calculate the allowable stress the column can sustain for this modulus. It is 32,197 psi. Since the calculated allowable stress for this value of the tangent modulus is less than the actual stress corresponding to the tangent modulus, the column could safely carry an axial load that produced a stress of 32,197 psi.

It is possible to improve this approximate solution. Try a slightly lower actual stress, say $\sigma = 34,000$ psi. The corresponding tangent modulus is $E_T = 13.3 \times 10^6$ psi and the calculated allowable stress is $\sigma_{ALL} = 38,929$ psi. The calculated allowable stress exceeds the actual stress corresponding to E_T. Therefore, the column cannot sustain this allowable stress.

As a third trial, choose $\sigma = 34,250$ psi for which $E_T = 12.37 \times 10^6$ psi. Equation (c) yields $\sigma_{ALL} = 36,207$ psi for the allowable stress. Since σ_{ALL} exceeds the actual stress corresponding to E_T used to calculate σ_{ALL}, the column cannot safely sustain a load corresponding to this stress.

As a fourth refinement, choose $\sigma = 34,500$ psi for which $E_T = 11.75 \times 10^6$ psi. Equation (c) then gives $\sigma_{ALL} = 34,392$ psi. Since $\sigma_{ALL} < \sigma$, the column can safely carry the load corresponding to $\sigma_{ALL} = 34,392$ psi. Attempts to further refine the critical stress are not feasible in view of the sensitivity of the tangent modulus to small changes in stress and the difficulty encountered in determining E_T accurately from the diagram. Accordingly, the critical load is

$$P_{CR} = \sigma_{CR} A = 34,392(15.60) = 536,515 \text{ lb} \tag{d}$$

It may be more convenient to first establish the column curve for the material as shown in Figure 9-23 and then simply pick off the critical stress corresponding to the actual $k\ell/r$. Consequently, since $k\ell/r = 58.1$, determine, from Figure 9-23, $\sigma_{CR} = 34,400$ psi.

PROBLEMS / Section 9-6

9-15 The stress-strain diagram for a certain material is approximated by straight lines as shown in Figure P9-15a. Construct the column curve for this material. That is, construct the curve of critical stress versus effective slenderness ratio. Then determine the load carrying capacity of a simply supported, concentrically loaded column with the cross section shown in Figure P9-15b. The length of the column is 30 inches.

9-16 A 1-in. diameter rod made of 2024-T3 aluminum alloy is to be used as a compression member in a mechanism in a manner that can be closely approximated as shown in Figure P9-16. The Ramberg–Osgood representation of the stress-strain diagram for this material is $\epsilon = (\sigma/E) + (\sigma/\sigma_0)^n$ where $E = 10 \times 10^6$ psi, $\sigma_0 = 101,200$ psi, and $n = 20$. Determine the maximum load that the member can support. Use a safety factor of 2 (50.5 k).

9-17 Stress-strain data for aluminum alloy 2024-T3 ($\sigma_{ULT} = 65$ ksi) is given in Figure P9-17. (a) Construct the stress-strain diagram. (b) Construct a graph of tangent modulus E_T versus stress. (c) Construct the column curves for pinned-end and fixed-end columns.

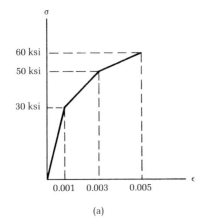

(a)

(b)

Figure P9-15

σ ksi	ϵ in./in.
0	0.
5	0.0005
10	0.0010
15	0.0015
20	0.0020
25	0.0025
30	0.0032
32	0.0035
34	0.0039
36	0.0045
38	0.0054
40	0.0067
42	0.0086
44	0.0114
46	0.0155
48	0.0214
50	0.0300
55	0.0704
60	0.1609
62	0.2212
65	0.3514

Figure P9-17

9-7 The Secant Formula

The Euler column analysis presented in Section 9-5 presumes that the column is initially straight and that the axial load is directed along its centroidal axis. That is, Euler column theory assumes a perfect column free from initial geometric imperfections and/or eccentricities of load. These hypotheses are extremely difficult to realize for actual columns. Moreover, situations arise where some intentional load eccentricity is necessary. Thus an analysis that accounts for initial geometric imperfections and load eccentricities is required. A rational approach to an analysis of this type is the so-called **secant formula**. To develop this formula, write the differential equation for the deflected centroidal line of a column with an eccentric load as shown in Figure 9-24.

$$EIv''(x) + Pv(x) = 0 \tag{9-54}$$

with boundary conditions

$$\left. \begin{aligned} v(0) &= e \\ v'\left(\frac{\ell}{2}\right) &= 0 \end{aligned} \right\} \tag{9-55}$$

Equation (9-54) is rewritten in the form

$$v''(x) + \lambda^2 v(x) = 0 \tag{9-56}$$

where

$$\lambda^2 = \frac{P}{EI} \tag{9-57}$$

is a load factor.

The general solution for Eq. (9-56) is

$$v(x) = C_1 \cos \lambda x + C_2 \sin \lambda x \tag{9-58}$$

Applying the boundary conditions given by Eqs. (9-55) yields

$$C_1 = e$$

$$\left. \begin{aligned} C_2 &= e \, \frac{\sin \dfrac{\lambda \ell}{2}}{\cos \dfrac{\lambda \ell}{2}} \end{aligned} \right\} \tag{9-59}$$

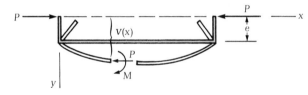

Figure 9-24

Consequently, the equation for the deflected centroidal line of the beam is

$$v(x) = e\left\{\cos \lambda x + \tan \frac{\lambda \ell}{2} \sin \lambda x\right\} \tag{9-60}$$

To determine a formula for the maximum normal stress on a cross section of the column, we need the maximum moment. The maximum moment occurs at midspan and has the value

$$M_{max} = Pv\left(\frac{\ell}{2}\right) \tag{9-61}$$

From Eq. (9-60) with $x = \ell/2$, determine

$$v_{max} = e \sec \frac{\lambda \ell}{2} \tag{9-62}$$

so that

$$M_{max} = Pe \sec \frac{\lambda \ell}{2} \tag{9-63}$$

The maximum stress on a cross section of the column is

$$\left.\begin{aligned}
\sigma_{max} &= \frac{P}{A} + \frac{M_{max}c}{I} = \frac{P}{A} + \frac{Pec}{Ar^2} \sec \frac{\lambda \ell}{2} \\
\sigma_{max} &= \frac{P}{A}\left\{1 + \frac{ec}{r^2} \sec \frac{\lambda \ell}{2}\right\}
\end{aligned}\right\} \tag{9-64}$$

The load factor λ is expressed in terms of P by Eq. (9-57); thus Eq. (9-64) assumes the form

$$\sigma_{max} = \frac{P}{A}\left\{1 + \frac{ec}{r^2} \sec \left(\sqrt{\frac{P}{4AE}}\frac{\ell}{r}\right)\right\} \tag{9-65}$$

Notice that r is the radius of gyration of the cross section with respect to the centroidal axis about which the column bends due to eccentricity. Since this may not be the minimum radius of gyration for the cross section, the column could buckle as a concentrically loaded column with respect to the centroidal axis for which the radius of gyration has its least value. Consequently, buckling loads should be calculated for both possibilities in this event. The smallest of these loads is the critical load for the column.

If a safety factor (S.F.) is to be applied to the load, P in Eq. (9-65) must be replaced by S.F. $\times P$. Application of the safety factor to the stress does not result in the same margin of safety that the application of the safety factor to the load does because of the nonlinear relation between stress and load.

Application of Eq. (9-65) is difficult and time consuming. First, it requires a trial-and-error procedure, and second, it requires that the eccentricity e be known. A design procedure can be established by constructing column curves such as the ones shown in Figure 9-25. An important observation to be made from this figure is that the secant formula approaches the Euler formula asymptotically for long slender columns regardless of the magnitude of the eccentricity. Euler's formula therefore predicts accurate critical loads for long slender columns even though initial deviations from straightness and eccentricities of loading may occur. It is because of this observation that almost every empirical column code incorporates the Euler formula for long slender columns.

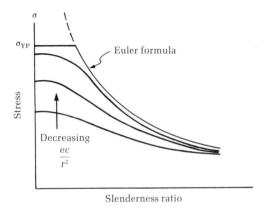

Figure 9-25

Finally, observe that the deflection at midspan is

$$\delta = e \sec \frac{\lambda \ell}{2} \tag{9-66}$$

Now, according to the imperfection stability criterion, the critical load for a column with an initial geometric imperfection corresponds to the smallest load for which the deflections of the column become infinite. Accordingly, δ becomes infinite at the points

$$\frac{\lambda \ell}{2} = \left(\frac{2n + 1}{2}\right)\pi \qquad (n = 0, 1, 2, \ldots) \tag{9-67}$$

Therefore, the critical load for a straight column according to the imperfection crition is

$$P_{CR} = \frac{\pi^2 EI}{\ell^2} \tag{9-68}$$

which is precisely the same value obtained by the adjacent equilibrium method.

PROBLEMS / Section 9-7

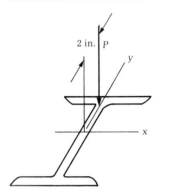

Figure P9-18

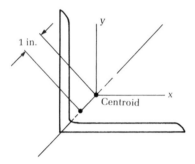

Figure P9-19

9-18   A W 8 × 21 structural steel section is used as a compression member required to carry a load that has an eccentricity along the minor axis as shown in Figure P9-18. If the compression member can be considered to be pinned at each end and is 14 ft long, calculate the maximum permissible value of P using the secant formula. Assume $\sigma_{YP} = 36$ ksi and $E = 29 \times 10^6$ psi (132 k). Calculate the Euler load for buckling about the y axis.

9-19   Select the most economical equal-leg structural steel angle to carry an eccentric load of 9.5 k on an unsupported length of 15 ft. Consider the ends of the column to be pinned, and assume that $\sigma_{YP} = 36$ ksi and $E = 29 \times 10^6$ psi. Use the secant formula. The eccentricity of loading is shown in Figure P9-19 $(2 \times 2 \times \frac{3}{8})$.

9-20   A 5-m length of aluminum pipe ($E = 70$ GPa, $\sigma_{YP} = 210$ MPa) is used as a compression member whose ends can be considered to be pinned. Calculate the maximum load this column can carry if the eccentricity is 40 mm as shown in Figure P9-20. Use the secant formula (14.8 kN).

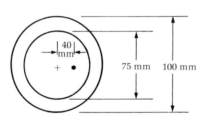

Figure P9-20

9-8   Column Design by Empirical Formulas

The Euler column formula is valid for concentrically loaded columns that become unstable for loads that induce direct stresses in the straight equilibrium configuration that do not exceed the proportional limit of the material from which the column is made. The Engesser column formula is valid for concentrically loaded columns that become unstable for loads associated with direct stresses that exceed the proportional limit. The Engesser and Euler column formulas cover the full range of $k\ell/r$ values for concentrically loaded columns. Because these formulas are developed from physical considerations, they are referred to as **rational formulas**, and a column design procedure that incorporates them is called a **rational design procedure**.

Many widely used procedures for the design of concentrically loaded columns are **semi-rational** in the sense that a rational design formula (Euler formula) is used for a prescribed range of slenderness ratios, and an empirical formula is used for slenderness ratios not covered by the rational formula.

For example, a parabolic equation of the general form

$$\sigma_{CR} = C_1 + C_2\left(\frac{k\ell}{r}\right)^2$$

(9-69)

is used in several design codes to predict critical loads for concentrically loaded columns that fall outside the Euler column range. The parameters C_1 and C_2 are adjusted so that the parabolic equation closely approximates experimental data in the short column range. Generally, the parameter C_1 is determined by prescribing the value of σ_{CR} for $k\ell/r = 0$. The parameter C_2 is determined by specifying the value of σ_{CR} at which it is desirable for the parabolic equation and the Euler formula to possess a common tangent. A safety factor may be included directly in the parameters C_1 and C_2, or it may be specified separately.

Some widely used procedures for the design of concentrically loaded columns employ a linear formula of the general form

$$\sigma_{CR} = C_1 + C_2\left(\frac{k\ell}{r}\right)$$

(9-70)

to predict critical loads for columns that fall outside the Euler column range. Again, the parameters C_1 and C_2 are adjusted so that this linear equation closely approximates the experimental data for columns of a given material.

Figures 9-26a and 9-26b indicate the manner in which the empirical formulas and Euler's column formula are used to form a column code. The least value of $k\ell/r$ for which the Euler formula remains valid is called the **limiting slenderness ratio** and designated by $(k\ell/r)_0$. Clearly, $(k\ell/r)_0$ separates the range of slenderness ratios for which the Euler formula applies from the range of slenderness ratios for which the empirical formula applies. Thus $(k\ell/r)_0$ determines which of the two formulas is to be used in calculating the critical load for a column.

AISC code. The American Institute of Steel Construction (AISC) recommends a semi-rational design procedure that incorporates the parabolic formula for short columns and the Euler formula for long columns.

The parameter C_1 appearing in Eq. (9-69) is determined so that $\sigma_{CR} = \sigma_{YP}$ when $k\ell/r = 0$. The parameter C_2 is determined by requiring that the parabolic equation and the Euler hyperbola possess a common tangent at $\sigma_{CR} = \sigma_{YP}/2$. The limiting slenderness ratio is obtained from the Euler formula with $\sigma_{CR} = \sigma_{YP}/2$. It is denoted by C_C and is given

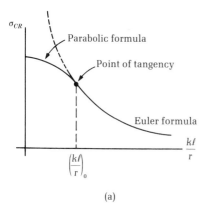

(a)

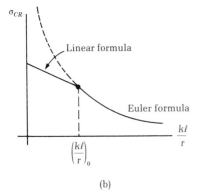

(b)

Figure 9-26

by the formula

$$C_C = \left(\frac{k\ell}{r}\right)_0 = \sqrt{\frac{2\pi^2 E}{\sigma_{YP}}}.$$

(9-71)

Moreover, it is no trouble to show that $C_1 = \sigma_{YP}$ and $C_2 = -\sigma_{YP}/2C_C^2$. Consequently, the AISC column design code stipulates that the following formulas are to be used to calculate critical loads for steel columns:

$$\sigma_{CR} = \frac{\sigma_{YP}}{\text{S.F.}}\left\{1 - \frac{\left(\frac{k\ell}{r}\right)^2}{2C_C^2}\right\} \quad \text{for } 0 \leq \frac{k\ell}{r} \leq C_C$$

(9-72)

with

$$\text{S.F.} = \frac{5}{3} + \frac{3}{8}\frac{\left(\frac{k\ell}{r}\right)}{C_C} - \frac{1}{8}\frac{\left(\frac{k\ell}{r}\right)^3}{C_C^3}$$

(9-73)

and

$$\sigma_{CR} = \frac{\pi^2 E}{\text{S.F.}\left(\frac{k\ell}{r}\right)^2} \quad \text{for } C_C \leq \frac{k\ell}{r} \leq 200$$

(9-74)

with

$$\text{S.F.} = 1.92$$

(9-75)

Observe that the safety factor to be used with the parabolic formula depends on the slenderness ratio of the column and varies between the limits of 1.67 at $k\ell/r = 0$ and 1.92 at the point of tangency with the Euler hyperbola. The safety factor to be used with the Euler formula is 1.92 over its complete range of validity. Finally, the AISC code specifies that $E = 29 \times 10^6$ psi. No column in excess of $k\ell/r = 200$ is permitted.

Column formulas for aluminum. The Aluminum Company of America (ALCOA) recommends the use of the linear formula for short columns and the Euler column formula for long columns. There are a large number of aluminum alloys available with a consequent large range of yield strengths. Since the modulus of elasticity is nearly constant for most alloys, ALCOA recommends that $E = 10.5 \times 10^6$ psi in the Euler formula for all aluminum columns.

As an example of a typical column code, the *ALCOA Structural Handbook* recommends that the following formulas be used for an extruded 2024-T4 alloy:

$$\sigma_{CR} = \frac{44{,}800 - 313\left(\dfrac{k\ell}{r}\right)}{\text{S.F.}} \qquad \text{for } 0 \leq \frac{k\ell}{r} \leq 64 \tag{9-76}$$

and

$$\sigma_{CR} = \frac{\pi^2 E}{\text{S.F.}\left(\dfrac{k\ell}{r}\right)^2} \qquad \text{for } 64 < \frac{k\ell}{r} \tag{9-77}$$

A suitable safety factor must be supplied.

EXAMPLE 9-9

Using the AISC column code, select the most economical wide flange section to carry a concentric load of 200,000 lb on an unsupported length of 15 ft. The column is simply supported with respect to bending in any plane, and it is made from A441 structural steel with $\sigma_{YP} = 50$ ksi.

SOLUTION

The selection of rolled sections to carry specified concentric loads is a trial-and-error procedure. The column code for A441 steel is

$$\sigma_{CR} = \frac{50}{\text{S.F.}}\left\{1 - \frac{1}{2}\left(\frac{\dfrac{k\ell}{r}}{C_C}\right)^2\right\} \text{ksi} \qquad \text{for } 0 \leq \frac{k\ell}{r} \leq C_C \tag{a}$$

and

$$\sigma_{CR} = \frac{149{,}000}{\left(\dfrac{k\ell}{r}\right)^2} \text{ksi} \qquad \text{for } C_C \leq \frac{k\ell}{r} \leq 200 \tag{b}$$

where

$$\text{S.F.} = \frac{5}{3} + \frac{3}{8}\left(\frac{\dfrac{k\ell}{r}}{C_C}\right) - \frac{1}{8}\left(\frac{\dfrac{k\ell}{r}}{C_C}\right)^3 \tag{c}$$

The limiting value of effective slenderness ratio for A441 structural steel is

$$C_C = \sqrt{\frac{2\pi^2 29 \times 10^6}{50 \times 10^3}} = 107 \qquad\qquad (d)$$

To start the trial and error procedure, note that the allowable stress for failure by yielding is $\sigma_{ALL} = 50{,}000/\frac{5}{3} = 30$ ksi. Consequently, a cross section with an area of at least $A = 200{,}000/30{,}000 = 6.67$ in^2 is required. Choose a W 8 × 24 as a first candidate. Pertinent properties for this section are obtained from a steel handbook and are listed in columns two and three in Table 9-2.

Table 9-2

Section designation	Area (in^2)	Radius of gyration (in.)	Effective slenderness ratio ($k\ell/r$)	Allowable stress, psi	Actual stress (psi)
W 8 × 24	7.06	1.61	112	11,880	28,300
W 8 × 58	17.10	2.10	85.8	17,760	11,696
W 8 × 40	11.70	2.04	88.2	17,320	17,100
W 8 × 35	10.30	2.03	88.7	17,183	19,417

Since the column is simply supported, $k = 1.0$. Consequently, for the W 8 × 24 section, the actual effective slenderness ratio is

$$\frac{k\ell}{r} = \frac{1.0(15 \times 12)}{1.61} = 112$$

It appears in column four of Table 9-2.

Since the actual $k\ell/r = 112$ is greater than the limiting value of 107, the allowable stress for this column must be calculated from Eq. (b). Thus

$$\sigma_{ALL} = \frac{149 \times 10^6}{(112)^2} = 11{,}880 \text{ psi}$$

This allowable stress is entered in the fifth column of Table 9-2.

The actual stress acting on a cross section of the column in its straight configuration is

$$\sigma_{ACT} = \frac{200{,}000}{7.06} = 28{,}300 \text{ psi}$$

and is entered in column six of Table 9-2.

The criterion that determines the adequacy of a candidate section is $\sigma_{ACT} \leq \sigma_{ALL}$. Accordingly, the W 8 × 24 is inadequate.

As a second candidate, select a W 8 × 58. The cross-sectional area and minimum radius of gyration are obtained from a handbook and are listed in Table 9-2. The effective slenderness ratio for this column is

$$\frac{k\ell}{r} = \frac{1.0(15 \times 12)}{2.10} = 85.8$$

Consequently, the parabolic formula applies, and we calculate

$$\text{S.F.} = \frac{5}{3} + \frac{3}{8}\left(\frac{85.8}{107}\right) - \frac{1}{8}\left(\frac{85.8}{107}\right)^3 = 1.91$$

and, hence,

$$\sigma_{ALL} = \frac{50,000}{1.91}\left\{1 - 0.5\left(\frac{85.8}{107}\right)^2\right\} = 17,760 \text{ psi}$$

Finally, the actual stress is calculated as

$$\sigma_{ACT} = \frac{200,000}{17.10} = 11,696 \text{ psi}$$

Now the actual stress is less than the allowable stress so that this candidate is adequate; that is, it will carry the stipulated concentric load. However, it is necessary to check other available wide flange sections to ascertain whether there are other sections that may be more economical as well as adequate.

Consider a W 8 × 40. The pertinent properties for this section are shown in Table 9-2. Since $k\ell/r = 88.2$ is less than C_C, the parabolic formula must be used to calculate the allowable stress. The allowable and actual stresses are listed in the appropriate columns. This section is clearly adequate and is more economical than the W 8 × 58.

Further checking shows that no other available wide flange beam is both more economical and adequate.

EXAMPLE 9-10

A standard 3-in. diameter steel pipe 10 ft long is to be used as a simply supported column. Using the AISC column code, determine the maximum concentric load P to which the column can be subjected. The pipe is made of structural steel for which $\sigma_{YP} = 35$ ksi.

SOLUTION

From Appendix D, the properties of a standard 3-in. diameter steel pipe are $A = 2.228 \text{ in}^2$ and $r = 1.16$ in. Thus the limiting value of effective slenderness ratio is

$$C_C = \sqrt{\frac{2\pi^2 29 \times 10^6}{35 \times 10^3}} = 127.9 \tag{a}$$

and the actual effective slenderness ratio is

$$\frac{k\ell}{r} = (1.0)\frac{12 \times 10}{1.16} = 103.4 \tag{b}$$

Since $(k\ell/r)_{ACT} < C_C$, the AISC column code requires that the allowable column stress be calculated from the parabolic formula. The required safety factor is

$$\text{S.F.} = \frac{5}{3} + \frac{3}{8}\left(\frac{103.4}{127.9}\right) - \frac{1}{8}\left(\frac{103.4}{127.9}\right)^3 = 1.90 \tag{c}$$

Consequently, the allowable stress for the column is

$$\sigma_{ALL} = \frac{35{,}000}{1.90}\left\{1 - 0.5\left(\frac{103.4}{127.9}\right)^2\right\} = 12{,}363 \text{ psi} \tag{d}$$

and the allowable concentric load is

$$P_{ALL} = 12{,}363(2.228) = 27{,}546 \text{ lb} \tag{e}$$

EXAMPLE 9-11

A steel rod (A441) 10 ft long is required to support a concentric load of 10,000 lb. If both ends of the rod are assumed to be clamped, determine the minimum diameter required to support the load according to the AISC column code ($\sigma_{YP} = 50{,}000$ psi for A441 steel).

SOLUTION

The limiting effective slenderness ratio is

$$C_C = \sqrt{\frac{2\pi^2 29 \times 10^6}{50 \times 10^3}} = 107 \tag{a}$$

Now the actual effective slenderness ratio can be expressed in terms of the diameter of the rod as

$$\left(\frac{k\ell}{r}\right)_{ACT} = \frac{0.5(120)}{\dfrac{d}{4}} = \frac{240}{d} \tag{b}$$

Since the actual slenderness ratio cannot be calculated explicitly, we do not know a priori which formula of the AISC code to use to calculate the critical stress for the column. Assume that one formula is valid and calculate a diameter based on that assumption. Once the diameter has been calculated, the actual $k\ell/r$ corresponding to it can be determined and compared with C_C. The most convenient assumption is that Euler's formula prevails. Thus

$$\frac{10{,}000}{\dfrac{\pi d^2}{4}} = \frac{\pi^2 29 \times 10^6}{1.92\left(\dfrac{240}{d}\right)^2}$$

or

$$40{,}000 = 8.131 \times 10^3 \, d^4 \tag{c}$$

Accordingly,

$$d = 1.489 \text{ in.} \tag{d}$$

Since steel rod usually is available only in fractional sizes, a 1.5-in. diameter rod would be specified. Consequently, the actual effective slenderness ratio is

$$\left(\frac{k\ell}{r}\right)_{\text{ACT}} = \frac{240}{1.5} = 160 \tag{e}$$

Since $(k\ell/r)_{\text{ACT}} > C_C$, use of the Euler formula is justified.

EXAMPLE 9-12

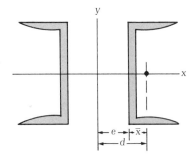

(a)

(b)

Figure 9-27

Two C 3 × 5 steel channels ($\sigma_{\text{YP}} = 35$ ksi) are placed back to back and latticed together to form a column required to carry a concentric load on a length of 10 ft. Both ends of the column are simply supported. Determine the maximum concentric load that can be sustained by the column when the channels are spaced so that the moments of inertia relative to the principal axes through the centroid of the cross section are equal. Use the AISC code (see Figure 9-27a).

SOLUTION

The properties for a C 3 × 5 steel channel are obtained from Appendix D. They are

$$\left. \begin{array}{l} A = 1.47 \text{ in}^2 \\ \bar{x} = 0.438 \text{ in.} \\ r_x = 1.12 \text{ in.} \\ r_y = 0.41 \text{ in.} \end{array} \right\} \tag{a}$$

The requirement that the principal moments of inertia of the composite section be equal leads to the expression

$$I_{xx} = I_{yy}$$

or

$$2Ar_x^2 = 2(Ar_y^2 + Ad^2)$$

or

$$r_x^2 = r_y^2 + d^2 \tag{b}$$

where A, r_x, and r_y are properties of a single channel, and d is shown in Figure 9-27a. Substituting the numerical values from Eq. (a) in Eq. (b) yields

$$d = 1.042 \text{ in.}$$

or

$$e = 1.042 - 0.438 = 0.604 \text{ in.} \tag{c}$$

The limiting effective slenderness ratio is

$$C_C = \sqrt{\frac{2\pi^2 29 \times 10^6}{35 \times 10^3}} = 127.9 \qquad\qquad (d)$$

and the actual effective slenderness ratio is

$$\left(\frac{k\ell}{r}\right)_{ACT} = \frac{1.0(120)}{1.12} = 107.1 \qquad\qquad (e)$$

Thus the parabolic formula yields the allowable stress

$$\sigma_{ALL} = \frac{35,000}{1.91}\left\{1 - 0.5\left(\frac{107.1}{127.9}\right)^2\right\} = 11,912 \text{ psi} \qquad\qquad (f)$$

where the safety factor has been calculated from the relation

$$\text{S.F.} = \frac{5}{3} + \frac{3}{8}\left(\frac{107.1}{127.9}\right) - \frac{1}{8}\left(\frac{107.1}{127.9}\right)^3 = 1.91 \qquad\qquad (g)$$

The maximum allowable load that the column can carry is

$$P_{ALL} = \sigma_{ALL}(2A) = 11,912(2 \times 1.47) = 35,021 \text{ lb} \qquad\qquad (h)$$

PROBLEMS / Section 9-8

9-21   A W 8 × 40 structural steel beam is to be used as a column required to support a concentric load on an unsupported length of 25 ft. Determine the maximum permissible load if the ends of the column can be considered to be pinned. Use the AISC code with $\sigma_{YP} = 36$ ksi.

9-22   Determine the load-carrying capacity of a concentrically loaded L 3 × 3 × $\frac{1}{4}$ structural steel angle with an unsupported length of 10 ft. Assume that the ends of the column are pinned with respect to bending in any plane. Use the AISC code with $\sigma_{YP} = 36$ ksi.

9-23   Two L 3 × 2 × $\frac{1}{2}$ structural steel angles are riveted so that their long sides are back to back. The fabricated beam is to be used as a column to carry a concentric load on an unsupported length of 10 ft. Each end of the column can be considered to be clamped. Determine the maximum load permitted by the AISC column code. Assume that $\sigma_{YP} = 36$ ksi.

9-24   Select the most economical wide flange structural steel ($\sigma_{YP} = 36$ ksi) beam to act as a column with pinned ends. The column is to carry a concentric load of 200 k on an unsupported length of 12 ft. Use the AISC code.

9-25   Four L 2 × 2 × $\frac{3}{8}$ equal-leg structural aluminum angles are latticed together to form the cross section of a beam as shown in Figure P9-25. The beam is to be used as a 12-ft. boom on a small crane in a lumber yard. The ends of the boom may be considered to be pinned with regard to bending about any axis. Determine the maximum concentric load that the boom can resist if a safety factor of 3 is required.

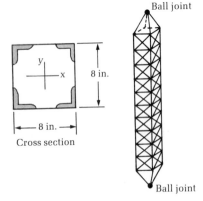

Cross section

Figure P9-25

9-9 SUMMARY

Three stability criteria commonly used to predict the critical loads that columns can carry are (a) the adjacent equilibrium criterion, (b) the imperfection criterion, and (c) the energy criterion. We used the adjacent equilibrium criterion to obtain the Euler loads for straight columns with various end supports, and we used the imperfection criterion to obtain the critical loads for eccentrically loaded columns.

Euler Columns:

$$\sigma_{CR} = \frac{\pi^2 E}{\left(\dfrac{k\ell}{r}\right)^2} \qquad k = \begin{cases} 1.0 \text{ pinned-pinned} \\ 0.7 \text{ pinned-clamped} \\ 0.5 \text{ clamped-clamped} \\ 2.0 \text{ clamped-free} \end{cases}$$

Engesser Columns:

$$\sigma_{CR} = \frac{\pi^2 E_T}{\left(\dfrac{k\ell}{r}\right)^2} \qquad \begin{array}{l} \text{Stress-strain diagram is required.} \\ \text{A } \sigma \text{ versus } E_T \text{ diagram is useful.} \end{array}$$

Secant Formula:

$$\sigma_{max} = \frac{P}{A}\left\{1 + \frac{ec}{r^2}\sec\left(\sqrt{\frac{P}{4AE}}\,\frac{\ell}{r}\right)\right\} \qquad \frac{ec}{r^2} = \text{eccentricity ratio}$$

Empirical Formulas:

$$\sigma_{CR} = C_1 + C_2\left(\frac{k\ell}{r}\right)^2 \qquad \text{Parabolic formula}$$

$$\sigma_{CR} = C_1 + C_2\left(\frac{k\ell}{r}\right) \qquad \text{Straight line formula}$$

The AISC code and the ALCOA code were described as examples of the parabolic and linear formulas.

PROBLEMS / CHAPTER 9

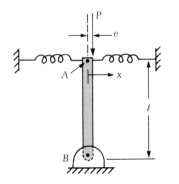

Figure P9-26

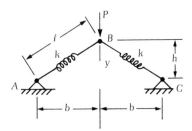

Figure P9-27

9-26 A rigid, weightless bar AB with a frictionless pin joint at B is connected to transverse springs as shown in Figure P9-26. A vertical force P is acting at point A with an eccentricity e. Assume that the force in each spring is given by

$$f = kx \left\{ 1 + \beta \left(\frac{x}{\ell}\right)^2 + \gamma \left(\frac{x}{\ell}\right)^3 \right\}$$

where x is the horizontal displacement of point A, and β and γ are parameters that fix the nature of the nonlinearity of the springs. The spring constant for a linear spring is k. Construct the nondimensional load $(P/2k\ell)$ versus nondimensional deflection (x/ℓ) curves when $\alpha = \beta = -10$, 0, and 10 for values of the non-dimensional imperfection parameter $e/\ell = 0.0$, 0.01, and 0.001.

9-27 The mechanism shown in Figure P9-27 consists of two linear springs, and its initial height is h. Plot the P versus y curves where P is the applied load and y is the deflection of point B. Assume that $h/\ell = 0.1$. Discuss the stability on the various branches of the load-deflection curve. How would you define the so-called snap-through buckling load? Is there a bifurcation load or a bifurcated path on the load-deflection curve? Discuss the reasons.

9-28 Compute the minimum radius of gyration for each of the cross sections shown in Figure P9-28. All dimensions are millimeters.

9-29 Two C 3 × 6 steel sections are latticed together by light bars to form a column whose cross section is shown in Figure P9-29. Determine the distance d if the centroidal moments of inertia with respect to the x and y axes are to be equal. Determine the radius of gyration for the cross section. Assume that the lattice bars do not contribute to the moment of inertia of the section.

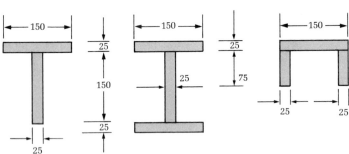

Figure P9-28

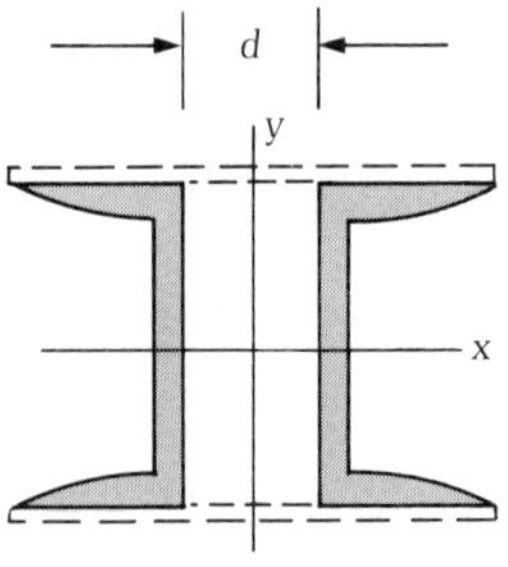

Figure P9-29

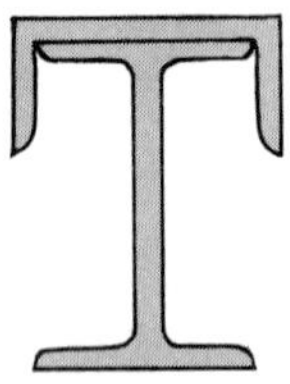

Figure P9-30

9-30 A C 6 × 13 and a S 4 × 9.5 are welded together to form a column whose cross section is shown in Figure P9-30. Compute the minimum radius of gyration for the cross section. If the length of the column is 18 ft and both ends are clamped, calculate its effective slenderness ratio.

9-31 Compute the minimum radius of gyration for the z section shown in Figure P9-31.

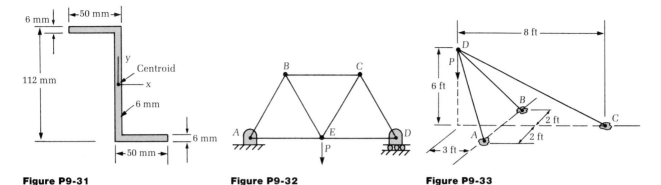

Figure P9-31   **Figure P9-32**   **Figure P9-33**

9-32   Each member of the pin-connected truss shown in Figure P9-32 is 50 in.
long and 1 in. in diameter. Determine the maximum force P that can be applied
to the truss if the members are made of steel with a yield point of 30 ksi (10 k).

9-33   Members AD and BD of the structure shown in Figure P9-33 are made from
2-in. standard steel pipe. If the ends of the pipes at points A and B are connected
to the supports by ball-and-socket joints, and if the common joint at D is also a
ball and socket, calculate the maximum force P that can be applied to the structure. Assume that Euler's formula is valid and use a safety factor of 2. Verify
the validity of the use of Euler's formula ($\sigma_{YP} = 30$ ksi and $E = 30 \times 10^6$ psi).

9-34   Members AB and DE in Figure P9-34 are made of aluminum ($E = 10 \times
10^6$ psi, $\alpha = 13 \times 10^6$ in./in./°F) and steel ($E = 30 \times 10^6$ psi, $\alpha = 6.5 \times 10^{-6}$ in./
in./°F), respectively, and they have diameters of 1 in. and 0.75 in., respectively.
Determine the maximum increase in temperature to which the mechanism can
be subjected. Consider the plate to which the bars are attached to be rigid, and
the pins at A, B, C, D, and E to be smooth and adequately designed. Assume
Euler's formula is valid.

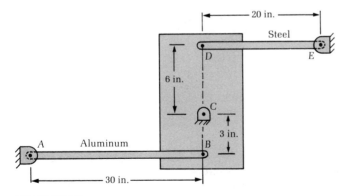

Figure P9-34

9-35   The frame and pulley assembly shown in Figure P9-35 is used to lift
supplies to craftsmen on a construction site. Member AB is an aluminum tube
whose inside and outside diameters are 50 mm and 35 mm, respectively. Pertinent mechanical properties for the aluminum are shown in the figure. Determine

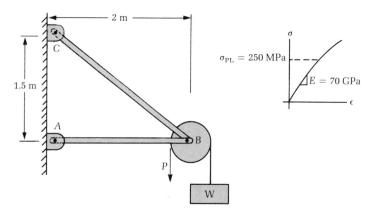

Figure P9-35

the largest load W that can be raised if it is assumed that member AB is the decisive factor. Justify the use of Euler's formula.

9-36 Determine the maximum force P that can be safely applied to the lever of the press shown in Figure P9-36 if it is assumed that the stability of member AB is the decisive factor. The clevis at A provides a simple support with respect to buckling in the "strong" direction and a clamped support with respect to buckling in the "weak" direction. The base of AB is a rigid disk. Properties of the material are given in the figure. Use a safety factor of 3. (2.5 kN)

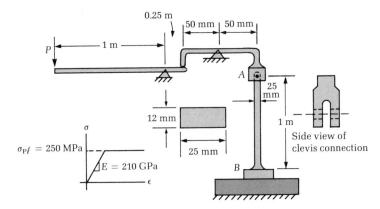

Figure P9-36

9-37 An 18-in. long aluminum rod (2024-T3) is to be used as a concentrically loaded compression member with both ends fixed. If the rod is required to support a concentric load of 25,000 lb, determine its minimum permissible diameter. The stress-strain data of Figure P9-17 may be used, or the Ramberg–Osgood equation with $E = 10 \times 10^6$ psi, $B = 72{,}300$ psi, and $n = 10$.

9-38 Use the AISC code to select a wide-flange beam to carry a concentric compressive load of 200 k on an unsupported length of 12 ft. Assume that each end of the column is pinned and that A441 steel with a yield point of 50 ksi is used. (W 10 × 39)

9-39 Select the most economical S shape to carry a concentric compressive load of 200 k on an unsupported length of 12 ft. One end of the column can be considered to be clamped, and the other end pinned. Use the AISC code and assume A36 steel with $\sigma_{YP} = 36$ ksi.

9-40 An S shape 2024-T4 aluminum beam with the dimensions shown in Figure P9-40 is to be used as a column 12 ft. long with simply supported end conditions. Using the ALCOA code, determine the maximum permissible concentric load for this column. Use a safety factor of 2.0.

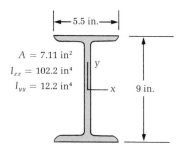

$A = 7.11$ in^2
$I_{xx} = 102.2$ in^4
$I_{yy} = 12.2$ in^4

Figure P9-40

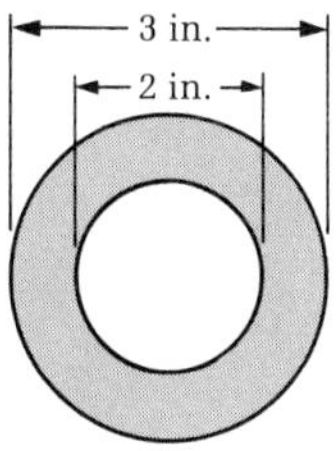

Figure P9-41

9-41 An aluminum pipe 48 in. long with simply supported ends is to support a concentric load. Use the ALCOA column code with a safety factor of 1.5 to determine the allowable load for the column. Dimensions of the cross section are shown in Figure P9-41 (73,690 lb).

CHAPTER 10

Structural Connections

10-1 Introduction

Members of a structure are frequently interconnected by rivets, bolts, or welds. The use of rivets in structural connections is not so common as the use of high-strength bolts or welds. However, the mechanics of riveted connections is developed in detail, with the understanding that the developments apply equally well to bolted connections. The mechanics of welded connections differs somewhat from the mechanics of riveted or bolted connections and, therefore, requires special attention.

The principal purpose of this chapter is to acquaint the reader with the fundamental mechanics that serves as the basis for the analysis and design of riveted, bolted, and welded connections. Design codes that are relevant to the design of riveted, bolted, and welded connections are not referred to in this text. Design codes are usually dealt with in more advanced, specialized publications.

When two structural members are overlapped and joined with rivets, the resulting connection is called a **lap joint**. The rivets (bolts) of a lap joint are said to act in **single shear** since only one cross-sectional area of each rivet (bolt) participates in the transmission of the applied load from one member to the other as shown in Figure 10-1a.

A **butt joint** occurs when the ends of two structural members are butted one against the other and the members are riveted (bolted) to common cover plates as shown in Figure 10-1b. The rivets (bolts) of a butt joint are said to act in **double shear** because two cross-sectional areas of each rivet (bolt) participate in the transmission of the applied load from one member to the other as can be seen in Figure 10-1b.

Additional nomenclature used in the design and analysis of riveted (bolted) connections are the pitch, back pitch, and repeating section. **Pitch** is defined as the distance between rivets (bolts) in the same row

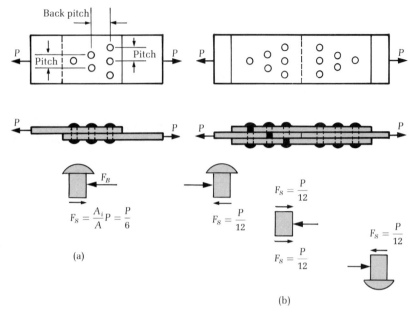

Figure 10-1

and is not necessarily the same for every row. **Back pitch** is the distance between adjacent rows of rivets (bolts), and it can have different values for different pairs of adjacent rows. Pitch and back pitch are depicted in Figure 10-1a. The smallest symmetric grouping of rivets (bolts) that repeats itself along the length of a joint is called a **repeating section**. The strength analysis of a riveted (bolted) connection is based on its repeating section.

The **full capacity** of a riveted, bolted, or welded connection is defined to be the maximum concentric force that either of the two connecting members can resist at a section where there has been no reduction in cross-sectional area because of rivet or bolt holes. This maximum force is based on a specified allowable normal stress for the material from which the connection is made.

The **efficiency** of a connection is defined as the ratio of the allowable load the connection can resist to the full capacity of the connection.

10-2 Concentrically Loaded, Riveted Connections

A riveted (or bolted) connection is said to be concentrically loaded when the force applied to it does not cause the connection to twist. We will show that the physical condition for a riveted connection to be concentrically loaded is that the action line of the applied force pass through the centroid of the rivet areas. This conclusion is a result of the following assumption regarding the manner in which the applied force is transmitted to the rivets of a repeating section. It is commonly assumed that the applied force P is distributed among the rivets in proportion to the shearing area of a rivet. Consequently, if A_i is the shearing area of the ith rivet at the interface of two plates, and if A is the sum of all such areas, then the shearing force F_i acting on the shearing area A_i is

$$F_i = \frac{A_i}{A} P \tag{10-1}$$

In determining the total shearing area A, we count all rivet areas that undergo rivet shear. Accordingly, the triple-riveted lab joint of Figure 10-1a has six shearing areas, and the triple-riveted butt joint of Figure 10-1b has twelve shearing areas. The shearing forces acting on the shearing areas of various portions of a rivet for a lap joint and a butt joint are shown in Figures 10-1a and 10-1b.

Consider a free-body diagram of a plate from either a lap joint or a butt joint such as the one shown in Figure 10-2. Let F_i be the force exerted on the plate by the ith rivet, y_i the distance of the action line of F_i from some reference point, P the applied load, and e the distance of the action line of P from the same reference point. If the force P is to be applied so that twisting of the plate is precluded, the resultant of the forces F_i must be equal, opposite, and collinear with P.

The assumption embodied in Eq. (10-1) assures that the magnitude of the resultant of the forces F_i is equal to the magnitude of P. If the action lines of this resultant and the applied force P are collinear, Newton's third law requires that the resultant be directed opposite to P. Collinearity is assured when the applied force P passes through the centroid of the rivet areas.

To show that collinearity of the rivet forces and the applied force P is assured when P passes through the centroid of the rivet areas, equate the moments of the forces F_i and the moment of P relative to a common point O. Thus

$$\sum_{i=1}^{N} F_i y_i = Pe \tag{10-2}$$

Because of Eq. (10-1), Eq. (10-2) can be rewritten as

$$\frac{\sum_{i=1}^{N} A_i y_i}{A} = e \tag{10-3}$$

The left-hand side of Eq. (10-3) defines the centroid of the rivet areas. Observe that Eq. (10-1) tacitly assumes that the rivets are all of the same material, which is commonly the case. The following examples illustrate the analysis of concentrically loaded riveted connections.

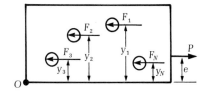

Figure 10-2

EXAMPLE 10-1

Two $\frac{1}{2}$-in. structural plates are joined together by $\frac{3}{4}$-in. rivets to form the lap joint shown in Figure 10-3a. The pitch of the rivets in the inner and outer rows is 3 in. and 6 in., respectively. The back pitch for each row is 3 in. If the allowable stresses for the material from which the plates and the rivets are made are 20 ksi in tension, 15 ksi in shear, and 32 ksi in bearing, determine the efficiency of the joint.

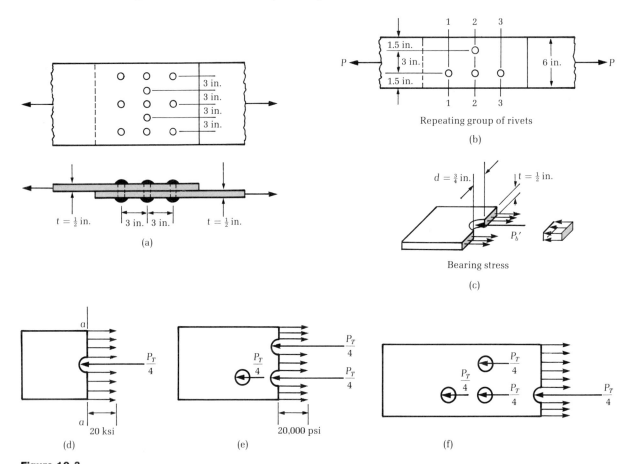

Figure 10-3

SOLUTION

The repeating group of rivets is shown in Figure 10-3b.

Rivet shear. First calculate the allowable load P_s based on the shearing stress in the rivets. The shearing force on any one rivet is

$$F_s = \frac{A_i}{A} P = \frac{P_s}{N} \tag{a}$$

where N is the number of rivet areas. The allowable shearing force F_s per rivet is equal to the allowable shearing stress times the rivet area. Consequently,

$$P_s = 4F_s = 4\left[15,000 \times \frac{\pi}{4}(\tfrac{3}{4})^2\right] = 26,507 \text{ lb}$$

Plate bearing.   The bearing stress is the same at each hole. A typical situation is shown in Figure 10-3c. The average bearing stress is calculated from the formula

$$(\sigma_b)_{\text{ave}} = \frac{P_b'}{td} \tag{b}$$

so that $P_b' = 32,000(\tfrac{1}{2} \times \tfrac{3}{4}) = 12,000$ lb. Here P_b' is the allowable force that each rivet can carry if bearing stress governs. Since there are four rivets in the group, the total allowable force for this case is

$$P_b = 4P_b' = 48,000 \text{ lb}$$

Tension.   The analysis is based on the net cross-sectional area of the plate at various sections. The allowable load P_T based on the tensile properties of the joint is calculated as follows. At section 1-1, of Figure 10-3b through the bottom plate, the free-body diagram of Figure 10-3d yields

$$\frac{P_T}{4} = 20,000 \left[\tfrac{1}{2} \times (6 - \tfrac{3}{4})\right] \qquad \text{or} \qquad P_T = 210,000 \text{ lb}$$

At section 2-2 through the bottom plate, the free-body diagram of Figure 10-3e yields

$$\frac{3P_T}{4} = 20,000 \left[\tfrac{1}{2}(6 - 2 \times \tfrac{3}{4})\right] \qquad \text{or} \qquad P_T = 60,000 \text{ lb}$$

At section 3-3, the free-body diagram of Figure 10-3f yields

$$P_T = 20,000 \left[\tfrac{1}{2} \times (6 - \tfrac{3}{4})\right] \qquad \text{or} \qquad P_T = 52,500 \text{ lb}$$

The maximum allowable force that the joint can safely resist is the smallest of the forces P_s, P_b, and P_T. Thus $P_{\text{ALL}} = 26,507$ lb. The **efficiency of a joint** is defined as the ratio of the maximum allowable force that a riveted plate can carry to that which can be carried by the same plate without rivets. For the present case,

$$\eta = \frac{26,507}{20,000(6 \times \tfrac{1}{2})} \times 100 = 44.2\%$$

EXAMPLE 10-2

The multiple-riveted structural butt joint shown in Figure 10-4a is to transmit a concentric load of 240 kN. The main plates are 10 mm thick by 250 mm wide, and the cover plates are each 6 mm thick by

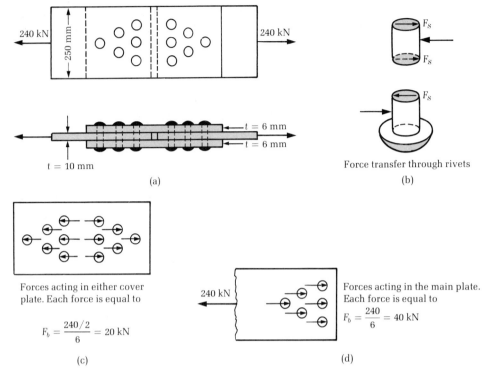

Force transfer through rivets

(a)   (b)

Forces acting in either cover
plate. Each force is equal to

$$F_b = \frac{240/2}{6} = 20 \text{ kN}$$

(c)

Forces acting in the main plate.
Each force is equal to

$$F_b = \frac{240}{6} = 40 \text{ kN}$$

(d)

Figure 10-4

250 mm wide. The diameter of each rivet is 20 mm. Assume that the holes have the same diameter as the rivets. Determine the maximum average shearing stress in the rivets, the maximum bearing stresses in the main plate and in the cover plates, and the maximum tensile stress in the main plate and the cover plates.

SOLUTION

Rivet shear. The force in the main plate is transmitted to the cover plates via six rivets. The shearing force associated with any rivet shearing area is

$$F_s = \frac{240}{(2)(6)} = 20 \text{ kN}$$

Here we use the fact that the rivets are acting in double shear. The average shearing stress in any rivet is

$$\tau_{ave} = \frac{20 \text{ kN}}{\frac{\pi}{4}(0.02)^2} = 63.66 \text{ MPa}$$

Note that this shearing stress acts on each shearing area because a force of the same magnitude acts on each shearing area (see Figure 10-4b).

Bearing stress in cover plates. Figure 10-4c shows that the average bearing stress in either cover plate (the thicknesses of the cover plates are equal) is

$$\sigma_b = \frac{20}{(0.02)(0.006)} = 166.7 \text{ MPa}$$

Bearing stress in the main plate. Figure 10-4d shows that the average bearing stress in the main plate is

$$\sigma_b = \frac{40}{(0.020)(0.010)} = 200 \text{ MPa}$$

Tension in the cover plates. Equilibrium of forces for the free-body diagrams shown in Figure 10-5a yields

$$\sigma_{1\text{-}1}(0.250 - 0.020)(0.006) = 20 \text{ kN}$$
$$\sigma_{2\text{-}2}(0.250 - 0.040)(0.006) = 60 \text{ kN}$$
$$\sigma_{3\text{-}3}(0.250 - 0.060)(0.006) = 120 \text{ kN}$$

The tensile stresses at sections 1-1, 2-2, and 3-3 are

$$\sigma_{1\text{-}1} = 14.5 \text{ MPa}$$
$$\sigma_{2\text{-}2} = 47.6 \text{ MPa}$$
$$\sigma_{3\text{-}3} = 105.3 \text{ MPa}$$

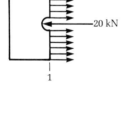
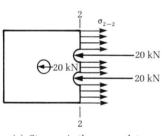
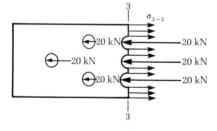

(a) Stresses in the cover plates

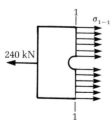
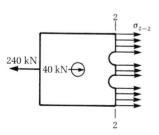

(b) Stresses in the main plates

Figure 10-5

Tension in the main plates.   The tensile stresses in the main plate are obtained from force equilibrium using Figure 10-5b. Thus

$$\sigma_{1-1}(0.250 - 0.020)(0.010) = 240$$
$$\sigma_{2-2}(0.250 - 0.040)(0.010) = 200$$
$$\sigma_{3-3}(0.250 - 0.060)(0.010) = 120$$

The corresponding tensile stresses in the main plates are

$$\sigma_{1-1} = 104.3 \text{ MPa}$$
$$\sigma_{2-2} = 95.2 \text{ MPa}$$
$$\sigma_{3-3} = 63.2 \text{ MPa}$$

As far as tension is concerned, the critical section in the cover plates is section 3-3; for the main plates, the critical section is 1-1.

EXAMPLE 10-3

The 100-in. diameter boiler shown in Figure 10-6a has plates $\frac{3}{4}$ in. thick and a longitudinal double-riveted butt joint whose cover plates are $\frac{1}{4}$ in. thick. The diameter of each rivet is $\frac{3}{4}$ in., the pitch in each row is 3 in., and the back pitch is also 3 in. If the allowable stresses in shear, bearing, and tension are 8800 psi, 19,000 psi, and 11,000 psi, respectively, determine the maximum pressure to which the boiler may be subjected if the induced stresses are not to exceed the corresponding allowable stresses.

SOLUTION

The repeated group of rivets is shown in Figure 10-6b.

Rivet shear.   If rivet shear is the controlling factor, the allowable load is

$$P_s = 2(4)\left[8800 \times \frac{\pi}{4}(\tfrac{3}{4})^2\right] = 31,100 \text{ lb} \tag{a}$$

Note that the rivets are acting in double shear.

Bearing stress in the cover plates.   If bearing stress in the cover plates is the controlling factor, the allowable load is

$$P_b = 2(4)\,[19,000(\tfrac{1}{4} \times \tfrac{3}{4})] = 28,500 \text{ lb} \tag{b}$$

Bearing in the main plates.   If bearing stress in the main plate is the controlling factor, the allowable load is

$$P_b = 4\,[19,000(\tfrac{3}{4} \times \tfrac{3}{4})] = 42,750 \text{ lb} \tag{c}$$

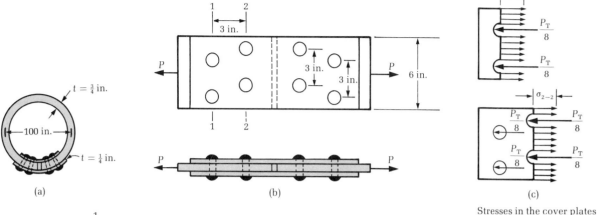

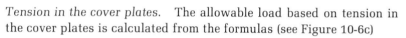

Stresses in the cover plates

(c)

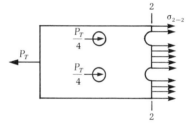

(d)

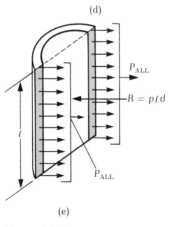

(e)

Figure 10-6

Tension in the cover plates. The allowable load based on tension in the cover plates is calculated from the formulas (see Figure 10-6c)

$$11,000\,[6 - 2(\tfrac{3}{4})]\tfrac{1}{4} = \frac{P_T}{4} \quad \text{or} \quad P_T = 49,500\ \text{lb}$$

$$11,000\,[6 - 2(\tfrac{3}{4})]\tfrac{1}{4} = \frac{P_T}{2} \quad \text{or} \quad P_T = 24,750\ \text{lb}$$ (d)

Tension in the main plates. The allowable load based on tension in the main plates can be calculated from the free-body diagrams of Figure 10-6d. Thus force equilibrium yields

$$P_T = 11,000\,[6 - 2(\tfrac{3}{4})]\tfrac{3}{4} = 37,125\ \text{lb}$$

$$P_T - \frac{P_T}{2} = 11\,000\,[6 - 2(\tfrac{3}{4})]\tfrac{3}{4} \quad \text{or} \quad P_T = 74,250\ \text{lb}$$ (e)

The maximum load P that the joint can be subjected to without exceeding any of the allowable stresses is

$$P_{\text{ALL}} = 24,750\ \text{lb}$$ (f)

Tension at section 2-2 of the cover plates governs the analysis.

The maximum allowable pressure can be calculated from the formula for the circumferential stress in a thin-walled cylinder or from a free-body diagram like the one shown in Figure 10-6e. From equilibrium of forces,

$$2P_{\text{ALL}} = p\ell d$$

or

$$p = \frac{2P_{\text{ALL}}}{\ell d} = \frac{2(24,750)}{6(100)} = 82.5\ \text{psi}$$ (g)

PROBLEMS / Section 10-2

10-1 Two $\frac{1}{2}$-in. × 12-in. steel plates are riveted together to form the lap joint shown in Figure P10-1. The diameter of each rivet is 1 in., and the allowable stresses are 20 ksi in tension, 15 ksi in shear, and 32 ksi in bearing. Determine the maximum permissible load P that the connection can carry. Determine the joint efficiency.

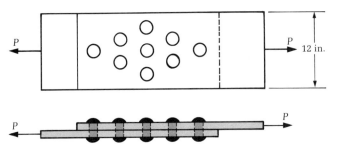

Figure P10-1

10-2 The diameter of each rivet of the lap joint shown in Figure P10-2 is 25 mm. Calculate the maximum tensile stress in the plates, the shearing stress in the rivets, and the maximum bearing stress.

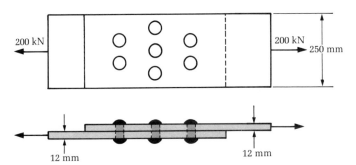

Figure P10-2

10-3 The diameter of each rivet of the butt joint shown in Figure P10-3 is 18 mm, and the thickness of each cover plate is 10 mm. Determine the maximum

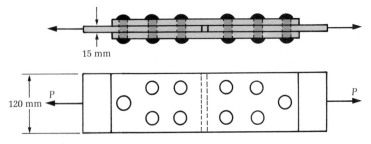

Figure P10-3

permissible load P the joint can resist if the allowable stresses in tension, shear, and bearing are 130 MPa, 100 MPa, and 200 MPa, respectively.

10-4   The diameter of each rivet of the triple-riveted butt joint shown in Figure P10-4 is 1 in. Calculate the shearing stress in the rivets, the maximum tensile stress in the main plate and in the cover plates, and the maximum bearing stress.

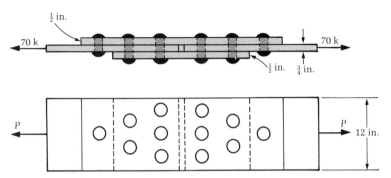

Figure P10-4

10-3   Pure Twisting of Riveted Connections

Rotating shafts are frequently connected by torsional couplings such as the one shown in Figure 10-7a. Moreover, eccentrically loaded riveted joints can be analyzed by separating the applied loads into concentric and torsional components. Unlike analyses of concentrically loaded connections, analyses of riveted connections due to pure torsional loads are considered to be elastic.

Consider a riveted (or bolted) connection that is subjected to a pure torque T. Figure 10-7b shows a free-body diagram of the plate. The forces $F_1, F_2, \ldots, F_n$ are the reactions of the rivets on the plate.

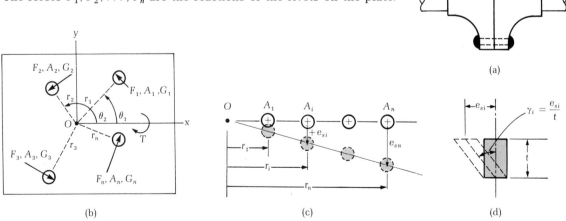

Figure 10-7

Denote the radial distance from point O, about which the plate tends to twist, to the ith rivet by r_i. Let θ_i be the corresponding polar angle as shown in the figure. Let A_i and G_i be the area and shearing modulus of elasticity of the ith rivet, respectively.

Force equilibrium in the x and y directions yields

$$\Sigma F_x = \sum_{i=1}^{n} F_i \sin \theta_i \equiv \sum_{i=1}^{n} F_i \left(\frac{y_i}{r_i} \right) = 0 \tag{10-4a}$$

and

$$\Sigma F_y = \sum_{i=1}^{n} F_i \cos \theta_i \equiv \sum_{i=1}^{n} F_i \left(\frac{x_i}{r_i} \right) = 0 \tag{10-4b}$$

Introduce the kinematic assumption that the plate is rigid; that is, straight lines in the plate remain straight and unstrained. Thus the shearing deformation of a rivet varies linearly with its distance from the center of twist O. Consequently (see Figure 10-7c),

$$\frac{e_{si}}{r_i} = \text{constant} \tag{10-5}$$

Because the plate has constant thickness, the shearing strains also vary linearly. Therefore,

$$\frac{\gamma_i}{r_i} = \text{constant} \tag{10-6}$$

Assume that the rivets deform elastically so that the shearing stress on a rivet is

$$\tau_i = G_i \gamma_i \tag{10-7}$$

and if, as is customary, the shearing stress is assumed to be uniformly distributed over the shearing area of the rivet,

$$F_i = \tau_i A_i \tag{10-8}$$

Equations (10-6), (10-7), and (10-8) combine to give

$$\frac{F_i}{r_i A_i G_i} = k \text{ (constant)} \tag{10-9}$$

Substituting F_i/r_i from Eq. (10-9) into Eqs. (10-4a) and (10-4b) yields

$$\sum_{i=1}^{n} A_i y_i G_i = 0 \tag{10-10a}$$

and

$$\sum_{i=1}^{n} A_i x_i G_i = 0 \tag{10-10b}$$

Assuming that all rivets are made from the same material, Eqs. (10-10) give

$$\sum_{i=1} A_i y_i = 0 \tag{10-11a}$$

$$\sum_{i=1}^{n} A_i x_i = 0 \tag{10-11b}$$

Equations (10-11) indicate that the center of twist for the connection coincides with the centroid of the rivet areas provided that all rivets are made from the same material.

Moment equilibrium about point O of Figure 10-7b yields

$$T = \sum_{i=1} r_i F_i \tag{10-12}$$

or, because of Eq. (10-9) (with $G_i = G$),

$$T = kG \sum_{i=1}^{n} r_i^2 A_i = kJG \tag{10-13}$$

where

$$J = \sum_{i=1}^{n} r_i^2 A_i$$

is the polar moment of inertia of the rivet areas relative to the centroid of rivet areas. (Note that the polar moment of inertia of each rivet relative to an axis through its centroid is discarded.) Equation (10-13) gives

$$k = \frac{T}{JG} \tag{10-14}$$

so that Eqs. (10-8) and (10-9) combine to give

$$\tau_i = \frac{T}{J} r_i \tag{10-15}$$

Thus the shearing stress associated with the ith rivet shearing area is given by the torsion formula, Eq. (10-15). The x and y components of the shearing stress are given by

$$\tau_x = \frac{T}{J} y_i \tag{10-16a}$$

and

$$\tau_y = \frac{T}{J} x_i$$

(10-16b)

EXAMPLE 10-4

Determine the maximum shearing stress in the rivets and the maximum bearing stress in the plates of the riveted joint shown in Figure 10-8a. The thickness of each plate is $\frac{1}{2}$ in. and the diameter of each rivet is 1 in.

SOLUTION

Rivet shear.   A free-body diagram of the top plate is shown in Figure 10-8b. The centroid of the rivet areas is located by the relations

$$A\bar{y} = \Sigma A_i y_i = 3A \qquad \text{or} \qquad \bar{y} = 1 \text{ in.} \tag{a}$$

and $\bar{x} = 0$ by symmetry.

The radial distances r_i of the rivet centers from the centroid of the rivet areas are

$$r_1 = r_2 = r_3 = 2 \text{ in.} \tag{b}$$

Consequently, the polar moment of inertia is

$$J = \sum_{i=1}^{3} A_i r_i^2 = 12A_1 = 12\left(\frac{\pi}{4} \times 1^2\right) = 3\pi \text{ in.}^4 \tag{c}$$

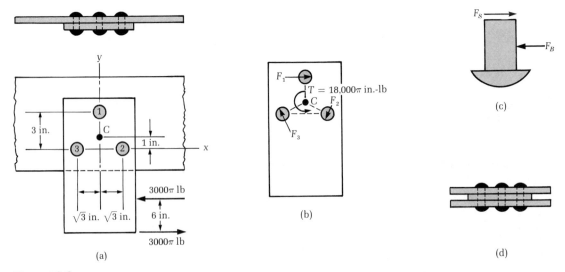

(a)

(b)

(c)

(d)

Figure 10-8

The shearing stress on each rivet is found from the formula

$$\tau_i = \frac{T}{J} r_i$$

or

$$\tau_1 = \tau_2 = \tau_3 = \frac{18{,}000\pi(2)}{3\pi} = 12 \text{ ksi} \tag{d}$$

Bearing stress. The bearing force at each rivet hole is the same since each rivet is the same radial distance from the centroid of the rivet areas. Figure 10-8c shows that

$$F_B = F_S = 12{,}000 \left[\frac{\pi}{4}(1)^2 \right] = 9425 \text{ lb} \tag{e}$$

Therefore, the bearing stress at each of the rivet holes is

$$\sigma_B = \frac{F_B}{t\,d} = \frac{9425}{(0.5)1} = 18{,}850 \text{ psi} \tag{f}$$

If the top plate had been sandwiched between two $\frac{1}{2}$-in. plates as shown in Figure 10-8d, the procedure to be used in calculating the shearing stresses in the rivets would be the same except that J should be multiplied by 2 to account for the fact that the rivets are in double shear.

Alternate method of solution Consider the free-body diagram of Figure 10-8b. From moment equilibrium about the centroid of rivet areas,

$$2F_1 + 2F_2 + 2F_3 = 18{,}000\pi \tag{g}$$

Also, since the plate is considered to be rigid, from Eq. (10-9)

$$F_1 = F_2 = F_3 \tag{h}$$

Therefore, solving Eqs. (g) and (h) simultaneously yields

$$F_1 = F_2 = F_3 = 3000\pi \text{ lb} \tag{i}$$

The average shearing stress on each rivet area is

$$\tau_{ave} = \frac{3000\pi}{\dfrac{\pi}{4}(1)^2} = 12{,}000 \text{ psi} \tag{j}$$

and the average bearing stress is

$$(\sigma_B)_{ave} = \frac{3000\pi}{0.5 \times 1} = 18{,}850 \text{ psi} \tag{k}$$

The latter method of solution is probably more straightforward, and it provides a clear indication of the procedure required to solve the same problem with the top view shown in Figure 10-8d.

EXAMPLE 10-5

Determine the maximum torque that can be applied to the riveted joint shown in Figure 10-9a. The cross-sectional area of rivets 1, 2, and 3 are 0.2, 0.4, and 0.2 in^2, respectively, and the allowable shearing stress is 15 ksi. Assume that rivet shear controls the magnitude of the largest torque that can be applied.

SOLUTION

The centroid of the rivet areas is shown in Figure 10-9a. The polar moment of inertia is

$$J = \Sigma A_i r_i^2 = 0.2(10) + 0.4(2) + 0.2(10) = 4.8 \text{ in}^4 \tag{a}$$

Consequently, the maximum torque that can be applied to the connection without exceeding the allowable shearing stress is

$$T_{ALL} = \frac{\tau_{ALL}J}{r_{max}} = \frac{15,000(4.8)}{\sqrt{10}} = 22,768 \text{ in.-lb} \tag{b}$$

Alternate method of solution   The forces exerted by the rivets on the plate vary according to Eq. (10-9) with distance from the centroid of the rivet areas. Thus

$$\frac{F_1}{A_1 r_1} = \frac{F_2}{A_2 r_2} = \frac{F_3}{A_3 r_3}$$

or

$$F_1 = F_3 \tag{c}$$

and

$$F_2 = \frac{2}{\sqrt{5}} F_3 \tag{d}$$

Moment equilibrium with respect to the centroid of rivet areas gives a third equation in F_1, F_2, and F_3;

$$\sqrt{10}F_1 + \sqrt{2}F_2 + \sqrt{10}F_3 = T \tag{e}$$

Solving Eqs. (c), (d), and (e) simultaneously gives

$$\left.\begin{array}{l} F_1 = F_3 = 0.1318T \\ F_2 = 0.1179T \end{array}\right\} \tag{f}$$

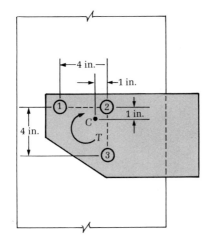

Top view

(a)

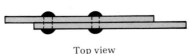

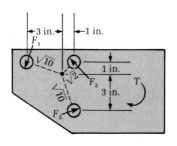

(b)

Figure 10-9

Consequently, either

$$15,000(0.2) = 0.1318T, \quad T = 22,762 \text{ in.-lb}$$
$$15,000(0.4) = 0.1179T, \quad T = 50,891 \text{ in.-lb}$$

(g)

The maximum allowable torque is clearly 22,762 in.-lb.

PROBLEMS / Section 10-3

10-5 Locate the rivet in the rivet connection of Figure P10-5 that experiences the largest shearing stress. The diameter of each rivet is 1 in. ($\tau_{max} = 12,155$ psi).

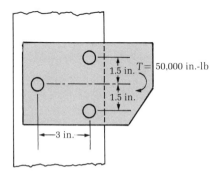

Figure P10-5

Figure P10-6

10-6 The allowable shearing stress for the rivet material of the rivet connection shown in Figure P10-6 is 70 MPa. If the diameter of each rivet is 25 mm, determine the maximum permissible torque that the connection can resist ($T = 11.8$ kN·m).

10-7 Locate the rivet in the rivet connection shown in Figure P10-7 that experiences the largest shearing stress, and calculate the magnitude of this stress. The diameter of each rivet is $\frac{1}{2}$ in. ($\tau_{max} = 10,188$ psi).

10-8 A machine requires 300 HP at 330 rpm in its operation. It is proposed that this power be supplied by a motor through a shaft with a coupling as shown schematically in Figure P10-8. If the diameter of each bolt is 1 in., and if the allowable shearing stress for the bolt material is 12 ksi, determine whether the proposed coupling is adequate.

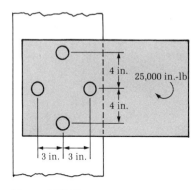

Figure P10-7

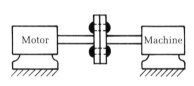

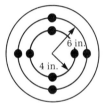

Figure P10-8

10-4 Eccentrically Loaded, Riveted Connections

Eccentrically loaded, riveted connections can be analyzed by separating the applied load into its concentric and torsional parts. Results for the combined loading are obtained by superposition. The procedures are illustrated by the following example.

EXAMPLE 10-6

Two structural steel plates are riveted together to form the connection shown in Figure 10-10a. The diameter of each rivet is 10 mm. Determine the maximum shearing stress in the rivets. The centroid of the rivet areas is denoted by C in the accompanying figures.

SOLUTION

The applied load is replaced by a statically equivalent force-couple system acting at the centroid of the rivet areas as shown in Figure 10-10b. Consider the three problems shown in Figures 10-10c, 10-10d, and 10-10e where free-body diagrams of one plate are subjected to each of the three loads shown in Figure 10-10b.

Direct shear in the x direction.   The concentric 15-kN force of Figure 10-10c divides equally among the rivets. Therefore,

$$F_h = \tfrac{15}{5} = 3 \text{ kN} \tag{a}$$

and the direct shearing stress on each rivet is

$$\tau_h = \frac{3000}{\dfrac{\pi}{4}(0.01)^2} = 38.2 \text{ MPa} \tag{b}$$

Direct shear in the y direction.   The concentric 20-kN force of Figure 10-10d divides equally among the rivets. Consequently,

$$F_v = \tfrac{20}{5} = 4 \text{ kN} \tag{c}$$

and the direct shearing stress on each rivet is

$$\tau_v = \frac{4000}{\dfrac{\pi}{4}(0.01)^2} = 50.9 \text{ MPa} \tag{d}$$

Shearing stress due to twisting.   The polar moment of inertia of the rivet areas relative to an axis through the centroid is

$$J = \Sigma A_i r_i^2 = 78.5 \times 10^{-6}\{(0.0541)^2 + (0.0304)^2 + (0.0626)^2$$
$$+ (0.0541)^2 + (0.0636)^2\}$$
$$= 1.1578 \times 10^{-6} \text{ m}^4 \tag{e}$$

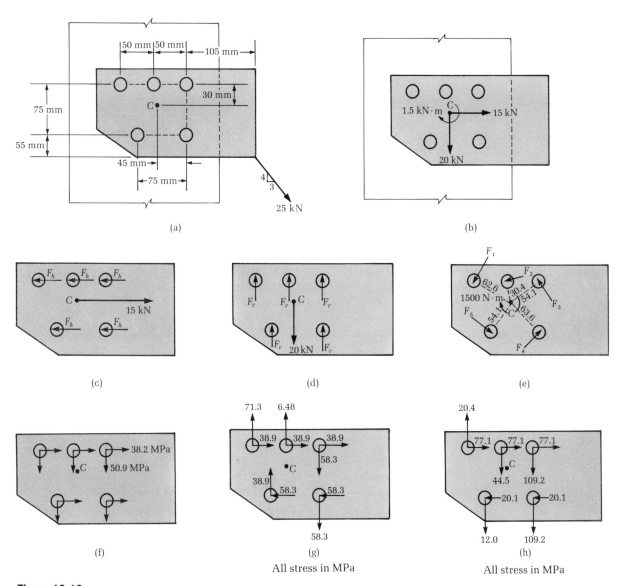

Figure 10-10

Consequently, the torsion formula yields as the shearing stress at the ith rivet

$$\tau_i = \frac{T}{J}\, r_i = \frac{1500\, r_i}{1.1578 \times 10^{-6}} = 1.2956 \times 10^9 r_i \qquad (f)$$

or, in terms of x and y components,

$$(\tau_i)_x = 1.2956 \times 10^9 y_i \qquad (g)$$

and

$$(\tau_i)_y = 1.2956 \times 10^9 x_i \qquad \text{(h)}$$

Figures 10-10f and 10-10g show the shearing stresses acting on each rivet due to the two concentric forces and the torsional moment, respectively. Figure 10-10h shows the combined effects of these stresses in component form.

The maximum shearing stress occurs at the rivet in the upper right-hand corner of Figure 10-10h. Its magnitude is

$$\tau_{max} = \sqrt{(109.2)^2 + (77.1)^2} = 133.7 \text{ MPa} \qquad \text{(i)}$$

Note that the addition of shearing stresses vectorially is valid here because the components act on the same shearing area. The shearing stresses could be multiplied by the shearing area to obtain the associated force components. These force components could then be added vectorially to obtain the resultant shearing force. Subsequent division by the shearing area yields the shearing stress. The operations of multiplication and division by the shearing area cancel each other.

PROBLEMS / Section 10-4

10-9 If the rivets of the eccentrically loaded connection shown in Figure P10-9 have equal diameters, and if the shearing stress in any rivet is not to exceed 12 ksi, determine the minimum rivet diameter required to resist the indicated load ($d = 0.65$ in. Use $\frac{3}{4}$-in. rivets.)

10-10 Determine the maximum permissible eccentric load P that the rivet connection shown in Figure P10-10 can resist if the allowable shearing stress for the rivet material is 12 ksi. The diameter of each rivet is 0.75 in.

10-11 Locate the rivet in the eccentrically loaded connection in Figure P10-11 that experiences the largest shearing stress. Determine the magnitude of this maximum shearing stress. The diameter of each rivet is a 25 mm ($\tau_{max} = 58.8$ MPa).

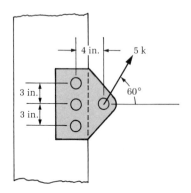

Figure P10-9

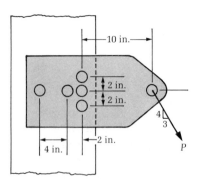

Figure P10-10

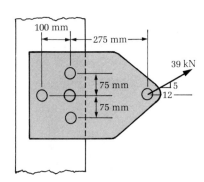

Figure P10-11

10-5 Welded Connections .

Welding is a method widely used to connect structural members. In this book, we describe only the mechanics of **butt welds** and **fillet welds**.

A **butt weld** occurs when the ends of two structural members are butted together and welded. There are numerous types of butt welds, but the basic purpose common to all butt welds is to connect structural members aligned in the same plane. The mechanics of butt welds is simple. The function of a butt weld is to transmit a force applied to one structural member to a connecting member by either a tensile or a compressive resistance. An allowable tensile or compressive stress is assigned to the weld, and the **effective thickness** of the weld is taken to be the thickness of the thinner of the connecting members. The allowable force that a butt weld can transmit is the product of the allowable stress, the effective thickness, and the length of the weld.

A **fillet weld** is used to connect two overlapping structural members, and it is placed in the corner made by the connecting structural members as shown in Figure 10-11a. The smallest dimension across a fillet weld is called the **throat** of the weld. The effective throat thicknesses t_e for equal and unequal leg fillet welds are expressed in terms of the lengths of the legs of the fillets as shown in Figure 10-11b.

The **effective area** of a fillet weld is defined as the product of its length and the effective throat thickness. All strength calculations for a fillet weld, regardless of the direction of the applied loads, are based on the effective area of the weld. In this manner, the analysis or design of a connection that involves a fillet weld is simplified considerably. Consequently, the allowable force that a fillet weld can carry is the product of the allowable shearing stress for the weld material and the effective area of the fillet weld.

Welded connections can be loaded concentrically or eccentrically. No twisting of the connection occurs when the resultant of the forces applied to the connection passes through the centroid of the weld areas. In this event, the fillet weld is said to be subjected to direct shear. If the resultant of the applied forces does not pass through the centroid of the weld areas, additional shearing stresses are induced in the weld due to twisting. Pure twisting of the welded connection occurs when the resultant of the applied forces is a couple.

The shearing stress induced in a fillet weld due to direct shear and torsional shear is obtained through the principle of superposition. Thus shearing stress due to direct shear and shearing stress due to pure twisting are computed separately, and, subsequently, added vectorially. The analysis is analogous to that employed to analyze riveted connections subjected to eccentric forces.

The coordinates of the centroid of the weld areas play an important role in the analysis of fillet welds. In determining the coordinates of the centroid of the weld areas, we assume that a fillet weld is concentrated along a line coincident with the root line of the weld. Moments of

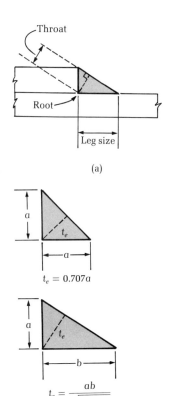

(a)

$t_e = 0.707a$

$t_e = \dfrac{ab}{\sqrt{a^2 + b^2}}$

(b)

Figure 10-11

inertia are calculated under the assumption that a weld area has a width equal to the effective throat thickness. These concepts are illustrated in the following examples.

EXAMPLE 10-7

A $4 \times 4 \times \frac{1}{2}$ in. equal leg angle is welded to a structural steel plate using $\frac{5}{16}$-in. fillet welds as shown in Figure 10-12a. If the angle is to carry a load P equal to the full capacity of the angle, determine (a) the length of side fillet welds required if the end fillet weld *is not* used and (b) the length of side fillet welds required if the end fillet weld *is* used. Assume that the allowable shearing stress on a section through the throat of the weld is 13.6 ksi and that the allowable tensile stress for steel is 20 ksi.

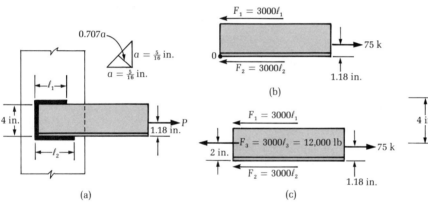

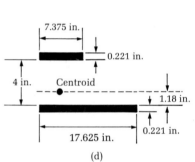

(a)   (b)   (c)   (d)

Figure 10-12

SOLUTION

The effective throat thickness, as depicted in Figure 10-12a, is calculated from the formula

$$t_e = 0.707a = (0.707)(\tfrac{5}{16}) = 0.221 \text{ in.} \tag{a}$$

Consequently, the allowable shearing resistance per linear inch of weld is

$$R = 13,600(0.221) = 3000 \text{ lb/in.} \tag{b}$$

Since the cross-sectional area of a $4 \times 4 \times \frac{1}{2}$ angle is 3.75 in², the maximum allowable force P that can be applied to the angle is

$$P = 20,000(3.75) = 75,000 \text{ lb} \tag{c}$$

Thus the total length of the fillet weld that is required is

$$\ell = \frac{P}{R} = \frac{75,000}{3,000} = 25 \text{ in.} \tag{d}$$

(a) If the applied force P is not to cause the connection to twist, its action line must be coincident with the action line of the resultant of the shearing forces acting in the two side fillet welds. Equating the moments of the weld forces F_1 and F_2 and the moment of the applied force P about point O gives (see Figure 10-12b)

$$1.18(75{,}000) = 3000\ell_1(4)$$

or

$$\ell_1 = 7.375 \text{ in.} \tag{e}$$

Consequently, since $\ell_1 + \ell_2 = 25$,

$$\ell_2 = 17.625 \text{ in.} \tag{f}$$

(b) When an end fillet weld is added, an appropriate free-body diagram of the angle appears as in Figure 10-12c. The pertinent equations become

$$\ell_1 + \ell_2 = 25 - 4 = 21 \tag{g}$$

and

$$1.18(75{,}000) = 3000\ell_1(4) + 12{,}000(2) \tag{h}$$

Equations (g) and (h) give

$$\left.\begin{array}{l} \ell_1 = 5.375 \text{ in.} \\ \ell_2 = 15.625 \text{ in.} \end{array}\right\} \tag{i}$$

Notice that the moment arms in Eqs. (e) and (h) are determined according to the assumption that the weld areas are concentrated along the root line of the fillet weld; that is, along the edges of the angle.

Finally, to check the calculations, observe that the action line of the applied force P should pass through the centroid of the weld areas. Thus, with the aid of Figure 10-12d, calculate the location of the centroid of the weld areas from the formula

$$A\bar{y} = A_1 y_1 + A_2 y_2 \tag{j}$$

Using ℓ_1 and ℓ_2 calculated in part (a), we find the effective areas for the two fillet areas to be

$$\left.\begin{array}{l} A_1 = (7.375)(0.221) = 1.6225 \text{ in}^2 \\ A_2 = (17.625)(0.221) = 3.8775 \text{ in}^2 \end{array}\right\} \tag{k}$$

Equations (j) and (k) yield

$$5.5\bar{y} = 1.6225(4) + 3.8775(0)$$

or

$$\bar{y} = 1.18 \text{ in.} \tag{l}$$

Consequently, the action line of the applied force passes through the centroid of the weld areas. Note that the width of a weld area is equal to its effective thickness and that the moment arm of a weld area is based on the root line of the fillet weld.

EXAMPLE 10-8

Two structural steel plates are welded to the flanges of a wide-flange beam that is being used as a column as shown in Figure 10-13a. If $\frac{5}{16}$-in. fillet welds are used, determine the largest load P that can be applied. Assume that the allowable shearing stress for the fillet weld is 13,600 psi.

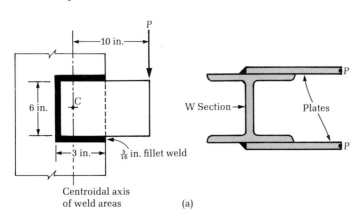

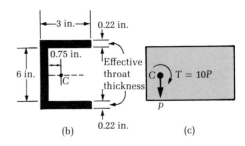

(a)   (b)   (c)

SOLUTION

Locate the centroid of the fillet welds first. Recall that a weld area is considered to be the product of the length of the weld and its effective throat thickness, and that moment arms are determined as if the weld area were concentrated in a line coincident with the edges of the plate along which the weld lies. Therefore, the weld area A and the x coordinate of the centroid of the weld areas can be calculated using Figure 10-13b. Thus

$$A = 2(3)(0.22) + 6(0.22) = 2.64 \text{ in}^2 \qquad (a)$$

and, hence,

$$2.64\bar{x} = (0.66)(1.5)2 + (1.32)(0)$$

or

$$\bar{x} = 0.75 \text{ in.} \qquad (b)$$

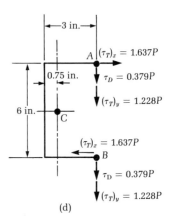

Figure 10-13

Also $\bar{y} = 0$ by symmetry.

The applied load P is replaced by an equivalent force-couple system with the force acting through the centroid of the weld areas as shown in Figure 10-13c. The shearing stresses in the weld due to the direct load P and the twisting moment T can now be calculated.

Direct shearing stress. The shearing stress at each point of the weld due to the concentric load P is the same and is equal to

$$\tau_D = \frac{P}{A} = \frac{P}{2.64} = 0.379P \tag{c}$$

Twisting shearing stress. To calculate the shearing stress due to the couple, the polar moment of inertia J of the weld areas relative to the centroid is needed. J can be calculated most conveniently from the relation

$$J = I_{xx} + I_{yy} \tag{d}$$

Accordingly, since

$$I_{xx} = 2[\tfrac{1}{12}(3)(0.221)^3 + 0.66(3)^2] + \tfrac{1}{12}(0.221)6^3$$
$$= 15.85 \text{ in}^4 \tag{e}$$

and

$$I_{yy} = 2[\tfrac{1}{12}(0.221)(3)^3 + 0.66(0.75)^2]$$
$$+ [\tfrac{1}{12}(6)(0.221)^3 + 1.32(0.75)^2]$$
$$= 2.48 \text{ in}^4 \tag{f}$$

it follows that

$$J = 15.85 + 2.48 = 18.33 \text{ in}^4 \tag{g}$$

The components of shearing stress due to the torque are given by the formulas

$$(\tau_T)_x = \frac{T}{J} y = \frac{10P}{18.33} y = 0.5456Py \tag{h}$$

and

$$(\tau_T)_y = \frac{T}{J} x = 0.5456Px \tag{i}$$

The stresses at points A and B of the weld are critical as can be seen from Figure 10-13d. The maximum stress is

$$\tau_{\max} = \sqrt{(1.637P)^2 + (0.379P + 1.228P)^2} = 2.294P \tag{j}$$

The maximum allowable load P that can be applied to the welded joint is

$$P_{\text{ALL}} = \frac{13,600}{2.294} = 5929 \text{ lb} \tag{k}$$

EXAMPLE 10-9

A structural steel plate is welded to another structural member by means of $\frac{5}{16}$-in. fillet welds as shown in Figure 10-14a. Determine the maximum shearing stress that occurs in the weld and state where it occurs.

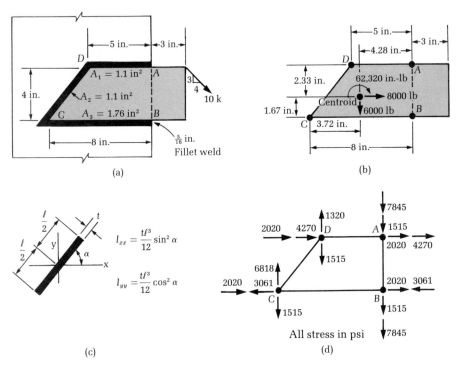

Figure 10-14

SOLUTION

The effective shearing areas are shown in Figure 10-14a. The centroid of the weld areas has the coordinates

$$\bar{x} = \frac{1.1(2.5) + 1.1(6.5) + 1.76(4)}{(1.1 + 1.1 + 1.76)} = 4.28 \text{ in.} \quad \text{(from line } AB) \quad \text{(a)}$$

and

$$\bar{y} = \frac{1.1(4) + 1.1(2)}{(1.1 + 1.1 + 1.76)} = 1.67 \text{ in.} \quad \text{(from line } BC) \quad \text{(b)}$$

The location of the centroid of the weld areas is shown in Figure 10-14b. Figure 10-14b shows an equivalent force-couple system with the forces applied at the centroid of the weld areas. The shearing stress at any point in the weld is obtained using the superposition principle.

Direct shearing stress.   Shearing stresses due to the concentric forces are

$$(\tau_D)_x = \frac{8000}{3.96} = 2{,}020 \text{ psi} \tag{c}$$

and

$$(\tau_D)_y = \frac{6000}{3.96} = 1515 \text{ psi} \tag{d}$$

Twisting shearing stress.   To obtain the shearing stresses due to twisting, it is necessary to calculate the polar moment of inertia of the weld areas with respect to the centroid. The moments of inertia of an inclined rectangular area about two orthogonal centroidal axes are given in Figure 10-14c in terms of the angle of inclination α. The parallel axis theorem is valid for such areas also. Thus the moments of inertia of the weld areas about the centroidal axes are

$$I_{xx} = [\tfrac{1}{12}(5)(0.221)^3 + 1.1(2.33)^2] + [\tfrac{1}{12}(8)(0.221)^3 + 1.76(1.67)^2]$$
$$+ [\tfrac{1}{12}(0.221)(5)^3(0.8)^2 + 1.1(0.33)^2]$$
$$= 12.477 \text{ in}^4 \tag{e}$$

and

$$I_{yy} = [\tfrac{1}{12}(0.221)(5)^3 + 1.1(1.78)^2] + [\tfrac{1}{12}(0.221)(8)^3 + 1.76(0.28)^2]$$
$$+ [\tfrac{1}{12}(0.221)(5)^3(0.6)^2 + 1.1(2.22)^2]$$
$$= 21.54 \text{ in}^4 \tag{f}$$

so that

$$J = I_{xx} + I_{yy} = 34.0 \text{ in}^4 \tag{g}$$

The components of the shearing stress at any point of the weld are given by the formulas

$$(\tau_T)_x = \frac{T}{J}\, y = \frac{62{,}320}{34.0}\, y = 1833y \tag{h}$$

and

$$(\tau_T)_y = \frac{T}{J}\, x = \frac{62{,}320}{34.0}\, x = 1833x \tag{i}$$

These components at various points of the weld are shown in Figure 10-14d. The maximum shearing stress occurs at point A as can be seen from an examination of Figure 10-14d. Its magnitude is

$$\tau_{\max} = \sqrt{(2020 + 4270)^2 + (1515 + 7845)^2}$$

or

$$\tau_{\max} = 11{,}277 \text{ psi} \tag{j}$$

This welded connection would be considered capable of carrying the designated load if the allowable shearing stress for the weld material is larger than τ_{max}.

PROBLEMS / Section 10-5

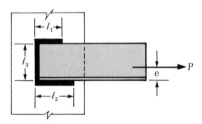

Figure P10-12

10-12 A 4 × 3 × $\frac{1}{2}$ in. unequal leg angle is welded to a structural steel plate using $\frac{5}{16}$-in. fillet welds as shown in Figure P10-12. Assume that it is the long leg that is welded to the plate. In this case, $\ell_3 = 4$ in. If the angle is to carry a load P equal to its full capacity ($e = 1.33$ in.), determine the length of side fillet welds required (a) if the end fillet weld *is not* used and (b) if the end fillet weld *is* used. The maximum allowable shearing stress on a section through the throat of the fillet weld is 13,600 psi and the allowable tensile stress for the plate material is 20,000 psi.

10-13 Solve Problem 10-12 assuming that the short leg of the angle is welded to the structural steel plate. In this case, $\ell_3 = 3$ in. and $e = 0.827$ in.

10-14 The bracket shown in Figure P10-14 is subjected to an 80-kN force. Equal leg fillet welds are used for the construction of the bracket. Assume that the same size welds are used on all sides. The allowable shearing stress of the weld is 150 MN/m^2. Determine the smallest size of fillet welds that can be used.

10-15 The bracket shown in Figure P10-15 is constructed using 10-mm fillet welds. Assuming that the allowable shearing stress of the weld is 150 MPa, determine the maximum allowable force P that can be applied to the bracket.

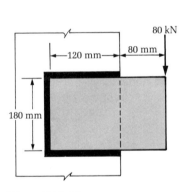

Figure P10-14

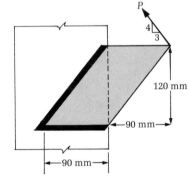

Figure P10-15

10-6 SUMMARY

In this chapter, we develop procedures used to analyze and design riveted and welded connections under concentric forces, pure twisting by couples, and eccentric forces.

	Riveted or bolted connections	Welded connections
Concentric loading	$F_i = \dfrac{A_i}{A} P, \; \tau_i = \dfrac{F_i}{A_i}$	$\tau = \dfrac{P}{A}$
	$A_i = i$th shearing area $A = \Sigma A_i$ $F_i =$ shearing force acting on the ith shearing area	$A =$ effective weld area which is equal to the total length of the weld times the effective throat thickness
Pure twisting	$(\tau_i)_x = \dfrac{T}{J} y$ $(\tau_i)_y = \dfrac{T}{J} x$	$\tau_x = \dfrac{T}{J} y$ $\tau_y = \dfrac{T}{J} x$
	$J =$ polar moment of inertia of the rivet areas with respect to the centroid of the rivet areas $x, y =$ coordinates of a rivet measured from the centroid of the rivet areas.	$J =$ polar moment of inertia of the effective weld areas with respect to the centroid of the weld areas $x, y =$ coordinates of a point on a weld measured from the centroid of the weld areas
Eccentric loading	Vectorial superposition of shearing stresses due to concentric loading and pure twisting	Vectorial superposition of shearing stresses due to concentric loading and pure twisting

PROBLEMS / CHAPTER 10

10-16 A 1.5-m diameter boiler is made of 12-mm steel plates that are joined by a double riveted butt joint. The cover plates for the joint are each 8 mm thick. The pitch of the rivets in the inner row is 100 mm, and the pitch of the rivets in the outer row is 75 mm. If the diameter of each rivet is 20 mm, and if the allowable stresses in tension, bearing, and shear are 80 MPa, 140 MPa, and 60 MPa, respectively, determine the efficiency of the joint and the maximum permissible internal pressure to which the boiler can be subjected.

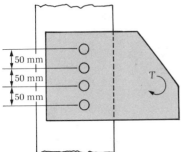

Figure P10-17

10-17   Calculate the maximum permissible torque T that can be applied to the rivet connection shown in Figure P10-17 if the shearing stress in the rivet material is not to exceed 70 MPa and if the diameter of each rivet is 25 mm ($T =$ 5.73 kN·m).

10-18   The mechanical coupling shown in Figure P10-18 is required to transmit 300 HP at 330 rpm. The diameter of each bolt is 10 mm. If there are four bolts in the interior bolt circle, and if the shearing stress in the bolt material is not to exceed 70 MPa, determine the number of bolts required in the exterior bolt circle ($N = 5$).

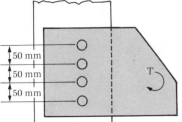

Figure P10-18

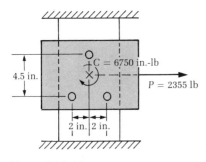

Figure P10-19

10-19   Two steel plates are connected with 1-in. diameter rivets as shown in Figure P10-19. The joint is subjected to a concentrated force P that passes through the centroid of the rivet areas and a couple C. Determine the maximum shearing stress that occurs in the rivets.

10-20   Locate the rivet in the eccentrically loaded rivet connection shown in Figure P10-20 that experiences the largest shearing stress. Calculate its magnitude if the diameter of each rivet is 25 mm.

10-21   Rivets A and B of Figure P10-21 each have a diameter of 1 in., and rivets C and D each have a diameter of $\frac{1}{2}$ in. Locate the rivet that experiences the largest shearing stress and calculate its magnitude ($\tau_{max} = 20{,}130$ psi).

10-22   The bracket shown in Figure P10-22 is constructed using $\frac{1}{4}$-in. fillet welds. The allowable shearing stress of the weld is 13.6 ksi. Determine the maximum allowable force P that can be applied as shown.

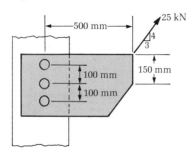

Figure P10-20

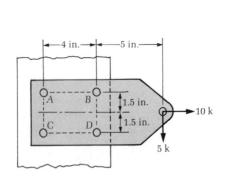

Figure P10-21

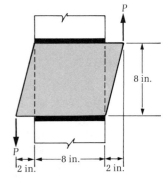

Figure P10-22

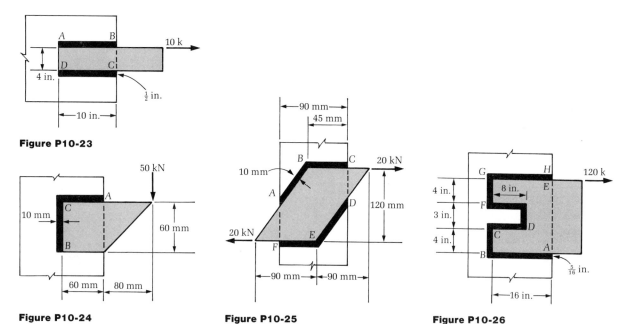

Figure P10-23

Figure P10-24

Figure P10-25

Figure P10-26

10-23–10-26 A structural steel plate is welded to another structural member by means of equal leg fillet welds as shown in the figure. Determine the shearing stresses that occur at the points designated. Determine the location and magnitude of the critical shearing stress in the connection.

CHAPTER 11

Energy Methods in Solid Mechanics

11-1 Introduction

The principle of total potential energy, Castigliano's second theorem, and the unit load method are corollaries of the more fundamental principle of virtual work. In this chapter, we take the principle of virtual work as the starting point for the development of the three corollary principles just mentioned. The principle of virtual work can be traced to Aristotle. John Bernoulli (1667–1748) gave a formal statement of the principle in a letter to P. Varignon in 1717, and J. L. Lagrange (1736–1813) adopted it as the principal axiom of mechanics in his famous treatise *Mechanique Analytique (1788).*

11-2 Principle of Virtual Work

Prior to stating the principle of virtual work, we need to define such terms as virtual displacement and virtual work.

Virtual displacement. Consider a single point i on the surface of the deformable mechanical system shown in Figure 11-1. A force **F** and a couple **C** act at point i. The deformed equilibrium configuration of the system is indicated by the solid curve in Figure 11-1.

Suppose point i is given an infinitesimal displacement $\delta \mathbf{u}$ that is consistent with the geometric constraints of the system, but which is otherwise arbitrary. A displacement of this type is called a **virtual displacement** because it does not necessarily coincide with the displacement of point i caused by the actual force **F** and the couple **C**. The symbol δ is customarily used to indicate a virtual quantity.

A virtual displacement can also be a rotation or an angular displacement. For example, consider an infinitesimal element of area surrounding point i. The area is shaded in Figure 11-1. The line normal

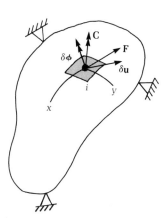

Figure 11-1

485

to this element of surface area can rotate about three mutually perpendicular axes x, y, and z as a result of the virtual deformation the surface undergoes, giving rise to the virtual rotation $\delta\boldsymbol{\varphi}$.

Virtual work.   The work done by force $\mathbf{F}$ on the virtual displacement $\delta\mathbf{u}$ is called the **virtual work of the force F**, and the work done by the couple $\mathbf{C}$ on the virtual rotation $\delta\boldsymbol{\varphi}$ is called the **virtual work of the couple C**. Virtual work is calculated assuming that the actual forces and/or couples that act on the system remain constant during the application of the virtual displacements. Consequently, the virtual work of $\mathbf{F}$ is $\delta W_{\mathbf{F}} = \mathbf{F} \cdot \delta\mathbf{u}$, and the virtual work of $\mathbf{C}$ is $\delta W_C = \mathbf{C} \cdot \delta\boldsymbol{\varphi}$. Note that the dot product can be interpreted either as the product of magnitudes of $\mathbf{F}$ and the component of $\delta\mathbf{u}$ parallel to $\mathbf{F}$, or the product of the magnitudes of $\delta\mathbf{u}$ and the component of $\mathbf{F}$ parallel to $\delta\mathbf{u}$. Similar observations can be made for δW_C.

The virtual work associated with several forces and/or couples is the sum of the virtual work associated with each force and/or couple.

Principle of Virtual Work.   If a deformable mechanical system is in equilibrium under a system of external forces P_i and remains in equilibrium when it is given a virtual displacement δq_i, the sum of the virtual work of the internal and external forces on this virtual displacement must vanish.

The terms force and displacement are used in a general sense; that is, force means either a force or a couple, and displacement means either a linear displacement or a rotation.

If δW_e represents the virtual work associated with the external forces, and δW_i represents the virtual work associated with the internal forces (stresses), then the principle of virtual work asserts that

$$\delta W_i + \delta W_e = 0 \tag{11-1}$$

for an equilibrium configuration of a mechanical system.

11-3 Principle of Stationary Potential Energy

The principle of stationary potential energy is a direct consequence of the principle of virtual work. To obtain this principle, we must restrict our attention to mechanical systems for which mechanical energy is conserved; that is, the principle remains valid only for conservative mechanical systems.

The principle of stationary potential energy is widely used in structural analysis, particularly in the development of large-scale finite-element and finite-difference structural computer programs.

Principle of Stationary Potential Energy. Equilibrium configurations of a conservative mechanical system are characterized by a stationary value of its total potential energy; that is,

$$\frac{\partial V}{\partial q_i} = 0, \, i = 1, 2, 3, \dots, n.$$

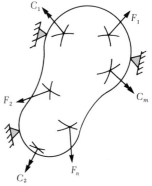

Figure 11-2

Consider the deformable mechanical system shown in Figure 11-2. Assume that the system is supported by a number of unyielding supports (support reactions therefore do no work) sufficient to prevent rigid body translations and/or rotations. The system is acted on by n concentrated forces F_i and m couples C_i. For convenience, these applied forces and couples are denoted collectively as generalized forces P_i. Consequently, a particular P_i can be either a force or a couple.

Let q_i denote the component of displacement of the point of application of the force P_i parallel to its action line, or the component of rotation parallel to the action line of the couple C_i. Note that the displacements q_i are associated with the deformed equilibrium configuration corresponding to forces P_i acting on the system. The q_i are referred to in this chapter as generalized coordinates because when they are known, the deformed configuration of the system is known. These coordinates are required to be independent. Also, their increments δq_i are required to be independent.

Now, imagine that the deformable system undergoes a virtual deformation so that the points of application of the forces and couples experience virtual displacements with the components δq_i parallel to the action lines of the P_i. The virtual work done by the forces and couples P_i on the virtual displacements δq_i is

$$\delta W_e = \Sigma P_i \, \delta q_i \tag{11-2}$$

Define a function $\Omega = \Omega(q_i)$ such that

$$P_i = -\frac{\partial \Omega}{\partial q_i} \tag{11-3}$$

Then, by Eqs. (11-2) and (11-3),

$$\delta W_e = -\Sigma \frac{\partial \Omega}{\partial q_i} \, \delta q_i \tag{11-4}$$

The virtual work done by the internal forces (stresses) that are associated with the equilibrium configuration shown in Figure 11-2 is

$$\delta W_i = -\delta U \tag{11-5}$$

where $U = U(q_i)$ in the strain energy stored in the system because of the forces P_i. Now, the increment in strain energy that results because of the virtual displacements δq_i is

$$\delta U = \Sigma \frac{\partial U}{\partial q_i} \, \delta q_i \tag{11-6}$$

so that

$$\delta W_i = -\Sigma \frac{\partial U}{\partial q_i} \delta q_i \tag{11-7}$$

According to the principle of virtual work,

$$\delta W_i + \delta W_e = 0 \tag{11-8}$$

so that

$$\Sigma \left(\frac{\partial U}{\partial q_i} + \frac{\partial \Omega}{\partial q_i} \right) \delta q_i = \Sigma \frac{\partial}{\partial q_i} (U + \Omega) \delta q_i = 0 \tag{11-9}$$

Now define another function $V = V(q_i)$ by the relation

$$V = U + \Omega \tag{11-10}$$

so that Eq. (11-9) becomes

$$\Sigma \frac{\partial V}{\partial q_i} \delta q_i = 0 \tag{11-11}$$

If attention is restricted to mechanical systems for which δq_i is independent, Eq. (11-11) leads to the equilibrium requirements

$$\frac{\partial V}{\partial q_i} = 0, i = 1, 2, 3, \dots, n \tag{11-12}$$

The functions $V(q_i)$, $U(q_i)$, and $\Omega(q_i)$ are referred to as the total potential energy, potential energy of the internal forces, and potential energy of the external forces. Notice that U is the strain energy associated with an elastic deformation process—not necessarily a **linear** elastic process. Also note that Ω is the negative of the work done by the external forces of the system. The use of the principle of stationary potential energy is illustrated in the following examples.

EXAMPLE 11-1

A rigid weightless beam is suspended by three cables as shown in Figure 11-3a. Before the forces P and Q are applied, the beam is horizontal. If the extensional rigidities EA of the cables are the same, determine the orientation of the beam after the forces have been applied. Derive formulas for the elongations and forces in the cables.

SOLUTION

The position of the beam after the loads have been applied is shown in Figure 11-3b. The elongations of the cables are seen to be

$$\left. \begin{array}{l} e_1 = x - l\varphi \\ e_2 = x + l\varphi \\ e_3 = x \end{array} \right\} \tag{a}$$

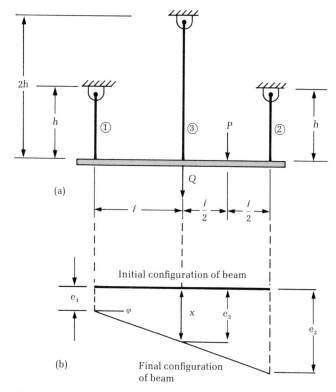

Figure 11-3

The strain energy for each cable is

$$U_1 = \frac{EA}{2h} e_1^2$$
$$U_2 = \frac{EA}{2h} e_2^2$$ (b)
$$U_3 = \frac{EA}{4h} e_3^2$$

and the potential energy of the external forces is

$$\Omega = -P\left(x + \frac{\ell}{2}\varphi\right) - Qx$$ (c)

Consequently, the total potential energy for the system is

$$V = \frac{EA}{2h}\{(x - \ell\varphi)^2 + (x + \ell\varphi)^2 + \tfrac{1}{2}x^2\} - P\left(x + \frac{\ell}{2}\varphi\right) - Qx$$ (d)

The equilibrium configuration for the system is characterized by the relations

$$\left.\begin{array}{l} \dfrac{\partial V}{\partial x} = \dfrac{EA}{2h} \{2(x - \ell\varphi) + 2(x + \ell\varphi) + x\} - Q - P = 0 \\[3mm] \dfrac{\partial V}{\partial \varphi} = \dfrac{EA}{2h} \{-2\ell\,(x - \ell\,\varphi) + 2\ell\,(x + \ell\,\varphi)\} - \dfrac{P\ell}{2} = 0 \end{array}\right\} \tag{e}$$

Equations (e) yield

$$\left.\begin{array}{l} x = \dfrac{2(P + Q)h}{5AE} \\[3mm] \varphi = \dfrac{Ph}{4EA\ell} \end{array}\right\} \tag{f}$$

From Eqs. (a) and (f), the elongations are found to be

$$\left.\begin{array}{l} e_1 = \dfrac{h}{20EA}\,(3P + 8Q) \\[3mm] e_2 = \dfrac{h}{20EA}\,(13P + 8Q) \\[3mm] e_3 = \dfrac{2h}{5EA}\,(P + Q) \end{array}\right\} \tag{g}$$

and, finally, the cable forces are

$$\left.\begin{array}{l} N_1 = \dfrac{EA}{h}\left[\dfrac{h}{20EA}\,(3P + 8Q)\right] = \dfrac{3P + 8Q}{20} \\[3mm] N_2 = \dfrac{EA}{h}\left[\dfrac{h}{20EA}\,(13P + 8Q)\right] = \dfrac{13P + 8Q}{20} \\[3mm] N_3 = \dfrac{EA}{2h}\left[\dfrac{2h}{5EA}\,(P + Q)\right] = \dfrac{P + Q}{5} \end{array}\right\} \tag{h}$$

Notice that these cable forces satisfy the vertical equilibrium condition for the beam.

EXAMPLE 11-2

A rigid rectangular plate is suspended by four cables attached at its corners as shown in Figure 11-4. The extensional stiffness of each cable is listed in the figure. Neglect the weight of the plate and determine the configuration of the plate after the two forces shown in the figure have been applied. Determine the elongations and forces in the cables. The plate is horizontal prior to the applications of the forces Q and $2Q$.

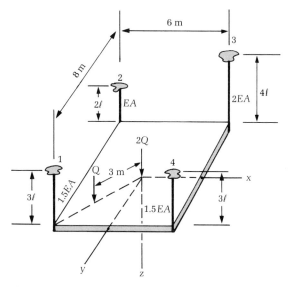

Figure 11-4

SOLUTION

The plate is rigid. Therefore the plate will remain a plane after the forces are applied. The equation of a plane is

$$z = ax + by + c \tag{a}$$

The constants a, b, and c are to be determined so that Eq. (a) represents the equation of the plate after the loads have been applied.

The elongations of the cables are found from Eq. (a) by substituting the coordinates of their attachment points on the plate. Thus

$$\left. \begin{aligned}
e_1 &= -3a + 4b + c \\
e_2 &= -3a - 4b + c \\
e_3 &= 3a - 4b + c \\
e_4 &= 3a + 4b + c
\end{aligned} \right\} \tag{b}$$

The strain energy for each cable is $[U_i = (E_i A_i / 2\ell_i)e_i^2]$

$$\left. \begin{aligned}
U_1 &= \frac{EA}{4\ell}\, e_1^2 \\[4pt]
U_2 &= \frac{EA}{4\ell}\, e_2^2 \\[4pt]
U_3 &= \frac{EA}{4\ell}\, e_3^2 \\[4pt]
U_4 &= \frac{EA}{4\ell}\, e_4^2
\end{aligned} \right\} \tag{c}$$

The potential energy of each external force is

$$\left.\begin{array}{l} \Omega_1 = -2Qc \\ \Omega_2 = -Q(-1.8a + 2.4b + c) \end{array}\right\} \tag{d}$$

Consequently, the total potential energy associated with the system is

$$V = \frac{EA}{4\ell} \{e_1^2 + e_2^2 + e_3^2 + e_4^2\} - 2Qc - Q(-1.8a + 2.4b + c) \tag{e}$$

Substituting the elongations into Eq. (e) from Eqs. (b) expresses V as a function of the parameters a, b, and c. Accordingly,

$$\begin{aligned} V(a, b, c) = \frac{EA}{4\ell} \{&(-3a + 4b + c)^2 + (-3a - 4b + c)^2 \\ &+ (3a - 4b + c)^2 + (3a + 4b + c)^2\} \\ &- Q(-1.8a + 2.4b + 3c) \end{aligned} \tag{f}$$

According to the principle of stationary potential energy, the equilibrium configuration of the plate is characterized by a stationary value for V. Therefore, by differentiation of Eq. (f),

$$\left.\begin{array}{lll} \dfrac{\partial V}{\partial a} = \dfrac{EA}{2\ell} \{36a\} + 1.8Q = 0 & \therefore a = -\dfrac{Q\ell}{10AE} \\[3mm] \dfrac{\partial V}{\partial b} = \dfrac{EA}{2\ell} (64b) - 2.4Q = 0 & \therefore b = \dfrac{3Q\ell}{40AE} \\[3mm] \dfrac{\partial V}{\partial c} = \dfrac{EA}{2\ell} (4c) - 3.0Q = 0 & \therefore c = \dfrac{3Q\ell}{2AE} \end{array}\right\} \tag{g}$$

The configuration of the plate after the loads have been applied is obtained by substituting a, b, and c, from Eqs. (g) into Eq. (a). Therefore,

$$z = \frac{Q\ell}{40AE} (-4x + 3y + 60) \tag{h}$$

The elongations of the cables are calculated from Eq. (h) by substituting the coordinates of their attachment points. Thus

$$\left.\begin{array}{l} e_1 = \dfrac{Q\ell}{40AE} [-4(-3) + 3(4) + 60] = \dfrac{84}{40} \dfrac{Q\ell}{AE} = 2.1 \dfrac{Q\ell}{AE} \\[3mm] e_2 = \dfrac{Q\ell}{40AE} [-4(-3) + 3(-4) + 60] = \dfrac{60}{40} \dfrac{Q\ell}{AE} = 1.5 \dfrac{Q\ell}{AE} \\[3mm] e_3 = \dfrac{Q\ell}{40AE} [-4(3) + 3(-4) + 60] = \dfrac{36}{40} \dfrac{Q\ell}{AE} = 0.9 \dfrac{Q\ell}{AE} \\[3mm] e_4 = \dfrac{Q\ell}{40AE} [-4(3) + 3(4) + 60] = \dfrac{60}{40} \dfrac{Q\ell}{AE} = 1.5 \dfrac{Q\ell}{AE} \end{array}\right\} \tag{i}$$

The cable forces are

$$N_1 = \frac{E_1 A_1}{\ell_1} e_1 = \frac{1.5EA}{3\ell} \left(2.1 \frac{Q\ell}{AE} \right) = 1.05Q$$

$$N_2 = \frac{E_2 A_2}{\ell_2} e_2 = \frac{EA}{2\ell} \left(1.5 \frac{Q\ell}{AE} \right) = 0.75Q$$

$$N_3 = \frac{E_3 A_3}{\ell_3} e_3 = \frac{2EA}{4\ell} \left(0.9 \frac{Q\ell}{AE} \right) = 0.45Q$$

$$N_4 = \frac{E_4 A_4}{\ell_4} e_4 = \frac{1.5EA}{3\ell} \left(1.5 \frac{Q\ell}{AE} \right) = 0.75 \, Q$$

(j)

PROBLEMS / Section 11-3

11-1 Determine the equilibrium configurations associated with a mechanical system whose total potential energy is

$$V(x) = \frac{1}{5} x^5 - \frac{a}{4} x^4 - \frac{2}{3} a^2 x^3$$

where a is system constant.

11-2 A thin, homogeneous rod of length ℓ and weight w rests in a smooth hemispherical cavity of radius R as shown in Figure P11-2. Using the method of total potential energy, determine a formula for the equilibrium position θ. $[(\cos 2\theta / \cos \theta) = \ell / 4R]$

11-3 The extensional rigidity EA is the same for each of the elastic rods shown in Figure P11-3. If the rigid crossbeam is horizontal prior to the application of the load P, determine its configuration after P has been applied.

11-4 The rigid triangular plate of negligible weight shown in Figure P11-4 lies in the xy plane prior to the application of force P. It is suspended by three

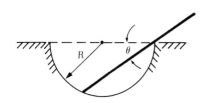

Figure P11-2

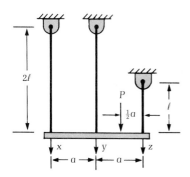

Figure P11-3

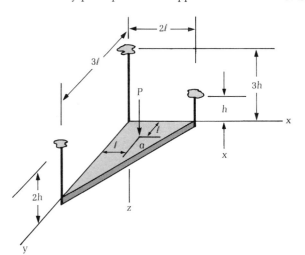

Figure P11-4

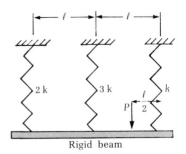

Figure P11-5

linearly elastic rods with identical extensional stiffnesses EA. Determine (a) the equation of the plate after the force P is applied, (b) the elongation of each rod, and (c) the force in each rod.

11-5 A weightless rigid beam is suspended by three linearly elastic springs as shown in Figure P11-5. The spring constant for each spring appears beside the springs in the figure. Determine the elongation and force in each spring.

11-4   Principle of Stationary Potential Energy Applied to Trusses

A plane truss, or a space truss, consists of straight members each of which is connected by smooth pins to the other members of the truss.

This definition is an idealization since members of real trusses are actually connected to gusset plates by welding or riveting or some similar method. However, since the members are usually long and slender, the bending effects that occur in the vicinity of the gusset plates are highly localized. The bending stresses decline rapidly with distance from the connection point so that the dominant action in a member a short distance from the gusset plate is a single axial force. Note that the loads applied to a truss are always applied at the joints. Consequently, this loading restriction, together with the pin-end assumption, make each member of a truss a two-force member.

The strain energy associated with the ith member of a truss made from a linearly elastic material is

$$U_i = \frac{\sigma_i^2}{2E_i} A_i \ell_i = \frac{N_i^2 \ell_i}{2E_i A_i} = \frac{E_i A_i}{2\ell_i} e_i^2 \qquad (11\text{-}13)$$

Here N_i, E_i, ℓ_i, A_i, and e_i are the axial force, modulus of elasticity, length, cross-sectional area, and elongation of the ith member of the truss, respectively.

The strain energy for the truss is the sum of the strain energies of its members. Thus

$$U = \sum_{i=1}^{n} U_i \qquad (11\text{-}14)$$

where n is the number of members of the truss.

The principle of stationary potential energy is valid for statically determinate and statically indeterminate trusses (both plane and space trusses). It remains valid if the material from which the members are made is nonlinear. Modifications of Eq. (11-13) must be made to account for the nonlinear material behavior. Moreover, the principle remains valid for geometric nonlinearities such as those that would occur if the displacement of the joints of the truss were large.

Elongations of truss members. Figure 11-5 shows a single member of a truss. Its end points in the undeformed configuration are denoted by A and B, and by A^* and B^* in the deformed configuration. The horizontal and vertical displacements of the ends are signified by u_A, v_A and u_B, v_B.

The length of the member in the deformed configuration is

$$\ell^* = \sqrt{(x_B^* - x_A^*)^2 + (y_B^* - y_A^*)} \tag{11-15}$$

and its length in the undeformed configuration is

$$\ell = \sqrt{(x_B - x_A)^2 + (y_B - y_A)^2} \tag{11-16}$$

From Figure 11-5, we see that

$$\left. \begin{array}{ll} x_A^* = x_A + u_A, & y_A^* = y_A + v_A \\ x_B^* = x_B + u_B, & y_B^* = y_B + v_B \end{array} \right\} \tag{11-17}$$

so that the elongation of the member can be expressed as

$$\begin{aligned} e &= \ell^* - \ell \\ &= \sqrt{\begin{array}{c} 1 + 2\left(\dfrac{u_B - u_A}{\ell}\right)\cos\theta + 2\left(\dfrac{v_B - v_A}{\ell}\right)\sin\theta \\ + \left(\dfrac{u_B - u_A}{\ell}\right)^2 + \left(\dfrac{v_B - v_A}{\ell}\right)^2 \end{array}} - \ell \end{aligned} \tag{11-18}$$

Note that θ is the angle between the axis of the member in the undeformed configuration and the x axis. This is an exact geometric relation. It remains valid for large deflections of the joints of a truss. Now, if the displacements of the ends of the member are small, the term

$$\left(\frac{u_B - u_A}{\ell}\right)^2 + \left(\frac{v_B - v_A}{\ell}\right)^2 \ll 1 \tag{11-19}$$

so that

$$e \approx \sqrt{1 + 2\left(\frac{u_B - u_A}{\ell}\right)\cos\theta + 2\left(\frac{v_B - v_A}{\ell}\right)\sin\theta} - \ell \tag{11-20}$$

Expanding the radical in Eq. (11-20) by means of the binomial expansion theorem leads to the expression

$$e = \ell\left\{1 + \left(\frac{u_B - u_A}{\ell}\right)\cos\theta + \left(\frac{v_B - v_A}{\ell}\right)\sin\theta - \cdots\right\} - \ell \tag{11-21}$$

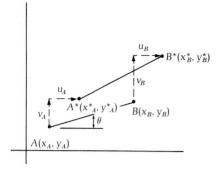

Figure 11-5

Consequently, for small displacements of the ends of the member, or the so-called linear theory of trusses,

$$e = (u_B - u_A) \cos \theta + (v_B - v_A) \sin \theta \qquad (11\text{-}22)$$

This relation shows that the elongations of the members of a truss are equal to the projections of the relative displacement components upon the **original direction** of the member. Consequently, to obtain the elongations of a member of a truss, we simply project the displacement components of its ends on the original direction of the axis of the member.

EXAMPLE 11-3

The extensional stiffness of each member of the pin-connected, plane truss of Figure 11-6a is the same. If the members are made from a linearly elastic material, determine (a) the horizontal and vertical displacements of the joints, (b) the elongation of each member, and (c) the internal force in each member.

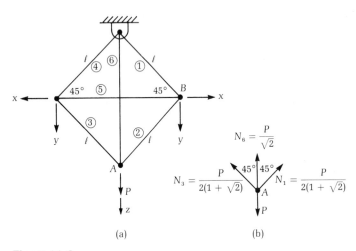

(a)   (b)

Figure 11-6

SOLUTION

(a) The members of the truss are numbered 1 through 6 as shown in Figure 11-6a. Formulate the potential energy associated with the linearly elastic truss in terms of the displacements of its joints. Let x, y, and z be components of displacements of the joints of the truss. Notice that symmetry of loading and geometry has been taken into account in selecting the displacement components.

The elongations of the various members are

$$
\left.
\begin{aligned}
e_1 = e_4 &= \frac{1}{\sqrt{2}}(x + y) \\[2mm]
e_2 = e_3 &= \frac{1}{\sqrt{2}}(x - y + z) \\[2mm]
e_5 &= 2x \\[2mm]
e_6 &= z
\end{aligned}
\right\}
\tag{a}
$$

The strain energy for the truss is equal to the sum of the strain energy of its members. Thus

$$
U = \frac{EA}{2\ell}\left(2e_1^2 + 2e_2^2 + \frac{e_5^2}{\sqrt{2}} + \frac{e_6^2}{\sqrt{2}}\right)
\tag{b}
$$

The potential energy of the external load is

$$
\Omega = -Pz
\tag{c}
$$

Equations (a), (b), and (c) lead to the total potential energy expression

$$
V = \frac{EA}{2\ell}\left\{(x + y)^2 + (x - y + z)^2 + 2\sqrt{2}x^2 + \frac{1}{\sqrt{2}}z^2\right\} - Pz
\tag{d}
$$

According to the principle of stationary potential energy, an equilibrium configuration for a conservative mechanical system is characterized by a stationary value of its total potential energy. Consequently,

$$
\left.
\begin{aligned}
\frac{\partial V}{\partial x} &= \frac{EA}{\ell}\{(x + y) + (x - y + z) + 2\sqrt{2}x\} = 0 \\[2mm]
\frac{\partial V}{\partial y} &= \frac{EA}{\ell}\{(x + y) - (x - y + z)\} = 0 \\[2mm]
\frac{\partial V}{\partial z} &= \frac{EA}{\ell}\left\{(x - y + z) + \frac{1}{\sqrt{2}}z\right\} - P = 0
\end{aligned}
\right\}
\tag{e}
$$

These equations are the conditions that characterize the equilibrium configuration of the truss under the load P. Equations (e) can be simplified and expressed as

$$
\left.
\begin{aligned}
2(1 + \sqrt{2})x \qquad\qquad\quad + z &= 0, \\[2mm]
2y \qquad\quad - z &= 0, \\[2mm]
x - y + \left(1 + \frac{1}{\sqrt{2}}\right)z &= \frac{P\ell}{AE}
\end{aligned}
\right\}
\tag{f}
$$

The solution of these equations is easily determined to be

$$\left.\begin{array}{l} x = -\dfrac{1}{2(1 + \sqrt{2})}\dfrac{P\ell}{AE} \\[4mm] y = \dfrac{1}{2}\dfrac{P\ell}{AE} \\[4mm] z = \dfrac{P\ell}{AE} \end{array}\right\} \qquad (g)$$

(b) The elongations of the various members are calculated from Eqs. (a) and (g). Accordingly,

$$\left.\begin{array}{l} e_1 = e_4 = \dfrac{P\ell}{2(1 + \sqrt{2})AE} \\[4mm] e_2 = e_3 = \dfrac{P\ell}{2(1 + \sqrt{2})AE} \\[4mm] e_5 = \dfrac{-P\ell}{(1 + \sqrt{2})AE} \\[4mm] e_6 = \dfrac{P\ell}{AE} \end{array}\right\} \qquad (h)$$

(c) The internal forces in the various members are calculated from the formula $N_i = (E_i A_i / \ell_i)e_i$. Consequently, with the aid of Eqs. (h),

$$\left.\begin{array}{l} N_1 = N_2 = N_3 = N_4 = \dfrac{P}{2(1 + \sqrt{2})} \\[4mm] N_5 = \dfrac{-P}{\sqrt{2}(1 + \sqrt{2})} \\[4mm] N_6 = \dfrac{P}{\sqrt{2}} \end{array}\right\} \qquad (i)$$

Figure 11-6b shows a free-body diagram of joint A. It is evident that the forces of Eqs. (i) satisfy equilibrium at this joint. Equilibrium is also satisfied at joint B.

Observe that the elongations (internal displacements) given in Eqs. (a) are compatible with the displacements x, y, and z. In accordance with the principle of stationary potential energy, displacements that also satisfy equilibrium requirements are assured.

EXAMPLE 11-4

Use the principle of stationary potential energy to determine the displacements of the joints of the pin-connected truss shown in Figure 11-7.

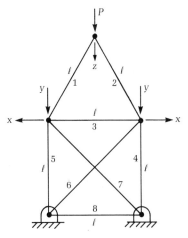

Figure 11-7

The length of each outside member is ℓ and the extensional rigidity EA is the same for each member. Also determine member elongations and forces.

SOLUTION

Because of the symmetry of loading, of support conditions, and of the extensional rigidities of the truss, only three generalized coordinates are required. These coordinates are denoted by x, y, and z in Figure 11-7. The labels appearing beside each member are identification labels.

The elongations of the various members that are compatible with the displacements x, y, and z, are

$$
\left.
\begin{aligned}
e_1 = e_2 &= \frac{1}{2} x + \frac{\sqrt{3}}{2} y - \frac{\sqrt{3}}{2} z \\
e_3 &= 2x \\
e_4 = e_5 &= -y \\
e_6 = e_7 &= \frac{1}{\sqrt{2}} (x - y)
\end{aligned}
\right\} \tag{a}
$$

The potential energy of the internal forces is

$$
U = \frac{EA}{2\ell} \left(e_1^2 + e_2^2 + e_3^2 + e_4^2 + e_5^2 + \frac{e_6^2}{\sqrt{2}} + \frac{e_7^2}{\sqrt{2}} \right) \tag{b}
$$

and the potential energy of the external force is

$$
\Omega = -Pz \tag{c}
$$

The total potential energy for the truss, expressed in terms of the joint displacements, is

$$
V(x, y, z) = \frac{EA}{2\ell} \left\{ \frac{1}{2} (x + \sqrt{3}y - \sqrt{3}z)^2 + 4x^2 + 2y^2 + \frac{(x-y)^2}{\sqrt{2}} \right\} - Pz \tag{d}
$$

According to the principle of stationary potential energy, the equilibrium configuration for the truss is given by the relations

$$
\left.
\begin{aligned}
\frac{\partial V}{\partial x} &= \frac{EA}{2\ell} \left\{ (x + \sqrt{3}y - \sqrt{3}z) + 8x + \sqrt{2}(x - y) \right\} = 0 \\
\frac{\partial V}{\partial y} &= \frac{EA}{2\ell} \left\{ (x + \sqrt{3}y - \sqrt{3}z)\sqrt{3} + 4y - \sqrt{2}(x - y) \right\} = 0 \\
\frac{\partial V}{\partial z} &= \frac{EA}{2\ell} \left\{ -(x + \sqrt{3}y - \sqrt{3}z)\sqrt{3} \right\} - P = 0
\end{aligned}
\right\} \tag{e}
$$

These equations can be simplified and arranged in standard form for linear algebraic equations. Thus

$$\left.\begin{array}{r} 5.207x + 0.159y - 0.866z = 0 \\ 0.159x + 4.207y - 1.500z = 0 \\ -0.866x - 1.500y + 1.500z = \dfrac{P\ell}{AE} \end{array}\right\} \quad \text{(f)}$$

Equations (f) have the solution

$$\left.\begin{array}{l} x = 0.187 \dfrac{P\ell}{AE} \\[2mm] y = 0.421 \dfrac{P\ell}{AE} \\[2mm] z = 1.199 \dfrac{P\ell}{AE} \end{array}\right\} \quad \text{(g)}$$

The member elongations are calculated from Eqs. (a). Therefore,

$$\left.\begin{array}{l} e_1 = e_2 = -0.581 \dfrac{P\ell}{AE} \\[2mm] e_3 = 0.373 \dfrac{P\ell}{AE} \\[2mm] e_4 = e_5 = -0.421 \dfrac{P\ell}{AE} \\[2mm] e_6 = e_7 = -0.165 \dfrac{P\ell}{AE} \end{array}\right\} \quad \text{(h)}$$

and the corresponding member forces are [from $N_i = (E_i A_i / \ell_i) e_i$]

$$\left.\begin{array}{r} N_1 = N_2 = -0.581P \\ N_3 = 0.373P \\ N_4 = N_5 = -0.421P \\ N_6 = N_7 = -0.117P \end{array}\right\} \quad \text{(i)}$$

EXAMPLE 11-5

The members of the simple truss shown in Figure 11-8 have the same cross-sectional area A and are both made from a nonlinear material whose stress-strain relation is $\sigma = K\epsilon^{1/3}$. Using the principle of stationary potential energy, determine formulas for the horizontal and vertical displacements of joint A.

SOLUTION

The principle of stationary potential energy remains valid for nonlinear **elastic** behavior. The strain energy density increment for a uniaxial

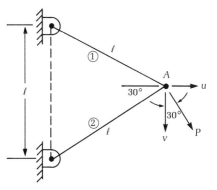

Figure 11-8

stress field is

$$\delta U_0 = \sigma \delta \epsilon \tag{a}$$

Substituting $\sigma = K\epsilon^{1/3}$ in Eq. (a) yields, after integration,

$$U_0 = \tfrac{3}{4}K\epsilon^{4/3} \tag{b}$$

Since the strain is uniform over the length of a member,

$$U_i = \tfrac{3}{4}K_i A_i \ell_i \epsilon_i^{4/3} \tag{c}$$

for the strain energy of the ith member.

The elongations of the two members are

$$\begin{aligned} e_1 &= \tfrac{1}{2}(\sqrt{3}u + v) \\ e_2 &= \tfrac{1}{2}(\sqrt{3}u - v) \end{aligned} \tag{d}$$

and the potential energy of the external force is

$$\Omega = -\frac{1}{2}Pu - \frac{\sqrt{3}}{2}Pv \tag{e}$$

Consequently, the total potential energy for the truss is

$$V = c\{(\sqrt{3}u + v)^{4/3} + (\sqrt{3}u - v)^{4/3}\} - \frac{1}{2}Pu - \frac{\sqrt{3}}{2}Pv \tag{f}$$

where

$$c = \frac{3}{8\sqrt[3]{2}}\frac{KA}{\sqrt[3]{\ell}} \tag{g}$$

The principle of stationary potential energy requires that $\partial V/\partial u = \partial V/\partial v = 0$ at an equilibrium configuration. Accordingly, we obtain from Eq. (f)

$$\left.\begin{aligned} (\sqrt{3}u + v)^{1/3} + (\sqrt{3}u - v)^{1/3} &= \frac{\sqrt{3}}{8}\frac{P}{c} \\ (\sqrt{3}u + v)^{1/3} - (\sqrt{3}u - v)^{1/3} &= \frac{3\sqrt{3}}{8}\frac{P}{c} \end{aligned}\right\} \tag{h}$$

The solution of these nonlinear equations is

$$\left.\begin{aligned} u &= \frac{7}{9}\left(\frac{P}{KA}\right)^3 \ell \\ v &= \sqrt{3}\left(\frac{P}{KA}\right)^3 \ell \end{aligned}\right\} \tag{i}$$

Equations (h) are tractable by direct techniques used to solve systems of linear algebraic equations. However, a numerical approach will almost always be necessary for the solution of nonlinear problems of practical significance. The Newton–Raphson iterative procedure can be used to advantage in the solution of these nonlinear problems.

Observe that the nonlinearity of the algebraic equations representing the equilibrium of the truss of the present problem arises out of the material nonlinearity. The principle of stationary potential energy also remains valid for nonlinearities that arise because of large changes in geometry.

PROBLEMS / Section 11-4

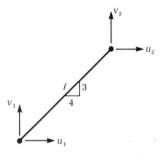

Figure P11-6

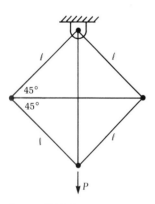

Figure P11-9

11-6 Determine an expression for the elongation of the truss member shown in Figure P11-6 in terms of the end displacements u_1, v_1 and u_2, v_2.

11-7 The extensional stiffness EA of each of the exterior members of the three-rod assembly shown in Figure P11-7 is twice that of the interior member. If the material of the rods is assumed to be linearly elastic, derive formulas for the displacements of point A in terms of P, ℓ, A, and E. Obtain formulas for the force in each member in terms of the applied load P.

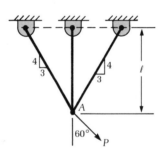

Figure P11-7

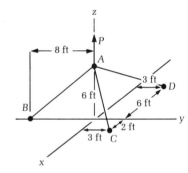

Figure P11-8

11-8 The members of the tripod shown in Figure P11-8 are connected by ball-and-socket joints at points A, B, C, and D. Use the method of total potential energy to determine the displacement components (u, v, w) of point A due to a force P acting along the z axis of the reference coordinate system. Assume that the extensional rigidity is the same for each member and that the material behaves elastically.

11-9 Determine the force in each member of the statically indeterminate truss shown in Figure P11-9. Use the method of total potential energy, consider the extensional stiffness EA to be the same for each member, and assume linearly elastic material behavior.

11-5 Castigliano's First Theorem

Consider a conservative mechanical system with a finite number of degrees of freedom N on which several concentrated forces F_i, several couples C_i, and several distributed loads act. Moreover, rigid body motions are prevented by a sufficient number of rigid supports. Such a system is shown in Figure 11-9.

For convenience, the forces F_i and the couples C_i are denoted by the generalized forces P_i. Displacements of the application points of forces F_i and the angular displacements about the axes of the couples C_i are denoted by the generalized coordinates q_i. Distributed loads enter the development implicitly through the strain energy U. Assume that U can be expressed in terms of the generalized coordinates q_i. Since the q_i are generalized coordinates, they are required to satisfy the constraints at the rigid supports. Thus it is implied that the internal displacements of the system are compatible with the displacements and rotations at the surface of the system.

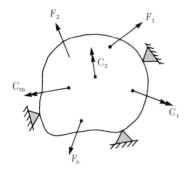

Figure 11-9

According to the principle of stationary potential energy, an equilibrium configuration of a conservative mechanical system is characterized by a stationary value of potential energy.

From Eqs. (11-3) and (11-9),

$$\sum_{i=1}^{N} \left(\frac{\partial U}{\partial q_i} - P_i \right) \delta q_i = 0 \qquad (11\text{-}23)$$

The δq_i are independent so that the coefficient of each δq_i must vanish. Consequently, we obtain

$$P_i = \frac{\partial U}{\partial q_i}, \ i = 1, 2, \ldots, N \qquad (11\text{-}24)$$

which is known as **Castigliano's first theorem**. If P_i is a force, Castigliano's first theorem asserts that the force in the direction of the displacement q_i is equal to the partial derivative of the strain energy with respect to q_i. If P_i is a couple, the theorem asserts that the couple associated with rotation q_i about the axis of the couple is equal to the partial derivative of the strain energy with respect to rotation q_i.

Observe that Eq. (11-24) is valid for geometric nonlinearities and nonlinear **elastic** material behavior because the principle of stationary potential energy does not preclude these nonlinear factors.

11-6 Castigliano's Second Theorem

To establish Castigliano's second theorem, it is necessary to introduce the **complementary energy** for a mechanical system. The complementary energy is defined through the relation

$$\gamma = -U + \sum_{i=1}^{N} P_i q_i \qquad (11\text{-}25)$$

This time it is assumed that U can be expressed in terms of an equilibrium system of external forces P_i. Otherwise, the stipulations already established as the basis for Castigliano's first theorem remain in effect.

By differentiation of Eq. (11-25) with respect to the generalized force P_k, we obtain

$$\frac{\partial \gamma}{\partial P_k} = -\left\{ \frac{\partial U}{\partial q_1}\frac{\partial q_1}{\partial P_k} + \frac{\partial U}{\partial q_2}\frac{\partial q_2}{\partial P_k} + \cdots + \frac{\partial U}{\partial q_N}\frac{\partial q_N}{\partial P_k} \right\}$$

$$+ \left\{ P_1\frac{\partial q_1}{\partial P_k} + P_2\frac{\partial q_2}{\partial P_k} + \cdots + P_N\frac{\partial q_N}{\partial P_k} \right\}$$

$$+ \left\{ \frac{\partial P_1}{\partial P_k}q_1 + \frac{\partial P_2}{\partial P_k}q_2 + \cdots + \frac{\partial P_N}{\partial P_k}q_N \right\} \qquad (11\text{-}26)$$

Now all the partial derivatives $\partial P_i/\partial P_k = 0$ when $i \neq k$ and $\partial P_k/\partial P_k = 1$. Therefore, the last line in Eq. (11-26) reduces to q_k. Collecting coefficients of the $\partial q_i/\partial P_k$ leads to the expression

$$\frac{\partial \gamma}{\partial P_k} = \left(P_1 - \frac{\partial U}{\partial q_1} \right)\frac{\partial q_1}{\partial P_k} + \left(P_2 - \frac{\partial U}{\partial q_2} \right)\frac{\partial q_2}{\partial P_k} + \cdots$$

$$+ \left(P_N - \frac{\partial U}{\partial q_N} \right)\frac{\partial q_N}{\partial P_k} + q_k \qquad (11\text{-}27)$$

The quantities in parentheses are identically zero according to Eqs. (11-24). Consequently, the mathematical expression of **Castigliano's second theorem** is

$$\frac{\partial \gamma}{\partial P_k} = q_k \qquad (11\text{-}28)$$

If P_k is a force, Castigliano's second theorem asserts that the displacement q_k in the direction of P_k is equal to the partial derivative of the complementary energy of the system with respect to the force P_k. If P_k is a couple, the theorem asserts that the rotation q_k about the axis of the couple P_k is equal to the partial derivative of the complementary energy of the system with respect to the couple P_k.

If no force or couple occurs at the points whose displacement or rotation is sought, a fictitious force or couple can be applied there. These fictitious quantities are set equal to zero after the differentiation indicated in Eq. (11-28) has been performed.

Linearly elastic system. For a linearly elastic system, the external forces and couples P_i are related linearly to the generalized displacements q_i. Thus

$$\left. \begin{aligned} P_1 &= k_{11}q_1 + k_{12}q_2 + \cdots + k_{1N}q_N \\ P_2 &= k_{21}q_1 + k_{22}q_2 + \cdots + k_{2N}q_N \\ &\;\;\vdots \\ P_N &= k_{N1}q_1 + k_{N2}q_2 + \cdots + k_{NN}q_N \end{aligned} \right\} \qquad (11\text{-}29)$$

It is convenient to write Eqs. (11-29) in the compact form

$$P_i = \sum_{j=1}^{N} k_{ij}q_j \qquad i = 1, 2, \ldots, N \qquad (11\text{-}30)$$

The coefficients k_{11}, k_{12} ... are called stiffness coefficients.

The increment of strain energy associated with virtual displacements δq_i is

$$\delta U = P_1\,\delta q_1 + P_2\,\delta q_2 + \cdots + P_N\,\delta q_N \qquad (11\text{-}31)$$

or, on using Eq. (11-29),

$$\begin{aligned}
\delta U = {}&(k_{11}q_1 + k_{12}q_2 + \cdots + k_{1N}q_N)\,\delta q_1 \\
&+ (k_{21}q_1 + k_{22}q_2 + \cdots + k_{2N}q_N)\,\delta q_2 \\
&\;\;\vdots \\
&+ (k_{N1}q_1 + k_{N2}q_2 + \cdots + k_{NN}q_N)\,\delta q_N \qquad (11\text{-}32)
\end{aligned}$$

Now regroup the terms in Eq. (11-32) to isolate exact differentials.

$$\begin{aligned}
\delta U = {}&k_{11}q_1\,\delta q_1 + k_{12}(q_2\,\delta q_1 + q_1\,\delta q_2) + k_{13}(q_3\,\delta q_1 + q_1\,\delta q_3) + \cdots + k_{1N}(q_N\,\delta q_1 + q_1\,\delta q_N) \\
&+ k_{22}q_2\,\delta q_2 \qquad\qquad\quad + k_{23}(q_3\,\delta q_2 + q_2\,\delta q_3) + \cdots + k_{2N}(q_N\,\delta q_2 + q_2\,\delta q_N) \\
&+ k_{33}q_3\,\delta q_3 \qquad\qquad\qquad\qquad\qquad\qquad\; + \cdots + k_{3N}(q_N\,\delta q_3 + q_3\,\delta q_N) \\
&\qquad\qquad\qquad\qquad\qquad\qquad\qquad\qquad\qquad\;\;\vdots \\
&\qquad\qquad\qquad\qquad\qquad\qquad\qquad\qquad\qquad\quad + k_{NN}q_N\,\delta q_N \qquad (11\text{-}33)
\end{aligned}$$

Note that it has been assumed that the k_{ij} are symmetric; that is, $k_{12} = k_{21}$, $k_{13} = k_{31}$, and so on. Also observe that each grouping of terms in Eq. (11-33) is an exact differential in the displacements. For example, $q_1\,\delta q_1 \equiv \delta(\tfrac{1}{2}q_1^2)$ and $q_2\,\delta q_1 + q_1\,\delta q_2 \equiv \delta(q_1 q_2)$. Consequently, Eq. (11-33) can be integrated over a specific set of q_i. The result is

$$\begin{aligned}
U = {}&\tfrac{1}{2}k_{11}q_1^2 + k_{12}q_1q_2 + k_{13}q_1q_3 + \cdots + k_{1N}q_1q_N \\
&+ \tfrac{1}{2}k_{22}q_2^2 + k_{23}q_2q_3 + \cdots + k_{2N}q_2q_N \\
&+ \tfrac{1}{2}k_{33}q_3^2 + \cdots + k_{3N}q_3q_N \\
&\qquad\qquad\qquad\quad\;\;\vdots \\
&\qquad\qquad\qquad + \tfrac{1}{2}k_{NN}q_N^2 \qquad (11\text{-}34)
\end{aligned}$$

Now write Eq. (11-34) in compact form.

$$U = \frac{1}{2}\sum_{i=1}^{N}\left(\sum_{j=1}^{N} k_{ij}q_j\right)q_i = \frac{1}{2}\sum_{i=1}^{N} P_i q_i \qquad (11\text{-}35)$$

The complementary energy for a linear system is obtained from Eqs. (11-25) and (11-35). It is

$$\gamma = -\frac{1}{2}\sum_{i=1}^{N} P_i q_i + \sum_{i=1}^{N} P_i q_i = \frac{1}{2}\sum_{i=1}^{N} P_i q_i \qquad (11\text{-}36)$$

or, after using Eq. (11-30),

$$\gamma = \frac{1}{2} \sum_{i=1}^{N} \sum_{j=1}^{N} k_{ij} q_i q_j \tag{11-37}$$

For a linear system, the strain energy and the complementary energy are equal, and Castigliano's second theorem becomes

$$\frac{\partial U}{\partial P_k} = q_k \tag{11-38}$$

11-7 Direct Application of Castigliano's Second Theorem

The use of Castigliano's theorem to determine deflections and rotations for straight beams, curved beams, and planar trusses is demonstrated in the examples contained in the following three sections. To implement Castigliano's theorem, we must express the complementary energy of the structural elements in terms of the internal reactions characteristic to them. Consequently, before proceeding to the examples, we must establish formulas for the bending and direct shearing energy for straight beams, for twisting energy for circular rods, and for the axial energy for axially loaded rods.

Linearly elastic systems are considered here. For nonlinear elastic systems, formulas for complementary energy must be derived directly.

Axially loaded rods. The strain energy per unit volume associated with a uniaxial stress field is

$$U_0 = \frac{\sigma^2}{2E} \tag{11-39}$$

for a linearly elastic material. The strain energy for the rod is

$$U = \int_{\text{vol}} U_0 \, dV = \int_{\text{vol}} \frac{\sigma^2}{2E} \, dA \, dx \tag{11-40}$$

and, since $\sigma = N/A$ is uniform on the cross section,

$$U = \int_0^l \frac{N^2}{2AE} \, dx \tag{11-41}$$

Torsionally loaded rods. The strain energy per unit volume associated with a state of pure shear is

$$U_0 = \frac{\tau^2}{2G} \tag{11-42}$$

for a linearly elastic material, or, since $\tau = Tr/J$, the strain energy for the rod is

$$U = \int_{\text{vol}} U_0 \, dV = \int_0^l \frac{T^2}{2J^2 G} \left(\int_{\text{area}} r^2 \, dA \right) dx$$

or

$$U = \int_0^l \frac{T^2}{2JG} \, dx \tag{11-43}$$

Bending of straight beams.   The strain energy associated with a straight beam due to bending is

$$U = \int_{\text{vol}} \frac{\sigma^2}{2E} \, dA \, dx \tag{11-44}$$

or, since $\sigma = My/I$,

$$U = \int_0^l \frac{M^2}{2EI^2} \left(\int_{\text{area}} y^2 \, dA \right) dx = \int_0^l \frac{M^2}{2EI} \, dx \tag{11-45}$$

In a similar manner, the strain energy due to transverse shearing stresses in a straight beam is

$$U = \int_{\text{vol}} \frac{\tau^2}{2G} \, dA \, dx \tag{11-46}$$

or, since $\tau = VQ/Ib$,

$$U = \int_0^l \frac{V^2}{2AG} \left(\frac{A}{I^2} \int_{\text{area}} \frac{Q^2}{b^2} \, dA \right) dx = \int_0^l \frac{\kappa V^2}{2AG} \, dx \tag{11-47}$$

where

$$\kappa = \frac{A}{I^2} \int_{\text{area}} \frac{Q^2}{b^2} \, dA \tag{11-48}$$

is a property of the cross section called a **shape factor**. The shearing strain energy is frequently small compared to bending or twisting energy so that κ is a small correction to a relative small quantity. Consequently, κ is frequently set equal to unity. However, there are situations where κ might be significant enough to warrant its calculation.

11-8 Application of Castigliano's Theorem to Straight Beams

The strain energy associated with a straight beam is the sum of the energies due to bending and direct shear. Thus

$$U = \int_0^l \frac{M^2}{2EI} \, dx + \int_0^l \frac{V^2}{2AG} \, dx \tag{11-49}$$

The application of Castigliano's principle is facilitated if the required differentiation is performed before the integrands in the integrals of Eq. (11-49) are formed. Accordingly, displacement and rotation of a point on the centroidal line of a beam are given by the general formulas

$$q_P = \frac{\partial U}{\partial P} = \int_0^l \frac{M}{EI} \frac{\partial M}{\partial P} \, dx + \int_0^l \frac{V}{AG} \frac{\partial V}{\partial P} \, dx \tag{11-50}$$

and

$$\theta_c = \frac{\partial U}{\partial C} = \int_0^l \frac{M}{EI} \frac{\partial M}{\partial C} \, dx + \int_0^l \frac{V}{AG} \frac{\partial V}{\partial C} \, dx \tag{11-51}$$

where P is a concentrated force applied at the point on the centroidal line whose displacement is sought, and C is a couple applied at the point whose rotation is sought.

Observe that M and V are internal reactions and can always be expressed in terms of the applied external loads and the support reactions. For statically determinate beams, the support reactions can be determined from equilibrium considerations so that M and V are known explicitly. For statically indeterminate beams, not all support reactions can be determined in terms of the applied loads by statical considerations. Thus formulas for M and V will contain certain support reactions as undetermined quantities. In this case, Castigliano's principle can be used to derive algebraic equations among the undetermined support reactions that express the manner in which the beam is supported. The following examples illustrate the use of Castigliano's principle as it is applied to statically determinate and statically indeterminate beams.

EXAMPLE 11-6

Use Castigliano's principle to determine the deflection and rotation of the free end of the uniformly loaded cantilever beam shown in Figure 11-10a. Assume that the cantilever is long compared to its depth so that deflection due to transverse shearing forces can be neglected.

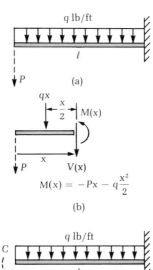

(a)

(b)

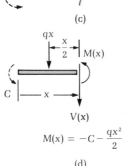

(c)

(d)

Figure 11-10

SOLUTION

To determine the deflection of the free end of the cantilever, introduce a fictitious force P at the point whose deflection is sought. The appropriate moment equation is given in Figure 11-10b. According to Castigliano's principle, the deflection of the free end of the cantilever is

$$\delta_P = \frac{\partial U}{\partial P}\bigg|_{P=0} = \int_0^\ell \left(\frac{-Px - \frac{qx^2}{2}}{EI}\right)(-x)\,dx \tag{a}$$

Since P is a fictitious force, it must be set equal to zero after the differentiation has been accomplished. Consequently,

$$\delta_P = \int_0^\ell \frac{qx^3}{2EI}\,dx = \frac{q\ell^4}{8EI} \tag{b}$$

The positive character of this deflection indicates that it has the same sense as the fictitious force P.

To determine the rotation of the free end of the cantilever, introduce a fictitious couple as shown in Figure 11-10c. The appropriate moment equation is obtained from Figure 11-10d as

$$M(x) = -C - \frac{qx^2}{2} \tag{c}$$

According to Castigliano's principle, the rotation at the end of the cantilever is

$$\theta = \frac{\partial U}{\partial C}\bigg|_{C=0} = \int_0^\ell \left(-\frac{qx^2}{2EI}\right)(-1)\,dx = \frac{q\ell^3}{6EI} \tag{d}$$

The positive character of this rotation indicates that its sense coincides with the sense of the fictitious couple C.

EXAMPLE 11-7

Determine the reaction at the roller for the statically indeterminate beam shown in Figure 11-11a. Assume that bending is the dominant deformational response and use Castigliano's principle.

SOLUTION

Free-body diagrams appropriate for establishing bending moment equations are shown in Figure 11-11b. These equations are

$$\left.\begin{array}{ll} M(x) = Rx & 0 \leq x \leq \ell \\ M(x) = Rx - M_0 & \ell \leq x \leq 2\ell \end{array}\right\} \tag{a}$$

where R is the unknown reaction at the roller.

According to Castigliano's principle, the deflection δ_R at the free end of a cantilever loaded by a couple M_0 at its center and a concentrated force R at its free end is

$$\delta_R = \frac{\partial U}{\partial R} = \int_0^l \left(\frac{Rx}{EI}\right) x\, dx + \int_l^{2l} \left(\frac{Rx - M_0}{EI}\right) x\, dx \qquad (b)$$

Integration and evaluation at the limits lead to the formula

$$\delta_R = \frac{8}{3}\frac{Rl^3}{EI} - \frac{3}{2}\frac{M_0}{EI}^2 \qquad (c)$$

Now the deflection at the free end of the cantilever is zero because of the rigid roller. Consequently, Eq. (c) yields

$$R = \frac{9}{16}\frac{M_0}{l} \qquad (d)$$

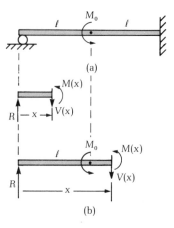

Figure 11-11

EXAMPLE 11-8

Determine the reactions at supports for the statically indeterminate beam shown in Figure 11-12a. Assume that bending is the dominant deformational response. Use Castigliano's principle.

SOLUTION

Appropriate moment equations are obtained from the free-body diagrams shown in Figure 11-12b,

$$\left.\begin{array}{ll} M(x) = V_l x + M_l & 0 \le x \le l \\ M(x) = V_l x + M_l - P(x - l) & l \le x \le 2l, \end{array}\right\} \qquad (a)$$

where V_l and M_l are the unknown support reactions at the left end.

Now use Castigliano's principle to establish formulas for the deflection δ_l and rotation θ_l at the left end of the beam if it was a cantilever with V_l and M_l applied as external forces. Accordingly,

$$\theta_l = \frac{\partial U}{\partial M_l} = \int_0^l \left(\frac{V_l x + M_l}{EI}\right) dx + \int_l^{2l}\left[\frac{V_l x + M_l - P(x - l)}{EI}\right] dx \qquad (b)$$

and

$$\delta = \frac{\partial U}{\partial V_l} = \int_0^l \left(\frac{V_l x + M_l}{EI}\right) x\, dx + \int_l^{2l}\left[\frac{V_l x + M_l - P(x - l)}{EI}\right] x\, dx \qquad (c)$$

Integration of Eqs. (b) and (c) and evaluation at the limits yields the two formulas

$$\theta_l = 2\frac{V_l l^2}{EI} + 2\frac{M_l l}{EI} - \frac{Pl^2}{2EI} \qquad (d)$$

and

$$\delta_l = \frac{8}{3}\frac{V_l l^3}{EI} + 2\frac{M_l l^2}{EI} - \frac{5}{6}\frac{Pl^3}{EI} \qquad (e)$$

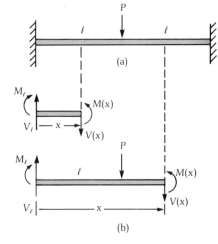

Figure 11-12

The rigid clamp at the left end of the beam requires that $\theta_l = \delta_l = 0$. Consequently, two algebraic equations for the unknown reactions V_l and M_l are

$$\left. \begin{array}{l} V_l \ell^2 + M_l \ell = \dfrac{P\ell^2}{4} \\[2ex] V_l \ell^3 + \frac{3}{4}M_l \ell^2 = \frac{5}{16} P\ell^3 \end{array} \right\}$$ (f)

Solving Eqs. (f) simultaneously yields

$$\left. \begin{array}{l} V_l = \dfrac{P}{2} \\[2ex] M_l = -\dfrac{P\ell}{4} \end{array} \right\}$$ (g)

The minus sign of M_l indicates that the direction of the moment is opposite to that shown in Figure 11-12b. The reactions at the right end of the cantilever can be obtained from a free-body diagram of the entire beam or by symmetry considerations. They are

$$\left. \begin{array}{l} V_r = \dfrac{P}{2} \\[2ex] M_r = -\dfrac{P\ell}{4} \end{array} \right\}$$ (h)

PROBLEMS / Section 11-8

11-10 Use Castigliano's principle to determine the deflection of the free end of the cantilever beam shown in Figure P11-10. Assume linearly elastic material behavior. Include the effect of transverse shear. Show that, for a slenderness ratio $\ell/r \geq 20$, the deflection due to shear is less than 2% of the total deflection. Use $\nu = 1/3$.

Figure P11-10

11-11 Use Castigliano's second theorem to determine the rotation at the left end of the linearly elastic beam shown in Figure P11-11.

11-12 Use Castigliano's second theorem to determine the reaction at the roller for the statically indeterminate beam shown in Figure P11-12.

11-13 Use Castigliano's second theorem to determine the reactions at the rollers for the statically indeterminate beam shown in Figure P11-13.

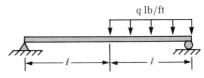

Figure P11-11

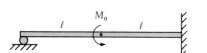

Figure P11-12

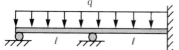

Figure P11-13

11-9   Application of Castigliano's Second Theorem to Curved Beams

The strain energy associated with the bending and stretching of a curved beam is given by the Winkler–Bach formula*

$$U = \int_0^l \left\{ \frac{N^2}{2AE} + \frac{MN}{EAR} + \left(1 + \frac{1}{Z}\right) \frac{M^2}{2EAR^2} \right\} ds \qquad (11\text{-}52)$$

Bending of a curved beam is usually accompanied by torsional deformations so that the total energy must be considered to be the sum of the energies due to bending, stretching, direct shear, and twisting.

The bending energy associated with curved beams can frequently be approximated with sufficient accuracy by the straight beam bending energy formula. This approximation usually reduces the difficulties encountered in applying Castigliano's principle. In many problems, the strain energy due to transverse shearing stresses is small compared to bending and/or twisting energy, and it can be neglected without introducing significant inaccuracies. This approximation also reduces the difficulties encountered in applying Castigliano's principle to curved beams.

The bending energy for a curved beam can usually be approximated by the straight beam bending energy when $R/c > 4$; that is, when the ratio of the local radius of curvature to the maximum dimension of the cross section is greater than 4. This observation is illustrated in one of the following examples.

The direct shearing strain energy can usually be neglected when the rod is slender; that is, when the radius of the rod is large compared to its depth. This is analogous to the known relative importance of direct shear in the calculation of deflections of straight beams.

Again, Castigliano's principle is facilitated if the required differentiation is performed before the integrands of the energy integrals are formed. If the **bending energy** is represented by the **bending energy of a straight beam**, the displacement and rotation of a point on the centroidal line of a curved beam are given by a formula of the form

$$\frac{\partial U}{\partial P} = \int_0^l \left\{ \frac{N}{AE} \frac{\partial N}{\partial P} + \frac{M}{EI} \frac{\partial M}{\partial P} + \frac{T}{JG} \frac{\partial T}{\partial P} + \frac{V}{AG} \frac{\partial V}{\partial P} \right\} ds \qquad (11\text{-}53)$$

and

$$\frac{\partial U}{\partial C} = \int_0^l \left\{ \frac{N}{AE} \frac{\partial N}{\partial C} + \frac{M}{EI} \frac{\partial M}{\partial C} + \frac{T}{JG} \frac{\partial T}{\partial C} + \frac{V}{AG} \frac{\partial V}{\partial C} \right\} ds \qquad (11\text{-}54)$$

* For a derivation of this formula, see Sealy, F. B. and J. O. Smith. 1957. *Advanced Mechanics of Materials*. Wiley and Sons, New York, 2nd Edition.

P and C have the same interpretations as they had for straight beam theory. Observe that N, M, T, and V are internal reactions, and they can always be expressed in terms of the applied external loads and support reactions. For statically determinate beams, the support reactions can be determined from equilibrium considerations so that N, M, T, and V are known explicitly in terms of the applied loads. For statically indeterminate beams, not all support reactions can be determined in terms of the applied loads by statical considerations. Consequently, formulas for N, M, T, and V, will contain certain support reactions as undetermined quantities. In this case, Castigliano's principle can be used to establish algebraic equations, relating the undetermined support reactions, that express the manner in which the curved beam is supported.

The following examples illustrate the use of Castigliano's principle as it is applied to statically determinate and statically indeterminate curved beams.

EXAMPLE 11-9

Determine a formula for the roller reaction at the end of the slender rod shown in Figure 11-13a. Consider bending energy only.

SOLUTION

Appropriate moment equations are obtained from the free-body diagrams shown in Figure 11-13b.

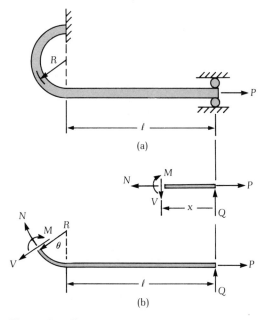

Figure 11-13

$$M(x) = Qx \qquad\qquad 0 \le x \le \ell$$
$$M(x) = Q(\ell + R \sin \theta) + PR(1 - \cos \theta) \qquad 0 \le \theta \le \pi$$

where Q is the unknown reaction at the roller.

The deflection that would occur at the right end of the rod if a concentrated force Q were applied in place of the roller is

$$\delta_Q = \frac{\partial U}{\partial Q} = \int_0^\ell \left(\frac{Qx}{EI}\right) x \, dx$$
$$+ \int_0^\pi \frac{[Q(\ell + R \sin \theta) + PR\,(1 - \cos \theta)]}{EI} (\ell + R \sin \theta)R \, d\theta$$

Integration and evaluation at the limits yield the formula

$$\delta_Q = \frac{Q\ell^3}{3EI} + \frac{QR}{EI}\left(\pi\ell^2 + 4R\ell + \frac{\pi}{2}R^2\right) + \frac{PR^2}{EI}(\pi\ell + 2R)$$

Since the deflection δ_Q is prevented by the roller,

$$Q = \frac{-PR^2(\pi\ell + 2R)}{\ell^3\left\{\dfrac{1}{3} + \dfrac{\pi R}{\ell} + 4\left(\dfrac{R}{\ell}\right)^2 + \dfrac{\pi}{2}\left(\dfrac{R}{\ell}\right)^3\right\}}$$

EXAMPLE 11-10

A rod with a circular cross section is bent into a quarter of a circular ring of radius R. The quarter ring is clamped at one end and is loaded by a concentrated force P perpendicular to its plane at the other end as shown in Figure 11-14. Assuming that the radius r of cross section is small compared to the radius of curvature R of the quarter ring, determine the displacement perpendicular to the free end of the rod. Assume linearly elastic material behavior.

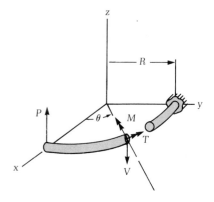

Figure 11-14

SOLUTION

A free-body diagram of a portion of the quarter ring is shown in Figure 11-14. The internal reactions at an arbitrary section are determined from equilibrium considerations. Thus

$$\left.\begin{aligned} V &= P \\ M &= -PR \sin \theta \\ T &= -PR(1 - \cos \theta) \end{aligned}\right\} \qquad\qquad (a)$$

Since it is assumed that $r \ll R$, the strain energy due to bending can be approximated through the straight beam formula. Also, the strain energy due to the direct shearing force V is approximated by setting $\kappa = 1$ in Eq. (11-47). Castigliano's theorem on deflections gives

$$\frac{\partial U}{\partial P} = \int_0^{\pi/2} \frac{M}{EI}\frac{\partial M}{\partial P} R \, d\theta + \int_0^{\pi/2} \frac{T}{JG}\frac{\partial T}{\partial P} R \, d\theta + \int_0^{\pi/2} \frac{V}{AG}\frac{\partial V}{\partial P} R \, d\theta \qquad (b)$$

Substitution of Eqs. (a) into Eq. (b) yields

$$\frac{\partial U}{\partial P} = \frac{PR^3}{EI} \int_0^{\pi/2} \sin^2 \theta \, d\theta + \frac{PR^3}{JG} \int_0^{\pi/2} (1 - \cos \theta)^2 \, d\theta + \frac{PR}{AG} \int_0^{\pi/2} d\theta$$

(c)

Integration with respect to θ yields the formula

$$\delta_P = \frac{\pi PR^3}{4EI} \left\{ 1 + \frac{4}{\pi} \frac{EI}{JG} \left(\frac{3\pi}{4} - 2 \right) + 2 \frac{EI}{GAR^2} \right\}$$

(d)

The first, second, and third terms in the brackets represent, respectively, the deflection in the direction of P due to bending, twisting, and direct shear. Note that $EI/JG = 1 + v$ so that the deflection due to twisting is of the same order of magnitude as that due to bending $\left(\text{the factor } \frac{4}{\pi} \frac{EI}{JG} \left(\frac{3\pi}{4} - 2 \right) = 0.567 \right)$. Also observe that the contribution due to direct shear is $\frac{2EI}{AGR^2} = 4(1 + v) \left(\frac{k_g}{R} \right)^2 = 5 \left(\frac{k_g}{R} \right)^2$ for $v = \frac{1}{4}$. Here k_g denotes the radius of gyration of the cross section. Thus when $k_g/R < \frac{1}{10}$, $2EI/AGR^2 < 0.05$. Consequently, the contribution to the total lateral deflection of the free end of the quarter ring due to direct shear is less than 5% when $k_g/R < \frac{1}{10}$. Deflection due to shear is frequently neglected when the rod is slender. In the present example, no difficulty arises by including direct shear effects; however, there may be cases where solutions are complicated enough by the inclusion of direct shear effects so that it is computationally expedient to ignore them.

EXAMPLE 11-11

Determine a formula for the rotation of the free end of the quarter ring of Example 11-10 about a line parallel to the x axis. Assume linearly elastic material behavior.

SOLUTION

Since no couple is applied at the free end, a fictitious couple must be applied there in the direction of the line about which the rotation is required to determine a formula for the required rotation. Figure 11-15 shows a free-body diagram of a portion of the quarter ring with the fictitious couple applied. The internal reactions at an arbitrary section are determined from equilibrium considerations. Thus

$$\left. \begin{array}{l} V = P \\ M = C \cos \theta - PR \sin \theta \\ T = C \sin \theta - PR(1 - \cos \theta) \end{array} \right\}$$

(a)

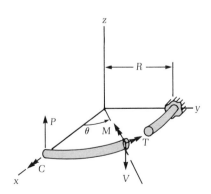

Figure 11-15

According to Castigliano's theorem on deflections

$$\theta = \frac{\partial U}{\partial C}\bigg|_{C=0} = \int_0^{\pi/2} \frac{M}{EI}\frac{\partial M}{\partial C} R\,d\theta + \int_0^{\pi/2} \frac{T}{JG}\frac{\partial T}{\partial C} R\,d\theta$$

$$+ \int_0^{\pi/2} \frac{V}{AG}\frac{\partial V}{\partial C} R\,d\theta \qquad (b)$$

Consequently, using Eqs. (a), Eq. (b) becomes

$$\theta = \int_0^{\pi/2} \frac{(-PR\sin\theta)}{EI}\cos\theta(R\,d\theta)$$

$$+ \int_0^{\pi/2} \frac{[-PR(1-\cos\theta)]}{JG}\sin\theta(R\,d\theta) \qquad (c)$$

or

$$\theta = -\frac{PR^2}{2EI} - \frac{PR^2}{2JG} = -\left(\frac{2+v}{2}\right)\frac{PR^2}{EI} = -\frac{9}{8}\frac{PR^2}{EI} \quad \text{for } v = \frac{1}{4} \quad (d)$$

EXAMPLE 11-12

A statically indeterminate steel ring with a mean diameter of 5 in. and a circular cross section 2-in. in diameter is subjected to a diametrally directed load of $2P$ (see Figure 11-16a). If the stress caused by the external load does not exceed the proportional limit of the material, determine the stress at points A and B of the ring. Calculate the decrease in the diameter along which the load acts. Assume that the straight beam formula for bending energy is sufficiently accurate (steel $E = 30 \times 10^6$ psi).

SOLUTION

Because of symmetry, only one quarter of the ring needs to be considered as shown in the free-body diagram of Figure (11-16b). Equilibrium of the shaded portion of the quarter ring gives

$$\left.\begin{aligned} V &= -P\sin\theta \\ N &= -P\cos\theta \\ M &= C - PR(1-\cos\theta) \end{aligned}\right\} \qquad (a)$$

Here C denotes the unknown moment at the section through point A.

Using the formula for bending energy associated with a straight beam, by Castigliano's principle, the rotation of the section at point A is

$$\theta_A = 4\int_0^{\pi/2}\left(\frac{N}{AE}\frac{\partial N}{\partial C} + \frac{M}{EI}\frac{\partial M}{\partial C} + \frac{V}{AG}\frac{\partial V}{\partial C}\right)R\,d\theta \qquad (b)$$

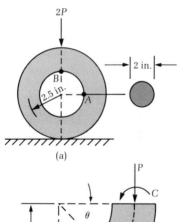

(a)

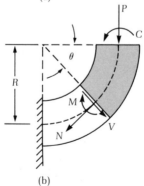

(b)

Figure 11-16

Now $\partial N/\partial C = 0$, $\partial V/\partial C = 0$, and $\partial M/\partial C = 1$ so that

$$\theta_A = 4 \int_0^{\pi/2} \frac{C - PR(1 - \cos \theta)}{EI} (1)R \, d\theta \tag{c}$$

Integration of Eq. (c) leads to the formula

$$\theta_A = \frac{4R}{EI} \left\{ \frac{\pi}{2} C - PR \left(\frac{\pi}{2} - 1 \right) \right\} \tag{d}$$

Consequently, since the section through point A does not rotate,

$$C = PR \left(1 - \frac{2}{\pi} \right) = 0.363PR \tag{e}$$

Notice that Eq. (e) is based on the assumption that the bending energy associated with the ring can be approximated by the straight beam bending energy.

The moment at an arbitrary section is given by the third of Eqs. (a). Therefore,

$$M(\theta) = (-0.637 + \cos \theta)PR \tag{f}$$

so that the moment at a section through B is (by setting $\theta = \pi/2$ in Eq. (f))

$$M_B = -0.637PR \tag{g}$$

The stresses at points A and B are calculated from the Winkler–Bach curved beam formula,*

$$\sigma = \frac{N}{A} + \frac{M}{AR} \left(1 + \frac{1}{Z} \frac{\xi}{R + \xi} \right) \tag{h}$$

Now $A = \pi$ in^2 and the Winkler–Bach curved beam factor is given by the formula

$$Z = -1 + 2 \left(\frac{R}{c} \right)^2 - 2 \left(\frac{R}{c} \right) \sqrt{\left(\frac{R}{c} \right)^2 - 1} \tag{i}$$

Since $R/c = 2.5$, $Z = 0.0436$.

Direct substitution into Eq. (h) leads to the stresses at points A and B. Thus

$$\left.\begin{aligned}
\sigma_A &= -\frac{P}{\pi} + \frac{0.363PR}{\pi R} \left\{ 1 + 22.94 \left(\frac{-1}{2.5 - 1} \right) \right\} = -1.973P \\[2mm]
\sigma_B &= -\frac{0.637PR}{\pi R} \left\{ 1 + 22.94 \left(\frac{-1}{2.5 - 1} \right) \right\} = 2.899P
\end{aligned}\right\} \tag{j}$$

---

* The Winkler–Bach curved beam formula is derived in chapter 12. The quantities A, R, and Z are, respectively, the cross-sectional area, the radius of curvature of the centroidal axis, and the Winkler–Bach curved beam factor. The coordinate ξ is the distance from the centroidal axis to the point in the cross section whose stress is to be calculated. It is positive when measured away from the center of curvature.

The Bernoulli theory of straight beams predicts the stresses

$$\left.\begin{array}{l}\sigma_A = -\dfrac{P}{\pi} - \dfrac{(0.363PR)r}{\dfrac{\pi r^4}{4}} = -\dfrac{P}{\pi}\left\{1 + 0.363(2.5)4\right\} = -1.475P \\[2em] \sigma_B = \dfrac{(0.637PR)r}{\dfrac{\pi r^4}{4}} = 2.026P\end{array}\right\} \qquad \text{(k)}$$

If we adopt the curved beam formula as a basis for a comparison, then the magnitudes of the stresses at points A and B as calculated from the straight beam formula, are 25.3% and 30.1% smaller, respectively, than those calculated from the Winkler–Bach curved beam formula. Note that $R/c = 2.5$ for this problem.

The deflection along the diameter coinciding with the action line of the external force can be calculated via Castigliano's principle. Thus

$$\delta_{2P} = 4 \int_0^{\pi/2} \left\{ \frac{N}{AE}\frac{\partial N}{\partial(2P)} + \frac{M}{EI}\frac{\partial M}{\partial(2P)} + \frac{V}{AG}\frac{\partial V}{\partial(2P)} \right\} R\, d\theta \qquad \text{(l)}$$

or, if the energy due to direct shear is neglected,

$$\begin{aligned}
\delta_{2P} &= 4 \int_0^{\pi/2} \frac{(-P\cos\theta)}{AE}(-\tfrac{1}{2}\cos\theta)R\, d\theta \\[1em]
&\quad + 4 \int_0^{\pi/2} \frac{[C - PR(1-\cos\theta)]}{EI}[-\tfrac{1}{2}R(1-\cos\theta)]R\, d\theta \\[1em]
&= \frac{2PR}{AE}\int_0^{\pi/2} \cos^2\theta\, d\theta \\[1em]
&\quad - \frac{2R^2}{EI}\int_0^{\pi/2} [C - PR(1-\cos\theta)](1-\cos\theta)\, d\theta \qquad \text{(m)}
\end{aligned}$$

Integration with respect to θ and evaluation at the limits yields

$$\begin{aligned}
\delta_{2P} &= \frac{\pi}{2}\frac{PR}{AE} - \frac{2PR^3}{E\ell}\left\{0.363\left(\frac{\pi}{2}-1\right) - \left(\frac{3\pi}{4}-2\right)\right\} \\[1em]
&= 1.571\frac{PR}{AE} + 0.298\frac{PR^3}{EI}
\end{aligned} \qquad \text{(n)}$$

Substituting the given material and geometric parameters

$$\begin{aligned}
\delta_{2P} &= (0.417 + 1.973) \times 10^{-7}P \\
&= 2.392 \times 10^{-7}P \qquad \text{(without shear effects)} \qquad \text{(o)}
\end{aligned}$$

If shear effects are included, add to Eq. (a) the contribution due to the term

$$4 \int_0^{\pi/2} \frac{V}{AG} \frac{\partial V}{\partial(2P)} ds = 4 \int_0^{\pi/2} \frac{(-P \sin \theta)}{AG} (-\tfrac{1}{2} \sin \theta) R \, d\theta$$

$$= \frac{2PR}{AG} \int_0^{\pi/2} \sin^2 \theta \, d\theta = \frac{\pi PR}{2AG}$$

$$= 1.042 \times 10^{-7} P \qquad \text{(p)}$$

Consequently, the decrease in the diameter coincident with the action line of 2P is

$$\delta_{2P} = (2.392 + 1.042) \times 10^{-7} P$$

$$= 3.434 \times 10^{-7} P \qquad \text{(with shear effects)} \qquad \text{(q)}$$

Clearly, the deflection due to direct shear is a significant contribution in this example.

EXAMPLE 11-13

Rework Example 11-12 using the curved beam formula to compute the bending energy of the ring.

SOLUTION

The rotation of the cross section at point A is given by the formula

$$\theta = 4 \int_0^{\pi/2} \frac{N}{AE} \frac{\partial N}{\partial C} R \, d\theta + 4 \int_0^{\pi/2} \frac{V}{AG} \frac{\partial V}{\partial C} R \, d\theta$$

$$+ \frac{4}{EAR} \int_0^{\pi/2} \left\{ N \frac{\partial M}{\partial C} + \frac{\partial N}{\partial C} M \right\} R \, d\theta \qquad \text{(a)}$$

$$+ \frac{4\left(1 + \dfrac{1}{Z}\right)}{EAR^2} \int_0^{\pi/2} M \frac{\partial M}{\partial C} R \, d\theta$$

Now $\partial N/\partial C = \partial V/\partial C = 0$ and $\partial M/\partial C = 1$ so that Eq. (a) reduces to

$$\theta = \frac{4}{EAR} \int_0^{\pi/2} NR \, d\theta + \frac{4\left(1 + \dfrac{1}{Z}\right)}{EAR^2} \int_0^{\pi/2} MR \, d\theta \qquad \text{(b)}$$

Substituting from Eqs. (a) of Example 11-12, we obtain

$$\theta = \frac{4}{EAR} \int_0^{\pi/2} (-P \cos \theta) R \, d\theta$$

$$+ \frac{4\left(1 + \dfrac{1}{Z}\right)}{EAR^2} \int_0^{\pi/2} [C - PR(1 - \cos \theta)] R \, d\theta \qquad \text{(c)}$$

Integration with respect to θ and evaluating at the limits gives

$$\theta = -\frac{4PR}{EAR} + \frac{4\left(1 + \dfrac{1}{Z}\right)}{EAR}\left[\frac{\pi}{2}C - \left(\frac{\pi}{2} - 1\right)PR\right] \tag{d}$$

Since the section through point A does not rotate, Eq. (d) gives

$$C = 0.390PR \tag{e}$$

The normal stresses at points A and B that correspond to this couple are

$$\left.\begin{aligned}
\sigma_A &= -\frac{P}{A} + \frac{0.390PR}{AR}\left(1 + \frac{1}{Z}\frac{-1}{2.5 - 1}\right) = -2.094P \\[2mm]
\sigma_B &= \frac{(-0.610PR)}{AR}\left(1 + \frac{1}{Z}\frac{-1}{2.5 - 1}\right) = 2.775P
\end{aligned}\right\} \tag{f}$$

The percentage differences between these stresses and those computed using the straight beam bending energy formula are 5.8 and 4.4 based on the more accurate stresses given by Eq. (f).

The deflection along the diameter coincident with the action line of the external force $2P$ is given by the formula

$$\begin{aligned}
\delta_{2P} = {}& 4\int_0^{\pi/2} \frac{N}{AE}\frac{\partial N}{\partial(2P)}R\,d\theta + 4\int_0^{\pi/2}\frac{V}{AG}\frac{\partial V}{\partial(2P)}R\,d\theta \\[2mm]
&+ \frac{4}{EAR}\int_0^{\pi/2}\left(N\frac{\partial M}{\partial(2P)} + M\frac{\partial N}{\partial(2P)}\right)R\,d\theta \\[2mm]
&+ \frac{4\left(1 + \dfrac{1}{Z}\right)}{EAR^2}\int_0^{\pi/2}M\frac{\partial M}{\partial(2P)}R\,d\theta
\end{aligned} \tag{g}$$

Substituting for N, V, and M from Eqs. (a) of Example 11-12, and for $\partial N/\partial(2P)$, $\partial V/\partial(2P)$, and $\partial M/\partial(2P)$ by means of the same formulas, yields

$$\begin{aligned}
\delta_{2P} = {}& \frac{2PR}{AE}\int_0^{\pi/2}\cos^2\theta\,d\theta + \frac{2PR}{AG}\int_0^{\pi/2}\sin^2\theta\,d\theta \\[2mm]
&+ \frac{2}{AE}\int_0^{\pi/2}\{PR\cos\theta(1 - \cos\theta) - [C - PR(1 - \cos\theta)]\cos\theta\}\,d\theta \\[2mm]
&- \frac{2\left(1 + \dfrac{1}{Z}\right)}{AE}\int_0^{\pi/2}\{C - PR(1 - \cos\theta)](1 - \cos\theta)\}\,d\theta
\end{aligned} \tag{h}$$

Integration and evaluation at the limits yields the formula

$$\begin{aligned}
\delta_{2P} = {}& \frac{\pi}{2}\frac{PR}{AE} + \frac{\pi}{2}\frac{PR}{AG} + \frac{2}{AE}\left\{\left(1 - \frac{\pi}{4}\right)PR - \left[C - \left(1 - \frac{\pi}{4}\right)PR\right]\right\} \\[2mm]
&- \frac{2\left(1 + \dfrac{1}{Z}\right)}{AE}\left\{\left(\frac{\pi}{2} - 1\right)C - \left(\frac{\pi}{2} - 2 + \frac{\pi}{4}\right)PR\right\}
\end{aligned} \tag{i}$$

Substituting Eq. (e) into Eq. (i) along with the values $A = \pi$ in^2, $Z = 0.0436$, $R = 2.5$ in., $E = 30 \times 10^6$ psi, $G = E/[2(1 + v)]$, and $v = \frac{1}{4}$ yields

$$\delta_{2P} = 3.17 \times 10^{-7}P \text{ in.} \tag{j}$$

The deflection calculated on the basis of straight beam bending energy is 8.4% greater than the deflection computed using the Winkler–Bach curved beam bending energy.

EXAMPLE 11-14

For the machine part shown in Figure 11-17a, determine a formula for the relative displacement of the points of application of the applied forces (a) by approximating the bending energy in the curved portion by the straight beam formula and (b) by the more accurate Winkler–Bach formula including the coupling of M and N. Compute numerical values for each case if $\ell = 10$ in., $R = 3.75$ in., $P = 3000$ lb, $E = 30 \times 10^6$ psi, and $v = \frac{1}{4}$.

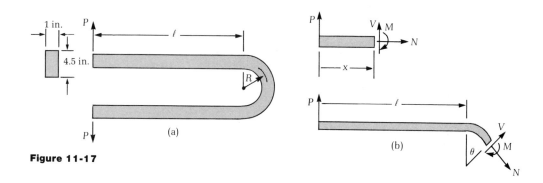

Figure 11-17

SOLUTION

The cross-sectional dimensions are shown in Figure 11-17a.
(a) From the free-body diagrams shown in Figure 11-17b,

$$\left.\begin{array}{l} N = 0 \\ V = -P \\ M = -Px \end{array}\right\} \quad 0 \le x \le \ell \tag{a}$$

and

$$\left.\begin{array}{l} N = P \sin \theta \\ V = -P \cos \theta \\ M = -P(\ell + R \sin \theta) \end{array}\right\} \quad 0 \le \theta \le \frac{\pi}{2} \tag{b}$$

The strain energy associated with the machine part is

$$U = 2 \int_0^\ell \left\{ \frac{M^2}{2EI} + \frac{N^2}{2AE} + \frac{V^2}{2AG} \right\} dx$$

$$+ 2 \int_0^{\pi/2} \left\{ \frac{M^2}{2EI} + \frac{N^2}{2AE} + \frac{V^2}{2AG} \right\} R \, d\theta$$ (c)

According to Castigliano's second theorem, the separation at P is

$$\frac{\partial U}{\partial P} = 2 \int_0^\ell \left\{ \frac{M}{EI} \frac{\partial M}{\partial P} + \frac{V}{AG} \frac{\partial V}{\partial P} \right\} dx$$

$$+ 2 \int_0^{\pi/2} \left\{ \frac{M}{EI} \frac{\partial M}{\partial P} + \frac{N}{AE} \frac{\partial N}{\partial P} + \frac{V}{AG} \frac{\partial V}{\partial P} \right\} R \, d\theta$$

$$= 2 \int_0^\ell \left(\frac{Px^2}{EI} + \frac{P}{AG} \right) dx$$

$$+ 2 \int_0^{\pi/2} \left\{ \frac{P(\ell + R \sin \theta)^2}{EI} + \frac{P \sin^2 \theta}{AE} + \frac{P \cos^2 \theta}{AG} \right\} R \, d\theta$$ (d)

Integration and evaluation at the limits yield the formula

$$\frac{\partial U}{\partial P} = 2 \left(\frac{P\ell^3}{3EI} + \frac{P\ell}{AG} \right) + 2 \frac{PR}{EI} \left(\frac{\pi\ell^2}{2} + 2\ell R + \frac{\pi R^2}{4} \right)$$

$$+ \frac{\pi}{2} \frac{PR}{AE} + \frac{\pi}{2} \frac{PR}{AG}$$ (e)

Now separate the terms in Eq. (e) into the contribution due to bending, shear, and axial deformations. Then

$$\frac{\partial U}{\partial P} = 2 \left\{ \frac{P\ell^3}{3EI} + \frac{\pi PR\ell^2}{2EI} + \frac{2PR^2\ell}{EI} + \frac{\pi PR^3}{4EI} \right\}$$

$$+ 2 \left\{ \frac{P\ell}{AG} + \frac{\pi PR}{4AG} \right\} + 2 \left\{ \frac{\pi PR}{4AE} \right\}$$ (f)

or

$$\frac{\partial U}{\partial P} = 2(0.00439 + \underline{0.007757 + 0.003704 + 0.000545})$$

$$+ 2(0.000555 + 0.000164)$$

$$+ 2(0.000065)$$

$$= 0.03279 + 0.00144 + 0.00013 = 0.03436 \text{ in.}$$ (g)

The bending, shearing, and axial effects constitute 95.4, 4.2, and 0.38%, respectively, of the total deflection of the points of application of the forces P.

(b) To obtain a formula for the deflection using the Winkler–Bach curved beam formula for bending energy, replace the underlined term in Eq. (e) with the term

$$Q = 2 \int_0^{\pi/2} \frac{M \dfrac{\partial N}{\partial P} + N \dfrac{\partial M}{\partial P}}{EAR} \, R \, d\theta$$

$$+ 2 \int_0^{\pi/2} \left(1 + \frac{1}{Z}\right) \frac{M}{EAR^2} \frac{\partial M}{\partial P} \, R \, d\theta$$

$$= 2 \int_0^{\pi/2} \left[\frac{-P(\ell + R \sin \theta) \sin \theta}{EAR} \right] R \, d\theta$$

$$+ \int_0^{\pi/2} \frac{-P \sin \theta (\ell + R \sin \theta)}{EAR} \, R \, d\theta$$

$$+ 2 \int_0^{\pi/2} \left(1 + \frac{1}{Z}\right) \frac{P(\ell + R \sin \theta)^2}{EAR^2} \, R \, d\theta \tag{h}$$

Integration and evaluation at the limits give

$$Q = -\frac{2P}{AE} \left(\ell + \frac{\pi R}{4} \right) - \frac{2P}{AE} \left(\ell + \frac{\pi R}{4} \right)$$

$$+ \frac{2P}{EAR} \left(1 + \frac{1}{Z}\right) \left(\frac{\pi \ell^2}{2} + 2R\ell + \frac{\pi R^2}{4} \right) \tag{i}$$

Substituting the given data into Eq. (i) gives

$$Q = -0.001151 + 0.021442 = 0.020291 \text{ in.} \tag{j}$$

Replacing the underlined term in Eq. (g) with the value given in Eq. (j) yields the deflection

$$\frac{\partial U}{\partial P} = 0.03064 \text{ in.} \tag{k}$$

The approximate formula predicts a deflection that is almost 11% greater than the deflection predicted by the more exact Winkler–Bach formula.

PROBLEMS / Section 11-9

11-14   A slender circular rod is formed into a semi-circular ring of radius R. The ring is clamped at one end as shown in Figure P11-14, and a concentrated force P perpendicular to the plane of the ring is applied at the other end. Derive a formula for the deflection of the free end of the ring in the direction of P. Also derive a formula for the rotation of the tangent line at the free end.

11-15   Determine the out-of-plane displacement of the free end of the semi-circular ring shown in Figure P11-15. Consider the radius of gyration of the cross section of the ring to be small compared to R.

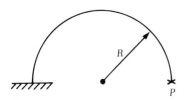

Figure P11-14

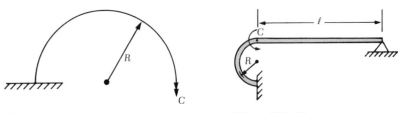

Figure P11-15   **Figure P11-16**

11-16 Use Castigliano's second theorem to determine the algebraic equations that determine the pin reactions of the member shown in Figure P11-16. Consider flexural energy only.

11-10 Application of Castigliano's Second Theorem to Trusses

By definition, the members of a truss are two-force members. Thus the strain energy associated with the ith member is

$$U_i = \frac{N_i^2 \ell_i}{2 A_i E_i} \tag{11-55}$$

The strain energy for the truss is the sum of the strain energy of its members. Consequently,

$$U = \Sigma \frac{N_i^2 \ell_i}{2 A_i E_i} \tag{11-56}$$

Now, according to Castigliano's principle, the deflection q_P of a joint at which a force P is applied, in the direction of the force P, is given by the relation

$$q_P = \frac{\partial U}{\partial P} = \Sigma \frac{N_i \ell_i}{A_i E_i} \frac{\partial N_i}{\partial P} \tag{11-57}$$

The rotation of a member is obtained by applying a couple whose forces act at the ends of the member and are perpendicular to the member. If P is the magnitude of the forces of the couple, and if ℓ is the length of the member whose rotation is sought, then the angular rotation of the member is

$$\theta = \frac{\partial U}{\partial C} = \frac{\partial U}{\partial (\ell P)} = \frac{1}{\ell} \Sigma \frac{N_i \ell_i}{A_i E_i} \frac{\partial N_i}{\partial P} \tag{11-58}$$

Only statically determinant trusses are considered in the following examples.

EXAMPLE 11-15

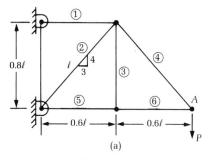

(a)

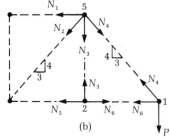

(b)

Figure 11-18

Determine the displacement of joint A of the truss of Figure 11-18a in the direction of the load P. The stress-strain relation for the material from which each member is made is $\sigma = K\epsilon^{1/3}$. Numbers near the middle of the members of the truss are member labels. The cross-sectional areas of the various members are $A_2 = A_4 = 2A_1$, $A_3 = 0.5A_1$, and $A_5 = A_6 = 1.5A_1$.

SOLUTION

The complementary energy density for the ith member of the truss is

$$\delta\gamma_{0i} = \epsilon_i\delta\sigma_i \tag{a}$$

Using the given stress-strain relation,

$$\gamma_{0i} = \frac{1}{4K^3}\left(\frac{N_i}{A_i}\right)^4 \tag{b}$$

so that the complementary energy for the ith member is

$$\gamma_i = \frac{A_i\ell_i}{4K^3}\left(\frac{N_i}{A_i}\right)^4 \tag{c}$$

Finally, the complementary energy for truss is

$$\gamma = \sum_{i=1}^{6}\gamma_i = \sum_{i=1}^{6}\frac{A_i\ell_i}{4K^3}\left(\frac{N_i}{A_i}\right)^4 \tag{d}$$

According to Castigliano's principle, the displacement of the point of application of the force P parallel to P is

$$q_P = \frac{\partial\gamma}{\partial P} = \sum_{i=1}^{6}\frac{\ell_i}{K^3}\left(\frac{N_i}{A_i}\right)^3\frac{\partial N_i}{\partial P} \tag{e}$$

The internal forces N_i are obtained from the free-body diagrams of the joints 1, 2, and 5 shown in Figure 11-18b. By the method of joints,

$$N_1 = \frac{3}{2}P$$

$$N_2 = -\frac{5}{4}P$$

$$N_3 = 0$$

$$N_4 = \frac{5}{4}P \tag{f}$$

$$N_5 = \frac{3}{4}P$$

$$N_6 = -\frac{3}{4}P$$

The partial derivatives appearing in Eq. (e) are calculated from Eqs. (f). Consequently,

$$\frac{\partial N_1}{\partial P} = \frac{3}{2}, \quad \frac{\partial N_2}{\partial P} = -\frac{5}{4}, \quad \frac{\partial N_3}{\partial P} = 0, \quad \frac{\partial N_4}{\partial P} = \frac{5}{4},$$

$$\frac{\partial N_5}{\partial P} = -\frac{3}{4}, \quad \text{and} \quad \frac{\partial N_6}{\partial P} = -\frac{3}{4} \tag{g}$$

Expanding Eq. (e) yields

$$q_P = \frac{1}{K^3} \left\{ \ell_1 \left(\frac{N_1}{A_1}\right)^3 \frac{\partial N_1}{\partial P} + \ell_2 \left(\frac{N_2}{A_2}\right)^3 \frac{\partial N_2}{\partial P} + \ell_3 \left(\frac{N_3}{A_3}\right)^3 \frac{\partial N_3}{\partial P} \right.$$

$$\left. + \ell_4 \left(\frac{N_4}{A_4}\right)^3 \frac{\partial N_4}{\partial P} + \ell_5 \left(\frac{N_5}{A_5}\right)^3 \frac{\partial N_5}{\partial P} + \ell_6 \left(\frac{N_6}{A_6}\right)^3 \frac{\partial N_6}{\partial P} \right\} \tag{h}$$

Now, $\ell_1 = \ell_5 = \ell_6 = 0.6\ell$, $\ell_2 = \ell_4 = \ell$, and $\ell_3 = 0.8\ell$. Also $A_2 = A_4 = 2A_1$, $A_3 = 0.5A_1$, and $A_5 = A_6 = 1.5A_1$. Therefore, Eq. (h) can be written as

$$q_P = \frac{\ell}{(KA_1)^3} \left\{ 0.6 N_1^3 \frac{\partial N_1}{\partial P} + \left(\frac{N_2}{2}\right)^3 \frac{\partial N_2}{\partial P} + 0.8 \left(\frac{N_3}{0.5}\right)^3 \frac{\partial N_3}{\partial P} \right.$$

$$\left. + \left(\frac{N_4}{2}\right)^3 \frac{\partial N_4}{\partial P} + 0.6 \left(\frac{N_5}{1.5}\right)^3 \frac{\partial N_5}{\partial P} + 0.6 \left(\frac{N_6}{1.5}\right)^3 \frac{\partial N_6}{\partial P} \right\} \tag{i}$$

Substituting for N_i and $\partial N_i / \partial P$ from eqs. (f) and (g) leads to the formula

$$q_P = \frac{P^3 \ell}{(K\Lambda_1)^3} \left\{ 0.6 \left(\frac{3}{2}\right)^3 \left(\frac{3}{2}\right) + 2 \left(-\frac{5}{8}\right)^3 \left(-\frac{5}{4}\right) \right.$$

$$\left. + 2(0.6) \left(-\frac{1}{2}\right)^3 \left(-\frac{3}{4}\right) \right\}$$

or

$$q_P = 3.76 \left(\frac{P}{KA_1}\right)^3 \ell \tag{j}$$

EXAMPLE 11-16

Determine the rotation of member 6 of the pin-connected truss of Example 11-15.

SOLUTION

To obtain the rotation of member 6, introduce equal and opposite forces Q at its ends as shown in Figure 11-19a. Then, according to Castigliano's principle, the angular displacement of this member is

$$\theta_6 = \frac{\partial \gamma}{\partial C} \tag{a}$$

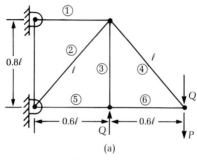

(a)

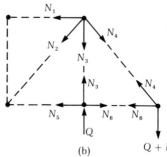

(b)

Figure 11-19

where $C = \ell_6 Q$. The complementary energy for the truss is

$$\gamma = \frac{1}{4} \sum_{i=1}^{6} \frac{A_i \ell_i}{K^3} \left(\frac{N_i}{A_i} \right)^4 \tag{b}$$

so that

$$\theta_6 = \sum_{i=1}^{6} \frac{\ell_i}{K^3} \left(\frac{N_i}{A_i} \right)^3 \frac{\partial N_i}{\partial C} \tag{c}$$

From the free-body diagrams of Figure 11-19b, by the method of joints,

$$\left. \begin{aligned}
N_1 &= \frac{3}{2} P + \frac{3}{4} Q \\[1em]
N_2 &= -\frac{5}{4} P \\[1em]
N_3 &= -Q \\[1em]
N_4 &= \frac{5}{4}(Q + P) \\[1em]
N_5 &= -\frac{3}{4}(Q + P) \\[1em]
N_6 &= -\frac{3}{4}(Q + P)
\end{aligned} \right\} \tag{d}$$

The derivatives required in Eq. (c) are

$$\frac{\partial N_1}{\partial C} = \frac{1}{\ell_6} \left(\frac{3}{4} \right), \quad \frac{\partial N_2}{\partial C} = 0, \quad \frac{\partial N_3}{\partial C} = \frac{1}{\ell_6}(-1), \quad \frac{\partial N_4}{\partial C} = \frac{1}{\ell_6} \left(\frac{5}{4} \right),$$

$$\frac{\partial N_5}{\partial C} = \frac{1}{\ell_6} \left(-\frac{3}{4} \right), \quad \frac{\partial N_6}{\partial C} = \frac{1}{\ell_6} \left(-\frac{3}{4} \right) \tag{e}$$

or, since $\ell_6 = \frac{3}{5}\ell$,

$$\left. \begin{aligned}
\ell \frac{\partial N_1}{\partial C} &= \frac{5}{4} \\[1em]
\ell \frac{\partial N_2}{\partial C} &= 0 \\[1em]
\ell \frac{\partial N_3}{\partial C} &= -\frac{5}{3} \\[1em]
\ell \frac{\partial N_4}{\partial C} &= \frac{25}{12} \\[1em]
\ell \frac{\partial N_5}{\partial C} &= -\frac{5}{4} \\[1em]
\ell \frac{\partial N_6}{\partial C} &= -\frac{5}{4}
\end{aligned} \right\} \tag{f}$$

Expansion of Eq. (c) yields

$$
\theta_6 = \frac{\ell}{(KA_1)^3} \left\{ 0.6N_1^3 \frac{\partial N_1}{\partial C} + \left(\frac{N_2}{2}\right)^3 \frac{\partial N_2}{\partial C} + 0.8\left(\frac{N_3}{0.5}\right)^3 \frac{\partial N_3}{\partial C} \right.
$$
$$
\left. + \left(\frac{N_4}{2}\right)^3 \frac{\partial N_4}{\partial C} + 0.6\left(\frac{N_5}{1.5}\right)^3 \frac{\partial N_5}{\partial C} + 0.6\left(\frac{N_6}{1.5}\right)^3 \frac{\partial N_6}{\partial C} \right\}
$$

(g)

Setting $Q = 0$ in Eqs. (d) and subsequently replacing the N_i and $\partial N_i/\partial C$ with their equivalents from Eqs. (d) and (e) yields

$$
\theta_6 = \left(\frac{P}{KA_1}\right)^3 \left\{ 0.6\left(\frac{3}{2}\right)^3 \left(\frac{5}{4}\right) + \left(\frac{5}{8}\right)^3 \left(\frac{25}{12}\right) \right.
$$
$$
\left. + 2(0.6)\left(-\frac{1}{2}\right)^3 \left(-\frac{5}{4}\right) \right\}
$$

(h)

or

$$
\theta_6 = 3.23 \left(\frac{P}{KA_1}\right)^3
$$

(i)

PROBLEMS / Section 11-10

11-17 Using Castigliano's principle for linearly elastic materials, establish formulas for the displacement components of point A of the two-link mechanism shown in Figure P11-17. Assume that the extensional rigidity EA is the same for each member.

11-18 The extensional stiffnesses for each member of the statically determinate truss shown in Figure P11-18 are identical to one another. Determine the horizontal and vertical components of displacement of joint A using Castigliano's principle for linearly elastic behavior.

11-19 The extensional rigidities of the members of the pin-connected, statically determine truss shown in Figure P11-19 are the same. Make use of symmetry and Castigliano's principle for linearly elastic materials to derive formulas for the displacement of point A.

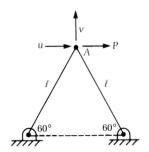

Figure P11-17

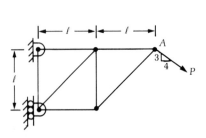

Figure P11-18

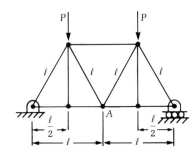

Figure P11-19

11-11 Unit Load Method

The unit load method common to structure analysis is actually Castigliano's second theorem in disguise. To confirm this assertion, recall that the internal reactions (M_x, M_y, M_z, V_x, V_y, N) at any section of a deformable body are related linearly to the forces and couples applied to the system. This observation is expressed analytically as

$$Q_i = \sum_{j=1}^{N} a_{ij}P_j \qquad i = 1, 2, \ldots, 6 \tag{11-59}$$

where Q_i denotes the ith internal reaction at a section, P_j is the jth applied force or couple, a_{ij} is independent of the forces or couples, and N is the number of applied forces and couples.

The partial derivatives required for the application of Castigliano's second theorem are given by

$$\frac{\partial Q_i}{\partial P_k} = \left(a_{i1} \frac{\partial P_1}{\partial P_k} + a_{i2} \frac{\partial P_2}{\partial P_k} + \cdots + a_{ik} \frac{\partial P_k}{\partial P_k} \right.$$
$$\left. + \cdots + a_{1N} \frac{\partial P_N}{\partial P_k} \right) \tag{11-60}$$

Now, $P_1, P_2, \ldots, P_N$ are independent so that all the derivatives appearing in Eq. (11-60) are zero except $\partial P_k/\partial P_k$ which is equal to unity. Thus

$$\frac{\partial Q_i}{\partial P_k} = a_{ik} \tag{11-61}$$

Notice that a_{ik} is the value that Q_i of Eq. (11-59) acquires when $P_k = 1$ and all other $P_i = 0$. Now a_{ik} is nothing more that the ith internal reaction at the section due to P_k when its value is unity. If there is no force (couple) at the point whose displacement (rotation) is sought, a unit force (couple) is applied there. a_{ik} is the internal reaction due to the fictitious unit load. Of course this force (couple) must be set equal to zero in Eq. (11-59) *after* a_{ik} has been determined.

The following examples will serve to illustrate the unit load method.

Straight beams. The unit load method for straight beams has the form

$$q_P = \int_0^t \frac{M}{EI} m\, dx + \int_0^t \frac{V}{AG} v\, dx + \int_0^t \frac{N}{AE} n\, dx \tag{11-62}$$

where m, v, and n are the internal moment, internal shear force, and internal normal force at a section of the beam due to a unit force or a unit couple applied at a point on the centroidal line of the beam whose displacement or rotation is required.

Observe that $(M/EI) dx = d\theta$, $(V/AG) dx = dv$, and $(N/AE) dx = du$; which are the rotation of a normal section of the beam, the increment of transverse displacement due to shear, and the increment of displacement parallel to the normal force, respectively.

Curved beams. The unit load method for curved beams is precisely the same as for straight beams, except that it is usually necessary to include twisting energy and, for sharply curved beams, the Winkler–Bach bending energy formula. If bending energy can be represented by the straight beam formula

$$q_P = \int_0^t \frac{M}{EI} m\, ds + \int_0^t \frac{T}{JG} t\, ds + \int_0^t \frac{N}{AE} n\, ds + \int_0^t \frac{V}{AG} v\, ds \qquad (11\text{-}63)$$

where m, n, and v have the same meanings as for straight beams, and t is the internal twisting moment at a section due to a unit force or a unit couple applied at a point on the centroidal line of the beam whose displacement or rotation is sought.

Trusses. For planar or spatial trusses, the displacement of the point of application of the force P_k in the direction of P_k, or the rotation about the action of the couple P_k, is given by Castigliano's formula

$$q_{P_k} = \frac{\partial \gamma}{\partial N_1} \frac{\partial N_1}{\partial P_k} + \frac{\partial \gamma}{\partial N_2} \frac{\partial N_2}{\partial P_k} + \cdots + \frac{\partial \gamma}{\partial N_N} \frac{\partial N_N}{\partial P_k} \qquad (11\text{-}64)$$

where γ is the complementary energy of the truss, N_i is the axial force in the ith member, and P_k is the externally applied force or couple at whose location the deflection or rotation is sought. Note that $\partial \gamma / \partial N_i = e_i$, the elongation of the ith member due to the actual forces that act on the truss. Also note that $\partial N_i / \partial P_k = n_i$ is the tension in the ith member due either to a unit load applied at the point and in the direction of the displacement that is sought, or to a unit couple whose forces are applied to the ends of the member whose angular rotation is sought. Consequently,

$$q_P = \sum_{i=1}^{N} e_i n_i \qquad (11\text{-}65)$$

The following examples illustrate this method.

EXAMPLE 11-17

Use the unit load method to determine the reaction R at the roller for the plane frame shown in Figure 11-20a. Assume that the material of the frame is linearly elastic.

SOLUTION

The internal reactions in the two legs of the frame due to the actual load C are obtained from the free-body diagrams shown in Figures 11-20b

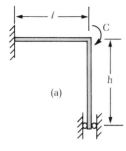

(a)

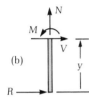

(b)

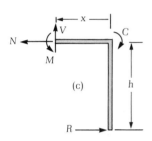

(c)

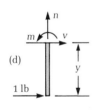

(d)

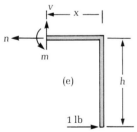

(e)

Figure 11-20

and 11-20c. Consequently,

$$\left.\begin{array}{l} V = -R \\ N = 0 \\ M = -Ry \end{array}\right\} \quad 0 \le y \le h \tag{a}$$

and

$$\left.\begin{array}{l} V = 0 \\ N = R \\ M = -Rh + C \end{array}\right\} \quad 0 \le x \le \ell \tag{b}$$

In a similar manner, the internal reactions in the two legs of the frame due to a unit force at the roller are obtained from the free-body diagrams shown in Figures 11-20d and 11-20e.

$$\left.\begin{array}{l} v = -1 \\ n = 0 \\ m = -y \end{array}\right\} \quad 0 \le y \le h \tag{c}$$

and

$$\left.\begin{array}{l} v = 0 \\ n = 1 \\ m = -h \end{array}\right\} \quad 0 \le x \le \ell \tag{d}$$

If the roller were replaced by a force R, the deflection at the lower end of the frame would be

$$\delta = \int_0^h \left(\frac{M}{EI} m + \frac{N}{AE} n + \frac{V}{AG} v\right) dy \\ + \int_0^\ell \left(\frac{M}{EI} m + \frac{N}{AE} n + \frac{V}{AG} v\right) dx \tag{e}$$

Substituting from Eqs. (a) through (d) in Eq. (e) yields

$$\delta = \int_0^h \left(\frac{Ry^2}{EI} + \frac{R}{AG}\right) dy + \int_0^\ell \left(\frac{Rh^2 - Ch}{EI} + \frac{R}{AE}\right) dx$$

or, upon integration and evaluation at the limits,

$$\delta = \frac{Rh^3}{3EI} + \frac{Rh}{AG} + \frac{Rh^2\ell - Ch\ell}{EI} + \frac{R\ell}{AE} \tag{f}$$

The roller prevents this deflection from occurring so that Eq. (f) yields

$$R = \frac{3C\ell}{h^2\left\{1 + 6(1 + v)\left(\dfrac{r}{h}\right)^2 + 3\left(\dfrac{\ell}{h}\right) + 3\left(\dfrac{\ell}{h}\right)\left(\dfrac{r}{h}\right)^2\right\}} \tag{g}$$

where v is Poisson's ratio for the material of the truss and r is the radius of gyration of the cross section of its members.

EXAMPLE 11-18

Determine (a) the vertical deflection and (b) the rotation of the free end of the semi-circular ring shown in Figure 11-21a. Use the unit load method and regard R as large in comparison with the dimensions of the cross section so that bending energy can be approximated by the straight beam formula for bending energy. Assume the material behavior is linearly elastic.

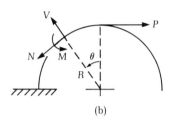

(a)

SOLUTION

(a) The complementary energy for a linearly elastic system is equal to its strain energy. Therefore, the vertical deflection of the free end of the ring is given by the formula

$$q = \int_0^{\pi/2} \frac{M}{EI} mR \, d\theta + \int_0^{\pi/2} \frac{N}{AE} nR \, d\theta + \int_0^{\pi/2} \frac{V}{AG} vR \, d\theta \qquad (a)$$

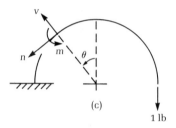

(b)

From the free-body diagram shown in Figure 11-21b, obtain, from equilibrium considerations,

$$\left.\begin{array}{l} V(\theta) = P \sin \theta \\ N(\theta) = P \cos \theta \\ M(\theta) = PR(1 - \cos \theta) \end{array}\right\} \quad 0 \le \theta \le \frac{\pi}{2} \qquad (b)$$

Note that the internal reactions V, N, and M are zero for the right-hand portion of the ring.

The internal reactions due to a vertical unit load at the free end of the ring are obtained from Figure 11-21c. Thus force and moment equilibrium require that

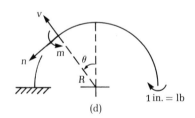

(c)

1 lb

$$\left.\begin{array}{l} v(\theta) = \cos \theta \\ n(\theta) = -\sin \theta \\ m(\theta) = R(1 + \sin \theta) \end{array}\right\} \qquad (c)$$

Substituting Eqs. (b) and (c) into Eq. (a) gives

$$q = \int_0^{\pi/2} \frac{PR^3}{EI} (1 - \cos \theta)(1 + \sin \theta) \, d\theta$$

$$+ \int_0^{\pi/2} \left(\frac{-PR}{AE}\right) \sin \theta \cos \theta \, d\theta + \int_0^{\pi/2} \frac{PR}{AG} \sin \theta \cos \theta \, d\theta \qquad (d)$$

(d)   **Figure 11-21**

Integration and evaluation of the integrals in Eq. (d) yields

$$q = (\pi - 1) \frac{PR^3}{2EI} - \frac{PR}{2AE} + \frac{PR}{2AG} \qquad (e)$$

(b) The rotation at the free end of the ring is given by the formula

$$\theta = \int_0^{\pi/2} \frac{M}{EI} mR \, d\theta + \int_0^{\pi/2} \frac{N}{AE} nR \, d\theta + \int_0^{\pi/2} \frac{V}{AG} vR \, d\theta \qquad (f)$$

where m, n, and v are internal reactions due to a couple of unit magnitude applied at the free end.

From the free-body diagram shown in Figure 11-21d, obtain, by equilibrium considerations,

$$\left.\begin{aligned} v(\theta) &= 0 \\ n(\theta) &= 0 \\ m(\theta) &= 1 \end{aligned}\right\} \tag{g}$$

Substituting from Eqs. (b) and (g) in Eq. (f) yields the formula

$$\theta = \int_0^{\pi/2} \frac{PR^2}{EI}(1 - \cos\theta)\,d\theta = (\pi - 2)\frac{PR^2}{2EI} \tag{h}$$

EXAMPLE 11-19

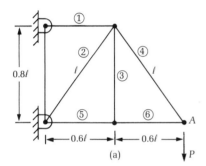

(a)

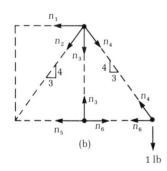

(b)

1 lb

Figure 11-22

Rework Example 11-15 using the unit load method. The structural configuration is reproduced in Figure 11-22a for convenient reference.

SOLUTION

According to the unit load method, the deflection of joint A in the direction of P is given by the formula

$$q_P = \Sigma e_i n_i \tag{a}$$

where e_i is elongation of the ith member of the truss due to the *actual* loads, and n_i is the force in the ith member due to a virtual force of unit magnitude applied at A in the direction of the deflection sought.

The member forces due to the actual loads were determined in Example 11-15 from the free-body diagrams shown in Figure 11-18b. They are

$$\left.\begin{aligned} N_1 &= \frac{3}{2}P \\[4pt] N_2 &= -\frac{5}{4}P \\[4pt] N_3 &= 0 \\[4pt] N_4 &= \frac{5}{4}P \\[4pt] N_5 &= -\frac{3}{4}P \\[4pt] N_6 &= -\frac{3}{4}P \end{aligned}\right\} \tag{b}$$

The elongation of the ith member is given by the formula

$$e_i = \epsilon_i \ell_i = \left(\frac{N_i}{A_i K_i}\right)^3 \ell_i \tag{c}$$

Thus the elongations of the various members of the truss are

$$e_1 = 0.6 \left(\frac{3}{2}\right)^3 \left(\frac{P}{A_1 K}\right)^3 \ell$$

$$e_2 = \left(-\frac{5}{8}\right)^3 \left(\frac{P}{A_1 K}\right)^3 \ell$$

$$e_3 = 0$$

$$e_4 = \left(\frac{5}{8}\right)^3 \left(\frac{P}{A_1 K}\right)^3 \ell \qquad\qquad (d)$$

$$e_5 = 0.6 \left(-\frac{1}{2}\right)^3 \left(\frac{P}{A_1 K}\right)^3 \ell$$

$$e_6 = 0.6 \left(-\frac{1}{2}\right)^3 \left(\frac{P}{A_1 K}\right)^3 \ell$$

The forces n_i due to the unit load at point A are determined from the free-body diagrams shown in Figure 11-22b. Accordingly, by the method of joints,

$$n_1 = \frac{3}{2}, \quad n_2 = -\frac{5}{4}, \quad n_3 = 0, \quad n_4 = \frac{5}{4}, \quad n_5 = n_6 = -\frac{3}{4} \qquad (e)$$

Direct substitution of Eqs. (d) and (e) into Eq. (a) yields

$$q_P = \left(\frac{P}{A_1 K}\right)^3 \ell \left\{ 0.6 \left(\frac{3}{2}\right)^3 \left(\frac{3}{2}\right) + \left(-\frac{5}{8}\right)^3 \left(-\frac{5}{4}\right) + \left(\frac{5}{8}\right)^3 \left(\frac{5}{4}\right) \right.$$
$$\left. + 0.6 \left(-\frac{1}{2}\right)^3 \left(-\frac{3}{4}\right) + 0.6 \left(-\frac{1}{2}\right)^3 \left(-\frac{3}{4}\right) \right\} \qquad (f)$$

or

$$q_P = 3.76 \left(\frac{P}{A_1 K}\right)^3 \ell \qquad\qquad (g)$$

Equation (g) is precisely the same result obtained by the direct application of Castigliano's principle in Example 11-15.

EXAMPLE 11-20

Use the unit load method to determine the rotation of member 6 of the truss of Example 11-15.

SOLUTION

According to the unit load method, the rotation of member 6 is given by the formula

$$\theta_6 = \sum_{i=1}^{6} e_i n_i \qquad\qquad (a)$$

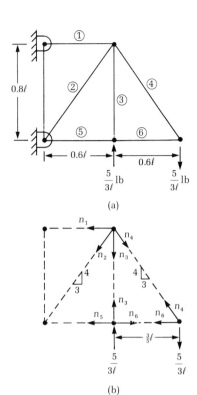

(a)

(b)

Figure 11-23

where e_i is the elongation of the ith member, and n_i is the force in the ith member due to a unit couple applied as shown in Figure 11-23a. The elongations due to the actual loads were calculated in Example 11-19 as Eqs. (d). They are repeated here for easy reference.

$$\left.\begin{array}{l} e_1 = 0.6 \left(\dfrac{3}{2}\right)^3 \left(\dfrac{P}{A_1 K}\right)^3 \ell \\[2mm] e_2 = \left(-\dfrac{5}{8}\right)^3 \left(\dfrac{P}{A_1 K}\right)^3 \ell \\[2mm] e_3 = 0 \\[2mm] e_4 = \left(\dfrac{5}{8}\right)^3 \left(\dfrac{P}{A_1 K}\right)^3 \ell \\[2mm] e_5 = 0.6 \left(-\dfrac{1}{2}\right)^3 \left(\dfrac{P}{A_1 K}\right)^3 \ell \\[2mm] e_6 = 0.6 \left(-\dfrac{1}{2}\right)^3 \left(\dfrac{P}{A_1 K}\right) \ell \end{array}\right\} \quad (b)$$

The forces n_i due to the unit couple are found from the free-body diagrams of Figure 11-23b. They are

$$\left.\begin{array}{l} n_1 = \dfrac{5}{4\ell} \\[2mm] n_2 = 0 \\[2mm] n_3 = -\dfrac{5}{3\ell} \\[2mm] n_4 = \dfrac{25}{12\ell} \\[2mm] n_5 = -\dfrac{5}{4\ell} \\[2mm] n_6 = -\dfrac{5}{4\ell} \end{array}\right\} \quad (c)$$

Equations (a), (b), and (c) yield

$$\theta_6 = \left(\frac{P}{A_1 K}\right)^3 \left\{ 0.6 \left(\frac{3}{2}\right)^3 \left(\frac{5}{4}\right) + \left(\frac{5}{8}\right)^3 \left(\frac{25}{12}\right) + 2(0.6) \left(-\frac{1}{2}\right)^3 \left(-\frac{5}{4}\right) \right\} \quad (d)$$

or

$$\theta_6 = 3.23 \left(\frac{P}{A_1 K}\right)^3 \quad (e)$$

This is obviously the same result that was obtained in Example 11-16 using the direct application of Castigliano's principle.

PROBLEMS / Section 11-11

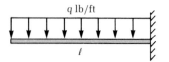

Figure P11-20

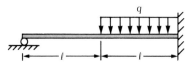

Figure P11-21

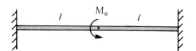

Figure P11-22

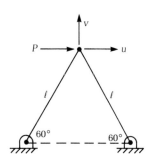

Figure P11-23

11-20   Use the unit load method to determine formulas for the slope and deflection of the free end of the cantilever beam shown in Figure P11-20. Consider bending to be the dominant action.

11-21   Use the unit load method to determine the roller reaction for the beam shown in Figure P11-21. Bending is the dominant action.

11-22   Use the unit load method to determine the reactions of the left support on the elastic beam shown in Figure P11-22. Neglect energy due to transverse shear.

11-23   Use the unit load method to establish formulas for the displacement components of point A of the two-link mechanism shown in Figure P11-23. Assume that the material of the members behaves elastically and that the extensional rigidity EA is the same for each member.

11-24   The extensional stiffness for each member of the statically determinate truss shown in Figure P11-24 is EA. Determine the horizontal and vertical displacement components of point A using the unit load method. Assume linearly elastic material behavior.

11-25   A slender semi-circular ring of radius R supports a load P as shown in Figure P11-25. Use the unit load method to determine the roller reaction. Assume linearly elastic behavior.

11-26   Use the unit load method to determine the horizontal deflection of point A of the slender elastic semi-circular ring shown in Figure P11-26. Determine a formula for the rotation of the tangent line at A.

11-27   A slender quarter ring is acted on by a force P perpendicular to its end as shown in Figure P11-27. Using the unit load method, determine the deflection of point A parallel to the action line of P and the angle of twist at point A.

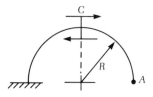

Figure P11-26

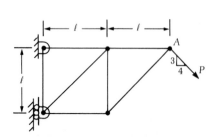

Figure P11-24

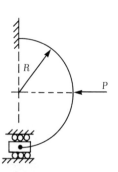

Figure P11-25

Figure P11-27

11-12 SUMMARY

In this chapter, we derived the principle of stationary potential energy using the principle of virtual work as a starting point. We obtained Castigliano's first and second theorems directly from the principle of stationary potential energy. We also showed that the unit load method is equivalent to a direct application of Castigliano's second theorem.

Principle of Stationary Potential Energy

$$V = U + \Omega \quad \text{and} \quad \frac{\partial V}{\partial q_i} = 0, \; i = 1, 2, \ldots, N \quad \text{at an equilibrium configuration}$$

Castigliano's Second Theorem

$$q_P = \frac{\partial \gamma}{\partial P}$$

γ = complementary energy of the system

q_P = displacement in the direction of P

$$\theta_C = \frac{\partial \gamma}{\partial C}$$

θ_C = rotation in the plane of the couple C

$\gamma = U$ for linear material behavior

Trusses:

$$q_P = \sum_{i=1}^{N} \frac{\partial \gamma_i}{\partial P} \quad \text{and} \quad \theta_C = \sum_{i=1}^{N} \frac{\partial \gamma_i}{\partial C}$$

Straight Beams:

$$q_P = \int_0^l \frac{M}{EI} \frac{\partial M}{\partial P} \, dx + \int_0^l \frac{V}{AG} \frac{\partial V}{\partial P} \, dx$$

$$\theta_C = \int_0^l \frac{M}{EI} \frac{\partial M}{\partial C} \, dx + \int_0^l \frac{V}{AG} \frac{\partial V}{\partial C} \, dx$$

$$q_P = \int_0^l \frac{M}{EI} \frac{\partial M}{\partial P} \, ds + \int_0^l \frac{N}{AE} \frac{\partial N}{\partial P} \, ds + \int_0^l \frac{V}{AG} \frac{\partial V}{\partial P} \, ds + \int_0^l \frac{T}{JG} \frac{\partial T}{\partial P} \, ds$$

$$\theta_C = \int_0^l \frac{M}{EI} \frac{\partial M}{\partial C} \, ds + \int_0^l \frac{N}{AE} \frac{\partial N}{\partial C} \, ds + \int_0^l \frac{V}{AG} \frac{\partial V}{\partial C} \, ds + \int_0^l \frac{T}{JG} \frac{\partial T}{\partial C} \, ds$$

Unit Load Method:

Trusses:

$$q_P,\ \theta_C = \Sigma e_i n_i \qquad e_i = \text{elongation of the } i\text{th member} \\ \text{due to the actual loads} \\ n_i = \text{tension in the } i\text{th member due to a} \\ \text{unit force or a unit couple}$$

Straight Beams:

$$q_P,\ \theta_C = \int_0^l \frac{M}{EI}\, m\, dx + \int_0^l \frac{V}{AG}\, v\, dx$$

Curved Beams:

$$q_P,\ \theta_C = \int_0^l \frac{M}{EI}\, m\, ds + \int_0^l \frac{N}{AE}\, n\, ds + \int_0^l \frac{V}{AG}\, v\, ds + \int_0^l \frac{T}{JG}\, t\, ds$$

Here m, n, v, and t are the internal bending moment, normal force, shear force, and twisting moment due to a unit force or a unit couple.

PROBLEMS / CHAPTER 11

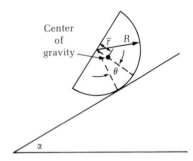

Figure P11-28

11-28 A homogeneous, semi-cylindrical drum rests on an inclined plane as shown in Figure P11-28. The drum may roll without slipping. Using the method of total potential energy, determine a formula for its equilibrium configuration θ.

11-29 The potential energy of a certain mechanical system is given as $V(x, y) = \frac{1}{2}(x - a)^2 + \frac{1}{2}(y - b)^2 + 2xy$ where a and b are constants. Locate the equilibrium position for the system using the method of total potential energy.

11-30 The extensional stiffness EA is the same for each member of the assembly of pin-connected rods shown in Figure P11-30. Using the principle of total potential energy, determine the force in each rod. Assume linearly elastic material behavior.

11-31 Consider the rotations of the various members of the plane truss shown in Figure P11-31 to be negligible. The extensional rigidity EA is the same for each member. Using the principle of stationary potential energy, determine the displacements of joints A, B, and C, and, subsequently, determine formulas for the force in each member.

11-32 The flexural rigidity of each segment of the engineering member shown in Figure P11-32 is EI. If bending is the dominant deformational response, determine the component of displacement of point A in the direction of P by

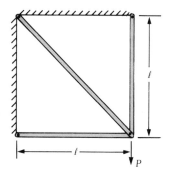

Figure P11-30

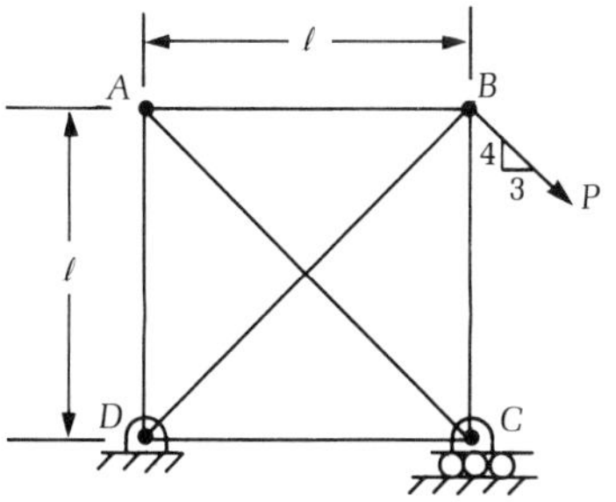

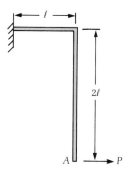

Figure P11-31

Figure P11-32

Castigliano's second theorem. Determine a formula for the rotation at point A of the centroidal line of the vertical segment of the member.

11-33   Use Castigliano's second theorem to determine the roller reaction for the slender elastic frame shown in Figure P11-33.

11-34   Use Castigliano's second theorem to determine the deflection and rotation of the free end of the cantilever shown in Figure P11-34. Assume linearly elastic material behavior.

11-35   Determine the vertical displacement of, and the rotation at, point A of the slender elastic frame shown in Figure P11-35. Use Castigliano's second theorem.

11-36   A slender semi-circular ring of radius R supports a load P as shown in Figure P11-36. Determine a formula for the roller reaction using Castigliano's principle for linearly elastic material.

11-37   The slender semi-circular ring shown in Figure P11-37 is subjected to a concentrated force P whose action line is perpendicular to the plane of the ring. Determine a formula for the bending moment at the midpoint of the ring. Use Castigliano's second theorem.

11-38   Using Castigliano's principle for linearly elastic material, determine the deflection and rotation of the free end of the cantilever spring shown in Figure P11-38 due to flexure only.

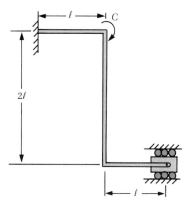

Figure P11-33

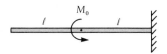

Figure P11-34

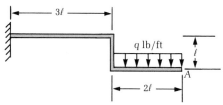

Figure P11-35

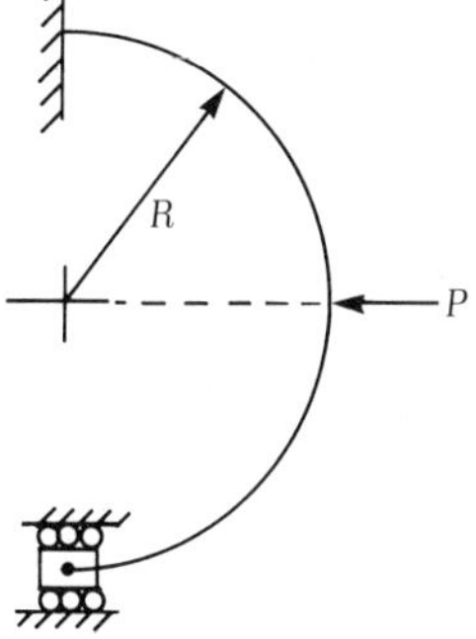

Figure P11-36

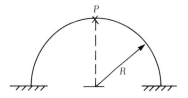

Figure P11-37

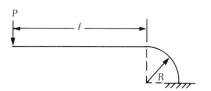

Figure P11-38

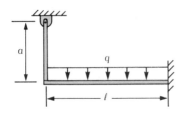

Figure P11-39

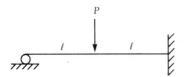

Figure P11-40

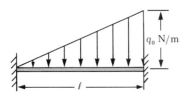

Figure P11-41

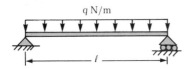

Figure P11-42

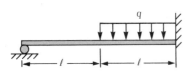

Figure P11-43

11-39 The uniformly loaded cantilever beam shown in Figure P11-39 is supported by an elastic rod of length a and cross-sectional area A at its end. The flexural rigidity EI for the beam is constant, and all material behavior is linear. Use Castigliano's principle to show that the force in the rod is

$$N = \frac{\frac{3}{8}q\ell}{1 + \dfrac{3E_B Ia}{E_R A\ell^3}}$$

11-40 Use Castigliano's principle to determine the reaction at the roller for the linearly elastic beam shown in Figure P11-40.

11-41 Use Castigliano's principle to determine the reactions at the left support for the beam shown in Figure P11-41. Draw shear and moment diagrams.

11-42 Use the unit load method to determine the deflection at the center of the simply supported beam shown in Figure P11-42 and the slope at its left end. Consider bending to be the dominant action.

11-43 Use the unit load method to determine the roller reaction for the beam shown in Figure P11-43. Consider bending to be the dominant action.

11-44 Use the unit load method to determine the forces in the various members of the linearly elastic truss shown in Figure P11-44. The extensional stiffnesses of the members are identical.

11-45 A slender quarter ring is clamped at one end and is free at the other end as shown in Figure P11-45. A couple C is applied so that the vector representing it is perpendicular to the plane of the ring. Use the unit load method to establish formulas for the horizontal and vertical components of the displacement of point A and the rotation of the tangent line at A.

11-46 Use the unit load method to determine the displacement of point A of the slender semi-circular ring shown in Figure P11-46. Determine the rotation of the tangent line at point A.

11-47 Use the unit load method to determine the displacement components of the free end of the slender quarter ring shown in Figure P11-47.

11-48 A slender rod is formed into the shape shown in Figure P11-48. If the material is linearly elastic, determine the roller reaction using the unit load method.

11-49 Derive formulas for the vertical deflection of joint A of the pin-connected assembly of rods shown in Figure P11-49 by (a) the principle of stationary

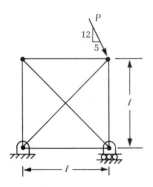

Figure P11-44

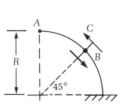

Figure P11-45

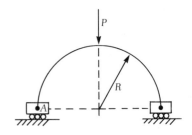

Figure P11-46

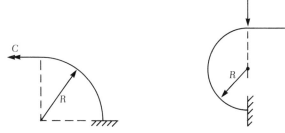

Figure P11-47   **Figure P11-48**

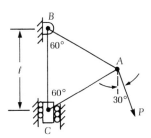

Figure P11-49

potential energy, (b) the principle of complementary energy, and (c) the unit load method. The extensional stiffnesses EA of the members are identical.

11-50 The simple truss, which is commonly known as the **Mises truss**, shown in Figure 11-50a, is made of a linearly elastic material. The deflection of the point of application of the vertical force P is denoted by f. The deformed configuration of the truss under the action of P is depicted by the heavy dashed lines in Figure P11-50a.

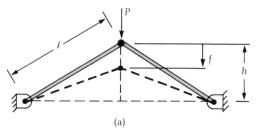

(a)   (b)

Figure P11-50

Show that the shortening of each member of the truss is

$$e = \ell\left\{1 - \sqrt{1 - 2\varphi\lambda + \varphi^2}\right\}\ell \qquad (a)$$

where $\varphi = f/\ell$ is the nondimensional deflection and $\lambda = h/\ell$ is the so-called shallowness ratio.

Using principle of stationary potential energy, show that the equilibrium configurations of the deformed truss are described by the equations

$$\psi = (\lambda - \varphi)\left\{(1 - 2\lambda\varphi + \varphi^2)^{-1/2} - 1\right\} \qquad (b)$$

where $\psi = P/2AE$ is the nondimensional load. For $\lambda = 0.1, 0.2$, and so on, plot the ψ versus φ curves similar to that given in Figure P11-50b. For one of the curves, discuss the behavior of the truss for increasing ψ starting from $\psi = 0$. Explain what kind of a phenomenon is taking place at point A, which is a so-called **limit point**. Also discuss the configurations of the truss at points B, C, and D.

CHAPTER 12

Special Topics

12-1 Introduction

This chapter contains information on procedures used to calculate: normal and shearing stresses for the bending of straight beams with unsymmetric cross sections, normal stresses for the bending of a curved beam in its plane, normal stresses in thick-walled cylinders under axisymmetric pressure, and normal stresses in general rotationally symmetric pressure vessels.

12-2 Bending of Beams with Unsymmetric Cross Sections

In the development of elementary straight beam theory, it is usually assumed that each cross section along the beam axis possesses an axis of symmetry. For convenience, this axis is assumed to be vertical. The aggregate of these axes of symmetry forms a plane. Elementary beam theory further requires this **plane of symmetry** to contain the applied loads. These requirements ensure that the loads will pass through the **axis of twist** of the beam, thereby allowing the beam to bend without twisting. Recall that the axis of twist for a straight beam is the axis parallel to the centroidal axis about which the cross sections tend to rotate under torsional loads. When the plane of the loads contains this axis, the beam bends without twisting. These requirements furthermore ensure that bending occurs about a **principal axis of inertia** of the cross section. Recall also that an axis of symmetry for any plane area is a principal axis of inertia. Figures 12-1 through 12-3 illustrate cross sections of beams that possess vertical planes of symmetry.

Straight beams having cross sections with one or two axes of symmetry are frequently required to resist loads whose action lines

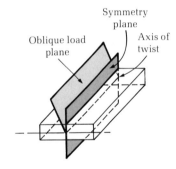

Figure 12-1

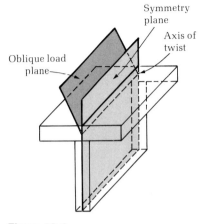

Figure 12-2

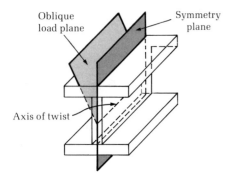

Figure 12-3

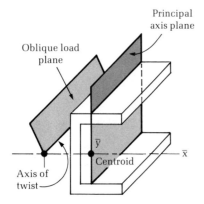

Figure 12-4

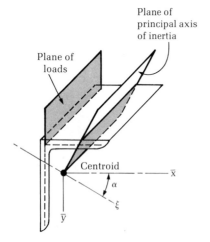

Figure 12-5

form a plane that does not coincide with a plane of symmetry. Furthermore, straight beams with unsymmetric cross sections are commonly used to resist transverse loads. The former are illustrated in Figures 12-1 through 12-4 and the latter in Figure 12-5.

We need a theory that permits the calculation of normal and shearing stresses for the more general case wherein the plane of the applied loads is not parallel to the plane formed by the principal axes of inertia. Two such cases are illustrated in Figures 12-4 and 12-5. To begin the development of such a theory, let the plane of the applied loads contain the axis of twist so that bending is not accompanied by twisting.

Figure 12-6 shows a typical beam cross section. The xy system of axes is an arbitrary set of rectangular axes whose origin coincides with the centroid of the cross-sectional area. The $\xi\eta$ system of axes coincides with the centroidal principal axes of inertia for the cross section. N, V_x, V_y, and M_x, M_y, M_z are the components, respectively, of a force-couple system statically equivalent to the system of distributed forces that act on the cross section. Recall that there are an infinite number of force-couple systems statically equivalent to the actual distributed system of forces. The resultant force associated with any of these equivalent force-couple systems is identical in magnitude, sense, and direction. However, the couple corresponding to a particular force-couple system depends on the point of application of the resultant force. In beam theory, it is customary to perceive the resultant force to act at the centroid of the cross-sectional area. Thus static equivalence requires that

$$M_x = \int_A \sigma_z y \, dA \tag{12-1a}$$

$$M_y = \int_A \sigma_z x \, dA \tag{12-1b}$$

$$M_z = \int_A (x\tau_{zy} - y\tau_{zx}) \, dA \tag{12-1c}$$

and

$$V_x = \int_A \tau_{zx} \, dA \tag{12-2a}$$

$$V_y = \int_A \tau_{zy} \, dA \tag{12-2b}$$

$$N = \int_A \sigma_z \, dA \tag{12-2c}$$

In these equations, M_x and M_y are bending moments exerted on a cross section by the portion of the beam contiguous with it. Positive senses for M_x and M_y are such that each produces a tensile stress in the first quadrant of the xy system as shown in Figure 12-6. According to this sign convention, positive M_x is in the **positive** x direction, whereas positive M_y is in the **negative** y direction. Since we required that the plane of the loads contains the axis of twist, the beam bends without twisting. Consequently, $M_z \equiv 0$.

The normal force N exerted on the cross section is considered to be positive when it produces a tensile stress. The transverse shearing forces V_x and V_y are discussed after the development of a formula for computing the normal stress σ_z.

Normal stress. The **Bernoulli theory** of straight beams hypothesizes that a plane cross section normal to the centroidal axis before deformation remains plane and normal to the centroidal axis in the deformed state. Figure 12-7 shows three mutually perpendicular line elements at an arbitrary point on a cross section of a straight beam. If the cross section must remain normal to the deformed centroidal axis of the beam, the shearing strain components $\gamma_{zx} = \gamma_{zy} = 0$. Moreover, since the normal stresses $\sigma_x = \sigma_y = 0$, it follows from the three-dimensional stress-strain relations given in Chapter 3 that

$$\epsilon_x = \epsilon_y = -\nu\epsilon_z \tag{12-3a}$$

Furthermore, since twisting of the cross section is prevented,

$$\gamma_{xy} = 0 \tag{12-3b}$$

Consequently, the Bernoulli hypothesis, which is based on geometric considerations, admits only the Poisson strains given by Eq. (12-3a) and the normal strain ϵ_z. In this case, the isotropic, three-dimensional, stress-strain relations show that

$$\sigma_x = \sigma_y = \tau_{xy} = \tau_{zx} = \tau_{zy} = 0$$

so that the only nonzero component of stress that is consistent with the Bernoulli hypothesis is the stress that is normal to the cross section.

Equations (12-2a) and (12-2b) show that when $\tau_{zx} = 0$ and $\tau_{zy} = 0$, $V_x = 0$ and $V_y = 0$. This result is unacceptable for any reasonable structural theory. Nevertheless, the Bernoulli hypothesis remains the basis for the derivation of the formula for computing the normal stress σ_z. Long experience with the flexure formula has shown that it can be used with confidence. Moreover, transverse shearing forces are reintroduced through equilibrium considerations.

The hypothesis that plane sections before deformation remain plane after deformation implies that the normal strain must vary linearly over the cross section. Thus

$$\epsilon_z = a'x + b'y + c' \tag{12-4}$$

The coefficients a', b', and c' are constants at a given cross section.

If it is assumed that the material from which the beam is made is isotropic and elastic, then the normal stress has the general form (from $\sigma_z = E\epsilon_z$)

$$\sigma_z = ax + by + c \tag{12-5}$$

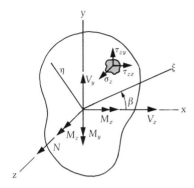

Figure 12-6

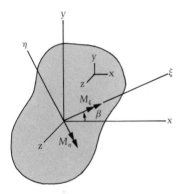

Figure 12-7

where $a = Ea'$, $b = Eb'$, and $c = Ec'$. Substituting Eq. (12-5) into Eqs. (12-1a), (12-1b), and (12-2c) yields three linear, algebraic equations for the constants a, b, and c;

$$M_x = aI_{xy} + bI_{xx} \tag{12-6a}$$

$$M_y = aI_{yy} + bI_{xy} \tag{12-6b}$$

$$N = cA \tag{12-6c}$$

In the derivation of Eqs. (12-6), we have made use of the fact that the origin of the coordinate system coincides with the centroid of the cross section.

Simultaneous solution of Eqs. (12-6) yields

$$a = \frac{M_y I_{xx} - M_x I_{xy}}{I_{xx} I_{yy} - I_{xy}^2} \tag{12-7a}$$

$$b = \frac{M_x I_{yy} - M_y I_{xy}}{I_{xx} I_{yy} - I_{xy}^2} \tag{12-7b}$$

$$c = \frac{N}{A} \tag{12-7c}$$

Substituting Eqs. (12-7) into Eq. (12-5), we obtain the flexure formula for unsymmetric bending.

$$\sigma_z = \left(\frac{M_y I_{xx} - M_x I_{xy}}{I_{xx} I_{yy} - I_{xy}^2} \right) x + \left(\frac{M_x I_{yy} - M_y I_{xy}}{I_{xx} I_{yy} - I_{xy}^2} \right) y + \frac{N}{A} \tag{12-8}$$

Equation (12-8) is applicable for any set of centroidal axes x, y. It can be simplified considerably by letting the xy system coincide with principal axes of inertia for the cross section. Let the principal axes of inertia be denoted by ξ and η. For the principal axes of inertia, $I_{\xi\eta} \equiv 0$ so that Eq. (12-8) reduces to

$$\sigma_z = \frac{M_\eta}{I_{\eta\eta}} \xi + \frac{M_\xi}{I_{\xi\xi}} \eta + \frac{N}{A} \tag{12-9}$$

Here M_ξ and M_η are the moments about the principal axes of inertia, and their positive senses are shown in Figure 12-7. Equation (12-9) shows that the normal stress can be calculated by the superposition of the bending stresses about the principal axes of inertia and the uniform axial stress due to stretching.

Equation (12-9) appears to be much less complicated than Eq. (12-8); however, appearance is misleading in this case. From an efficiency-of-calculation standpoint, Eq. (12-8) may be more attractive than Eq. (12-9) if normal stresses are the only goal of the analysis. This situation is true because $I_{\xi\xi}$ and $I_{\eta\eta}$ must also be calculated, M_ξ and M_η

must be determined, and the coordinates of the points in the cross section at which critical values of stress are desired may be difficult to determine in the $\xi\eta$ system. Formulas that facilitate the determination of the ξ, η coordinates of a point on the cross section are

$$\xi = x \cos \beta + y \sin \beta \tag{12-10a}$$

$$\eta = -x \sin \beta + y \cos \beta \tag{12-10b}$$

The $\xi\eta$ system and the angle β are shown in Figure 12-7. Note, in passing, that the neutral axis associated with the unsymmetric bending and stretching of straight beams can be obtained from Eq. (12-8) or (12-9) by setting the normal stress σ_z equal to zero. In either case, we obtain the equation for the neutral axis of the cross section.

EXAMPLE 12-1

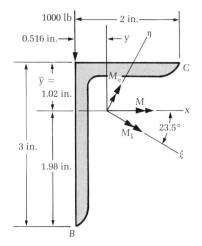

Figure 12-8

A $3 \times 2 \times \frac{5}{16}$ in. unequal leg angle section is being used as a cantilever beam. It supports a concentrated force of 1000 lb at one end as shown in Figure 12-8. Determine the normal stresses at points B and C of the cross section at the support if the beam span is 30 in.

SOLUTION

The cross-sectional properties are found from Appendix D to be:

$$I_{xx} = 1.32 \text{ in}^4, \ I_{yy} = 0.47 \text{ in}^4$$

$$I_{xy} = 0.456 \text{ in}^4, \ r_{min} = 0.432 \text{ in}.$$

$$A = 1.46 \text{ in}^2$$

$$\tan \alpha = 0.435$$

Here r_{min} is the minimum radius of gyration of the cross section.

The stresses at points B and C are calculated from Eq. (12-8).

$$\sigma_z = \left(\frac{M_y I_{xx} - M_x I_{xy}}{I_{xx} I_{yy} - I_{xy}^2}\right) x + \left(\frac{M_x I_{yy} - M_y I_{xy}}{I_{xx} I_{yy} - I_{xy}^2}\right) y \tag{a}$$

Since the external force passes through the shear center, there is no twisting of the cross section. Now, for the present problem, $M_x = 1000(30) = 30,000$ in.-lb and $M_y = 0$. Consequently, since $I_{xx} I_{yy} - I_{xy}^2 = 0.412$ in^4,

$$\sigma_z = \frac{-30,000(0.456)}{0.412} x + \frac{30,000(0.47)}{0.412} y$$

or

$$\sigma_z = -33,204x + 34,223y \tag{b}$$

Equation (b) is valid for all points x and y of the cross section at the wall. In particular, at point C whose coordinates are (1.484, 1.02) and at point B whose coordinates are $(-0.516, -1.98)$, the normal stresses are

$$\sigma_C = -33{,}240(1.484) + 34{,}223(1.02) = 14{,}421 \text{ psi(C)}$$

and

$$\sigma_B = -33{,}240(-0.516) + 34{,}223(-1.98) = 50{,}510 \text{ psi(C)}$$

ALTERNATIVE SOLUTION

The principal axes of inertia $\xi\eta$ are located as shown in Figure 12-8. The minimum radius of gyration as found from Appendix D is 0.432 in.; therefore, the minimum moment of inertia for the cross section is

$$I_{\eta\eta} = r_{\min}^2 A = (0.432)^2(1.46) = 0.272 \text{ in}^4$$

Thus, from the identity $I_{xx} + I_{yy} = I_{\xi\xi} + I_{\eta\eta}$,

$$I_{\xi\xi} = 1.32 + 0.47 - 0.272 = 1.518 \text{ in}^4$$

The resisting moments at the section at the wall are seen to be

$$M_\xi = 30{,}000 \cos 23.5° = 27{,}512 \text{ in.-lb}$$

and

$$M_\eta = -30{,}000 \sin 23.5° = -11{,}963 \text{ in.-lb}$$

Note that M_η as shown in Figure 12-8 is negative according to the stated sign convention. Equation (12-9) for the present case becomes

$$\sigma_z = \frac{M_\eta}{I_{\eta\eta}} \xi + \frac{M_\xi}{I_{\xi\xi}} \eta = -\frac{11{,}963}{0.272} \xi + \frac{27{,}512}{1.518} \eta$$

or

$$\sigma_z = -43{,}982\xi + 18{,}124\eta \tag{c}$$

Equation (c) is valid for all points ξ and η of the cross section at the support. To obtain the stresses at points B and C, it is necessary to calculate the ξ and η coordinates of these points. Equations (12-10) are used. Note that the rotation of the $\xi\eta$ system is clockwise and, thus, the angle β is negative. Therefore, at point C:

$$\left. \begin{array}{l} \xi = 1.484 \cos (-23.5°) + 1.02 \sin (-23.5°) = 0.954 \text{ in.} \\ \eta = -1.484 \sin (-23.5°) + 1.02 \cos (-23.5°) = 1.527 \text{ in.} \end{array} \right\} \tag{d}$$

so that

$$\sigma_C = -43{,}982(0.954) + 18{,}124(1.527) = 14{,}283 \text{ psi(C)}$$

At point B:

$$\left. \begin{array}{l} \xi = -0.516 \cos (-23.5°) - 1.98 \sin (-23.5°) = 0.316 \text{ in.} \\ \eta = +0.516 \sin (-23.5°) - 1.98 \cos (-23.5°) = -2.022 \text{ in.} \end{array} \right\} \tag{e}$$

so that

$$\sigma_B = -43{,}982(0.316) + 18{,}124(-2.022) = 50{,}545 \text{ psi(C)}$$

The stresses at B and C, as calculated by the alternative methods, are not numerically equal; however, the negligible difference can be attributed to rounding errors associated with the numerical procedures employed.

EXAMPLE 12-2

The cross section of a simply supported beam is built from an S 12 × 31.8, a C 12 × 20.7, and a cover plate 12 in. wide by $\frac{3}{8}$ in. thick, riveted together as shown in Figure 12-9a. The built-up section is to be used as a simply supported beam with an 18-ft span. The beam is to carry a uniformly distributed load of 1500 lb/ft which includes the weight of the beam. If the plane of the applied loads passes through the axis of twist of the beam, determine the equation of the neutral axis and the maximum tensile and compressive stresses in the beam.

SOLUTION

Properties of the S 12 × 31.8 section are shown in Figure 12-9b, and those of the C 12 × 20.7 section are shown in Figure 12-9c. First determine the coordinates of the centroid of the composite section. Since $A\overline{x}_0 = \sum_{i=1}^{3} A_i \overline{x}_i$,

$$(9.35 + 6.09 + 4.5)\overline{x}_0 = 4.5(6) + 9.35(2.5) + 6.09(8.5 - 0.698)$$

$$19.94\overline{x}_0 = 97.889$$

$$\overline{x}_0 = 4.909 \text{ in.} \tag{a}$$

Likewise, the $\overline{y}$ coordinate of the centroid of the composite section is determined from the relation $A\overline{y}_0 = \sum_{i=1}^{3} A_i y_i$ or

$$19.94\overline{y}_0 = 4.5(\tfrac{3}{16}) + 9.35(6.375) + 6.09(6.375) = 99.273,$$

$$\overline{y}_0 = 4.979 \text{ in.} \tag{b}$$

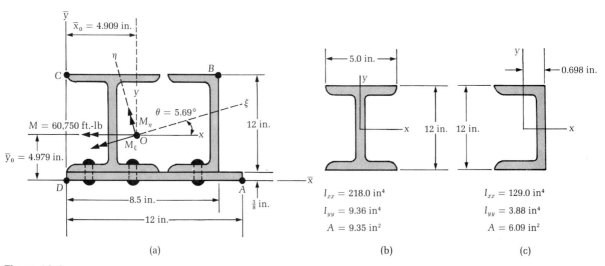

$I_{xx} = 218.0 \text{ in}^4$
$I_{yy} = 9.36 \text{ in}^4$
$A = 9.35 \text{ in}^2$

$I_{xx} = 129.0 \text{ in}^4$
$I_{yy} = 3.88 \text{ in}^4$
$A = 6.09 \text{ in}^2$

(a)   (b)   (c)

Figure 12-9

The rectangular moments of inertia with respect to a set of centroidal axes parallel to the $\bar{x}\bar{y}$ axes are required. First calculate the moments of inertia $I_{\bar{x}\bar{x}}$, $I_{\bar{y}\bar{y}}$, and $I_{\bar{x}\bar{y}}$ and transfer these values to the centroidal axes x and y. Since the moment of inertia of a composite section is equal to the sum of the moments of inertia of its parts with respect to the same line,

$$
\begin{aligned}
I_{xx} &= \tfrac{1}{3}(12)(0.375)^3 + [218.0 + 9.35(6.375)^2] \\
&\quad + [129.0 + 6.09(6.375)^2] = 974.70 \text{ in}^4 \\
I_{yy} &= \tfrac{1}{3}(0.375)(12)^3 + [9.36 + 9.35(2.5)^2] \\
&\quad + [3.88 + 6.09(8.5 - 0.698)^2] = 658.38 \text{ in}^4 \\
I_{xy} &= 4.5(6)(\tfrac{3}{16}) + 9.35(2.5)(6.375) + 6.09(7.8)(6.375) \\
&= 456.90 \text{ in}^4
\end{aligned}
\tag{c}
$$

The corresponding centroidal quantities are obtained by means of the parallel axis theorem,

$$
\begin{aligned}
I_{xx} &= 974.70 - 19.94(4.979)^2 = 480.38 \text{ in}^4 \\
I_{yy} &= 658.38 - 19.94(4.909)^2 = 177.86 \text{ in}^4 \\
I_{xy} &= 456.90 - 19.94(4.909)(4.979) = -30.47 \text{ in}^4
\end{aligned}
\tag{d}
$$

Also, $I_{xx}I_{yy} - I_{xy}^2 = 84{,}512.$

The shear and moment diagrams for the beam are shown in Figure 12-10. Using Eq. (12-8), the expression for the stress on the cross section at which the maximum moment occurs is

$$
\begin{aligned}
\sigma_z &= -\frac{M_x I_{xy}}{I_{xx}I_{yy} - I_{xy}^2} x + \frac{M_x I_{yy}}{I_{xx}I_{yy} - I_{xy}^2} y \\
&= -\frac{[-60{,}750(12)](-30.47)}{84{,}512} x + \frac{[-60{,}750(12)](176.65)}{84{,}512} y
\end{aligned}
$$

or

$$
\sigma_z = -262.83x - 1523.78y
\tag{e}
$$

The stresses at points A, B, C, and D of Figure 12-9a are:

$$
\begin{aligned}
\sigma_A &= -262.83(12 - 4.909) - 1523.78(-4.979) \\
&= 5723 \text{ psi}(T) \\
\sigma_B &= -262.83(8.5 - 4.909) - 1523.78(12.375 - 4.979) \\
&= 12{,}214 \text{ psi}(C) \text{ (maximum compressive stress)} \\
\sigma_C &= -262.83(-4.909) - 1523.78(12.375 - 4.979) \\
&= 9980 \text{ psi}(C) \\
\sigma_D &= -262.83(-4.909) - 1523.78(-4.979) \\
&= 8877 \text{ psi}(T) \text{ (maximum tensile stress)}
\end{aligned}
\tag{f}
$$

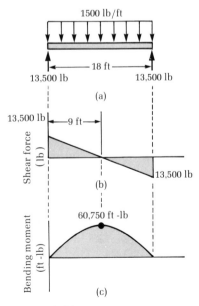

1500 lb/ft

18 ft

13,500 lb   13,500 lb

(a)

13,500 lb   9 ft

Shear force (lb)

13,500 lb

(b)

60,750 ft-lb

Bending moment (ft-lb)

(c)

Figure 12-10

ALTERNATIVE SOLUTION

The same results should be obtained when the principal axes of inertia are employed. First calculate the principal moments of inertia. The magnitudes of the principal moments of inertia are

$$\begin{Bmatrix} I_{max} \\ I_{min} \end{Bmatrix} = \frac{480.38 + 177.86}{2} \pm \sqrt{\left(\frac{480.38 - 177.86}{2}\right)^2 + (30.47)^2}$$

$$= \begin{Bmatrix} 483.42 \\ 174.82 \end{Bmatrix} \text{in}^4 \tag{g}$$

and the orientation of the principal axes is

$$\tan 2\theta = \frac{+30.47}{151.26} = 0.2014$$

$$2\theta = 11.39° \quad \text{and} \quad \theta = 5.69° \tag{h}$$

A correlation calculation shows that $I_{\xi\xi} = 483.42 \text{ in}^4$ and, hence, $I_{\eta\eta} = 177.82 \text{ in}^4$. The principal axes $\xi\eta$ are shown in Figure 12-9a. Decomposition of the resisting moment at the section into components along the ξ and η axes gives

$$\left.\begin{aligned} M_\xi &= -60,750(12) \cos (5.69°) = -725,408 \text{ in.-lb} \\ M_\eta &= -60,750(12) \sin (5.69°) = -72,277 \text{ in.-lb} \end{aligned}\right\} \tag{i}$$

The formula for the normal stress in the cross section is, from Eq. (12-9),

$$\sigma_z = -\frac{72,277}{177.82}\xi - \frac{725,408}{483.42}\eta = -406.5\xi - 1500.6\eta \tag{j}$$

To calculate the stresses at points A, B, C, and D of Figure 12-9a, we need the ξ and η coordinates of the points. Thus, from Eqs. (12-10a) and (12-10b), for example,

$$\left.\begin{aligned} \xi_A &= (12 - 4.909) \cos (5.69°) + (-4.979) \sin (5.69°) \\ &= 6.562 \text{ in.} \\ \eta_A &= -(12 - 4.909) \sin (5.69°) + (-4.979) \cos (5.69°) \\ &= -5.658 \text{ in.} \end{aligned}\right\} \tag{k}$$

Equation (j) yields, for the normal stress at point A,

$$\sigma_A = -406.5(6.562) - 1500.6(-5.658) = 5823 \text{ psi}(T) \tag{l}$$

In an identical manner, the coordinates and the normal stresses at the other points are: $B(4.307, 7.004)$, $\sigma_B = 12,261 \text{ psi}(C)$; $C(-4.152, 7.846)$, $\sigma_C = 10,086 \text{ psi}(C)$; and $D(-5.378, -4.468)$, $\sigma_D = 8891 \text{ psi}(T)$. Compare these results with those calculated from Eq. (f). Also observe that you need to use significantly more numerical computations with the principal axes of inertia approach than you do when you use a natural set of rectangular centroidal axes.

Transverse shearing stress.   In this section, we consider the transverse shearing stresses in straight beams due to unsymmetric bending. Figure 12-11a depicts a cross section of arbitrary shape. The xy coordinate axes are centroidal axes, but they are not necessarily coincident with the principal axes of inertia. The z coordinate is directed along the centroidal line of the beam. Plane aa is parallel to the centroidal line of the beam, but it is otherwise arbitrary.

A segment of the beam contained between the sections at z and z + Δz is shown in Figure 12-11b. Consider the portion of this segment that lies above the plane aa. The forces $R(z + \Delta z)$ and $R(z)$ generally have different magnitudes because the stresses $\sigma_z(z + \Delta z)$ and $\sigma_z(z)$ have different magnitudes when the moments $M_x(z + \Delta z)$, $M_y(z + \Delta z)$, and $M_x(z)$, $M_y(z)$ are different. Thus equilibrium of forces along the axis of the beam can be satisfied for the portion contained between sections z and z + Δz and the plane aa only if a shearing force acts in the plane aa. This shearing force will be denoted by $q\,\Delta z$, where q is force per unit of length and is called the shear flow at the plane aa. Consequently,

$$R(z + \Delta z) - R(z) - q\,\Delta z = 0 \tag{12-11}$$

Dividing Eq. (12-11) by Δz and passing to the limit yields

$$\frac{dR}{dz} = q \tag{12-12}$$

Now the force $R(z)$ is given by the relation

$$R(z) = \int_{A^*} \sigma_z(z)\,dA \tag{12-13}$$

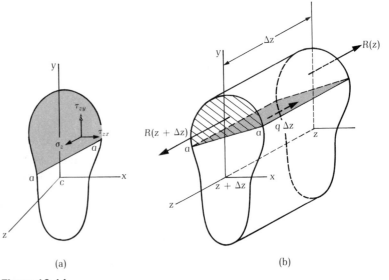

(a)   (b)

Figure 12-11

where A^* is the area of the end of the portion of the beam considered. This area is crosshatched in Figure 12-11b. The normal stress on the cross section located at z is denoted by $\sigma_z(z)$.

Equations (12-5) and (12-13) give

$$R(z) = \int_{A^*} [a(z)x + b(z)y + c] \, dA \tag{12-14}$$

from which

$$\frac{dR}{dz} = \frac{da}{dz} Q_y + \frac{db}{dz} Q_x = q \tag{12-15}$$

where

$$\left.\begin{aligned} Q_x &= \int_{A^*} y \, dA \\ Q_y &= \int_{A^*} x \, dA \end{aligned}\right\} \tag{12-16}$$

It is assumed that the normal force N acting on the cross section of the beam does not vary continuously along it length and, thus,

$$\frac{dc}{dz} = 0$$

The quantities Q_x and Q_y are the first moments of the area A^* with respect to the x and y axes, respectively.

From Eqs. (12-7a) and (12-7b),

$$\frac{da}{dz} = \frac{\dfrac{dM_y}{dz} I_{xx} - \dfrac{dM_x}{dz} I_{xy}}{I_{xx}I_{yy} - I_{xy}^2} \tag{12-17a}$$

and

$$\frac{db}{dz} = \frac{\dfrac{dM_x}{dz} I_{yy} - \dfrac{dM_y}{dz} I_{xy}}{I_{xx}I_{yy} - I_{xy}^2} \tag{12-17b}$$

At this point, we digress briefly to derive some important differential relations applicable for unsymmetric bending of beams. Consider a segment of a beam between the sections at z and $z + \Delta z$ as shown in Figures 12-12a and 12-12b. In Figure 12-12a, only the forces that enter into a force balance along the y direction are shown. Likewise, only moments that enter into a moment balance with respect to the x axis are shown. Figure 12-12b shows only those forces that enter into a force balance parallel to the x axis and moments that enter into moment balance relative to the y axis. Thus equilibrium of forces parallel to the x and y directions gives

$$\frac{dV_x}{dz} = -p_x \tag{12-18a}$$

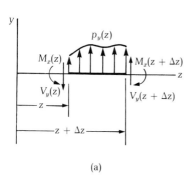

(a)

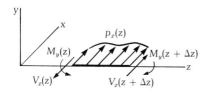

(b)

Figure 12-12

and

$$\frac{dV_y}{dz} = -p_y \tag{12-18b}$$

while moment equilibrium with respect to the same axes gives

$$\frac{dM_x}{dz} = V_y \tag{12-19a}$$

and

$$\frac{dM_y}{dz} = V_x \tag{12-19b}$$

Equations (12-18a) and (12-19a) are the differential relations that serve as the basis for constructing shear and moment diagrams by the summation technique for loads applied parallel to the y axis. Equations (12-18b) and (12-19b) are similar relations for loads applied parallel to the x axis. Consequently, the shear and moment diagrams for a beam subjected to unsymmetric bending can be obtained by the familiar summation technique developed in elementary beam theory. Two sets of diagrams are required: one set for bending about the x axis only and a second set for bending about the y axis only.

Substituting Eqs. (12-19) into Eqs. (12-17) gives

$$\frac{da}{dz} = \frac{V_x I_{xx} - V_y I_{xy}}{I_{xx}I_{yy} - I_{xy}^2} \tag{12-20a}$$

and

$$\frac{db}{dz} = \frac{V_y I_{yy} - V_x I_{xy}}{I_{xx}I_{yy} - I_{xy}^2} \tag{12-20b}$$

so that Eq. (12-15) can be written as

$$q = \frac{dR}{dz} = \left(\frac{V_x I_{xx} - V_y I_{xy}}{I_{xx}I_{yy} - I_{xy}^2} \right) Q_y + \left(\frac{V_y I_{yy} - V_x I_{xy}}{I_{xx}I_{yy} - I_{xy}^2} \right) Q_x \tag{12-21}$$

Observe that q(lb/in. or N/m) is the shear flow in the aa plane parallel to the z axis. Since shearing stresses must occur in pairs at right angles to each other, it follows that q of Eq. (12-21) is equal to the shear flow in the cross section normal to the beam axis at z.

Equation (12-21) is simplified considerably (in appearance) if the xy system coincides with the principal axes of inertia. Thus, if ξ, η denote a set of axes that coincide with the principal axes of inertia,

$$q = \frac{V_\xi Q_\eta}{I_{\eta\eta}} + \frac{V_\eta Q_\xi}{I_{\xi\xi}} \tag{12-22}$$

Equation (12-22) shows that the shear flow associated with any plane parallel to the axis of a straight beam can be considered to be the **sum of the shear flows due to bending about the two principal axes of inertia**. Note that this superposition of shear flows permits beams undergoing unsymmetric bending to be analyzed for shearing stresses by considering the bending about each principal axis separately. This procedure may not, however, be the most computationally effective one.

Finally, the shearing stress corresponding to a shear flow q is assumed to be uniformly distributed over the width of plane aa. Consequently,

$$\tau = \frac{q}{b} \tag{12-23}$$

where b is the width of the plane aa. **The uniform distribution of shearing stress may not be realistic when the width of plane <u>aa</u> is large compared to other cross-sectional dimensions**. For example, the transverse shearing stresses in the flanges of an S section are not accurately predicted by Eq. (12-23), while the transverse shearing stresses in the web are predicted with reasonable accuracy.

EXAMPLE 12-3

A straight beam 30 in. long with an unsymmetric cross section is loaded at the third point by forces $P = 50$ lb as shown in Figure 12-13a. The dimensions of the cross section are shown in Figure 12-13b. Determine the shearing stress induced by the applied loads at the plane aa in a region containing the first 10 in. of the beam.

SOLUTION

From the dimensions of the cross section given in Figure 12-13b, the section properties are calculated as

$$\left.\begin{array}{l} A = 0.275 \text{ in}^2 \\ \bar{x}_C = 0.0773 \text{ in.} \\ \bar{y}_C = 0.5391 \text{ in.} \\ I_{xx} = 0.0576 \text{ in}^4 \\ I_{yy} = 0.0150 \text{ in}^4 \\ I_{xy} = 0.0117 \text{ in}^4 \end{array}\right\} \tag{a}$$

The moments of inertia given in Eq. (a) are centroidal moments of inertia.

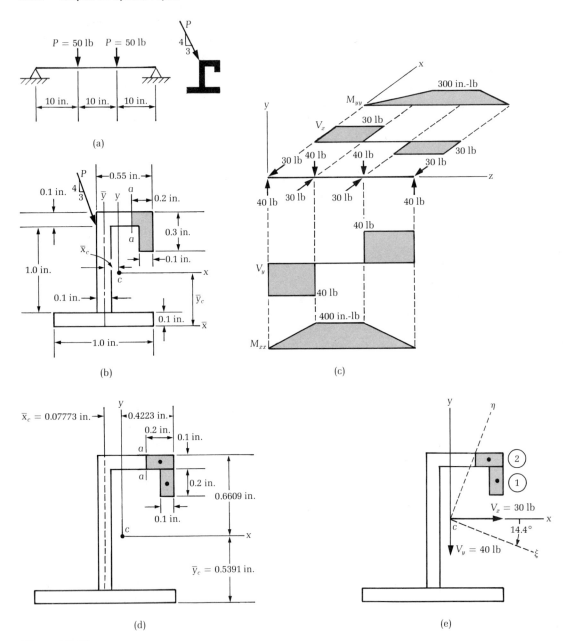

(a)

(b)

(c)

(d)

(e)

Figure 12-13

Shear and moment diagrams for the beam are constructed in Figure 12-13c. It is seen from the shear diagrams that, for the specified region of interest,

$$V_x = 30 \text{ lb}$$
$$V_y = -40 \text{ lb}$$

(b)

The factors Q_x and Q_y are the first moments of the shaded area of Figure 12-13b with respect to the x and y axes, respectively. Therefore, from Figure 12-13d,

$$
\left.
\begin{aligned}
Q_x &= (0.2 \times 0.1)(0.6609 - 0.05) + (0.2 \times 0.1)(0.6609 - 0.20) \\
&= 0.0214 \text{ in}^3 \\
Q_y &= (0.2 \times 0.1)(0.4223 - 0.1) + (0.2 \times 0.1)(0.4223 - 0.05) \\
&= 0.0139 \text{ in}^3
\end{aligned}
\right\}
\quad \text{(c)}
$$

The shear flow at section aa is calculated from Eq. (12-21).

$$
\begin{aligned}
q &= \frac{30(0.0576) - (-40)(0.0117)}{0.000726}(0.0139) \\
&+ \frac{(-40)(0.0150) - 30(0.0117)}{0.000726}(0.0214)
\end{aligned}
$$

or

$$q = 14.01 \text{ lb/in.} \qquad \text{(d)}$$

From Eq. (12-23), the shearing stress at the section

$$\tau = \frac{q}{b} = \frac{14.01}{0.1} = 140.1 \text{ psi} \qquad \text{(e)}$$

ALTERNATIVE SOLUTION

Solve this problem using principal axes of inertia. The magnitudes of the principal moments of inertia are

$$
\begin{aligned}
\begin{Bmatrix} I_{max} \\ I_{min} \end{Bmatrix} &= \frac{0.0576 + 0.015}{2} \pm \sqrt{\left(\frac{0.0576 - 0.015}{2}\right)^2 + (-0.0117)^2} \\
&= 0.0363 \pm 0.0243 \\
&= 0.0606 \text{ in}^4, 0.0120 \text{ in}^4 \qquad \text{(f)}
\end{aligned}
$$

and the orientation of the principal axes is

$$\tan 2\theta_p = \frac{(-0.0117)}{0.0213} = -0.55$$

or

$$\theta_p = -14.40° \qquad \text{(g)}$$

A correlation calculation shows that $I_{\xi\xi} = 0.0606 \text{ in}^4$ and $I_{\eta\eta} = 0.0120 \text{ in}^4$, where the ξ and η axes are shown in Figure 12-13e.

The shearing forces along the principal axes are

$$
\left.
\begin{aligned}
V_\xi &= 30 \cos (14.4°) + 40 \sin (14.4°) = 39.00 \text{ lb} \\
V_\eta &= 30 \sin (14.4°) - 40 \cos (14.4°) = -31.28 \text{ lb}
\end{aligned}
\right\}
\quad \text{(h)}
$$

The quantities Q_ξ and Q_η are determined from the expressions

$$Q_\xi = \sum_{i=1}^{2} A_i \bar{\eta}_i \quad \text{and} \quad Q_\eta = \sum_{i=1}^{2} A_i \bar{\xi}_i$$

where $\bar{\xi}_i$ and $\bar{\eta}_i$ are the centroidal coordinates of the ith part. The coordinates of the centroids of the two parts shown in Figure 12-13e are

$$\begin{aligned}
\bar{\xi}_1 &= 0.3723 \cos{(-14.40°)} + 0.4609 \sin{(-14.40°)} \\
&= 0.2460 \text{ in.} \\
\bar{\eta}_1 &= -0.3723 \sin{(-14.40°)} + 0.4609 \cos{(-14.40°)} \\
&= 0.5390 \text{ in.} \\
\bar{\xi}_2 &= 0.3223 \cos{(-14.40°)} + 0.6109 \sin{(-14.40°)} \\
&= 0.1602 \text{ in.} \\
\bar{\eta}_2 &= -0.3223 \sin{(-14.40°)} + 0.6109 \cos{(-14.4°)} \\
&= 0.6719 \text{ in.}
\end{aligned} \right\} \quad \text{(i)}$$

Consequently,

$$\begin{aligned}
Q_\xi &= 0.02(0.5390) + 0.02(0.6719) = 0.0242 \text{ in}^3 \\
Q_\eta &= 0.02(0.2460) + 0.02(0.1602) = 0.0081 \text{ in}^3
\end{aligned} \right\} \quad \text{(j)}$$

Finally, the shear flow at section aa is found to be

$$q = \frac{V_\xi}{I_{\eta\eta}} Q_\eta + \frac{V_\eta}{I_{\xi\xi}} Q_\xi = \frac{39}{0.0120}(0.0081) + \frac{(-31.28)}{0.0606}(0.0242)$$

$$q = 13.83 \text{ lb/in.} \quad \text{(k)}$$

The discrepancy ($\approx 1.28\%$) between this result and the one calculated previously can be accredited to numerical rounding errors.

This example illustrates the computation effort required for an analysis using principal axes of inertia beyond that required when a natural set of axes is used. Nevertheless, the principal axes theory is a convenient conceptual tool.

PROBLEMS / Section 12-2

12-1 Calculate the maximum tensile and compressive stresses in the cantilever beam shown in Figure P12-1. Assume that the plane of the loads contains the axis of twist for the beam (112 MPa).

12-2 A simply supported aluminum beam with the cross section shown in Figure P12-2a supports a concentrated force at midspan. Calculate the normal stress at points A, B, and C of the cross section. Assume that the plane of the loads contains the axis of twist of the beam ($\sigma_A = 10.4$ psi(C)).

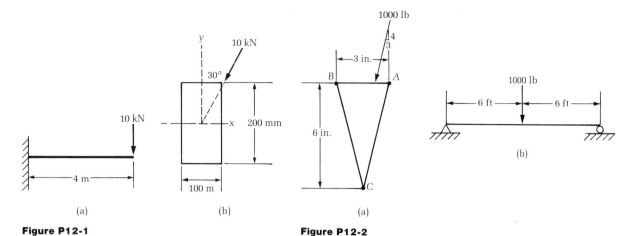

(a)   (b)   (a)

Figure P12-1   **Figure P12-2**

12-3 A steel beam having the cross section shown in Figure P12-3 is subjected to a maximum internal bending moment of 1000 N·m as shown in the figure. Calculate the normal stress at point A of the cross section.

12-4 An aluminum strut having the cross section shown in Figure P12-4 is subjected to a maximum internal bending moment of 1000 N·m as shown in the figure. Calculate the normal stress at point A of the cross section. Pertinent cross-sectional properties are: $\bar{I}_{xx} = 7.79 \times 10^{-6}$ m^4, $\bar{I}_{yy} = 2.30 \times 10^{-6}$ m^4, $\bar{I}_{xy} = 3.17 \times 10^{-6}$ m^4 (7.76 MPa(C)).

12-5 A cantilever beam 6 ft long with the cross section shown in Figure P12-5 supports a concentrated force of 1000 lb at its free end. Calculate the normal stress at points A and B of the cross section near the support. Verify that $\bar{I}_{xx} = 104.8$ in^4, $\bar{I}_{yy} = 12.2$ in^4, and $\bar{I}_{xy} = 25.5$ in^4.

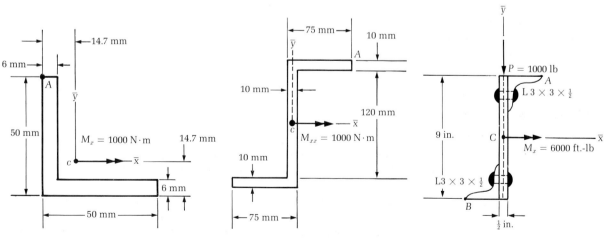

Figure P12-3   **Figure P12-4**   **Figure P12-5**

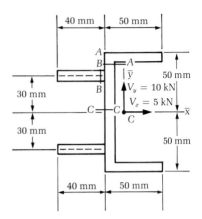

Figure P12-6

12-6 The transverse shearing forces that act at a certain cross section of a straight beam are $V_x = 5$ kN and $V_y = 10$ kN as shown in Figure P12-6. If the beam bends without twisting, determine the shearing stress at locations AA, BB, and CC. The thickness of each segment of the cross section is 4 mm.

12-7 The cross section of an aluminum beam has the shape and dimensions shown in Figure P12-7. Assume that the beam bends without twisting. Determine the shearing stresses at the four locations AA, BB, CC, and DD due to the shearing forces $V_x = 5$ kN and $V_y = 10$ kN, directed as shown in the figure. The thickness of each segment of the cross section is 3 mm.

12-8 Calculate the shearing stress at locations AA and BB of the cross section shown in Figure P12-8. The thickness of each segment is 4 mm and the shearing forces are $V_x = +5$ kN and $V_y = 10$ kN. Assume that bending occurs without twisting.

12-9 Pertinent properties for the cross section shown in Figure P12-9 are: $A = 560$ mm^2, $\bar{I}_{xx} = 0.755 \times 10^{-6}$ m^4, $\bar{I}_{yy} = 0.095 \times 10^{-6}$ m^4, $\bar{I}_{xy} = 0.109 \times 10^{-6}$ m^4. Determine shearing stress at the locations AA, BB, and CC when $V_x = 10$ kN and $V_y = 5$ kN. The thickness of each segment of the cross section is 4 mm ($\tau_{AA} = 38.1$ MPa).

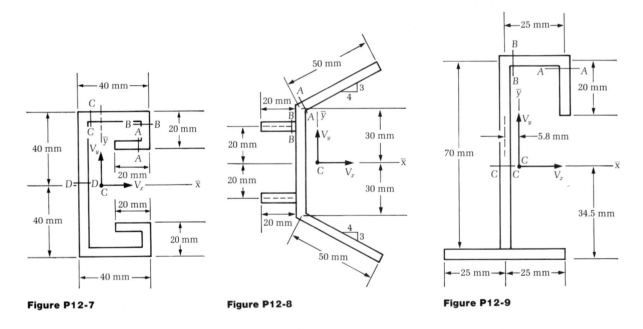

Figure P12-7   **Figure P12-8**   **Figure P12-9**

12-3   Normal Stresses in Curved Beams

Winkler–Bach curved beam theory. Consider a curved beam whose cross section is symmetric with respect to the plane containing its centroidal line. The centroidal line for such a curved beam is shown by the dash-dot line in Figure 12-14a.

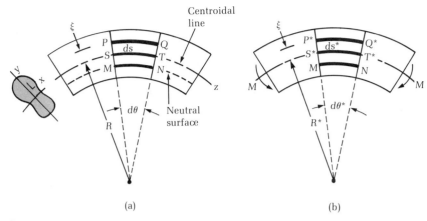

Figure 12-14

Assume that the external forces are applied in this plane of symmetry so that bending of the beam occurs in the plane of symmetry. That is, the beam does not displace perpendicular to its plane of symmetry. These restrictions ensure that $\gamma_{xz} = \gamma_{xy} = 0$.

The Winkler–Bach curved beam theory is an extension of the Bernoulli theory for straight beams. Kinematically, it is assumed that plane sections perpendicular to the centroidal line of the beam before deformation remain perpendicular to the centroidal line in the deformed configuration of the beam. This assumption implies that $\gamma_{zy} = 0$. Since the normal stresses that act on the lateral surfaces of the beam are either zero or small in comparison to the flexural stress due to bending, these stresses (σ_x and σ_y) are assumed to be zero throughout the cross section. The three-dimensional stress-strain relations show that $\epsilon_x = \epsilon_y = -\nu\epsilon_z$.

Observe that, since $\gamma_{xy} = \gamma_{xz} = \gamma_{yz} = 0$, the three-dimensional shearing stress-strain relations show that $\tau_{xy} = \tau_{xz} = \tau_{yz} = 0$. The Winkler–Bach curved beam theory admits only a normal strain ϵ_z parallel to the centroidal line of the beam and the Poisson strains $\epsilon_x = \epsilon_y = -\nu\epsilon_z$.

Figures 12-14a and 12-14b show a curved beam in its undeformed and deformed configurations. PQ is a line of particles in the undeformed configuration a distance ξ from the centroidal line, ST is a line of particles coincident with the centroidal line, and MN is a line of particles coincident with the neutral surface for the beam. Note that the neutral surface for a curved beam is not generally coincident with its centroidal line. The local radius of curvature of the centroidal line is designated by R, and the angle between two cross-sectional planes separated by an infinitesimal distance ds is signified by $d\theta$. Corresponding quantities for the deformed configuration are indicated by an asterisk, ( )*. Note that the line of particles MN does not change its length during the deformation process because it lies in the neutral surface.

The object of the Winkler–Bach curved beam theory is to establish a formula for the normal stress σ_z acting on a cross-sectional plane.

Kinematic considerations.   The strain of the fiber PQ as it passes into the deformed state is

$$\epsilon_z = \lim_{Q \to P} \frac{P^*Q^* - PQ}{PQ} \tag{12-24}$$

From Figures 12-14a and 12-14b,

$$PQ = \frac{R + \xi}{R}\, ds \tag{12-25a}$$

and

$$P^*Q^* = \frac{R^* + \xi}{R^*}\, ds^* \tag{12-25b}$$

so that

$$\epsilon_z = \left(\frac{ds^* - ds}{ds} \right) \frac{R}{R + \xi} + \left(\frac{\dfrac{ds^*}{R^*} - \dfrac{ds}{R}}{\dfrac{ds}{R}} \right) \frac{\xi}{R + \xi} \tag{12-26}$$

The quantities in parentheses are the normal strain at the centroid of the cross section and the angular strain, respectively. These quantities are denoted by

$$\epsilon_0 = \frac{ds^* - ds}{ds} \qquad \text{and} \qquad \omega = \frac{\dfrac{ds^*}{R^*} - \dfrac{ds}{R}}{\dfrac{ds}{R}} \tag{12-27}$$

Thus the normal strain for the Winkler–Bach theory of curved beams is

$$\epsilon_z = \frac{R}{R + \xi}\left(\epsilon_0 + \frac{\xi}{R}\,\omega \right) \tag{12-28}$$

Observe that the normal strain associated with the Winkler–Bach theory of curved beams does not vary linearly over the depth of the beam as it does for the Bernoulli theory of straight beams. The strain distribution in a curved beam follows a hyperbolic pattern.

Equilibrium considerations.   Let N and M constitute a force-couple system equivalent to the distributed force system represented by the normal stress σ_z. This system is illustrated in Figure 12-15.

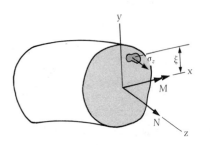

Figure 12-15

This equivalency is expressed by the relations

$$N = \int_A \sigma_z \, dA \tag{12-29a}$$

and

$$M = \int_A \sigma_z \xi \, dA \tag{12-29b}$$

Material behavior. Assuming a linearly elastic material ($\sigma_z = E\epsilon_z$), Eqs. (12-28) and (12-29) lead to two algebraic equations for the parameters ϵ_0 and ω,

$$ER\epsilon_0 \int_A \frac{dA}{R + \xi} + E\omega \int_A \frac{\xi \, dA}{R + \xi} = N \tag{12-30a}$$

and

$$ER\epsilon_0 \int_A \frac{\xi \, dA}{R + \xi} + E\omega \int_A \frac{\xi^2 \, dA}{R + \xi} = M \tag{12-30b}$$

The integrals appearing in Eqs. (12-30a) and (12-30b) are related to the geometric properties of the cross section as follows. Let

$$\int_A \frac{\xi \, dA}{R + \xi} = -AZ \tag{12-31a}$$

where Z is the so-called Winkler–Bach curved beam factor and A is the area of the cross section. From this definition, it follows that

$$\int_A \frac{\xi^2 \, dA}{R + \xi} = \int_A \left(\xi - \frac{R\xi}{R + \xi} \right) dA = -R \int_A \frac{\xi \, dA}{R + \xi} = RAZ \tag{12-31b}$$

and

$$\int_A \frac{\xi \, dA}{R + \xi} = \int_A \left(1 - \frac{R}{R + \xi} \right) dA = A - R \int_A \frac{dA}{R + \xi} = -AZ \tag{12-31c}$$

from which

$$\int_A \frac{dA}{R + \xi} = \frac{A}{R} (1 + Z) \tag{12-31d}$$

Equations (12-30a) and (12-30b) can now be written in the form

$$(1 + Z)\epsilon_0 - Z\omega = \frac{N}{AE} \tag{12-32a}$$

and

$$-\epsilon_0 + \omega = \frac{M}{EARZ} \tag{12-32b}$$

Solving these equations simultaneously yields

$$\epsilon_0 = \frac{M}{EAR} + \frac{N}{AE} \tag{12-33a}$$

and

$$\omega = \frac{N}{AE} + \frac{M}{EAR}\left(1 + \frac{1}{Z}\right) \tag{12-33b}$$

Substituting these formulas for ϵ_0 and ω into Eq. (12-28) yields the Winkler–Bach formula for the normal strain in a curved beam

$$\epsilon_z = \frac{N}{AE} + \frac{M}{EAR}\left[1 + \frac{1}{Z}\left(\frac{\xi}{R + \xi}\right)\right] \tag{12-34}$$

which leads to the normal stress

$$\sigma_z = \frac{N}{A} + \frac{M}{AR}\left[1 + \frac{1}{Z}\left(\frac{\xi}{R + \xi}\right)\right] \tag{12-35}$$

It should be emphasized that Eq. (12-35) is valid only for linearly elastic material behavior. Observe that the normal force N and the bending moment M produce stresses that can be obtained by the superposition principle.

The sign convention employed in the preceding development for the normal stress in a curved beam is that (a) a tensile value of N is positive, (b) M is positive when it tends to increase the curvature of the beam (decrease the radius of curvature), and (c) ξ is positive when measured away from the center of curvature of the beam.

Location of the neutral axis. The neutral axis for a Winkler–Bach curved beam is obtained from Eq. (12-35) by setting $\sigma_z = 0$. In this manner, we find the location of the neutral axis relative the centroidal axis to be

$$\xi = -RZ\left[\frac{1 + \dfrac{NR}{M}}{1 + Z\left(1 + \dfrac{NR}{M}\right)}\right] \tag{12-36}$$

which reveals that in the absence of the normal force N, the neutral axis always lies on the side of the centroidal axis nearest to the center of curvature for the cross section.

Figure 12-16 lists formulas for the Winkler–Bach curved beam factor Z for some typical cross sections. Ordinarily Z must be calculated with a reasonably high degree of accuracy. Calculation of the appropriate Z factor is a simple operation for a hand calculator.

Winkler-Bach curved beam factors

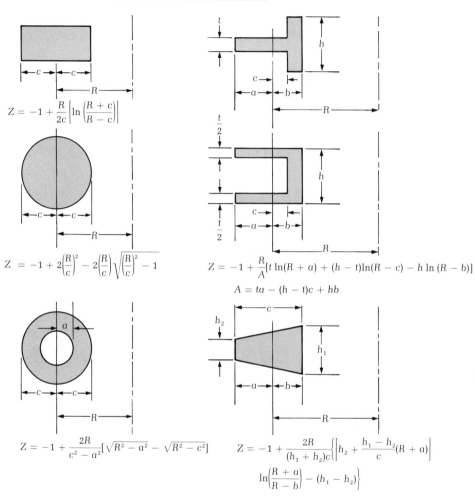

$$Z = -1 + \frac{R}{2c}\left[\ln\left(\frac{R+c}{R-c}\right)\right]$$

$$Z = -1 + 2\left(\frac{R}{c}\right)^2 - 2\left(\frac{R}{c}\right)\sqrt{\left(\frac{R}{c}\right)^2 - 1}$$

$$Z = -1 + \frac{R}{A}[t\ln(R+a) + (h-t)\ln(R-c) - h\ln(R-b)]$$

$$A = ta - (h-t)c + hb$$

$$Z = -1 + \frac{2R}{c^2 - a^2}[\sqrt{R^2 - a^2} - \sqrt{R^2 - c^2}]$$

$$Z = -1 + \frac{2R}{(h_1 + h_2)c}\left\{\left[h_2 + \frac{h_1 - h_2}{c}(R+a)\right]\right.$$

$$\left.\ln\left(\frac{R+a}{R-b}\right) - (h_1 - h_2)\right\}$$

Figure 12-16

EXAMPLE 12-4

For the structural member shown in Figure 12-17a, determine the normal stresses at points A and B of the curved region.

SOLUTION

Calculate the normal stresses at points A and B. From Figure 12-17b,

$$N = 400 \text{ lb}$$

and

$$M = -1800 \text{ in.-lb}$$

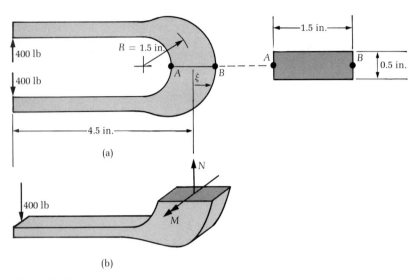

(a)

(b)

Figure 12-17

Also $A = 1.5(0.5) = 0.75$ in^2, $R = 1.5$ in., and $Z = -1 + 1.5/3$

$$\ln\left(\frac{1.5 + 0.75}{1.5 - 0.75}\right) = 0.0986,$$

which is obtained from Figure 12-16.

The normal stress at a generic point ξ of the cross section AB is, from Eq. (12-35),

$$\sigma_z = \frac{400}{0.75} + \frac{(-1800)}{0.75(1.5)}\left[1 + \frac{1}{0.0986}\frac{\xi}{1.5 + \xi}\right]$$

or

$$\sigma_z = 533 - 1600\left[1 + \frac{1}{0.0986}\left(\frac{\xi}{1.5 + \xi}\right)\right] \tag{a}$$

Consequently, the normal stresses at points A and B are calculated from Eq. (a) by setting $\xi = -0.75$ and $\xi = 0.75$, respectively. Thus

$$\left.\begin{array}{l} \sigma_A = 15{,}160 \text{ psi(T)} \\ \sigma_B = 6476 \text{ psi(C)} \end{array}\right\} \tag{b}$$

The stresses at A and B, when calculated from straight beam theory, are

$$\left.\begin{array}{l} \sigma_A = \dfrac{1800(0.75)}{0.14063} + 533 = 10{,}133 \text{ psi(T)} \\[4mm] \sigma_B = -\dfrac{1800(0.75)}{0.14063} + 533 = 9067 \text{ psi(C)} \end{array}\right\} \tag{c}$$

where

$$I = \tfrac{1}{12}(0.5)(1.5)^3 = 0.14063 \text{ in}^4 \qquad\qquad (d)$$

The stresses at points A and B as calculated from the Winkler–Bach curved beam theory are, respectively, 49.6% higher and 28.6% lower than the stresses calculated at the same points via the straight beam formula. Use of the straight beam formula to calculate normal stresses in curved beams can clearly lead to significant inaccuracies. Note that the ratio $R/c = 2.0$ for this example, which represents a sharply curved member. When $R \gg c$, the straight beam formula provides excellent approximate values for the normal stresses in curved beams.

The neutral axis occurs at the value of ξ calculated from Eq. (12-36). Thus

$$\xi = -1.5(0.0986)\left[\dfrac{1 + \dfrac{400(1.5)}{-1800}}{1 + 0.0986\left(1 + \dfrac{400(1.5)}{-1800}\right)}\right] = -0.0925$$

$$\xi = -0.0925 \text{ in.}$$

toward the center of curvature.

EXAMPLE 12-5

The ring shown in Figure 12-18a has a circular cross section 75 mm in diameter. For the dimensions shown and a load $P = 25$ kN, determine the normal stresses at points A and B.

SOLUTION

From Figure 12-18b, it follows that

$$N = -25 \text{ kN}, \ M = 0.0875(25,000) = 2187.5 \text{ N·m}$$

$$A = \frac{\pi}{4}(0.075)^2 = 0.00442 \text{ m}^2 \qquad \text{and} \qquad R = 0.0875 \text{ m}$$

The curved beam factor is

$$Z = -1 + 2\left(\frac{0.0875}{0.0375}\right)^2 - 2\left(\frac{0.0875}{0.0375}\right)\sqrt{\left(\frac{0.0875}{0.0375}\right)^2 - 1} = 0.0507$$

Equation (12-35) becomes

$$\sigma_z = \frac{-25,000}{0.00442} + \frac{2187.5}{0.00442(0.0875)}\left[1 + \frac{1}{0.0507}\left(\frac{\xi}{0.0875 + \xi}\right)\right]$$

$$= \left\{-5.656 + 5.656\left[1 + 19.76\left(\frac{\xi}{0.0875 + \xi}\right)\right]\right\} \times 10^6 \text{ N/m}^2$$

$$(a)$$

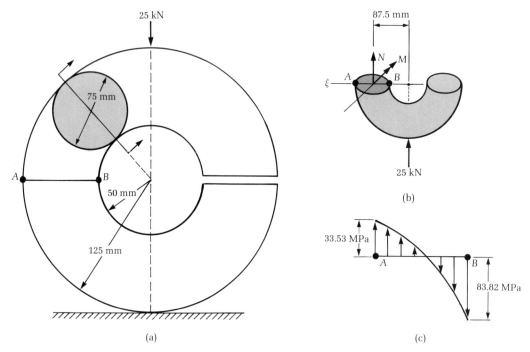

25 kN

75 mm

50 mm

A   B

125 mm

(a)

87.5 mm

N   M

ξ   A   B

25 kN

(b)

33.53 MPa

A   B

83.82 MPa

(c)

Figure 12-18

The normal stresses at points A and B are calculated from Eq. (a) by setting $\xi = 0.0375$ m and $\xi = -0.0375$ m, respectively. Thus

$$\left.\begin{array}{l} \sigma_A = 33.53 \text{ MPa}(T) \\ \sigma_B = 83.82 \text{ MPa}(C) \end{array}\right\} \tag{b}$$

The normal stress distribution on the cross section AB is shown in Figure 12-18c. Note that an analysis based on elastic behavior has been used to arrive at this stress distribution. Consequently, a stress-strain diagram for the material from which the ring is made should be consulted to ensure that the calculated stresses do not exceed the proportional limit of the material.

PROBLEMS / Section 12-3

12-10   Calculate the maximum compressive and tensile stresses on section aa of the crane hook shown in Figure P12-10 using the Winkler–Bach curved beam theory. Plot the stress distribution. Determine the location of the neutral axis. Compare the maximum compressive and tensile stresses calculated by the Winkler–Bach theory with the corresponding stresses calculated from the straight beam theory ($\sigma_{\max} = 115$ MPa(T), 123 MPa(C)).

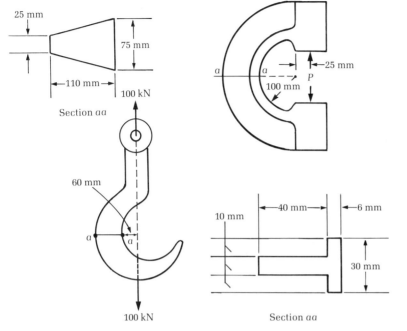

Figure P12-10   **Figure P12-11**

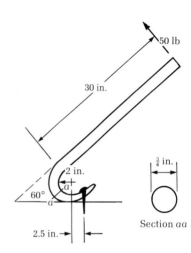

Figure P12-12

12-11 The allowable stress in tension and compression for the material of the clamping device shown in Figure P12-11 is 100 MPa. Use the Winkler–Bach curved beam theory to determine the maximum permissible load P that the device can resist (3652 N).

12-12 A 50-lb force is required at the end of the crowbar shown in Figure P12-12 to pull a nail from a timber. Calculate the maximum tensile and compressive stresses on section aa.

12-4 Thick-Walled Cylinders

We discussed thin-walled cylindrical and spherical pressure vessels in Section 6-5. The salient features of the developments in Section 6-5 are that circumferential normal stress is constant through the thickness of the container, and the radial stress is negligibly small. In this section, we consider thick-walled cylinders that are subjected to internal or external pressures. For **thick-walled cylinders**, we must abandon the assumptions noted here for thin-walled cylinders.

Figure 12-19a shows a cross section of a thick-walled cylinder that is subjected to internal and external pressures p_i and p_o. The internal and external radii of curvature are denoted by r_i and r_o, respectively. A differential element, shown shaded in Figure 12-19a, is isolated from the cylinder as shown in Figure 12-19b. The quantities

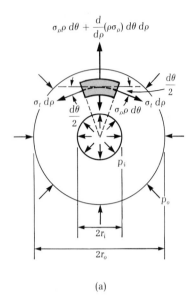

$$\sigma_\rho\rho\, d\theta + \frac{d}{d\rho}(\rho\sigma_\rho)\, d\theta\, d\rho$$

$$\frac{d\theta}{2}$$

$$\sigma_t\, d\rho \qquad \frac{d\theta}{2} \qquad \sigma_\rho\rho\, d\theta \qquad \sigma_t\, d\rho$$

$$p_i$$

$$2r_i$$

$$2r_o$$

(a)

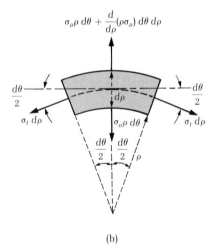

$$\sigma_\rho\rho\, d\theta + \frac{d}{d\rho}(\rho\sigma_\rho)\, d\theta\, d\rho$$

$$\frac{d\theta}{2} \qquad\qquad\qquad d\rho \qquad\qquad \frac{d\theta}{2}$$

$$\sigma_t\, d\rho \qquad \sigma_\rho\rho\, d\theta \qquad \sigma_t\, d\rho$$

$$\frac{d\theta}{2}\ \Big|\ \frac{d\theta}{2}\ \Big/\rho$$

(b)

Figure 12-19

σ_ρ and σ_t denote radial and circumferential stress components that are induced by the pressures p_i and p_o.

Note that the pressure loads cause deformations that are independent of the circumferential coordinate θ. Indeed, stress components and strain components, as well as displacements, are independent of θ.

Radial and circumferential stresses.   Consider the equilibrium of forces acting on the differential element shown in Figure 12-19b. Force equilibrium in the circumferential direction is satisfied identically. Equilibrium of forces in the radial direction requires that

$$\frac{d}{d\rho}(\rho\sigma_\rho)\, d\theta\, d\rho - \sigma_t\, d\theta\, d\rho = 0$$

or

$$\frac{d\sigma_\rho}{d\rho} + \frac{\sigma_\rho - \sigma_t}{\rho} = 0 \tag{12-37}$$

Equation (12-37) contains two unknown stresses, σ_ρ and σ_t. Therefore, these stress components cannot be determined from equilibrium considerations alone. The equilibrium Eq. (12-37) must be augmented with an additional equation involving σ_ρ and σ_t. This additional equation is obtained through the use of the strain-displacement relations (geometrically compatible displacements) and the stress-strain relations for the material from which the cylinder is made.

The stress-strain relations for linearly elastic material behavior are

$$\left.\begin{aligned} \epsilon_\rho &= \frac{1}{E}(\sigma_\rho - v\sigma_t - v\sigma_z) \\ \epsilon_t &= \frac{1}{E}(\sigma_t - v\sigma_\rho - v\sigma_z) \\ \epsilon_z &= \frac{1}{E}(\sigma_z - v\sigma_\rho - v\sigma_t) \end{aligned}\right\} \tag{12-38}$$

In the following derivation, a state of plane stress is assumed to exist in the cylinder. Consequently, the longitudinal normal stress σ_z is zero. The linearly elastic stress-strain relations (Eqs. (12-38)) reduce to

$$\left.\begin{aligned} \sigma_\rho &= \frac{E}{1 - v^2}(\epsilon_\rho + v\epsilon_t) \\ \sigma_t &= \frac{E}{1 - v^2}(\epsilon_t + v\epsilon_\rho) \end{aligned}\right\} \tag{12-39}$$

The required strain-displacement relations are obtained from the basic definitions of ϵ_ρ and ϵ_t. Thus, if $u(\rho)$ is the radial displacement of an arbitrary point in the cylinder,

$$\epsilon_\rho(\rho) = \lim_{\Delta\rho \to 0} \frac{\Delta u}{\Delta\rho} = \frac{du}{d\rho}$$

Moreover, the circumferential strain

$$\epsilon_t(\rho) = \frac{2\pi(\rho + u) - 2\pi\rho}{2\pi\rho} = \frac{u(\rho)}{\rho}$$

Thus

$$\left.\begin{aligned}
\epsilon_\rho &= \frac{du}{d\rho}\\[2mm]
\epsilon_t &= \frac{u}{\rho}
\end{aligned}\right\} \tag{12-40}$$

The strain-displacement relations (Eqs. (12-40)) together with the equilibrium equation (Eq. (12-37)) and the stress-strain relations (Eqs. (12-39)) can be combined to obtain a single differential equation for the radial displacement $u(\rho)$. Having determined $u(\rho)$, we can calculate the strains ϵ_ρ and ϵ_t from Eqs. (12-40) and the stress components σ_ρ and σ_t from Eqs. (12-39).

Substituting Eqs. (12-40) into the stress-strain relations (Eqs. (12-39)) yields

$$\left.\begin{aligned}
\sigma_\rho &= \frac{E}{1 - v^2}\left(\frac{du}{d\rho} + v\,\frac{u}{\rho}\right)\\[2mm]
\sigma_t &= \frac{E}{1 - v^2}\left(\frac{u}{\rho} + v\,\frac{du}{d\rho}\right)
\end{aligned}\right\} \tag{12-41}$$

Equations (12-37) and (12-41) are combined to yield the equilibrium equation in terms of the displacement $u(\rho)$. Thus

$$\frac{d^2 u}{d\rho^2} + \frac{1}{\rho}\frac{du}{d\rho} - \frac{u}{\rho^2} = 0 \tag{12-42}$$

The general solution of this ordinary, second-order differential equation is

$$u(\rho) = C_1\rho + \frac{C_2}{\rho} \tag{12-43}$$

where C_1 and C_2 are constants of integration to be determined from the boundary conditions. Substituting Eq. (12-43) into Eqs. (12-41) leads to the relations

$$\left.\begin{aligned}
\sigma_\rho &= \frac{E}{1 - v^2}\left\{(1 + v)C_1 - \frac{(1 - v)C_2}{\rho^2}\right\}\\[2mm]
\sigma_t &= \frac{E}{1 - v^2}\left\{(1 + v)C_1 + \frac{(1 - v)C_2}{\rho^2}\right\}
\end{aligned}\right\} \tag{12-44}$$

At the internal and external surfaces, we must satisfy the following boundary conditions:

$$\left.\begin{array}{l} \sigma_\rho(r_i) = -p_i \\ \sigma_\rho(r_o) = -p_o \end{array}\right\} \tag{12-45}$$

so that

$$\left.\begin{array}{l} -p_i = \dfrac{E}{1-v^2}\left[(1+v)C_1 - \dfrac{(1-v)C_2}{r_i^2}\right] \\[4mm] -p_o = \dfrac{E}{1-v^2}\left[(1+v)C_1 - \dfrac{(1-v)C_2}{r_o^2}\right] \end{array}\right\} \tag{12-46}$$

Equations (12-46) are algebraic equations that determine the constants of integration C_1 and C_2. Thus

$$\left.\begin{array}{l} C_1 = -\dfrac{1-v}{E}\dfrac{p_o r_o^2 - p_i r_i^2}{(r_o^2 - r_i^2)} \\[4mm] C_2 = -\dfrac{1+v}{E}\dfrac{(p_o - p_i)r_o^2 r_i^2}{(r_o^2 - r_i^2)} \end{array}\right\} \tag{12-47}$$

The radial displacement and the radial and circumferential stresses are readily obtained from Eqs. (12-43) and (12-44) with the aid of Eqs. (12-47). Consequently,

$$u(\rho) = -\frac{(1-v)}{E}\frac{\rho}{(r_o^2 - r_i^2)}\left\{(p_o r_o^2 - p_i r_i^2) + \frac{1+v}{1-v}(p_o - p_i)\frac{r_o r_i^2}{\rho^2}\right\} \tag{12-48a}$$

$$\left\{\begin{array}{l}\sigma_\rho \\ \sigma_t\end{array}\right\} = \frac{(p_i r_i^2 - p_o r_o^2) \pm (p_o - p_i)\dfrac{r_o^2 r_i^2}{\rho^2}}{(r_0^2 - r_i^2)} \tag{12-48b} \tag{12-48c}$$

It is convenient to have analogous formulas for the special cases for which either the internal or the external pressure is zero. From Eqs. (12-48), we obtain the following formulas:

External Pressure Only ($p_i \equiv 0$)

$$u(\rho) = -\frac{1-v}{E}\frac{p_o r_o^2}{(r_o^2 - r_i^2)}\rho\left\{1 + \frac{1+v}{1-v}\left(\frac{r_i}{\rho}\right)^2\right\} \tag{12-49a}$$

$$\left\{\begin{array}{l}\sigma_t \\ \sigma_\rho\end{array}\right\} = \frac{-p_o r_o^2}{(r_o^2 - r_i^2)}\left\{1 \pm \left(\frac{r_i}{\rho}\right)^2\right\} \tag{12-49b} \tag{12-49c}$$

The maximum circumferential stress occurs at $\rho = r_i$, and the maximum radial stress occurs at $\rho = r_o$. Consequently,

$$(\sigma_t)_{max} = \frac{-2p_o r_o^2}{r_o^2 - r_i^2}$$ and $$(\sigma_\rho)_{max} = -p_o$$ (12-49d)

Internal Pressure Only ($p_o \equiv 0$)

$$u(\rho) = \frac{1 - v}{E} \frac{p_i r_i^2}{(r_o^2 - r_i^2)} \rho \left\{ 1 + \frac{1 + v}{1 - v} \left(\frac{r_o}{\rho} \right)^2 \right\}$$ (12-50a)

$$\left\{ \begin{array}{c} \sigma_t \\ \sigma_\rho \end{array} \right\} = \frac{p_i r_i^2}{(r_o^2 - r_i^2)} \left\{ 1 \pm \left(\frac{r_o}{\rho} \right)^2 \right\}$$ (12-50b)
(12-50c)

The maximum circumferential stress and the maximum radial stress each occur at $\rho = r_i$. Thus

$$(\sigma_t)_{max} = \frac{r_o^2 + r_i^2}{r_o^2 - r_i^2} p_i$$ and $$(\sigma_\rho)_{max} = -p_i$$ (12-50d)

The relations given in Eqs. (12-48), (12-49), and (12-50) have been derived on the assumption that a state of plane stress prevails. Notice that these equations remain valid when ϵ_z is independent of ρ. We validate this assertion in the problems at the end of this chapter.

Longitudinal stresses.   A formula for the longitudinal stress σ_z in thick-walled cylinders when the ends are capped can be obtained through the consideration of a free-body diagram of a portion of the cylinder as shown in Figure 12-20. Equilibrium along the axis of the cylinder requires that

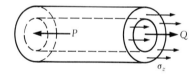

Figure 12-20

$$P = Q$$ (12-51)

P is the resultant force due to the pressure distribution acting on the end of the cylinder, and Q is the resultant force due to longitudinal stress acting over the thickness of the cylinder. If σ_z is assumed to be uniformly distributed over the thickness, then Eq. (12-51) can be written as

$$\pi r_i^2 p_i - \pi r_o^2 p_o = \pi (r_o^2 - r_i^2) \sigma_z$$

from which

$$\sigma_z = \frac{r_i^2 p_i - r_o^2 p_o}{r_o^2 - r_i^2}$$ (12-52)

The general stress-strain relations given in Eq. (12-38), under a plane state of strain $\epsilon_z = 0$, lead to

$$
\left.\begin{aligned}
\sigma_\rho &= \frac{E}{(1+v)(1-2v)} \{(1-v)\epsilon_\rho + v\epsilon_t\} \\
&= \frac{E}{(1+v)(1-2v)} \left\{(1-v)\frac{du}{d\rho} + v\frac{u}{\rho}\right\} \\
\sigma_t &= \frac{E}{(1+v)(1-2v)} \{(1-v)\epsilon_t + v\epsilon_\rho\} \\
&= \frac{E}{(1+v)(1-2v)} \left\{(1-v)\frac{u}{\rho} + v\frac{du}{d\rho}\right\}
\end{aligned}\right\}
\tag{12-53}
$$

Substitution of Eqs. (12-53) into Eq. (12-37) leads to the differential equation given in Eq. (12-42). Application of the boundary conditions given in Eqs. (12-45) leads to the algebraic equations

$$
\left.\begin{aligned}
-p_o &= \frac{E}{(1+v)(1-2v)} \left\{C_1 - (1-2v)\frac{C_2}{r_o^2}\right\} \\
-p_i &= \frac{E}{(1+v)(1-2v)} \left\{C_1 - (1-2v)\frac{C_2}{r_i^2}\right\}
\end{aligned}\right\}
\tag{12-54}
$$

Thus, for plane strain,

$$
\left.\begin{aligned}
C_1 &= -\frac{(1+v)(1-2v)}{E}\frac{p_o r_o^2 - p_i r_i^2}{(r_o^2 - r_i^2)} \\
C_2 &= -\frac{(1+v)}{E}\frac{(p_o - p_i)r_o^2 r_i^2}{(r_o^2 - r_i^2)}
\end{aligned}\right\}
\tag{12-55}
$$

Notice that C_2 is precisely the same as for the plane stress condition. Consequently, the radial displacement for the plane strain condition is

$$
u(\rho) = -\frac{(1+v)}{E}\frac{\rho}{(r_o^2 - r_i^2)}\left\{(1-2v)(p_o r_o^2 - p_i r_i^2) + (p_o - p_i)\frac{r_o^2 r_i^2}{\rho^2}\right\}
\tag{12-56}
$$

The radial and tangential stresses are given by the same formulas as for the plane stress. The radial displacement for the special cases of pressure considered previously are given here for plane strain.

External Pressure Only ($p_i \equiv 0$)

$$
u(\rho) = -\frac{(1+v)}{E}\frac{p_o r_o^2 \rho}{(r_o^2 - r_i^2)}\left\{(1-2v) + \left(\frac{r_i}{\rho}\right)^2\right\}
\tag{12-57}
$$

Internal Pressure Only ($p_o \equiv 0$)

$$u(\rho) = \frac{1 + v}{E} \frac{p_i r_i^2 \rho}{(r_o^2 - r_i^2)} \left\{ (1 - 2v) + \left(\frac{r_o}{\rho} \right)^2 \right\} \tag{12-58}$$

EXAMPLE 12-6

Compute the maximum circumferential stress for a thick-walled cylinder under an internal pressure p_i for the ratios $r_o/r_i = 1.0$, 1.1, 1.2, 1.4, 1.6, 1.8, and 2.0. Also determine the circumferential stress for these ratios using thin-walled cylinder theory.

SOLUTION

Equation (12-50d) is used to calculate the maximum circumferential stress for the thick-walled cylinder, and

$$\sigma_t = \frac{p_i r_i}{t_i} = \frac{p_i}{\dfrac{r_o}{r_i} - 1}$$

is used to calculate the circumferential stress according to thin-walled cylinder theory. It is convenient to tabulate the ratio $(\sigma_t)_{\text{thick}}/(\sigma_t)_{\text{thin}}$ as a function of the ratio r_o/r_i. Thus

$$\frac{(\sigma_t)_{\text{thick}}}{(\sigma_t)_{\text{thin}}} = \frac{1 + \left(\dfrac{r_o}{r_i} \right)^2}{1 + \left(\dfrac{r_o}{r_i} \right)} \tag{a}$$

Table 12-1 lists the results for the specified ratios of r_o/r_i. Table 12-1 shows that the thin-walled formula gives results that differ by less than 5.2% from those given by the more accurate thick-walled theory when the inner radius is equal to or greater than about 90% of the outer radius. Substantial inaccuracies result from the use of the thin-walled formula as r_o/r_i increases.

Table 12-1

$\dfrac{r_o}{r_i}$ = 1.0	1.1	1.2	1.4	1.6	1.8	2.0
$\dfrac{(\sigma_t)_{\text{thick}}}{(\sigma_t)_{\text{thin}}}$ = 1.0	1.052	1.109	1.233	1.369	1.514	1.667

EXAMPLE 12-7

A hollow cylinder with closed ends has an 8-in. internal diameter and a 24-in. external diameter. Determine the maximum circumferential, radial, and longitudinal stresses at a section away from the ends of the cylinder so that any end effects can be neglected for (a) an internal pressure of 10,000 psi and for (b) an external pressure of 10,000 psi.

SOLUTION

(a) Equations (12-50d) give the maximum values for the circumferential and radial stresses for an internal pressure. Thus

$$(\sigma_t)_{max} = \frac{(12)^2 + (4)^2}{(12)^2 - (4)^2} \, 10{,}000 = 12{,}500 \text{ psi}(T) \qquad \text{(a)}$$

and

$$(\sigma_\rho)_{max} = 10{,}000 \text{ psi}(C) \qquad \text{(b)}$$

Each of these stresses occurs at the inner surface of the cylinder.

The longitudinal stress is uniformly distributed over the cross section of the cylinder. Thus, from Eq. (12-52),

$$\sigma_z = \frac{(4)^2}{(12)^2 - (4)^2} \, 10{,}000 = 1250 \text{ psi}(T) \qquad \text{(c)}$$

These stresses are shown on a differential element at the inner surface of the cylinder in Figure 12-21.

(b) The maximum circumferential and radial stresses are given by Eqs. (12-49d) for external pressure only. The maximum circumferential stress occurs at the inner surface, and the maximum radial stress occurs at the outer surface. Consequently,

$$(\sigma_t)_{max} = \frac{-2(12)^2}{(12)^2 - (4)^2} \, 10{,}000 = 22{,}500 \text{ psi}(C) \qquad \text{(d)}$$

and

$$(\sigma_\rho)_{max} = 10{,}000 \text{ psi}(C) \qquad \text{(e)}$$

The longitudinal stress is given by Eq. (12-53).

$$\sigma_z = \frac{-(12)^2}{(12)^2 - (4)^2} \, 10{,}000 = 11{,}250 \text{ psi}(C) \qquad \text{(f)}$$

Observe the significant difference between the magnitudes of the circumferential stress and between the magnitudes of the longitudinal stress for the two cases.

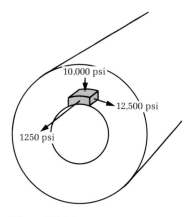

Figure 12-21

EXAMPLE 12-8

A thick-walled cylinder has an internal diameter of 75 mm and an external diameter of 125 mm, and it is acted on by an internal pressure of 21 MPa and an external pressure of 70 MPa. Calculate the maximum circumferential, radial, and longitudinal stresses in the cylinder. Assume that the elastic strength of the material is not exceeded and that end effects are negligible.

SOLUTION

The circumferential and radial stress distributions are calculated from Eqs. (12-48b) and (12-48c). The longitudinal stress is calculated from Eq. (12-52). Thus, for the present problem,

$$\begin{Bmatrix} \sigma_\rho \\ \sigma_t \end{Bmatrix} = \frac{[21(37.5)^2 - 70(62.5)^2] \pm (70 - 21)\dfrac{(37.5)^2(62.5)^2}{\rho^2}}{(62.5)^2 - (37.5)^2}$$

or

$$\begin{Bmatrix} \sigma_\rho \\ \sigma_t \end{Bmatrix} = -97.56 \pm \frac{107,666}{\rho^2} \tag{a, b}$$

The maximum values of the circumferential and radial stresses are observed from Eqs. (a) and (b) to be

$$(\sigma_t)_{max} = 174.1 \text{ MPa(C)} \quad \text{(at the inner surface)} \tag{c}$$

and

$$(\sigma_\rho)_{max} = 70 \text{ MPa(C)} \quad \text{(at the outer surface)} \tag{d}$$

Finally, by Eq. (12-52), the longitudinal stress is

$$\sigma_z = \frac{(37.5)^2 21 - (62.5)^2 70}{2500} = 97.56 \text{ MPa}(C) \tag{e}$$

EXAMPLE 12-9

A gun barrel is assembled by shrinking an outer barrel over an inner barrel so that the maximum principal stress equals 70% of the yield strength of the material. Both members are made of steel for which $\sigma_{YP} = 78$ ksi, $v = 0.292$, and $E = 30 \times 10^6$ psi. The nominal radii of the barrels are $\frac{3}{16}, \frac{6}{16}, \frac{9}{16}$ in. (a) Plot the circumferential stress distribution for each barrel. (b) What should be the difference in radii if the maximum principal stress is to exist? (c) When the gun is fired, an internal pressure of 40 ksi is created. Plot the circumferential stress distribution for the peak internal pressure.

SOLUTION

(a) First calculate the interface pressure p_s that results from the shrink-fit process. From Eq. (12-49d),

$$-0.7(78,000) = \frac{-2p_s r_2^2}{(r_2^2 - r_1^2)} = \frac{-2p_s\left(\dfrac{6}{16}\right)^2}{\left(\dfrac{6}{16}\right)^2 - \left(\dfrac{3}{16}\right)^2} = -2.6667\, p_s$$

or

$$p_s = 20,475 \text{ psi} \tag{a}$$

The circumferential stress distributions for the inner and outer barrels are obtained from Eqs. (12-49b) and (12-50b). Thus

$$(\sigma_t)_{\text{inner}} = -27,300\left[1 + \left(\frac{\frac{3}{16}}{\rho}\right)^2\right] \tag{b}$$

and

$$(\sigma_t)_{\text{outer}} = 16,380\left[1 + \left(\frac{\frac{9}{16}}{\rho}\right)^2\right] \tag{c}$$

These stress distributions are shown in Figures 12-22a and 12-22b, respectively.

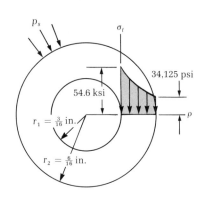

Inner cylinder

(a)

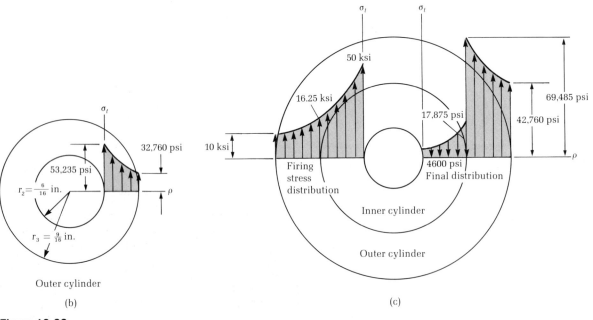

Outer cylinder

(b)

Firing stress distribution

Inner cylinder

Outer cylinder

Final distribution

(c)

Figure 12-22

(b) The difference between the external radius of the inner barrel and the internal radius of the outer barrel required to induce a shrink-fit pressure of 20,475 psi is the sum of the absolute values of the radial displacements of the common surfaces of the two cylinders. Consequently, the outer surface of the inner cylinder has a radial displacement given by Eq. (12-49a) with $\rho = r_2$. Thus

$$u(r_2)_{\text{inner cylinder}} = \frac{-r_2 p_s}{E}\left(\frac{r_2^2 + r_1^2}{r_2^2 - r_1^2} - \nu\right)$$

$$= \frac{-\dfrac{6}{16}(20,475)}{30 \times 10^6}\left(\frac{5}{3} - 0.292\right)$$

$$= -0.000352 \text{ in.} \tag{d}$$

The inner surface of the outer cylinder has a radial displacement given by Eq. (12-50a) with $\rho = r_2$. Thus

$$u(r_2)_{\text{outer cylinder}} = \frac{r_2 p_s}{E}\left(\frac{r_3^2 + r_2^2}{r_3^2 - r_2^2} + \nu\right)$$

$$= \frac{\dfrac{6}{16}(20,475)}{30 \times 10^6}\left(\frac{13}{5} + 0.292\right)$$

$$= 0.000740 \text{ in.} \tag{e}$$

The radial interference required to induce the shrink-fit maximum stress $(0.7\sigma_{\text{YP}})$ is

$$\delta = \left|u(r_2)_{\text{inner}}\right| + \left|u(r_2)_{\text{outer}}\right| = 0.001092 \text{ in.} \tag{f}$$

(c) The firing stresses are calculated from Eq. (12-50b) using $r_i = \frac{3}{16}$ in. and $r_o = \frac{9}{16}$ in. Thus, the distribution of the circumferential stresses in the composite barrel due to firing pressure only is

$$\sigma_t = \frac{p r_1^2}{(r_3^2 - r_1^2)}\left[1 + \left(\frac{r_3}{\rho}\right)^2\right] = 5000\left[1 + \left(\frac{\frac{9}{16}}{\rho}\right)^2\right] \tag{g}$$

This distribution is shown on the left portion of the cross section of the composite cylinder in Figure 12-22c.

The stress distribution at peak pressure during firing is obtained by adding, algebraically, the stresses due to the shrink-fit process (see Figures 12-22a and 12-22b) and those due to the firing process shown in Figure 12-22c. The result of this superposition of circumferential stresses is shown on the right portion of the cross section of the composite cylinder in Figure 12-22c.

PROBLEMS / Section 12-4

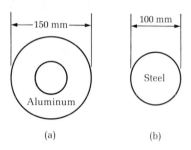

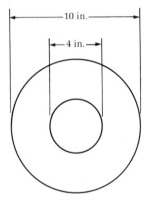

(a)

(b)

Figure P12-15

Figure P12-16

12-13 A steel cylinder is subjected to an internal pressure of 30 MPa and an external pressure of 60 MPa. If the inner and outer diameters of the cylinder are 75 mm and 150 mm, respectively, determine analytical expressions for the radial and circumferential stress distributions. Construct graphs of σ_ρ versus ρ and σ_t versus ρ. Calculate the radial displacements at the inner and outer surfaces ($E = 200$ GPa and $\nu = 0.25$).

12-14 Calculate the radial and circumferential stress distributions in a 12-in. standard steel pipe when the external pressure is the equivalent of 100 ft of water. Construct graphs of σ_ρ versus ρ and σ_t versus ρ using the thick-walled cylinder formulas. Plot the radial and circumferential stress distributions predicted by thin-walled cylinder theory.

12-15 The aluminum cylinder shown in Figure P12-15a is to be shrunk on the steel cylinder shown in Figure P12-15b. If the stresses in the steel and the aluminum are not to exceed 120 MPa and 140 MPa, respectively, determine an approximate value for the internal diameter of the aluminum cylinder. How could this approximate value be improved? (99.758 mm).

12-16 A steel ($E = 30 \times 10^6$ psi, $\nu = 0.25$) cylinder with the dimensions shown in Figure P12-16 is heated to an appropriate temperature and slipped over a steel shaft whose nominal diameter is 4 in. If the radial interference is 0.004 in., determine the shrink-fit pressure and the normal stress of greatest magnitude in the cylinder and in the shaft.

12-5 Pressure Vessels of General Shape

The cylindrical and spherical pressure vessels considered in Section 6-5 are special cases of surfaces of revolution. In this section, we consider a more general surface of revolution.

Figure 12-23a depicts a differential element *ABCD* of a surface of revolution that has been obtained by rotating an arbitrary plane curve *OA* about the z axis. Let s be an arc length along a meridian of the surface of revolution and θ be the polar coordinate that locates a meridian relative to the xz plane. Let n represent a direction normal to the surface. Let $\mathbf{F}_s$ and $\mathbf{F}_\theta$ be the forces and σ_s and σ_θ the corresponding stresses acting on the negative faces of the element. Let **p** be the normal force acting on its surface. The side of the differential element along the meridional coordinate s is ds. This length is the length *BA* in Figure 12-23a. The side along the circumferential direction is $r\,d\theta$, and it is denoted by *BC* in the same figure. The constant thickness of

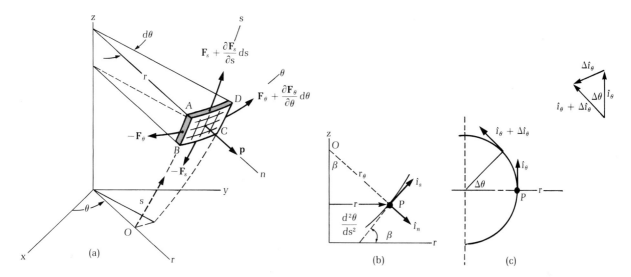

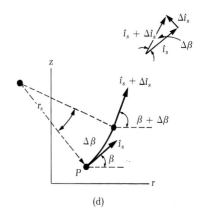

Figure 12-23

the element is t. Thus the area of the face on which σ_θ acts is $t\,ds$; σ_s acts on the area $tr\,d\theta$. The surface area of the differential element is $r\,d\theta\,ds$. The forces acting on the coordinate faces AB and BC and on the surface $ABCD$ are

$$
\begin{aligned}
\mathbf{F}_s &\equiv (\sigma_s tr\,d\theta)\mathbf{i}_s \\
\mathbf{F}_\theta &\equiv (\sigma_\theta t\,ds)\mathbf{i}_\theta \\
\mathbf{p} &\equiv (pr\,d\theta\,ds)\mathbf{i}_n
\end{aligned}
\tag{12-59}
$$

where $\mathbf{i}_s$, $\mathbf{i}_\theta$, and $\mathbf{i}_n$ are unit vectors as shown in Figures 12-23b and 12-23c. Note that $\mathbf{F}_s$ and $\mathbf{F}_\theta$ are the forces acting on the BC and BA faces of the differential element as shown in Figure 12-23a. The corresponding forces acting on the AD and CD faces are slightly different and are obtained by adding differential increments to the forces acting on the coordinate faces BC and AB, respectively. These incremental forces are shown in Figure 12-23a.

Now, for the element $ABCD$, force equilibrium requires that

$$
\left(\mathbf{F}_s + \frac{\partial \mathbf{F}_s}{\partial s}\,ds\right) - \mathbf{F}_s + \left(\mathbf{F}_\theta + \frac{\partial \mathbf{F}_\theta}{\partial \theta}\,d\theta\right) - \mathbf{F}_\theta + \mathbf{p} = 0
$$

or

$$
\frac{\partial \mathbf{F}}{\partial s}\,ds + \frac{\partial \mathbf{F}}{\partial \theta}\,d\theta + \mathbf{p} = 0
\tag{12-60}
$$

Substituting from Eqs. (12-59) yields

$$
\frac{\partial}{\partial s}(\sigma_s tr\mathbf{i}_s) + \frac{\partial}{\partial \theta}(\sigma_\theta t\mathbf{i}_\theta) + pr\mathbf{i}_n = 0
\tag{12-61}
$$

Carrying out the indicated differentiations leads to the expression

$$\frac{\partial}{\partial s}(\sigma_s tr)\mathbf{i}_s + \sigma_s tr \frac{\partial \mathbf{i}_s}{\partial s} + \frac{\partial}{\partial \theta}(\sigma_\theta t)^0 \mathbf{i}_\theta + \sigma_\theta t \frac{\partial \mathbf{i}_\theta}{\partial \theta} + pr\mathbf{i}_n = 0 \qquad (12\text{-}62)$$

Now $(\partial/\partial \theta)(\sigma_\theta t) \equiv 0$ because of the geometric symmetry and the assumed symmetry of the loading.

The derivative of the unit vector $\mathbf{i}_s$ is obtained with the aid of Figure 12-23d. The magnitude of $\Delta \mathbf{i}_s$ is seen to be

$$|\Delta \mathbf{i}_s| = 2 \sin \frac{\Delta \beta}{2} \approx \Delta \beta \qquad (12\text{-}63)$$

Consequently, the magnitude of the derivative $d\mathbf{i}_s/ds$ is

$$\left|\frac{d\mathbf{i}_s}{ds}\right| = \lim_{\Delta \beta \to 0} \frac{\Delta \beta}{\Delta s} = \frac{d\beta}{ds} = \frac{1}{r_s} \qquad (12\text{-}64)$$

where r_s is the radius of curvature of the meridian at the point P.

The direction of the derivative is the same as the direction of $d\mathbf{i}_s$. Thus

$$\frac{d\mathbf{i}_s}{ds} = -\frac{\mathbf{i}_n}{r_s} \qquad (12\text{-}65)$$

Similarly, the magnitude of $d\mathbf{i}_\theta/d\theta$ is seen from Figure 12-23c to be

$$\left|\frac{d\mathbf{i}_\theta}{d\theta}\right| = 1 \qquad (12\text{-}66)$$

and its direction is parallel to the radius r of a cross-sectional circle as shown in Figure 12-23b. Thus

$$\frac{d\mathbf{i}_\theta}{d\theta} = -\cos \beta \mathbf{i}_s - \sin \beta \mathbf{i}_\theta \qquad (12\text{-}67)$$

Substituting Eqs. (12-65) and (12-67) into Eq. (12-61) yields the vector equation

$$\mathbf{i}_s \left\{ \frac{\partial}{\partial s}(rt\sigma_s) - \sigma_\theta t \cos \beta \right\} + \mathbf{i}_n \left\{ \frac{-rt\sigma_s}{r_s} - t\sigma_\theta \sin \beta + rp \right\} = 0 \qquad (12\text{-}68)$$

Since $\mathbf{i}_s$ and $\mathbf{i}_\theta$ are orthogonal and, thus, independent vectors, two scalar equilibrium equations result.

$$\frac{\partial}{\partial s}(r\sigma_s) - \sigma_\theta \cos \beta = 0 \qquad (12\text{-}69)$$

and

$$\frac{\sigma_s}{r_s} + \frac{\sigma_\theta}{r} \sin \beta = \frac{p}{t} \qquad (12\text{-}70)$$

It is shown in differential geometry that the curvature of a normal section in the direction of a cross-sectional circle is the distance OP

shown in Figure 12-23b. This distance is denoted by r_θ. Note that the center of curvature for the normal section in the direction of a cross-sectional circle lies on the axis of revolution. Notice also that r_θ is *not* the radius of a cross-sectional circle.

From Figure 12-23b,

$$r = r_\theta \sin \beta \qquad (12\text{-}71)$$

so that Eq. (12-70) becomes

$$\frac{\sigma_s}{r_s} + \frac{\sigma_\theta}{r_\theta} = \frac{p}{t} \qquad (12\text{-}72)$$

Usually, the meridional stress σ_s can be obtained by taking a free-body diagram of the pressure vessel to either side of a plane perpendicular to the axis of revolution. The circumferential stress σ_θ is determined directly from Eq. (12-72). The following examples illustrate the use of Eq. (12-72).

EXAMPLE 12-10

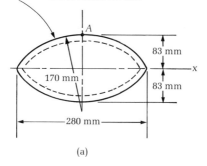

Circular arc of radius 170 mm

170 mm

280 mm

83 mm

83 mm

(a)

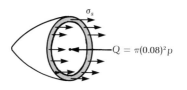

σ_s

$Q = \pi(0.08)^2 p$

(b)

Figure 12-24

Determine the stresses at point A of the football shown in Figure 12-24a. The internal pressure is 0.1 MPa and the skin thickness is 3 mm.

SOLUTION

The surface of the football is assumed to be generated by rotating the circular arc of outer radius 170 mm about the x axis. The radii of curvature at point A of the meridian and a cross-sectional circle are

$$r_s = 168.5 \text{ mm} \qquad \text{and} \qquad r_\theta = 81.5 \text{ mm} \qquad (a)$$

From the free-body diagram shown in Figure 12-24b, the meridional stress is calculated from the equilibrium equation

$$2\pi(0.0815)\sigma_s(0.003) = \pi(0.08)^2 p$$

or, for $p = 0.1$ MPa,

$$\sigma_s = 1.31 \text{ MPa} \qquad (b)$$

Then, from Eq. (12-72),

$$\frac{\sigma_\theta}{0.0815} = \frac{0.1}{0.003} - \frac{1.31}{0.1685} = 25.56$$

or

$$\sigma_\theta = 2.1 \text{ MPa}(T) \qquad (c)$$

EXAMPLE 12-11

A thin-walled toroidal pressure vessel with the dimensions shown in Figure 12-25a is subjected to an internal pressure $p = 30$ psi. Calculate the normal stresses σ_s and σ_θ that occur at points A and B on the outer and the inner surfaces of the toroid.

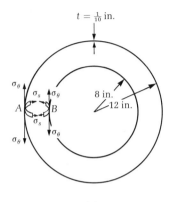

(a)

SOLUTION

At point A, the radii of curvature to be used in formula (12-72) are $r_s = 2$ in. and $r_\theta = 12$ in. Equation (12-72) gives the formula

$$\frac{30}{0.10} = \frac{\sigma_s}{2} + \frac{\sigma_\theta}{12} \tag{a}$$

A second relationship between σ_s and σ_θ is required. σ_s can be calculated from the free-body diagram shown in Figure 12-25b. On this free-body diagram, the resultant force R is

$$R = (2\pi r_\theta t)\sigma_s$$
$$= 2\pi(12)t\sigma_s = 24\pi t\sigma_s \tag{b}$$

Q is the resultant pressure force in a direction parallel to R. Its value is

$$Q = \pi(12^2 - 10^2)p = 44\pi p \tag{c}$$

Equating Eqs. (b) and (c) yields

$$\sigma_s = \frac{11}{6}\frac{p}{t} = \frac{11}{6}\frac{30}{0.10} = 550 \text{ psi}(T) \tag{d}$$

σ_θ can now be calculated from Eq. (a). Therefore,

$$\frac{30}{0.10} = \frac{550}{2} + \frac{\sigma_\theta}{12} \tag{e}$$

or

$$\sigma_\theta = 300 \text{ psi}(T) \tag{f}$$

At point B, the radii of curvature to be used in Eq. (12-72) are $r_\theta = -8$ in. and $r_s = 2$ in. Consequently, Eq. (12-72) gives the formula

$$300 = \frac{\sigma_s}{2} + \frac{\sigma_\theta}{-8} \tag{g}$$

σ_s is obtained from the free-body diagram shown in Figure 12-25c. The resultants of the stress distribution σ_s and the pressure are, respectively,

$$\left.\begin{array}{l} R = 2\pi(8)t\sigma_s \\ Q = \pi(10^2 - 8^2)p \end{array}\right\} \tag{h}$$

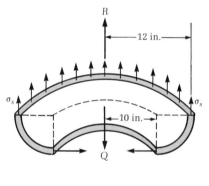

(b)

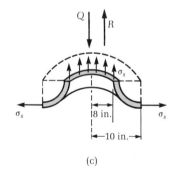

(c)

Figure 12-25

Equating Eqs. (h) gives

$$\sigma_s = \frac{9}{4}\frac{p}{t} = \frac{9}{4}\frac{30}{0.10} = 675 \text{ psi(T)} \tag{i}$$

Now σ_θ can be calculated from Eq. (g). Consequently,

$$300 = \frac{675}{2} - \frac{\sigma_\theta}{8}$$

or

$$\sigma_\theta = 300 \text{ psi(T)} \tag{j}$$

PROBLEMS / Section 12-5

12-17 The thin-walled truncated tank shown in Figure P12-17 is used to store a gas under a pressure of 100 psi. Determine the circumferential and longitudinal stresses at point A shown in the figure if the wall thickness is 0.1 in.

12-18 A thin toroidal pressure vessel is to be used to store a gas at a pressure of 1.25 MPa. The outer and inner diameters of the toroid are 3 m and 2 m, respectively. Determine the minimum thickness of the pressure vessel if the yield point of the material from which it is made is 350 MPa and a safety factor of 2.0 is required. (2 mm).

12-19 Show that the maximum normal stress in a toroidal pressure vessel occurs at point A shown in Figure P12-19.

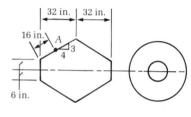

Figure P12-17

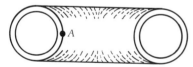

Figure P12-19

12-6 SUMMARY

In this chapter, we developed formulas used to determine flexural stress and shearing stress in beams with unsymmetric cross sections.

Unsymmetric Bending
Normal Stress:

$$\sigma_z = \left(\frac{M_y I_{xx} - M_x I_{xy}}{I_{xx}I_{yy} - I_{xy}^2}\right)x$$
$$+ \left(\frac{M_x I_{yy} - M_y I_{xy}}{I_{xx}I_{yy} - I_{xy}^2}\right)y + \frac{N}{A} \text{ (arbitrary centroidal axes)}$$

$$\sigma_z = \frac{M_\eta}{I_{\eta\eta}}\xi + \frac{M_\xi}{I_{\xi\xi}}\eta + \frac{N}{A} \text{ (principal axes of inertia)}$$

Shearing Stress:

$$q = \left(\frac{V_x I_{xx} - V_y I_{xy}}{I_{xx} I_{yy} - I_{xy}^2}\right) Q_y$$

$$+ \left(\frac{V_y I_{yy} - V_x I_{xy}}{I_{xx} I_{yy} - I_{xy}^2}\right) Q_x \text{ (arbitrary centroidal axes)}$$

$$q = \frac{V_\xi}{I_{\eta\eta}} Q_\eta + \frac{V_\eta}{I_{\xi\xi}} Q_\xi \text{ (principal centroidal axes)}$$

Transformation of Coordinates:

$$\xi = x \sin \beta + y \cos \beta$$

$$\eta = -x \cos \beta + y \sin \beta$$

Curved Beams

The normal stress in a curved beam is given by the Winkler–Bach formula:

$$\sigma_z = \frac{N}{A} + \frac{M}{AR}\left\{1 + \frac{1}{Z}\left(\frac{\xi}{R + \xi}\right)\right\}$$

where M is positive if it tends to increase the curvature of the beam, N is positive in tension, and ξ is positive when measured away from the center of curvature.

Thick-Walled Cylinders

We developed formulas for the normal stresses and radial displacement for a thick-walled cylinder subjected to internal pressure p_i and external pressure p_o in this chapter.

General Thick-Walled Cylinder Formulas:

$$\begin{Bmatrix} \sigma_\rho \\ \sigma_t \end{Bmatrix} = \frac{(p_i r_i^2 - p_o r_o^2) \pm (p_o - p_i)\dfrac{r_o^2 r_i^2}{\rho^2}}{(r_o^2 - r_i^2)} \begin{array}{l} \text{(plane stress or} \\ \text{plane strain)} \end{array}$$

$$u(\rho) = \begin{cases} -\dfrac{(1-v)}{E}\dfrac{\rho}{(r_o^2 - r_i^2)}\left\{(p_o r_o^2 - p_i r_i^2) + \dfrac{1+v}{1-v}(p_o - p_i)\dfrac{r_o^2 r_i^2}{\rho^2}\right\} \\ \qquad\qquad\qquad\qquad\qquad\qquad\qquad \text{(plane stress)} \\ \\ -\dfrac{1+v}{E}\dfrac{\rho}{(r_o^2 - r_i^2)}\left\{(1-2v)(p_o r_o^2 - p_i r_i^2) + (p_o - p_i)\dfrac{r_o^2 r_i^2}{\rho^2}\right\} \\ \qquad\qquad\qquad\qquad\qquad\qquad\qquad \text{(plane strain)} \end{cases}$$

External Pressure Only $(p_i = 0)$:

$$\begin{Bmatrix} \sigma_\rho \\ \sigma_t \end{Bmatrix} = -\frac{p_o r_o^2}{(r_o^2 - r_i^2)} \left\{ 1 \pm \left(\frac{r_i}{\rho} \right)^2 \right\} \quad \text{(plane stress or plane strain)}$$

$$u(\rho) = \begin{cases} -\dfrac{(1-v)}{E} \dfrac{p_o r_o^2}{(r_o^2 - r_i^2)} \rho \left\{ 1 + \dfrac{1+v}{1-v} \left(\dfrac{r_i}{\rho} \right)^2 \right\} \quad \text{(plane stress)} \\[3mm] -\dfrac{(1+v)}{E} \dfrac{p_o r_o^2}{(r_o^2 - r_i^2)} \rho \left\{ (1 - 2v) + \left(\dfrac{r_i}{\rho} \right)^2 \right\} \quad \text{(plane strain)} \end{cases}$$

The maximum circumferential stress occurs at $\rho = r_i$, and the maximum radial stress occurs at $\rho = r_o$. They are

$$(\sigma_t)_{max} = -\frac{2p_o r_o^2}{(r_o^2 - r_i^2)} \quad \text{and} \quad (\sigma_\rho)_{max} = -p_o$$

Internal Pressure Only $(p_o = 0)$:

$$\begin{Bmatrix} \sigma_\rho \\ \sigma_t \end{Bmatrix} = \frac{p_i r_i^2}{(r_o^2 - r_i^2)} \left\{ 1 \pm \left(\frac{r_o}{\rho} \right)^2 \right\} \quad \text{(plane stress and plane strain)}$$

$$u(\rho) = \begin{cases} \dfrac{1-v}{E} \dfrac{p_i r_i^2}{(r_o^2 - r_i^2)} \rho \left\{ 1 + \dfrac{1+v}{1-v} \left(\dfrac{r_o}{\rho} \right)^2 \right\} \quad \text{(plane stress)} \\[3mm] \dfrac{1+v}{E} \dfrac{p_i r_i^2}{(r_o^2 - r_i^2)} \rho \left\{ (1 - 2v) + \left(\dfrac{r_o}{\rho} \right)^2 \right\} \quad \text{(plane strain)} \end{cases}$$

The maximum circumferential and radial stresses occur at $\rho = r_i$. They are

$$(\sigma_t)_{max} = \frac{r_o^2 + r_i^2}{(r_o^2 - r_i^2)} p_i \quad \text{and} \quad (\sigma_\rho)_{max} = -p_i$$

Longitudinal Stress:

$$\sigma_z = \frac{r_i^2 p_i - r_o^2 p_o}{(r_o^2 - r_i^2)}$$

Thin-Walled Pressure Vessels

We extended thin-walled pressure vessel theory to general rotationally symmetric shapes. The additional formula required is

$$\frac{\sigma_s}{r_s} + \frac{\sigma_\theta}{r_\theta} = \frac{p}{t}$$

PROBLEMS / CHAPTER 12

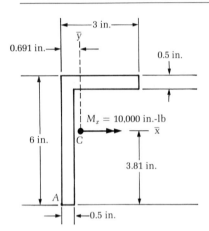

Figure P12-20

12-20   A beam with the cross section shown in Figure P12-20 experiences a maximum internal bending moment of 10,000 in.-lb directed along the positive x axis as shown. Determine the normal stress at point A without using the principal axes of inertia. Repeat the calculations using the principal axes approach. Pertinent properties of the cross section are: $\bar{I}_{xx} = 15.66$ in^4, $\bar{I}_{yy} = 2.70$ in^4, and $\bar{I}_{xy} = 3.64$ in^4 (2677 psi(C)).

12-21   The cross section of a 6-ft long cantilever beam is fabricated by nailing two 2-in. × 6-in. wood planks together to form the T shape shown in Figure P12-21. If the beam is used as a cantilever supporting a concentrated load at its free end as indicated in the figure, calculate the normal stress at point A of the cross section (759.2 psi(T)).

12-22   An S 5 × 10 structural steel beam is used as a cantilever to support an end load as shown in Figure P12-22. (a) Calculate the normal stress at point A of the cross section at the fixed support. (b) Assume that the same load acts in the vertical plane of symmetry and recalculate the normal stress at point A.

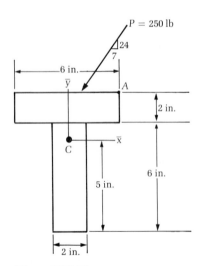

Figure P12-21

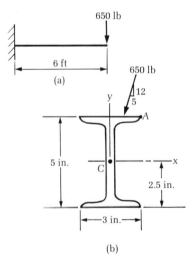

Figure P12-22

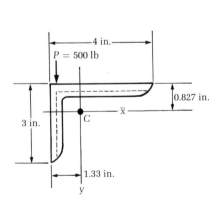

Figure P12-23

Compare the two results (30,912 psi, 9512 psi, 225% greater for the inclined force).

12-23 An L 4 × 3 × $\frac{1}{2}$ structural steel angle is used as a simply supported beam and carries a concentrated force of 500 lb at the midpoint of a 12-ft span. Determine the maximum tensile and compressive normal stresses that are present in the beam (see Figure P12-23).

12-24 A beam is fabricated from a C 6 × 13 structural steel channel and an L 3 × 3 × $\frac{1}{2}$ structural steel angle to form the cross section shown in Figure P12-24. The centroid of the angle section is situated on the horizontal center line of the channel section. Show that (a) $d = 0.09$ in. for the composite section and (b) $\bar{I}_{xx} = 19.62$ in.^{4}, $\bar{I}_{yy} = 6.62$ in.^{4}, and $\bar{I}_{xy} = -1.28$ in.^{4}. Calculate the normal stress at point A if the section is subjected to a bending moment along the positive x axis of 60,000 in.-lb.

12-25 Calculate the shearing stress at locations AA and BB for the cross section shown in Figure P12-25 when $V_x = 2000$ lb and $V_y = 1000$ lb. The thickness of each segment of the cross section is 0.1 in. Assume that bending occurs without twisting.

12-26 The cross section shown in Figure P12-26 is 4 mm thick at each point. Calculate the shearing stress at the locations AA and BB due to shearing forces $V_x = 10$ kN and $V_y = 5$ kN. Pertinent section properties are: $\bar{I}_{xx} = 140.5 \times 10^{-9}$ m^4, $\bar{I}_{yy} = 36.53 \times 10^{-9}$ m^4, $\bar{I}_{xy} = 11.34 \times 10^{-9}$ m^4, and $A = 448$ mm^2 ($\tau_{AA} = 70.5$ MPa).

12-27 Calculate the normal stress on section aa of the toggle pliers shown in Figure P12-27. Assume that the center line of a jaw is the arc of a circle of radius 3.1875 in. with the center at point O. Use the Winkler–Bach theory of curved beams ($\sigma_{max} = 24,904$ psi(C), 21,122 psi(T)).

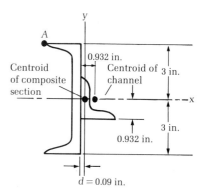

Figure P12-24

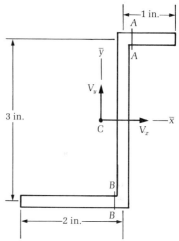

Figure P12-25

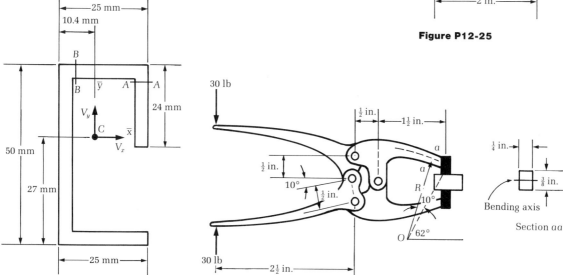

Figure P12-26

Figure P12-27

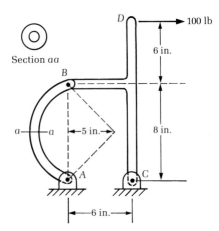

Section aa

Figure P12-28

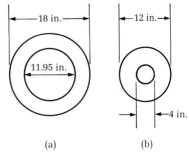

(a) (b)

Figure P12-29

12-28 Member AB of the mechanism shown in Figure P12-28 has the shape of the arc of a circle, and its cross section is tubular with interior and exterior diameters of 0.25 in. and 0.5 in., respectively. Calculate the maximum tensile and compressive stresses acting on section aa using the Winkler–Bach curved beam theory.

12-29 The steel collar shown in Figure P12-29a is heated to an appropriate temperature and fitted over a steel tube whose dimensions are shown in Figure P12-29b. Assuming elastic behavior, determine the shrink-fit pressure p_s. Plot the tangential stress distribution across the thickness of each component. $E = 30 \times 10^6$ psi and $v = 0.25$ ($p_s = 32{,}468$ psi, $(\sigma_t)_{max} = 73{,}053$; $40{,}585$ psi(C) for the inner cylinder, $(\sigma_t)_{max} = 84{,}417$; $51{,}949$ psi(T) for the outer cylinder).

12-30 Two steel cylinders ($E = 30 \times 10^6$ psi, $v = 0.25$, $\sigma_{YP} = 60$ ksi) of nominal diameters 1.0, 3.0, and 5.0 in. are to be subjected to a shrink-fit process. The maximum stress in either cylinder is not to exceed 75% of the yield point of the material. (a) Determine the maximum permissible radial interference and the corresponding shrink-fit pressure. (b) Plot the circumferential stress distribution for each cylinder. (c) Determine the resulting stress distribution in each cylinder when an additional internal pressure of 8000 psi is applied to the assembly. Note that the superposition principle is valid ($\delta = 0.003375$ in., $p_s = 20$ ksi).

12-31 Consider a thick-walled cylinder for which the longitudinal strain ϵ_z is independent of ρ but is not zero. This is the case of a cylinder with end caps. The strain-stress relations are

$$
\left.
\begin{aligned}
\epsilon_\rho &= \frac{1}{E}\left(\sigma_\rho - v\sigma_t - v\sigma_z\right) \\[4pt]
\epsilon_t &= \frac{1}{E}\left(\sigma_t - v\sigma_\rho - v\sigma_z\right) \\[4pt]
\epsilon_z &= \frac{1}{E}\left(\sigma_z - v\sigma_\rho - v\sigma_t\right)
\end{aligned}
\right\}
\tag{a}
$$

Inversion yields

$$
\left.
\begin{aligned}
\sigma_\rho &= \frac{E}{(1+v)(1-2v)}\left\{(1-v)\epsilon_\rho + v\epsilon_t + v\epsilon_z\right\} \\[4pt]
&= \frac{E}{(1+v)(1-2v)}\left\{(1-v)\frac{du}{d\rho} + v\frac{u}{\rho} + v\epsilon_z\right\} \\[6pt]
\sigma_t &= \frac{E}{(1+v)(1-2v)}\left\{(1-v)\epsilon_t + v\epsilon_\rho + v\epsilon_z\right\} \\[4pt]
&= \frac{E}{(1+v)(1-2v)}\left\{(1-v)\frac{u}{\rho} + v\frac{du}{d\rho} + v\epsilon_z\right\}
\end{aligned}
\right\}
\tag{b}
$$

(a) Show that one arrives at the same governing differential equation for the radial displacement as for the case of plane strain or plane stress and, consequently, that the general solutions are identical.

(b) Show that the radial and tangential stresses can be expressed in terms of the constants of integration C_1 and C_2 in the form

$$\sigma_\rho = \frac{E}{(1+v)(1-2v)}\left\{C_1 - \frac{(1-2v)C_2}{\rho^2} + v\epsilon_z\right\}$$

$$\sigma_t = \frac{E}{(1+v)(1-2v)}\left\{C_1 + \frac{(1-2v)C_2}{\rho^2} + v\epsilon_z\right\}$$

(c)

(c) Show that the boundary conditions $\sigma_\rho(r_i) = -p_i$ and $\sigma_\rho(r_o) = -p_o$ lead to the algebraic equations

$$-p_o = \frac{E}{(1+v)(1-2v)}\left\{C_1 - \frac{(1-2v)C_2}{r_o^2} + v\epsilon_z\right\}$$

$$-p_i = \frac{E}{(1+v)(1-2v)}\left\{C_1 - \frac{(1-2v)C_2}{r_i^2} + v\epsilon_z\right\}$$

(d)

which, in turn, lead to the formulas,

$$C_1 = -\frac{(1+v)(1-2v)}{E}\left\{\frac{p_o r_o^2 - p_i r_i^2}{r_o^2 - r_i^2}\right\} - v\epsilon_z$$

$$C_2 = -\frac{(1+v)}{E}\frac{(p_o - p_i)r_o^2 r_i^2}{(r_o^2 - r_i^2)}$$

(e)

(d) By substituting Eqs. (e) into Eqs. (c), show that the formulas obtained for the radial and circumferential stresses in the present case ($\epsilon_z = $ constant) are precisely the same as those for the plane stress condition ($\sigma_z = 0$).

12-32 The aluminum collar shown in Figure P12-32a is heated to an appropriate temperature and fitted over the solid steel cylinder shown in Figure P12-32b. Assuming elastic behavior, determine the shrink-fit pressure p_s. Plot the tangential stress distribution across the thickness of each component. Take $E_{ST} = 210$ GPa and $v_{ST} = 0.25$, and $E_{AL} = 70$ GPa and $v_{AL} = 0.30$ ($p_s = 50.3$ MPa).

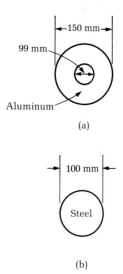

Aluminum

(a)

Steel

(b)

Figure P12-32

CHAPTER 13

Three-Dimensional Stress and Strain Transformations

13-1 Introduction

Chapter 6 dealt with the transformations of stress and strain in two dimensions. We developed formulas that permitted the determination of the magnitudes and orientations of the principal normal stresses and strains and the magnitudes and orientations of the principal shearing stresses and strains. In this chapter, we develop analogous formulas for normal and shearing stresses and normal and shearing strains for three dimensions.

13-2 General Stress Transformations

Consider the three-dimensional state of stress shown in Figure 13-1. Only the stress components that act on the positive coordinate faces are shown in the figure. A positive coordinate face is a face of the differential element for which the outwardly directed normal has the same sense as the positive coordinate axis perpendicular to the face. For example, face $ABCD$ of Figure 13-1 is the positive x coordinate face. The face that lies in the yz plane is the negative x coordinate face.

Instead of labeling the coordinate axes x, y, and z as depicted in Figure 13-1, let the x axis be denoted by x_1, the y axis by x_2, and the z axis by x_3 as shown in Figure 13-2. Then the stress components on the x_1 coordinate face become σ_{11}, σ_{12}, and σ_{13}. In a similar manner, the stress components on the x_2 face are σ_{21}, σ_{22}, and σ_{23}, and those on the x_3 face are σ_{31}, σ_{32}, and σ_{33}. The stress components associated with the x_1, x_2, and x_3 coordinate faces are shown in Figure 13-2.

It is customary to define the **stress vector** that acts on the x_1 coordinate face by $\mathbf{T}_1$, while $\mathbf{T}_2$ and $\mathbf{T}_3$ denote the stress vectors that act on the x_2 and x_3 coordinate faces, respectively.

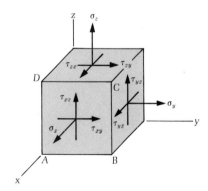

Figure 13-1

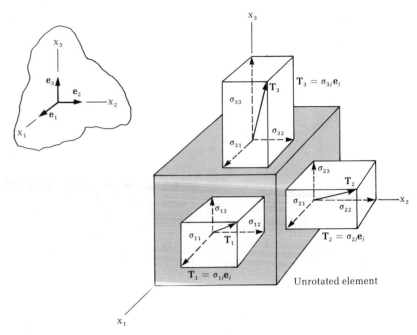

Figure 13-2

Now let $\mathbf{e}_1$, $\mathbf{e}_2$, and $\mathbf{e}_3$ be unit vectors parallel to the x_1, x_2, and x_3 coordinate axes, respectively. Then the stress vectors $\mathbf{T}_1$, $\mathbf{T}_2$, and $\mathbf{T}_3$ can be written as

$$
\left.
\begin{aligned}
\mathbf{T}_1 &= \sigma_{11}\mathbf{e}_1 + \sigma_{12}\mathbf{e}_2 + \sigma_{13}\mathbf{e}_3 \equiv \sum_{j=1}^{3} \sigma_{1j}\mathbf{e}_j \\
\mathbf{T}_2 &= \sigma_{21}\mathbf{e}_1 + \sigma_{22}\mathbf{e}_2 + \sigma_{23}\mathbf{e}_3 \equiv \sum_{j=1}^{3} \sigma_{2j}\mathbf{e}_j \\
\mathbf{T}_3 &= \sigma_{31}\mathbf{e}_1 + \sigma_{32}\mathbf{e}_2 + \sigma_{33}\mathbf{e}_3 \equiv \sum_{j=1}^{3} \sigma_{3j}\mathbf{e}_j
\end{aligned}
\right\}
\tag{13-1}
$$

These formulas can be written in the more compact form,

$$
\mathbf{T}_i = \sum_{j=1}^{3} \sigma_{ij}\mathbf{e}_j, \qquad i = 1, 2, 3
\tag{13-2}
$$

Customarily, the summation sign is discarded; the repeated index *implies* a summation on the index j. Thus Eqs. (13-1) are written as

$$
\mathbf{T}_i = \sigma_{ij}\mathbf{e}_j \qquad i, j = 1, 2, 3
\tag{13-3}
$$

Figure 13-2 shows the stress vectors for the three positive coordinate faces. From this point forward, an index that is repeated in a term will imply a summation on that index over the integers 1, 2, and 3.

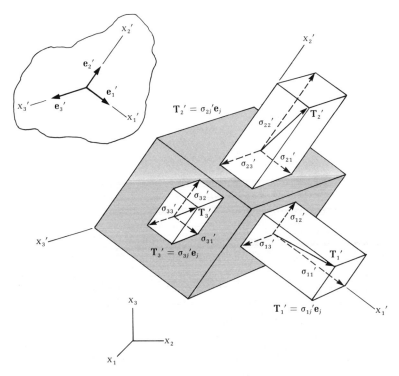

Rotated axes

Figure 13-3

Figure 13-3 shows the stress vectors $\mathbf{T}'_1$, $\mathbf{T}'_2$, and $\mathbf{T}'_3$ associated with the positive coordinate faces of a differential element in a **rotated orientation**. The coordinate system whose axes are perpendicular to the coordinate faces of the rotated element is denoted by x'_1, x'_2, and x'_3. It should be understood that the orientations shown in Figures 13-2 and 13-3 are different ways of representing the same state of stress at the same point.

If $\mathbf{e}'_1$, $\mathbf{e}'_2$, and $\mathbf{e}'_3$ denote unit vectors parallel to the x'_1, x'_2, and x'_3 axes, then the stress vectors that act on the positive coordinate faces of the differential element shown in Figure 13-3 can be written as

$$\mathbf{T}'_1 = \sigma'_{11}\mathbf{e}'_1 + \sigma'_{12}\mathbf{e}'_2 + \sigma'_{13}\mathbf{e}'_3 = \sum_{j=1}^{3} \sigma'_{1j}\mathbf{e}'_j$$

$$\mathbf{T}'_2 = \sigma'_{21}\mathbf{e}'_1 + \sigma'_{22}\mathbf{e}'_2 + \sigma'_{23}\mathbf{e}'_3 = \sum_{j=1}^{3} \sigma'_{2j}\mathbf{e}'_j \qquad (13\text{-}4)$$

$$\mathbf{T}'_3 = \sigma'_{31}\mathbf{e}'_1 + \sigma'_{32}\mathbf{e}'_2 + \sigma'_{33}\mathbf{e}'_3 = \sum_{j=1}^{3} \sigma'_{3j}\mathbf{e}'_j$$

or, using the compact form,

$$\mathbf{T}'_i = \sigma'_{ij}\mathbf{e}'_j, \qquad i, j = 1, 2, 3 \tag{13-5}$$

Suppose that the stress components σ_{ij} are known in the unprimed coordinate system, and that we must calculate the stress components σ'_{ij} for the primed coordinate system. To establish a law of transformation for the stress components, we must establish a law of transformation for the stress vectors $\mathbf{T}_i$ first.

To establish the transformation law for the stress vectors, isolate an infinitesimal tetrahedron of material surrounding point P by passing a cutting plane through the differential parallelogram of Figure 13-2 in the manner shown in Figure 13-4. Denote the area perpendicular to the x_i coordinate axis by ΔA_i and the area of the cutting plane ABC by ΔA. The forces acting on the negative x_i coordinate faces are denoted by $-\mathbf{T}_1 \Delta A_1$, $-\mathbf{T}_2 \Delta A_2$, and $-\mathbf{T}_3 \Delta A_3$ as shown in Figure 13-4. Moreover, let the x'_1 axis of the rotated system shown in Figure 13-3 be perpendicular to the area ABC exposed by the cutting plane. Then the stress vector acting on this plane is $\mathbf{T}'_1$.

According to Newton's law of motion,

$$\mathbf{T}' \Delta A - \mathbf{T}_1 \Delta A_1 - \mathbf{T}_2 \Delta A_2 - \mathbf{T}_3 \Delta A_3 = \rho \Delta V\mathbf{a} \tag{13-6}$$

where ΔV is the volume of the differential element, ρ is its mass density, and $\mathbf{a}$ is the acceleration vector for point P.

Dividing Eq. (13-6) by ΔA and letting $\Delta A \rightarrow 0$ yields

$$\mathbf{T}'_1 = \frac{dA_1}{dA}\mathbf{T}_1 + \frac{dA_2}{dA}\mathbf{T}_2 + \frac{dA_3}{dA}\mathbf{T}_3 \tag{13-7}$$

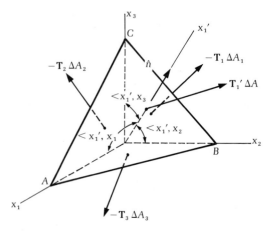

Differential tetrahedron

Figure 13-4

The acceleration term vanishes in the limit because $\lim_{\Delta A \to 0} (\Delta V/\Delta A) \to 0$. Furthermore, the differential quantities

$$\left.\begin{aligned}
\frac{dA_1}{dA} &\equiv \cos(x'_1, x_1) = \ell_{11}\\[2mm]
\frac{dA_2}{dA} &\equiv \cos(x'_1, x_2) = \ell_{12}\\[2mm]
\frac{dA_3}{dA} &\equiv \cos(x'_1, x_3) = \ell_{13}
\end{aligned}\right\} \qquad (13\text{-}8)$$

Observe that ℓ_{11}, ℓ_{12}, and ℓ_{13} are the **direction cosines** of a line perpendicular to the cutting plane ABC. The direction angles to which these direction cosines correspond are shown in Figure 13-4.

Using the definitions given in Eq. (13-8), Eq. (13-7) can be written as

$$\mathbf{T}'_1 = \ell_{11}\mathbf{T}_1 + \ell_{12}\mathbf{T}_2 + \ell_{13}\mathbf{T}_3 = \ell_{1j}\mathbf{T}_j \qquad (13\text{-}9\text{a})$$

By orienting the cutting plane so that first the x'_2 axis and then the x'_3 axis is perpendicular to it, and proceeding as before, we find that

$$\mathbf{T}'_2 = \ell_{2j}\mathbf{T}_j \qquad (13\text{-}9\text{b})$$

and

$$\mathbf{T}'_3 = \ell_{3j}\mathbf{T}_j \qquad (13\text{-}9\text{c})$$

Finally, the three stress vectors $(\mathbf{T}'_1, \mathbf{T}'_2, \mathbf{T}'_3)$ can be written collectively as

$$\mathbf{T}'_i = \ell_{ij}\mathbf{T}_j \qquad i, j = 1, 2, 3 \qquad (13\text{-}10)$$

Note that $\ell_{21} \equiv \cos(x'_2, x_1)$, $\ell_{22} \equiv \cos(x'_2, x_2)$, and $\ell_{23} \equiv \cos(x'_2, x_3)$; and $\ell_{31} \equiv \cos(x'_3, x_1)$, $\ell_{32} \equiv \cos(x'_3, x_2)$, and $\ell_{33} \equiv \cos(x'_3, x_3)$ are the direction cosines of the x'_2 and the x'_3 axes with respect to the x_i coordinate system.

Substituting Eqs. (13-3) and (13-5) into Eq. (13-10) yields

$$\sigma'_{ij}\mathbf{e}'_j = \ell_{ip}\sigma_{pq}\mathbf{e}_q \qquad (13\text{-}11)$$

Now dot both sides of Eq. (13-11) with the unit vector $\mathbf{e}'_k$ and obtain

$$\sigma'_{ij}\mathbf{e}'_j \cdot \mathbf{e}'_k = \ell_{ip}\sigma_{pq}\mathbf{e}_q \cdot \mathbf{e}'_k \qquad (13\text{-}12)$$

Observe that, because of the definition of the scalar product,*

$$\mathbf{e}'_j \cdot \mathbf{e}'_k = \begin{cases} 1 & j = k \\ 0 & j \neq k \end{cases} \qquad (13\text{-}13)$$

---

* For example, $\mathbf{e}'_1 \cdot \mathbf{e}'_1 = \cos 0° = 1$, $\mathbf{e}'_1 \cdot \mathbf{e}'_2 = \cos(\pi/2) = 0$, and so on. Also, see the accompanying figure, $\mathbf{e}'_1 \cdot \mathbf{e}_1 = \cos(x'_1, x_1) = \ell_{11}$, $\mathbf{e}'_1 \cdot \mathbf{e}_2 = \cos(x'_1, x_2) = \ell_{12}$, and so on. Similar reasoning can be used to verify the remaining components of the relations given in Eq. (13-14).

and

$$\mathbf{e}'_k \cdot \mathbf{e}_q = \cos{(x'_k, x_q)} = \ell_{kq} \tag{13-14}$$

Consequently, Eq. (13-12) leads to the transformation law for the stress components,

$$\sigma'_{ij} = \ell_{ip}\ell_{jq}\sigma_{pq}, \qquad i, j, p, q = 1, 2, 3 \tag{13-15}$$

Normal stress formula. The normal stress acting on the area perpendicular to the x'_i coordinate axis is obtained by setting $i = 1$, $j = 1$ in Eq. (13-15). Thus

$$\sigma'_{11} = \ell_{1p}\ell_{1q}\sigma_{pq} \tag{13-16a}$$

or, after performing the implied summations,

$$\begin{aligned}
\sigma'_{11} = &\ \ell_{11}\ell_{11}\sigma_{11} + \ell_{11}\ell_{12}\sigma_{12} + \ell_{11}\ell_{13}\sigma_{13} \\
&+ \ell_{12}\ell_{11}\sigma_{21} + \ell_{12}\ell_{12}\sigma_{22} + \ell_{12}\ell_{13}\sigma_{23} \\
&+ \ell_{13}\ell_{11}\sigma_{31} + \ell_{13}\ell_{12}\sigma_{32} + \ell_{13}\ell_{13}\sigma_{33}
\end{aligned} \tag{13-16b}$$

A more convenient expression for the normal stress σ'_{11} can be established by setting

$$\left. \begin{aligned}
\ell_{11} &\equiv \cos{(x'_1, x_1)} = n_1 \\
\ell_{12} &\equiv \cos{(x'_1, x_2)} = n_2 \\
\ell_{13} &\equiv \cos{(x'_1, x_3)} = n_3
\end{aligned} \right\} \tag{13-17}$$

The angles used in Eq. (13-17) are shown in Figure 13-5. Thus Eq. (13-16b) becomes

$$\begin{aligned}
\sigma'_{11} = &\ n_1 n_1 \sigma_{11} + n_1 n_2 \sigma_{12} + n_1 n_3 \sigma_{13} \\
&+ n_2 n_1 \sigma_{21} + n_2 n_2 \sigma_{22} + n_2 n_3 \sigma_{23} \\
&+ n_3 n_1 \sigma_{31} + n_3 n_2 \sigma_{32} + n_3 n_3 \sigma_{33}
\end{aligned} \tag{13-18a}$$

or, using the compact notation,

$$\sigma'_{11} = \sigma_{ij} n_i n_j, \qquad i, j = 1, 2, 3 \tag{13-18b}$$

This formula is valid for calculating the normal stress on *any* plane whose unit normal vector has the direction cosines n_i. Consequently, if the normal stress on the plane whose normal vector n_i is denoted by σ_n then

$$\sigma_n = \sigma_{ij} n_i n_j, \qquad i, j = 1, 2, 3 \tag{13-19}$$

Since this formula can be used to calculate the normal stress on any plane whose normal vector is n_i, it can be used to calculate the magnitudes of the normal stresses associated with the coordinate faces of a rotated differential volume element. Only the direction cosines as-

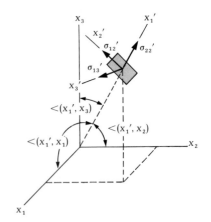

Figure 13-5

sociated with the normal vectors of the coordinate surfaces are required.

Shearing stress formula.   The shearing stress on the coordinate surface perpendicular to the x'_1 axis that acts in the direction of the x'_2 axis is obtained from Eq. (13-15) by setting $i = 1$ and $j = 2$. Consequently,

$$\sigma'_{12} = \ell_{1p}\ell_{2q}\sigma_{pq} \tag{13-20a}$$

or, after carrying out the implied summations,

$$\begin{aligned}
\sigma'_{12} = {} & \ell_{11}\ell_{21}\sigma_{11} + \ell_{11}\ell_{22}\sigma_{12} + \ell_{11}\ell_{23}\sigma_{13} \\
& + \ell_{12}\ell_{21}\sigma_{21} + \ell_{12}\ell_{22}\sigma_{22} + \ell_{12}\ell_{23}\sigma_{23} \\
& + \ell_{13}\ell_{21}\sigma_{31} + \ell_{13}\ell_{22}\sigma_{32} + \ell_{13}\ell_{23}\sigma_{23}
\end{aligned} \tag{13-20b}$$

Now, in addition to the definitions for n_i, denote the direction cosines corresponding to the x'_2 axis by

$$\left.\begin{aligned}
t_1 &\equiv \cos(x'_2, x_1) = \ell_{21} \\
t_2 &\equiv \cos(x'_2, x_2) = \ell_{22} \\
t_3 &\equiv \cos(x'_2, x_3) = \ell_{23}
\end{aligned}\right\} \tag{13-21}$$

so that Eq. (13-20b) can be written as

$$\begin{aligned}
\sigma'_{12} = {} & n_1 t_1 \sigma_{11} + n_1 t_2 \sigma_{12} + n_1 t_3 \sigma_{13} \\
& + n_2 t_1 \sigma_{21} + n_2 t_2 \sigma_{22} + n_2 t_3 \sigma_{23} \\
& + n_3 t_1 \sigma_{31} + n_3 t_2 \sigma_{32} + n_3 t_3 \sigma_{33},
\end{aligned} \tag{13-22a}$$

or, in the compact form,

$$\sigma'_{12} = \sigma_{ij} n_i t_j, \qquad i, j = 1, 2, 3 \tag{13-22b}$$

Equation (13-22b) can be used to calculate the shearing stress in any direction t_i in the plane whose normal vector has the direction cosines n_i. Thus, if σ_t designates the shearing stress in the direction t_i of the plane whose normal vector is n_i, then Eq. (13-22b) becomes

$$\sigma_t = \sigma_{ij} n_i t_j, \qquad i, j = 1, 2, 3 \tag{13-23}$$

Note that when n_i and t_i correspond to the direction cosines of the x'_1 and x'_2 axes, respectively, then $\sigma_t = \sigma'_{12}$, the shearing stress on the positive x'_1 coordinate face that acts in the direction of the positive x'_2 coordinate axis.

EXAMPLE 13-1

The state of stress at a point in a material is shown in Figure 13-6a. The direction cosines of the x'_i coordinate system relative to the x_i

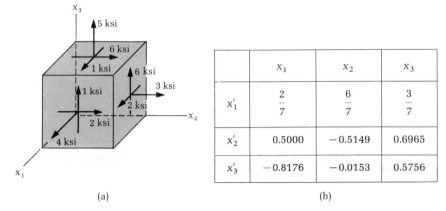

	x_1	x_2	x_3
x_1'	$\dfrac{2}{7}$	$\dfrac{6}{7}$	$\dfrac{3}{7}$
x_2'	0.5000	−0.5149	0.6965
x_3'	−0.8176	−0.0153	0.5756

(a)   (b)

Figure 13-6

coordinate system are given in Figure 13-6b. (a) Show that the direction cosines listed in Figure 13-6b are valid direction cosines. (b) Determine the components of stress that act on the plane that is perpendicular to the x_1' axis.

SOLUTION

(a) The direction cosines must satisfy the condition $n_i n_i = n_1^2 + n_2^2 + n_3^2 = 1$. Moreover, they must also satisfy the orthogonality condition $n_i t_i = n_1 t_1 + n_2 t_2 + n_3 t_3 = 0$.

Consequently, to satisfy the normality conditions, the sum of the squares of the direction cosines appearing in any row or column of Figure 13-6b must be equal to unity. For example, in the first row

$$\left(\frac{2}{7}\right)^2 + \left(\frac{6}{7}\right)^2 + \left(\frac{3}{7}\right)^2 = 1 \qquad\qquad (a)$$

or, in the first column,

$$\left(\frac{2}{7}\right)^2 + (0.5000)^2 + (-0.8176)^2 = 1.0001 \approx 1 \qquad\qquad (b)$$

Moreover, to satisfy the orthogonality conditions, the sum of the products of corresponding direction cosines in any two rows or any two columns must be equal to zero. For example, the direction cosines in the first two rows yield

$$\frac{2}{7}(0.5000) + \frac{6}{7}(-0.5149) + \frac{3}{7}(0.6965) = 0 \qquad\qquad (c)$$

and the direction cosines in the first two columns satisfy the condition

$$\frac{2}{7}\left(\frac{6}{7}\right) + 0.5000(-0.5149) + (-0.8176)(-0.0153) = 0 \qquad\qquad (d)$$

(b) The normal stress on the plane perpendicular to the x_1' axis can be calculated from Eq. (13-19) using the direction cosines for the x_1' axis (first row). Thus

$$\sigma'_{11} = 4\left(\frac{2}{7}\right)\left(\frac{2}{7}\right) + 2\left(\frac{2}{7}\right)\left(\frac{6}{7}\right) + 1\left(\frac{2}{7}\right)\left(\frac{3}{7}\right)$$

$$+ 2\left(\frac{6}{7}\right)\left(\frac{2}{7}\right) + 3\left(\frac{6}{7}\right)\left(\frac{6}{7}\right) + 6\left(\frac{6}{7}\right)\left(\frac{3}{7}\right) \tag{e}$$

$$+ 1\left(\frac{3}{7}\right)\left(\frac{2}{7}\right) + 6\left(\frac{3}{7}\right)\left(\frac{6}{7}\right) + 5\left(\frac{3}{7}\right)\left(\frac{3}{7}\right) = 9.082$$

The components of shearing stress on the same plane that act in the directions of the x'_2 and x'_3 axes are calculated from Eq. (13-23) using the direction cosines for the x'_2 and x'_3 axes, respectively. Accordingly,

$$\sigma'_{12} = \frac{2}{7}[4(0.5000) + 2(-0.5149) + 0.6965]$$

$$+ \frac{6}{7}[2(0.5000) + 3(-0.5149) + 6(0.6965)] \tag{f}$$

$$+ \frac{3}{7}[1(0.5000) + 6(-0.5149) + 5(0.6965)] = 3.974$$

and

$$\sigma'_{13} = \frac{2}{7}[4(-0.8176) + 2(-0.01528) + 1(0.5756)]$$

$$+ \frac{6}{7}[2(-0.8176) + 3(-0.01528) + 6(0.5756)] \tag{g}$$

$$+ \frac{3}{7}[1(-0.8176) + 6(-0.01528) + 5(0.5756)] = 1.584$$

The normal stress and shearing stress components that act on the planes perpendicular to the x'_2 and x'_3 axes can be calculated in exactly the same manner. Consequently, the complete state of stress can be established for a volume element whose orientation is described by the direction cosines listed in Figure 13-6a.

EXAMPLE 13-2

For a plane state of stress (the stress components lie in the same plane, say the x_1x_2 plane), show that the transformation laws for the normal and shearing stress are

$$\sigma_n = \sigma_{\alpha\beta} n_\alpha n_\beta$$

and

$$\sigma_t = \sigma_{\alpha\beta} n_\alpha t_\beta, \qquad \alpha, \beta = 1, 2$$

Show that these formulas are equivalent to the ones derived in Chapter 7.

SOLUTION

For a plane state of stress where the components of stress lie in the x_1x_2 plane, $\sigma_{13} = \sigma_{23} = \sigma_{33} = 0$. Equations (13-18a) and (13-22a) give

$$\sigma_n = n_1 n_1 \sigma_{11} + n_1 n_2 \sigma_{12}$$
$$+ n_2 n_1 \sigma_{21} + n_2 n_2 \sigma_{22} = \sigma_{\alpha\beta} n_\alpha n_\beta \qquad \text{(a)}$$

and

$$\sigma_t = n_1 t_1 \sigma_{12} + n_1 t_2 \sigma_{12}$$
$$+ n_2 t_1 \sigma_{21} + n_2 t_3 \sigma_{22} = \sigma_{\alpha\beta} n_\alpha t_\beta, \qquad \alpha, \beta = 1, 2 \qquad \text{(b)}$$

To show that Eqs. (a) and (b) are equivalent to the transformation equations for normal stress and shearing stress derived in Chapter 7, consider a biaxial state of stress referred to the x_i coordinate system shown in Figure 13-7a. The triangular element shown in Figure 13-7b is obtained from the rectangular element of Figure 13-7a by passing a cutting plane perpendicular to the x_1' axis through the rectangular element. This procedure isolates the stress components $\sigma_n = \sigma_{11}'$ and $\sigma_t = \sigma_{12}'$.

Observe from Figure 13-7b that

$$\left.\begin{array}{l} n_1 \equiv \cos(x_1', x_1) = \cos\theta \\[2mm] n_2 \equiv \cos(x_1', x_2) = \cos\left(\dfrac{\pi}{2} - \theta\right) = \sin\theta \\[4mm] t_1 \equiv \cos(x_2', x_1) = \cos\left(\dfrac{\pi}{2} + \theta\right) = -\sin\theta \\[4mm] t_2 \equiv \cos(x_2', x_2) = \cos\theta \end{array}\right\} \qquad \text{(c)}$$

so that, from Eqs. (a) and (b),

$$\sigma_n = \sigma_{11} \cos^2\theta + \sigma_{22} \sin^2\theta + 2\sigma_{12} \sin\theta \cos\theta$$

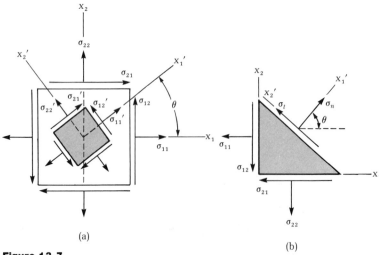

(a)

(b)

Figure 13-7

and

$$\sigma_t = -(\sigma_{11} - \sigma_{22}) \sin \theta \cos \theta + \sigma_{12}(\cos^2 \theta - \sin^2 \theta)$$

These equations are identical to those derived in Chapter 7.

PROBLEMS / Section 13-2

13-1 The direction cosines associated with two distinct lines relative to an x_i coordinate system are listed in Figure P13-1. Show that the direction cosines for each line are valid direction cosines and that the two lines are perpendicular.

Line	x_1	x_2	x_3
A	0.2857	0.4286	0.8571
B	0.3268	−0.8845	0.3333

Figure P13-1

Line	x_1	x_2	x_3
A	0.4444	0.7778	0.4444
B	−0.8435	0.1963	0.5000
C	0.3016	−0.5971	0.7433

Figure P13-2

13-2 The direction cosines associated with three distinct lines relative to an x_i coordinate system are listed in Figure P13-2. Show that the direction cosines listed for each line are valid direction cosines and that each line is perpendicular to the other two lines.

13-3 The state of stress at a point in an engineering member is shown in Figure P13-3. Calculate the normal stress acting on the plane whose normal vector has the direction cosines $\frac{2}{7}, \frac{3}{7}, \frac{6}{7}$. Calculate the shearing stress that acts on this plane in the direction parallel to the line whose direction cosines are 0.3268, −0.8845, and 0.3333 ($\sigma_n = 133.7$ MPa, $\sigma_t = 4.2$ MPa).

13-4 The state of stress at a point in an engineering material is shown in Figure P13-4a. The edges of a differential volume element at this point are

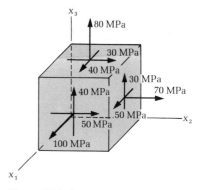

Figure P13-3

	x_1	x_2	x_3
x_1'	0.4444	0.7778	0.4444
x_2'	−0.8435	0.1963	0.5000
x_3'	0.3016	−0.5971	0.7433

(a)

(b)

Figure P13-4

oriented along the x'_1, x'_2, and x'_3 axes, whose direction cosines are listed in Figure P13-4b. Determine the components of stress that act on the face of the element perpendicular to the x'_1 axis ($\sigma'_{11} = 27.7$ ksi, $\sigma'_{12} = -0.924$ ksi, $\sigma'_{13} = -0.798$ ksi).

13-3 Principal Normal Stresses

The normal stress σ_n that acts on the differential area with normal vector n_i assumes different magnitudes as the direction of the normal vector changes. Consequently, σ_n can obtain extreme values for certain sets of direction cosines n_i. These extreme values of σ_n are called **principal normal stresses**, and the orientations of the lines along which they act are called **principal directions**. Principal normal stresses for biaxial states of stress were discussed in Chapter 7.

The shearing stresses on a principal plane of normal stress are zero. Consequently, the stress vector acting on a principal plane is perpendicular to the plane. Accordingly, assume that the x'_1 axis in Figure 13-4 is a principal axis, and then set $\mathbf{T}'_1 \equiv \sigma\hat{n}$. Here σ is a scalar multiplier and $\hat{n}$ is parallel to the x'_1 axis. Equation (13-9a) becomes, for this case,

$$\hat{n}\sigma = n_1\mathbf{T}_1 + n_2\mathbf{T}_2 + n_3\mathbf{T}_3 \tag{13-24a}$$

or

$$\sigma(n_1\mathbf{e}_1 + n_2\mathbf{e}_2 + n_3\mathbf{e}_3) = n_1(\sigma_{11}\mathbf{e}_1 + \sigma_{12}\mathbf{e}_2 + \sigma_{12}\mathbf{e}_3)$$
$$+ n_2(\sigma_{21}\mathbf{e}_1 + \sigma_{22}\mathbf{e}_2 + \sigma_{23}\mathbf{e}_3)$$
$$+ n_3(\sigma_{31}\mathbf{e}_1 + \sigma_{32}\mathbf{e}_2 + \sigma_{33}\mathbf{e}_3) \tag{13-24b}$$

Now collect coefficients of $\mathbf{e}_1$, $\mathbf{e}_2$, and $\mathbf{e}_3$,

$$\mathbf{e}_1\{(\sigma_{11} - \sigma)n_1 + \sigma_{12}n_2 + \sigma_{13}n_3\}$$
$$+ \mathbf{e}_2\{\sigma_{12}n_1 + (\sigma_{22} - \sigma)n_2 + \sigma_{23}n_3\}$$
$$+ \mathbf{e}_3\{\sigma_{13}n_1 + \sigma_{23}n_2 + (\sigma_{33} - \sigma)n_3\} = 0 \tag{13-24c}$$

Because the unit vectors are independent, the quantities in parentheses must vanish. Thus the following set of homogeneous algebraic equations is obtained.

$$\left.\begin{array}{l} (\sigma_{11} - \sigma)n_1 + \sigma_{12}n_2 + \sigma_{13}n_3 = 0 \\ \sigma_{12}n_1 + (\sigma_{22} - \sigma)n_2 + \sigma_{23}n_3 = 0 \\ \sigma_{13}n_1 + \sigma_{23}n_2 + (\sigma_{33} - \sigma)n_3 = 0 \end{array}\right\} \tag{13-25}$$

Equations (13-25) possess nonzero solutions for n_1, n_2, and n_3 if and only if the determinant of the coefficients is zero:

$$\begin{vmatrix} (\sigma_{11} - \sigma) & \sigma_{12} & \sigma_{13} \\ \sigma_{12} & (\sigma_{22} - \sigma) & \sigma_{23} \\ \sigma_{13} & \sigma_{23} & (\sigma_{33} - \sigma) \end{vmatrix} = 0 \tag{13-26}$$

Expansion of the determinant of Eq. (13-26) leads to a third-order polynomial that determines specific values of σ that correspond to nonzero values of n_i—that is, to the orientation of the planes on which the principal values of σ_n act. This polynomial is

$$\sigma^3 - I_1\sigma^2 + I_2\sigma - I_3 = 0 \qquad (13\text{-}27)$$

where

$$I_1 = \sigma_{11} + \sigma_{22} + \sigma_{33} \qquad (13\text{-}28a)$$

$$\begin{aligned} I_2 &= \sigma_{11}\sigma_{22} + \sigma_{22}\sigma_{33} + \sigma_{33}\sigma_{11} \\ &\quad - \sigma_{12}^2 - \sigma_{23}^2 - \sigma_{31}^2 \end{aligned} \qquad (13\text{-}28b)$$

$$I_3 = \begin{vmatrix} \sigma_{11} & \sigma_{12} & \sigma_{13} \\ \sigma_{21} & \sigma_{22} & \sigma_{23} \\ \sigma_{31} & \sigma_{32} & \sigma_{33} \end{vmatrix} \qquad (13\text{-}28c)$$

The direction cosines n_1, n_2, and n_3 that correspond to each root of Eq. (13-27) can be determined from Eqs. (13-25) and the normalizing condition $n_1^2 + n_2^2 + n_3^2 = 1$.

Because the coefficients of the polynomial of Eq. (13-27) are real and the stress components are symmetric, it can be shown that the roots σ of the polynomial are real. Three different sets of roots can be encountered: three distinct roots, two equal roots, and three equal roots.

If the roots are distinct, Eqs. (13-25)—along with the condition $n_1^2 + n_2^2 + n_3^2 = 1$—determine a distinct set of direction cosines for each root. Moreover, the three directions defined by these direction cosines are mutually perpendicular.

If two roots are equal, Eqs. (13-25) and the normalizing condition determine a distinct set of direction cosines for the single distinct root only. Because the principal directions are always orthogonal, the remaining principal stresses must lie in a plane perpendicular to the direction corresponding to the distinct principal stress. Any direction in this plane is a principal direction. Consequently, a convenient direction is selected for one of the equal principal stresses, and the direction of the third principal stress is chosen perpendicular to the second.

If all roots are equal, every direction is a principal direction. This state is the so-called **spherical state of stress**.

EXAMPLE 13-3

The state of stress at a point in a material is shown in Figure 13-8. Determine the magnitude of the principal normal stresses and the orientations of the planes on which they act.

SOLUTION

The coefficients that appear in the characteristic polynomial, Eq. (13-27), are

$$
\left.
\begin{aligned}
I_1 &= 1 + 2 + 3 = 6 \\
I_2 &= 1(2) + 2(3) + 3(1) - 2^2 - 2^2 = 3 \\
I_3 &= \begin{vmatrix} 1 & 2 & 0 \\ 2 & 2 & 2 \\ 0 & 2 & 3 \end{vmatrix} = -10
\end{aligned}
\right\}
\tag{a}
$$

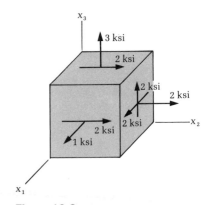

Figure 13-8

Consequently, the characteristic polynomial is

$$
\sigma^3 - 6\sigma^2 + 3\sigma + 10 = 0
\tag{b}
$$

which can be factored into the form

$$
(\sigma + 1)(\sigma - 2)(\sigma - 5) = 0
\tag{c}
$$

The magnitudes of the principal normal stresses are therefore

$$
\left.
\begin{aligned}
\sigma_{\mathrm{I}} &= 5 \text{ ksi} \\
\sigma_{\mathrm{II}} &= 2 \text{ ksi} \\
\sigma_{\mathrm{III}} &= -1 \text{ ksi}
\end{aligned}
\right\}
\tag{d}
$$

Generally, it is not possible to factor the characteristic polynomial, and a more general method of solution is required.

To determine the orientation of the plane on which each principal normal stress acts, we must solve Eqs. (13-25) for each value of σ. Consider $\sigma_{\mathrm{I}} = 5$ ksi first. Equations (13-25) become

$$
\left.
\begin{aligned}
-4n_1 + 2n_2 \quad\;\; &= 0 \\
2n_1 - 3n_2 + 2n_3 &= 0 \\
2n_2 - 2n_3 &= 0
\end{aligned}
\right\}
\tag{e}
$$

Note that these equations must be linearly dependent because the determinant of their coefficients is zero. This means that unique values for the n_i cannot be determined from these relations alone. Only two ratios involving n_1, n_2, and n_3 can be obtained from Eqs. (d).

Solve the first equation for n_1 in terms of n_2, and solve the third equation for n_3 in terms of n_2. In this way,

$$
\left.
\begin{aligned}
\frac{n_1}{n_2} &= \frac{1}{2} \\
\frac{n_2}{n_3} &= 1
\end{aligned}
\right\}
\tag{f}
$$

Using the normalizing condition, we obtain a distinct set of direction cosines for the vector normal to the plane on which $\sigma_{\mathrm{I}} = 5$ ksi acts.

Hence,

$$(\tfrac{1}{2}n_2)^2 + n_2^2 + n_2^2 = 1 \tag{g}$$

which leads to

$$\left.\begin{aligned} n_1^{\mathrm{I}} &= \pm\frac{1}{3} \\[6pt] n_2^{\mathrm{I}} &= \pm\frac{2}{3} \\[6pt] n_3^{\mathrm{I}} &= \pm\frac{2}{3} \end{aligned}\right\} \qquad (\sigma_{\mathrm{I}} = 5 \text{ ksi}) \tag{h}$$

Second root $\sigma_{\mathrm{II}} = 2$ *ksi.* To determine the direction cosines for the normal to the plane on which $\sigma_{\mathrm{II}} = 2$ ksi acts, repeat the foregoing analysis using $\sigma = 2$ in Eqs. (13-25). Accordingly,

$$\left.\begin{aligned} -n_1 + 2n_2 \qquad\;\; &= 0 \\ 2n_1 \qquad + 2n_3 &= 0 \\ 2n_2 + \; n_3 &= 0 \end{aligned}\right\} \tag{i}$$

From the first and third of Eqs. (i),

$$\left.\begin{aligned} \frac{n_1}{n_2} &= \frac{1}{2} \\[6pt] \frac{n_3}{n_2} &= -2 \end{aligned}\right\} \tag{j}$$

so that, from the normalizing condition,

$$(2n_2)^2 + (n_2)^2 + (-2n_2)^2 = 1 \tag{k}$$

Consequently,

$$\left.\begin{aligned} n_1^{\mathrm{II}} &= \pm\frac{2}{3} \\[6pt] n_2^{\mathrm{II}} &= \pm\frac{1}{3} \\[6pt] n_3^{\mathrm{II}} &= \mp\frac{2}{3} \end{aligned}\right\} \qquad (\sigma_{\mathrm{II}} = 2 \text{ ksi}) \tag{l}$$

Third root $\sigma_{\mathrm{III}} = -1$ *ksi.* The direction cosines for the vector normal to the plane on which $\sigma_{\mathrm{III}} = -1$ ksi acts are obtained in the same manner as those for σ_{I} and σ_{II}. Therefore, substituting $\sigma = -1$ ksi into Eqs. (13-25) yields

$$\left.\begin{aligned} 2n_1 + 2n_2 \qquad\;\; &= 0 \\ 2n_1 + 3n_2 + 2n_3 &= 0 \\ 2n_2 + 4n_3 &= 0 \end{aligned}\right\} \tag{m}$$

from which

$$\left.\begin{array}{l} \dfrac{n_1}{n_2} = -1 \\[2ex] \dfrac{n_3}{n_2} = -\dfrac{1}{2} \end{array}\right\} \tag{n}$$

The normalizing condition yields

$$(-n_2)^2 + (n_2)^2 + (-\tfrac{1}{2}n_2)^2 = 1 \tag{o}$$

or

$$\left.\begin{array}{l} n_1^{\text{III}} = \mp\dfrac{2}{3} \\[2ex] n_2^{\text{III}} = \pm\dfrac{2}{3} \\[2ex] n_3^{\text{III}} = \mp\dfrac{1}{3} \end{array}\right\} \quad (\sigma_{\text{III}} = -1 \text{ ksi}) \tag{p}$$

13-4   Tartaglia's Method

In practice, the roots of the characteristic polynomial are rarely integer valued. This situation requires a solution procedure capable of accommodating decimal valued roots. The Scientific Subroutine Package (SSP) contains computer programs designed to determine roots of polynomials of a higher order than the cubic polynomial encountered in three-dimensional stress calculations. The SSP also contains programs that determine the direction cosines. In mathematical terminology, σ_{I}, σ_{II}, and σ_{III} are referred to as **eigenvalues** or **characteristic values**, and the corresponding direction cosines n_i are referred to as **eigen-modes**, **eigen-functions**, or **characteristic functions**.

A direct method that can be used to determine the roots of third-order polynomials when computer facilities are unavailable was devised by Tartaglia.* Tartaglia's method for finding the roots of third-degree polynomials with real coefficients is similar to the quadratic formula used to determine the roots of quadratic equations (second-order polynomials).

Let x_1, x_2, and x_3 be roots of the polynomial

$$x^3 + b_1 x^2 + b_2 x + b_3 = 0 \tag{13-29}$$

---

* A proof of Tartaglia's method can be found in *College Algebra* by Raymond W. Brink, Second Edition, 1951, Appleton-Century-Crofts, Inc.

Then, according to Tartaglia, the three roots are

$$x_1 = z_1 - \frac{c_2}{z_1} - \frac{1}{3}b_1 \qquad\qquad c_2 = \frac{1}{3}b_2 - \frac{1}{9}b_1^2$$

$$x_2 = z_2 - \frac{c_2}{z_2} - \frac{1}{3}b_1 \quad \text{where} \quad c_3 = \frac{2}{27}b_1^3 - \frac{1}{3}b_1b_2 + b_3 \qquad (13\text{-}30)$$

$$x_3 = z_3 - \frac{c_2}{z_3} - \frac{1}{3}b_1 \qquad\qquad z^3 = \frac{-c_3 + \sqrt{c_3^2 + 4c_2^3}}{2}$$

EXAMPLE 13-4

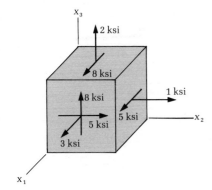

Figure 13-9

The state of stress at a point in a material is shown in Figure 13-9. Determine the magnitude of the principal normal stresses and the orientation of the plane on which the maximum principal normal stress acts.

SOLUTION

The characteristic polynomial is found to be

$$\sigma^3 - 6\sigma^2 - 78\sigma + 108 = 0 \qquad\qquad (a)$$

Consequently, $b_1 = -6$, $b_2 = -78$, and $b_3 = 108$. Accordingly, Eqs. (13-30) give

$$c_2 = \frac{1}{3}(-78) - \frac{1}{9}(-6)^2 = -30$$

$$c_3 = \frac{2}{27}(-6)^3 - \frac{1}{3}(-6)(-78) + 108 = -64 \qquad (b)$$

$$z^3 = \frac{64 + \sqrt{(-64)^2 + 4(-30)^3}}{2} = 32 + 161.14i$$

Here i is the imaginary unit, $\sqrt{-1}$. The three values of z are determined as the roots of z^3. This determination can be accomplished using **DeMoire's theorem**. First write z^3 in its polar form

$$z^3 = re^{i\theta} = 164.3e^{0.439\pi i} \qquad\qquad (c)$$

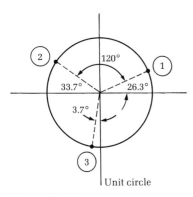

Figure 13-10

The modulus $r = \sqrt{(32)^2 + (161.14)^2} = 164.3$, and the argument $\theta =$ arc tan (161.14/32) rad. Next, construct a unit circle and locate points on its circumference that correspond to the roots of z^3. According to DeMoire's theorem, the first root is located at the angle $\theta/3 = 79°/3 = 26.3°$, and the second root is located at $360/3 = 120$ degrees from the first root. The third root is located at the same angular displacement from the second root. This construction is shown in Figure 13-10.

The three roots Z_i can now be determined from the unit circle:

$$\left.\begin{array}{l} z_1 = 5.48(\cos 26.3° + i \sin 26.3°) = 5.48(0.89649 + 0.44307i) \\ z_2 = 5.48(-\cos 33.7° + i \sin 33.7°) = 5.48(-0.83533 + 0.54975i) \\ z_3 = 5.48(-\sin 3.7° - i \cos 3.7°) = 5.48(-0.06453 - 0.99792i) \end{array}\right\}$$
(d)

Finally, the first root of the characteristic polynomial is obtained as follows from Eqs. (13-30).

$$\sigma_1 = 5.48(0.89649 + 0.44307i) - \frac{-30}{5.48(0.89649 + 0.44307i)} - \frac{1}{3}(-6)$$

$$= 5.48(0.89649 + 0.44307i) + 5.48(0.89649 - 0.44307i) + 2$$

$$= 5.48(0.89649)2 + 2 = 11.8255 \text{ ksi}$$
(e)

The other two roots are determined in precisely the same manner. Hence

$$\sigma_{II} = -7.1552 \text{ ksi}$$
(f)

and

$$\sigma_{III} = 1.2928 \text{ ksi}$$
(g)

A check on the accuracy of these roots can be done by substituting each root into the characteristic polynomial and examining the residual. If the residual is nearly zero, the roots are accurate.

The orientations of the planes on which the principal normal stresses act are obtained in precisely the same manner as they were determined for the example in which the roots were integers.

For principal stress $\sigma_I = 11.8255$ ksi, Eqs. (13-25) become

$$\left.\begin{array}{rl} -8.8255n_1 + 5.0000n_2 + 8.0000n_3 &= 0 \\ 5.0000n_1 - 10.8255n_2 &= 0 \\ 8.0000n_1 - 9.8255n_3 &= 0 \end{array}\right\}$$
(h)

The second and third equations yield $n_2 = 0.4619n_1$ and $n_3 = 0.8142n_2$. The normalization condition is

$$(n_1)^2 + (0.4619n_1)^2 + (0.8142n_2)^2 = 1$$
(i)

which leads to the set of direction cosines

$$\left.\begin{array}{l} n_1^I = \pm 0.7300 \\ n_2^I = \pm 0.3372 \\ n_3^I = \pm 0.5944 \end{array}\right\} \quad (\sigma_I = 11.8255 \text{ ksi})$$
(j)

In precisely the same manner, the direction cosine corresponding to $\sigma_{II} = 1.2928$ ksi and $\sigma_{III} = -7.1552$ ksi are found to be

$$\left.\begin{array}{ll} n_1^{II} = \pm 0.0488 & n_1^{III} = \pm 0.6837 \\ n_2^{II} = \pm 0.8327 \quad \text{and} & n_2^{III} = \mp 0.4192 \\ n_3^{II} = \mp 0.5516 & n_3^{III} = \mp 0.5974 \end{array}\right\}$$
(k)

Table 13-1

	x_1	x_2	x_3
x_1'	± 0.7300	± 0.3372	± 0.5944
x_2'	± 0.0488	± 0.8327	∓ 0.5516
x_3'	± 0.6837	∓ 0.4192	∓ 0.5974

To check the correctness of the direction cosines corresponding to each principal normal stress, recall the condition for the orthogonality of two lines in space; that is, the direction cosines associated with the principal directions must satisfy the conditions $n_i^I n_i^{II} = 0$, $n_i^I n_i^{III} = 0$, and $n_i^{II} n_i^{III} = 0$. These computations can be facilitated by constructing a table of direction cosines like Table 13-1. The orthogonality of the principal axes can be checked conveniently with a hand calculator using the direction cosines shown in Table 13-1.

13-5   Octahedral Normal and Shearing Stress

The normal stress and the shearing stress that act on the eight planes whose normal vectors form equal angles with the **principal axes of normal stress** are called octahedral stresses.

A formula for the octahedral normal stress is obtained by from Eq. (13-19) by letting the x_i' coordinate system coincide with the principal axes of normal stress and setting $n_1 = n_2 = n_3 = 1/\sqrt{3}$. Because the shearing stresses on the principal planes of normal stress are zero, Eq. (13-19) yields

$$\sigma_{\text{oct}} = \frac{1}{3}(\sigma_I + \sigma_{II} + \sigma_{III}) \tag{13-31}$$

A formula for the octahedral shearing stress is obtained with the aid of Figure 13-11. From Eq. (13-9a), the stress vector that acts on the plane whose normal vector has the direction cosines n_i is

$$\mathbf{T}' = n_i \mathbf{T}_i \tag{13-32}$$

In particular, the stress vector that acts on the octahedral plane $(n_i = 1/\sqrt{3}, 1/\sqrt{3}, 1/\sqrt{3})$ is

$$\mathbf{T}_{\text{oct}} = \frac{1}{\sqrt{3}}(\sigma_I \mathbf{e}_1 + \sigma_{II}\mathbf{e}_2 + \sigma_{III}\mathbf{e}_3) \tag{13-33}$$

From Figure (13-11),

$$\tau_{\text{oct}}^2 = \mathbf{T}_{\text{oct}} \cdot \mathbf{T}_{\text{oct}} - \sigma_{\text{oct}}^2 \tag{13-34}$$

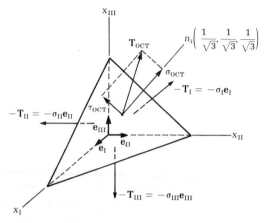

Figure 13-11

or

$$\tau_{oct}^2 = \frac{1}{3}(\sigma_I^2 + \sigma_{II}^2 + \sigma_{III}^2) - \frac{1}{9}(\sigma_I + \sigma_{II} + \sigma_{III})^2 \tag{13-35}$$

A little algebraic manipulation leads to the preferred form

$$\tau_{oct} = \frac{1}{3}\sqrt{(\sigma_I - \sigma_{II})^2 + (\sigma_{II} - \sigma_{III})^2 + (\sigma_{III} - \sigma_I)^2} \tag{13-36}$$

Equations (13-31) and (13-36) express the components of stress that act on the octahedral planes in terms of the principal normal stresses.

EXAMPLE 13-5

Determine the octahedral normal and shearing stresses associated with the state of stress shown in Figure 13-9.

SOLUTION

The principle normal stresses for this state of stress were determined in Example 13-4. They are $\sigma_I = 11.8255$ ksi, $\sigma_{II} = -7.1552$ ksi, and $\sigma_{III} = 1.2928$ ksi. Consequently,

$$\sigma_{oct} = \frac{1}{3}(11.8255 - 7.1552 + 1.2928) = 1.9877 \text{ ksi}$$

and

$$\tau_{oct} = \frac{1}{3}\sqrt{[11.8255 - (-7.1552)]^2 + [-7.1552 - 1.2928]^2}$$

$$\overline{+ [1.2928 - 11.8255]^2} = 7.7644 \text{ ksi}$$

13-6 Principal Shearing Stresses

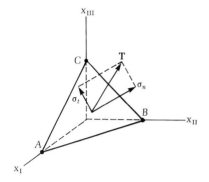

Figure 13-12

To determine the magnitudes of the principal shearing stresses and the orientation of the planes on which they act, it is convenient to let the x_i coordinate system coincide with the principal axes for normal stress as shown in Figure 13-12.

The shear stress acting on the plane ABC of Figure 13-12 whose normal vector has the direction cosines n_i is denoted by σ_t, and the normal stress on this plane is σ_n. $\mathbf{T}$ is the stress vector that acts on the plane ABC. Using the Pythagorean theorem,

$$\sigma_t^2 = \mathbf{T} \cdot \mathbf{T} - \sigma_n^2 \tag{13-37}$$

Equations (13-32), (13-19), and (13-37) yield

$$\sigma_t^2 = (\sigma_I^2 n_1^2 + \sigma_{II}^2 n_2^2 + \sigma_{III}^2 n_3^2) - (\sigma_I n_1^2 + \sigma_{II} n_2^2 + \sigma_{III} n_3^2)^2 \tag{13-38}$$

Also recall that the direction cosines must satisfy

$$n_1^2 + n_2^2 + n_3^2 = 1 \tag{13-39}$$

The determination of the magnitudes of the principal shearing stresses and the orientation of the planes on which they act consists of determining the extreme values of σ_t subject to the constraint given by Eq. (13-39).

Elimination of n_3 from Eq. (13-38) and, subsequently, differentiating the resulting expression with respect to n_1 and then n_2 yields the following two independent relations:

$$2n_1(\sigma_I - \sigma_{III})\{(\sigma_I - \sigma_{III}) - 2[(\sigma_I - \sigma_{III})n_1^2 + (\sigma_{II} - \sigma_{III})n_2^2]\} = 0 \tag{13-40a}$$

and

$$2n_2(\sigma_{II} - \sigma_{III})\{(\sigma_{II} - \sigma_{III}) - 2[(\sigma_I - \sigma_{III})n_1^2 + (\sigma_{II} - \sigma_{III})n_2^2]\} = 0 \tag{13-40b}$$

One solution of these equations is $n_1 = n_2 = 0$ and $n_3 \pm 1$. These are the direction cosines of the x_{III} axis of Figure 13-12 and, therefore, they correspond to the principal plane that is normal to the x_{III} axis. But the shearing stress acting on a principal plane is known to be zero. Consequently, other solutions must be sought.

Another solution is obtained by noting that $n_1 = 0$ and $n_2 = \pm 1/\sqrt{2}$ satisfies Eqs. (13-40), and then $n_i n_i = 1$ yields $n_3 = \pm 1/\sqrt{2}$. Substituting $n_1 = 0$ and $n_2 = n_3 = \pm 1/\sqrt{2}$ into Eq. (13-38) gives the magnitude of the shearing stress associated with the plane defined by these direction cosines. Thus

$$\tau_I = \pm \frac{1}{2}(\sigma_{II} - \sigma_{III}), \quad \left(n_1 = 0, n_2 = n_2 = \pm \frac{1}{\sqrt{2}}\right) \tag{13-41}$$

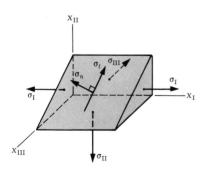

Figure 13-13

This principal shearing stress is shown pictorially in Figure 13-13. Note that τ_I acts on the plane that bisects the planes on which the

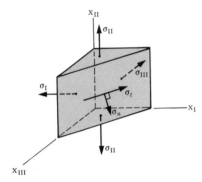

Figure 13-14

principal normal stresses σ_{II} and σ_{III} act. Also note that the normal stress associated with the principal shearing plane is given by Eq. (13-19), so that

$$\sigma_n = \frac{1}{2}(\sigma_{II} + \sigma_{III}), \left(n_1 = 0, n_2 = n_3 = \pm\frac{1}{\sqrt{2}}\right) \tag{13-42}$$

A third solution of Eqs. (13-40) is $n_2 = 0$ and $n_1 = \pm 1/\sqrt{2}$. The constraint relation $n_i n_i = 1$ yields $n_3 = \pm 1/\sqrt{2}$. The normal and shearing stresses that act on this plane are

$$\left.\begin{aligned} \tau_{II} &= \pm\frac{1}{2}(\sigma_I - \sigma_{III}) \\[2mm] \sigma_n &= \frac{1}{2}(\sigma_I + \sigma_{III}) \end{aligned}\right\} \quad n_1 = \pm\frac{1}{\sqrt{2}}, \; n_2 = 0, \quad n_3 = \pm\frac{1}{\sqrt{2}} \tag{13-43}$$

Note that this principal shearing stress acts on a plane that bisects the planes on which σ_I and σ_{III} act as shown in Figure 13-14.

To get the third principal shearing stress and the orientation of the plane on which it acts, eliminate n_2 from Eq. (13-38) and, subsequently, differentiate with respect to n_1 and then n_3. This procedure leads to

$$2n_1(\sigma_I - \sigma_{II})\{(\sigma_I - \sigma_{II}) - 2[(\sigma_I - \sigma_{II})n_1^2 + (\sigma_{III} - \sigma_{II})n_3^2]\} = 0 \tag{13-44a}$$

$$2n_3(\sigma_{III} - \sigma_{II})\{(\sigma_{III} - \sigma_{II}) - 2[(\sigma_I - \sigma_{II})n_1^2 + (\sigma_{III} - \sigma_{II})n_3^2]\} = 0 \tag{13-44b}$$

One set of direction cosines that satisfies Eqs. (13-44) is $n_1 = n_3 = 0$ and $n_2 = \pm 1$. This set of direction cosines is associated with the plane normal to the x_{II} axis. This principal normal stress plane is one for which the shearing stress is known to be zero.

Another set of direction cosines that satisfy Eqs. (13-44) is $n_1 = n_2 = \pm 1/\sqrt{2}$ and $n_3 = 0$. The shear and normal stresses that act on the plane associated with these direction cosines are

$$\left.\begin{aligned} \tau_{III} &= \pm\frac{1}{2}(\sigma_I - \sigma_{II}) \\[2mm] \sigma_n &= \frac{1}{2}(\sigma_I + \sigma_{II}) \end{aligned}\right\} \quad \left(n_1 = n_2 = \pm\frac{1}{\sqrt{2}}, \; n_3 = 0\right) \tag{13-45}$$

Note that this principal shear plane bisects the planes on which σ_I and σ_{II} act. This result is shown in Figure 13-15.

Normally, calculations of the principal shearing stresses and the determination of the planes on which they act are preceded by calculations for the principal normal stresses and the orientations of the

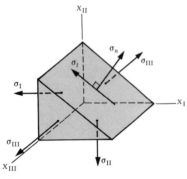

Figure 13-15

planes on which they act. The magnitudes of the principal shearing stresses can then be calculated with relative ease, and the planes on which they act are known to bisect the planes on which the principal normal stresses act.

EXAMPLE 13-6

Determine the magnitudes of the principal shearing stresses and the orientation of the planes on which they act for the state of stress shown in Figure 13-8.

SOLUTION

The principal normal stresses were determined in Example 13-3 to be $\sigma_I = 5$ ksi, $\sigma_{II} = 2$ ksi, and $\sigma_{III} = -1$ ksi. According to Eqs. (13-41), (13-43), and (13-45), the magnitudes of the principal shearing stresses and the accompanying normal stresses are

$$\tau_I = \frac{1}{2}\,[2 - (-1)] = 1.5 \text{ ksi} \qquad \sigma_n = \frac{1}{2}\,(2 - 1) = 0.5 \text{ ksi}$$

$$\tau_{II} = \frac{1}{2}\,[5 - (-1)] = 3.0 \text{ ksi} \qquad \sigma_n = \frac{1}{2}\,(5 - 1) = 2.0 \text{ ksi}$$

$$\tau_{III} = \frac{1}{2}\,[5 - 2] = 1.5 \text{ ksi} \qquad \sigma_n = \frac{1}{2}\,(5 + 2) = 3.5 \text{ ksi}$$

The orientations of the planes on which the principal shearing stress act are shown in Figure 13-16.

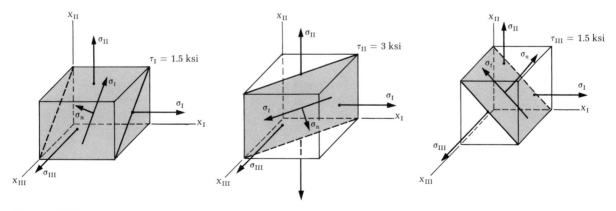

Figure 13-16

PROBLEMS / Sections 13-3, 13-4, 13-5, and 13-6

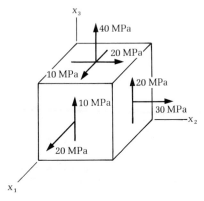

Figure P13-5

13-5 The state of stress at a point P in an engineering member is shown in Figure P13-5. (a) Determine the principal normal stresses and the orientation of the planes on which they act. (b) Determine the principal shearing stresses and the orientation of the planes on which they act. (c) Determine the magnitudes of the octahedral normal and the octahedral shearing stresses.

13-6–13-8 Provide solutions as stipulated in Problem 13-5 for the states of stress shown in Figures P13-6, P13-7, and P13-8.

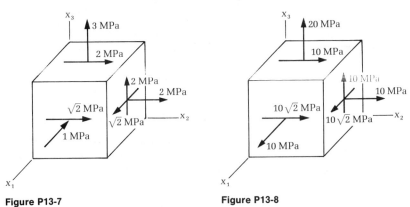

Figure P13-7   **Figure P13-8**

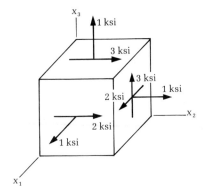

Figure P13-6

13-9 Show that the normal stress that acts on a plane on which a principal shearing stress acts is the average of the two principal normal stresses that lie in the plane containing the normal vector to this plane.

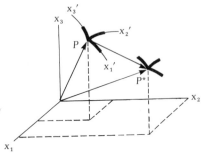

Figure 13-17

13-7   Transformations of Strains

Consider three mutually perpendicular infinitesimal line elements at a point P in a material. Let the x_1' axis coincide with one line element, the x_2' axis coincide with a second line element, and the x_3' axis coincide with the third line element. This situation is depicted in Figure 13-17.

After the material is deformed, P will reside at P^* and the triad of line elements will have undergone a rigid body translation and a rigid body rotation, neither of which produces strains in the material. Additionally, each element of the triad will have had its length altered. The change in length results in a normal strain of a line element. Moreover, the initial right angle between any pair of line elements can also be altered during the deformation process. The angle change between any pair of line elements is shearing strain. Thus three normal strains ϵ'_{11}, ϵ'_{22}, and ϵ'_{33}, and three shearing strains $2\epsilon'_{12}$, $2\epsilon'_{13}$, and $2\epsilon'_{23}$ are required to describe the state of strain at a point.

It can be shown that these normal and shearing strains transform in precisely the same manner as normal and shearing stress. Indeed, the transformation equation for the normal strain of a line element situated along a line whose direction cosines are n_1, n_2, and n_3 is

$$\epsilon_n = \epsilon_{ij} n_i n_j, \qquad i, j = 1, 2, 3 \tag{13-46}$$

and the transformation equation for shearing strain between a pair of line elements whose directions are defined by the direction cosines n_1, n_2, and n_3, and t_1, t_2, and t_3 is

$$\epsilon_{nt} = \epsilon_{ij} n_i t_j \qquad i, j = 1, 2, 3 \tag{13-47}$$

These are precisely analogous to the transformation equations for normal stress and shearing stress, respectively. Accordingly, the procedures followed for determining the stress components acting on the faces of a rotated element can be used to calculate the components of strain associated with a set of rotated line elements. The following examples illustrate the procedures.

EXAMPLE 13-7

Determine the normal and shearing strains associated with line elements whose orientations are given by the list of direction cosines shown in Figure 13-18a when the state of strain relative to the x_1, x_2, and x_3 axes is as shown in Figure 13-18b.

SOLUTION

Expanding Eq. (13-46) yields

$$\epsilon'_n = n_1^2 \epsilon_{11} + n_2^2 \epsilon_{22} + n_3^2 \epsilon_{33} + 2n_1 n_2 \epsilon_{12} + 2n_1 n_3 \epsilon_{13} + 2n_2 n_3 \epsilon_{23}$$

	x_1	x_2	x_3
x'_1	$\dfrac{2}{7}$	$\dfrac{6}{7}$	$\dfrac{3}{7}$
x'_2	0.5000	−0.5149	0.6965
x'_3	−0.8176	−0.0153	0.5756

(a)

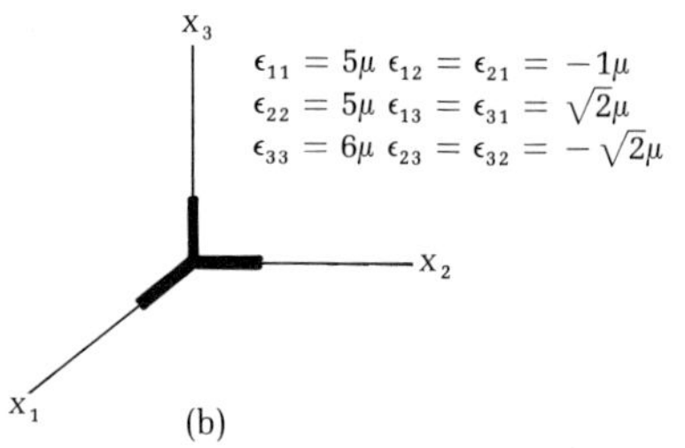

$\epsilon_{11} = 5\mu$ $\epsilon_{12} = \epsilon_{21} = -1\mu$
$\epsilon_{22} = 5\mu$ $\epsilon_{13} = \epsilon_{31} = \sqrt{2}\mu$
$\epsilon_{33} = 6\mu$ $\epsilon_{23} = \epsilon_{32} = -\sqrt{2}\mu$

(b)

Figure 13-18

Consequently,

$$\epsilon'_{11} = 5\left(\frac{2}{7}\right)^2 + 5\left(\frac{6}{7}\right)^2 + 6\left(\frac{3}{7}\right)^2 + 2(-1)\left(\frac{2}{7}\right)\left(\frac{6}{7}\right)$$
$$+ 2\sqrt{2}\left(\frac{2}{7}\right)\left(\frac{3}{7}\right) + 2(-\sqrt{2})\left(\frac{6}{7}\right)\left(\frac{3}{7}\right)$$
$$= 4.0012\mu$$

and, in a similar way,

$$\epsilon'_{22} = 8.0\mu$$

and

$$\epsilon'_{33} = 4.0\mu$$

The shearing strains are given by Eq. (13-47). Consequently, expansion of Eq. (13-47) gives

$$\epsilon'_{nt} = n_1 t_1 \epsilon_{11} + n_2 t_2 \epsilon_{22} + n_3 t_3 \epsilon_{33} + (n_1 t_2 + n_2 t_1)\epsilon_{12}$$
$$+ (n_1 t_3 + n_3 t_1)\epsilon_{13} + (n_2 t_3 + n_3 t_2)\epsilon_{23}$$

Thus

$$\epsilon'_{12} = 5\left(\frac{2}{7}\right)(0.5000) + 5\left(\frac{6}{7}\right)(-0.5149) + 6\left(\frac{3}{7}\right)(0.6965)$$
$$+ \left[\left(\frac{2}{7}\right)(-0.5149) + (0.5000)\left(\frac{6}{7}\right)\right](-1)$$
$$+ \left[\left(\frac{2}{7}\right)(0.6965) + (0.5000)\left(\frac{3}{7}\right)\right]\sqrt{2}$$
$$+ \left[\left(\frac{6}{7}\right)(0.6965) + (-0.5149)\left(\frac{3}{7}\right)\right](-\sqrt{2})$$
$$= 0.0694\mu$$

and, in a similar way,

$$\epsilon'_{13} = 0.00029\mu$$
$$\epsilon'_{23} = 0.02336\mu$$

Observe that the shearing strains are quite small, indicating that the normal strains that have been calculated here are very nearly principal normal strains; that is, the x_i axes of this problem lie very close to the principal strain axes.

EXAMPLE 13-8

For a plane state of strain (the strain components all lie in the same plane, say the $x_1 x_2$ plane), show that the transformation laws for the

normal strain and shearing strain are

$$\epsilon_n = \epsilon_{\alpha\beta} n_\alpha n_\beta$$
$$\epsilon_{nt} = \epsilon_{\alpha\beta} n_\alpha t_\beta \qquad \alpha, \beta = 1, 2$$

Show that these formulas are equivalent to the ones derived in Chapter 7.

SOLUTION

For a plane state of strain, $\epsilon_{13} = \epsilon_{23} = \epsilon_{33} = 0$. Equations (13-46) and (13-47) yield

$$\epsilon_n = n_1 n_1 \epsilon_{11} + n_1 n_2 \epsilon_{12} + n_2 n_1 \epsilon_{21} + n_2 n_2 \epsilon_{22}$$
$$= \epsilon_{\alpha\beta} n_\alpha n_\beta \tag{a}$$

and

$$\epsilon_{nt} = n_1 t_1 \epsilon_{11} + n_1 t_2 \epsilon_{12} + n_2 t_1 \epsilon_{21} + n_2 t_2 \epsilon_{22}$$
$$= \epsilon_{\alpha\beta} n_\alpha t_\beta \qquad \alpha, \beta = 1, 2 \tag{b}$$

To show that Eqs. (a) and (b) are equivalent to the transformation equations derived in Chapter 7 for normal strain and shearing strain, observe from Figure 13-19 that

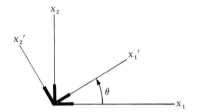

Figure 13-19

$$\left. \begin{array}{l} n_1 = \cos \theta \\ n_2 = \sin \theta \\ t_1 = -\sin \theta \\ t_2 = \cos \theta \end{array} \right\} \tag{c}$$

Consequently, Eqs. (a) and (b) become

$$\left. \begin{array}{l} \epsilon_n = \epsilon_{11} \cos^2 \theta + \epsilon_{22} \sin^2 \theta + 2\epsilon_{12} \sin \theta \cos \theta \\ \epsilon_{nt} = -(\epsilon_{11} - \epsilon_{22}) \sin \theta \cos \theta + \epsilon_{12}(\cos^2 \theta - \sin^2 \theta) \end{array} \right\} \tag{d}$$

Equations (d) are identical to the transformation equations derived in Chapter 7. Note that $2\epsilon_{nt} = \gamma_{\xi\eta}$.

PROBLEMS / Section 13-7

In the following problems, the components of strain are given as elements of the array:

$$\begin{vmatrix} \epsilon_{11} & \epsilon_{12} & \epsilon_{13} \\ \epsilon_{21} & \epsilon_{22} & \epsilon_{23} \\ \epsilon_{31} & \epsilon_{32} & \epsilon_{33} \end{vmatrix}$$

13-10 The state of strain at a point in a structural member is given in Figure P13-10. Calculate the normal strain that a line element undergoes when its direction is defined by direction cosines $\frac{2}{7}$, $\frac{3}{7}$, and $\frac{6}{7}$. Calculate the shearing

$$\begin{vmatrix} 400 & 300 & 200 \\ 300 & 500 & 100 \\ 200 & 100 & 600 \end{vmatrix} \times 10^{-6}$$

Figure P13-10

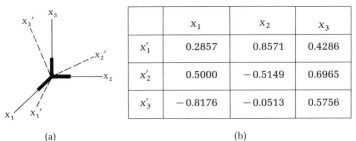

	x_1	x_2	x_3
x_1'	0.2857	0.8571	0.4286
x_2'	0.5000	-0.5149	0.6965
x_3'	-0.8176	-0.0513	0.5756

$$\begin{vmatrix} 600 & 250 & 0 \\ 250 & 600 & 0 \\ 0 & 0 & 800 \end{vmatrix} \times 10^{-6}$$

(a)   (b)   (c)

Figure P13-11

strain between this line element and a second line element whose direction cosines are 0.3268, −0.8845, and 0.3333.

13-11 The state of strain at a certain point in a structural member referred to the x_i coordinate system shown in Figure P13-11a is given in Figure P13-11c. Determine the components of strain referred to the x_i' coordinate system. The orientation of the x_i' coordinate system is established by the direction cosines listed in Figure P13-11b.

13-12 The state of strain relative to an x_i coordinate system is listed in Figure P13-12a. Determine the components of strain referred to the x_i' axes. The orientation of the x_i' coordinate axes relative to the x_i coordinate axes is established by the direction cosines listed in Figure P13-12b.

$$\begin{vmatrix} 100 & 300 & 0 \\ 300 & 200 & 100\sqrt{5} \\ 0 & 100\sqrt{5} & 300 \end{vmatrix} \times 10^{-6}$$

	x_1	x_2	x_3
x_1'	0.4444	0.7778	0.4444
x_2'	-0.8435	0.1963	0.5000
x_3'	0.3016	-0.5971	0.7433

(a)   (b)

Figure P13-12

13-8 Principal Normal Strains

Equation (13-46) shows that the normal strain of a line element through a point P in a material is a function of the direction of the line element. It is of considerable interest to be able to determine both the directions of the line elements that undergo the maximum and minimum normal strains and the magnitudes of the strains associated with these line elements. The problem of determining the magnitudes and directions of the principal normal strains is mathematically identical to the problem of determining the magnitudes and directions of the principal normal stresses. In the present case, the extreme values of the function

$$\epsilon_n = \epsilon_{ij} n_i n_j \qquad \text{(13-48a)}$$

are sought, subject to the relation

$$n_i n_i = 1 \qquad \text{(13-48b)}$$

Because of the analogy between the principal strain problem and the principal stress problem, the equations and procedures developed for principal normal stresses can be used in the solution of principal normal strain problems by setting $\sigma_{11} = \epsilon_{11}$, $\sigma_{22} = \epsilon_{22}$, $\sigma_{33} = \epsilon_{33}$, $\sigma_{12} = \epsilon_{12}$, $\sigma_{13} = \epsilon_{13}$, and $\sigma_{23} = \epsilon_{23}$. Thus the characteristic determinant becomes

$$\begin{vmatrix} (\epsilon_{11} - \epsilon) & \epsilon_{12} & \epsilon_{13} \\ \epsilon_{21} & (\epsilon_{22} - \epsilon) & \epsilon_{23} \\ \epsilon_{31} & \epsilon_{32} & (\epsilon_{33} - \epsilon) \end{vmatrix} = 0 \qquad \text{(13-49)}$$

where ϵ represents the magnitude of a principal normal strain. Expansion of Eq. (13-49) leads to the characteristic polynomial

$$\epsilon^3 - J_1 \epsilon^2 + J_2 \epsilon - J_3 = 0 \qquad \text{(13-50)}$$

where

$$J_1 = \epsilon_{11} + \epsilon_{22} + \epsilon_{33} \qquad \text{(13-51a)}$$

$$J_2 = \epsilon_{11}\epsilon_{22} + \epsilon_{22}\epsilon_{33} + \epsilon_{33}\epsilon_{11} - \epsilon_{12}^2 - \epsilon_{13}^2 - \epsilon_{23}^2 \qquad \text{(13-51b)}$$

$$J_3 = \begin{vmatrix} \epsilon_{11} & \epsilon_{12} & \epsilon_{13} \\ \epsilon_{21} & \epsilon_{22} & \epsilon_{23} \\ \epsilon_{31} & \epsilon_{32} & \epsilon_{33} \end{vmatrix} \qquad \text{(13-51c)}$$

The quantities J_1, J_2, and J_3 are called strain invariants.

Having determined the roots of Eq. (13-50), we can now determine the direction cosines corresponding to *each* root from the equations

$$\left. \begin{array}{l} (\epsilon_{11} - \epsilon)n_1 + \epsilon_{12}n_2 + \epsilon_{13}n_3 = 0 \\ \epsilon_{21}n_1 + (\epsilon_{22} - \epsilon)n_2 + \epsilon_{23}n_3 = 0 \\ \epsilon_{31}n_1 + \epsilon_{32}n_2 + (\epsilon_{33} - \epsilon)n_3 = 0 \end{array} \right\} \qquad \text{(13-52)}$$

and the normalizing relation

$$n_1^2 + n_2^2 + n_3^2 = 1 \qquad \text{(13-53)}$$

It can be shown that the principal axes of strain are mutually perpendicular. Moreover, the roots of the characteristic polynomial (13-50) are real because the strains are symmetric and the coefficients in Eq. (13-50) are real. This property suggests three different types of roots: three distinct roots, two equal roots, or three equal roots.

If the roots are distinct, Eqs. (13-52) and (13-53) determine a distinct set of direction cosines for each of the three roots. Moreover, as noted previously, these three directions are mutually orthogonal.

If only one root is distinct, say ϵ_{III}, Eqs. (13-52) and (13-53) determine a distinct set of direction cosines for ϵ_{III} only. The principal directions corresponding to $\epsilon_{I} = \epsilon_{II}$ lie in the plane perpendicular to the principal direction associated with ϵ_{III}. Furthermore, any direction in this plane is a principal direction. Consequently, a convenient direction is chosen for, say, ϵ_{I}, and the direction for ϵ_{II} is required to be perpendicular to it.

If all roots are equal, every direction is a principal direction. This state is the so-called **spherical state of strain**.

Finally, the line elements that experience the principal normal strains are characterized by zero shearing strains.

EXAMPLE 13-9

Determine the magnitudes of the principal normal strains and the orientations of the line elements that experience these strains for the state of strain shown in Figure 13-20.

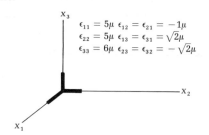

$$\epsilon_{11} = 5\mu \quad \epsilon_{12} = \epsilon_{21} = -1\mu$$
$$\epsilon_{22} = 5\mu \quad \epsilon_{13} = \epsilon_{31} = \sqrt{2}\mu$$
$$\epsilon_{33} = 6\mu \quad \epsilon_{23} = \epsilon_{32} = -\sqrt{2}\mu$$

Figure 13-20

SOLUTION

The strain invariants are

$$J_1 = 16\mu, \ J_2 = 80\mu^2, \ \text{and} \ J_3 = 128\mu^3 \tag{a}$$

so that the characteristic polynomial becomes

$$\epsilon^3 - 16\mu\epsilon^2 + 80\mu^2\epsilon - 128\mu^3 = 0 \tag{b}$$

To solve this equation set

$$\epsilon = \mu\eta \tag{c}$$

and obtain

$$\eta^3 - 16\eta^2 + 80\eta - 128 = 0 \tag{d}$$

The roots of Eq. (d) are

$$\eta_I = 8 \quad \text{and} \quad \eta_{II} = \eta_{III} = 4 \tag{e}$$

so that, by Eq. (c), the magnitudes of the principal normal strains are

$$\epsilon_I = 8\mu \quad \text{and} \quad \epsilon_{II} = \epsilon_{III} = 4\mu \tag{f}$$

Notice that Eq. (b) possesses two equal roots. Therefore, the sets of direction cosines corresponding to these roots are arbitrary except that they must define directions that are perpendicular to the principal strain axis corresponding to ϵ_I.

The orientations of the axes of principal strain are determined as follows.

Principal strain $\epsilon_\text{I} = 8\mu$.   Eqs. (13-52) become

$$\left. \begin{array}{rrr} -3n_1 - & n_2 + \sqrt{2}n_3 = 0 \\ -n_1 - & 3n_2 - \sqrt{2}n_3 = 0 \\ \sqrt{2}n_1 - & \sqrt{2}n_2 - & 2n_3 = 0 \end{array} \right\} \qquad (g)$$

Eliminating n_1 from the first two equations gives $n_2 = -(1/\sqrt{2})n_3$, and, using this result in the first equation gives $n_1 = (1/\sqrt{2})n_3$. The normalization condition yields

$$\left(\frac{1}{\sqrt{2}} n_3 \right)^2 + \left(-\frac{1}{\sqrt{2}} n_3 \right)^2 + n_3^2 = 1$$

or

$$n_3 = \pm \frac{1}{\sqrt{2}}$$

Consequently, the direction cosines associated with the element that undergoes the principal normal strain $\epsilon_\text{I} = 8\mu$ are

$$\left. \begin{array}{l} n_1^\text{I} = \pm \dfrac{1}{2} \\[10pt] n_2^\text{I} = \mp \dfrac{1}{2} \qquad (\epsilon_\text{I} = 8\mu) \\[10pt] n_3^\text{I} = \pm \dfrac{1}{\sqrt{2}} \end{array} \right\} \qquad (h)$$

Principal strain $\epsilon_\text{II} = \epsilon_\text{III} = 4\mu$.   Equations (13-52) become

$$\left. \begin{array}{rrr} n_1 - & n_2 + \sqrt{2}n_3 = 0 \\ -n_1 + & n_2 - \sqrt{2}n_3 = 0 \\ \sqrt{2}n_1 - & \sqrt{2}n_2 + & 2n_3 = 0 \end{array} \right\} \qquad (i)$$

Only one equation in this set is linearly independent. Denote the direction cosines for ϵ_I, ϵ_II, and ϵ_III by n_i^I, n_i^II, and n_i^III, respectively. The line elements that experience the normal strains ϵ_II and ϵ_III are perpendicular to ϵ_I, the distinct normal strain. Consequently, two additional equations are obtained from among the three sets of direction cosines because of the orthogonality requirements.

$$\left. \begin{array}{l} n_1^\text{I} n_1^\text{II} + n_2^\text{I} n_2^\text{II} + n_3^\text{I} n_3^\text{II} = 0 \\ n_1^\text{I} n_1^\text{III} + n_2^\text{I} n_2^\text{III} + n_3^\text{I} n_3^\text{III} = 0 \end{array} \right\} \qquad (j)$$

Moreover, the normality conditions corresponding to the principal strains ϵ_{II} and ϵ_{III} provide two additional equations. They are

$$\left. \begin{array}{l} (n_1^{II})^2 + (n_2^{II})^2 + (n_3^{II})^2 = 1 \\ (n_1^{III})^2 + (n_2^{III})^2 + (n_3^{III})^2 = 1 \end{array} \right\} \tag{k}$$

Finally, note that any direction perpendicular to the principal strain ϵ_I is a principal direction. It is customary to select a convenient axis in the plane perpendicular to the ϵ_I axis for, say, the direction of ϵ_{II}, and then determine the ϵ_{III} direction to be perpendicular to both the directions of ϵ_I and ϵ_{II}. Therefore, a sixth equation among the direction cosines n_i^{II} and n_i^{III} is

$$n_1^{II} n_1^{III} + n_2^{II} n_2^{III} + n_3^{II} n_3^{III} = 0 \tag{l}$$

Now

$$n_1^I = \pm \frac{1}{2}, \, n_2^I = \mp \frac{1}{2}, \quad \text{and} \quad n_3^I = \pm \frac{1}{\sqrt{2}}$$

Consequently, six equations among the direction cosines n_i^{II} and n_i^{III} result.

A solution of these equations is

$$\left. \begin{array}{l} n_1^{II} = \mp \dfrac{1}{3\sqrt{2}}, \, n_2^{II} = \pm \dfrac{1}{\sqrt{2}}, \, n_3^{II} = \pm \dfrac{2}{3} \\[3mm] n_1^{III} = \pm \dfrac{5}{\sqrt{34}}, \, n_2^{III} = \mp \dfrac{1}{\sqrt{34}}, \, n_3^{III} = \pm \dfrac{2}{\sqrt{17}} \end{array} \right\} \tag{m}$$

Check to see that each set of n_i satisfies the normality requirement individually and that the orthogonality conditions are satisfied by each possible combination of sets of direction cosines.

PROBLEMS / Section 13-8

13-13 Determine the principal normal strains and the orientations of the elements that undergo these strains for the state of strain shown in Figure P13-13.

13-14 Determine the principal normal strains and the orientations of the elements that undergo these strains for the state of strain shown in Figure P13-14.

13-15 Supply the information requested in Problem 13-14 for the state of strain shown in Figure P13-15.

$$\begin{vmatrix} 200 & 0 & 100 \\ 0 & 300 & 200 \\ 100 & 200 & 400 \end{vmatrix} \times 10^{-6}$$

Figure P13-13

$$\begin{vmatrix} 600 & 250 & 0 \\ 250 & 600 & 0 \\ 0 & 0 & 800 \end{vmatrix} \times 10^{-6}$$

Figure P13-14

$$\begin{vmatrix} 100 & 300 & 0 \\ 300 & 200 & 100\sqrt{5} \\ 0 & 100\sqrt{5} & 300 \end{vmatrix} \times 10^{-6}$$

Figure P13-15

13-9 Principal Shearing Strains

Equation (13-47) shows that the shearing strain between any two initially perpendicular line elements depends on the initial directions of these line elements. Consequently, the magnitude of this shearing strain assumes different values as the initial directions of the pair are varied. The magnitude of the shearing strain ϵ_{nt} acquires extreme values for particular initial directions of the pair of elements. The extreme values of shearing strain are called **principal shearing strains**, and the directions of the line elements that undergo these strains are called **principal directions of shearing strains**.

Mathematically, the magnitudes of the principal shearing strains and the orientations of the line elements that experience them are obtained by determining the extreme values of

$$\epsilon_{nt} = \epsilon_{ij} n_i t_j \tag{13-54}$$

subject to the orthogonality and normalizing conditions

$$n_i n_i = 1, \; t_i t_i = 1, \quad \text{and} \quad n_i t_i = 0 \tag{13-55}$$

Customarily, the solution to this problem is facilitated by letting the x_1, x_2, x_3 coordinate system coincide with the principal axes of normal strain as shown in Figure 13-21. Because the shearing strains are zero for line elements situated along the principal axes of normal strain, Eq. (13-54) becomes

$$\epsilon_{nt} = n_1 t_1 \epsilon_{\mathrm{I}} + n_2 t_2 \epsilon_{\mathrm{II}} + n_3 t_3 \epsilon_{\mathrm{III}} \tag{13-56}$$

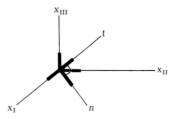

Figure 13-21

Notice that ϵ_{I}, ϵ_{II}, and ϵ_{III} are principal normal strains and that n_i and t_i are the direction cosines of an arbitrary pair of line elements at P. The directions of these line elements are indicated by the n and t directions in Figure 13-21. Also observe that the direction cosines n_i and t_i are measured relative to the principal axes of normal strain.

The mathematical procedure used to obtain the orientations of the line elements that undergo the principal shearing strains is tedious and, therefore, is not presented here. Instead, we present the results of such a process so that we can calculate the orientations and magnitudes of the principal shearing strains.

There are three pairs of line elements that undergo extreme values of shearing strain. The direction cosines associated with each pair of line elements are

$$\left.\begin{aligned}
n_1 &= \pm\frac{1}{\sqrt{2}}, \; n_2 = \pm\frac{1}{\sqrt{2}}, \; n_3 = 0 \\[2mm]
t_1 &= \pm\frac{1}{\sqrt{2}}, \; t_2 = \mp\frac{1}{\sqrt{2}}, \; t_3 = 0
\end{aligned}\right\} \tag{13-57a}$$

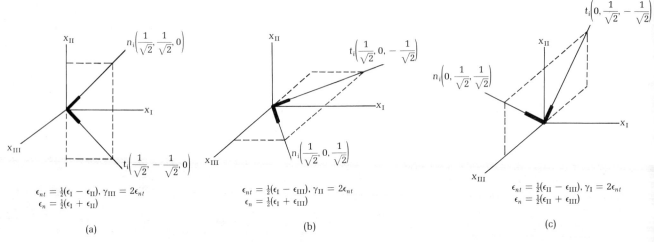

Figure 13-22

$$n_1 = \pm\frac{1}{\sqrt{2}}, \, n_2 = 0, \qquad n_3 = \pm\frac{1}{\sqrt{2}}$$
$$t_1 = \pm\frac{1}{\sqrt{2}}, \, t_2 = 0 \qquad t_3 = \mp\frac{1}{\sqrt{2}} \qquad (13\text{-}57\text{b})$$

$$n_1 = 0, \qquad n_2 = \pm\frac{1}{\sqrt{2}}, \, n_3 = \pm\frac{1}{\sqrt{2}}$$
$$t_1 = 0, \qquad t_2 = \pm\frac{1}{\sqrt{2}}, \, t_3 = \mp\frac{1}{\sqrt{2}} \qquad (13\text{-}57\text{c})$$

Figures 13-22a through 13-22c depict the orientation of the line elements that undergo the shearing strains associated with the directions given by Eqs. (13-57a), (13-57b), and (13-57c). The magnitudes of the principal shearing strains and the magnitudes of the accompanying normal strains are easily obtained from Eqs. (13-56) and (13-48a), respectively. Formulas for these magnitudes are shown in Figures 13-22a through 13-22c.

Observe that the pair of line elements that experience the principal shearing strain bisects the directions of the elements that experience the principal normal strains. Also note that the total shearing strain between a pair of initially perpendicular line elements is $\gamma_{nt} = 2\epsilon_{nt}$.

EXAMPLE 13-10

Determine the principal shearing strains and the corresponding normal strains for the state of strain shown in Figure 13-20.

SOLUTION

The principal normal strains were determined in Example 13-9 to be: $\epsilon_I = 8\mu$ and $\epsilon_{II} = \epsilon_{III} = 4\mu$. Thus, from the formulas shown in Figures 13-22a through 13-22c,

$$\gamma_I = \left|\epsilon_{II} - \epsilon_{III}\right| = |4 - 4| = 0, \quad \epsilon_n = \frac{1}{2}(4 + 4) = 4\mu$$

$$\gamma_{II} = \left|\epsilon_I - \epsilon_{III}\right| = |8 - 4| = 4\mu, \, \epsilon_n = \frac{1}{2}(8 + 4) = 6\mu$$

$$\gamma_{III} = \left|\epsilon_I - \epsilon_{II}\right| = |8 - 4| = 4\mu, \, \epsilon_n = \frac{1}{2}(8 + 4) = 6\mu$$

PROBLEMS / Section 13-9

13-16 For the state of strain given in Figure P13-13, determine the principal shearing strains. Establish the orientation of the elements that experience these strains with reference to the principal axes of normal strain. Calculate the normal strains associated with these elements.

13-17 For the state of strain given in Figure P13-14, determine the principal shearing strains. Establish the orientation of the elements that experience these strains with reference to the principal axes of normal strain. Calculate the normal strains associated with these elements.

13-18 Solve Problem 13-17 for the state of strain listed in Figure P13-15.

13-10 SUMMARY

In this chapter, we established transformation equations for **normal and shearing stresses** in three dimensions. We also established procedures for determining the **magnitudes** and **orientations** of the **principal normal and shearing stresses**. In addition we established analogous transformation equations for **normal and shearing strains**, as well as procedures for the determination of the **magnitudes** and **orientations** of the **principal normal and shearing strains**.

Stress Transformations

General Stress Transformations

Normal Stress:

$$\sigma_n = \sigma_{ij}n_i n_j$$

Shearing Stress:

$$\sigma_t = \sigma_{ij}n_i t_j$$

Principal Normal Stresses

Magnitudes:

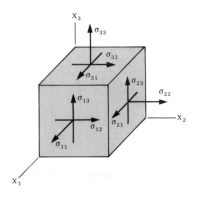

$$\sigma^3 - I_1\sigma^2 + I_2\sigma - I_3 = 0$$
$$I_1 = \sigma_{11} + \sigma_{22} + \sigma_{33}$$
$$I_2 = \sigma_{11}\sigma_{22} + \sigma_{22}\sigma_{33} + \sigma_{33}\sigma_{11} - \sigma_{12}^2 - \sigma_{23}^2 - \sigma_{31}^2$$
$$I_3 = \begin{vmatrix} \sigma_{11} & \sigma_{12} & \sigma_{13} \\ \sigma_{21} & \sigma_{22} & \sigma_{23} \\ \sigma_{31} & \sigma_{32} & \sigma_{33} \end{vmatrix}$$

Orientations:

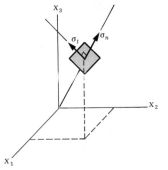

$$(\sigma_{11} - \sigma)n_1 + \sigma_{12}n_2 + \sigma_{13}n_3 = 0$$
$$\sigma_{21}n_1 + (\sigma_{22} - \sigma)n_2 + \sigma_{23}n_3 = 0$$
$$\sigma_{31}n_1 + \sigma_{32}n_2 + (\sigma_{33} - \sigma)n_3 = 0$$
$$n_1^2 + n_2^2 + n_3^2 = 1$$

Notes:

1. The directions of the principal normal stresses are mutually perpendicular.
2. The planes on which the principal normal stresses act are free of shearing stresses.
3. When the principal normal stresses are unequal, a distinct direction is defined for each principal normal stress.
4. When two principal normal stresses are equal, any direction in the plane containing them is a principal direction.
5. When all three principal normal stresses are equal, every direction is a principal direction.

Principal Shearing Stresses

Magnitudes:

Shearing Stress   Normal Stress

$$\tau_I = \frac{1}{2}(\sigma_{II} - \sigma_{III}) \qquad \sigma_n = \frac{1}{2}(\sigma_{II} + \sigma_{III})$$

$$\tau_{II} = \frac{1}{2}(\sigma_I - \sigma_{III}) \qquad \sigma_n = \frac{1}{2}(\sigma_I + \sigma_{III})$$

$$\tau_{III} = \frac{1}{2}(\sigma_I - \sigma_{II}) \qquad \sigma_n = \frac{1}{2}(\sigma_I + \sigma_{II})$$

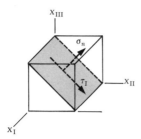

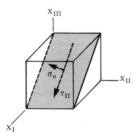

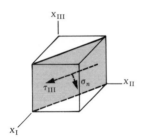

Orientations:

The principal shearing stresses act on planes that bisect the planes on which the principal normal stresses act.

Notes:

1. The orientations of the principal shearing stresses are usually given relative to a set of coordinate axes that coincide with the directions of the principal normal stresses.
2. The determination of the principal normal stresses usually precedes the determination of the principal shearing stresses.

Octahedral Stresses

Normal Stresses:

$$\sigma_{\text{oct}} = \frac{1}{3}(\sigma_\text{I} + \sigma_\text{II} + \sigma_\text{III})$$

Shearing Stresses:

$$\tau_{\text{oct}} = \frac{1}{3}\sqrt{(\sigma_\text{I} - \sigma_\text{II})^2 + (\sigma_\text{II} - \sigma_\text{III})^2 + (\sigma_\text{III} - \sigma_\text{I})^2}$$

Note: The octahedral normal stress and the octahedral shearing stress act on the eight planes whose normal vectors make equal angles with the directions of the principal normal stresses.

Strain Transformations

General Strain Transformations

Normal Strain:

$$\epsilon_n = \epsilon_{ij} n_i n_j$$

Shearing Strain:

$$\epsilon_{nt} = \epsilon_{ij} n_i t_j$$

Principal Normal Strains

Magnitudes:

$$\epsilon^3 - J_1\epsilon^2 + J_2\epsilon - J_3 = 0$$
$$J_1 = \epsilon_{11} + \epsilon_{22} + \epsilon_{33}$$
$$J_2 = \epsilon_{11}\epsilon_{22} + \epsilon_{22}\epsilon_{33} + \epsilon_{33}\epsilon_{11} - \epsilon_{12}^2 - \epsilon_{23}^2 - \epsilon_{31}^2$$
$$J_3 = \begin{vmatrix} \epsilon_{11} & \epsilon_{12} & \epsilon_{13} \\ \epsilon_{21} & \epsilon_{22} & \epsilon_{23} \\ \epsilon_{31} & \epsilon_{32} & \epsilon_{33} \end{vmatrix}$$

Orientations:

$$(\epsilon_{11} - \epsilon)n_1 + \epsilon_{12}n_2 \quad\;\; + \epsilon_{13}n_3 \quad\;\; = 0$$
$$\epsilon_{21}n_1 + (\epsilon_{22} - \epsilon)n_2 + \epsilon_{23}n_3 \quad\;\; = 0$$
$$\epsilon_{31}n_1 + \epsilon_{32}n_2 \quad\;\; + (\epsilon_{33} - \epsilon)n_3 = 0$$
$$n_1^2 + n_2^2 + n_3^2 = 1$$

Notes:

1. The directions of the line elements that undergo the principal normal strains are mutually perpendicular.
2. The line elements that experience the principal normal strains do not experience shearing strains.
3. When the principal normal strains are unequal, a distinct direction is defined for each principal normal strain.
4. When two principal normal strains are equal, any direction in the plane containing the two line elements that experience these strains is a principal direction.
5. When all three principal normal strains are equal, every direction is a principal direction.

Principal Shearing Strains

Magnitudes:

$$\gamma_I = \epsilon_{II} - \epsilon_{III} \qquad \epsilon_n = \frac{1}{2}(\epsilon_{II} + \epsilon_{III})$$

$$\gamma_{II} = \epsilon_I - \epsilon_{III} \qquad \epsilon_n = \frac{1}{2}(\epsilon_I + \epsilon_{III})$$

$$\gamma_{III} = \epsilon_I - \epsilon_{II} \qquad \epsilon_n = \frac{1}{2}(\epsilon_I + \epsilon_{II})$$

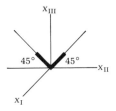

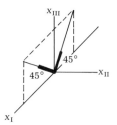

Orientations:

The line elements that undergo the principal shearing strains bisect the directions corresponding to the principal normal strains.

Notes:

1. The orientations of the line elements that experience the principal shearing strains are usually given with respect to a set of coordinate axes that coincide with the line elements that experience the principal normal strains.

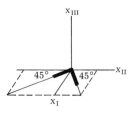

2. The determination of the principal shearing strains is usually preceded by the determination of the magnitudes and orientations of the principal normal strains.

PROBLEMS / CHAPTER 13

In the problems that follow, the state of stress at a point in a material is specified by the following two-dimensional array.

$$\begin{Vmatrix} \sigma_{11} & \sigma_{12} & \sigma_{13} \\ \sigma_{21} & \sigma_{22} & \sigma_{23} \\ \sigma_{31} & \sigma_{32} & \sigma_{33} \end{Vmatrix}$$

Note that the elements of the first row in this array are the components of stress that act on the coordinate plane perpendicular to the x_1 coordinate axis. Similarly, the elements of the second and third rows are the components of stress that act, respectively, on the x_2 and x_3 coordinate planes.

13-19–13-23 For the state of stress shown in the appropriate figure: (a) Determine the magnitudes of the principal normal stresses. (b) Determine the direction cosines associated with the maximum principal normal stress. (c) Determine the magnitudes of the principal shearing stresses. (d) Determine the normal stress that acts on the plane whose normal vector has the direction cosines $1/\sqrt{3}$, $1/\sqrt{3}$, $1/\sqrt{3}$. (e) Determine the octahedral normal and shearing stresses.

13-24 Using the stress invariants, show that the octahedral shearing stress is expressed in terms of a general state of stress by the formula

$$\tau_{oct} = \frac{1}{3}\sqrt{(\sigma_x - \sigma_y)^2 + (\sigma_y - \sigma_z)^2 + (\sigma_z - \sigma_x)^2 + 6(\tau_{xy}^2 + \tau_{yz}^2 + \tau_{zx}^2)}$$

13-25 Use the result of Problem 13-24 to show that the octahedral shearing stress for a plane state of stress ($\sigma_z = \tau_{zx} = \tau_{zy} \equiv 0$) is

$$\tau_{oct} = \frac{\sqrt{2}}{3}\sqrt{\sigma_x^2 + \sigma_y^2 - \sigma_x\sigma_y + 3\tau_{xy}^2}$$

13-26 Show that the energy of distortion can be expressed in general stress components as

$$U_0^d = \frac{1+v}{6E}\{(\sigma_x - \sigma_y)^2 + (\sigma_y - \sigma_z)^2 + (\sigma_z - \sigma_x)^2 + 6(\tau_{xy}^2 + \tau_{yz}^2 + \tau_{zx}^2)\}$$

13-27–13-29 For the state of strain shown in the appropriate figure: (a) Determine the magnitudes of the principal normal strains. (b) Determine the direction cosines associated with the maximum normal strain. (c) Determine the magnitudes of the principal shearing strains and the magnitudes of the accompanying normal strains. (d) Determine the normal strain associated with a line element defined by the direction cosines $\frac{1}{3}, \frac{2}{3}, \frac{2}{3}$. (e) Determine the shearing strain associated with the directions $(2/\sqrt{6}, 1/\sqrt{6}, 1/\sqrt{6})$ and $(1/\sqrt{3}, -1/\sqrt{3}, -1/\sqrt{3})$.

$$\begin{Vmatrix} 1 & 2 & 0 \\ 2 & 1 & 3 \\ 0 & 3 & 1 \end{Vmatrix} \text{ksi}$$

Figure P13-19

$$\begin{Vmatrix} -1 & \sqrt{2} & 0 \\ \sqrt{2} & 2 & 2 \\ 0 & 2 & 3 \end{Vmatrix} \text{ksi}$$

Figure P13-20

$$\begin{Vmatrix} 1 & \sqrt{2} & 0 \\ \sqrt{2} & 1 & 0 \\ 0 & 0 & 2 \end{Vmatrix} \text{ksi}$$

Figure P13-21

$$\begin{Vmatrix} 100 & 300 & 0 \\ 300 & 200 & 100\sqrt{5} \\ 0 & 100\sqrt{5} & 300 \end{Vmatrix} \text{MPa}$$

Figure P13-22

$$\begin{Vmatrix} 20 & 0 & 10 \\ 0 & 30 & 20 \\ 10 & 20 & 40 \end{Vmatrix} \text{MPa}$$

Figure P13-23

$$\begin{Vmatrix} 200 & -100 & 300 \\ -100 & 400 & 200 \\ 300 & 200 & 500 \end{Vmatrix} \mu \text{ in./in.}$$

Figure P13-27

$$\begin{Vmatrix} 100 & 200 & 0 \\ 200 & 100 & 300 \\ 0 & 300 & 100 \end{Vmatrix} \mu \text{ mm/mm}$$

Figure P13-28

$$\begin{Vmatrix} 0 & 300 & 400 \\ 300 & 0 & 200 \\ 400 & 200 & 0 \end{Vmatrix} \mu \text{ in./in.}$$

Figure P13-29

13-30 A state of strain associated with line elements situated along the x_1, x_2, x_3 coordinate axes is shown in Figure P13-30a. Determine the normal and shearing strain components associated with line elements situated along the x_1', x_2', x_3' coordinate axes whose direction cosines are given in Figure P13-30b.

$$\begin{Vmatrix} 500 & -200 & 100 \\ -200 & 600 & -300 \\ 100 & -300 & 700 \end{Vmatrix} \mu \text{ in./in.}$$

(a)

	x_1	x_2	x_3
x_1'	$\dfrac{2}{\sqrt{6}}$	$\dfrac{1}{\sqrt{6}}$	$\dfrac{1}{\sqrt{6}}$
x_2'	$\dfrac{1}{\sqrt{3}}$	$-\dfrac{1}{\sqrt{3}}$	$-\dfrac{1}{\sqrt{3}}$
x_3'	0	$\dfrac{1}{\sqrt{2}}$	$-\dfrac{1}{\sqrt{2}}$

(b)

Figure P13-30

CHAPTER 14

Failure Theories for Isotropic Materials

14-1 Introduction

The design of any structural element requires that a decision be made, prior to the actual design, regarding the circumstances which, when violated, constitute a failure of the element to perform its intended purpose. Thus a design criterion can be of varying complexity depending on the intended function of the structural element, the environment in which it functions, the nature of the applied loads (static, impact, fatigue), esthetic requirements, economic and environmental constraints, fabrication limitations, and material behavior (elastic, inelastic, time-dependent).

Static theories of failure address the need to establish criteria that determine when a material from which a structural element is made ceases to perform its static load carrying function. Consequently, the static theories of failure address only one aspect of the total design procedure. In general, the static theories of failure attempt to correlate either the initiation of yielding or fracture of a material subjected to a multiaxial state of stress with certain mechanical properties that accompany the initiation of yielding or the occurrence of fracture in a simple uniaxial tension test.

The practical value of such theories of failure becomes evident when we realize that much experimental work can be avoided if it can be shown that yielding or fracture of a material under a multiaxial state of stress is controlled by the same mechanical property that controls yielding or fracture in a uniaxial tension test. If such a correlation did not exist, it would be necessary to conduct an experiment for each combination of stress components to determine the value of the mechanical property that controls the initiation of yielding or the occurrence of fracture. Only then could a rational design of the structural element proceed. This procedure would clearly be impractical, not only because of the number of combinations of the stress components likely to occur in a multiaxial state of stress, but also because of the complexity of the testing apparatus required.

Theories of failure are phenomenological in the sense that they attempt to correlate the mechanism believed to be active when yielding is initiated, or when fracture occurs, with combinations of stresses, strains, or strain energy that are most likely to maximize the effect of the mechanism. Thus it is not surprising that several static theories of failure have been proposed. It is even less surprising that no single failure theory is universally applicable for materials that behave in either a ductile or a brittle manner for all possible states of stress. Moreover, given the difficulty of establishing a theory of failure that is universally valid for **isotropic** materials, it is not surprising that failure theories for **anisotropic** materials (fiber-reinforced materials) are even more difficult to establish.

The lack of universal applicability of any of the proposed theories of failure for isotropic materials should not obscure their widespread use. Consequently, a designer should be thoroughly acquainted with them. Moreover, many of the concepts embodied in these failure theories provide useful intellectual launching points from which failure criteria for anisotropic materials can be developed.

Much experimental evidence has been gathered to indicate the range of applicability of several of the more durable failure theories for isotropic materials. This evidence, as will be seen later, clearly shows the lack of universal applicability of any of the proposed theories. It also provides reassurance that some of them can provide realistic predictions for the initiation of yielding in certain materials for restricted combinations of stress components, while others provide realistic predictions of fracture for restricted combinations of stress components.

14-2 Uniaxial Tension Tests

Mechanical properties that play prominent roles in the static failure theories described in this chapter are obtained from a standard tension test. Six failure theories are described:

1. The **maximum principal normal stress** theory of failure.
2. The **maximum principal shearing stress** theory of failure.
3. The **maximum energy of distortion** theory of failure.
4. The **octahedral shearing stress** theory of failure.
5. The **maximum principal normal strain** theory of failure.
6. The **total strain energy** theory of failure.

Each theory of failure attempts to predict when yielding will be initiated by a multiaxial state of stress using information concerning the initiation of yielding in a standard tension test. Before stating each theory of failure, we describe the mechanical property associated with a standard tension test used in the statement of the failure theory.

Maximum principal normal stress theory of failure. In a standard tension test, yielding is initiated when the axial stress (maximum normal stress) reaches the yield stress of the material. Therefore, $\sigma_{max} = \sigma_{YP}$ when yielding is initiated in a standard tension test.

Maximum principal shearing stress theory of failure. The maximum shearing stress in a standard tension test is equal to one-half the axial stress, as can be seen from Mohr's circle for a uniaxial state of stress. Therefore, $\tau_{max} = \frac{1}{2}\sigma_{YP}$ when yielding is initiated in a standard tension test.

Maximum energy of distortion theory of failure. A formula for the energy of distortion associated with a standard tension test is obtained from Eq. (3-58) by setting $\sigma_{II} = \sigma_{III} = 0$. Thus,

$$U_0^d = \frac{1+\nu}{3E}\,\sigma_I^2$$

Therefore,

$$U_0^d = \frac{1+\nu}{3E}\,\sigma_{YP}^2$$

when yielding is initiated in a standard tension test.

Octahedral shearing stress theory of failure. A formula for the octahedral shearing stress associated with a standard tension test is obtained from Eq. (13-36) by setting $\sigma_{II} = \sigma_{III} = 0$. Thus $\tau_{oct} = (\sqrt{2/3})\sigma_I$. Therefore, $\tau_{oct} = (\sqrt{2/3})\sigma_{YP}$ when yielding is initiated in a simple tension test.

Maximum principal normal strain theory of failure. In a standard tensile test, yielding is initiated when the axial strain (maximum normal strain) reaches the yield strain of the material. Therefore, $\epsilon_{max} = \epsilon_{YP}$ when yielding is initiated in a standard tensile test.

Total strain energy theory of failure. A formula for the strain energy associated with a standard tension test is obtained from Eq. (3-57) or by setting $\sigma_{II} = \sigma_{III} = 0$ from the area under the linear region of the stress-strain diagram. Accordingly, $U_0 = \sigma_I^2/2E$. Therefore, $U_0 = \sigma_{YP}^2/2E$ when yielding is initiated in a standard tension test.

14-3 Maximum Principal Normal Stress Theory of Failure

The maximum principal normal stress theory of failure asserts that yielding will occur for a multiaxial state of stress whenever the maximum principal normal stress reaches a value equal to the tensile yield stress (σ_{YP}^t) for a uniaxial tension test, or whenever the minimum principal normal stress reaches a value equal to the compressive yield stress (σ_{YP}^c) for a uniaxial compression test.

The controlling mechanical property associated with this theory of failure is either the tensile yield stress σ_{YP}^t or the compressive yield stress σ_{YP}^t.

The maximum principal normal stress theory of failure possesses two notable flaws. First, yielding of material is fundamentally a shearing process. Therefore, it is reasonable to expect that a failure theory that would be valid for materials that behave in a ductile manner should be based on a shearing quantity. Secondly, this failure theory completely ignores the other two principal normal stresses. There is ample experimental evidence to indicate that a material may sustain a hydrostatic pressure of high intensity without yielding even though its compressive yield strength may be significantly less than the hydrostatic pressure to which it is subjected. This observation indicates that intermediate and minimum principal normal stresses effect the initiation of yielding.

Materials that behave in a brittle manner fail by fracture. In this case, it seems reasonable that a failure criterion should be controlled by the maximum principal normal stress. Experimental evidence substantiates this conclusion.

For many engineering materials, the tensile and compressive yield stresses are so nearly equal that it is impractical to make a distinction between them. Therefore, it is assumed that they are equal; that is, $\sigma_{YP}^t = \sigma_{YP}^c = \sigma_{YP}$.

For a plane state of stress ($\sigma_{III} = 0$, or $\sigma_{13} = \sigma_{23} = \sigma_{33} = 0$) and a safety factor (S.F.) applied to the stresses, this theory takes the form

$$\sigma_I = \frac{\sigma_{11} + \sigma_{22}}{2} + \sqrt{\left(\frac{\sigma_{11} - \sigma_{22}}{2}\right)^2 + \sigma_{12}^2} = \pm\frac{\sigma_{YP}}{\text{S.F.}} \quad (14\text{-}1a)$$

or

$$\sigma_{II} = \frac{\sigma_{11} + \sigma_{22}}{2} - \sqrt{\left(\frac{\sigma_{11} - \sigma_{22}}{2}\right)^2 + \sigma_{12}^2} = \pm\frac{\sigma_{YP}}{\text{S.F.}} \quad (14\text{-}1b)$$

Without the safety factor, these relations can be written in the normalized form

$$\frac{\sigma_I}{\sigma_{YP}} = \pm 1 \quad \text{or} \quad \frac{\sigma_{II}}{\sigma_{YP}} = \pm 1 \quad (14\text{-}2)$$

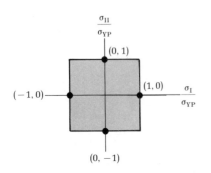

Figure 14-1

The curves represented by Eq. (14-2) are depicted in Figure 14-1.

Combinations of stresses σ_I and σ_{II} that lie on the interior of the square represent states of stress for which no yielding occurs. Points that lie on the boundary of the square represent states of stress for which yielding is initiated.

EXAMPLE 14-1

A circular bar is made from a material whose tensile yield stress is $\sigma_{YP}^t = 340$ MPa. The bar is subjected to a combined load consisting of a bending moment $M = 9$ kN·m and a twisting moment $T = 27$ kN·m as shown in Figure 14-2a. Determine the minimum diameter of the bar if a safety factor of 2.0 with respect to the yield stress is required. Use the maximum principal normal stress theory of failure.

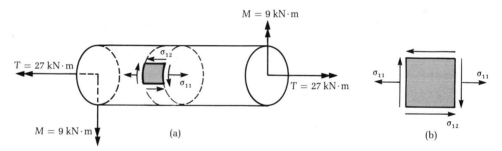

Figure 14-2

SOLUTION

The critical state of stress occurs at a point on the surface of the bar. Figure 14-2a shows the loading situation and Figure 14-2b shows an element at the surface of the bar. The normal stress and the shearing stress are given by the elementary formulas

$$\sigma_{11} = \frac{Mc}{I} = \frac{32}{\pi}\frac{M}{d^3} \qquad (a)$$

and

$$\sigma_{12} = \frac{Tc}{J} = \frac{16}{\pi}\frac{T}{d^3} \qquad (b)$$

where d is the diameter of the bar.

According to the maximum principal normal stress theory of failure, $\sigma_{max} = \sigma_{YP}$ when yielding is initiated. The maximum normal stress for the state of stress shown in Figure 14-2 is

$$\sigma_{max} = \frac{\sigma_{11}}{2} + \sqrt{\left(\frac{\sigma_{11}}{2}\right)^2 + \sigma_{12}^2} = \frac{16M}{\pi d^3} + \sqrt{\left(\frac{16M}{\pi d^3}\right)^2 + \left(\frac{16T}{\pi d^3}\right)^2}$$

$$= \frac{16}{\pi}\frac{M}{d^3}\left\{1 + \sqrt{1 + \left(\frac{T}{M}\right)^2}\right\} = 21.2\frac{M}{d^3} \qquad (c)$$

Then, by the maximum principal normal stress theory of failure, the minimum diameter is calculated from the relation

$$21.2 \frac{M}{d^3} = \frac{\sigma_{YP}}{S.F.} = \frac{340 \times 10^6}{2} \qquad \text{(d)}$$

Therefore,

$$d = 0.104 \text{ m}$$

EXAMPLE 14-2

Determine the internal pressure to which the thin-walled cylindrical pressure vessel shown in Figure 14-3 may be subjected if a safety factor of 2.0 is desired. Use the maximum principal normal stress theory of failure. The cylinder has a wall thickness of 0.25 in. and a mean radius of 20 in. The yield stress for the material of the cylinder is 30 ksi.

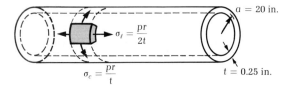

Figure 14-3

SOLUTION

The state of stress on the surface of the cylinder is shown in Figure 14-3. The maximum principal normal stress is the circumferential stress σ_c.

According to the maximum principal normal stress theory of failure, yielding is initiated when $\sigma_c = \sigma_{YP}/S.F.$ Therefore, the maximum internal pressure to which the cylinder can be subjected is given by the condition

$$\frac{pr}{t} = \frac{30,000}{2}$$

$$p = 15,000 \left(\frac{0.25}{20} \right) = 187.5 \text{ psi}$$

EXAMPLE 14-3

Using the maximum principal normal stress theory of failure, determine the maximum allowable load P that can be applied to the engineering member shown in Figure 14-4a. The tensile and compressive

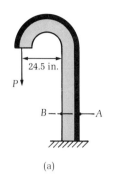

24.5 in.

P↓

B——A

(a)

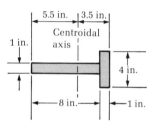

5.5 in.   3.5 in.

Centroidal axis

1 in.

4 in.

8 in.   1 in.

$I_{NA} = 97\ in^4$ and $A = 12\ in^2$

(b)

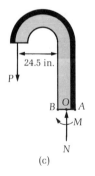

24.5 in.

P↓

B | O | A

M

N

(c)

Figure 14-4

yield stresses for the material of which the member is made, with a suitable safety factor, are 5 ksi and 20 ksi, respectively. The cross section of the member is shown in Figure 14-4b. Pertinent properties of the cross section are $I_{NA} = 97\ in^4$ and $A = 12\ in^2$.

SOLUTION

A free-body diagram that isolates the portion of the member above the plane AB is shown in Figure 14-4c. Equilibrium requirements yield

$$\Sigma F_y = 0: \quad N = P$$
$$\Sigma M_0 = 0: \quad M = 30P$$

According to the maximum principal normal stress theory of failure, either

$$|\sigma_I| = \frac{\sigma_{YP}^t}{S.F.} \tag{a}$$

or

$$|\sigma_{II}| = \frac{\sigma_{YP}^c}{S.F.} \tag{b}$$

The principal normal stresses σ_I and σ_{II} occur at points A and B of Figure 14-4c, respectively. They are

$$\sigma_I = -\frac{P}{A} + \frac{Mc_A}{I} = -\frac{P}{12} + \frac{30P(3.5)}{97} = 0.99914P \tag{c}$$

and

$$\sigma_{II} = -\frac{P}{A} - \frac{Mc_B}{I} = -\frac{P}{12} - \frac{30P(5.5)}{97} = -1.78436P \tag{d}$$

The maximum allowable load P is the smaller of the two loads calculated from Eqs. (a) and (b). Thus

$$0.99941P = 5000 \quad \text{or} \quad P_t = 5004\ lb$$

and

$$-1.78436P = -20,000 \quad \text{or} \quad P_c = 11,209\ lb$$

Thus the tensile stress governs the allowable load that the member can carry.

PROBLEMS / Section 14-3

14-1   The engineering member shown in Figure P14-1 is made from a material for which $\sigma_{YP} = 45$ ksi. Use the maximum principal normal stress theory of failure to determine the minimum permissible diameter for the circular shaft if

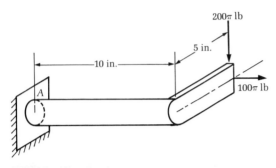

Figure P14-1

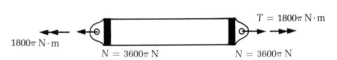

Figure P14-3

the design is governed by the state of stress at point A. Neglect stress concentrations at the support and use a safety factor of 2.

14-2 A thin-walled cylinder is used to store gas under a pressure of 250 psi. The mean radius of the cylinder is 20 in. and the wall thickness is 0.25 in. If the tensile yield stress for the material of the cylinder is 60 ksi, determine the safety factor used in the design of the pressure vessel. Assume that the maximum principal normal stress theory of failure served as the basis of the design.

14-3 A thin-walled tube with a mean radius of 120 mm and a wall thickness of 3 mm is subjected to an internal pressure p, an axial force $N = 3600\pi$ N, and a torque $T = 1800\pi$ N·m as shown in Figure P14-3. If the tensile yield stress for the material of the tube is 140 MPa, determine the magnitude of the pressure that will initiate yielding according to the maximum principal normal stress theory of failure.

14-4 The state of stress at a point in a material whose yield stress is 32 ksi is shown in Figure P14-4. Determine whether yielding has occurred according to the maximum principal normal stress theory of failure.

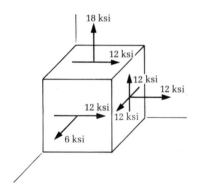

Figure P14-4

14-5 A 50-mm diameter cast iron shaft is subjected to equal and opposite torques T at its ends. Determine the maximum torque T that can be applied to the shaft if the ultimate strength of cast iron in tension is 210 MPa. Use the maximum principal normal stress theory of failure as the basis for the analysis.

14-4 Maximum Principal Shearing Stress Theory of Failure

The maximum principal shearing stress theory of failure asserts that yielding will occur for a multiaxial state of stress whenever the maximum shearing stress reaches a value equal to the maximum shearing stress associated with the initiation of yielding in a uniaxial tension test.

The controlling quantity associated with this failure theory is $\tau_{\max} = \frac{1}{2}\sigma_{YP}$ as can be seen from a Mohr's circle for a uniaxial state of stress. The general form for the maximum principal shearing stress theory of failure is

$$-\sigma_{YP} \leq (\sigma_{I} - \sigma_{II}) \leq \sigma_{YP} \qquad (14\text{-}3a)$$

or

$$-\sigma_{YP} \le (\sigma_{II} - \sigma_{III}) \le \sigma_{YP} \qquad \text{(14-3b)}$$

or

$$-\sigma_{YP} \le (\sigma_{III} - \sigma_{I}) \le \sigma_{YP} \qquad \text{(14-3c)}$$

Dividing each of the inequalities in Eqs. 14-3a, 14-3b, and 14-3c by σ_{YP} yields the normalized form for the maximum principal shearing stress theory of failure. Thus

$$\frac{\sigma_{I}}{\sigma_{YP}} - \frac{\sigma_{II}}{\sigma_{YP}} = \pm 1 \qquad \text{(14-4a)}$$

or

$$\frac{\sigma_{II}}{\sigma_{YP}} - \frac{\sigma_{III}}{\sigma_{YP}} = \pm 1 \qquad \text{(14-4b)}$$

or

$$\frac{\sigma_{III}}{\sigma_{YP}} - \frac{\sigma_{I}}{\sigma_{YP}} = \pm 1 \qquad \text{(14-4c)}$$

These relations constitute six intersecting planes that form the sides of a regular hexagonal prism whose geometric axis is the line

$$\frac{\sigma_{I}}{\sigma_{YP}} = \frac{\sigma_{II}}{\sigma_{YP}} = \frac{\sigma_{III}}{\sigma_{YP}}$$

This hexagonal prism is infinite in length and is referred to as the **yield surface** for the maximum principal shearing stress theory of failure. It is shown in Figure 14-5 and is called **Tresca's hexagonal**.

Observe that points on the geometric axis of the hexagonal prism fall within a region for which no yielding can occur. Also note that these points correspond to spherical states of stress (either tension or compression). Consequently, the maximum principal shearing stress theory of failure implies that yielding is independent of the spherical state of stress. Another way to see that yielding is independent of the spherical state of stress according to this theory of failure is to observe that equal increments of stress can be added to each relation in Eqs. (14-3) or (14-4) without altering them.

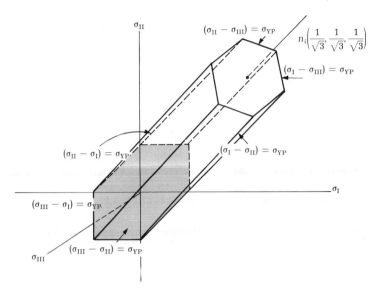

Figure 14-5   Tresca's hexagonal in principal stress space. Shaded area in $\sigma_I\sigma_{II}$ plane is Tresca's hexagon for plane stress.

For plane stress conditions, Eqs. (14-4) become

$$\frac{\sigma_I}{\sigma_{YP}} - \frac{\sigma_{II}}{\sigma_{YP}} = \pm 1 \tag{14-5a}$$

or

$$\frac{\sigma_{II}}{\sigma_{YP}} = \pm 1 \tag{14-5b}$$

or

$$\frac{\sigma_I}{\sigma_{YP}} = \pm 1 \tag{14-5c}$$

These relations represent the line segments corresponding to the intersection of the hexagonal prism with the $\sigma_I\sigma_{II}$ coordinate plane. This plane is the shaded area in Figure 14-5 and is repeated for clarity as Figure 14-6. The hexagon shown in Figure 14-6 is known as Tresca's hexagon. For comparison, the square yield surface associated with the maximum principal normal stress theory of failure is superimposed

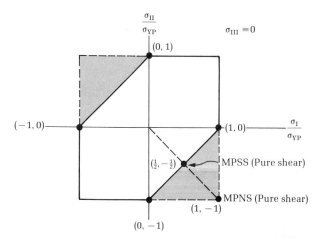

Figure 14-6

on the yield surface corresponding to the maximum principal shearing stress theory of failure. Notice that in the first and third quadrants where the principal normal stresses have the **same sign**, the maximum principal normal stress and the maximum principal shearing stress theories of failure predict identical results for the initiation of yielding. However, in the second and fourth quadrants, where the principal normal stresses have opposite signs, a large discrepancy exists between the predictions for the onset of yielding by the two theories. The shaded area in Figure 14-6 indicates the region where the two failure theories disagree. Note that the greatest disagreement (nearly 71%) occurs for a state of pure shear $(\sigma_I = -\sigma_{II})$.

EXAMPLE 14-4

Solve Example 14-1 using the maximum principal shearing stress theory of failure.

SOLUTION

According to the maximum principal shearing stress theory of failure, yielding will occur when the maximum shearing stress in the material reaches a value equal to $\frac{1}{2}\sigma_{YP}$. The maximum shearing stress is given by the formula

$$\tau_{max} = \sqrt{\left(\frac{\sigma_{11}}{2}\right)^2 + (\sigma_{12})^2} \tag{a}$$

so that yielding is initiated when

$$\sqrt{\left(\frac{\sigma_{11}}{2}\right)^2 + (\sigma_{12})^2} = \frac{\sigma_{YP}}{\text{S.F.}(2)} \tag{b}$$

Substituting the given quantities from Example 14-1 yields

$$\frac{16}{\pi}\frac{M}{d^3}\sqrt{1+\left(\frac{T}{M}\right)^2}=\frac{340\times10^6}{4} \tag{c}$$

so that

$$d = 0.119 \text{ m}$$

This result is 14.4% larger than the diameter ($d = 0.104$ m) predicted by the maximum principal normal stress theory of failure.

EXAMPLE 14-5

Solve Example 14-2 using the maximum principal shearing stress theory of failure.

SOLUTION

The maximum shearing stress in the cylinder occurs at any point on the interior surface of the cylinder and is given by the formula

$$\tau_{\max} = \frac{1}{2}(\sigma_c + p) \tag{a}$$

According to the MPSS theory of failure, yielding will occur when $\tau_{\max} = \frac{1}{2}\sigma_{YP}$. Therefore, with a safety factor of 2, yielding is initiated at points on the interior of the cylinder when

$$\frac{1}{2}(\sigma_c + p) = \frac{1}{2}\left(\frac{pr}{t} + p\right) = \frac{1}{2}\frac{\sigma_{YP}}{\text{S.F.}} \tag{b}$$

Accordingly, $p = 185.2$ psi, which differs by only 1.2% from the pressure predicted by the maximum principal normal stress theory of failure.

PROBLEMS / Section 14-4

14-6 A ductile aluminum tube with an inside diameter of 1 in. and an outside diameter of 2 in. is to be used to transmit a torque T between a motor and a pump. If $\sigma_{YP} = 20$ ksi for the aluminum, determine the maximum torque that can be transmitted according to the maximum principal shearing stress theory of failure.

14-7 A thin-walled square tube with the cross section shown in Figure P14-7b is subjected to the forces shown in Figure P14-7a. The tensile yield stress of the material of the tube is 140 MPa. Determine whether the loads shown will cause yielding in the material according to the maximum principal shearing stress theory of failure. The tube thickness is 2 mm.

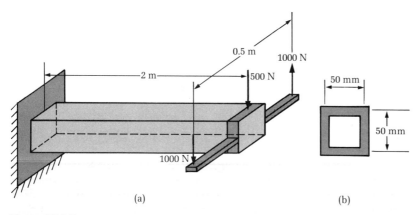

Figure P14-7

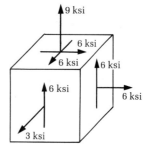

Figure P14-8

14-8 The state of stress at a point in a material whose tensile yield stress is 15 ksi is shown in Figure P14-8. Determine whether this state of stress is sufficient to initiate yielding according to the maximum principal shearing stress theory of failure.

14-9 A thin-walled sphere with a mean radius of 500 mm and wall thickness of 5 mm is subjected to an internal pressure of 2 MPa. The material from which the sphere is made has a tensile yield stress of 140 MPa. Determine whether this internal pressure is sufficient to cause yielding according to the maximum principal shearing stress theory of failure.

14-5 Maximum Energy of Distortion Theory of Failure

The maximum energy of distortion theory of failure asserts that yielding will occur for a multiaxial state of stress when the maximum energy of distortion reaches a value equal to the energy of distortion associated with the initiation of yielding in a uniaxial tension test.

The controlling quantity associated with the maximum energy of distortion theory of failure is $U_0^d = [(1 + v)/3E]\sigma_{YP}^2$. The general form for the maximum energy of distortion theory of failure is

$$U_0^d \leqq \frac{1 + v}{3E}\left(\frac{\sigma_{YP}}{S.F.}\right)^2 \tag{14-6}$$

An expression for the energy of distortion for an isotropic material under a multiaxial state of stress was derived in Chapter 3. The maximum energy of distortion theory of failure is

$$(\sigma_I - \sigma_{II})^2 + (\sigma_{II} - \sigma_{III})^2 + (\sigma_{III} - \sigma_I)^2 = 2\left(\frac{\sigma_{YP}}{S.F.}\right)^2 \tag{14-7}$$

In terms of the dimensionless stress ratios, without the safety factor,

$$\left(\frac{\sigma_{\mathrm{I}}}{\sigma_{\mathrm{YP}}} - \frac{\sigma_{\mathrm{II}}}{\sigma_{\mathrm{YP}}}\right)^2 + \left(\frac{\sigma_{\mathrm{II}}}{\sigma_{\mathrm{YP}}} - \frac{\sigma_{\mathrm{III}}}{\sigma_{\mathrm{YP}}}\right)^2 + \left(\frac{\sigma_{\mathrm{III}}}{\sigma_{\mathrm{YP}}} - \frac{\sigma_{\mathrm{I}}}{\sigma_{\mathrm{YP}}}\right)^2 = 2 \qquad (14\text{-}8)$$

Equation (14-8) represents a right circular cylinder whose geometric axis coincides with the geometric axis of the regular hexagonal prism of the maximum principal shearing stress failure theory and whose radius is $\sqrt{\frac{2}{3}}$. Consequently, points lying on the geometric axis of this cylinder correspond to spherical states of stress. Equation (14-8) shows that equal increments of stress can be added to the principal normal stresses without altering the validity of the yield condition. Consequently, this theory of failure implies that yielding is closely associated with the shearing stresses in the material. We expect that this theory is more reliable when it is applied to materials that behave in a ductile manner. Figure 14-7 shows the right circular cylinder represented by Eq. (14-8). The shaded area in this figure corresponds to the intersection of the cylinder with the $\sigma_{\mathrm{I}}\sigma_{\mathrm{II}}$ coordinate plane and, therefore, represents the yield locus for a plane state of stress. The right circular cylinder of Figure 14-7 is known as the **Mises cylinder**.

For plane stress conditions ($\sigma_{\mathrm{III}} = 0$), Eq. (14-8) becomes

$$\left(\frac{\sigma_{\mathrm{I}}}{\sigma_{\mathrm{YP}}}\right)^2 + \left(\frac{\sigma_{\mathrm{II}}}{\sigma_{\mathrm{YP}}}\right)^2 - \frac{\sigma_{\mathrm{I}}}{\sigma_{\mathrm{YP}}}\frac{\sigma_{\mathrm{II}}}{\sigma_{\mathrm{YP}}} = 1 \qquad (14\text{-}9)$$

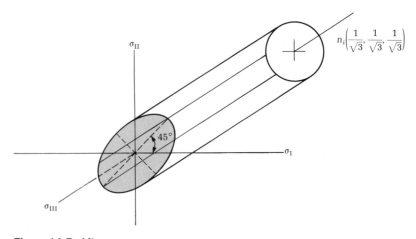

Figure 14-7 Mises cylinder in principal stress space. Shaded area in $\sigma_{\mathrm{I}}\sigma_{\mathrm{II}}$ plane is the Mises ellipse for plane stress.

This equation represents the intersection of the cylinder with the $\sigma_I\sigma_{II}$ plane. Consequently, the yield locus for a plane state of stress is an ellipse with its major axis coinciding with the bisector of the σ_I and σ_{II} axes. Figure 14-8 shows this yield locus superimposed on the yield loci associated with the maximum principal normal stress (dot-dash lines) and the maximum principal shearing stress (dashed lines). The ellipse shown in Figure 14-8 is called the Mises ellipse. Major differences between the predictions made by the Tresca and Mises yield criteria are indicated on the figure.

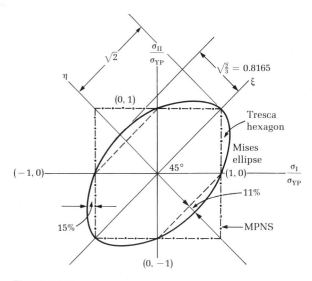

Figure 14-8

EXAMPLE 14-6

Show that Eq. (14-7) represents a right circular cylinder of radius $\sqrt{\frac{2}{3}}$ whose geometric axis has the direction cosines $(1/\sqrt{3},\ 1/\sqrt{3},\ 1/\sqrt{3})$.

SOLUTION

First write the equation for a right circular cylinder in the coordinate system shown in Figure 14-9a. Thus

$$(x_1')^2 + (x_2')^2 = a^2 \tag{a}$$

where a is the radius of the cross section of the cylinder.

Now let the x_3' axis—the geometric axis of the cylinder—coincide with the direction $1/\sqrt{3},\ 1/\sqrt{3},\ 1/\sqrt{3}$ in the x_i coordinate system as shown in Figure 14-9b.

Moreover, for convenience choose the x_1' axis to lie in the plane containing the x_3' axis and the x_3 axis. Consequently, as can be seen

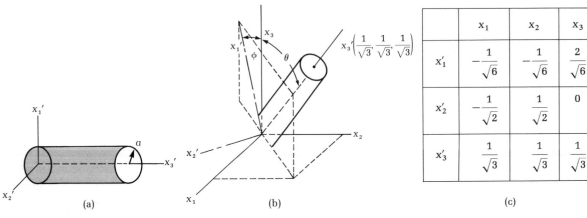

Figure 14-9

from Figure 14-9c,

$$\ell_{13} = \cos \varphi = \cos \left(\frac{\pi}{2} - \theta \right) = \sqrt{\frac{2}{3}} \qquad \text{(b)}$$

The direction cosines associated with the x_1' axis and the x_2' axis can be calculated from the orthogonality and normality requirements.

Direction cosines for the x_1' axis. The condition that the x_1' axis be orthogonal to the x_3' axis is

$$\ell_{11}\ell_{31} + \ell_{12}\ell_{32} + \ell_{13}\ell_{33} = 0$$

or, since $\ell_{31} = \ell_{32} = \ell_{33} = 1/\sqrt{3}$ and $\ell_{13} = \sqrt{\frac{2}{3}}$,

$$\ell_{11} + \ell_{12} + \sqrt{\frac{2}{3}} = 0 \qquad \text{(c)}$$

The normality condition for the x_1' axis is

$$\ell_{11}^2 + \ell_{12}^2 + \ell_{13}^2 = 1 \qquad \text{(d)}$$

Equations (c) and (d) yield

$$\ell_{11} = \ell_{12} = -\frac{1}{\sqrt{6}} \qquad \text{(e)}$$

The direction cosines for the x_1' axis are entered as the first row in Figure 14-9c.

Direction cosines for the x_2' axis. The condition that the x_1' axis be orthogonal to the x_2' axis is

$$\ell_{11}\ell_{21} + \ell_{12}\ell_{22} + \ell_{13}\ell_{23} = 0 \qquad \text{(f)}$$

or, since $\ell_{11} = \ell_{12} = -1/\sqrt{6}$ and $\ell_{13} = \sqrt{\frac{2}{3}}$, Eq. (f) becomes

$$\ell_{21} + \ell_{22} - 2\ell_{23} = 0 \qquad \text{(g)}$$

Similarly, the condition that the x_2' axis be orthogonal to the x_3' axis is

$$\ell_{21} + \ell_{22} + \ell_{23} = 0 \tag{h}$$

Finally, the normality condition for the direction cosines corresponding to the x_2' axis is

$$\ell_{21}^2 + \ell_{22}^2 + \ell_{23}^2 = 1 \tag{i}$$

Equations (g), (h), and (i) yield

$$\ell_{21} = \pm\frac{1}{\sqrt{2}}, \ell_{22} = \mp\frac{1}{\sqrt{2}}, \ell_{23} = 0 \tag{j}$$

These direction cosines are entered as the second row in Figure 14-9c.

The coordinate transformation corresponding to these direction cosines $(x_i' = \ell_{ij}x_j)$ is

$$\left.\begin{array}{l} x_1' = \dfrac{1}{\sqrt{6}}(-x_1 - x_2 + 2x_3) \\[2mm] x_2' = \dfrac{1}{\sqrt{2}}(-x_1 + x_2) \\[2mm] x_3' = \dfrac{1}{\sqrt{3}}(x_1 + x_2 + x_3) \end{array}\right\} \tag{k}$$

Substituting Eqs. (k) in Eq. (a) yields, after a little algebraic manipulation,

$$\frac{2}{3}(x_1^2 + x_2^2 + x_3^2 - x_1x_2 - x_1x_3 - x_2x_3) = a^2 \tag{l}$$

which can easily be shown to have the equivalent form

$$\frac{1}{3}\{(x_1 - x_2)^2 + (x_2 - x_3)^2 + (x_3 - x_1)^2\} = a^2 \tag{m}$$

Equation (m) is the equation of a right circular cylinder whose geometric axis coincides with the line through the origin with direction cosines $(1/\sqrt{3}, 1/\sqrt{3}, 1/\sqrt{3})$. Since Eq. (14-8) is of the same form as Eq. (m), it follows that Eq. (14-8) is the equation of a right circular cylinder with geometric axis coincident with the line through the origin with direction cosines $(1/\sqrt{3}, 1/\sqrt{3}, 1/\sqrt{3})$.

EXAMPLE 14-7

Show that the major and minor axes of the Mises ellipse coincide with the bisectors of the σ_I and σ_{II} axes. Also show that the principal radii of the ellipse are $\sqrt{2}$ and $\sqrt{\frac{2}{3}}$.

SOLUTION

Let the ξ, η coordinate system coincide with the major and minor axes of the Mises ellipse as shown in Figure 14-8. Then

$$\frac{\sigma_I}{\sigma_{YP}} = \frac{1}{\sqrt{2}} (\xi - \eta) \quad \text{and} \quad \frac{\sigma_{II}}{\sigma_{YP}} = \frac{1}{\sqrt{2}} (\xi + \eta) \qquad (a)$$

Substituting Eqs. (a) into Eq. (14-9) yields

$$\frac{\xi^2}{(\sqrt{2})^2} + \frac{\eta^2}{\left(\sqrt{\frac{2}{3}}\right)^2} = 1 \qquad (b)$$

Equation (b) is the equation of the Mises yield criterion in the $\xi\eta$ coordinate system. The yield locus is clearly an ellipse with major and minor axes coinciding with the ξ and η axes, respectively. The principal radii are seen to be $\sqrt{2}$ and $\sqrt{\frac{2}{3}}$.

EXAMPLE 14-8

Solve Example 14-1 using the maximum energy of distortion theory of failure.

SOLUTION

According to the maximum energy of distortion theory of failure, yielding will occur when the strain energy of distortion reaches a value equal to $U_0^d = [(1 + v)/3E]\sigma_{YP}^2$. Therefore, the yield condition for this failure theory becomes

$$\sigma_{11}^2 + 3\sigma_{12}^2 = \left(\frac{\sigma_{YP}}{\text{S.F.}}\right)^2$$

Substituting the data given in Example 14-1,

$$\left(\frac{32}{\pi}\frac{M}{d^3}\right)^2 + 3\left(\frac{16}{\pi}\frac{T}{d^3}\right)^2 = \left(\frac{\sigma_{YP}}{\text{S.F.}}\right)^2$$

or

$$d = 0.114 \text{ m}$$

This result is about 4.2% less than the result obtained using the maximum principal shearing stress failure theory, and about 9.6% greater than the result predicted by the maximum principal normal stress failure theory.

EXAMPLE 14-9

Solve Example 14-2 using the maximum energy of distortion theory of failure.

SOLUTION

From Eq. (14-9), the yield locus for the maximum energy of distortion failure theory under plane stress conditions is

$$\sigma_I^2 + \sigma_{II}^2 - \sigma_I\sigma_{II} = \left(\frac{\sigma_{YP}}{S.F.}\right)^2 \tag{a}$$

Now with $\sigma_I = 2\sigma_{II} = pr/t$, Eq. (a) reduces to

$$\frac{3}{4}\left(\frac{pr}{t}\right)^2 = \left(\frac{\sigma_{YP}}{S.F.}\right)^2 \tag{b}$$

or

$$p = \sqrt{\frac{4}{3}\frac{t}{r}\frac{\sigma_{YP}}{S.F.}} = \sqrt{\frac{4}{3}\frac{0.25}{20}\frac{30,000}{2}} = 216.5 \text{ psi} \tag{c}$$

The pressure to cause yielding in the thin-walled pressure vessel according to the maximum energy of distortion failure theory is 15.47% greater than the 187.5 psi predicted by the maximum principal normal stress failure theory.

EXAMPLE 14-10

Suppose that the tensile and compressive yield stresses are equal in Example 14-3. Let $\sigma_{YP} = 20$ ksi. Use the maximum energy of distortion failure theory to determine the maximum allowable load P.

SOLUTION

Observe that the maximum energy of distortion failure theory requires that the tensile and compressive yield stresses be equal.

The states of stress at points A and B are shown in Figures 14-10a and 14-10b, respectively. According to the maximum energy of distortion failure theory, the yield condition should be

$$\sigma_{II}^2 = \left(\frac{\sigma_{YP}}{S.F.}\right)^2$$

The critical condition occurs at point B and, hence, the maximum allowable load P is calculated from the formula

$$(-1.78436P)^2 = \left(\frac{20,000}{1}\right)^2$$

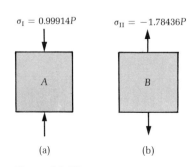

$\sigma_I = 0.99914P$ $\sigma_{II} = -1.78436P$

(a)   (b)

Figure 14-10

or

$$P = 11,209 \text{ lb}$$

This value is exactly the one that would have been predicted by either the maximum principal normal stress or the maximum principal shearing stress failure theories. This fact can be seen by examining Figure 14-9.

14-6 Octahedral Shearing Stress Theory of Failure

The octahedral shearing stress (OSS) theory of failure asserts that yielding will occur for a multiaxial state of stress when the octahedral shearing stress reaches a value equal to the octahedral shearing stress associated with the initiation of yielding in a simple tension test.

The controlling quantity associated with the octahedral shearing stress failure theory is $\tau_{\text{oct}} = (\sqrt{2}/3)\sigma_{\text{YP}}$. Eqs. (13-36) and (14-7) show that the OSS and the MED failure theories are equivalent. It should be noted that while the two theories are equivalent, they originate from substantially different concepts. Since the mechanism that serves as the basis for the two theories is the shearing process, it is not surprising that they are in best agreement with available experimental data for materials that behave in a ductile manner.

EXAMPLE 14-11

Show that the octahedral shearing stress theory of failure is equivalent to the maximum energy of distortion theory of failure.

SOLUTION

From Eq. (13-36), the octahedral shearing stress is

$$\tau_{\text{oct}} = \frac{1}{3}\sqrt{(\sigma_{\text{I}} - \sigma_{\text{II}})^2 + (\sigma_{\text{II}} - \sigma_{\text{III}})^2 + (\sigma_{\text{III}} - \sigma_{\text{I}})^2} \qquad \text{(a)}$$

According to the OSS theory of failure, yielding will be initiated when

$$\frac{1}{3}\sqrt{(\sigma_{\text{I}} - \sigma_{\text{II}})^2 + (\sigma_{\text{II}} - \sigma_{\text{III}})^2 + (\sigma_{\text{III}} - \sigma_{\text{I}})^2} = \frac{\sqrt{2}}{3}\sigma_{\text{YP}} \qquad \text{(b)}$$

Equation (b) can be written as

$$(\sigma_{\text{I}} - \sigma_{\text{II}})^2 + (\sigma_{\text{II}} - \sigma_{\text{III}})^2 + (\sigma_{\text{III}} - \sigma_{\text{I}})^2 = 2\sigma_{\text{YP}}^2 \qquad \text{(c)}$$

which is precisely the condition given by the maximum energy of distortion theory of failure for the initiation of yielding.

PROBLEMS / Sections 14-5 and 14-6

14-10 A solid rectangular bar made from a material whose tensile yield stress is 30 ksi is subjected to the forces shown in Figure P14-10a. If the depth of the cross section is to be twice its width, determine the minimum cross-sectional dimensions according to the maximum energy of distortion theory of failure. A safety factor of 2.33 is required.

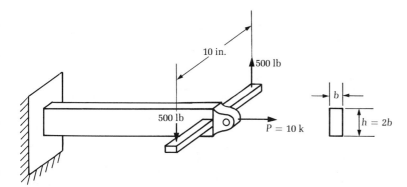

Figure P14-10

14-11 Solve Problem 14-3 using the maximum energy of distortion theory of failure.

14-12 Solve Problem 14-4 using the maximum energy of distortion theory of failure.

14-13 The state of stress at a point in a material whose tensile yield stress is 210 MPa is shown in Figure P14-13. Determine whether this state of stress causes yielding according to the maximum energy of distortion theory of failure.

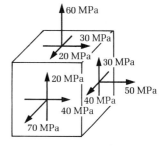

Figure P14-13

14-7 Maximum Principal Normal Strain Theory of Failure

The maximum principal normal strain theory of failure asserts that yielding will occur for a multiaxial state of stress when the maximum principal normal strain reaches a value equal to the tensile yield strain (ϵ_{YP}^t) for a uniaxial tension test, or when the minimum principal normal strain reaches a value equal to the compressive yield strain (ϵ_{YP}^c) for a uniaxial compression test.

The controlling quantity associated with the maximum principal normal strain failure theory is either the tensile or the compressive yield strain. This theory has two principal flaws that are physically the same flaws attached to the maximum principal normal stress failure theory. First, failure in a ductile material is a shearing process and, therefore, we would expect that a failure theory for such a material would depend on shearing strains. Secondly, two principal normal strains are ignored, which implies that these strains do not influence when yielding will occur. This implication is contrary to the experimental evidence.

For many engineering materials, the tensile and compressive yield strains are so nearly equal that it is impractical to make a distinction between them. Consequently, we assume that $\epsilon_{YP}^t = \epsilon_{YP}^c = \epsilon_{YP}$.

For plane stress conditions ($\sigma_{III} = 0$) and a safety factor applied to the strains, this theory takes the form

$$\epsilon_I = \frac{\epsilon_{11} + \epsilon_{22}}{2} + \sqrt{\left(\frac{\epsilon_{11} - \epsilon_{22}}{2}\right)^2 + \epsilon_{12}^2} = \pm\frac{\epsilon_{YP}}{S.F.} \qquad (14\text{-}10a)$$

or

$$\epsilon_{II} = \frac{\epsilon_{11} + \epsilon_{22}}{2} - \sqrt{\left(\frac{\epsilon_{11} - \epsilon_{22}}{2}\right)^2 + \epsilon_{12}^2} = \pm\frac{\epsilon_{YP}}{S.F.} \qquad (14\text{-}10b)$$

In terms of the principal stresses σ_I and σ_{II}, this theory can be written as

$$\frac{\sigma_I}{\sigma_{YP}} - v\frac{\sigma_{II}}{\sigma_{YP}} = \pm 1 \qquad \text{for} \qquad |\sigma_I| \geq |\sigma_{II}| \qquad (14\text{-}11a)$$

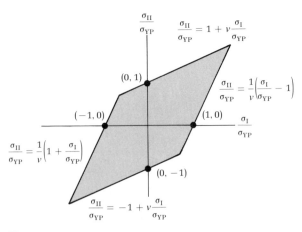

Figure 14-11

or

$$\frac{\sigma_{\text{II}}}{\sigma_{\text{YP}}} - v\frac{\sigma_{\text{I}}}{\sigma_{\text{YP}}} = \pm 1 \qquad \text{for} \qquad |\sigma_{\text{II}}| \geq |\sigma_{\text{I}}| \qquad (14\text{-}11\text{b})$$

The yield locus corresponding to these relations is shown in Figure 14-11. This theory does not agree well with experimental data, but it has been used successfully to design gun barrels.

EXAMPLE 14-12

Solve Example 14-1 using the maximum principal normal strain failure theory.

SOLUTION

The maximum principal normal strain can be written in terms of the principal stresses as

$$\epsilon_{\text{I}} = \frac{1}{E}(\sigma_{\text{I}} - v\sigma_{\text{II}}) \qquad (a)$$

and, then, in terms of the coordinate stresses,

$$\epsilon_{\text{I}} = \frac{1}{E}\left\{\left(\frac{\sigma_{11} + \sigma_{22}}{2} + \sqrt{\left(\frac{\sigma_{11} - \sigma_{22}}{2}\right)^2 + \sigma_{12}^2}\right)\right.$$
$$\left. - v\left(\frac{\sigma_{11} + \sigma_{22}}{2} - \sqrt{\left(\frac{\sigma_{11} - \sigma_{22}}{2}\right)^2 + \sigma_{12}^2}\right)\right\}$$
$$= \frac{1}{E}\left\{(1-v)\frac{\sigma_{11} + \sigma_{22}}{2} + (1+v)\sqrt{\left(\frac{\sigma_{11} - \sigma_{22}}{2}\right)^2 + \sigma_{12}^2}\right\} \qquad (b)$$

Taking $v = 0.3$, $\sigma_{22} = 0$, $\sigma_{11} = (32/\pi)/(M/d^3)$, and $\sigma_{12} = (16/\pi)/(T/d^3)$, Eq. (b) becomes

$$\epsilon_{\text{I}} = \frac{1}{E}\left\{0.35\sigma_{11} + 0.65\sqrt{\sigma_{11}^2 + 4\sigma_{12}^2}\right\}$$
$$= \frac{1}{E}\frac{32}{\pi}\frac{M}{d^3}\left(0.35 + 0.65\sqrt{1 + \left(\frac{T}{M}\right)^2}\right)$$

or

$$\epsilon_{\text{I}} = \frac{0.2205 \times 10^6}{Ed^3} \qquad (c)$$

According to the maximum principal normal strain failure theory,

$$\epsilon_{\text{I}} = \frac{\epsilon_{\text{YP}}}{\text{S.F.}} = \frac{\sigma_{\text{YP}}}{E(\text{S.F.})}$$

Consequently,

$$\frac{0.2205 \times 10^6}{Ed^3} = \frac{340 \times 10^6}{2E}$$

from which

$$d = 0.109 \text{ m} \tag{d}$$

This result is about 4.8% greater than the result obtained using the maximum principal normal stress failure theory and about 8.4% less than the result obtained using the maximum principal shearing stress failure theory.

EXAMPLE 14-13

Solve Example 14-2 using the maximum principal normal strain failure theory. Assume Poisson's ratio to be 0.3.

SOLUTION

The maximum normal strain in the thin-walled pressure vessel is

$$\epsilon_I = \frac{1}{E} (\sigma_I - v\sigma_{II}) = \frac{1}{E} \left(\frac{pr}{t} - v\frac{pr}{2t} \right) = \frac{pr}{2t}\frac{2-v}{E} \tag{a}$$

According to the maximum principal normal strain failure theory, $\epsilon_I = \sigma_{YP}/E(\text{S.F.})$ so that

$$\frac{pr}{2t}\frac{2-v}{E} = \frac{\sigma_{YP}}{E(\text{S.F.})} \tag{b}$$

Substituting the data given in Example 14-2 yields

$$p = \frac{\sigma_{YP}}{(2-v)}\frac{t}{r} = \frac{30,000}{2-0.3}\frac{0.25}{20} = 220.6 \text{ psi} \tag{c}$$

This result is about 17.6% greater than result predicted by the maximum principal normal stress failure theory.

EXAMPLE 14-14

Solve Example 14-3 using the maximum principal normal strain failure theory.

SOLUTION

Since the member experiences a uniaxial state of stress, the maximum permissible normal strain is

$$|\epsilon_I| = \left|\frac{\sigma_I}{E}\right| = \left|\frac{0.99914P}{E}\right| \tag{a}$$

Thus

$$0.99914P = \frac{\sigma_{\text{YP}}}{\text{S.F.}} = 5000 \qquad (b)$$

which gives the same result as the maximum principal normal stress failure theory or $P_t = 5004$ lb.

PROBLEMS / Section 14-7

Solve the following problems using the maximum principal normal strain theory of failure. Assume $E = 30 \times 10^6$ psi or 210 GPa as the case may require and $v = 0.25$.

14-14 Problem 14-1.

14-15 Problem 14-3.

14-16 Problem 14-7.

14-17 The state of plane strain at a point in a material whose yield stress $\sigma_{\text{YP}} = 32$ ksi is shown in Figure P14-17. Determine whether yielding has been initiated.

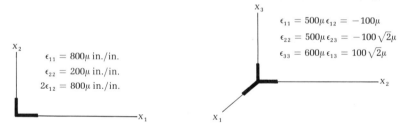

Figure P14-17   **Figure P14-18**

14-18 The state of strain at a point in a material whose tensile yield stress $\sigma_{\text{YP}} = 210$ MPa is shown in Figure P14-18. Determine whether yielding has occurred.

14-19 Problem 14-9.

14-8 Comparison of Theories of Failure—Experimental Results

Observe that all theories of failure agree for the uniaxial state of stress since the simple tension test is the basis for comparison. All theories should be expected to give essentially the same results whenever one principal normal stress is much larger than the other principal normal stress.

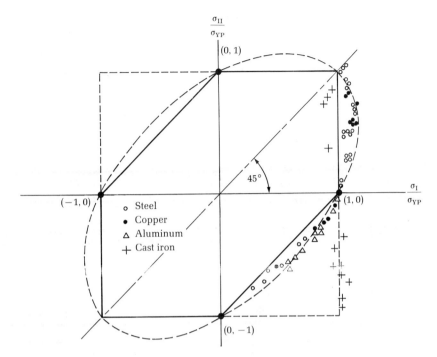

Figure 14-12 Comparison of yield and fracture criteria with experimental data

The results of experiments on materials that behave in a ductile manner—such as steel, copper, and aluminum—are shown in Figure 14-12*. Experimental results for cast iron, which behaves in a brittle manner in tension, are also shown in Figure 14-12.

Observe that all theories agree reasonably well with the experimental data for both brittle and ductile behavior whenever both principal normal stresses are tensile stresses. For ductile behavior, the maximum principal normal stress and the maximum principal shearing stress theories of failure predict the same results. Also note that these predictions are on the conservative side. For states of stress where the principal normal stresses have different signs, the maximum principal normal stress theory of failure is inadequate. However, the maximum principal shearing stress theory of failure agrees well with experiments. It should also be observed that the maximum energy of distortion theory of failure is in excellent agreement with experiments for ductile behavior. Since the maximum shearing stress theory of failure is usually easier to apply, and because it tends to give conservative results, it has gained widespread use for design of members made from materials that behave in a ductile manner.

* The experimental data shown in this figure are based on classical experiments by several investigators. This figure is based on data compiled by G. Murphy, 1964. *Advanced Mechanics of Materials*. McGraw-Hill, New York.

The maximum principal normal stress theory of failure agrees well with the experimental results over the full range of stresses for materials that behave in a brittle manner. Because of this observation and because of its extreme simplicity of application, it has gained widespread use for design of members made from materials that behave in a brittle manner.

The discussion of the physical bases for the various theories of failure and the observation of experiments indicate some of the difficulties involved in arriving at any single theory of failure that would be universally applicable for all states of stress and for materials that behave in both a ductile and a brittle manner. Recall that these theories have been developed for isotropic materials. With the development of modern composite materials, we need theories of failure that permit designers to use these new materials. It is interesting to reflect on the ingredients of the theories of failure and attempt to formulate failure theories for fiber-reinforced composite materials. Considerable research effort is presently being given to the development of such theories of failure.

14-9 SUMMARY

In this chapter, we discussed six theories of failure for the static loading of engineering members made of ductile materials. These six theories of failure are (1) maximum principal normal stress, (2) maximum principal shearing stress, (3) energy of distortion, (4) octahedral shearing stress, (5) maximum principal normal strain, and (6) total strain energy.

Each theory of failure attempts to predict when yielding will be initiated by a multiaxial state of stress using information concerning the initiation of yielding in a standard tension test.

Maximum Principal Normal Stress:

$$\sigma_{max} \leq \sigma_{YP}$$

Maximum Principal Shearing Stress:

$$\tau_{max} \leq \frac{1}{2}\sigma_{YP}$$

Maximum Energy of Distortion:

$$U_0^d \leq \frac{1+v}{3E}\sigma_{YP}^2$$

Octahedral Shearing Stress:

$$\tau_{oct} \leq \frac{\sqrt{2}}{3}\sigma_{YP}$$

Maximum Principal Normal Strain:

$$\epsilon_{max} \leq \epsilon_{YP} = \frac{\sigma_{YP}}{E}$$

Total Strain Energy:

$$U_0 \leq \frac{\sigma_{YP}^2}{2E}$$

For materials that behave in a brittle manner, the yield stress is replaced by the ultimate strength of the material. The theories of failure then attempt to predict fracture instead of yielding.

PROBLEMS / CHAPTER 14

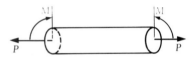

Figure P14-20

14-20   A 4-in. diameter steel bar is subjected to combined bending and axial loads at its ends as shown in Figure P14-20. The material from which the bar is made is isotropic and has the properties $\sigma_{YP} = 32$ ksi, $E = 30 \times 10^6$ psi, and $v = 0.3$. If failure occurs by general yielding, and if the applied moment $M = 150,000$ in.-lb, determine, according to the octahedral shearing stress theory of failure, the maximum axial load P that can be applied without causing permanent deformation.

14-21   A spherical pressure vessel 20 in. in diameter and 0.25 in. thick is made from an isotropic material that behaves in a ductile manner. If the mechanical properties of this material are $\sigma_{YP} = 30$ ksi, $E = 10.5 \times 10^6$ psi, and $v = 0.3$, determine the maximum permissible pressure to which the vessel can be subjected if yielding is assumed to be mode of failure. Use the maximum principal shearing stress and the maximum energy of distortion theories of failure ($p = 732$ psi or 517 psi).

14-22   A steel cylindrical pressure vessel 20 in. in diameter and 0.25 in. thick also acts as a cantilever beam as shown in Figure P14-22. Determine the maximum internal pressure that can be sustained by the vessel if the maximum shearing stress theory of failure is adopted as the criterion for predicting the initiation of yielding. To what internal pressure could the vessel be subjected if the maximum energy of distortion criterion was adopted to predict the initiation of yielding? Assume the yield point for the material is 40 ksi in tension and the stress state at point A governs.

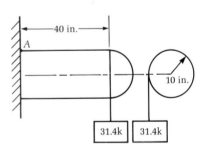

Figure P14-22

14-23   The thin cylinder shown in Figure P14-23 has a mean diameter of 200 mm and a wall thickness of 3 mm. If the axial load $P = 100$ kN and the yield point for the material is 210 MPa, determine the torque T that will cause yielding to be initiated according to the energy of distortion theory of failure ($T = 10.7$ kN·m).

Figure P14-23

14-24   A sheet of material having a yield point of 280 MPa is subjected to the plane state of stress shown in Figure P14-24. Determine the value of the shearing stress required to initiate yielding according to (a) the maximum principal shearing stress theory of failure and (b) the maximum energy of distortion theory of failure.

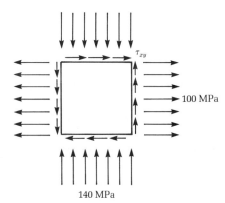

Figure P14-24

14-25   The state of stress at a point in an engineering member is characterized on the differential element shown in Figure P14-25. If the shearing stresses τ_{xz} and τ_{yz} are equal, and if the yield point for the material is 210 MPa, determine the magnitude of the shearing stress required to initiate yielding as predicted by the energy of distortion theory of failure.

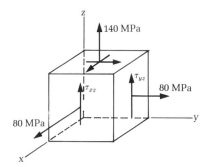

Figure P14-25

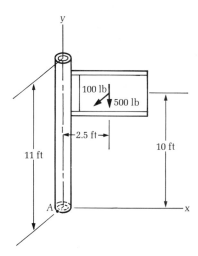

Figure P14-26

14-26   A 500-lb sign is supported by a $3\frac{1}{2}$-in. standard steel pipe as shown in Figure P14-26. The maximum horizontal wind force acting on the sign is approximately 100 lb. Determine the safety factor with respect to yielding at point A that prevails according to the octahedral shearing stress theory of failure and according to the maximum normal stress theory of failure. Assume that the system is stable and that $\sigma_{\text{YP}} = 30$ ksi.

14-27   The state of stress at a point in a structural member at which yielding was initiated is given in Figure P14-27a. Determine the value of the uniaxial tensile yield strength of the material that each of the following failure theories predict: (a) octahedral shearing stress, (b) maximum normal stress, (c) maximum

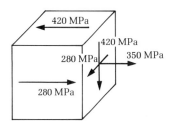

Figure P14-27

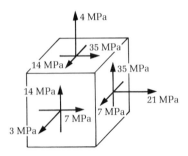

Figure P14-28

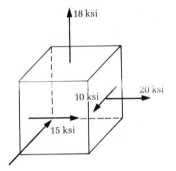

Figure P14-29

shearing stress, and (d) maximum normal strain. The Poisson ratio and modulus of elasticity for the material are 0.25 and 210 GPa, respectively.

14-28 The state of stress at a point is given in Figure P14-28. If $\sigma_{YP} = 140$ MPa, $\nu = 0.3$, $E = 70$ GPa, and a safety factor of 2.2 is required, determine whether yielding has occurred according to (a) the maximum normal strain theory of failure and (b) the maximum energy of distortion theory of failure.

14-29 The state of stress at a point in a machine element is shown in Figure P14-29. If $\sigma_{YP} = 36$ ksi and $\nu = 0.285$, determine whether yielding has been initiated according to (a) the energy of distortion theory of failure and (b) the maximum shearing stress theory of failure.

14-30 An engineering member is subjected to a force $P = 10,000$ lb as shown in Figure P14-30. It is made from an isotropic material that behaves in a ductile manner and has the following mechanical properties: $\sigma_{YP} = 40$ ksi, $E = 30 \times 10^6$ psi, and $\nu = 0.25$. Determine the minimum permissible diameter d according to each of the failure theories if yielding is to be avoided. A safety factor of 2 is required.

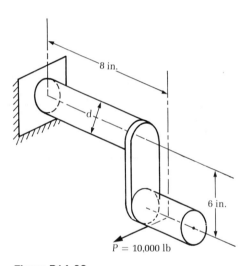

Figure P14-30

Appendices

* Data for Appendix D are based on the eighth edition of the AISC Manual of Steel Construction. Courtesy of the American Institute of Steel Construction, Chicago, Illinois.

Appendix A

Typical Properties for Some Common Materials

Material	Ultimate Strength ksi (MPa)			Yield Strength[a] ksi (MPa)	Elastic Moduli × 10^6 psi (GPa)		Coefficient of Thermal Expansion × 10^-6 F^-1 (μC^-1)	Percent Elongation in 2-in. Gage Length	Unit Weight lb/in^3 (kN/m^3)
	Tension	Comp.	Shear	Tension	Tension or Compression	Shear			
Aluminum									
2014-T4[b]	62 (427)	...	38 (262)	42 (290)	10.6 (73)	4.0 (27.6)	12.8 (23.0)	20	0.101 (27.67)
2024-T4	68 (469)	...	41 (283)	47 (324)	10.6 (73)	4.0 (27.6)	12.9 (23.2)	19	0.100 (27.40)
6061-T6	45 (310)	...	30 (207)	40 (276)	10.0 (69)	3.75 (25.8)	13.1 (23.6)	17	0.098 (26.85)
7075-T6	83 (572)	...	48 (331)	73 (503)	10.4 (72)	3.90 (26.9)	13.1 (23.6)	11	0.101 (27.67)
Brass									
RED BRASS, COLD DRAWN	75 (518)	...	...	72 (496)	15.0 (103)	5.6 (38.6)	9.8 (18.7)	55	0.316 (85.76)
RED BRASS, ANNEALED	37 (255)	...	...	14 (96)	15.0 (103)	5.6 (38.6)	9.8 (18.7)	18	0.316 (85.76)
Magnesium Alloy									
AM 100A	40 (276)	...	21 (145)	22 (152)	6.5 (44.8)	2.4 (16.5)	14.0 (25.2)	...	0.065 (17.81)
Gray Cast Iron No.									
25[c]	26 (179)	97 (668)	...	...	13.0 (89.5)	5.3 (36.5)			0.260 (71.24)
35	36 (248)	124 (854)	...	...	16.0 (110)	6.4 (44.1)			0.260 (")
50	52 (358)	164 (1130)	...	...	21.0 (145)	7.6 (52.4)			0.260 (")
Steels									
AISI[d] 1010 HR[e]	47 (324)	47 (324)	...	26 (179)	30.0[g] (207)	11.5[g] (79.3)	11.6 (20.9)	28	0.282 (77.27)
AISI 1040 HR	76 (524)	76 (524)	...	42 (289)	same	same	11.6 (")	18	0.282 (")
AISI 1045 HR	82 (565)	82 (562)	...	45 (310)	same	same	11.6 (")	16	0.282 (")
AISI 4140 HR	90 (620)	90 (620)	...	63 (434)	same	same	11.6 (")	27	0.282 (")
AISI 4140 CD[f]	102 (703)	102 (703)	...	90 (620)	same	same	11.6 (")	18	0.282 (")
ASTM[h] A7	60 (413)	60 (413)	...	...	same	same	11.6 (")		
ASTM A141	52 (358)	52 (358)	...	...	same	same	11.6 (")		
Wood									
DOUGLAS FIR, AIR DRY	...	7.4 (51)	1.1 (7.6)	...	1.9 (13.1)				0.020 (5.48)
RED OAK, AIR DRY	...	6.9 (48)	1.8 (12.4)	...	1.8 (12.4)				0.025 (6.85)
Concrete									
MEDIUM STRENGTH	...	3.0 (21)	...	1.2 (8.3)	3.0 (21)		6.0 (10.8)	...	0.087 (23.84)
HIGH STRENGTH	...	5.0 (35)	...	2.0 (13.8)	4.5 (31)		6.0 (10.8)	...	0.087

[a] The values for yield strength are for 0.2% offset.
[b] *Aluminum Standards and Data*, The Aluminum Association, Inc., 1976.
[c] Gray cast iron is designated by a number which represents close to its minimum ultimate tensile strength.
[d] American Iron and Steel Institute.
[e] Hot rolled
[f] Cold drawn
[g] Typical values for the Young's moduli and shearing moduli for carbon steels.
[h] American Society for Testing Materials.

Appendix B

Useful Properties of Plane Areas

Areas and Centroidal Moments of Inertia	
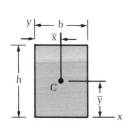 $A = bh$ $\bar{x} = b/2$ $\bar{y} = h/2$ $\bar{I}_{xx} = \frac{1}{12}bh^3$ $\bar{I}_{yy} = \frac{1}{12}hb^3$ $\bar{J} = \bar{I}_{xx} + \bar{I}_{yy}$ **Rectangle**	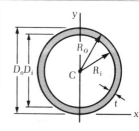 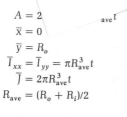 $A = 2\pi R_{ave}t$ $\bar{x} = 0$ $\bar{y} = R_o$ $\bar{I}_{xx} = \bar{I}_{yy} = \pi R_{ave}^3 t$ $\bar{J} = 2\pi R_{ave}^3 t$ $R_{ave} = (R_o + R_i)/2$ **Thin Tube**
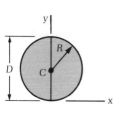 $A = \pi R^2 = \dfrac{\pi D^2}{4}$ $\bar{x} = 0$ $\bar{y} = R$ $\bar{I}_{xx} = \bar{I}_{yy} = \dfrac{\pi R^4}{4} = \dfrac{\pi D^4}{64}$ $\bar{J} = 2\bar{I}_{xx} = \dfrac{\pi D^4}{32}$ **Circle**	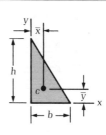 $A = \frac{1}{2}bh$ $\bar{x} = b/3$ $\bar{y} = h/3$ $\bar{I}_{xx} = \frac{1}{36}bh^3$ $\bar{I}_{yy} = \frac{1}{36}hb^3$ $\bar{J} = \bar{I}_{xx} + \bar{I}_{yy}$ $\quad = \frac{1}{36}bh(b^2 + h^2)$ **Triangle**
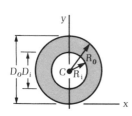 $A = \pi(R_o^2 - R_i^2) = \dfrac{\pi}{4}(D_o^2 - D_i^2)$ $\bar{x} = 0$ $\bar{y} = R_o$ $\bar{I}_{xx} = \bar{I}_{yy} = \dfrac{\pi}{4}(R_o^4 - R_i^4)$ $\quad = \dfrac{\pi}{64}(D_o^4 - D_i^4)$ $\bar{J} = 2\bar{I}_{xx} = \dfrac{\pi}{32}(D_o^4 - D_i^4)$ **Thick Tube**	$A = \dfrac{\pi R^2}{4} = \dfrac{\pi D^2}{16}$ $\bar{x} = \dfrac{4R}{3\pi}$ $\bar{y} = \dfrac{4R}{3\pi}$ $\bar{I}_{xx} = \bar{I}_{yy} = \left(\dfrac{\pi}{16} - \dfrac{4}{9\pi}\right)R^4$ $\quad = 0.0549R^4$ **Quarter Circle**
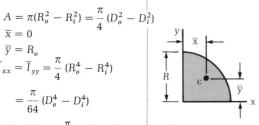 **AREA I** $A = \frac{1}{3}bh$ $\bar{x} = \frac{3}{4}b$ $\bar{y} = \frac{3}{10}h$ **AREA II** $A = \frac{2}{3}bh$ $\bar{x} = \frac{3}{8}b$ $\bar{y} = \frac{3}{5}h$ **Parabolic Spandrel**	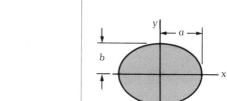 $A = \pi ab$ $\bar{x} = \bar{y} = 0$ $\bar{I}_{xx} = \dfrac{\pi}{4}ab^3$ $\bar{I}_{yy} = \dfrac{\pi}{4}ba^3$ $\bar{J} = \dfrac{\pi}{4}ab(a^2 + b^2)$ **Ellipse**

Centroids and Moments of Inertia of Composite Areas

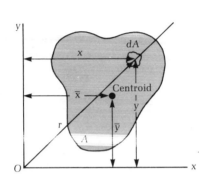

Definitions:

Rectangular Moments of Inertia

$$I_{xx} = \int_A y^2 \, dA$$

$$I_{yy} = \int_A x^2 \, dA$$

Product of Inertia

$$I_{xy} = \int_A xy \, dA$$

Polar Moment of Inertia

$$J_0 = \int_A r^2 \, dA$$

Coordinates of Centroid

$$A\bar{x} = \int_A x \, dA$$

$$A\bar{y} = \int_A y \, dA$$

Sum Theorem: The moment of inertia of a composite area with respect to a line $\ell\ell$ is equal to the sum of the moments of inertia of its parts with respect to the same line.

Proof: For a plane area composed of N parts,

$$I_{xx} = \int_{A_1} y^2 \, dA + \int_{A_2} y^2 \, dA + \cdots + \int_{A_N} y^2 \, dA$$

$$I_{xx} = \sum_{i=1}^{N} I_{xx}^{(i)}$$

Similar theorems can be stated for products of inertia and polar moments of inertia. Thus

$$I_{xy} = \sum_{i=1}^{N} I_{xy}^{(i)}$$

and

$$J_0 = \sum_{i=1}^{N} J_0^{(i)}$$

Parallel Axis Theorem: The moment of inertia of a plane area with respect to a line $\ell\ell$ is equal to the moment of inertia of the area with respect to a parallel line through the centroid plus the product of the area and the square of the perpendicular distance between the two lines. Thus

$$I_{\ell\ell} = \bar{I}_{\ell\ell} + Ad^2$$

Parallel axis theorems for products and polar moments of inertia are

$$I_{xy} = \bar{I}_{xy} + A\overline{xy}$$

and

$$J_0 = \bar{J} + Ad^2$$

Centroids: The coordinates of the centroid of a plane area composed of N parts are given by the formulas

$$A\bar{x} = \sum_{i=1}^{N} A_i x_i$$

and

$$A\bar{y} = \sum_{i=1}^{N} A_i y_i$$

where x_i and y_i are the coordinates of the centroid of the ith part and A_i is the area of the ith part.

Appendix D Properties of Structural Shapes

W Shapes

(Wide-Flange Shapes)

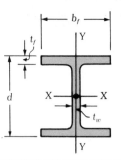

Designation*	Area A, in^2	Depth d, in.	Flange Width b_f, in.	Flange Thickness t_f, in.	Web Thickness t_w, in.	Axis XX I_x, in^4	Axis XX S_x, in^3	Axis XX r_x, in.	Axis YY I_y, in^4	Axis YY S_y, in^3	Axis YY r_y, in.
W 24 × 62	18.2	24.74	7.040	0.590	0.430	1830	154	9.55	70.4	15.7	1.87
× 55	16.2	23.57	7.005	0.505	0.395	1350	114	9.11	29.1	8.30	1.34
W 21 × 147	43.2	22.06	12.510	1.150	0.720	3630	329	9.17	376	60.1	2.95
× 122	35.9	21.68	12.390	0.960	0.600	2960	273	9.09	305	49.2	2.92
× 111	32.7	21.51	12.340	0.875	0.550	2670	249	9.05	274	44.5	2.90
× 101	29.8	21.36	12.290	0.800	0.500	2420	227	9.02	248	40.3	2.89
× 93	27.3	21.62	8.420	0.930	0.580	2070	192	8.70	92.9	22.1	1.84
× 83	24.3	21.43	8.355	0.835	0.515	1830	171	8.67	81.4	19.5	1.83
× 73	21.5	21.24	8.295	0.740	0.455	1600	151	8.64	70.6	17.0	1.81
× 68	20.0	21.13	8.270	0.685	0.430	1480	140	8.60	64.7	15.7	1.80
× 62	18.3	20.99	8.240	0.615	0.400	1330	127	8.54	57.5	13.9	1.77
× 57	16.7	21.06	6.555	0.650	0.405	1170	111	8.36	30.6	9.35	1.35
× 50	14.7	20.83	6.530	0.535	0.380	984	94.5	8.18	24.9	7.64	1.30
× 44	13.0	20.66	6.500	0.450	0.350	843	81.6	8.06	20.7	6.36	1.26
W 18 × 119	35.1	18.97	11.265	1.060	0.655	2190	231	7.90	253	44.9	2.69
× 106	31.1	18.73	11.200	0.940	0.590	1910	204	7.84	220	39.4	2.66
× 97	28.5	18.59	11.145	0.870	0.535	1750	188	7.82	201	36.1	2.65
× 86	25.3	18.39	11.090	0.770	0.480	1530	166	7.77	175	31.6	2.63
× 76	22.3	18.21	11.035	0.680	0.425	1330	146	7.73	152	27.6	2.61
W 18 × 71	20.8	18.47	7.635	0.810	0.495	1170	127	7.50	60.3	15.8	1.70
× 65	19.1	18.35	7.590	0.750	0.450	1070	117	7.49	54.8	14.4	1.69
× 60	17.6	18.24	7.555	0.695	0.415	984	108	7.47	50.1	13.3	1.69
× 55	16.2	18.11	7.530	0.630	0.390	890	98.3	7.41	44.9	11.9	1.67
× 50	14.7	17.99	7.495	0.570	0.355	800	88.9	7.38	40.1	10.7	1.65
W 18 × 46	13.5	18.06	6.060	0.605	0.360	712	78.8	7.25	22.5	7.43	1.29
× 40	11.8	17.90	6.015	0.525	0.315	612	68.4	7.21	19.1	6.35	1.27
× 35	10.3	17.70	6.000	0.425	0.300	510	57.6	7.04	15.3	5.12	1.22
W 16 × 100	29.4	16.97	10.425	0.985	0.585	1490	175	7.10	186	35.7	2.51
× 89	26.2	16.75	10.365	0.875	0.525	1300	155	7.05	163	31.4	2.49
× 77	22.6	16.52	10.295	0.760	0.455	1110	134	7.00	138	26.9	2.47
× 67	19.7	16.33	10.235	0.665	0.395	954	117	6.96	119	23.2	2.46
W 16 × 57	16.8	16.43	7.120	0.715	0.430	758	92.2	6.72	43.1	12.1	1.60
× 50	14.7	16.26	7.070	0.630	0.380	659	81.0	6.68	37.2	10.5	1.59
× 45	13.3	16.13	7.035	0.565	0.345	586	72.7	6.65	32.8	9.34	1.57
× 40	11.8	16.01	6.995	0.505	0.305	518	64.7	6.63	28.9	8.25	1.57
× 36	10.6	15.86	6.985	0.430	0.295	448	56.5	6.51	24.5	7.00	1.52
W 16 × 31	9.12	15.88	5.525	0.440	0.275	375	47.2	6.41	12.4	4.49	1.17
× 26	7.68	15.69	5.500	0.345	0.250	301	38.4	6.26	9.59	3.49	1.12

Appendix D   Properties of Structural Shapes

W Shapes* (English Units)

(Wide-Flange Shapes)

Designation *	Area A, in^2	Depth d, in.	Flange Width b_f, in.	Flange Thickness t_f, in.	Web Thickness t_w, in.	Axis XX I_x, in^4	Axis XX S_x, in^3	Axis XX r_x, in.	Axis YY I_y, in^4	Axis YY S_y, in^3	Axis YY r_y, in.
W 14 × 82	24.1	14.31	10.130	0.855	0.510	882	123	6.05	148	29.3	2.48
× 74	21.8	14.17	10.070	0.785	0.450	796	112	6.04	134	26.6	2.48
× 68	20.0	14.04	10.035	0.720	0.415	723	103	6.01	121	24.2	2.46
× 61	17.9	13.89	9.995	0.645	0.375	640	92.2	5.98	107	21.5	2.45
× 53	15.6	13.92	8.060	0.660	0.370	541	77.8	5.89	57.7	14.3	1.92
W 14 × 48	14.1	13.79	8.030	0.595	0.340	485	70.3	5.85	51.4	12.8	1.91
× 43	12.6	13.66	7.995	0.530	0.305	428	62.7	5.82	45.2	11.3	1.89
× 38	11.2	14.10	6.770	0.515	0.310	385	54.6	5.87	26.7	7.88	1.55
× 34	10.0	13.98	6.745	0.455	0.285	340	48.6	5.83	23.3	6.91	1.53
× 30	8.85	13.84	6.730	0.385	0.270	291	42.0	5.73	19.6	5.82	1.49
W 12 × 58	17.0	12.19	10.010	0.640	0.360	475	78.0	5.28	107	21.4	2.51
× 53	15.6	12.06	9.995	0.575	0.345	425	70.6	5.23	95.8	19.2	2.48
× 50	14.7	12.19	8.080	0.640	0.370	394	64.7	5.18	56.3	13.9	1.96
× 45	13.2	12.06	8.045	0.575	0.335	350	58.1	5.15	50.0	12.4	1.94
× 40	11.8	11.94	8.005	0.515	0.295	310	51.9	5.13	44.1	11.0	1.93
W 12 × 35	10.3	12.50	6.560	0.520	0.300	285	45.6	5.25	24.5	7.47	1.54
× 30	8.79	12.34	6.520	0.440	0.260	238	38.6	5.21	20.3	6.24	1.52
× 26	7.65	12.22	6.490	0.380	0.230	204	33.4	5.17	17.3	5.34	1.51
W 10 × 45	13.3	10.10	8.020	0.620	0.350	248	49.1	4.32	53.4	13.3	2.01
× 39	11.5	9.92	7.985	0.530	0.315	209	42.1	4.27	45.0	11.3	1.98
× 33	9.71	9.73	7.960	0.435	0.290	170	35.0	4.19	36.6	9.20	1.94
× 30	8.84	10.47	5.810	0.510	0.300	170	32.4	4.38	16.7	5.75	1.37
× 26	7.61	10.33	5.770	0.440	0.260	144	27.9	4.35	14.1	4.89	1.36
× 22	6.49	10.17	5.750	0.360	0.240	118	23.2	4.27	11.4	3.97	1.33
W 8 × 67	19.7	9.00	8.280	0.935	0.570	272	60.4	3.72	88.6	21.4	2.12
× 58	17.1	8.75	8.220	0.810	0.510	228	52.0	3.65	75.1	18.3	2.10
× 48	14.1	8.50	8.110	0.685	0.400	184	43.3	3.61	60.9	15.0	2.08
× 40	11.7	8.25	8.070	0.560	0.360	146	35.5	3.53	49.1	12.2	2.04
× 35	10.3	8.12	8.020	0.495	0.310	127	31.2	3.51	42.6	10.6	2.03
W 8 × 31	9.13	8.00	7.995	0.435	0.285	110	27.5	3.47	37.1	9.27	2.02
× 28	8.25	8.06	6.535	0.465	0.285	98.0	24.3	3.45	21.7	6.63	1.62
× 24	7.08	7.93	6.495	0.400	0.245	82.8	20.9	3.42	18.3	5.63	1.61
× 21	6.16	8.28	5.270	0.400	0.250	75.3	18.2	3.49	9.77	3.71	1.26
× 18	5.26	8.14	5.250	0.330	0.230	61.9	15.2	3.43	7.97	3.04	1.23
W 8 × 15	4.44	8.11	4.015	0.315	0.245	48.0	11.8	3.29	3.41	1.70	0.876
× 13	3.84	7.99	4.000	0.255	0.230	39.6	9.91	3.21	2.73	1.37	0.843
× 10	2.96	7.89	3.940	0.205	0.170	36.8	7.81	3.22	2.09	1.06	0.841
W 6 × 25	7.34	6.38	6.080	0.455	0.320	53.4	16.7	2.70	17.1	5.61	1.52
× 20	5.87	6.20	6.020	0.365	0.260	41.4	13.4	2.66	13.3	4.41	1.50
× 15	4.43	5.99	5.990	0.260	0.230	29.1	9.72	2.56	9.32	3.11	1.46
× 16	4.74	6.28	4.030	0.405	0.260	32.1	10.2	2.60	4.43	2.20	0.966
× 12	3.55	6.03	4.000	0.280	0.230	22.1	7.31	2.49	2.99	1.50	0.918
× 9	2.68	5.90	3.940	0.215	0.170	16.4	5.56	2.47	2.19	1.11	0.905

*A wide flange shape is designated by the letter W followed by the nominal depth in inches and the weight in pounds per foot.

(Table continued on page 670)

Appendix D Properties of Structural Shapes

S Shapes * (English Units)

(American Standard Shapes)

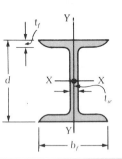

Designation	Area A, in^2	Depth d, in.	Flange Width b_f, in.	Flange Thickness t_f, in.	Web Thickness t_w, in.	Axis XX I_x, in^4	Axis XX S_x, in^3	Axis XX r_x, in.	Axis YY I_y, in^4	Axis YY S_y, in^3	Axis YY r_y, in.
S 24 × 121	35.6	24.50	8.050	1.090	0.800	3160	258	9.43	83.3	20.7	1.53
× 106	31.2	24.50	7.870	1.090	0.620	2940	240	9.71	77.1	19.6	1.57
× 100	29.3	24.00	7.245	0.870	0.745	2390	199	9.02	47.7	13.2	1.27
× 90	26.5	24.00	7.125	0.870	0.625	2250	187	9.21	44.9	12.6	1.30
× 80	23.5	24.00	7.000	0.870	0.500	2100	175	9.47	42.2	12.1	1.34
S 20 × 96	28.2	20.30	7.200	0.920	0.800	1670	165	7.71	50.2	13.9	1.33
× 86	25.3	20.30	7.060	0.920	0.660	1580	155	7.89	46.8	13.3	1.36
× 75	22.0	20.00	6.385	0.795	0.635	1280	128	7.62	29.8	9.32	1.16
× 66	19.4	20.00	6.255	0.795	0.505	1190	119	7.83	27.7	8.85	1.19
S 18 × 70	20.6	18.00	6.251	0.691	0.711	926	103	6.71	24.1	7.72	1.08
× 54.7	16.1	18.00	6.001	0.691	0.461	804	89.4	7.07	20.8	6.94	1.14
S 15 × 50	14.7	15.00	5.640	0.622	0.550	486	64.8	5.75	15.7	5.57	1.03
× 42.9	12.6	15.00	5.501	0.622	0.411	447	59.6	5.95	14.4	5.23	1.07
S 12 × 50	14.7	12.00	5.477	0.659	0.687	305	50.8	4.55	15.7	5.74	1.03
× 40.8	12.0	12.00	5.252	0.659	0.462	272	45.4	4.77	13.6	5.16	1.06
× 35	10.3	12.00	5.078	0.544	0.428	229	38.2	4.72	9.87	3.89	0.980
× 31.8	9.35	12.00	5.000	0.544	0.350	218	36.4	4.83	9.36	3.74	1.00
S 10 × 35	10.3	10.00	4.944	0.491	0.594	147	29.4	3.78	8.36	3.38	0.901
× 25.4	7.46	10.00	4.661	0.491	0.311	124	24.7	4.07	6.79	2.91	0.954
S 8 × 23	6.77	8.00	4.171	0.426	0.441	64.9	16.2	3.10	4.31	2.07	0.798
× 18.4	5.41	8.00	4.001	0.426	0.271	57.6	14.4	3.26	3.73	1.86	0.831
S 7 × 20	5.88	7.00	3.860	0.392	0.450	42.4	12.1	2.69	3.17	1.64	0.734
× 15.3	4.50	7.00	3.662	0.392	0.252	36.7	10.5	2.86	2.64	1.44	0.766
S 6 × 17.25	5.07	6.00	3.565	0.359	0.465	26.3	8.77	2.28	2.31	1.30	0.675
× 12.5	3.67	6.00	3.332	0.359	0.232	22.1	7.37	2.45	1.82	1.09	0.705
S 5 × 14.75	4.34	5.00	3.284	0.326	0.494	15.2	6.09	1.87	1.67	1.01	0.620
× 10	2.94	5.00	3.004	0.326	0.214	12.3	4.92	2.05	1.22	0.809	0.643
S 4 × 9.5	2.79	4.00	2.796	0.293	0.326	6.79	3.39	1.56	0.903	0.646	0.569
× 7.7	2.26	4.00	2.663	0.293	0.193	6.08	3.04	1.64	0.764	0.574	0.581
S 3 × 7.5	2.21	3.00	2.509	0.260	0.349	2.93	1.95	1.15	0.586	0.468	0.516
× 5.7	1.67	3.00	2.330	0.260	0.170	2.52	1.68	1.23	0.455	0.390	0.522

*An American Standard Beam is designated by the letter S followed by the nominal depth in inches and the weight in pounds per foot.

Appendix D Properties of Structural Shapes

C Shapes* (English Units)

(American Standard Channels)

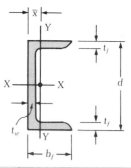

Designation	Area A, in^2	Depth d, in.	Flange Width b_f, in.	Flange Thickness t_f, in.	Web Thickness t_w, in.	Axis XX I_x, in^4	S_x, in^3	r_x, in.	Axis YY I_y, in^4	S_y, in^3	r_y, in.	$\bar{X}$, in.
C 15 × 50	14.7	15.00	3.716	0.650	0.716	404	53.8	5.24	11.0	3.78	0.867	0.798
× 40	11.8	15.00	3.520	0.650	0.520	349	46.5	5.44	9.23	3.37	0.886	0.777
× 33.9	9.96	15.00	3.400	0.650	0.400	315	42.0	5.62	8.13	3.11	0.904	0.787
C 12 × 30	8.82	12.00	3.170	0.501	0.510	162	27.0	4.29	5.14	2.06	0.763	0.674
× 25	7.35	12.00	3.047	0.501	0.387	144	24.1	4.43	4.47	1.88	0.780	0.674
× 20.7	6.09	12.00	2.942	0.501	0.282	129	21.5	4.61	3.88	1.73	0.799	0.698
C 10 × 30	8.82	10.00	3.033	0.436	0.673	103	20.7	3.42	3.94	1.65	0.669	0.649
× 25	7.35	10.00	2.886	0.436	0.526	91.2	18.2	3.52	3.36	1.48	0.676	0.617
× 20	5.88	10.00	2.739	0.436	0.379	78.9	15.8	3.66	2.81	1.32	0.692	0.606
× 15.3	4.49	10.00	2.600	0.436	0.240	67.4	13.5	3.87	2.28	1.16	0.713	0.634
C 9 × 20	5.88	9.00	2.648	0.413	0.448	60.9	13.5	3.22	2.42	1.17	0.642	0.583
× 15	4.41	9.00	2.485	0.413	0.285	51.0	11.3	3.40	1.93	1.01	0.661	0.586
× 13.4	3.94	9.00	2.433	0.413	0.233	47.9	10.6	3.48	1.76	0.962	0.669	0.601
C 8 × 18.75	5.51	8.00	2.527	0.390	0.487	44.0	11.0	2.82	1.98	1.01	0.599	0.565
× 13.75	4.04	8.00	2.343	0.390	0.303	36.1	9.03	2.99	1.53	0.854	0.615	0.553
× 11.5	3.38	8.00	2.260	0.390	0.220	32.6	8.14	3.11	1.32	0.781	0.625	0.571
C 7 × 14.75	4.33	7.00	2.299	0.366	0.419	27.2	7.78	2.51	1.38	0.779	0.564	0.532
× 12.25	3.60	7.00	2.194	0.366	0.314	24.2	6.93	2.60	1.17	0.703	0.571	0.525
× 9.8	2.87	7.00	2.090	0.366	0.210	21.3	6.08	2.72	0.968	0.625	0.581	0.540
C 6 × 13	3.83	6.00	2.157	0.343	0.437	17.4	5.80	2.13	1.05	0.642	0.525	0.514
× 10.5	3.09	6.00	2.034	0.343	0.314	15.2	5.06	2.22	0.866	0.564	0.529	0.499
× 8.2	2.40	6.00	1.920	0.343	0.200	13.1	4.38	2.34	0.693	0.492	0.537	0.511
C 5 × 9	2.64	5.00	1.885	0.320	0.325	8.90	3.56	1.83	0.632	0.450	0.489	0.478
× 6.7	1.97	5.00	1.750	0.320	0.190	7.49	3.00	1.95	0.479	0.378	0.493	0.484
C 4 × 7.25	2.13	4.00	1.721	0.296	0.321	4.59	2.29	1.47	0.433	0.343	0.450	0.459
× 5.4	1.59	4.00	1.584	0.296	0.184	3.85	1.93	1.56	0.319	0.283	0.449	0.457
C 3 × 6	1.76	3.00	1.596	0.273	0.356	2.07	1.38	1.08	0.305	0.268	0.416	0.455
× 5	1.47	3.00	1.498	0.273	0.258	1.85	1.24	1.12	0.247	0.233	0.410	0.438
× 4.1	1.21	3.00	1.410	0.273	0.170	1.66	1.10	1.17	0.197	0.202	0.404	0.436

*An American Standard Channel is designated by the letter C followed by the nominal depth in inches and the weight in pounds per foot.

Appendix D   Properties of Structural Shapes

Unequal Leg Angles (English Units)

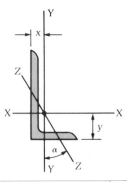

Size and Thickness	Weight per foot	Area A	Axis XX				Axis YY				Axis ZZ	
			I_x	S_x	r_x	y	I_y	S_y	r_y	x	r_z	Tan
in.	lb	in^2	in^4	in^3	in.	in.	in^4	in^3	in.	in.	in.	α
L 6 × 4 × $\frac{7}{8}$	27.2	7.98	27.7	7.15	1.86	2.12	9.75	3.39	1.11	1.12	0.857	0.421
$\frac{3}{4}$	23.6	6.94	24.5	6.25	1.88	2.08	8.68	2.97	1.12	1.08	0.860	0.428
$\frac{5}{8}$	20.0	5.86	21.1	5.31	1.90	2.03	7.52	2.54	1.13	1.03	0.864	0.435
$\frac{9}{16}$	18.1	5.31	19.3	4.83	1.90	2.01	6.91	2.31	1.14	1.01	0.866	0.438
$\frac{1}{2}$	16.2	4.75	17.4	4.33	1.91	1.99	6.27	2.08	1.15	0.987	0.870	0.440
$\frac{7}{16}$	14.3	4.18	15.5	3.83	1.92	1.96	5.60	1.85	1.16	0.964	0.873	0.443
$\frac{3}{8}$	12.3	3.61	13.5	3.32	1.93	1.94	4.90	1.60	1.17	0.941	0.877	0.446
$\frac{5}{16}$	10.3	3.03	11.4	2.79	1.94	1.92	4.18	1.35	1.17	0.918	0.882	0.448
L 6 × 3$\frac{1}{2}$ × $\frac{1}{2}$	15.3	4.50	16.6	4.24	1.92	2.08	4.25	1.59	0.972	0.833	0.759	0.344
$\frac{3}{8}$	11.7	3.42	12.9	3.24	1.94	2.04	3.34	1.23	0.988	0.787	0.767	0.350
$\frac{5}{16}$	9.8	2.87	10.9	2.73	1.95	2.01	2.85	1.04	0.996	0.763	0.772	0.352
L 4 × 3$\frac{1}{2}$ × $\frac{5}{8}$	14.7	4.30	6.37	2.35	1.22	1.29	4.52	1.84	1.03	1.04	0.719	0.745
$\frac{1}{2}$	11.9	3.50	5.32	1.94	1.23	1.25	3.79	1.52	1.04	1.00	0.722	0.750
$\frac{7}{16}$	10.6	3.09	4.76	1.72	1.24	1.23	3.40	1.35	1.05	0.978	0.724	0.753
$\frac{3}{8}$	9.1	2.67	4.18	1.49	1.25	1.21	2.95	1.17	1.06	0.955	0.727	0.755
$\frac{5}{16}$	7.7	2.25	3.56	1.26	1.26	1.18	2.55	0.994	1.07	0.932	0.730	0.757
$\frac{1}{4}$	6.2	1.81	2.91	1.03	1.27	1.16	2.09	0.808	1.07	0.909	0.734	0.759
L 4 × 3 × $\frac{5}{8}$	13.6	3.98	6.03	2.30	1.23	1.37	2.87	1.35	0.849	0.871	0.637	0.534
$\frac{1}{2}$	11.1	3.25	5.05	1.89	1.25	1.33	2.42	1.12	0.864	0.827	0.639	0.543
$\frac{7}{16}$	9.8	2.87	4.52	1.68	1.25	1.30	2.18	0.992	0.871	0.804	0.641	0.547
$\frac{3}{8}$	8.5	2.48	3.96	1.46	1.26	1.28	1.92	0.866	0.879	0.782	0.644	0.551
$\frac{5}{16}$	7.2	2.09	3.38	1.23	1.27	1.26	1.65	0.734	0.887	0.759	0.647	0.554
$\frac{1}{4}$	5.8	1.69	2.77	1.00	1.28	1.24	1.36	0.599	0.896	0.736	0.651	0.558
L 3$\frac{1}{2}$ × 3 × $\frac{1}{2}$	10.2	3.00	3.45	1.45	1.07	1.13	2.33	1.10	0.881	0.875	0.621	0.714
$\frac{7}{16}$	9.1	2.65	3.10	1.29	1.08	1.10	2.09	0.975	0.889	0.853	0.622	0.718
$\frac{3}{8}$	7.9	2.30	2.72	1.13	1.09	1.08	1.85	0.851	0.897	0.830	0.625	0.721
$\frac{5}{16}$	6.6	1.93	2.33	0.954	1.10	1.06	1.58	0.722	0.905	0.808	0.627	0.724
$\frac{1}{4}$	5.4	1.56	1.91	0.776	1.11	1.04	1.30	0.589	0.914	0.785	0.631	0.727
L 3$\frac{1}{2}$ × 2$\frac{1}{2}$ × $\frac{1}{2}$	9.4	2.75	3.24	1.41	1.09	1.20	1.36	0.760	0.704	0.705	0.534	0.486
$\frac{7}{16}$	8.3	2.43	2.91	1.26	1.09	1.18	1.23	0.677	0.711	0.682	0.535	0.491
$\frac{3}{8}$	7.2	2.11	2.56	1.09	1.10	1.16	1.09	0.592	0.719	0.660	0.537	0.496
$\frac{5}{16}$	6.1	1.78	2.19	0.927	1.11	1.14	0.939	0.504	0.727	0.637	0.540	0.501
$\frac{1}{4}$	4.9	1.44	1.80	0.755	1.12	1.11	0.777	0.412	0.735	0.614	0.544	0.506

(Continued)

Size and Thickness	Weight per foot	Area A	Axis XX				Axis YY				Axis ZZ	
			I_x	S_x	r_x	y	I_y	S_y	r_y	x	r_z	Tan
in.	lb	in²	in⁴	in³	in.	in.	in⁴	in³	in.	in.	in.	α
L $2\frac{1}{2} \times 2 \times \frac{3}{8}$	5.3	1.55	0.912	0.547	0.768	0.831	0.514	0.363	0.577	0.581	0.420	0.614
$\frac{5}{16}$	4.5	1.31	0.788	0.466	0.776	0.809	0.446	0.310	0.584	0.559	0.422	0.620
$\frac{1}{4}$	3.62	1.06	0.654	0.381	0.784	0.787	0.372	0.254	0.592	0.537	0.424	0.626
$\frac{3}{16}$	2.75	0.809	0.509	0.293	0.793	0.764	0.291	0.196	0.600	0.514	0.427	0.631
L $3 \times 2\frac{1}{2} \times \frac{1}{2}$	8.5	2.50	2.08	1.04	0.913	1.00	1.30	0.744	0.722	0.750	0.520	0.667
$\frac{7}{16}$	7.6	2.21	1.88	0.928	0.920	0.978	1.18	0.664	0.729	0.728	0.521	0.672
$\frac{3}{8}$	6.6	1.92	1.66	0.810	0.928	0.956	1.04	0.581	0.736	0.706	0.522	0.676
$\frac{5}{16}$	5.6	1.62	1.42	0.688	0.937	0.933	0.898	0.494	0.744	0.683	0.525	0.680
$\frac{1}{4}$	4.5	1.31	1.17	0.561	0.945	0.911	0.743	0.404	0.753	0.661	0.528	0.684
$\frac{3}{16}$	3.39	0.996	0.907	0.430	0.954	0.888	0.577	0.310	0.761	0.638	0.533	0.688
L $3 \times 2 \times \frac{1}{2}$	7.7	2.25	1.92	1.00	0.924	1.08	0.672	0.474	0.546	0.583	0.428	0.414
$\frac{7}{16}$	6.8	2.00	1.73	0.894	0.932	1.06	0.609	0.424	0.553	0.561	0.429	0.421
$\frac{3}{8}$	5.9	1.73	1.53	0.781	0.940	1.04	0.543	0.371	0.559	0.539	0.430	0.428
$\frac{5}{16}$	5.0	1.46	1.32	0.664	0.948	1.02	0.470	0.317	0.567	0.516	0.432	0.435
$\frac{1}{4}$	4.1	1.19	1.09	0.542	0.957	0.993	0.392	0.260	0.574	0.493	0.435	0.440
$\frac{3}{16}$	3.07	0.902	0.842	0.415	0.966	0.970	0.307	0.200	0.583	0.470	0.439	0.446

Angles in shaded rows may not be readily available.

Appendix D   Properties of Structural Shapes

Equal Leg Angles (English Units)

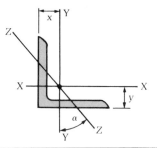

Size and Thickness	Weight per foot	Area A	Axis XX				Axis YY				Axis ZZ	
			I_x	S_x	r_x	y	I_y	S_y	r_y	x	r_z	Tan
in.	lb	in²	in⁴	in³	in.	in.	in⁴	in³	in.	in.	in.	α
L 8 × 8 × 1⅛	56.9	16.7	98.0	17.5	2.42	2.41	98.0	17.5	2.42	2.41	1.56	1.000
1	51.0	15.0	89.0	15.8	2.44	2.37	89.0	15.8	2.44	2.37	1.56	1.000
⅞	45.0	13.2	79.6	14.0	2.45	2.32	79.6	14.0	2.45	2.32	1.57	1.000
¾	38.9	11.4	69.7	12.2	2.47	2.28	69.7	12.2	2.47	2.28	1.58	1.000
⅝	32.7	9.61	59.4	10.3	2.49	2.23	59.4	10.3	2.49	2.23	1.58	1.000
9/16	29.6	8.68	54.1	9.34	2.50	2.21	54.1	9.34	2.50	2.21	1.59	1.000
½	26.4	7.75	48.6	8.36	2.50	2.19	48.6	8.36	2.50	2.19	1.59	1.000
L 6 × 6 × 1	37.4	11.0	35.5	8.57	1.80	1.86	35.5	8.57	1.80	1.86	1.17	1.000
⅞	33.1	9.73	31.9	7.63	1.81	1.82	31.9	7.63	1.81	1.82	1.17	1.000
¾	28.7	8.44	28.2	6.66	1.83	1.78	28.2	6.66	1.83	1.78	1.17	1.000
⅝	24.2	7.11	24.2	5.66	1.84	1.73	24.2	5.66	1.84	1.73	1.18	1.000
9/16	21.9	6.43	22.1	5.14	1.85	1.71	22.1	5.14	1.85	1.71	1.18	1.000
½	19.6	5.75	19.9	4.61	1.86	1.68	19.9	4.61	1.86	1.68	1.18	1.000
7/16	17.2	5.06	17.7	4.08	1.87	1.66	17.7	4.08	1.87	1.66	1.19	1.000
⅜	14.9	4.36	15.4	3.53	1.88	1.64	15.4	3.53	1.88	1.64	1.19	1.000
5/16	12.4	3.65	13.0	2.97	1.89	1.62	13.0	2.97	1.89	1.62	1.20	1.000
L 5 × 5 × ⅞	27.2	7.98	17.8	5.17	1.49	1.57	17.8	5.17	1.49	1.57	0.973	1.000
¾	23.6	6.94	15.7	4.53	1.51	1.52	15.7	4.53	1.51	1.52	0.975	1.000
⅝	20.0	5.86	13.6	3.86	1.52	1.48	13.6	3.86	1.52	1.48	0.978	1.000
½	16.2	4.75	11.3	3.16	1.54	1.43	11.3	3.16	1.54	1.43	0.983	1.000
7/16	14.3	4.18	10.0	2.79	1.55	1.41	10.0	2.79	1.55	1.41	0.986	1.000
⅜	12.3	3.61	8.74	2.42	1.56	1.39	8.74	2.42	1.56	1.39	0.990	1.000
5/16	10.3	3.03	7.42	2.04	1.57	1.37	7.42	2.04	1.57	1.37	0.994	1.000
L 4 × 4 × ¾	18.5	5.44	7.67	2.81	1.19	1.27	7.67	2.81	1.19	1.27	0.778	1.000
⅝	15.7	4.61	6.66	2.40	1.20	1.23	6.66	2.40	1.20	1.23	0.779	1.000
½	12.8	3.75	5.56	1.97	1.22	1.18	5.56	1.97	1.22	1.18	0.782	1.000
7/16	11.3	3.31	4.97	1.75	1.23	1.16	4.97	1.75	1.23	1.16	0.785	1.000
⅜	9.8	2.86	4.36	1.52	1.23	1.14	4.36	1.52	1.23	1.14	0.788	1.000
5/16	8.2	2.40	3.71	1.29	1.24	1.12	3.71	1.29	1.24	1.12	0.791	1.000
¼	6.6	1.94	3.04	1.05	1.25	1.09	3.04	1.05	1.25	1.09	0.795	1.000
L 3½ × 3½ × ½	11.1	3.25	3.64	1.49	1.06	1.06	3.64	1.49	1.06	1.06	0.683	1.000
7/16	9.8	2.87	3.26	1.32	1.07	1.04	3.26	1.32	1.07	1.04	0.684	1.000
⅜	8.5	2.48	2.87	1.15	1.07	1.01	2.87	1.15	1.07	1.01	0.687	1.000
5/16	7.2	2.09	2.45	0.976	1.08	0.990	2.45	0.976	1.08	0.990	0.690	1.000
¼	5.8	1.69	2.01	0.794	1.09	0.968	2.01	0.794	1.09	0.968	0.694	1.000

(Continued)

Size and Thickness	Weight per foot	Area A	Axis XX				Axis YY				Axis ZZ	
			I_x	S_x	r_x	y	I_y	S_y	r_y	x	r_z	Tan
in.	lb	in^2	in^4	in^3	in.	in.	in^4	in^3	in.	in.	in.	α
L 3 × 3 × $\frac{1}{2}$	9.4	2.75	2.22	1.07	0.898	0.932	2.22	1.07	0.898	0.932	0.584	1.000
$\frac{7}{16}$	8.3	2.43	1.99	0.954	0.905	0.910	1.99	0.954	0.905	0.910	0.585	1.000
$\frac{3}{8}$	7.2	2.11	1.76	0.833	0.913	0.888	1.76	0.833	0.913	0.888	0.587	1.000
$\frac{5}{16}$	6.1	1.78	1.51	0.707	0.922	0.865	1.51	0.707	0.922	0.865	0.589	1.000
$\frac{1}{4}$	4.9	1.44	1.24	0.577	0.930	0.842	1.24	0.577	0.930	0.842	0.592	1.000
$\frac{3}{16}$	3.71	1.09	0.962	0.441	0.939	0.820	0.962	0.441	0.939	0.820	0.596	1.000
L $2\frac{1}{2}$ × $2\frac{1}{2}$ × $\frac{1}{2}$	7.7	2.25	1.23	0.724	0.739	0.806	1.23	0.724	0.739	0.806	0.487	1.000
$\frac{3}{8}$	5.9	1.73	0.984	0.566	0.753	0.762	0.984	0.566	0.753	0.762	0.487	1.000
$\frac{5}{16}$	5.0	1.46	0.849	0.482	0.761	0.740	0.849	0.482	0.761	0.740	0.489	1.000
$\frac{1}{4}$	4.1	1.19	0.703	0.394	0.769	0.717	0.703	0.394	0.769	0.717	0.491	1.000
$\frac{3}{16}$	3.07	0.902	0.547	0.303	0.778	0.694	0.547	0.303	0.778	0.694	0.495	1.000
L 2 × 2 × $\frac{3}{8}$	4.7	1.36	0.479	0.351	0.594	0.636	0.479	0.351	0.594	0.636	0.389	1.000
$\frac{5}{16}$	3.92	1.15	0.416	0.300	0.601	0.614	0.416	0.300	0.601	0.614	0.390	1.000
$\frac{1}{4}$	3.19	0.938	0.348	0.247	0.609	0.592	0.348	0.247	0.609	0.592	0.391	1.000
$\frac{3}{16}$	2.44	0.715	0.272	0.190	0.617	0.569	0.272	0.190	0.617	0.569	0.394	1.000
$\frac{1}{8}$	1.65	0.484	0.190	0.131	0.626	0.546	0.190	0.131	0.626	0.546	0.398	1.000

Angles in shaded rows may not be readily available.

Appendix D Properties of Structural Shapes

Steel Pipe

Pipe Dimensions and Properties								
Dimensions				Weight per Foot Lbs Plain Ends	Properties			
Nominal Diameter in.	Outside Diameter in.	Inside Diameter in.	Wall Thickness in.		A in^2	I in^4	S in^3	r in.
Standard Weight								
$\frac{1}{2}$	0.840	0.622	0.109	0.85	0.250	0.017	0.041	0.261
$\frac{3}{4}$	1.050	0.824	0.113	1.13	0.333	0.037	0.071	0.334
1	1.315	1.049	0.133	1.68	0.494	0.087	0.133	0.421
$1\frac{1}{4}$	1.660	1.380	0.140	2.27	0.669	0.195	0.235	0.540
$1\frac{1}{2}$	1.900	1.610	0.145	2.72	0.799	0.310	0.326	0.623
2	2.375	2.067	0.154	3.65	1.07	0.666	0.561	0.787
$2\frac{1}{2}$	2.875	2.469	0.203	5.79	1.70	1.53	1.06	0.947
3	3.500	3.068	0.216	7.58	2.23	3.02	1.72	1.16
$3\frac{1}{2}$	4.000	3.548	0.226	9.11	2.68	4.79	2.39	1.34
4	4.500	4.026	0.237	10.79	3.17	7.23	3.21	1.51
5	5.563	5.047	0.258	14.62	4.30	15.2	5.45	1.88
6	6.625	6.065	0.280	18.97	5.58	28.1	8.50	2.25
8	8.625	7.981	0.322	28.55	8.40	72.5	16.8	2.94
10	10.750	10.020	0.365	40.48	11.9	161	29.9	3.67
12	12.750	12.000	0.375	49.56	14.6	279	43.8	4.38
Extra Strong								
$\frac{1}{2}$	0.840	0.546	0.147	1.09	0.320	0.020	0.048	0.250
$\frac{3}{4}$	1.050	0.742	0.154	1.47	0.433	0.045	0.085	0.321
1	1.315	0.957	0.179	2.17	0.639	0.106	0.161	0.407
$1\frac{1}{4}$	1.660	1.278	0.191	3.00	0.881	0.242	0.291	0.524
$1\frac{1}{2}$	1.900	1.500	0.200	3.63	1.07	0.391	0.412	0.605
2	2.375	1.939	0.218	5.02	1.48	0.868	0.731	0.766
$2\frac{1}{2}$	2.875	2.323	0.276	7.66	2.25	1.92	1.34	0.924
3	3.500	2.900	0.300	10.25	3.02	3.89	2.23	1.14
$3\frac{1}{2}$	4.000	3.364	0.318	12.50	3.68	6.28	3.14	1.31
4	4.500	3.826	0.337	14.98	4.41	9.61	4.27	1.48
5	5.563	4.813	0.375	20.78	6.11	20.7	7.43	1.84
6	6.625	5.761	0.432	28.57	8.40	40.5	12.2	2.19
8	8.625	7.625	0.500	43.39	12.8	106	24.5	2.88
10	10.750	9.750	0.500	54.74	16.1	212	39.4	3.63
12	12.750	11.750	0.500	65.42	19.2	362	56.7	4.33
Double Extra Strong								
2	2.375	1.503	0.436	9.03	2.66	1.31	1.10	0.703
$2\frac{1}{2}$	2.875	1.771	0.552	13.69	4.03	2.87	2.00	0.844
3	3.500	2.300	0.600	18.58	5.47	5.99	3.42	1.05
4	4.500	3.152	0.674	27.54	8.10	15.3	6.79	1.37
5	5.563	4.063	0.750	38.55	11.3	33.6	12.1	1.72
6	6.625	4.897	0.864	53.16	15.6	66.3	20.0	2.06
8	8.625	6.875	0.875	72.42	21.3	162	37.6	2.76

Appendix E

Straight Beam

Slope and Deflection Formulas—Shear and Moment Diagrams

Load and Support Conditions	Support Reactions	Slope and Deflection Equations	Important Slope and Deflection Formulas
	$R_A = 0$ $M_A = M_0$	$\theta(x) = \dfrac{M_0 x}{EI}$ $v(x) = \dfrac{M_0 x^2}{2EI}$	$\theta(\ell) = \dfrac{M_0 \ell}{EI}$ $v(\ell) = \dfrac{M_0 \ell^2}{2EI}$
	$R_A = P$ $M_A = P\ell$	$\theta(x) = \dfrac{Px}{2EI}(x - 2\ell)$ $v(x) = \dfrac{Px^2}{6EI}(x - 3\ell)$	$\theta(\ell) = -\dfrac{P\ell^2}{2EI}$ $v(\ell) = -\dfrac{P\ell^3}{3EI}$

(Continued)

Load and Support Conditions	Support Reactions	Slope and Deflection Equations	Important Slope and Deflection Formulas
	$R_A = q\ell$ $M_A = \dfrac{q\ell^2}{2}$	$\theta(x) = \dfrac{qx}{6EI}(-x^2 + 3\ell x - 3\ell^2)$ $v(x) = \dfrac{qx^2}{24EI}(-x^2 + 4\ell x - 6\ell^2)$	$\theta(\ell) = -\dfrac{q\ell^3}{6EI}$ $v(\ell) = -\dfrac{q\ell^4}{8EI}$
	$R_A = P$ $M_A = Pa$	$\underline{0 \le x \le a}$ $\theta(x) = \dfrac{Px}{2EI}(x - 2a)$ $v(x) = \dfrac{Px^2}{6EI}(x - 3a)$ $\underline{a \le x \le \ell}$ $\theta(x) = -\dfrac{Pa^2}{2EI}$ $v(x) = \dfrac{Pa^2}{6EI}(a - 3x)$	$\theta(a) = -\dfrac{Pa^2}{2EI}$ $v(a) = -\dfrac{Pa^3}{3EI}$ $\theta(\ell) = -\dfrac{Pa^2}{2EI}$ $v(\ell) = -\dfrac{Pa^2}{6EI}(3\ell - a)$
	$R_A = qa$ $M_A = \dfrac{qa^2}{2}$	$\underline{0 \le x \le a}$ $\theta(x) = \dfrac{qx}{6EI}(-x^2 + 3ax - 3a^2)$ $v(x) = \dfrac{qx^2}{24EI}(-x^2 + 4ax - 6a^2)$ $\underline{a \le x \le \ell}$ $\theta(x) = -\dfrac{qa^3}{6EI}$ $v(x) = -\dfrac{qa^3}{24EI}(4x - a)$	$\theta(a) = -\dfrac{qa^3}{6EI}$ $v(a) = -\dfrac{qa^4}{8EI}$ $\theta(\ell) = -\dfrac{qa^3}{6EI}$ $v(\ell) = -\dfrac{qa^3}{24EI}(4\ell - a)$

(Continued)

Load and Support Conditions	Support Reactions	Slope and Deflection Equations	Important Slope and Deflection Formulas
	$R_A = R_B = \dfrac{q\ell}{2}$	$\theta(x) = \dfrac{q}{24EI}(-4x^3 + 6\ell x^2 - \ell^3)$ $v(x) = \dfrac{qx}{24EI}(-x^3 + 2\ell x^2 - \ell^3)$	$\theta(0) = -\theta(\ell) = -\dfrac{q}{24EI}\ell^3$ $v\left(\dfrac{\ell}{2}\right) = \dfrac{5q\ell^4}{384EI}$
	$R_A = R_B = \dfrac{M_0}{\ell}$	$\underline{0 \le x \le a}$ $\theta(x) = \dfrac{M_0}{6EI\ell}(3x^2 + 3a^2 - 6a\ell + 2\ell^2)$ $v(x) = \dfrac{M_0 x}{6EI\ell}(x^2 + 3a^2 - 6a\ell + 2\ell^2)$ $\underline{a \le x \le b}$ $\theta(x) = \dfrac{M_0}{6EI\ell}(3x^2 - 6\ell x + 2\ell^2 + 3a^2)$ $v(x) = \dfrac{M_0}{6EI\ell}\{x^3 - 3\ell x^2$ $\quad + x(2\ell^2 + 3a^2) - 3a^2\ell\}$	$\theta(0) = \dfrac{M_0}{6EI\ell}(3a^2 - 6a\ell + 2\ell^2)$ $\theta(\ell) = \dfrac{M_0}{6EI\ell}(3a^2 - \ell^2)$
	$R_A = R_B = \dfrac{M_0}{\ell}$	$\theta(x) = \dfrac{M_0}{6EI\ell}(3x^2 - 6x\ell + 2\ell^2)$ $v(x) = \dfrac{M_0 x}{6EI\ell}(x^2 - 3x\ell + 2\ell^2)$	$\theta(0) = \dfrac{M_0\ell}{3EI}$ $\theta(\ell) = -\dfrac{M_0\ell}{6EI}$

(Continued)

Load and Support Conditions	Support Reactions	Slope and Deflection Equations	Important Slope and Deflection Formulas
	$R_A = \dfrac{b}{\ell} P$ $R_B = \dfrac{a}{\ell} P$	$\underline{0 \le x \le a}$ $\theta(x) = \dfrac{Pb}{6EI\ell}(3x^2 + b^2 - \ell^2)$ $v(x) = \dfrac{Pbx}{6EI\ell}(x^2 + b^2 - \ell^2)$ $\underline{a \le x \le \ell}$ $\theta(x) = \dfrac{Pa}{6EI\ell}(-3x^2 + 6\ell x - a^2 - 2\ell^2)$ $v(x) = \dfrac{Pa(\ell - x)}{6EI\ell}(x^2 - 2\ell x + a^2)$	$\theta(0) = \dfrac{Pb(b^2 - \ell^2)}{6EI\ell}$ $\theta(\ell) = \dfrac{Pa(\ell^2 - a^2)}{6EI\ell}$
	$R_A = \dfrac{Pb}{2\ell^3}(3\ell^2 - b^2)$ $R_B = \dfrac{Pa^2}{2\ell^3}(3\ell - a)$ $M_A = \dfrac{Pb}{3\ell^2}(\ell^2 - b^2)$	$\underline{0 \le x \le a}$ $\theta(x) = \dfrac{Pbx}{4EI\ell^3}\{2\ell(b^2 - \ell^2)$ $\qquad + x(3\ell^2 - b^2)\}$ $v(x) = \dfrac{Pbx^2}{12EI\ell^3}\{3\ell(b^2 - \ell^2)$ $\qquad + x(3\ell^2 - b^2)\}$ $\underline{a \le x \le \ell}$ $\theta(x) = [\theta(x)]_{AB} - \dfrac{P(x - a)^2}{2EI}$ $v(x) = [v(x)]_{AB} - \dfrac{P(x - a)^3}{6EI}$	$\theta(\ell) = \dfrac{Pb}{4EI\ell}(\ell - b)^2$
	$R_A = \dfrac{5}{8} q\ell$ $R_B = \dfrac{3}{8} q\ell$ $M_A = \dfrac{q\ell^2}{8}$	$\theta(x) = \dfrac{qx}{48EI}(-8x^2 + 15\ell x - 6\ell^2)$ $v(x) = \dfrac{qx^2}{48EI}(\ell - x)(2x - 3\ell)$	$\theta(\ell) = \dfrac{q\ell^3}{48EI}$

(Continued)

Load and Support Conditions	Support Reactions	Slope and Deflection Equations	Important Slope and Deflection Formulas
	$R_A = \dfrac{Pb^2}{\ell^3}(3a + b)$ $R_B = \dfrac{Pa^2}{\ell^3}(3b + a)$ $M_A = \dfrac{Pab^2}{\ell^2}$ $M_B = \dfrac{Pa^2b}{\ell^2}$	$0 \le x \le a$ $\theta(x) = \dfrac{Pb^2x}{2EI\ell^3}[x(3a + b) - 2a\ell]$ $v(x) = \dfrac{Pb^2x^2}{6EI\ell^3}[x(3a + b) - 3a\ell]$ $a \le x \le \ell$ $\theta(x) = \dfrac{Pa^2(\ell - x)}{2EI\ell^3}[2b\ell -$ $(\ell - x)(3a + b)]$ $v(x) = \dfrac{Pa^2(\ell - x)^2}{6EI\ell^3}[(\ell - x)(3b + a)$ $- 3b\ell]$	
	$R_A = R_B = \dfrac{q\ell}{2}$ $M_A = M_B = \dfrac{q\ell^2}{12}$	$\theta(x) = -\dfrac{qx}{12EI}(\ell - x)(\ell - 2x)$ $v(x) = -\dfrac{qx^2}{24EI}(\ell - x)^2$	

Appendix F

Answers to Selected Problems

Chapter 1

1-1 $N = P \sin \theta$, $V = -P \cos \theta$, $M = PR \sin \theta$

1-2 $N = \frac{3}{5}P$, $V = -\frac{4}{5}P$, $M_x = \frac{3}{5}Pb$, $M_y = \frac{3}{5}Pa - \frac{4}{5}Pc$, $T = \frac{4}{5}Pb$

1-3 $M_y = -Pe_x$, $M_x = Qh + Pe_y$, $N = -P$, $V_y = -Q$

1-5 $T = 10$ kN·m, all other reactions are zero

1-7 $N_{ED} = 8.32$ k(T), $N_{EF} = 7.68$ k(C), $N_{FG} = 13.68$ k(C), $N_{FC} = 7.81$ k(T), $N_{DC} = 9.6$ k(T)

1-9 $A_y = 0$, $A_x = 60$ kN, $D = 48.33$ kN, $C = 31.67$ kN, $N = 60$ kN(T), $V = 31.67$ kN, $M = 31.65$ kN·m, $V = 31.67$ N

1-11 $\sigma_{AB} = 308$ MPa(C), $\sigma_{AE} = 404.1$ MPa(T), $\sigma_{BC} = 19.4$ MPa(C), $\sigma_{BE} = 269.4$ MPa(C), $\sigma_{CE} = 269.4$ MPa(T), $\sigma_{CD} = 269.4$ MPa(C), $\sigma_{DF} = 134.7$ MPa(T)

1-13 $\sigma_{AB} = 13,248$ psi(C), $\sigma_{AC} = 6369$ psi(T)

1-15 $\tau_A = 40.83$ MPa, $\tau_D = 91.1$ MPa

1-17 $\tau_D = 18.04$ MPa, $\sigma_{AB} = 40$ MPa(T)

1-19 $\theta = 0°: \tau_{ave} = 0$, $\sigma_{ave} = 283$ MPa(T); $\theta = 30°: \tau_{ave} = 141.5$ MPa, $\sigma_{ave} = 245$ MPa(T); $\theta = 90°: \tau_{ave} = 283$ MPa, $\sigma_{ave} = 0$

1-21 $\sigma_{ED} = 1000$ psi(C), $\tau_C = 2223.6$ psi

1-22 $\sigma_{AB} = 10,000$ psi(T), $\tau_C = 5694$ psi

1-23 $\tau_{ave} = 23$ MPa

1-25 $N = 50$ N, $V = 100$ N, $T = 30$ N·m, all other reactions are zero

1-27 $\sigma_{cables} = 10,186$ psi(T), $\sigma_{bar} = 3000$ psi(C)

1-29 $T = 25$ kN·m

1-31 Section AA: $V_y = 50$ lb, $T = 600$ in.-lb, $M_x = 500$ in.-lb; section BB: $V_x = 30$ lb, $V_y = 50$ lb, $M_x = -1500$ in.-lb, $M_y = 300$ in.-lb, $T = 240$ in.-lb

Chapter 2

2-1 (a) $E = 10 \times 10^6$ psi   (b) $\sigma_{PL} = 67,000$ psi, $\epsilon_{PL} = 0.0067$ in./in. (c) $\sigma_{YS} = 76,500$ psi   (d) $\sigma_{ULT} = 87,000$ psi (e) M.R. $= 224.5$ in.-lb/in^3 (f) Toughness $\simeq 13,533$ in.-lb/in^3 (g) Percent elongation $= 16\%$

2-3 (a) $E = 200$ GPa   (b) $\sigma_{PL} = 280$ MPa   (c) $\sigma_{YP} = 285$ MPa (lower yield point)   (d) $\sigma_{ULT} = 480$ MPa   (e) M.R. $= 196$ kN·m/m^3 (f) Toughness $\simeq 126$ MN·m/m^3 (g) Percent elongation $= 28\%$ (h) Percent reduction in area $= 53\%$

2-5 $P_{all} = 131.25$ kN

2-7 (a) 14.3 ksi   (b) 13.5 ksi   (c) 8.25 ksi

2-9 $\ell_x = 0.1000000857$ m, $\ell_y = 0.2000001714$ m, $\ell_z = 0.2999991400$ m, $\Delta V = 6.9$ mm^3

2-13 0.00625 mm/mm

2-15 $G = 72.7$ GPa

2-17 2.9 mm

2-19 $Q = 1573$ lb

2-21 $P_{all} = 5000$ lb

2-23 $\sigma_{ST} = 2250$ psi(C), $\sigma_{CONC} = 225.2$ psi(C)

2-25 $\sigma_{AL} = 1137.5$ psi(C), $\sigma_{ST} = 2275$ psi(C)

2-27 $\sigma_{max} = 226.9$ MPa (right hole), $\sigma_{max} = 126$ MPa (left fillet), $\sigma_{max} = 149.4$ MPa (left hole), $\sigma_{max} = 118.5$ MPa (right fillet)

2-29 (a) $E = 14.3 \times 10^6$ psi   (b) $\sigma_{PL} = 28$ ksi   (c) $\sigma_{YS} = 35.5$ ksi   (d) $\sigma_{ULT} = 54$ ksi, M.R. ≈ 25.2 in.-lb/in^3, Toughness $\simeq 20{,}009$ in.-lb/in^3

2-31 (a) $E = 16 \times 10^6$ psi   (b) $v = 0.317$ (c) $\sigma_{PL} = 25{,}000$ psi, $\epsilon_{PL} = 0.0015625$ in./in.

2-33 $a = 8.89$ in., $b = 1.33$ in.

2-35 $\sigma_{AB} = 100$ MPa(T), $\sigma_{CB} = 173.2$ MPa(C)

2-37 $d = 0.153$ in. at A, $d = 0.121$ in. at B

2-39 (a) $\ell_{axial} = 14.9850$ in., $\ell_{lat} = 2.0006$ in.,   (b) $\dfrac{\Delta V}{V_{initial}} = -0.00040$

2-41 $\tau = \begin{cases} 4000\epsilon \ \text{(ksi)} & 0 \le \epsilon \le 0.002 \\ 8 \quad \text{(ksi)} & 0.002 \le \epsilon \le 0.010 \end{cases}$

2-43 $e_{A/E} = -0.5459$ mm, $(\sigma_{ST})_{max} = 100$ MPa(C), $\sigma_{AL} = 35.7$ MPa(C), $\sigma_{BRASS} = 71.4$ MPa(C)

2-45 $\sigma_{ST} = 38.2$ MPa(T), $\sigma_{AL} = 76.4$ MPa(C)

2-47 Reaction at right support $= 2917$ N(C), reaction at left support $= 15.083$ kN(T), $\sigma_{AB} = 5.8$ MPa(C) (to right of applied force), $\sigma_{BC} = 30.2$ MPa(T) (to left of applied force)

2-49 $N_3 = 2026$ N, $N_1 = N_2 = 1296$ N, $\sigma_1 = \sigma_2 = 129.6$ kN/m^2, $\sigma_3 = 202.6$ kN/m^2

Chapter 3

3-1 $\epsilon_{axial} = 0.000238$ mm/mm, $\epsilon_{circ} = 0.000834$ mm/mm, $\gamma = 0$, $\sigma_{axial} = 100$ MPa(T), $\sigma_{circ} = 200$ MPa(T)

3-3 $\gamma = 990\mu$ in./in., $\tau = 3762$ psi

3-5 $\sigma_r = 1978$ psi(T), $\sigma_\theta = 3407$ psi(C)

3-7 $U_1 = U_2 = \frac{1}{2}c\theta^2$, $U_3 = 2c\theta^2$, $U = 3c\theta^2$

3-9 $U_1 = \frac{1}{2}c\theta^2$, $U_4 = \frac{1}{2}c\theta^2$, $U_2 = \frac{1}{2}c(2\theta - \phi)^2$, $U_3 = \frac{1}{2}c(2\phi - \theta)^2$, $U = c(3\theta^2 + 3\phi^2 - 4\theta\phi)$

3-11 $U = \frac{1}{4}kx^4$

3-13 $U = \dfrac{AK\ell}{n+1}\left(\dfrac{P}{AK}\right)^{(n+1)/n}$

3-15 $h = 0.789$ in.

3-17 $\Delta\ell_x = -0.000292$ in., $\Delta\ell_y = 0.000542$ in., $\Delta\ell_z = -0.000083$ in.

3-19 $\epsilon_x = 476\mu$, $\epsilon_y = 571\mu$, $\epsilon_z = -857\mu$, $\gamma_{xy} = 760\mu$, $\gamma_{xz} = 0$, $\gamma_{yz} = 0$

3-21 $U = 9.15$ in.-lb

Chapter 4

4-1 $\tau_{BC} = 61.1$ MPa

4-3 $d_{min} = 87.9$ mm

4-5 $\phi = 0.087$ rad, $(\tau_{max})_{ST} = 16{,}000$ psi, $(\tau_{max})_{AL} = 8000$ psi

4-7 $\tau_{max} = 48.7$ MPa, $\phi = 1.07$ degrees

4-9 $T_R = 349$ in.-lb, $T_L = 2792.5$ in.-lb, $\tau_{max} = 1778$ psi in the steel, $\tau_{max} = 1778$ psi in the aluminum

4-11 $T_s = \dfrac{\dfrac{b}{\ell} T_0}{1 - \dfrac{JG}{k\ell}}, \quad T_R = \dfrac{a}{\ell} T_0 \left(\dfrac{1 - \dfrac{JG}{kA}}{1 - \dfrac{JG}{k\ell}} \right)$

4-13 $(\tau_{ST})_{max} = 3444$ psi, $(\tau_{AL})_{max} = 2295$ psi, $\phi = 0.789$ degrees

4-15 (a) $a = 25$ mm   (b) $\tau_\rho = \begin{cases} 5.6\rho \ \text{(MPa)} & 0 \le \rho \le 25 \ \text{mm} \\ 140 \ \text{MPa} & 25 \le \rho \le 50 \ \text{mm} \end{cases}$

(c) $\dfrac{d\phi}{dz} = 0.2$ rad

4-17

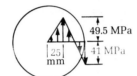

Residual stress distribution

4-19 $T = 89,427$ in.-lb, $d\phi/dz = 0.00018$ rad/in.

4-21 $\tau_{max} = 56.6$ MPA at the ends of the minor axis, $d\phi/dz = 0.0305$ rad/m

4-23 $\tau_{max} = 131.5$ MPa at midpoints of long sides, $d\phi/dz = 0.00215$ rad/m

4-25 $T = 2.88$ N·m

4-27 $\tau_{max} = 500.7$ MPa, $\phi = 0.328$ rad

4-29 $\phi = \dfrac{32}{3\pi} \dfrac{T}{G} \dfrac{\ell}{(d_\ell - d_s)} \left\{ \dfrac{1}{(2d_s)^3} - \dfrac{1}{d_\ell^3} \right\}$

4-31 $d = 58.8$ mm

4-33 $T_0 = 1329$ N·m

4-35 $d = 1.388$ in.

4-37 $d = 4.47$ in., $d = 3.097$ in.

4-39 $\tau_{max} = 99.8$ MPa, $\phi = 2.48$ degrees

4-40 (a) 1 in.   (b) $\tau_\rho = \begin{cases} 16\rho & 0 \le \rho \le 1 \\ 16 & 1 \le \rho \le 2 \end{cases}$ (c) 259,705 in.-lb

(d) 0.00142 rad/in.

4-41 (a)

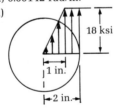

(b) 10.31 degrees

(c)

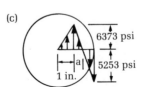

(d) 3.65 degrees

4-43 $T = \dfrac{2\pi c^3 n}{3n + 1}\, \tau_{max}, \quad \dfrac{d\phi}{dz} = \dfrac{1}{c}\left(\dfrac{\tau_{max}}{K}\right)^n$

4-45 (a) $T_{FE} = 25{,}771$ N·m (b) $T_{FP} = 32{,}070$ N·m (c) 24.5% increase

4-46 $T_{PP} = 285{,}307$ in.-lb

4-47 $T = 1304$ N·m

Chapter 5

5-1 R (left) $= 5$ k, R (right) $= 7$ k, $V = 3.4$ k, $M = 16{,}000$ ft-lb

5-3 $V = 5$ k, $M = 28{,}000$ ft-lb, section bb: $V = 1$ k, $M = -1000$ ft-lb,
section aa: $V = 4.25$ k, $M = 13{,}750$ ft-lb

5-5 $V = 5 - 0.4x$ (k), $M = 5x - 0.2x^2$ (k-ft), $0 \le x \le 10$;
$V = 1$ k, $M = 5x - 4(x - 5)$, $10 \le x \le 15$;
$V = -7$ k, $M = 5x - 4(x - 5) - 8(x - 15)$, $15 \le x \le 20$

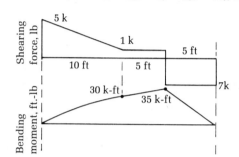

5-7

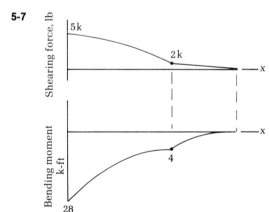

$\left.\begin{array}{l} V(x) = 5 - \frac{1}{12}x^2 \\ M(x) = -28 + 5x - \frac{1}{36}x^3 \end{array}\right\} \quad 0 \le x \le 6$

$\left.\begin{array}{l} V(x) = 2 - 0.5(x - 6) \\ M(x) = -28 + 5x - 3(x - 4) - 0.25(x - 6)^2 \end{array}\right\} \quad 6 \le x \le 10$

5-8 $N = -4.956$ kN, $V = 0.842$ kN, $M = 0.842$ kN·m (slanting member)
$N = -2.3$ kN, $V = 4.47$ kN, $M = 1.0526 + 4.47x$ kN·m
$0 \le x \le 1$ (left portion of horizontal member)
$N = -2.3$ kN, $V = -5.53$ kN, $M = 1.0526 - 10(x - 1) + 4.47x$
$1 \le x \le 2$ (right portion of horizontal member)

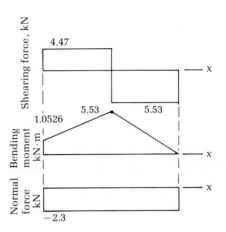

5-9

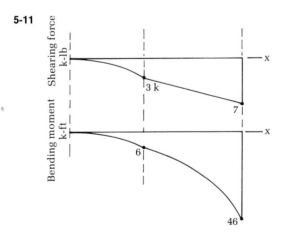

5-11

5-13 $\bar{y} = 5$ in. (neutral axis 5 in. from bottom), $\bar{I}_{xx} = 136$ in^4,
$(\sigma_T)_{max} = 5$ ksi(T), $(\sigma_C)_{max} = 3$ ksi(C)
$\bar{I}_{xx} = 428$ in^4 (neutral axis 5 in. from bottom or top), $\sigma_T = \sigma_C = 1589$ psi

5-15 (b) $\bar{I}_{xx} = 10.86$ in^4,
(c) $(\sigma_T)_{max} = 6814$ psi(T) top, $(\sigma_C)_{max} = 20,810$ psi(C) bottom

5-17 $b = 0.1$ m, $h = 0.2$ m

5-19 $a = 25.4$ mm

5-21 $M_{FP} = 19,687.5$ N·m, $M_{FP}/M_{FE} = 1.765$

5-23

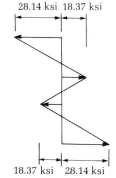

28.14 ksi 18.37 ksi

18.37 ksi   28.14 ksi
Residual stress distribution

5-25 $\tau_{AA} = 17.8$ MPa, $\tau_{BB} = 35.6$ MPa

5-27

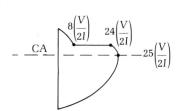
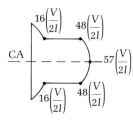
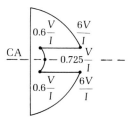

5-29 $s = 289$ mm   or $s = 433$ mm

5-31 $e = \dfrac{ht_2 c_2^3}{12I}$ from centerline of left flange

5-33 $e = 2R\left(\dfrac{\sin\beta - \beta\cos\beta}{\beta - \sin\beta\cos\beta}\right)$

5-35 $M = 861{,}480$ in.-lb

5-37 $q = 80$ lb/ft

5-39 $R_B = 8$ k, $R_A = 8$ k, $V = -8$ k, $M = 16{,}000$ ft-lb

5-41 $B = 48$ kN, $C = 30$ kN, $A_y = 6$ kN, $A_x = 18$ kN, $N = 0$, $V = 0$,
$M = 30$ kN·m at section bb

5-43 (a) $0 \le x \le a$: $V(x) = -M_0/\ell$, $M(x) = -(M_0/\ell)x$;
  $a \le x \le \ell$: $V(x) = -M_0/\ell$, $M(x) = -(M_0/\ell)x + M_0$
  (b) $0 \le x \le a$: $V(x) = bP/\ell$, $M(x) = (bP/\ell)x$;
  $a \le x \le \ell$: $V(x) = -aP/\ell$, $M(x) = (bP/\ell)x - P(x - a)$
  (c) $0 \le x \le \ell$: $V(x) = q\ell/2 - qx$, $M(x) = (q\ell/2)x - qx^2/2$
  (d) $0 \le x \le \ell$: $V(x) = q\ell/6 - \frac{1}{2}q_0(x^2/\ell)$, $M(x) = (q_0\ell/6)x - \frac{1}{6}q_0(x^3/\ell)$

5-45 (a) $0 \le x \le 2\ell/3$: $V(x) = -\frac{4}{3}M_0/\ell$, $M(x) = -\frac{4}{3}(M_0/\ell)x + M_0/3$;
  $2\ell/3 \le x \le \ell$: $V(x) = -\frac{4}{3}M_0/\ell$, $M(x) = -\frac{4}{3}(M_0/\ell)x + \frac{4}{3}M_0$
  (b) $0 \le x \le 2\ell/3$: $V(x) = -\frac{7}{27}P$, $M(x) = -\frac{2}{27}P\ell - \frac{7}{27}Px$;
  $2\ell/3 \le x \le \ell$: $V(x) = -\frac{34}{27}P$, $M(x) = -\frac{2}{27}P\ell - \frac{7}{27}Px - P(x - 2\ell/3)$
  (c) $0 \le x \le \ell$: $V(x) = q\ell/2 - qx$, $M(x) = (q\ell/2)x - q\ell^2/12 - qx^2/2$

5-47 $0 \le x \le 3$: $V(x) = 32 - 10x$, $M(x) = 32x - 5x^2$;
  $3 \le x \le 6$: $V(x) = 32 - 10x - \frac{4}{3}(x - 3)^2$, $M(x) = 32x - 5x^2 - \frac{4}{9}(x - 3)^3$

5-49 $0 \le x \le 5$: $V(x) = 8$ k, $M(x) = 8x$;
  $5 \le x \le 10$: $V(x) = -2$ k, $M(x) = 8x - 10(x - 5)$;
  $10 \le x \le 15$: $V(x) = -2$ k, $M(x) = 8x - 10(x - 5) + 40$;
  $15 \le x \le 20$: $V(x) = -12$ k, $M(x) = 8x - 10(x - 5) + 40 - 10(x - 15)$

5-51

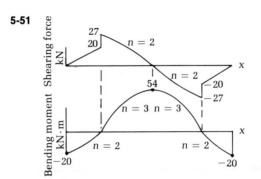

5-53

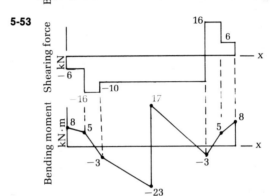

5-55

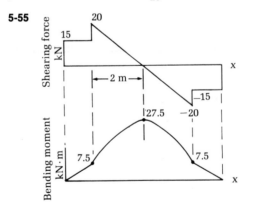

5-57

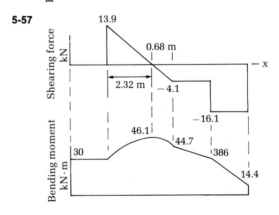

5-59 $I_{xx} = 96.9$ in^4, $\begin{bmatrix} \sigma_{\text{tension}} \\ \sigma_{\text{comp}} \end{bmatrix}_{\max} = \pm 11{,}868$ psi

5-61 $b = 50$ mm, $h = 100$ mm

5-63 $\dfrac{M_{\text{FP}}}{M_{\text{FE}}} = \dfrac{16 R_o}{3\pi} \dfrac{(R_o^3 - R_i^3)}{(R_o^4 - R_i^4)}$

5-65   (a) Partially plastic section: $M_{\text{PP}} = 3.125 \times 10^{-4}\, \sigma_{\text{YP}}$

(b)

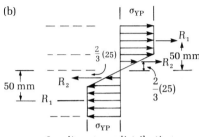

Loading stress distribution

(c)

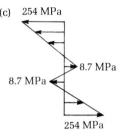

Residual stress distribution

5-67   (a) $M_{\text{PP}} = 15{,}819$ N·m,

(b)

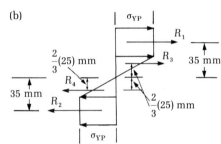

Loading stress distribution

(c)

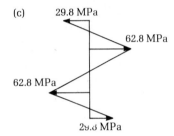

Residual stress distribution

5-69 $M = \dfrac{2bc^2}{n+2}\, \sigma_{\max}$

5-71 $P = 241.7$ lb

5-73 $\tau_{AA} = 1976$ psi, $R = 2631.6$ lb

5-74 $\tau_{\max} = \dfrac{4V}{3A}$

5-75 $\tau_{\max} = 2\ V/A$

5-77 $s = 11.4$ in.

5-79 $e = \pi R/2$

5-81 $M = 2.537 \times 10^5$ in.-lb

5-83 $q_{\max} = 574$ N/m

Chapter 6

6-1   Section AA: $\sigma_{\max} = 48.88$ MPa(T), $\sigma_{\max} = 47.28$ MPa(C);
     section BB: $\sigma_{\max} = 0.8$ MPa(T)

6-3 $\sigma_{\max} = 28.45$ MPa(T)

6-5 $\tau_A = 5600$ psi

6-7

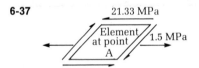

24 MPa

$6.4 + 57.6$
$= 64$ MPa

Element at point A

6-9

3.63 MPa

Element at point A

52.2 MPa

6-11 $\sigma_C = 75$ MPa

6-13 $\ell = 140$ mm

6-15 $P = 2.5$ lb, $P_{max} = 30$ lb, $\delta_{max} = 95$ mm

6-17 $\sigma_{max} = 54.93$ MPa(T), $\sigma_{max} = 25$ MPa(C)

6-19 $\sigma_{max} = 4260$ psi(T), $\sigma_{max} = 4380$ psi(C)

6-21 $P = 24{,}123.8$ N

6-23 $\sigma_{max} = 8.07$ MPa(T), $\sigma_{max} = 5.99$ MPa(C)

6-25 $\sigma_{max} = 76.7$ MPa(T), $\sigma_{max} = 78.3$ MPa(C)

6-27 $\sigma_A = 0.4$ MPa(C), $\sigma_B = 3.6$ MPa(C), $\sigma_C = 1.2$ MPa(C), $\sigma_D = 2.0$ MPa(T)

6-29 $\sigma_{(x,y)} = -1350 - 3600\sqrt{1 - x^2/4}$, $-1350x = 5850$ psi(C)

6-31 $h = 14.34$ m

6-33 $\tau_{ave} = 93.2$ MPa at top and bottom rivets

6-35 $\sigma(x, y) = -4.08 \times 10^7 \sqrt{1 - x^2/(0.025)^2} - 14.123 \times 10^9 x$,
$\sigma_{max} = 353$ MPa(C)

6-37

21.33 MPa

Element at point A

1.5 MPa

6-39 $\sigma_C = 16$ ksi < 20 ksi; the tank is adequate

6-41 $p_{max} = 166.7$ psi

6-43 $P = 7562$ N, $\delta = 283$ mm, $k = 26{,}721$ N/m

Chapter 7

7-1 (a) $\begin{Bmatrix} \sigma_{max} \\ \sigma_{min} \end{Bmatrix} = 7 \pm 3\sqrt{2}$ ksi, $\theta_p = 22.5°$ c.w.

(b) $\begin{Bmatrix} \tau_{max} \\ \tau_{min} \end{Bmatrix} = \pm 3\sqrt{2}$ ksi, $\theta_s = 22.5°$ c.c.w., $\sigma_\xi = \sigma_\eta = 7$ ksi

(c) $\sigma_\xi = 8.098$ ksi, $\sigma_\eta = 5.9$ ksi, $\tau_{\xi\eta} = -4.098$ ksi

7-3 (a) $\begin{Bmatrix} \sigma_{max} \\ \sigma_{min} \end{Bmatrix} = \pm 10$ ksi, $\theta_p = 45°$ c.c.w.

(b) $\begin{Bmatrix} \tau_{max} \\ \tau_{min} \end{Bmatrix} = \pm 10$ ksi, $\theta_s = 0°$, $\sigma_\xi = \sigma_\eta = 0$

(c) $\sigma_\xi = 5$ ksi, $\sigma_\eta = -5$ ksi, $\tau_{\xi\eta} = 8.66$ ksi

7-5   (a) $\begin{Bmatrix}\sigma_{max}\\\sigma_{min}\end{Bmatrix} = 1059$ psi(T), 59 psi(C), $\theta_p = 13.28°$ c.c.w.

    (b) $\begin{Bmatrix}\tau_{max}\\\tau_{min}\end{Bmatrix} = \pm 559$ psi, $\theta_s = 31.72°$ c.w., $\sigma_\xi = \sigma_\eta = 500$ psi

7-7 $P = 7{,}472.5$ N

7-9   (a) $\begin{Bmatrix}\epsilon_{max}\\\epsilon_{min}\end{Bmatrix} = -46\mu$ mm/mm, -654μ mm/mm, $\theta_p = 40.27°$ c.w.

    (b) $\begin{Bmatrix}\gamma_{max}\\\gamma_{min}\end{Bmatrix} = \pm 608\mu$ mm/mm, $\theta_s = 4.73°$ c.c.w.

    (c) $\epsilon_\xi = -456.7\mu$ mm/mm, $\epsilon_\eta = -243.3\mu$ mm/mm,
    $\gamma_{\xi\eta}/2 = -284.8\mu$ mm/mm

7-11 (a) $\gamma_{xy} = 600\mu$ in./in.

    (b) $\begin{Bmatrix}\epsilon_{max}\\\epsilon_{min}\end{Bmatrix} = 300\mu$ in./in., -700μ in./in., $\theta_p = 18.43°$ c.c.w.

    $\begin{Bmatrix}\gamma_{max}\\\gamma_{min}\end{Bmatrix} = \pm 1000\mu$ in./in., $\theta_s = 26.57°$ c.w.

    (c) $\sigma_{max} = 1778$ psi(T), $\sigma_{min} = 6667$ psi(C), $\tau_{max} = 4000$ psi

7-13 $\begin{Bmatrix}I_{max}\\I_{min}\end{Bmatrix} = 1.922 \times 10^{-5}$ m^4, 0.1942×10^{-5} m^4, $\theta_p = 21.35°$ c.c.w.

7-15 See solution for Problem 7-1.
7-17 See solution for Problem 7-3.
7-19 See solution for Problem 7-9.

7-21 $\begin{Bmatrix}I_{max}\\I_{min}\end{Bmatrix} = 252.5$ in^4, 29.96 in^4, $\theta_p = 12.26°$ c.w.

7-23 $\begin{Bmatrix}I_{max}\\I_{min}\end{Bmatrix} = 54$ in^4, 18 in^4, $\theta_p = 45°$ c.c.w.

7-25 (a) $\begin{Bmatrix}\sigma_{max}\\\sigma_{min}\end{Bmatrix} = 9$ ksi, -1 ksi, $\theta_p = 18.43°$ c.w.

    (b) $\begin{Bmatrix}\tau_{max}\\\tau_{min}\end{Bmatrix} = \pm 5$ ksi, $\theta_s = 26.57°$ c.c.w., $\sigma_\xi = \sigma_\eta = 4$ ksi

    (c) $\sigma_\xi = 5.964$ ksi, $\sigma_\eta = 2.036$ ksi, $\tau_{\xi\eta} = -4.598$ ksi

7-27 (a) $\begin{Bmatrix}\sigma_{max}\\\sigma_{min}\end{Bmatrix} = 11.66$ ksi, 0.34 ksi, $\theta_p = 22.5°$ c.w.

    (b) $\begin{Bmatrix}\tau_{max}\\\tau_{min}\end{Bmatrix} = \pm 5.66$ ksi, $\theta_s = 22.5°$ c.c.w., $\sigma_\xi = \sigma_\eta = 6$ ksi

    (c) $\sigma_\xi = 7.46$ ksi, $\sigma_\eta = 4.54$ ksi, $\tau_{\xi\eta} = -5.46$ ksi

7-29 (a) $\begin{Bmatrix}\sigma_{max}\\\sigma_{min}\end{Bmatrix} = \pm 50$ MPa, $\theta_p = 18.43°$ c.w.

    (b) $\begin{Bmatrix}\tau_{max}\\\tau_{min}\end{Bmatrix} = \pm 50$ MPa, $\theta_s = 26.57°$ c.c.w., $\sigma_\xi = \sigma_\eta = 0$

    (c) $\sigma_\xi = 19.64$ MPa, $\sigma_\eta = -19.64$ MPa, $\tau_{\xi\eta} = -45.98$ MPa

7-31 (a) $\begin{Bmatrix} \sigma_{max} \\ \sigma_{min} \end{Bmatrix}$ = 220 MPa, $-$ 280 MPa, θ_p = 36.87° c.c.w.

(b) $\begin{Bmatrix} \tau_{max} \\ \tau_{min} \end{Bmatrix}$ = $\pm$ 250 MPa, θ_s = 8.13° c.w., $\sigma_\xi = \sigma_\eta = -30$ MPa

(c) $\sigma_\xi = -210.6$ MPa, $\sigma_\eta = 150.6$ MPa, $\tau_{\xi\eta} = -172.8$ MPa

7-33 The given state of stress is the principal state of normal stress since no shearing stresses are present.

(a) σ_{max} = 140 MPa, σ_{min} = 0, θ_p = 0

(b) $\begin{Bmatrix} \tau_{max} \\ \tau_{min} \end{Bmatrix}$ = $\pm$ 70 MPa, θ_s = 45° c.c.w., $\sigma_\xi = \sigma_\eta = 70$ MPa

(c) $\sigma_\xi = 9.38$ MPa, $\sigma_\eta = 130.62$ MPa, $\tau_{\xi\eta} = 35$ MPa

7-35 (a) $\begin{Bmatrix} \sigma_{max} \\ \sigma_{min} \end{Bmatrix}$ = 22 ksi, 12 ksi, θ_p = 26.57° c.c.w.

(b) $\begin{Bmatrix} \tau_{max} \\ \tau_{min} \end{Bmatrix}$ = $\pm$ 5 ksi, θ_s = 18.43° c.w., $\sigma_\xi = \sigma_\eta = 17$ ksi

(c) $\sigma_\xi = 21.598$ ksi, $\sigma_\eta = 12.402$ ksi, $\tau_{\xi\eta} = 1.964$ ksi

7-37 See solution for Problem 7-25.

7-39 See solution for Problem 7-27.

7-41 See solution for Problem 7-29.

7-43 See solution for Problem 7-31.

7-45 See solution for Problem 7-33.

7-47 See solution for Problem 7-35.

7-48 (a) $\begin{Bmatrix} \sigma_{max} \\ \sigma_{min} \end{Bmatrix}$ = 9 ksi, $-$ 1 ksi, θ_p = 18.43° c.w.

(b) $\begin{Bmatrix} \tau_{max} \\ \tau_{min} \end{Bmatrix}$ = $\pm$ 5 ksi, θ_s = 26.57° c.c.w., $\sigma_\xi = \sigma_\eta = 4$ ksi

7-49 (a) $\begin{Bmatrix} \sigma_{max} \\ \sigma_{min} \end{Bmatrix}$ = 144 MPa, $-$ 16 MPa, θ_p = 18.43° c.c.w.

(b) $\begin{Bmatrix} \tau_{max} \\ \tau_{min} \end{Bmatrix}$ = $\pm$ 80 MPa, θ_s = 26.57° c.w., $\sigma_\xi = \sigma_\eta = 64$ MPa

7-51 (a) $\begin{Bmatrix} \epsilon_{max} \\ \epsilon_{min} \end{Bmatrix}$ = 145μ in./in., $-$115μ in./in., θ_p = 33.69° c.c.w.

(b) $\begin{Bmatrix} \dfrac{\gamma_{max}}{2} \\ \dfrac{\gamma_{min}}{2} \end{Bmatrix}$ = $\pm$ 130μ in./in., θ_s = 11.31° c.w., $\epsilon_\xi = \epsilon_\eta = 15\mu$ in./in.

(c) $\epsilon_\xi = 118.3\mu$ in./in., $\epsilon_\eta = -88.3\mu$ in./in., $\gamma_{\xi\eta}/2 = 78.9\mu$ in./in.

7-53 (a) The given state of strain is the principal normal state of strain since the shearing strains are zero.

(b) $\begin{Bmatrix} \dfrac{\gamma_{max}}{2} \\ \dfrac{\gamma_{min}}{2} \end{Bmatrix}$ = $\pm$ 70μ in./in., θ_s = 45° c.c.w., $\epsilon_\xi = \epsilon_\eta = -10\mu$ in./in.

(c) $\epsilon_\xi = 50.62\mu$ in./in., $\epsilon_\eta = -70.62\mu$ in./in., $\gamma_{\xi\eta}/2 = -35\mu$ in./in.

7-55 (a) $\begin{Bmatrix} \epsilon_{max} \\ \epsilon_{min} \end{Bmatrix} = 700\mu$ in./in., -1900μ in./in., $\theta_p = 11.31°$ c.w.

(b) $\begin{Bmatrix} \dfrac{\gamma_{max}}{2} \\ \dfrac{\gamma_{min}}{2} \end{Bmatrix} = \pm 1300\mu$ in./in., $\theta_s = 33.69°$ c.w.

(c) $\epsilon_\xi = -1389.2\mu$ in,/in., $\epsilon_\eta = 189.2\mu$ in./in., $\gamma_{\xi\eta}/2 = 1033\mu$ in./in.

7-57 (a) $\begin{Bmatrix} \epsilon_{max} \\ \epsilon_{min} \end{Bmatrix} = 300\mu$ mm/mm, -2300μ mm/mm, $\theta_p = 33.69°$ c.w.

(b) $\begin{Bmatrix} \dfrac{\gamma_{max}}{2} \\ \dfrac{\gamma_{min}}{2} \end{Bmatrix} = \pm 1300\mu$ mm/mm, $\theta_s = 11.31°$ c.c.w., $\epsilon_\xi = \epsilon_\eta = -1000\mu$ in./in.

(c) $\epsilon_\xi = -833\mu$ mm/mm, $\epsilon_\eta = -1167\mu$ mm/mm, $\gamma_{\xi\eta}/2 = 1289.2\mu$ mm/mm

7-59 (a) $\begin{Bmatrix} \epsilon_{max} \\ \epsilon_{min} \end{Bmatrix} = 324.3\mu$ mm/mm, -524.3μ mm/mm, $\theta_p = 22.5°$ c.w.

(b) $\begin{Bmatrix} \dfrac{\gamma_{max}}{2} \\ \dfrac{\gamma_{min}}{2} \end{Bmatrix} = \pm 424.3\mu$ mm/mm, $\theta_s = 22.5°$ c.c.w., $\epsilon_\xi = \epsilon_\eta = -100\mu$ mm/mm

(c) $\epsilon_\xi = -209.8\mu$ mm/mm, $\epsilon_\eta = 9.8\mu$ mm/mm, $\gamma_{\xi\eta}/2 = 409.8\mu$ mm/mm

7-61 (a) $\begin{Bmatrix} \epsilon_{max} \\ \epsilon_{min} \end{Bmatrix} = 42.68\mu$ mm/mm, -77.32μ mm/mm, $\theta_p = 15°$ c.w.

(b) $\begin{Bmatrix} \dfrac{\gamma_{max}}{2} \\ \dfrac{\gamma_{min}}{2} \end{Bmatrix} = 60\mu$ mm/mm, $\theta_s = 30°$ c.c.w., $\epsilon_\xi = \epsilon_\eta = -10\sqrt{3}\mu$ mm/mm

(c) $\epsilon_\xi = -47.32\mu$ mm/mm, $\epsilon_\eta = 12.68\mu$ mm/mm, $\gamma_{\xi\eta}/2 = 51.96\mu$ mm/mm

7-63 See solution for Problem 7-51.
7-65 See solution for Problem 7-53.
7-67 See solution for Problem 7-55.
7-69 See solution for Problem 7-57.
7-71 See solution for Problem 7-59.
7-73 See solution for Problem 7-61.

7-75 $\begin{Bmatrix} \epsilon_{max} \\ \epsilon_{min} \end{Bmatrix} = 711\mu$ mm/mm, -511μ mm/mm, $\theta_p = 24.55°$ c.c.w.,

$\begin{Bmatrix} \dfrac{\gamma_{max}}{2} \\ \dfrac{\gamma_{min}}{2} \end{Bmatrix} = \pm 611\mu$ mm/mm, $\theta_p = 20.45°$ c.w., $\epsilon_\xi = \epsilon_\eta = 100\mu$ mm/mm

7-77 $\begin{Bmatrix} I_{max} \\ I_{min} \end{Bmatrix} = 7.55 \times 10^{-6} \text{ m}^4, 1.26 \times 10^{-6} \text{ m}^4, \theta_p = 16.78° \text{ c.c.w.}$

$(I_{\xi\xi} = I_{max}, I_{\eta\eta} = I_{min})$

7-79 $\begin{Bmatrix} I_{max} \\ I_{min} \end{Bmatrix} = 80.2 \text{ in}^4, 17.3 \text{ in}^4, \theta_p = 11.71° \text{ c.w. } (I_{\xi\xi} = I_{max}, I_{\eta\eta} = I_{min})$

7-81 See solution for Problem 7-77.

Chapter 8

8-1 $\sigma = 50 \text{ ksi}$

8-3 $0 \le x \le \ell: EIv'(x) = \dfrac{M_0}{2\ell}\dfrac{x^2}{2} + C_1, EIv(x) = \dfrac{M_0}{2\ell}\dfrac{x^3}{6} + C_1 x + C_2;$

$\ell \le x \le 2\ell: EIv'(x) = \dfrac{M_0}{2\ell}\dfrac{x^2}{2} - M_0 x + C_3,$

$EIv(x) = \dfrac{M_0}{2\ell}\dfrac{x^3}{6} - M_0\dfrac{x^2}{2} + C_3 x + C_4;$

$C_1 = -\dfrac{M_0\ell}{12}, C_2 = 0, C_3 = \dfrac{11}{12}M_0\ell, C_4 = -\dfrac{1}{2}M_0\ell^2$

8-5 $\delta = 3.527 \text{ mm/graduation}$

8-7

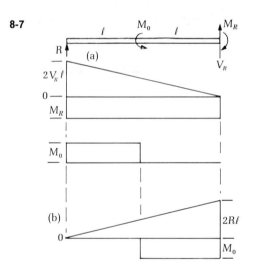

8-9   (a) $\theta = P\ell^2/2EI, \delta = -P\ell^3/3EI$ (b) $\delta = -0.0976 \text{ in.}$

8-11 (a) $R = \frac{7}{64}q\ell$ (b) $\theta = -\frac{5}{96}q\ell^2/EI$
(c) $x_0 = 0.9759\,\ell, \delta_0 = -0.0339 q\,\ell^4/EI$

8-13 $\theta = q\ell^3/6EI, \delta = \frac{11}{24}q\ell^4/EI$

8-15 $\theta = -P\ell^2/16EI - q\ell^3/24EI, \delta = -P\ell^3/48EI - 5q\ell^4/384EI$

8-17 $R = \dfrac{\frac{5}{4}q\ell}{1 + \dfrac{6EI}{k\ell^3}}$

8-19 $EIv'(x) = -\dfrac{q_0 x^3}{6} - \dfrac{q_0 x^4}{24\ell} + \dfrac{C_1 x^2}{2} + C_2 x + C_3,$

$EIv(x) = -\dfrac{q_0 x^4}{24} - \dfrac{q_0 x^5}{120\ell} + \dfrac{C_1 x^3}{6} + \dfrac{C_2 x^2}{2} + C_3 x + C_4,\ C_1 = \tfrac{2}{3} q_0 \ell,$

$C_3 = -\tfrac{11}{180} q \ell^3,\ C_2 = C_4 = 0$

8-21 $\theta(0) = \dfrac{2P\ell^2 + C\ell}{EI},\ v(0) = \dfrac{-\tfrac{8}{3}P\ell^3 - \tfrac{3}{2}C\ell^2}{EI}$

8-23 $\theta(4\ell) = \dfrac{\tfrac{31}{10} q_0 \ell^3 - \tfrac{13}{8} C\ell}{EI},\ v(4\ell) = \dfrac{\tfrac{85}{24} q \ell^4 - \tfrac{7}{6} C\ell^2}{EI}$

8-25 $V_\ell = \tfrac{3}{16} q \ell,\ M_\ell = -\tfrac{5}{48} q \ell^2$

8-27 $\theta_A = -M_0 \ell/3EI,\ \theta_B = M_0 \ell/6EI,\ \theta_A = -q\ell^3/24EI,\ \theta_B = q\ell^3/24EI$

8-29 $R = \tfrac{3}{8} q \ell,$

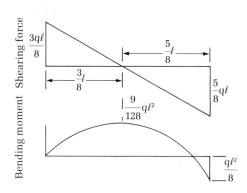

$M_R = -q\ell^2/8,\ V_R = \tfrac{5}{8} q \ell$

8-31 $R = \tfrac{3}{2} M_0/\ell,\ x_0 = \ell/3,\ v_{max}(\ell/3) = M_0 \ell^2/27EI$

8-33 $M_\ell = \tfrac{1}{3} M_0,\ V_\ell = -\tfrac{4}{9} M_0/\ell,\ x_0 = \tfrac{3}{2}\ell,\ v_{max} = M_0 \ell^2/8EI$

Shear and Moment Diagrams

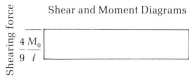

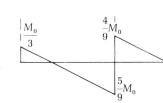

8-35 $\delta = 0.0922$ in.

8-37 $\theta_A = -\dfrac{11}{6}\dfrac{q\ell^4}{EI},\ \theta_B = +\dfrac{11}{6}\dfrac{q\ell^4}{EI};\ \delta_C = \dfrac{57}{24}\dfrac{q\ell^4}{EI}$

8-39 $\theta_A = \dfrac{M_0 \ell}{3EI};\ \delta_C = \dfrac{7}{6}\dfrac{M_0 \ell^2}{EI}$

8-41   (a) $\theta_A = \dfrac{M_0\ell}{12EI}$ (b) $x_0 = \dfrac{\ell}{\sqrt{3}}$, $v_{max} = +\dfrac{M_0\ell^2}{18\sqrt{3}EI}$ at $x = \dfrac{\ell}{\sqrt{3}}$ from A,

$$v_{max} = -\dfrac{M_0\ell^2}{18\sqrt{3}EI} \text{ at } x = \dfrac{\ell}{\sqrt{3}} \text{ from B}$$

8-43 $\theta = -\frac{3}{2}(M_0\ell/EI);\ \delta = \frac{5}{4}(M_0\ell^2/EI)\uparrow$

8-45 $\theta_A = M_0\ell/8EI;\ \delta_C = M_0\ell^2/24EI$

8-47 $R = \frac{3}{4}(M_0/\ell);\ \theta_A = \frac{3}{2}(M_0\ell/EI);\ \delta_A = -\frac{15}{8}(M_0\ell^2/EI)$

8-49 $R = \frac{7}{4}P;\ \theta_A = P\ell^2/EI;\ \delta_A = \frac{5}{6}(P\ell^3/EI)$

8-51 $R_A = \frac{3}{7}q\ell + \frac{30}{7}(EI/\ell^3)\Delta,\ R_B = \frac{19}{28}q\ell - \frac{32}{7}(EI/\ell^3)\Delta,$
$V_C = -\frac{3}{28}q\ell + \frac{2}{7}(EI\Delta/\ell^2),\ M_C = q\ell^2/28 + 4(EI\Delta/\ell^2)$

8-53 $V_\ell = \frac{7}{27}P,\ M_\ell = -\frac{2}{9}P\ell;\ M_R = -\frac{4}{9}P\ell,\ V_R = \frac{20}{27}P$
$x_0 = \frac{12}{7}\ell = 1.714\ell,\ \delta_{max} = -0.2021P\ell^3/EI$

8-55 $R = \dfrac{\frac{3}{8}q\ell}{1 + \dfrac{3EI}{k\ell^2}}$

8-57 $R = \dfrac{\dfrac{q\ell^+}{8E_BI}}{\dfrac{a}{E_RA} + \dfrac{\ell^3}{3E_B\ell}}$

8-59 $R = \dfrac{q\ell}{4} - \dfrac{2EI}{\ell^3}\Delta,\ \delta_{s.s.} = \dfrac{q\ell^4}{24EI} - \dfrac{1}{3}\Delta,\ \delta_{cant} = \dfrac{q\ell^4}{24EI} + \dfrac{2}{3}\Delta$

8-61 $R_A = \frac{11}{28}q\ell,\ R_B = \frac{8}{7}q\ell,\ V_C = -\frac{13}{28}q\ell,\ M_C = \frac{2}{28}q\ell^2$

8-63 $V(x) = \dfrac{q_0\ell}{15}\left[1 - 5\left(\dfrac{x}{\ell}\right)^2\right],\ M(x) = -\dfrac{q_0\ell^2}{60}\left[5\left(\dfrac{x}{\ell}\right)^4 - 4\left(\dfrac{x}{\ell}\right) + 1\right],$

$V\ell = \dfrac{q_0\ell}{15},\ M_\ell = -\dfrac{q_0\ell^2}{60}$

8-65 $v(0) = -\dfrac{67}{3}\dfrac{P\ell^3}{EI},\ v(\ell) = -\dfrac{25}{2}\dfrac{P\ell^3}{EI},\ v(2\ell) = -\dfrac{4P\ell^3}{EI}$

8-66 $R = \frac{9}{320}q_0\ell$

8-67 $R_A = \frac{3}{8}q\ell,\ R_B = \frac{5}{4}q\ell,\ R_C = \frac{3}{8}q\ell$

Chapter 9

9-1 $P_{CR} = \dfrac{c}{\ell},$

$$\dfrac{P\ell}{c} = \dfrac{\theta}{\sin\theta + \dfrac{e}{\ell}\cos\theta}$$

9-3 $\dfrac{P\ell}{c} = \dfrac{2\theta}{\sin\theta} + \dfrac{k\ell^2}{c}\cos\theta,\ P_{CR} = 75\text{ kN}$

9-5 $\dfrac{P\ell}{c} = \dfrac{2\theta + \dfrac{k\ell^2}{c}[\sin(\theta + \theta_0) - \sin\theta_0]\cos(\theta + \theta_0)}{\sin(\theta + \theta_0)}$

9-7 (a) $r_{xx} = \dfrac{h}{2\sqrt{3}}$, $r_{yy} = \dfrac{b}{2\sqrt{3}}$, $\left(\dfrac{\ell}{r}\right)_{xx} = 103.9$, $\left(\dfrac{\ell}{r}\right)_{yy} = 208$

(b) $r = \dfrac{d}{4}$, $\dfrac{\ell}{r} = 240$

(c) $r_{xx} = \dfrac{h}{3\sqrt{2}}$, $r_{yy} = \dfrac{b}{3\sqrt{2}}$, $\left(\dfrac{\ell}{r}\right)_{xx} = 127.3$, $\left(\dfrac{\ell}{r}\right)_{yy} = 254.8$

(d) $r = \dfrac{\sqrt{d_i^2 + d_o^2}}{4}$, $\ell/r = 90.3$

9-9 $r_{\min} = 6$ mm

9-11 $P = 7034$ N

9-13 $k = 0.7$

9-15 $P_{CR} = 25{,}782$ lb

9-17

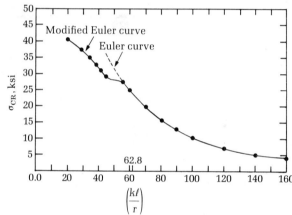

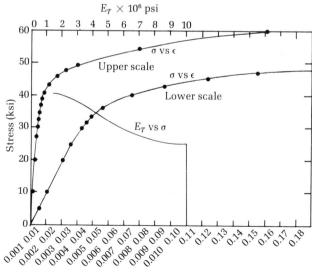

9-19 L $2 \times 2 \times \frac{3}{8}$

9-21 $P_{CR} = 80.7$ k

9-23 $P_{all} = 42.92$ k

9-25 $P_{all} = 61,300$ lb

9-27 $\dfrac{P}{2k\ell} = \left\{ \dfrac{1}{\sqrt{1 - 2\dfrac{h}{\ell}\dfrac{y}{\ell} + \left(\dfrac{y}{\ell}\right)^2}} - 1 \right\} \left(\dfrac{h}{\ell} - \dfrac{y}{\ell} \right)$

9-29 $d = 1.993$ in., $r_{xx} = r_{yy} = 1.08$ in.

9-31 $r_{min} = 12$ mm

9-33 $P = 13,834$ lb

9-35 $W = 15,074$ N

9-37 $d = 0.981$ in.

9-39 S 18×54.7

9-41 $P_{all} = 73,690$ lb

Chapter 10

10-1 $P_{max} = 105,975$ lb, $\eta = 0.883$

10-3 $P_{max} = 20.5$ kN

10-5 $\tau_{max} = 12,155$ psi, rivet at left of group

10-7 $\tau_{max} = 10,183$ psi, top and bottom rivets

10-9 $d = 0.65$ in.

10-11 $\tau_{max} = 58.5$ MPa, lowermost rivet in the group

10-13 (a) $\ell_1 = 5.97$ in., $\ell_2 = 15.70$ in.
    (b) $\ell_1 = 4.47$ in., $\ell_2 = 14.20$ in.

10-15 $P_{max} = 34,204$ N

10-17 $T = 5.73$ kN·m

10-19 $\tau_{max} = 2200$ psi, uppermost rivet in the group

10-21 $\tau_{max} = 20,150$ psi, rivet C

10-23 $\tau_A = \tau_B = \tau_{max} = 2196$ psi, $\tau_C = \tau_D = 1492$ psi

10-25 $\tau_{max} = 40.5$ MPa at points C and F

Chapter 11

11-1 $x = 0, 0, 2a, -a$

11-3 $x = \dfrac{21}{22}\dfrac{P\ell}{AE}, y = \dfrac{13}{22}\dfrac{P\ell}{AE}, z = \dfrac{5}{22}\dfrac{P\ell}{AE}$

11-5 $N_1 = -\dfrac{P}{17}, N_2 = \dfrac{21}{34}P, N_3 = \dfrac{15}{34}P, x = \dfrac{7}{34}\dfrac{P}{k}, \phi = \dfrac{4}{17}\dfrac{P}{k\ell}$ (rotation of cross beam)

11-7 $u = 0.7518\ P\ell/AE, v = 0.1641\ P\ell/AE, N_1 = 0.9318P, N_2 = 0.1641P,$
    $N_3 = -0.5117P$

11-9 $N_1 = N_2 = 0.2072P, N_3 = N_4 = 0.2072P, N_5 = 0.7071P, N_6 = -0.2927P$

11-11 $\theta = \dfrac{q\ell^3}{8EI}$

11-13 $R_1 = \frac{11}{28}q\ell, R_2 = \frac{32}{28}q\ell$

11-15 $\delta = \dfrac{\pi}{2}\dfrac{CR^2}{EI} + \dfrac{\pi}{2}\dfrac{CR^2}{JG}$

11-17 $u = 2P\ell/AE, v = 0$

11-19 $\delta_A = \frac{7}{3}(P\ell/AE)$

11-21 $R = \frac{7}{64}(q\ell/EI)$

11-23 $u = 2P\ell/AE, v = 0$

11-25 $Q = -\dfrac{P}{\pi}\left\{1 - \dfrac{(3 + v)}{AR^2}I\right\}$

11-27 $u_P = \dfrac{\pi}{4}\dfrac{PR^3}{EI} + \dfrac{PR^3}{JG}\left(\dfrac{3\pi}{4} - 2\right) + \dfrac{\pi}{2}\dfrac{PR}{AG}, \theta_A = \dfrac{\pi}{4}\dfrac{PR^2}{EI} - \dfrac{PR^2}{JG}\left(1 - \dfrac{\pi}{4}\right)$

11-29 $x = \dfrac{2b - a}{3}, y = \dfrac{2a - b}{3}$

11-31 $N_{AD} = 0.321P, N_{AB} = 0.321P, N_{BC} = -1.079P, N_{DC} = 0.321P,$
$N_{DB} = 0.395P, N_{CA} = -0.453P$

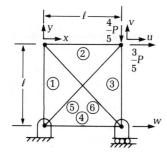

11-33 $Q = \dfrac{\frac{9}{28}c/\ell}{1 + (2 + v)(r/\ell)^2}$

11-35 $\phi = \dfrac{63q\ell^4}{EI}\left\{1 + \dfrac{2}{63}(9 + 8v)\left(\dfrac{r}{\ell}\right)^2\right\}, \theta = \dfrac{55}{3}\dfrac{q\ell^3}{EI}$

11-37 $M_0 = PR$

11-39 $R = \dfrac{\frac{3}{8}q\ell}{1 + \dfrac{3E_B I_B}{E_R A_R}\dfrac{a}{\ell^3}}$

11-41 $V_\ell = \dfrac{3}{20}q_0\ell, M_\ell = \dfrac{-q_0\ell^2}{30}$

11-43 $R = \dfrac{7}{64}q\ell$

11-45 Horizontal deflection $u_A = 0.493\, CR^2/EI$, vertical deflection $v_A = -0.707\, CR^2/EI$, rotation $\phi = (\pi/4)(CR/EI)$

11-47 $u = 0, v = 0, w = -\dfrac{\pi}{4}\dfrac{CR^2}{EI} + \dfrac{CR^2}{JG}\left(1 - \dfrac{\pi}{4}\right)$

11-49 $u = \dfrac{P\ell}{4AE}, v = \dfrac{13}{4\sqrt{3}}\dfrac{P\ell}{AE}, w = \dfrac{1}{2\sqrt{3}}\dfrac{P\ell}{AE}$

Chapter 12

12-1 $\sigma = 112$ MPa(T) (upper right hand corner),
$\sigma = 112$ MPa(C) (lower left hand corner)

12-3 $\sigma_A = 310.3$ MPa(T)

12-5 $\sigma_A = 3367.7$ psi(C)

12-7 $\tau_{AA} = 12.16$ MPa, $\tau_{BB} = 39.6$ MPa, $\tau_{DD} = 52.1$ MPa

12-9 $\tau_{AA} = 38.1$ MPa, $\tau_{BB} = 53$ MPa, $\tau_{CC} = 17.1$ MPa

12-11 $P_{all} = 3652$ N

12-13 $\left\{ \begin{matrix} \sigma_\rho \\ \sigma_t \end{matrix} \right\} = -70 \pm \dfrac{56,250}{\rho^2}$, u(37.5) = 0.0192 mm, u(75) = 0.0244 mm

12-15 $(d_i)_{al} = 99.758$ mm

12-17 $\sigma_\theta = 19,500$ psi(T), $\sigma_s = 9749$ psi(T)

12-21 $\sigma_A = 759.2$ psi(T)

12-23 $\sigma = 18,305$ psi(T) (lowermost point on cross section),
$\sigma = 15,143$ psi (C) (left uppermost corner of cross section)

12-25 $\tau_{AA} = 2580$ psi, $\tau_{BB} = 19,145$ psi

12-27 $\sigma = 24,904$ psi(C), $\sigma = 21,122$ psi (T)

12-29 $\left\{ \begin{matrix} \sigma_t \\ \sigma_\rho \end{matrix} \right\} = -36,526.5 \left\{ 1 \pm \dfrac{4}{\rho^2} \right\}$ for the internal cylinder,

$p_s = 32,468$ psi, $\left\{ \begin{matrix} \sigma_t \\ \sigma_\rho \end{matrix} \right\} = 25,974.4 \left\{ 1 \pm \dfrac{81}{\rho^2} \right\}$ for the external cylinder

Chapter 13

13-1 $n_i n_i = 1$, $n_i t_j = 0$ (direction cosines must satisfy these relations)

13-3 $\sigma_n = 133.7$ MPa, $\sigma_t = 4.2$ MPa

13-5   (a) $\sigma_{\text{I}} = 57.312$ MPa: $n_1 = \pm 0.2113$, $n_2 = \pm 0.5775$, $n_3 = \pm 0.7886$;
$\sigma_{\text{II}} = 10$ MPa: $n_1 = \pm 0.5774$, $n_2 = \pm 0.5774$, $n_3 = \mp 0.5774$;
$\sigma_{\text{III}} = 10$ MPa: $n_1 = \pm 0.7887$, $n_2 = \mp 0.5773$, $n_3 = \pm 0.2114$
(b) $\tau_{\text{I}} = 0$, $\sigma_n = 10$ MPa, $\tau_{\text{II}} = \tau_{\text{III}} = 23.66$ MPa, $\sigma_n = 33.66$ MPa.
The principal shearing stresses bisect the planes on which the principal normal stresses act.
(c) $\sigma_{\text{oct}} = 25.77$ MPa(T), $\tau_{\text{oct}} = 22.30$ MPa

13-7   (a) $\sigma_{\text{I}} = 1$ MPa: $n_2 = \pm 0.6325$, $n_1 = \pm 0.4472$, $n_3 = \mp 0.6325$;
$\sigma_{\text{II}} = 4.702$ MPa: $n_2 = \pm 0.6399$, $n_1 = \pm 0.1587$, $n_3 = \pm 0.7519$;
$\sigma_{\text{III}} = -1.702$ MPa: $n_2 = \pm 0.4368$, $n_1 = \mp 0.8802$, $n_3 = \mp 0.1858$
(b) $\tau_{\text{I}} = 3.202$ MPa, $\sigma_n = 1.5$ MPa(T); $\tau_{\text{II}} = 1.351$ MPa,
$\sigma_n = 0.351$ MPa(C); $\tau_{\text{III}} = 1.851$ MPa, $\sigma_n = 2.851$ MPa(T)
(c) $\sigma_{\text{oct}} = 1.333$ MPa(T), $\tau_{\text{oct}} = 7.88$ MPa

13-9   Normal stress on any plane referred to the principal axes x_{I}, x_{II}, x_{III} is
$\sigma_n = \sigma_{\text{I}} n_1^2 + \sigma_{\text{II}} n_2^2 + \sigma_{\text{III}} n_3^2$. Consider the principal shear plane $n_1 = 0$,
$n_2 = n_3 = \pm 1/\sqrt{2}$, then $\sigma_n = \sigma_{\text{II}}(\frac{1}{2}) + \sigma_{\text{III}}(\frac{1}{2}) = \frac{1}{2}(\sigma_{\text{II}} + \sigma_{\text{III}})$. Similar analyses are valid for $n_1 = \pm 1/\sqrt{2}$, $n_2 = 0$, $n_3 = \pm 1/\sqrt{2}$, and $n_1 = n_2 = \pm 1/\sqrt{2}$,
$n_3 = 0$

13-11 $\epsilon_{11} = 759.1\mu$, $\epsilon_{22} = 568.4\mu$, $\epsilon_{33} = 672.5\mu$, $\epsilon_{12} = 130.1\mu$, $\epsilon_{13} = -126.9\mu$,
$\epsilon_{23} = -27.0\mu$

13-13 $\epsilon_{\text{I}} = 573.2\mu$: $n_3 = \pm 0.7887$, $n_2 = \pm 0.5774$, $n_1 = \pm 0.2114$;
$\epsilon_{\text{II}} = 226.8\mu$: $n_3 = \pm 0.2113$, $n_2 = \pm 0.7886$, $n_1 = \mp 0.5774$;
$\epsilon_{\text{III}} = 100\mu$: $n_3 = \pm 0.5774$, $n_2 = \mp 0.5774$, $n_1 = \mp 0.5774$

13-15 $\epsilon_{\text{I}} = 593.4\mu$: $n_2 = \pm 0.7003$, $n_1 = \pm 0.4739$, $n_3 = \pm 0.5338$;
$\epsilon_{\text{II}} = -180.0\mu$: $n_2 = \pm 0.6503$, $n_1 = \mp 0.6967$, $n_3 = \mp 0.3029$;
$\epsilon_{\text{III}} = 186.7\mu$: $n_2 = \pm 0.2435$, $n_1 = \pm 0.8425$, $n_3 = \mp 0.4805$

13-16 $\gamma_{\text{III}} = 346.4\mu$, $\epsilon_n = 400\mu$; $\gamma_{\text{II}} = 473.2\mu$, $\epsilon_n = 336.6\mu$; $\gamma_{\text{I}} = 126.8\mu$,
$\epsilon_n = 163.4\mu$

13-17 $\gamma_{III} = 50\mu$, $\epsilon_n = 825\mu$; $\gamma_{II} = 500\mu$, $\epsilon_n = 600\mu$; $\gamma_I = 450\mu$, $\epsilon_n = 575\mu$

13-19 (a) $\sigma_I = 4.61$ ksi, $\sigma_{II} = 1$ ksi, $\sigma_{III} = -2.61$ ksi
  (b) $n_2 = \pm 0.7076$, $n_1 = \pm 0.3920$, $n_3 = \pm 0.5880$
  (c) $\tau_I = 1805$ psi, $\tau_{II} = 3610$ psi, $\tau_{III} = 1805$ psi
  (d) $\sigma_n = 4.33$ ksi(T)   (e)$\sigma_{oct} = 1000$ psi(T), $\tau_{oct} = 2.948$ ksi

13-21 (a) $\sigma_I = (1 + \sqrt{2})$ ksi, $\sigma_{II} = 2$ ksi, $\sigma_{III} = (1 - \sqrt{2})$ ksi
  (b) $n_1 = \pm 0.7071$, $n_2 = \pm 0.7071$, $n_3 = 0.0$
  (c) $\tau_I = 1.207$ ksi, $\tau_{II} = 1.414$ ksi, $\tau_{III} = 0.2071$ ksi
  (d) $\sigma_n = 2.276$ ksi(T)   (e) $\sigma_{oct} = 0.667$ ksi(T), $\tau_{oct} = 1.247$ ksi

13-23 (a) $\sigma_I = 5.732$ MPa, $\sigma_{II} = 22.68$ MPa, $\sigma_{III} = 10$ MPa
  (b) $n_3 = \pm 0.7887$, $n_2 = \pm 0.5774$, $n_1 = \pm 0.2113$
  (c) $\tau_I = 6.34$ MPa, $\tau_{II} = 23.66$ MPa, $\tau_{III} = 17.32$ MPa
  (d) $\sigma_n = 50$ MPa(T)   (e) $\sigma_{oct} = 30$ MPa(T), $\tau_{oct} = 20$ MPa

13-27 (a) $\epsilon_I = 734.09\mu$, $\epsilon_{III} = -66.15\mu$, $\epsilon_{II} = 432.09\mu$
  (b) $n_2 = \pm 0.2650$, $n_1 = \pm 0.7696$, $n_3 = \pm 0.5809$
  (c) $\gamma_I = 498.24\mu$ in./in., $\epsilon_n = 182.97\mu$ in./in.; $\gamma_{II} = 800.24\mu$ in./in.,
  $\epsilon_n = 333.97\mu$ in./in.; $\gamma_{III} = 302.0\mu$ in./in., $\epsilon_n = 583.09\mu$ in./in.
  (d) $\epsilon_n = 688.9\mu$ in./in. (e) $\gamma_{nt} = -518.6\mu$ in./in.

13-29 (a) $\epsilon_I = 607.4\mu$ mm/mm, $\epsilon_{II} = -188.6\mu$ mm/mm, $\epsilon_{III} = -442.4\mu$ mm/mm
  (b) $n_2 = \pm 0.5058$, $n_3 = \pm 0.5844$, $n_1 = \mp 0.6346$
  (c) $\gamma_I = 253.8\mu$ mm/mm, $\epsilon_n = -315.5\mu$ mm/mm; $\gamma_{II} = 524.9\mu$ mm/mm,
  $\epsilon_n = 82.5\mu$ mm/mm; $\gamma_{III} = 796.0\mu$ mm/mm, $\epsilon_n = 209.4\mu$ mm/mm
  (d) $\epsilon_n = 488.9\mu$ mm/mm   (e) $\epsilon_{nt} = -259.3\mu$ mm/mm

Chapter 14

14-1 $d = 1.45$ in.
14-3 $p = 2.9$ MPa
14-5 $T_{max} = 5154$ N·m
14-7 Yielding has occurred
14-9 Yielding has not occurred
14-11 $p = 2.7$ MPa
14-13 Yielding has not occurred
14-15 3.274 MPa
14-17 Yielding has not been initiated
14-19 Yielding has not occurred
14-21 $p = 1428.6$ psi
14-23 $T = 9861$ N·m
14-25 $\tau_{xz} = 82.2$ MPa
14-27 (a) $\sigma_{YP} = 1001$ MPa   (b) $\sigma_{YP} = 709.3$ MPa   (c)$\sigma_{YP} = 1068.6$ MPa
  (d) $\sigma_{YP} = 799$ MPa
14-29 (a) Yielding has occurred   (b) The maximum shearing stress theory of
  failure predicts that yielding has occurred

Index

Standard SI Prefixes[1,2]

Name	Symbol	Factor
tera	T	$1\ 000\ 000\ 000\ 000 = 10^{12}$
giga	G	$1\ 000\ 000\ 000 = 10^{9}$
mega	M	$1\ 000\ 000 = 10^{6}$
kilo	k	$1\ 000 = 10^{3}$
hecto*	h	$100 = 10^{2}$
deka*	da	$10 = 10^{1}$
deci*	d	$0.1 = 10^{-1}$
centi*	c	$0.01 = 10^{-2}$
milli	m	$0.001 = 10^{-3}$
micro	μ	$0.000\ 001 = 10^{-6}$
nano	n	$0.000\ 000\ 001 = 10^{-9}$
pico	p	$0.000\ 000\ 000\ 001 = 10^{-12}$
femto	f	$0.000\ 000\ 000\ 000\ 001 = 10^{-15}$
atto	a	$0.000\ 000\ 000\ 000\ 000\ 001 = 10^{-18}$

1. Use multiple and submultiple prefixes in increments of 1000 whenever possible. Specify length, for example, in mm, m, or km. Avoid using standard SI prefixes in the denominator of a combination of units. Use MN/m^2 (meganewton per square meter), but not N/cm^2 (newtons per square centimeter).

2. To avoid confusion with the practice in some foreign countries of using commas for decimal points, spaces are used to group numbers in the SI system.

* Use of these prefixes is not recommended but they are sometimes encountered.